Operations Research Proceedings 1993

DGOR / NSOR

Papers of the 22nd Annual Meeting of DGOR in Cooperation with NSOR
Vorträge der 22. Jahrestagung der DGOR zusammen mit NSOR

Edited by / Herausgegeben von
H. Dyckhoff, U. Derigs, M. Salomon
H. C. Tijms

With 103 Figures / Mit 103 Abbildungen

Springer-Verlag
Berlin Heidelberg New York
London Paris Tokyo
Hong Kong Barcelona
Budapest

Prof. Dr. Harald Dyckhoff
RWTH Aachen
Templergraben 64
D-52056 Aachen

Prof. Dr. Dr. Ulrich Derigs
Universität zu Köln
Albertus-Magnus-Platz
D-50923 Köln

Dr. Marc Salomon
Erasmus Universiteit Rotterdam
P.O. Box 1738
NL-3000 DR Rotterdam

Prof. Dr. Henk C. Tijms
Vrije Universiteit Amsterdam
De Boelelaan 1105
NL-1081 HV Amsterdam

ISBN-13: 978-3-540-57862-8 e-ISBN-13: 978-3-642-78910-6
DOI: 10.1007/978-3-642-78910-6

42/2202-5 4 3 2 1 0 – Printed on acid-free paper

Vorwort

Die 22. Jahrestagung der Deutschen Gesellschaft für Operations Research wurde 1993 zum zweiten Mal gemeinsam mit der Nederlandse Stichting voor Operations Research veranstaltet. Sie fand vom 25. bis 27. August an der Freien Universität in Amsterdam statt. Möglicherweise wegen dieses frühen, in der Ferienzeit liegenden Termins war sie mit ca. 330 Teilnehmern nur relativ schwach besucht. Dagegen bot die Tagung den Teilnehmern mit einem attraktiven wissenschaftlichen und sozialen Programm ein bewährtes Forum zum Austausch von Erfahrungen zwischen Wissenschaftlern und Praktikern aus Hochschulen, Wirtschaft und anderen Bereichen. Der hiermit vorgelegte Proceedings-Band gibt einen Überblick über das wissenschaftliche Programm.

Mit über 190 Fachvorträgen war das wissenschaftliche Programm eines der umfangreichsten und vielseitigsten aller bisherigen Jahrestagungen, wobei mit Blick auf den europäischen Binnenmarkt der mögliche Beitrag des Operations Research zur Verbesserung der internationalen Distribution und Logistik einen Schwerpunkt bildete. Besondere Höhepunkte stellten die sich mit der Praxis des Operations Research auseinandersetzenden drei Plenarvorträge international angesehener Wissenschaftler dar, die einen generell feststellbaren Trend zu mehr anwendungsspezifischen Beiträgen bestätigen. Neben den Plenarvorträgen und den drei prämierten Arbeiten der Preisträger des Studentenwettbewerbs waren die übrigen Beiträge in dreizehn Sektionen gegliedert, welche von fünfzehn Sektionsleitern weitgehend eigenverantwortlich ausgestaltet und betreut wurden.

Der Gesamtumfang des Tagungsbands hat Beschränkungen in den Seitenzahlen der abgedruckten Beiträge erforderlich gemacht, so daß der Inhalt vieler Beiträge nur durch eine einseitige Kurzfassung skizziert werden konnte. Bei der Auswahl der „Langfassungen" von ca. 7 Seiten (bzw. 12 Seiten bei Übersichtsvorträgen) hat sich der Programmausschuß inhaltlich wesentlich auf das Urteil des jeweiligen Sektionsleiters gestützt. Für die - zum Teil auch aus formalen Gründen - notwendigen Ablehnungen bitte ich die Referenten um Verständnis und darf den interessierten Leser zugleich darauf hinweisen, daß er gegebenenfalls eine ausführliche Version direkt beim Autor anfordern kann. Außerdem sind manche hier kurz gefaßten Beiträge zur Veröffentlichung an anderer Stelle, insbesondere in der Zeitschrift Operations Research-Spektrum, vorgesehen.

Eine erfolgreiche Tagung bedarf der Unterstützung durch viele Personen und Institutionen, die hier nicht einzeln genannt werden können. Ihnen allen sei herzlich gedankt, so den Referenten für ihre Beiträge, den Sitzungsleitern für die Moderation der Vorträge und der anschließenden Diskussion sowie den Mitgliedern und Mitarbeitern der verschiedenen mitwirkenden Kommittees, Ausschüsse, Vorständen und

Geschäftsstellen. Aus Sicht des Programmausschußvorsitzenden besonders hervorheben möchte ich zum ersten die Sektionsleiter, die zum Erfolg der Tagung und zum Erscheinen des Proceedings-Bandes wesentlich beigetragen haben, zum zweiten die Mitglieder des Programmausschusses für ihre aktive und fruchtbare Kooperation, allen voran Prof. Dr. Henk C. Tijms und seine Mitarbeitern für die hervorragende örtliche Tagungsorganisation und das attraktive Rahmenprogramm und last but not least die Helfer an meinem Lehrstuhl, hier besonders Herr Dipl. Wi-Ing. Rolf Soukal, der mit großem Engagement und Zeitaufwand die erfolgreiche Arbeit des Programmausschusses unterstützt und überhaupt erst ermöglicht hat.

Aachen, im Dezember 1993

Harald Dyckhoff

Programmausschuß

Sektionen und Sektionsleiter

1. *OR in der Praxis (Anwendungsberichte)/ Practical OR (application reports)*

(a) Karl-Peter Schuster
GUF GmbH Hamburg

(b) Prof. Dr. Gerrit Timmer
Vrije Universiteit Amsterdam

2. *Projektmanagement und Ablaufplanung/ Project Management and Scheduling*

Prof. Dr. Andreas Drexl
Christian-Albrechts-Universität zu Kiel

3. *Produktionsplanung und -steuerung/ Production Planning and Control*

Prof. Dr. Horst Tempelmeier
TU Braunschweig

4. *Distribution und Logistik/ Distribution and Logistics*

Prof. Dr. ir. Jo van Nunen
Erasmus Universiteit Rotterdam

5. *Marketing und Handel/ Marketing and Trade*

Prof. Dr. Klaus Ambrosi
Universität Hildesheim

6. *Finanzen, Banken und Versicherungen/ Finance, Banking and Insurance*

Prof. Dr. Bert Kersten
Rabobank Nederland Utrecht

7. *Öffentlicher Sektor und natürliche Umwelt/ Public Sector and Environment*

Prof. Dr. Günter Altrogge
Universität Hamburg

8. *Datenanalyse und Statistik/ Data Analysis and Statistics*

Prof. Dr. Karl Clemens Mosler
Universität der Bundeswehr Hamburg

9. *Daten- und Wissensverarbeitung/ Data and Knowledge Processing*

Prof. Dr. Michael Bastian
RWTH Aachen

10. *Unscharfe Mengen/ Fuzzy Sets*

Prof. Dr. Brigitte Werners
Ruhr-Universität Bochum

11. *Entscheidungsanalyse und Organisation/ Decision Analysis and Organization*

Prof. Dr. Rudolf Vetschera
Universität Konstanz

12. *Kombinatorische Optimierung/ Combinatorial Optimization*

Prof. Dr. Peter Brucker
Universität Osnabrück

13. *Stochastische Modelle/ Stochastic Models*

(a) Prof. Dr. Rommert Dekker
Erasmus Universiteit Rotterdam

(b) Prof. Dr. Lodewijk Kallenberg
Rijksuniversiteit Leiden

Sponsoren

Für die finanzielle Unterstützung der Tagung danken wir folgenden Unternehmen und Institutionen

Beyers & Partners Nederland

Centrum voor Informatieverwerking Utrecht

CQM Consultants Eindhoven

Dash Associates Ltd. Blisworth, U.K.

Fysisch en Elektronisch Laboratorium TNO

IBM Nederland NV

KNAW

Landbouwuniversiteit Wageningen

ORTEC Consultants Gouda

Shell Nederland BV

Vrije Universiteit Amsterdam

Inhaltsverzeichnis

Sektion 1 (b): OR in der Praxis/Practical OR

Sektion 2: Projektmanagement und Ablaufplanung/Project Management and Scheduling

Sektion 3: Produktionsplanung und -steuerung/Production Planning and Control

Sektion 4: Distribution und Logistik/Distribution and Logistics

Sektion 7: Öffentlicher Sektor und natürliche Umwelt/Public Sector and Environment

Sektion 8: Datenanalyse und Statistik/Data Analysis and Statistics

Sektion 9: Daten- und Wissensverarbeitung/Data and Knowledge Processing

Sektion 10: Unscharfe Mengen/Fuzzy Sets

Sektion 11: Entscheidungsanalyse und Organisation/Decision Analysis and Organization

Sektion 12: Kombinatorische Optimierung/Combinatorial Optimization

Sektion 13 (a): Stochastische Modelle/Stochastic Models

Sektion 13 (b): Stochastische Modelle/Stochastic Models

VEHICLE ROUTING IN PRACTICE: NEW SOLUTIONS AND CURRENT CHALLENGES

Achim Bachem, Universität zu Köln

In this talk we shall discuss two case studies of real life vehicle routing problems from different industry branches. We shall report on new solutions which solve online vehicle dispatch problems with time windows, truck dispatch problems in a network environment with many dispatchers and vehicle routing problems which are connected to a production planning system.

In addition we report on computational results of new parallel algorithms for solving complex vehicle routing problems, which we have implemented on a Parsytee 1024 processor system and on a workstation cluster using PVM.

The calculation of the distance resp. time matrix has to be done by using fast hierarchical shortest path algorithms on very large digitized street networks. We also have to take into account time depending travel times and traffic jams. We will present a model for simulating traffic which we have implemented on various supercomputer platforms.

Finally we shall present some of the needs for further research to develope computer aided dispatching systems in an industrial environment.

References:

1. A. Bachem, Hochstättler, and M. Malich, "The simulated trading heuristic for solving vehicle routing problems," Tech. Rep. 93-139, Mathematisches Institut, Universität zu Köln, Weyertal 86-90, 50931 Köln, Germany, 1993.
2. K. Nagel, "Life-times of simulated traffic jams," *to be published in: International Journal of Modern Physics C*, no. 141, 1993.
3. K. Nagel, "Fast low fidelity microsimulation of vehicle traffic on supercomputers," Tech. Rep. 93-140, Mathematisches Institut, Universität zu Köln, 1993.
4. G. Laporte, "The vehicle routing problem: An overview of exact and approximate algorithms," *European Journal of Operational Research*, vol. 59, pp. 345-358, 1992.
5. B. V. Cherkassky, A. V. Goldberg, and T. Radzik, "Shortest paths algorithms: Theory and experimental evaluation," tech. rep., Computer Science Department, Stanford University Stanford, CA 94305, USA, 1993.

New Directions in Modelling Production and Distribution Problems

Uday S. Karmarkar, William E.Simon Graduate School of Business Administration , University of Rochester, Rochester NY 14627, USA

Summary: Production and distribution problems have historically been a major category of research in Operations Research. Yet, applications of OR/MS methods in production and distribution practice remain limited and OR/MS groups are not thriving in industry. The reason may be in the way problems are approached in practice, which then has consequences for modelling and solution methods. Perhaps a slightly different perspective for OR/MS could lead to significantly greater successes in applications to production and distribution problems.

Introduction

Work on production and distribution problems remains the largest category of publication and research in Operations Research and Management Science (OR/MS). A very substantial proportion of all the papers submitted and published in OR/MS related journals deal with these topics. The application of OR/MS like methods to production problems could perhaps be said to start with the Economic Order Quantity model of Harris, Wilson and Camp more than seventy years ago, when the term OR/MS did not yet exist. The subsequent years of research on production and distribution problems have produced some successful applications, and a great number of research papers. Unfortunately, too many of the latter cannot be described as having had any impact on the practice of manufacturing. Apart from the low effectiveness of formal work, it is also apparent that practitioners in manufacturing and distribution environments do not necessarily turn first to OR/MS analysts to find answers to their problems. The most widely applied formal techniques in production control are Material Requirement Planning (MRP) and simple release rules, that do not make very much use of OR/MS techniques, modelling or optimization. Distribution management has been a somewhat better arena for OR/MS models. Many commercial distribution management software packages and practical applications do make use of modelling techniques. However, even here, the proportion of research output that finds actual application is small.

I do not mean to suggest that OR/MS has little to offer to practice. On the contrary, the opportunities for OR/MS applications are greater than ever due to the twin trends of the increasing costs of management and the ubiquitousness of cheap computing power. Nor am I arguing for a cessation of formal theoretical research. Clearly, such research is a part of the "supply chain" in that it provides the groundwork for applied techniques. However, I believe that the potential for application of OR/MS methods is not being fulfilled. The reality is that in industry as well as in academia, the role for OR/MS is declining. Many firms have closed down their OR/MS groups completely. There are fewer departments in academia, and courses on these topics are less popular, noticeably in schools of management.

What then of the opportunities for modelling and analysis due to cheap computing and expensive labor. The issue is that these opportunities will be garnered by other disciplines such as Artificial Intelligence (AI) and Knowledge-based Systems (KBS), or by spreadsheet and simulation models. While these methods ultimately are not really that different from OR/MS, it appears that many valuable opportunities are being lost to the field. Part of the reason for this is that OR/MS education and research has tended to emphasize formal academic research and rigorous publication standards at the expense of good applications. Many well-known authorities have already inveighed against excessive formalism and theoretical research in OR/MS and

I will not add more noise to that clamor. Rather, I propose to provide some observations regarding the obstacles encountered in applied work, and some practical suggestions on making progress in that area. Let me also hasten to admit that my qualifications to pronounce on these issues come as much from failures as from successes.

In the following, I first identify some broad inconsistencies between standard OR/MS paradigms and the needs of practice. I then discuss two cases from my own experience that illustrate some of issues I have raised. Next, I address in turn the issues of modelling approaches, solution strategies and solution methods. Finally, I list what I perceive as some of the outstanding opportunities for OR/MS and modelling methods in production and distribution management. These necessarily reflect my own perceptions, and there are undoubtedly many others that have not occurred to me.

The Limits of OR/MS Paradigms

Perhaps the most prevalent paradigm of OR/MS is that of the optimization model. In the paradigm, decision problems are phrased in terms of the familiar form of the extremization of an objective function, subject to a set of constraints on the decision variables. Constraint sets are the main tool for describing the context of the decision problem. One of the major contributions of OR/MS has been the identification of certain standard optimization problems, together with the construction of algorithms that can solve these problems to optimality. Outstanding examples include LP and the Dantzig and Karmarkar algorithms; Integer Linear Programming and Gomory's cutting plane algorithm and branch-and-bound; and various network flow and graph theoretic problems. One consequence of these successes has been a strong bent in the profession towards thinking in terms of abstract optimization problems and provably optimal algorithms. Yet the needs of practice can be quite different. Even the basic notion of optimality can really only make sense in the context of a formal model. Given that any such formal model is typically a limited and approximate representation of reality, to talk of optimal solutions to a real problem is often not meaningful.

The optimization formalism tends to foster a world view in which abstract models become a framework for representing real problems. The next step is to develop an algorithm which has desirable properties of convergence and optimality. The task of application and problem solution is then seen as a rather mechanical exercise of acquiring data and processing it through the available machinery. Implementation of the solution is even less interesting from this perspective. Practitioners however, do not view the world through the optimization framework. What they tend to see are streams of events and observations that are recorded as data. Practitioners also act within the context of management processes that are triggered by these events and observations. The resulting data stream is the primary view of the world for practitioners. Schematically, the OR/MS paradigm might be represented as:

World ----> Model ---> Algorithm ---> (+data) ---> Solution --> (implement?)

On the other hand the natural perspective for the practitioner might be more like:

World ----> Data ---> Analysis ----> Process ----> (model?) ---> Resolution

Perhaps the differences have been emphasized here more than the similarities, but the role of data and implementation in the former paradigm is underplayed as much as the role of models and algorithms is in the latter.

It might help to see the differences more clearly if these schemes are examined as models for a first course in OR/MS. The typical OR/MS course in a management school tends to follow the first scheme. The standard formulations are quickly introduced as templates for real problems. This is followed by algorithms with toy examples to illustrate the models and algorithms. Real world data is rarely seen. But imagine a course designed around the second scheme. It would present the students at the start of the course with data sets (numerical and otherwise) drawn from real situations. The students would first be taught to organize and analyze the data. Then they would be taught how to extract descriptive models of relationships, dynamics and feasibility. The course might describe how data and information drive management processes. Now models could be introduced as powerful tools to structure the decision points in the management process. The approximate nature of models would be very clear at this point, and concerns regarding algorithmic convergence and optimality would perhaps be muted.

The Primacy of Data and Process

I will hereafter use the terms "information" and "data" interchangeably simply to avoid a discussion of the differences. Information availability is a key limitation in developing solutions. It is true that specifying information requirements is part of the modelling process. However, it is a useful discipline to start the modelling process by first modelling existing information flows and processes. If these are so ineffective as to require re-engineering of business processes, then modelling is premature. Reengineering and simplification of business processes is generally a pre-requisite for modelling and building decision systems.

It should be made clear that when I refer to modelling information and data flows, I do not mean the development of data structures for algorithms, or even the organization of data base management systems (DBMS). While the latter may be more relevant, the real issue is to understand the interplay between information flows and management processes. Some critical information may not be (as yet) captured in formal DBM systems, and simply identifying data and information flows will not convey a sense of the management process by which actions and decisions are actually executed. Thus, a shop data collection system, and the associated data bases cannot convey the process of order management on the shop floor. The structure of an MRP system does not determine the management process by which the system is actually used to develop a consensus plan between marketing and manufacturing functions. The latter agreement requires a sequence of meetings and "sign-offs" which is usually very clearly defined in terms of process steps as well as responsibilities. A failure to understand these management processes will doom any modelling effort aimed at application of the results.

The practitioner's emphasis on data, information and process leads to significant differences with the world-view of the OR/MS specialist. My perception of some of the differences is described in Table I.

The view in practice reflects the realities of management and decision processes, where communication, common understanding, agreement or consensus, responsibility and clear action plans are crucial issues. The OR/MS world on the other hand, tends to be driven by structure and rigor.

The treatment of time is a good example of some of the distinctions in Table I. OR/MS often treat capacity in terms of loading time buckets. However, in practice, it is much easier to think in terms of time

Table I: A comparison of different modelling perspectives in OR/MS and practice

OR/MS	Practice
Optimization	Feasible improvement
Single best solution	Many good solutions
Single agent/objective	Multi agent/objective
Convergence = optimality	Convergence = consensus
Decision making	Management process
Standard engines (e.g., LP)	Robust computation
Structure oriented	Data oriented
Homogeneous structure	Mix of different structures
Abstract variables	Actual events, objects
Decision oriented	Action/event oriented
Formal model	Virtual representation

lines and events and intervals, much as in a calendar, diary or simulation model. For example, if we were decomposing a process across several stages in practice, we would identify the decomposition in terms of points in time at which certain events took place. Thus a trip might be thought of as arrivals and departures at locations, with the intervals to be scheduled. Similarly, delivery schedules and lead time intervals are characteristic of production and distribution co-ordination. In some models, such as scheduling with Gantt charts, we use this kind of modelling, but we lack ways of dealing with decomposition of composition of these models. As a result, capacity and planning models are often formulated very differently. Perhaps this is one reason that time interval and release oriented methods like MRP and DRP are used in practice for planning even though they are completely unable to actually deal with resource allocation decisions.

To reiterate, focussing on information flows and management processes, can create an entirely different view of modelling and solution methods. The emphasis shifts from rigor and homogeneous structure, to accurate representation and robust feasible solutions which are acceptable to all parties. The case examples of the following sections may serve to illustrate.

A Production Scheduling Example

The first of my examples is drawn from the development of a large scale scheduling system (Dobson et al., 1992). This system represented a deliberate attempt to span the range of OR/MS techniques from theoretical development of models and algorithms to a practical implementation of models and heuristics that could be used to solve large realistic problems drawn from actual factory data sets. In this effort we discovered that while we indeed did many things right, we did not go far enough, at least in our initial design.

Among the things we did right were:

The development of new theory to address real issues. The technical scheduling literature, while vast, offers little useful guidance on issues such as scheduling with multiple objectives. We were able to make new contributions in the area of understanding the novel issues of combining batching and sequencing, in scheduling parallel heterogeneous machines, and in scheduling resources that are required simultaneously (Karmarkar, 1987; Dobson et al., 1987, 1989). Work continues on other such models such as scheduling, batching and loading with batch processors (such as furnaces or kilns), and scheduling with finite buffer capacity.

- The development of "plug compatible" models. One of the problems with much of scheduling research is that very few results are amenable to combination with others. We emphasized the development of heuristics that worked in a common compatible framework, even at the expense of giving up on rigor and optimality. Our approach is based on concepts from Lagrangean decomposition (Dobson and Karmarkar, 1986).
- The scheduling system was designed to interface with MRP systems and with shop floor data collection systems. The data models used conformed to standard practice, but were extensible to enable richer modelling of specific processes.
- The system can work with incomplete data on costs and objectives, and objectives can be flexibly specified to include costs, and priorities.
- The system has some high level "knobs" or parameters which can be used to change the behavior of heuristics at an aggregate level, and which can be used to generate alternative solutions and scenarios. Internally, these settings are used to change heuristic parameters and relative weights on conflicting objectives such as capacity utilization and flow times or work-in-process.
- Optimality is sacrificed for speed. The system is extremely fast but is nevertheless able to significantly outperform other practical systems in tests with actual factory data (Dobson et al. 1992).
- As a "proof-of-concept", the system has been interfaced to a real-time shop floor system that also connects the system to certain MRP systems.

Despite these efforts to be realistic we have discovered that we need to go much further:

- We assumed that a standard DBMS would suffice for data management. It turns out that this approach is too slow, and we had to develop special procedures to work in a real time environment.
- Certain data elements that we assumed would be available in most plants, were not there in a plant that justifiably claimed to be the most data rich in the US.
- There were some standard fabrication processes that we could not model compatibly within our range of scheduling models.
- Certain industry practices in resource loading and scheduling could not be modelled even within our heuristic decision framework, and we doubt our ability to do this in the future.
- Practitioners felt much more comfortable with a simulation model, and in fact could not really distinguish between a feasible simulation run and an optimization approach. Fortunately, our scheduling model could essentially simulate as well.
- The scheduler solved difficult problems that practitioners did not think were feasible. However, on simple problems the solutions often did not look simple, and were hard to explain. Trying for every possible improvement sometimes led to complicated schedules that the practitioners did not understand.
- Practitioners were often more concerned with the user interface and reporting tools than the quality of the solutions produced.
- The system did not have the ability to refine and improve a user defined schedule.

The next version of this scheduling system will be an object oriented, simulation like model that will be able to utilize multiple solution techniques from OR/MS and AI/KBS. The system will accommodate "best practice" constraints and user defined rules. Any schedule, whether user or system generated, will be treated as a proposal that could be refined or modified through improvement heuristics and constraint satisfaction techniques. Graphical interfaces will be integral to the system. Nevertheless, I am sure that we will discover a number of new limitations to our approach.

A Distribution System Example

This "example" is really a composite of several different experiences with seemingly different distribution environments. They included a national distributor of industrial equipment, a manufacturer and national distributor of household cleaners, and a manufacturer and global distributor of medical products. In each case, the issue was the redesign of distribution systems including location and role of facilities, and transportation policies.

We discovered rapidly that we were unable to use traditional modelling methods for several reasons:

- These firms were facing competitive markets and were not interested in simply cost minimization.
- The distribution system design thus required significant mediation between marketing, distribution and manufacturing functions addressing basic company strategy issues.
- Changing the system entailed different contractual and business relationships with other entities in the distribution channels including independent distributors, freight forwarders, and agents.
- Strategic positioning was often defined in terms of a service-time profile that was affected not only by location, but also by inventory deployment and transportation policy. All these therefore had to be addressed.
- The roles of locations changed and had to be clearly defined. Some of these roles were: consolidation, breakbulk, seasonal storage, JIT warehouse, backup shipment warehouse, etc.

The net result was that we used simple heuristics, spreadsheet analyses and simulation to construct and evaluate a small set of design options within the limited time available to us. This was a stop-gap solution which nevertheless did the job, especially in terms of strategic positioning and revenue vs. cost tradeoffs. Our longer term aim is to develop models and solution methods that can address these complex problems in a way that communicates with practitioners.

Our approach at present is to use traditional methods to formulate the problem, and to analyze very small instances (Saxena, 1993). The formulation includes costs as well as service time requirements. The decision variables include locations, shipment paths, ship modes and frequencies, and inventory deployment including seasonal movements and safety stocks. To solve realistic versions of the problem, we are developing a combination of genetic algorithms and heuristics (Karmarkar and Saxena, 1992).

So far we have found that on very small problems, with 3 items, 3-6 retailers, and 2-6 potential sites, the genetic algorithms (GA's) produce optimal solutions in over 85% of the examples tried, with small gaps in other cases. In somewhat larger problems, with 4 supplying plants, 2-10 items, 8-12 retailers, and 12 potential warehouse and DC sites, the GA solutions average within 8-10% of the best lower bound. These are non-linear problems with up to 2500 constraints, 280 0-1 variables and 2000 continuous variables. We cannot presently compute optimal solutions.

We feel confident that GA's are a reasonable tool for large realistic problems, but we will have not ability (bounds or otherwise) to benchmark the quality of the solutions. We also do not have any other alternatives. We still have some way to go on modelling competitive situations, and on incorporating issues related to business relationships, contracting and facility roles.

Modelling and Representation

Examples like the above – and I am sure the reader has encountered similar situations -- lead me to reconsider our modelling approaches. There has been a tendency in OR/MS to identify homogeneous standardized modelling structure, and then to force problems into them. The problem is that real applications do not categorize that easily. The structured approach tends to be limiting in terms of the objectives as much as the constraints. There is an observable tendency to cast problems in terms of cost minimization. There are certainly many problems that can be cast purely in terms of cost. However, there are many that cannot. The current interest in industry for programs such as JIT and TQM might be interpreted as a recognition that the traditional cost and revenue decomposition of objectives is proving to be inadequate for today's competitive markets. Many companies in the US have made real shifts in their management emphasis from traditional purely financial measures to issues of lead time, quality and customer service. It does not appear that we have kept up in terms of model development. It is worth noting that after many years of research, few of our models address the time and quality impacts of decisions. It is only in the last five or ten years that we see a few models of production and inventory management decisions capable of analyzing time related issues, and I have reviewed some of this work in Karmarkar (1992). The situation with respect to other objectives such as quality and flexibility is even poorer, though I cannot claim to have authoritative knowledge of its current state.

An analogous situation exists with respect to the modelling of constraints and process dynamics. Since I do not have any ready made answers, let me at least pose some questions:

- *Do we have any way of modelling existing data and information flows?* There are information systems methodologies such as Data Flow Diagrams (DFD) to record and represent such flows but they do not have a symbolic representation, and in fact have no special methods for capturing decision points. Control theory methods have tended to stay with real time execution and control, and have not reached application to resource allocation problems. Petri nets may still have some promise in this regard, but do not seem to have accumulated analysis and solution techniques to accompany the representational method.
- *Can we model existing procedures and business processes?* MRP is one obvious example that I have mentioned earlier. There are indeed some models of MRP that can be included in decision formulations. There are increasingly effective models of production release procedures such as Kanban. However, we do not have an understanding of many key management processes even at the level of a taxonomy. For example, how are marketing and manufacturing needs reconciled in issues such as making forecasts, setting production and sales targets, or setting quality levels? How are process and communication issues to be included in multi-stage multi-facility manufacturing and distribution management?
- *Can we develop models of multi-agent decisions?* Game theory gives us some tools to represent both co-operative multi-agent decisions as well as non-cooperative or competitive situations. However, the methods though rigorous, are limited to simple examples. We need to be able to tackle issues such as contracting in manufacturing networks, in sourcing and supply chains and in distribution channels. Such analyses today are too often completely disjoint from the underlying production and inventory models,

which in turn usually do not consider these aspects.

- *How do we handle issues of competition?* Consider that the location literature is dominated by cost minimization formulations. Yet in many service industry examples, location is a competitive exercise, since customers choose the sites they will visit for service. Similarly, product design, service response, quality levels are all subject to competitive action. We often formulate models that assume that recourse can be made to subcontracting for say, transportation vehicles. However, it is often the case that such demands are seasonal and result in competition among demanders for scarce resources. Similarly, acquiring capacity or technology, as with a large flexible manufacturing system, may have to be analyzed in the light of competitors' responses. Again in addition to game theory, there are economic models that address these issues, but largely at a conceptual level.
- *Can we build modular, virtual, plug compatible models?* In a sense, an LP provides a framework for modular modelling, since many different aspects of a firm can be represented in an LP model. However, the linear structure that provides the standardization is too limiting for many problems. Perhaps formal methods of decomposition can suggest such modelling approaches. It becomes apparent that part of the problem is in the limitations of solution methods that require a very homogeneous structure to work. As examples of modular schemes let me point to queuing networks of both the open and closed variety, and Petri nets. We do not have an analogous methodology in say, scheduling or planning models. The ideal would be close to simulation in its modularity: it is relatively easy to add more to a simulation or to change the model without affecting other aspects; furthermore, the analysis process is unaffected by such changes. The process of representation is transparent to the practitioner and as a result the model has high credibility. Of course, simulation is a lot easier than decision making, but perhaps there is some common ground. Again, the availability of solution techniques is likely to be the limiting factors
- *Can we accommodate "best practice" requirements, and heuristics?* This is relatively easy to do in terms of constraints, except that the structure of the constraints may not be very convenient. So-called expert systems are a natural way of modelling such situations, where a method, guideline or practice has been found to work by trial and error over years or even decades and centuries. It is typically difficult or impossible to extract and state the underlying optimization problem. Such examples are common in process industries.
- *Can we model learning, and process improvement?* In the past, a typical shop floor control system would have focussed largely on material tracking and job scheduling. Today, quality control and process improvement are likely to be as if not more important. We have many models that address the evolution of systems in terms of production scheduling events and material movement. We are much poorer in terms of models of process improvement and learning. Our models tend to treat these phenomena as static, with given probability distributions. In fact, knowledge about processes changes over time, and the rate depends on how processes are managed. That is to say, process scheduling and process improvement interact with each other. Just for example: shorter production lead times aid learning and process improvement. On the other hand, complex scheduling of processes with many changes, may obscure cause and effect relationships, and hinder learning. In such areas we are weak even in terms of simulation methods.

I do not think that the questions above are that difficult to address. However, for whatever reason, OR/MS has not actively pursued such modelling approaches. I would argue that as one becomes interested in applications, and starts with the observation of real world problems, processes and data, the need for such methods will be apparent very quickly. On the other hand, if the direction in OR/MS is towards abstract research, such developments will not seem attractive because of the difficulty of solving problems optimally or establishing analytical results definitively. In other words, the difficulties of rigorous analysis and solution may have been inhibiting factors in the past.

Solution Strategies and Methods

By now my drift with respect to solution techniques should be apparent. Some of the great successes of OR/MS have come in the area of optimal solution techniques for significant problem classes. Progress in many areas continues to occur, and indeed there are many problems which can be modelled and solved by these methods. However, we have also come to recognize that there are also large classes of problem where optimal solutions are very difficult to obtain, even assuming large improvements in computing capability. Many of the best opportunities open for OR/MS fall in this latter category; most of the standard homogeneous problems have probably been identified already. It is likely that we will hereafter increasingly be in an environment not unlike that of the PC: no sooner is an application available than there is a demand for an extension or an enhancement. The most successful applications will be robust, easy to use, easy to modify and easy to build upon.

If our models must be flexible, extensible, and modular, we need appropriate solution strategies and solution methods. If we are willing to give up on rigor and optimality to some extent, many useful techniques already exist. Let me reiterate here that rigor remains important in basic research. However, in real problems where models are often poor approximations to reality, rigor and optimality can be misplaced objectives.

There have been many solution strategies and techniques that have appeared recently from the domain of artificial intelligence and knowledge based systems. Many of these techniques are not dissimilar to traditional OR methods. However, they tend to be less concerned with rigor, often because they were created to deal with extremely difficult problems such as chess playing, theorem proving and pattern recognition. They include search heuristics such as simulated annealing, tabu search and beam search, which are very closely allied to traditional gradient search methods and tree search methods. Other techniques like genetic algorithms, neural nets, blackboard methods, constraint satisfaction, learning systems, and production systems or expert systems may seem further afield. In these cases too, I would argue that the differences and novelties are much less than the similarities and in fact there are significant gains to be had from combining techniques. This is not the appropriate place to delve into these opportunities however, and I will limit myself to addressing what we might expect in the specific area of production and distribution modelling.

Many of these techniques are already being widely used, as a glance at the research papers in this conference shows. However, there is still a tendency to think of a model as having one type of structure with one solution technique. Alternative approaches are seen as competing rather than complementary. I would assert that this is not the right strategy to accompany a robust modular modelling approach. Rather I would suggest the following:

- *There may be many alternative acceptable solutions, rather than one optimum solution.* And it may require different methods to generate them. Some methods such as genetic algorithms naturally generate and maintain multiple solutions.

- *Any "solution" should be treated as a proposal to be changed or improved.* Even realistic models fall short of reality. Modifications to solutions will be necessary as new constraints and objectives are identified. It should be an intrinsic part of a solution strategy that solutions can constructed and/or improved. The solutions may come from algorithms or from a user.
- *An implication is that constructive methods must be supported by improvement methods and heuristics.* Traditional methods are especially weak on improvement. Even current improvement methods such as exchange heuristics can be too local and limited. For example, it may be required that a solution to a scheduling problem be modified to be more intuitively acceptable, or to cleave to some practical guideline.
- *Expert systems may provide methods for accommodating best practice modelling.* Expert system may also provide methods for improving starting solutions obtained by more structured methods, to then meet complex constraints on the form of the solution.
- *Alternative solution techniques can gain by trading solutions.* For example, simulated annealing can be used to enhance solutions generated by genetic algorithms. The results from GA's can be used as starting points for other methods.
- *There must be a mechanism for transferring information between different methods.* In my opinion, the GA framework can serve as a central "blackboard" or clearing house for alternative solutions generated by different methods.
- *Parallel computation will be an essential part of solution strategy.* Future computing advances are likely to rest on parallel computing. We need to develop intrinsically parallel methods in addition to those already available such as production systems and genetic algorithms.
- *General algorithmic engines could be equipped to learn about specific problem domains.* Techniques such as neural nets and genetic algorithms have been used to improve solution heuristics by learning about specific parameter domains that occur in particular applications.
- *We need a common language for models that can be understood by alternative solution techniques.* Current efforts in developing modelling languages are still aimed at a math programming framework. It should actually be simpler to develop a language that will be understandable by robust techniques.

Opportunities for the Future

I reiterate that the trends of increasing labor costs and cheap computing present the OR/MS professional with an unparalleled opportunity. Companies are trying as never before to improve and automate their management processes and to increase the productivity of personnel in all areas. However, the opportunity could be lost, if OR/MS remains limited to traditional decision modelling. This was perhaps appropriate 20 years ago, but is not in today's environment. It is necessary to recognize that to have an impact on applications, we need a paradigm that is founded in modelling at the level of data acquisition and management process. Furthermore, we also need modelling methods that are as robust and modular as simulations, spreadsheets or other modern PC applications. Changes in both modelling and solution strategies appear to be necessary.

Let me briefly list some of the application opportunities and trends that I perceive; I am sure the reader will be able to list many more:

- Applications are moving from generic models to specific areas. So-called *vertical applications* are the prevailing trend. That is to say, production control systems or vehicular routing models are focussing on particular industries rather than on generic methods. A scheduling system must address say metal cutting or chemical batch processes, instead of claiming to be a general purpose tool. It is already very clear that the models that can realistically solve one kind of problem can be worthless on another, because of the specialized heuristics, model structures and management processes that are used. Interestingly, commercial software packages also suffer from the problem of being too generic. MRP and scheduling systems are a good example where many of these systems are still sold as though they can work for any plant. A few software companies are moving rapidly to exploit this opportunity by creating differentiated products.

- From the research viewpoint, there are some very interesting problems in modelling specific manufacturing and distribution environments. Some examples are:

  Hot metal processing (e.g. steel casting and hot rolling)

  Batch chemical processing

  Semi-continuous chemical processes.

In each of these areas, there are differences across plants and products.

- The entire subject of managing operations for *profitability* in a *competitive* environment has been little addressed in either the OR/MS or any other literature. Just for example, capacity decisions or location decisions are significantly different in a competitive framework than in a cost minimization, single firm demand independent framework. In our traditional models, the shadow price of capacity only depends on internal factors. In the real world, if a competitor builds capacity, the value of every other company's capacity drops. I have already mentioned other examples such as location models and the need to look at objectives such as service, time response, innovation etc.
- There is a need to develop solution engines that can utilize a variety of algorithms from OR/MS and AI/KBS. The first issue to be addressed is probably the development of a suitable architecture in which multiple techniques can be accommodated in parallel.
- There is a great deal to be gained from combining ideas from different areas. For example, current GA techniques are weak on maintaining feasibility with complex constraints. Developing GA's that can exploit problem structure would be extremely useful in improving performance, while maintaining robustness.

It may seem that it is easier to talk about the issues that I have raised in this article, than to actually do something. However, I would argue that many research groups are pursuing every topic that I have mentioned. To emphasize my own interests in these topics let me end with a brief list what I am doing along these lines in addition to the two examples given earlier. I am working (with R. Jaikumar, Pierre Lefrancois, Benoit Montreuil) on the development of a shop floor system that will integrate data collection, simulation and decision support, for both material management and process learning or improvement. The aim is to design a platform that will support a variety of decision support tools, which might be developed by others in the future. In the area of scheduling, I would like to explore the differences between batch chemical processes and metal fabrication, and to develop modular modelling techniques that work like simulation modelling. I have already mentioned some ongoing work in distribution systems. I invite the reader to correspond with me either to share ideas or experience in any and all of these areas.

References

Dobson, G. and U. S. Karmarkar (1986). "Large Scale Shop Scheduling: Formulation and Decomposition". W. E. Simon Graduate School of Business Administration, Working Paper Series No. QM 86-31, University of Rochester. To appear in Optimization Models and Concepts in Production Management, Villa A. and P. Brandimarte (eds.).

Dobson, G., U. S. Karmarkar and J. L. Rummel (1987). "Batching to Minimize Flow Time on One Machine". *Management Science*, **33**, 784-799.

Dobson, G., U. S. Karmarkar and J. L. Rummel (1989). "Batching to Minimize Flow Times on Parallel Heterogeneous Machines". *Management Science*, **35**:5, 607-613.

Dobson, G., U. S. Karmarkar and J. L. Rummel (1992). "A Closed Loop Automatic Scheduling System (CLASS)". *Production Planning and Control*, **3**:2, 130-140.

Karmarkar, U. S. (1987). "Lot-Sizing and Sequencing Delays". *Management Science*, **33**:3, 419-423.

Karmarkar, U. S. and R. Saxena (1993). "Distribution Systems Design Using Genetic Algorithms". William E. Simon Graduate School of Business Administration, Working Paper CMOM 91-03, University of Rochester.

Saxena. R. (1993). "Multi-Item, Multi-Echelon Distribution Systems Design". Ph.D. Thesis, William E. Simon Graduate School of Business Administration, University of Rochester.

THE PRACTICE OF OR :
Between Art and Science

Gerrit T. Timmer
Vrije Universiteit Amsterdam
ORTEC Consultants, Gouda

Summary

Many application area's exist where an OR approach can be effective. This is especially true for many strategic and operational problems that arise in the field of distribution and logistics, due to major political and economical changes in Europe.
However, in order to play its role, it is necessary but not sufficient for the OR society to have access to the science aspect (knowledge of sound mathematical models and techniques) of OR. Equally important is the 'art' aspect of OR. That is the art to decompose a problem in sub-problems, the choice of the level of detail and the role and the type of the techniques (including OR techniques) to be used.

1. Introduction

In this paper, I outline a view on the way that Operations Research (OR) should be used, and the consequences of this view for OR practitioners and theorists.
This view is based on my experience both with the theory and the practice of OR. Practical experience comes from my work during the past 12½ years with ORTEC Consultants, a Dutch firm that uses OR techniques to solve practical problems for government and business organizations. Apart from ORTEC, I have always been active in the academic world, currently at the Free University of Amsterdam.

A central point will be, that I view OR both as a science and as an art. The science aspect focuses on mathematical models and optimization techniques. However, applying OR is also very much an art. It is an art, and demands good taste, to judge whether or not OR techniques are useful in specific cases, how the complex reality should be modelled, and how OR techniques should be mingled with possible other solutions and users knowledge and intuition. This art aspect of OR is crucial in the useful employment of the models and techniques offered by the science of OR.

To begin with, a short sketch of the history of OR is given. Although still very much alive, the status of OR is currently not what it could and should be. This has led to various discussions on how to proceed. Subsequently, I present my view on the role of OR and highlight on the implications of this view for both the universities and the OR consultants. Examples from the area of distribution and logistics will be used as an illustration. In the final section, I will give some more emphasis to this application area , which is a field of great opportunities for the OR society.

2. History

Building and manipulating mathematical models started centuries ago. Sets of linear inequalities were already solved by Fourier in 1826. However, the formulation of optimization methods only started in the 50's of this century. To quote Dantzig (1991) "Initially there was no objective function; broad goals were never stated explicitly in those days, because practical planners simply had no way to implement such a concept. Non computability was the chief reason, I believe for the total lack of interest in optimization prior to 1947".

In the first decennia after the second world war, a variety of models and techniques were developed. Applications were on an ad hoc basis. Applied mathematicians were confronted with practical problems, looked to its basic structure, and proposed solutions or solution techniques not thought possible before. It became apparent how amazingly powerful mathematics can be (c.f. Rinnooy Kan 1992).

OR was on his way to a bright future. This tendency was further accelerated by the extraordinary development of computer hardware and software. Due to the computerization of administrative processes, data to be used for optimization became available. Larger models could be solved due to increased computer power. The number of applications grew rapidly in the 60's and 70's. OR departments were established both in industry and universities. In the academic world integer programming, non-linear programming, stochastic programming techniques were developed, and larger and larger models could be solved. Standard packages for linear and integer programming became available.

Although one can speak of a triumph of mathematics, the current image of Operations Research is not overwhelmingly positive. This is indicated by a quote from an article about the mathematizing of economics (see Rinnooy Kan 1992) : "It is possible that economics today is following the same path as Operations Research (OR) did in the 1960's and 70's, leading to its heavily reduced status in industry. OR started life as a discipline for creative minds seeking original solutions to mathematical optimization problems. But within twenty to thirty years, the search for mathematical rigour and the solution to mathematical optimization problems was allowed to become an end in itself. Today, there is little demand for its services to help solve big problems of government and industry. It persists there among isolated specialists not involved with strategic decisions of the firm but, more typically, reporting to a 'production manager' whose problems are conceptually simple and for which a mathematically obtained optimum can reduce cost".

Apart from the fact that there is nothing wrong in solving operational problems for production managers, I think this quote exaggerates. However, we cannot deny it to be at least partly true. I guess that there are several reasons for the reduced status of OR. One reason has to do with the quality of most OR professionals. OR consultancy does not provide career possibilities within most companies. Persons who just come from the university use OR techniques for a few years, after which they are supposed to move to different kind of jobs, never to return. As a result the OR consultants generally have little experience in applying OR and hardly have any experience with the type of problems they have to tackle. There is a clear need for companies to increase the career possibilities of OR

consultants within the OR profession.
Another reason for the reduced status of OR is that the rapid improvements in computer hardware, apart from the obvious positive effects, also worked as a boomerang. Because computers became so fast, people were inclined not to think enough. Instead problems were formulated in the first model the modeller could think of and dumped into a computer. In doing so the role of a computer is severely overestimated. One creative thought, one spark of mathematical inspiration can make years of computer calculations redundant. Because of the increased computer possibilities people are also inclined to formulate models with a span of control much larger then the span of control of individual decision makers. Larger and larger models were used and these models and corresponding solution techniques, in contrast to the problems behind the models, became an end in itself.
Because of the tremendous number of details involved in huge computer implementations and because of the inadequate background and attitude of the consultants, strategic decision makers lost interest, and applications became more of an operational nature.

The negative trend led to confusion and different movements on how to proceed. Some people started talking about the gap between theory and practice. More and more sophisticated models and techniques are developed where nobody in practice is waiting for. Sound, hard mathematical techniques are said not to be useful. Instead more soft techniques are called for. That is a more business administrative kind of approach where relatively simple quantitative methods suffice, and where the focus is on social and consultancy skills.

3. Role of OR

I do not believe in the gap between theory and practice, nor in the usefulness of the distinction between hard and soft OR. The problem is not due to the theory. The problem is that the art of applying OR is not mastered well enough.

I believe sound mathematical techniques to be useful. The science aspect of OR should not be neglected. I agree with Boender (1993) in his statement that 'successful applied OR is not possible without a thorough and broad theoretical background'.
However, this is not the end of the story. How useful OR techniques may be, they should never become an end in themselves. Far too often the grasp and the interest of OR practitioners is too much confined to models, algorithms and software, whereas the focus should be on the client and his problem. The introduction of OR systems almost always touches other parts and persons of the clients' organization, and meets with several difficulties for which the OR consultant should have a skilful eye. A vehicle routing system requires an automated processing of order data, and threatens the responsibility, freedom of movement or even the employment of the planners and drivers.
But it is not only that the OR specialist should foresee the organizational or personnel effects of his work. From the start, he should be problem oriented, instead of discipline oriented.
That this is not always the case, can be seen from the fact that junior consultants often think in terms of simple and complex problems. Where the problem he has to solve is generally thought to be very

complex. Notions of simple and hard problems are relevant in complexity theory (Garey and Johnson 1979) and useful when thinking about algorithms to solve problems optimally. In a problem and user oriented context, however, the notion of simple and complex problems is not very relevant. The role of the OR specialist is to help solve a problem. That is to improve the situation in comparison to how it was before he arrived. This may be much harder for so-called simple problems, (which are probably treated well already), than it is for complex problems. Here, solutions which are far from optimal may already lead to huge improvements.

To begin with, one should determine what the problem is, and what solving the problem means. This is not as trivial as it may seem. Far too often the focus is on 'solving a puzzle'. Especially in strategic applications, the aim should generally be on enlarging the insight and understanding of a complex process.

Another important aspect is to determine the position of the client. What is his span of control. That is, from what kind and level of decisions are the clients' responsibilities. There is no sense in developing models or systems where decision variables are not grouped to match the responsibilities of the various decision makers.

Generally, a very difficult problem to solve is the level of detail required in treating the problem. Here, special care is required. If vital details are omitted, the model may divert too much from reality. However, more often there is a tendency of taking into account too many details. This will distract the attention from crucial factors, and makes it more difficult to interpret results. Supposing that results are obtained, and the project has not been killed before because of the impossibility to collect the necessary data.

If after careful consideration the problem to be solved is detected, it should be determined how to solve the problem. It is seldomly true that the entire problem can or should be solved using OR techniques only.

E.g. a nice example has been described by Ackoff (1956), about the problem of controlling elevators. If several elevators are next to each other in a building it is a nice OR problem to decide how the elevators should move in order to minimize the waiting times of the various persons to be transported up and down. However it turns out that, independent of the efficiency of the elevators, people always have the impression that they are waiting far too long. Far better results in terms of peoples satisfaction can be obtained by totally different approaches. E.g. if mirrors are placed, people do not mind waiting. Women look to see how they look, and men can look discretely to the women besides them.

This may be an extreme example, but in generally it is necessary to decompose the problem in sub problems and select the most adequate approach for every sub problem. In operational systems it is necessary to decide what aspects should be solved using OR techniques, on what aspects users intuition and knowledge should be used, and where to use other technologies. It is seldomly obvious how the user interaction should be treated. Too often OR specialists just try to solve a problem entirely using OR techniques, and reserve the user interaction only where they fail in this objective. They then parameterize their solution technique, and let the user specify some unknown parameter

values. E.g. if the optimal step length of a search strategy appears to be strongly problem dependent, just make it an input parameter of your program and let the user decide. In general, users have no intuition for these type of parameters, since they do not think in terms of the OR solution techniques. Instead user interaction is needed to let the user take care of the things he can handle adequately, like interpreting the importance and firmness of constraints and goals.
To optimize the usefulness of the user interaction, information technology should be used to present the relevant information in the best possible form. If it is possible to present all relevant information on a single screen, the user may be able to solve the problem himself, and OR techniques may become redundant. To come up with such a solution is not to be ashamed of. The client will be satisfied. However, in general many sub problems remain which can best be solved using more or less complex OR techniques. Combining the strength of human decision maker, mathematics and information technology then leads to the best possible result. Art and science of OR have played their role.

4. An application

As an illustration, I will shortly describe a project ORTEC carried out for the Suiker Unie (van Vliet et al. 1992). Suiker Unie is a farming co-operative that processes about 60 percent of the Dutch sugar beet crop. The bulk delivery problem differs in several aspects from the standard multi-depot vehicle-routing problem; trucks deliver full loads to each customer and can reload at any of five factories. Loading the trucks is very time consuming, and at each factory trucks can only be loaded one at a time.
Due to the combined routing (with time windows etc.) and scheduling characteristics, the problem is very hard from a mathematical point of view. The problem was divided in sub problems for several reasons. One reason being, that it would be very hard to solve all problems simultaneously anyhow. But an even stronger reason is that seldomly an overall feasible solution exists. The user is then needed to decide what to do, given his knowledge of the different clients. Modules are organized such that the user can re-optimize by activating the modules of his choice.

In solving the sub problems sophisticated techniques are used. A general assignment problem is solved using Lagrangean relaxation and the sub gradient method, assignment problems, improvement heuristics, etc.

However, apart from the OR techniques which proved to give good results, a main characteristic of the system we developed is an interaction window consisting of four communicating sub windows. Here, the user can simultaneously view the present orders, the planned routes of the trucks, the schedule for the loading facilities and the corresponding overall cost and other statistics. Always one of the windows (chosen by the user) is active. On the other windows the orders, trucks, or loading facilities are displayed relevant for the situation the user is viewing on the active window. This enables the user to make sensible changes in the active window, and after his change he can view the consequences on the other processes on the other windows.

This user interface is useful both as a tool for the user to adapt the schedules whenever needed, as for his willingness to accept the system. As a result the system is now being used on a daily basis to interactively determine and adjust the daily schedules. In addition to the improvement of the efficiency and quality of the planning process, the use of the system turned out to yield about seven percent savings in operating cost.

5. Academic world

Because sound mathematical techniques are also relevant in practice, I do not see a strong reason to dramatically change the research policy of the universities. Research should continue on a wide range of models. Models and techniques may be studied in great depth.
Research will probably produce more results if the activities are better bundled. Initiatives to do so have started in Holland. If several people work together on similar topics, more useful and better results may be expected.

To educate students, that is future consultants, more changes may be required.
Currently on most Dutch universities it is common that a few topics, that is the research topic of the teacher, are treated in relatively great depth, while others are hardly treated. This is not the way to educate consultants. They should be given an overview of a wide range of models. Solution techniques do not have to be examined in extreme detail. I do not believe that certain techniques can be used extremely often and others only seldomly. Inquiries stating that certain techniques (like DP) are hardly used, are not very informative. It does not mean that DP could never have been used successfully, it merely indicates that students were not trained in a way that they are able to do so. I myself for instance have used DP several times, with very good results.

For most techniques it is not enough to just teach them, and maybe force the students to do some exercises with them. Many students who have followed a course on LP cannot even formulate the simplest problem as a LP model. Therefore, a great deal of the curriculum must be devoted to case studies, where the students must model and solve more or less realistic problems. Only after having done that a few times there arises a change that they will also have the courage and ability to use mathematical models in their future lives. During these cases the art of applying OR can be shown to the students. Skills like giving presentations can also be trained. Finally, a sense of entrepeneurship, and the desire to jump on a problem and solve it can be stimulated.
The case studies can hardly be guided by someone who has no experience in practice. Therefore OR departments should at least recruit some personnel with consultancy experience.

6. OR Consultants

Below some of the remarks of section 3 are reformulated as guidelines to OR consultants.

- It is important to know the environment you are modelling. Do not just carry out what someone is telling you to do. Determine what the problem really is. What are the kind of solutions people will be happy with. It does not have to be enough to solve a problem in a mathematical sense. Fairly often the numerical outcome of the calculations is not the issue. Instead, a model is needed as a way of structuring thoughts.
- Determine the scope of the problem, do not introduce more decision variables than necessary. The model, including the decision variables, should be in harmony with the organization of the client.
- Decompose the problem. And be problem oriented, not discipline oriented.
  If sub problems remain which can or should be solved using mathematics, do not divert to simple heuristics too quickly. First try to make use of the mathematical structure.
  Decide carefully what can be gained by user interaction.
- During the project, always keep sight of the numbers. Fairly often people make ridiculous mistakes, which become evident much too late, because they did not use real numbers all the time to check whether their models were still sensible.
- In the relation to the client it is advantageous to produce some results quickly to whet the clients' appetite and to clarify the really important issues.
  Generally the gain of insight of the user when using the model is more important than the numbers that come out. Therefore the system as a whole should not be a black box. People should be able to understand the logic of the solutions.
- Finally, note that people are not enthusiastic about a project because it improves the world or their company. People should benefit themselves (e.g. with improved status due to the successful project), or at least should not be influenced negatively by the project. If someone apposes a project it is hardly impossible to make it a success. Any project needs at least one in favour, and preferably none against.

7. Distribution and logistics

Now let us finally turn to the area of distribution and logistics. Everything said before is also relevant for this area. Here the opportunities for OR are currently extremely well.

Due to drastic political and economical changes, eastern Europe, an economically unified Europe, many distribution networks and logistic processes are redesigned.
The political and economic changes come at a good time for OR. The past decennia many administrative processes have been automated, such that data are available for strategic studies, and companies are ready for the introduction of planning systems.

In the field of distribution and logistics both strategic and operational applications are possible in great numbers.

On strategic level ad hoc studies are possible because of the rearranging of distribution and logistic processes. For instance, we have been working for a large oil company who swapped their gasoline stations in country A with another oil companies' gasoline stations in country B. Such that both only operate in one country. Here the total distribution network had to be redesigned.

However it is also useful to use strategic models to improve the insight of managers in their complex world. This can be in the form of management games and simulation programs.
E.g. for the Concrete industry we made several models. There are about 200 companies producing concrete in Holland. They can be clustered in approximately 20 regions which are more or less independent. The distribution of the concrete to the clients is relatively expensive compared to the value of the concrete. The companies are small and only have a few trucks. To be able to meet demand at peak hours they all have an over capacity of these trucks on the average. Lately we guided them in their talks about using some trucks together. Clearly this is advantageous since peak hours of the different companies do not fall at the same time. However, we could never have convinced them by making whatever calculation and showing them the answer. Instead, we modelled the relation between the truck capacity per company and the corresponding service level (more trucks mean higher service levels) and the way that this was influenced by some joint trucks which could be used wherever necessary. The model was run using different datasets of real demands for a 14 days period. Results were discussed extensively with the company owners, and the different dependencies became clear.

In operational applications one can think about a huge number of planning systems, for routing and scheduling etc. Here, better use can and should be made of the rapid changes of information technology. Graphical interfaces for instance can be extremely powerful. More and more possibilities arise to present information to a planner, such that he can play a major role in the planning process. This does not make OR techniques redundant. Many sub problems can be solved mathematically. Moreover it is always advantageous to let the user start his session with an automatically generated schedule, which he can than further adapt.

The use of board computers and Global Position Systems will also have a major impact on the planning tools. Routing will become a real time application, where at any point in time truck positions will be known. As a result orders may be reallocated during the day on the basis of new information about delays, new orders, or other disturbances in the current schedule. This will also have its impact on the relevant OR models. Re-optimizing a current plan (in action) based on new information will be at least as important as the creation of the basic schedule from scratch. Objective functions may be influenced by these developments, in the sense that robust schedules which can be easily updated later, may become more attractive.
Currently we are involved in a project in Paris were a courier of valuables (mainly money) is introducing a GPS. Regulations state that a truck may not contain more than a certain amount of valuables. The customers to be visited each day are known in advance, but during the day the amounts of money to be collected or delivered are due to change. Therefore trucks have to be re-routed all the time. Still the process should evolve in a well controlled manner because of the high value of the loads. This again leads to an interesting challenge to OR.

To conclude I am convinced that the future of OR can be bright. More and more data become available and the companies are ready for the introduction of planning systems. Especially in the area of international distribution and logistics there are tremendous challenges. However, the OR-society will only be successful in meeting this challenge, if the art and science aspect of OR are combined properly.

References

Ackoff , R.L. (1956), "The Development of Operations Research as a Science", Operations Research, 4.

Boender, C.G.E. (1993), Decision Support of Asset-Liability Management, in: van P. Beek, L Fortuin and L. van Wassenhoven(eds.), Consult OR Perish, to appear.

Garey M.R., and D.S. Johnson (1979), Computers and Intractability, A Guide to the Theory of NP-completeness, Freeman, San Francisco. A.H.G., Rinnooy Kan (1992),

Dantzig, G.B. (1991), Linear programming: The story how it began, in: J.K. Lenstra, A.H.G. Rinnooy Kan and A. Schrijver (eds), History of Mathematical Programming, CWI, Amsterdam.

Rinnooy Kan , A.H.G. (1992), Opening speaker of the annual meeting of the 'Vereniging Voor Statistiek'.

van Vliet, A., C.G.E. Boender, and A.H.G. Rinnooy Kan, (1992), Interactive Optimization of Bulk Sugar Deliveries, Interfaces 22,3.

ZWEI- BZW. DREIDIMENSIONALE RÜSTUNGS- BZW. WETTBEWERBSMODELLE

Doris Anita Behrens, Klagenfurt

1. Einführung

Die Theorie nichtlinearer dynamischer Systeme stellt auf dem Gebiet angewandter Mathematik eine der großen Errungenschaften dieses Jahrhunderts dar. Auf naturwissenschaftlichem Sektor gibt es viele interessante und breit gefächterte Anwendungsgebiete dieser Theorie. Dennoch beschränkt sich der Einfluß der Lehre von den nichtlinearen Systemen in den Wirtschaftswissenschaften auf ein Minimum. Um den renommierten Ökonomen Michele Boldrin [1] zu zitieren.: "...bis vor kurzem war die nichtlineare Dynamik entweder ein ungebräuchliches Hilfsmittel für den mathematischen Ökonomen oder ihre Anwendung wurde oftmals, als für den Forschungszweck irrelevant, außer acht gelassen." In letzter Zeit aber erschienen einige Publikationen, die auch dem Leser mit mathematischem Standardwissen Einblick in die Chaostheorie ermöglichen [2,7]. Chaos wird meist mit der sensitiven Abhängigkeit von den Anfangswerten identifiziert, da die anderen hinreichenden Bedingungen mathematischer Natur und nur schwer empirisch nachweisbar sind. Jeder noch so winzige Effekt erreicht rasch makroskopische Dimensionen. Obwohl chaotische Systeme beschränkt sind, erzeugen sie stochastisches Verhalten aus sich selbst heraus.

Das Ziel meiner Nachforschungen war, zu beweisen, wie entsprechende Nichtlinearität chaotisches Verhalten eines zwei- bzw. dreidimensionalen Rüstungs- bzw. Wettbewerbsmodells verursacht. Weiters stellte sich heraus, daß auch Asymmetrie von entscheidender Bedeutung für das Auftreten von Chaos ist. Der Grundgedanke des Rüstungswettlaufes zweier und dreier konkurrierender Staaten im technischen Bereich basiert auf einem von Lewis Fry Richardson [4,5] in den 40er Jahren entwickelten Modell, das besagt, daß die militärische Rüstung konkurrierender Länder, die weder symmetrisch noch linear ist, im wesentlichen nicht nur von der Größe des eigenen, sondern von der Differenz der Rüstungspotentiale abhängt. Werden Nationen mit "feig-aggressivem" und "mutig-defensivem" Verhalten betrachtet, bedeutet "feig-aggressiv", daß im Fall rüstungsmäßiger Unterlegenheit keine Investitionen in die Vergrößerung des militärischen Potentials getätigt werden, um den zur Zeit überlegenen Gegner nicht zu reizen. Befindet sich der "feig-aggresive" Staat jedoch im Vorteil, so trachtet er danach, diesen noch weiter zu vergrößern und erhöht seine Ausgaben zur Anschaffung von Waffen. Das "mutig-defensive" Land handelt genau konträr. Es vergrößert seine Ausgaben im Fall der Unterlegenheit und tätigt bei größerem Rüstungspotential keine Investitionen. Beteiligen sich drei Staaten an einem Wettrüsten, ist die Anzahl der möglichen Reaktionen bezüglich des militärischen Budgets höher. Die Interaktionen werden durch Systeme natürlicher Differenzengleichungen beschrieben, die, wie bereits erwähnt, eine Entscheidungsfunktion beinhalten, welche von der Differenz der Militärbudgets abhängt.

Die auf strategischen Überlegungen beruhende mathematische Methode der Entscheidungsfindung, die es ermöglichen soll, aus mehreren Alternativen die beste auszuwählen, läßt sich auch auf Oligopole übertragen. Die Unternehmer trachten nach einer optimalen Marketingstrategie, einer am Absatzmarkt orientierten, systematischen Planung, Koordinierung und Durchführung aller Maßnahmen eines Betriebes mit dem Ziel einer langfristig optimalen Rendite. Da im Oligopol ein Stillhalten des einzelnen Unternehmers wahrscheinlich ist, solange keiner der Konkurrenten versucht, seinen Marktanteil durch Preissenkung zu vergrößern, zeigt sich typischerweise eine deutliche Preisstarrheit und ein Ausweichen auf Nicht-Preis-Wettbewerb mittels Werbung oder Produktveränderung. Betrachtet man also ein Duopol, in dem sich die beiden Kontrahenten mittels Werbung "bekriegen", so lassen sich die strategischen, dynamischen Interaktionen mit eingangs erwähntem zweidimensionalen Rüstungsmodell adäquat beschreiben. Folglich können die Aussagen über Rüstung denen über die Auswirkungen von Werbekampagnen eines spezialisierten Kleinbetriebes mit vorsichtigem Management und einem Großkonzern gleichgesetzt werden.

2. Die Modellierung des zweidimensionalen Falles

Wenn wir mit x_t bzw. y_t die Rüstungsausgaben bzw. Werbungskosten der Nationen bzw. Firmen x bzw. y zum Zeitpunkt t=0,1,2,... bezeichnen, so werden die Wechselwirkungen zwischen x und y von folgendem System zweier nichtlinearer Differenzengleichungen beschrieben:

$$x_{t+1} = (1-\alpha)x_t + f(x_t-y_t)$$
$$y_{t+1} = (1-\beta)y_t + g(x_t-y_t)$$

α und β repräsentieren jeweils das Maß für die Aufrechterhaltung des Status quo. Ist ein Beteiligter des Wettbewerbs nicht gezwungen auf die Aktionen seines Kontrahenten zu reagieren, ist die Größe des Rüstungspotentials bzw. des Werbebudgets rückläufig, d.h. $0<\alpha, \beta<1$. Der Einfluß der Differenz der beiden Potentiale kommt in den Entscheidungsfunktionen f und g zum Tragen. Diese haben folgende Gestalt:

$$f(x_t-y_t) = \varphi_a(x_t-y_t) = a\,[1 + \exp\{-c\,(x_t-y_t)\}]^{-1}\ ; f(\,) > 0,\ f'(\,) > 0$$
$$g(x_t-y_t) = \varphi_b(x_t-y_t) = b\,[1 + \exp\{-c\,(x_t-y_t)\}]^{-1}\ ; g(\,) > 0,\ g'(\,) > 0,$$

wobei die Parameter a, b und c positiv gewählt werden. a und b repräsentieren das Maß für die Aufrüstungskosten bzw. Investitionen. Der Parameter c mißt die "Steilheit" der Entscheidungsfunktionen φ_a und φ_b, wobei a und b die Asymptoten für den rechten Zweig der Graphen liefern. Je größer der Parameter c gewählt wird, desto ähnlicher werden φ_a und φ_b der Heavisidefunktion, mit der Sprunghöhe a bzw. b im Punkt 0, d.h. im Punkt x=y.

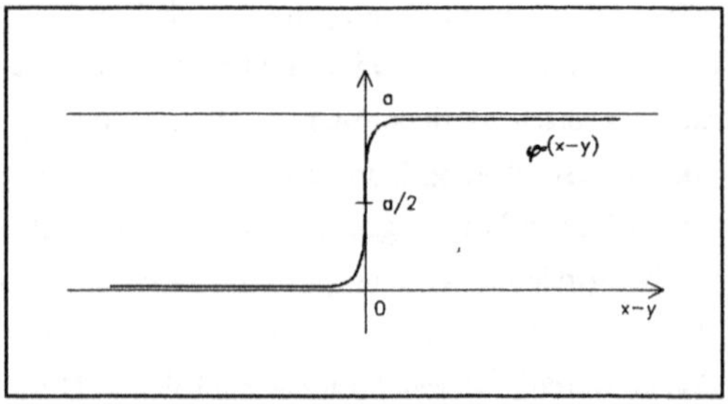

Abb.1: φ_a gezeichnet für a=0.2 und c=100

Da bei Verwendung der Entscheidungsfunktionen f und g Chaos nachgewiesen werden konnte und sie somit von zentraler Bedeutung für die Struktur des Systems sind, kommt ihren Eigenschaften besonders Interesse zu:

(i) Die Funktionen sind positiv und monoton wachsend.

(ii) $\varphi_a(x\text{-}y) \rightarrow \begin{cases} 0 & \text{für } x\text{-}y \rightarrow -\infty \\ a & \text{für } x\text{-}y \rightarrow +\infty \end{cases}$

$\varphi_a(0) = \frac{a}{2}$

$\varphi_b(x\text{-}y)$ analog.

(iii) für $c \rightarrow 0$: $\quad \varphi_a(x\text{-}y) \rightarrow \frac{a}{2} \qquad \forall\, x\text{-}y$

für $c \rightarrow \infty$: $\quad \varphi_a(x\text{-}y) = H_a(x\text{-}y) = \begin{cases} 0 & \text{für } x<y \\ a & \text{für } x>y \end{cases}$

$\varphi_b(x\text{-}y)$ analog.

Stehen drei Staaten oder Firmen im Wettbewerb, ist die Anzahl der möglichen Reaktionen bezüglich der Budgetgestaltung natürlich viel höher. Eine dieser Wechselwirkungen im dreidimensionalen Fall gestaltet sich wie folgt:

$$x_{t+1} = (1-\alpha)x_t + \psi_a(x_t\text{-}\max[y_t,z_t])$$

$$y_{t+1} = (1-\beta)y_t + \psi_b(\min[x_t,z_t]\text{-}y_t)$$

$$z_{t+1} = (1-\gamma)z_t + \psi_d(\max[x_t,y_t]\text{-}z_t)$$

wobei die Entscheidungsfunktionen ψ_a, ψ_b und ψ_d die Erweiterung des zwei- auf den dreidimensionalen Fall darstellen.

3. Numerische Untersuchungsmethoden auf Chaos

Ist es nun möglich, eine sinnvolle Werbestrategie zu erstellen, oder erweisen sich die Interaktionen in unserem Duopol als unvorhersagbar? Numerische Testverfahren, wie Bifurkationsdiagramme und die Berechnung von Lyapunovexponenten eines Systems diskreter Differenzengleichungen dienen zum Nachweis von Chaos im Rüstungswettlauf bzw. im "Werbekrieg". Interessant ist die Abfolge asymptotischer Zustände nichtlinearer Systeme, die auftreten können, wenn die Systemparameter variiert werden. Diese Übergänge im qualitativen Verhalten der Lösungen nennt man Bifurkationen. Bifurkation bedeutet Zweiteilung, Abspaltung bzw. Veränderung. Gegenstand der Bifurkationstheorie in der Lehre von den dynamischen Systemen ist die Betrachtung von Veränderungen, denen Abbildungen unterworfen werden, sobald Parameter variieren.

Interpretiert man Abb.2 im Sinne eines Duopols, dessen Teilnehmer im "Werbekrieg" leben, so veranschaulicht das Bifurkationsdiagramm die Auswirkungen der Obsoleszenzrate der Investitionen in Werbekampagnen von seiten des Großbetriebes auf die langfristigen Werbekosten des Kleinbetriebes. Anfangs sind die Werbekosten des Kleinunternehmers genau berechenbar. Bei Erhöhung der Obsoleszenzrate für Investitionen des Großbetriebes werden sie aber unvorhersagbar. Eine infinitesimale Änderung des Parameters in der Umgebung eines Bifurkationspunktes hat somit Unvorhersagbarkeit der Werbekosten und den Zusammenbruch des Investitionsplanes des Kleinunternehmers zur Folge. Wirft man nun noch einen Blick auf das Phasenportrait (Abb.3), in dem jeder Punkt (x,y) der Grafik die Werbekosten des spezialisierten Kleinbetriebes und des Großkonzerns zu einem festen Zeitpunkt symbolisiert, erkennt man, daß sich diese Unvorhersagbarkeit keineswegs nur auf den Kleinbetrieb beschränkt, sondern den Zusammenbruch der Werbestrategien beider Firmen bewirkt. Wie man sieht, pendeln sich die Kosten niemals bei einem festen Wert ein. Wie lange die beiden Firmen ihren Wettbewerb in Sachen Öffentlichkeitsarbeit auch weiterführen, ist es unmöglich, eine entsprechende Werbestrategie zu finden, geschweige denn sie vorherzusagen. Die Form und Lage des Attraktors läßt nur den Schluß zu, daß beide Firmen dazu tendieren, die Kosten gleichhoch zu halten, und daß die Investitionen des Großkonzerns so gut wie nie von denen des Kleinbetriebes überflügelt werden, da sich der Attraktor größtenteils von unten an die erste Mediane schmiegt.

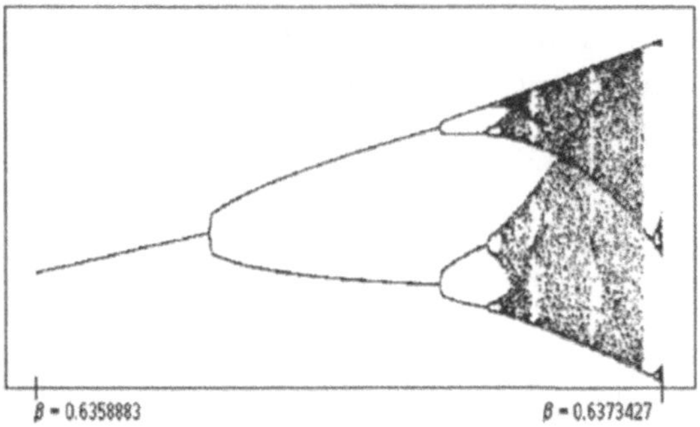

Abb.2: Bifurkationsdiagramm:
Ausgaben des Teilnehmers x aufgetragen gegen die Obsoleszenzrate β des Teilnehmers y

Abb.3: Phasenportrait gezeichnet für 60000 Iterationsschritte

Alle Indizien weisen auf Chaos hin. Was kann aber nun als Maß für die Unvorhersagbarkeit eines Systems angesehen werden? Beschränken wir unsere Betrachtungen vorerst auf eindimensionale Systeme bzw. Abbildungen, so liefert der Lyapunovexponent die durchschnittliche Rate von Konvergenz und Divergenz im Phasenraum. Der Lyapunovexponent (LCE) kann als verallgemeinerter Eigenwert angesehen werden. Ein negativer LCE steht für periodisches Verhalten des Systems, der Wert null für Stabilität und ein positiver LCE definiert chaotisches Verhalten der eindimensionalen Abbildung. Die Größe des positiven Lyapunovexponenten reflektiert das Zeitmaß, mit dem die Systemdynamiken unvorhersagbar werden. Liegt der LCE extrem nahe beim Wert null, so ist keine Aussage über die Existenz von Chaos möglich. Im mehrdimensionalen Fall ist die Berechnung dieses Exponenten ungleich schwieriger und erfordert einiges an mathematischen Kenntnissen; das zugrundeliegende Prinzip aber bleibt das gleiche. Da mathematisch keine einheitliche Definition von Chaos existiert, wird es in der Praxis durch das Auftreten eines positiven LCE verifiziert. Errechnet man für ein höherdimensionales System mit Hilfe von Softwarepaketen [3,6] ein Lyapunovspektrum, welches zumindest einen positiven LCE beinhaltet, so kann dieses als chaotisch angesehen werden.

Für das zweidimensionale System nichtlinearer, asymmetrischer Differenzengleichungen besteht das Lyapunovspektrum aus zwei Werten: LCE1 = 0.398807 und LCE2 = -0.371957. Der positive Wert beweist das Auftreten chaotischen Verhaltens. Das dreidimensionale System erforderte aus technischen Gründen die Verwendung eines anderen Softwarepakets. Für die Unkosten jeder Firma x, y und z wurde der LCE separat berechnet und alle Werte überstiegen 0.014495.

4. Schlußbemerkungen

Da in der praktischen Anwendung Parameter und Anfangsbedingungen Meßfehlern unterliegen, ist die Kenntnis über das qualitative Verhalten eines Systems von wesentlicher Bedeutung. In nicht chaotischen Systemen bleiben benachbarte Trajektorien im Zeitverlauf nahe beieinander. In diesem Sinne bezeichnet man solche Systeme als vorhersagbar. Messungen der Anfangswerte enthalten Informationen, dank denen sich künftiges Verhalten vorhersagen läßt. Vorhersagbare Systeme sind gegenüber Meßfehlern nicht sehr empfindlich. Chaotische Attraktoren hingegen wirken wie Pumpen, die vorhandene Informationen immer wieder durcheinanderwirbeln. Unter diesem Gesichtspunkt ist auch klar, daß keine exakte Lösung, keine Abkürzung bei der Vorhersage der Zukunft existieren kann. Trotz ihrer Beschränktheit gibt es in chaotischen Systemen keinen beschreibbaren Zusammenhang zwischen Vergangenheit und Zukunft.

Referenzen

[1] Boldrin, Michele. 1988. Persistent oscillations and chaos in dynamic economic models: notes for a survey. In *The economy as an envolving system*, edited by P.W. Anderson, K.J. Arrow, and D. Pines, 49-75. Redwood-City, CA: Addison-Wesley.

[2] Devaney, Robert L. 1989. *An introduction to chaotic dynamical systems.* Second edition. Redwood-City, CA: Addison-Wesley.

[3] Medio, Alfredo and Giampaolo Gallo. 1989. *DMC. Dynamical system analysis. Software, theory and applications to economics.* Department of Economics, University of Venice.

[4] Richardson, Lewis Fry. 1948. War moods. *Bychometrica* I: 147-174 and II: 197-232.

[5] Richardson, Lewis Fry. 1960. *Arms and insecurity.* Pittsburg: The Boxwood Press.

[6] Schaffer, W.M., Truty, G.L., and Fulmer, S.L. 1988. *Dynamical Software. User's manual and introduction to chaotic systems.* 2 Vols. Tucson, Arizona: Dynamical Systems, Inc.

[7] Wiggins, S. 1990. *Introduction to applied nonlinear dynamical systems and chaos.* New York: Springer.

Effiziente Algorithmen zur Bestimmung optimaler Losgrößen

Jan W. Höter

Zusammenfassung: Das 1958 von Wagner und Whitin entwickelte Verfahren zur Bestimmung optimaler Losgrößen kann, auch bei einer Dynamik sämtlicher Produktionsparameter, mit minimalem programmiertechnischen Möglichkeiten wesentlich effizienter implementiert werden, als es gemeinhin - zumindest im deutschen Sprachraum - angenommen wird. Durch die Verwendung erst kürzlich veröffentlichter Algorithmen, z.B. von Wagelmanns, van Hoesel und Kolen (1992), kann das mit steigendem Planungshorizont überproportionale Ansteigen des Rechenaufwandes für beliebige Datenkonstellationen erheblich reduziert werden ($O(n \cdot \log n)$-Komplexität). Trotzdem ist die Verwendung einer optimierten Version der von Evans (1985) vorgeschlagenen Implementierung für Problemgrößen bis 15 Perioden weiterhin sinnvoll und stellt, in Kombination mit den Planungshorizont-Theoremen von Eppen, Gould und Pashigian (1969), trotz $O(n^2)$-Komplexität, die wohl zur Zeit effizienteste Möglichkeit zur Losgrößenbestimmung in der betrachteten Planungssituation dar.

Abstract: The algorithm to determine optimal lot sizes, proposed by Wagner and Whitin in 1958, may be implemented, employing rather modest programming techniques and even allowing all production parameter to vary, far more efficiently than is commonly seen - at least in German literature. By using recently published algorithms, e.g. by Wagelmanns, van Hoesel und Kolen (1992), an increase in the planning horizon results for arbitrary data merely in a slightly over proportional enhancement of running time ($O(n \cdot \log n)$). However, an improved version of the implentation suggested by Evans (1985) is still useful for problem sizes up to 15 periods and, in combination with the planning horizon theorems of Eppen, Gould und Pashigian (1969), although still of $O(n^2)$-type, seems at the moment to be the most efficient way to solve lot sizing problems of the considered type.

1. Problemformulierung

Für einen n gleich lange[1] Perioden umfassenden Planungshorizont[2] sind für jede Periode j=1,2,...,n alle relevanten Daten - Rüstkosten $c1_j \geq 0$ GE, Lagerkosten $c2_j \geq 0$ GE/(ME·ZE), Herstellkosten $c3_j \geq 0$ GE/ME und Nettobedarf $d_j \geq 0$ ME mit $d_1 > 0$ - bekannt. Gesucht ist für j=1,2,...,3 die Produktionsmenge in der Periode j $x_j \geq 0$ ME bzw. der Lagerbestand am Ende der Periode j $y_j \geq 0$ ME (Fehlmengenausschluß) mit $y_j := 0$, so daß die Gesamtkosten innerhalb des Planungshorizontes n

$$g_n = \sum_{j=1}^{n}(c1_j \cdot f_j + c2_j \cdot y_j + c3_j \cdot x_j) \quad \text{mit} \quad f_j := \begin{cases} 1, \text{ falls } & x_j > 0 \\ 0, \text{ falls } & x_j = 0 \end{cases} \tag{1}$$

minimiert werden. Die Produktion erfolgt jeweils nur zu Beginn einer Periode und auf einmal (unendliche Produktionsgeschwindigkeit[3]).

---

[1] Sind die Perioden nicht gleich lang, so muß lediglich der Lagerkostensatz für jede Periode individuell und nicht pro ZE festgelegt werden.

[2] Die implizierte Prämisse $d_{n+i} = 0,\ i \in \mathbb{N}\backslash\{0\}$, kann sich bis zurück auf die optimale Produktionsmenge der ersten Periode auswirken und damit das gewählte Produktionsprogramm verfälschen, sofern (etwa bei rollierende Planung) unzutreffend (vgl. Zoller, 1990, S. 16).

[3] Der Durchschnittsvorrat, der sich ggf. aus einem nicht sofortigen Verzehr im Zeitpunkt j bzw. aus einer endlichen Produktionsgeschwindigkeit ergibt, ist nicht beeinflußbar. Intraperiodische Lagerungskosten bleiben somit als unvermeidlich außer Betracht. Weiterhin hat auch eine endliche Produktionsgeschwindigkeit selber keinen Einfluß auf die Planungsaufgabe (vgl. Robrade, 1990, S. 19).

2. Lösung mit $O(n^2)$-Algorithmen

Wagner und Whitin bewiesen 1958 (S.91) den folgenden Satz:

(S1) Es existiert ein optimales Produktionsprogramm, so daß für alle j=1,2,...,n gilt:

$$x_j = 0 \text{ oder } x_j = \sum_{i=j}^{t} d_i \quad \text{für einige } t,\ j \le t \le n.$$

(S1) impliziert, daß auf der Suche nach einem optimalen Produktionsprogramm lediglich die Kosten von $n \cdot (n+1)/2$ Produktionsstrategien ausgewertet werden müssen. Diese können mittels folgender Rekursionsformel bestimmt werden:

$$g_j = \min_{1 \le i \le j} \underbrace{\{c1_i + \sum_{t=i}^{j} (c3_i \cdot d_t + c2_t \cdot \sum_{h=t+1}^{j} d_h) + g_{i-1}\}}_{k_{i,j}} \quad \text{für } j=1,2,\dots,n \tag{2}$$

wobei

g_j	[GE]	die Kosten eines optimalen Produktionsprogramms für den die Periode 1 bis j umfassenden Planungszeitraum des betrachteten Losgrößenproblems mit $g_0 := 0$,
$k_{i,j}$	[GE]	die Gesamtkosten eines optimalen Produktionsprogramms für die Perioden 1 bis i-1 und der Produktion der Bedarfe der Perioden i bis j in der Periode i und
q_j	$\in \{1,2,\dots,j\}$	die optimale Produktionsperiode des Bedarfes der Periode j bei einem aus den Perioden 1 bis j bestehenden Planungszeitraum ($k_{i,j}$ wird für $i = q_j$ minimal)

bezeichnen, und sich zusammenfassend ergibt:

$$g_j = \min_{1 \le i \le j}(k_{i,j}) = k_{i^*,j} \quad \wedge \quad q_j := i^* . \tag{3}$$

Das optimale Produktionsprogramm ist durch die Variablen q_j eindeutig bestimmt und kann mit (S1) leicht rekonstruiert werden.

2.1. Der derzeitige Standard

Zumindest im deutschen Sprachraum werden die Kosten $k_{i,j}$ leider noch viel zu häufig exakt wie in (2) berechnet. Hierzu sind zwei ineinander verschachtelte Summationen für alle i und j ($1 \le i \le j$ für j=1,2,...,n) durchzuführen, was zu einem Algorithmus mit $O(n^4)$- Komplexität führt. Eine unter diesen Voraussetzungen noch recht gute "LOST"[4]-Implementierung benötigt, unter Vernachlässigung der, hinsichtlich optimaler Produktionsperiode und minimalen Kosten auch datenabhängigen Zuweisungen, zur Lösung eines n-Perioden-Problems $\chi_n^{LOST} = \frac{1}{12} \cdot n^4 + \frac{2}{3} \cdot n^3 + \frac{35}{12} \cdot n^2 - \frac{2}{3} \cdot n$ Operationen[5].

[4] Von "*lo*w *st*orage"; man beachte auch die Laufzeit dieser Implementierung, die sich mit steigender Problemgröße - insbesondere im Vergleich mit effizienteren Implementierungen - im Unendlichen 'verliert'.

[5] Für eine eingehendere Betrachtung sei hier und im folgenden auf das Literaturverzeichnis verwiesen.

2.2. Eine optimierte Version der Implementierung von Evans (1985)

*E*vans zeigte 1985 (S. 231), daß das vorliegende Problem auch durch einen Algorithmus mit $O(n^2)$-Komplexität gelöst werden kann. Seine Version ist jedoch insgesamt verbesserungswürdig, so daß im folgenden die *opt*imierte Version "EVOPT" behandelt wird. Es kann gezeigt werden, daß diese Implementierung der von Evans für beliebige Problemgrößen sowohl bezüglich Speicherplatzbedarf als auch hinsichtlich Laufzeit überlegen ist. Die Laufzeitverbesserungsrate liegt minimal bei 13% und für praxisrelevante Probemstellungen um 20%.

```
for j:=1 to n-1 do
  c2sum[j]:=0
g[1]:=c1[1]+d[1]*c3[1];
q[1]:=1;
k[1]:=g[1];
for j:=2 to n do begin
  g[j]:=c1[j]+d[j]*c3[j]+g[j-1];
  q[j]:=j;
  k[j]:=g[j];
  for i:=1 to j-1 do begin
    c2sum[i]:=c2sum[i]+c2[j-1];
    k[i]:=k[i]+d[j]*(c3[i]+c2sum[i]);
    if k[i]<g[j] then begin
      g[j]:=k[i];
      q[j]:=i
    end
  end
end;
```

Abb. 1: EVOPT-Implementierung

Wie bekannt gilt:

$$k_{i,j} = c1_i + \sum_{t=i}^{j}(c3_i \cdot d_t + c2_t \cdot \sum_{h=t+1}^{j} d_h) + g_{i-1} \text{ und damit auch } k_{i,j-1} = c1_i + \sum_{t=i}^{j-1}(c3_i \cdot d_t + c2_t \cdot \sum_{h=t+1}^{j-1} d_h) + g_{i-1}$$

$$\text{woraus folgt } k_{i,j} - k_{i,j-1} = d_j \cdot \sum_{t=i}^{j-1} c2_t + c3_i \cdot d_j = d_j \cdot (c3_i + \sum_{t=i}^{j-1} c2_t) \text{ und somit}$$

$$k_{i,j} = \begin{cases} k_{i,j-1} + d_j \cdot (c3_i + \sum_{t=i}^{j-1} c2_t), & \text{für } i = 1,2,\dots,j-1 \\ c1_j + c3_j \cdot d_j + g_{i-1} & \text{, für } i = j \end{cases} \qquad \text{für } j=1,2,\dots,n.$$

Wird weiterhin die Summation der Lagerkosten genau betrachtet, so wird deutlich, daß der zur Berechnung von $k_{i,j}$ benötigte Term, $c2sum_i^{(j)} = \sum_{t=i}^{j-1} c2_t$, günstig aus dem für die Berechnung von $k_{i,j-1}$ benötigtem Term, $c2sum_i^{(j-1)} = \sum_{t=i}^{j-2} c2_t$, ermittelt werden kann: $c2sum_i^{(j)} = c2sum_i^{(j-1)} + c2_{j-1}$. Hieraus ergibt sich eine Formel zur Berechnung der $k_{i,j}$, die mit $O(n^2)$-Komplexität implementiert werden kann:

$$k_{i,j} = \begin{cases} k_{i,j-1} + d_j \cdot (c3_i + c2sum_i^{(j-1)} + c2_{j-1}), & \text{für } i = 1,2,\dots,j-1 \\ c1_j + c3_j \cdot d_j + g_{i-1} & \text{, für } i = j \end{cases} \qquad \text{für } j=1,2,\dots,n. \tag{4}$$

Im Vergleich mit der Implementierung LOST benötigt die EVOPT-Implementierung zwei zusätzliche Vektoren der Länge n und (n-1), $k_i^{(j)}$ und $c2sum_i^{(j)}$, zeichnet sich jedoch aufgrund der geringeren Komplexität durch eine drastisch verringerte Laufzeit aus: $\chi_n^{EVOPT} = \frac{5}{2} \cdot n^2 + \frac{1}{2} \cdot n - 1$.

2.3. Zum Einsatz der Theoreme von Eppen Gould und Pashigian (1969)

Die Suche nach einer effizienten Möglichkeit zur Bestimmung der aus den zu überprüfenden Produktionsstrategien resultierenden Kosten scheint mit der Implementierung EVOPT vorläufig abgeschlossen. Weitere Einsparungspotentiale können nun dadurch erzielt werden, die Anzahl der auf der Suche nach einem optimalen Produktionsprogramm überhaupt zu überprüfenden Produktionsstrategien weiter zu begrenzen: Bereits 1958 beschrieben Wagner und Whitin (S. 92) ein sogenanntes "Planning Horizon Theorem" zur frühzeitigen Erkennung suboptimaler Strategien, dessen Anwendung jedoch auf den Fall konstanter (oder, wie Wagner 1975 (S. 321-322) zeigte, zumindest konkaver) Herstellkosten im Zeitablauf beschränkt ist. Auch für den Fall beliebig variabler Herstellkosten konnten Eppen, Gould und Pashigian (1969, S. 273-275) Planungshorizont-Theoreme herleiten, die zum Zwecke einer effizienten Implementierung in die folgenden Sätze überführt werden können (vgl. auch Zäpfel, 1971, S. 271-272):

```
for j:=1 to n-1 do
  c2sum[j]:=0
g[1]:=c1[1]+c3[1]*d[1];
q[1]:=1;
k[1]:=g[1];
iu:=1;
for j:=2 to n do begin
  q[j]:=q[j-1];
  id:=q[j];
  c2sum[id]:=c2sum[id]+c2[j-1];
  cq:=c2sum[id]+c3[id];
  cb:=cq;
  k[id]:=k[id]+d[j]*cq;
  g[j]:=k[id];
  if c3[j]<cq then begin
    cb:=c3[j];
    id:=j;
    k[j]:=c1[j]+d[j]*c3[j]+g[j-1];
    if k[j]<g[j] then begin
      g[j]:=k [j];
      q[j]:=j
    end
  end;
  for i:=iu to j-1 do begin
    if i<>q[j-1] then begin
      c2sum[i]:=c2sum[i]+c2[j-1];
      ci:=c2sum[i]+c3[i];
      if ci<cq then begin
        c2sum[i]:=ci-c3[i];
        if ci<cb then begin
          id:=i;
          cb:=ci
        end;
        k[i]:=k[i]+d[j]*(c2sum[i]+c3[i]);
        if k[i]<g[j] then begin
          g[j]:=k[i];
          q[j]:=i
        end
      end
    end
  end;
  if id=q[j] then iu:=j+1
end;
```

Abb. 2: EVEP-Implementierung

(S2) Gilt für ein i, $1 \le i \le j+1$, $k_{i,j+1} < k_{q_j,j+1}$ dann ist $i \in m(q_j, j+1)$ mit

$$m(q_j,j+1) = \{i \mid \; 1 \le i \le j+1 \;\wedge\; c3_i + \sum_{t=i}^{j} c2_t < c3_{q_j} + \sum_{t=q_j}^{j} c2_t\}.$$

(S3) Gilt in einem j-Perioden-Problem $\underset{1\le i\le j}{\mathrm{Min}}(c3_i + \sum_{t=i}^{j-1} c2_t) = c3_{q_j} + \sum_{t=q_j}^{j-1} c2_t$ dann

gilt in einem w-Perioden-Problem, $w > j$, $k_{i,w} \ge k_{q_j,w}$ für $1 \le i \le j$.

Die gemäß diesen Sätzen für eine Verkürzung der Anzahl der zu überprüfenden Produktionsstrategien erforderlichen Abprüfungen lassen sich verhältnismäßig einfach in die EVOPT-Implementierung integrieren und führen - abgesehen von besonders ungünstigen (und völlig unrealistischen) Datenkonstellationen, für die keiner der beiden Sätze Anwendung findet - zu einer erheblichen Verminderung von Rechenaufwand bzw. Laufzeit. Die vorstehend abgebildete Implementierung EVEP stellt eine Kombination der Vorteile der Ansätze von *Evans* (bzw. der Implementierung EVOPT) und *Eppen*, Gould und Pashigans dar: Berechnung der Kosten $k_{i,j}$ mit geringem zusätzlichen Rechenaufwand aus den Kosten $k_{i,j-1}$ mittels (4), sowie frühzeitigem Erkennen der Suboptimalität bestimmter Produktionsstrategien durch die Verwendung von (S2) und (S3). Neben den bereits bekannten, werden die Variablen ci, cb, cq, iu, und id verwendet:
In einer Periode $j \in \{2,3,...,n\}$ bezeichnet ci [cq] GE\ME die zusätzlichen Kosten, die für eine Einheit des Bedarfes der Periode j aufzuwenden sind, wenn in der Periode $i \in \{1,2,...j\}$ [Periode q_j] eine Einheit dieses Bedarfes mitproduziert wird ($c1_i$ somit nicht zusätzlich anfällt):

$$ci := c3_i + \sum_{t=i}^{j-1} c2_t \quad [\, cq := c3_{q_j} + \sum_{t=q_j}^{j-1} c2_t \,] .$$

Es bezeichnet id in einer Periode $j \in \{2,3,...,n\}$ die Periode $i \in \{1,2,...j\}$, in der die minimalen Kosten cb entstehen:

$$id := i^* \quad \text{mit} \quad cb := \underset{1\le i\le j}{\mathrm{Min}}(c3_i + \sum_{t=i}^{j-1} c2_t) = c3_{i^*} + \sum_{t=i^*}^{j-1} c2_t .$$

Die Variable iu stellt in einer Periode $j \in \{2,3,...,n\}$ die kleinste Periode $i \in \{1,2,...,j\} \setminus \{q_{j-1}\}$ dar, für die $k_{i,j}$ noch nicht gemäß (S3) als suboptimal identifiziert wurde: $k_{i,j} \not< k_{q_{j-1},j}$ für alle $i \in \{1,2,...,iu-1\}$.

3. Lösung mit O(n log n)-Algorithmen

Im "ASA" benannten Algorithmus von *A.* Wagelmanns, *S.* van Hoesel und *A.* Kolen (1992, S. S150-S151) schließlich werden je Periode nur noch die Kosten eines Produktionsprogramms, nämlich die des bis zu dieser Periode optimalen, ermittelt. Dies ist deshalb möglich, da im ASA Algorithmus weniger nach minimalen Kosten, als nach optimalen Produktionsperioden gesucht wird, aus denen sich die minimalen Kosten quasi als Kuppelprodukt ergeben. Die eigentliche Errungenschaft dieses Algorithmus besteht nun darin, daß aus einem geordneten Feld zuvor als 'brauchbar' ("efficient") identifizierten Perioden die optimale Produktionsperiode der jeweiligen Iteration mittels des Verfahrens der Binären Suche bestimmt werden kann, dessen Komplexität bekanntlich $O(\log_2 n)$ beträgt. Hierdurch kann, im Gegensatz zur Implementierung EVEP, für beliebige Datenkonstellationen eine Reduktion der explizit auf Optimalität zu prüfenden Produktionsstrategien (-perioden) erreicht werden, selbst wenn keine suboptimalen Strategien (unbrauchbare Perioden) von vornherein identifiziert werden können. Die in der Implementierung ASA durchzuführenden Vergleiche auf der Suche nach der optimalen Produktionsperiode sind jedoch ebenso wie die Prüfbedingungen zur Identifizierung unbrauchbarer Periode recht aufwendig. Auch benötigt sie einen hohen Intitialisierungsaufwand, weshalb die Implementierung EVOPT erst ab einer Problemgröße von 15 Perioden dominiert wird.

Der Vollständigkeit halber sei erwähnt, daß Aggarwal und Park (1990) und Federgruen und Tzur (1991) ebenfalls Algorithmen zur Bestimmung optimaler Losgrößen im hier betrachteten Fall entwickelten, deren Komplexität $O(n \log n)$ beträgt. Nach Federgruen und Tzur (1991, S. 911) besteht die Idee des noch unveröffentlichten Ansatzes von Aggarwal und Park in einer Aufspaltung des Losgrößenproblems in Teilprobleme, deren Datenkonstellationen bestimmten Anforderungen genügen und als Spezialfälle besonders effizient zu lösen sind. Dadurch soll der "constant factor in the time bounds" sehr hoch sein, so daß zumindest für kleinere Losgrößenprobleme keine Verbesserung gegenüber den bereits vorgestellten Implementierungen zu erwarten ist. Für die Implementierung ihres eigenen Algorithmus geben Federgruen und Tzur (1991, S. 910 und 922-923) eine Überlegenheit gegenüber der von Evans (1985, S. 235) für Problemgrößen $n \geq 20$ sowie eine durchschnittlich etwa um den Faktor 3 schnellere Lösungsfindung für $n = 500$ an. Da die ASA-Implementierung die für sie ungünstigste Datenkonstellation (alle Perioden sind brauchbar) bereits für n=15 schneller löst als die EVANS-Implementierung die für sie günstigste (die jeweils zuerst betrachtete Produktionsperiode ist optimal), und zwar um annähernd den Faktor 3 bereits für $n = 50$ (Faktor 5 für $n = 100$), erscheint auch der Algorithmus von Federgruen und Tzur (1991, S. 918) dem von Wagelmans, van Hoesel und Kolen (1992, S. S150-S151) klar unterlegen[6].

Anmerkungen

Aus Platzgründen war eine vollständige Wiedergabe der Ergebnisse aus Höter (1992) unmöglich. Die nebenstehende Tabelle liefert dennoch eine Kurzcharakteristik der beschriebenen Implementierungen inklusive ihrer Laufzeit für das auf die hier behandelte Problemstellung und Periodenanzahl erweiterte, und im Ergebnis als typisch zu bezeichnende Losgrößenproblem aus Wagner und Whitin (1958, S. 94).

Implementierung/ Charakteristika	LOST	EVANS	EVOPT	EVEP	ASA
Komplexität	n^4	n^2	n^2	n^2	$n \log n$
Speicherplatzbedarf[7]	$2 \cdot n + 9$	$4 \cdot n + 6$	$4 \cdot n + 2$	$4 \cdot n + 7$	$5 \cdot n + 10$
Laufzeit[8] $[10^{-3}$ sec$]$					
10 Perioden	54	9	7	5	12
50 Perioden	24330	227	182	25	81
100 Perioden	$\approx 385 \cdot 10^3$	923	738	50	194

Abb. 3: Kurzcharakteristik der betrachteten Implementierungen

6 Man beachte weiterhin, daß Federgruen und Tzur (1991, S. 917) zur Implementation ihres Algorithmus die Verwendung von Binären Bäumen vorschlagen, die ohne dynamische Datentypen nur sehr speicherplatzintensiv zu verwirklichen sind.

7 Exklusive der für die Eingabedaten notwendigen $4 \cdot n$ Speicherplätze.

8 Exklusive aller Ein- und Ausgabeoperationen, jedoch inklusive der zur Rekonstruktion des optimalen Produktionsprogramms notwendigen Anweisungen auf einem 386 SX 16 PC; Median aus jeweils drei Testläufen gerundet auf volle Millisekunden (die Abweichungen lagen jeweils im Mikrosekundenbereich).

Die Implementierung sämtlicher Algorithmen wurde aus Kompatibilitätsgründen in Standard Pascal, ISO 7185-1983 Stufe 0, vorgenommen. Die hierdurch ausgeschlossenen Verwendung dynamischer Datentypen birgt, insbesondere für die Implementierungen EVEP und ASA, beträchtliche Einsparungspotentiale in sich.

Die in der Problemformulierung gemachte Einschränkung $d_1 > 0$ ist, obwohl in der Literatur vielfach ignoriert, durchaus nicht nebensächlich: Perioden ohne Bedarf am Anfang eines Losgrößenproblems können zu suboptimalen Lösungen führen. Beispiele hierfür, ein kurzer Algorithmus zur Aufhebung dieser Einschränkung sowie vollständige Programme zu den betrachteten Implementierungen sind beim Verfasser erhältlich.

Literaturverzeichnis

Aggarwal, A. und J. K. Park. "Improved Algorithms for Economic Lot-Size Problems". Working Paper, IBM Thomas J. Watson Research Center. Yorktown Heights/New York, 1990. Zitiert nach Federgruen und Tzur (1991, S. 911/923).

Eppen, Gary D., F. J. Gould und B. Peter Pashigian. "Extensions of the Planning Horizon Theorem in the Dynamic Lot-Size Model". Management Science 15.5 (Januar 1969): 268-277.

Evans, James R. "An Efficient Implementation of the Wagner-Whitin Algorithm for Dynamic Lot-Sizing". Journal of Operations Management 5.2 (Februar 1985): 229-235.

Federgruen, Awi and Michal Tzur. "A Simple Forward Algorithm to Solve General Dynamic Lot Sizing Models With n Periods in O(n log n) or O(n) Time". Management Science 37.8 (August 1991): 909-925.

Höter, Jan W. "Effiziente Algorithmen zur Bestimmung optimaler Losgrößen". Diplomarbeit zur Erlangung des Grades eines Diplom-Kaufmanns am Fachbereich Wirtschafts- und Organisationswissenschaften der Universität der Bundeswehr in Hamburg. November 1992.

Robrade, Andreas Dieter. Dynamische Einprodukt-Lagerhaltungsmodelle bei periodischer Bestandsüberwachung. Physica-Schriften zur Betriebswirtschaft Bd. 34. Diss. Hamburg, 1990. Heidelberg: Physica, 1990.

Wagner, Harvey M. Principles of Operations Research with Applications to Managerial Decisions. 2. Aufl. Englewood Cliffs/New Jersey: Prentice-Hall, Inc., 1975.

Wagner, Harvey M. und Thomson M. Whitin. "Dynamic Version of the Economic Lot Size Model". Management Science 5.1 (1958): 89-96.

Wagelmans, Albert, Stan van Hoesel und Antoon Kolen. "Economic Lot Sizing: An O(n log n) Algorithm that Runs in Linear Time in the Wagner-Whitin Case". Operations Research Vol. 40 Supp. No. 1 (Januar-Februar 1992): S145-S156.

Zäpfel, Günther. "Exakte Verfahren der dynamischen Bestellrechnung". Angewandte Informatik 6 (1971): 269-275.

Zoller, Klaus. "Vorlesung Logistik I: Lagerhaltung". Arbeitspapier des Instituts für Betriebliche Logistik und Organisation an der Universität der Bundeswehr Hamburg. 1990.

Large Deviations and Queueing Applications

Michel Mandjes
Department of Econometrics
Vrije Universiteit Amsterdam

The theory of **Large Deviations** (LD) is concerned with the probability that certain stochastic processes attain extreme values, i.e., values far away from their mean value. This theory is founded by Cramér, Chernoff, and Sanov in the 1940's and 1950's. In particular, they applied it to stochastic problems in statistics and physics. The publications of Donsker, Varadhan, and Ellis in the 1970's and 1980's can also be regarded as important contributions to the development of the theory.

In recent years stochastic problems in the domain of Operations Research were tackled using techniques from LD, in particular problems in inventory and queueing. For instance Bucklew [1990] is an excellent summary of LD applications in engineering. The analysis of **rare events** is an important issue in stochastic Operations Research [overflows in queueing, stock outs in inventory, system failures in reliability, etc.].

LD provides us asymptotic expressions for the probability of certain rare events. Furthermore, it enables us to obtain variance reduction in Monte Carlo simulations. We call this method **Quick Simulation**.

1 An introduction: Large Deviations of sample averages

Consider a sequence of independent, identically distributed random variables $X_1, X_2, \ldots$ with finite mean μ. Denote by S_n the nth partial sum of the random variables: $S_n := \sum_{i=1}^{n} X_i$. As we know from probability [Weak Law of Large Numbers, Khinchine] the sample average S_n/n converges in probability to μ, i.e., for all positive ϵ $P(|S_n/n - \mu| \geq \epsilon) \to 0$ as $n \to \infty$. It is often the case that this convergence is also exponentially fast, i.e., there exist an $I > 0$ and $\alpha > 0$ such that $P(|S_n/n - \mu| \geq \epsilon) \approx \alpha \exp[-nI]$.

Now we look more precisely at one of the main results of the Theory of Large Deviations, Cramér's theorem. Before we state this theorem, we introduce the next two important functions:

(i) The **Moment Generating function**, also known as the free energy function, defined by:

$$M(\theta) := E\exp[\theta X_1] = \int_{-\infty}^{\infty} e^{\theta x} dP_{X_1}(x).$$

Using Hölder's inequality it can be shown that $\log M(\theta)$ is convex.

(ii) The **Large Deviations Rate function**, also known as the entropy function, defined as the Legendre-Fenchel transform [or convex conjugated] of $\log M(.)$:

$$I(x) := \sup_{\theta}[\theta x - \log M(\theta)].$$

In fact, the function $I(.)$ measures the discrepancy between the point x and mean μ. It can be proved that $I(x)$ is non-negative and attains its minimal value in $x = \mu$ and that $I(\mu) = 0$.

Now we can state **Cramér's theorem**. Assume $M(\theta) < \infty$ for all real θ. Then

(i) For every closed subset F on the real line:

$$\limsup_{n\to\infty} \frac{1}{n} \log P(S_n/n \in F) \leq - \inf_{x\in F} I(x).$$

(ii) For every open subset G on the real line:

$$\liminf_{n\to\infty} \frac{1}{n} \log P(S_n/n \in G) \geq - \inf_{x\in G} I(x).$$

For a proof of Cramér's theorem, see for instance Bucklew [1990] and Ellis [1985]. Let $\bar{K}$ denote the closure of the set K and K^o its interior. Note that if a set K is such that the infimum of $I(.)$ over $\bar{K}$ equals the infimum over K^o it may be derived that

$$\lim_{n\to\infty} \frac{1}{n} \log P(S_n/n \in K) = - \inf_{x\in K} I(x).$$

When we are dealing with for instance an interval K, it is immediately clear that we can replace the two inequalities in Cramér's theorem by one equality. Then we have that

$$P(S_n/n \in K) = \alpha(n) \exp[-n \inf_{x\in K} I(x)],$$

for a certain non negative function $\alpha(n)$ satisfying $\log \alpha(n) = o(n)$ where $n \to \infty$. We see that in some sense, the probability $P(S_n/n \in K)$ is asymptotically exponential, if $\mu \notin K$. Where laws of large numbers only tell us that certain probabilities go to zero, the LD provides us insight in the rate of convergence! It is clear now that LD can be regarded as an extension of the laws of large numbers.

2 Large Deviations of sample paths

Whereas we investigate the rate of convergence of sample averages in the introduction, we now focus on whole sample paths. Consider a sequence of independent, identically distributed (iid) random variables $\xi_1, \xi_2, \ldots$. Define S_k by $\sum_{i=1}^{k} \xi_i$. Next we scale the horizontal as well as the vertical axis by a factor $\frac{1}{n}$. To define the process for all t we interpolate linearly between $\frac{k}{n}$ and $\frac{k+1}{n}$, where $k \in \{0, 1, \ldots\}$. We denote the value of the obtained process at time t by $X_n(t)$, where $t \in [0, \infty)$. We say that $X_n(.)$ is the **Slow Markov Walk** belonging to the process $(S_k)_k$ and n.

Defining $\mu(t) := E\xi_1 \, t$, it is intuitively clear that the Slow Markov Walk $X_n(.)$ converges to $\mu(.)$. Just as in the case of the sample averages, LD bounds can be derived:

$$\limsup_{n\to\infty} \frac{1}{n} \log P(X_n(.) \in F) \leq - \inf_{f \in F} \int_0^T I(f'(t))dt = - \inf_{f \in F} I^*(f)$$

for closed sets F in the space of functions on $[0, T]$ equipped with the supremum metric, and

$$\liminf_{n\to\infty} \frac{1}{n} \log P(X_n(.) \in G) \geq - \inf_{f \in G} \int_0^T I(f'(t))dt = - \inf_{f \in G} I^*(f)$$

for open sets G. Here $I(x) := \sup_\theta(\theta x - \log E \exp[\theta\xi_1])$. Again, under weak conditions, the two inequalities may be replaced by equalities. In general, it is very difficult to perform the minimization over a given set of functions. However, it appeared that for a large class of sets, the level crossing sets, the minimization above can be calculated easily.

Fix an $r > 0$. Define A_r as the set of functions, starting in 0, who reach level r. As can be found in Bucklew [1990],

$$- \inf_{f \in A_r} \int_0^T I(f'(t))dt = -\theta^* r,$$

where θ^* is the (usually uniquely determined) positive solution of $E \exp[\theta\xi_1] = 1$. We can also find the element of A_r for which the minimum is attained: a linear function with slope $E\xi_1 \exp[\theta^*\xi_1]$. We conclude ($K$ large):

$$P\big((S_n)_n \text{ hits } [K, \infty)\big) = P\big(X_K(.) \in A_1\big) \approx \alpha \exp[-K\theta^*].$$

3 Quick Simulation in conjunction with LD

Consider the same random walk model with negative drift as in the last section. Suppose that we want to estimate the probability of $(S_n)_n$ hitting (M, ∞) before attaining a negative value by executing Monte Carlo simulation. Defining $dQ_{\xi_1}(x) := \exp[\theta^* x] dP_{\xi_1}(x)$, we see that ξ_1 distributed according to probability measure Q yields a random walk with a positive drift! Apart from that, it can be seen that the Slow Markov Walk under Q converges to a straight line with slope $E\xi_1 \exp[\theta^*\xi_1]$, which is also the slope of the optimum trajectory to a high level under the original distribution P!

This gives rise to simulating under Q instead of P. Calling the desired probability α, some details are given in the following recipe.

- Draw $\xi_1, \xi_2, \ldots$ under Q until for some k $\sum_{i=1}^{k} \xi_i = S_k > M$ or $\sum_{i=1}^{k} \xi_i = S_k < 0$. Let this be a run.

- Perform N_Q runs. If in run j $S_k = \sum_{i=1}^{k} \xi_i < 0$, then let $X_j := 0$. If on the contrary $S_k > M$ [with path $(S_1 = \omega_1, S_2 = \omega_2, .., S_k = \omega_k)$, say], we want to know how more likely such a realization is under Q than under P. Let f denote the density of ξ_1 under P. For the likelihood or Radon Nikodým derivative $dP(\omega)/dQ(\omega)$ of run j we have that

$$L_j := \frac{f(\omega_1)}{e^{\theta^* \omega_1} f(\omega_1)} \frac{f(\omega_2 - \omega_1)}{e^{\theta^*(\omega_2 - \omega_1)} f(\omega_2 - \omega_1)} \cdots \frac{f(\omega_k - \omega_{k-1})}{e^{\theta^*(\omega_k - \omega_{k-1})} f(\omega_k - \omega_{k-1})}$$

  Note that this equation can be simplified: $L_j = \exp[-\theta^* \omega_k]$.

- Defining $\hat{\alpha}_Q(n) := \sum_{j=1}^{n} (L_j X_j)/n$, an estimate of α is $\hat{\alpha}_Q(N_Q)$. A confidence interval can be derived easily.

This method is known as **Importance Sampling**. As can be found in Walrand [1988], we obtain variance reduction (since $L_j < 1$) in case of simulating under Q instead of P.

4 Applications in Queueing

In Queueing Theory we can use the Theory of Large Deviations to obtain approximations of the probability that the number of customers in the system exceeds some extreme value. Using Importance Sampling, we have also a tool to speed up the simulation of loss probabilities: stable queueing systems are modified into unstable systems in order to get better estimates. In the remainder of this section we treat these two items for some simple queueing systems.

- The **M/M/**c queue is a queueing system which assumes a Poisson arrival process [rate λ], exponentially distributed service times [with mean $\frac{1}{\mu}$] and c identical servers. To guarantee stability of the system, we let λ be smaller than $c\mu$.

  When a customer arrives or a service is completed, we say that the process makes a transition. Let X_j denote the number of customers in the queueing system immediately after the jth transition. It is not difficult to verify that $(X_j)_j$ is a [discrete time] Markov Chain. Its transition probabilities are given by:

$$\begin{aligned} P(X_j = i+1 | X_{j-1} = i) &= \frac{\lambda}{\lambda + \min(c,i)\mu} \qquad (i = 0, 1, \ldots) \\ P(X_j = i-1 | X_{j-1} = i) &= \frac{\min(c,i)\mu}{\lambda + \min(c,i)\mu} \qquad (i = 1, 2, \ldots) \end{aligned}$$

Now we want to examine the probability that, starting with an empty system, the number of customers in the system reaches n before returning to zero [for large n]. It can be seen that it approximately holds that $X_j := \sum_{i=1}^{j} \xi_i$, where $\xi_1, \xi_2, \ldots$ are iid random variables such that

$$\begin{aligned} \xi_j &:= \quad 1 \text{ with probability } \frac{\lambda}{\lambda + c\mu} \\ \xi_j &:= -1 \text{ with probability } \frac{c\mu}{\lambda + c\mu} \end{aligned}$$

With the theory of Large Deviations, we obtain

$$\lim_{n\to\infty} \frac{1}{n} \log P(n \text{ is reached before returning to } 0) = -\theta^*,$$

where θ^* denotes the positive solution of the equation $Ee^{\theta\xi_1} = 1$. Since this moment generating function equals $(\lambda e^{\theta} + c\mu e^{-\theta})/(\lambda + c\mu)$, $\theta^* = \log(c\mu/\lambda)$. The optimum trajectory is

$$E\xi_1 e^{\theta\xi_1}\, t = \left(\frac{c\mu - \lambda}{c\mu + \lambda}\right) t.$$

Performing Quick Simulation, the new transition probabilities are

$$\begin{aligned} \xi_j &:= \quad 1 \text{ with probability } e^{\theta\cdot(+1)} \frac{\lambda}{\lambda + c\mu} = \frac{c\mu}{\lambda + c\mu} \\ \xi_j &:= -1 \text{ with probability } e^{\theta\cdot(-1)} \frac{c\mu}{\lambda + c\mu} = \frac{\lambda}{\lambda + c\mu}, \end{aligned}$$

when $X_j \geq c$. In fact, we derived that the arrival rate λ and the service rate $c\mu$ interchange!

- An **M/E$_k$/1** queue is a stochastic process, in which customers arrive according to a Poisson process [rate λ]. One server is available and his service times are independent Erlang(k) distributed random variables with mean μ. We assume stability: $\lambda < \mu$.

The Erlang(k) distributed service time can be regarded as the sum of k independent exponentially distributed random variables, each with mean $1/k\mu$. So, the service time consists of k exponentially distributed phases. If a service phase is completed or a new customer has arrived the system makes a transition. Define X_j as the number of service phases yet to be completed, immediately after the jth transition. It is clear that the following relation holds:

$$\left\{ \begin{array}{l} \text{the number of customers in the system,} \\ \text{immediately after the } j^{\text{th}} \text{ transition, is } i \end{array} \right\} \Leftrightarrow X_j \text{ satisfies } (i-1)k < X_j \leq ik.$$

For more details, concerning this method of phases, see Tijms [1986]. $(X_j)_j$ is a Markov Chain. The transition probabilities are given by

$$\begin{aligned} P(X_j = i + k | X_{j-1} = i) &= \frac{\lambda}{\lambda + k\mu} \quad (i = 0, 1, \ldots) \\ P(X_j = k | X_{j-1} = 0) &= 1 \\ P(X_j = i - 1 | X_{j-1} = i) &= \frac{k\mu}{\lambda + k\mu} \quad (i = 1, 2, \ldots) \end{aligned}$$

Now our goal is to obtain an expression for

$$\lim_{n\to\infty}\frac{1}{n}\log P(n \text{ customers in the system before returning to } 0) =$$

$$\lim_{n\to\infty}\frac{1}{n}\log P((X_j)_j \text{ exceeds } nk \text{ before returning to } 0)$$

Analogously to the M/M/c queue it is easy to show that the expression above equals $-\theta_k^* k$, where θ_k^* denotes a positive solution of

$$M(\theta_k) := \frac{\lambda}{\lambda+k\mu}e^{\theta_k k} + \frac{k\mu}{\lambda+k\mu}e^{-\theta_k} = 1,$$

which appears to be uniquely determined.

Take for instance $k = 2, \mu = 6, \lambda = 1$, thus the arrival proces is a Poisson process with rate 1 and the services are Erlang(2) distributed [with mean $\frac{1}{6}$]. Solving $M(\theta) = 1$ we obtain $\theta = \log 3$. This yields

$$P(n \text{ customers in the system before returning to } 0) \approx \alpha(e^{-2\log 3})^n = \alpha\left(\frac{1}{9}\right)^n.$$

Performing Quick Simulation the new transition probabilities are

$$\begin{aligned} P(X_j = i+2 | X_{j-1} = i) &= e^{2\log 3}\frac{1}{13} = \frac{9}{13} \\ P(X_j = i-1 | X_{j-1} = i) &= e^{-\log 3}\frac{12}{13} = \frac{4}{13} \end{aligned}$$

Again, we obtain an unstable queueing system, just as we expected. After some calculations, we find that the optimum trajectory is given by $\frac{14}{13}t$.

- The $\mathbf{E}_k$**/M/1** model deals with interarrival times which are Erlang(k) distributed with mean $1/\lambda$. The service time of each customer is exponentially distributed. The expected service time is $1/\mu$. One server is available. In order to obtain a stable queueing system we assume $\lambda < \mu$.

  Analogously to the M/E_k/1 queue we obtain

$$\lim_{n\to\infty}\frac{1}{n}\log P(n \text{ customers in the system before returning to } 0)$$

$$= -\theta_k^* k \text{ , with } \theta_k^* > 0 \text{ satisfying } M(\theta_k) := \frac{k\lambda}{k\lambda+\mu}e^{\theta_k} + \frac{\mu}{k\lambda+\mu}e^{-k\theta_k} = 1.$$

- In the **M/D/c** queueing system we suppose a station with c servers at which customers arrive according to a Poisson process with rate λ. The service times of the customers are deterministic and equal $\frac{1}{\mu}$. The server utilization $\frac{\lambda}{c\mu}$ is smaller than 1. It is possible to deduce the asymptotics for the M/D/1 queue from the M/E_k/1 queue letting $k \to \infty$. However, we will use another technique.

When a customer arrives and server 1 is available, then we assume that the customer chooses server 1. Denote by X_j the number of customers in the system, immediately after the jth departure from server 1. $(X_j)_j$ is a Markov Chain. If $X_{j-1} < c$ then we have that $X_j = \text{Pois}(\lambda/\mu)$, where $\text{Pois}(\lambda)$ denotes a Poisson distributed random variable with mean λ; if $X_{j-1} \geq c$ then $X_j = \text{Pois}(\lambda/\mu) + X_{j-1} - c$. If, for large n, the Markov Chain exceeds n, X_j can be approximated by $\sum_{i=1}^{j} \xi_i$, where $\xi_1, \xi_2, \ldots$ is a sequence of iid random variables satisfying $\xi_1 \stackrel{d}{=} \text{Pois}(\lambda/\mu) - c$. Applying LD we obtain:

$$\lim_{n\to\infty} \frac{1}{n} \log P(n \text{ customers in the system before returning to } 0) = -\theta^*,$$

where θ^* denotes a positive of the equation $Ee^{\theta\xi_1} = 1$. Since

$$M(\theta) = Ee^{\theta\xi_1} = e^{-\theta c} \sum_{i=0}^{\infty} \frac{e^{\theta i} e^{-(\lambda/\mu)} (\lambda/\mu)^i}{i!} = \exp[-\theta c - \frac{\lambda}{\mu} + \frac{\lambda}{\mu} e^{\theta}],$$

we see immediately that we have to solve $\lambda e^{\theta} - \lambda - \mu\theta c = 0$. The most likely trajectory is given by: $\left(\lambda/\mu + (\theta^* - 1)c\right) t$, with θ^* satisfying $Ee^{\theta\xi_1} = 1$. Under the new distribution :

$$P(\xi_1 = i) = e^{\theta i} P\left(\text{Pois}(\lambda/\mu) - c = i\right).$$

- In the **D/M/**c queueing system the interarrival times are constant, say $\frac{1}{\lambda}$. The service times are independent exponentially distributed random variables [with mean $\frac{1}{\mu}$] and c servers. Again we assume $\lambda < c\mu$. Let X_j denote the number of customers in the system just before $\frac{j}{\lambda}$, i.e., the time of the jth arrival. Note that it approximately holds that $X_j = \sum_{i=1}^{j} \xi_i$, where the ξ_i are iid and $\xi_1 \stackrel{d}{=} 1 - \text{Pois}\,(c\mu/\lambda)$, when the process reaches a high level n. We obtain after some algebra,

$$\lim_{n\to\infty} \frac{1}{n} \log P(n \text{ customers in the system before returning to } 0) = -\theta^*,$$

where θ^* is a positive root of the equation $c\mu(e^{-\theta} - 1) + \lambda\theta = 0$. After some calculus we find that the optimum trajectory is given by $\left(1 + \theta^* - \frac{c\mu}{\lambda}\right) t$, with θ satisfying the equation above. Quick Simulation can be performed in a similar way as in the M/D/c case.

References

[1] J.A. Bucklew [1990]. *Large Deviation techniques in Decision, Simulation, and Estimation.* Wiley, New York.

[2] R.S. Ellis [1985]. *Entropy, Large Deviations, and Statistical Mechanics.* Springer Verlag, Berlin.

[3] H.C. Tijms [1986]. *Stochastic Modelling and Analysis, a Computational Approach.* Wiley, New York.

[4] J. Walrand [1988]. *An Introduction to Queueing Networks.* Prentice-Hall, New Jersey.

Entwicklung eines PC-gestützten Expertise-Modells für die Pflanzenschutz-Beratung

Angelika Adner, Hermann-Josef Kruse
Schering AG, 13342 Berlin, Germany

In einem Industriebetrieb stellt sich immer wieder die Aufgabe, für komplizierte und komplexe Sachverhalte, die sich noch dazu immer wieder wandeln, DV-gestützte Modelle zu entwickeln. Der folgende Beitrag handelt davon, wie in der Pflanzenschutz-Sparte der Schering AG mit einem solchen konkreten Problem praktisch umgegangen wurde.

Ein Pflanzenschutz-Berater benötigt für seine Empfehlung zur Pilzbekämpfung an Getreide eine Vielzahl von Informationen, denn eine gezielte Bekämpfungsmaßnahme (sprich: ein effektiver Fungizid-Einsatz) ist von vielen Faktoren abhängig. Es stellt sich nämlich die komplexe Aufgabe, aus einer Vielzahl von Fungiziden mit unterschiedlichen Wirkungen auf verschiedene Pilzkrankheiten ein solches Fungizid (oder eine Kombination) auszuwählen, mit dem ein bestimmter Pilzbefall an einer bestimmten Getreidesorte wirkungsvoll, ökonomisch sinnvoll und ökologisch vertretbar bekämpft werden kann. Hierbei sind viele Nebenbedingungen zu beachten (z.B. Zulassung, Wasserschutzauflage, Einsatztemperatur, erwarteter wirtschaftlicher Schaden).

Daraus ergab sich die Aufgabe, ein PC-Programm zu entwickeln, das den Pflanzenschutz-Beratern bei der Erstellung ihrer Expertisen über Maßnahmen zur Pilzbekämpfung Unterstützung liefern kann. Das Programm sollte anhand feldspezifischer Rahmenbedingungen und Angaben über Art und Stärke des Pilzbefalls eine Bekämpfungsempfehlung unterbreiten. Hierzu war es notwendig, alle für eine Expertise relevanten Informationen zu identifizieren und deren Verknüpfung in einem "Expertise-Modell" abzubilden.

In unserem Vortrag werden wir vor allem darauf eingehen, wie wir an das Projekt herangegangen, auf welche Schwierigkeiten wir gestoßen und wie wir damit umgegangen sind. Die Hauptschwierigkeit bestand im ständigen Ringen um den "optimalen Grad des Pragmatismus". Denn es ging zum einen um die Modellierung eines naturwissenschaftlich-technischen Sachverhalts, zum anderen mußte aber stets die praktische Beratungssituation im Auge behalten werden.

Die Akzeptanz unserer "Kompromißlösung" beruht u.E. auf folgenden drei Erfolgsfaktoren:

1. Beteiligung aller Know how-Bereiche im Projektteam (Vermeidung von "not invented here"-Effekten).
2. Frühzeitige und ständige Einbindung der Benutzer ("Kunden").
3. Eine schrittweise Programmrealisierung, bei der alle relevanten Unternehmensbereiche konkret an einer Weiterentwicklung des Modells beteiligt sind ("Prototyping").

DIE ANWENDUNG STATISTISCHER VERFAHREN ZUR RISIKOFRÜHERKENNUNG BEI DISPOSITIONSKREDITEN

Thomas Michael Bretzger, Deutsche Bank AG, 71522 Backnang

Der Dispositionskredit ist die wertmäßig wichtigste kurzfristige Kreditform im Privatkundenbereich. Aus diesem Grund kommt auch einer regelmäßigen Überwachung der eingeräumten Linien auf Ausfallrisiken besondere Bedeutung zu. Da die Stückgrößen dieser Kreditart vergleichsweise gering sind, bieten sich hierfür automatisierte Überwachungsverfahren an.

Beschrieben werden die Erfahrungen mit unterschiedlichen Diskriminanzanalyseverfahren beim Aufbau eines automatischen Risikofrüherkennungsverfahrens. Betrachtet werden neben der linearen und quadratischen Diskriminanzanalyse die in der Praxis häufig verwendeten Punktebewertungsverfahren sowie unter anderen nichtparametrische Dichteschätzungen nach der Potentialfunktions- und Nächste-Nachbarn Methode.

Von besonderer Bedeutung ist die geeignete Auswahl von Variablen im Zusammenhang mit verwendetem Verfahren und vorgegebenen Kosten der Fehlklassifikation.

Daneben sollen die einzelnen Schritte zum Aufbau eines derartigen Risikofrüherkennungssystems anhand eines praktischen Beispiels dargestellt, sowie einzelne Ergebnisse und Schlußfolgerungen erläutert werden.

GEMISCHT-GANZZAHLIGE MODELLE DER PRODUKTIONSOPTIMIERUNG IN DER CHEMISCHEN INDUSTRIE

Beate Brockmüller, BASF AG, Ludwigshafen
Josef Kallrath, BASF AG, Ludwigshafen

In diesem Vortrag werden einige in der chemischen Industrie auftretende Probleme aus den Bereichen Produktionsprogrammplanung, Anlagenbelegungsplanung und Losgrößenoptimierung vorgestellt, die mathematisch auf gemischt-ganzzahlige Modelle abgebildet werden können.

Diese Modelle berücksichtigen in den Nebenbedingungen unter anderem Produktionskapazitäten, Rezepturen, Lagerbilanzen, Sicherheitsbestände, Zukaufmöglichkeiten, Transporte, Verkäufe und in den Zielfunktionen insbesondere Herstellkosten, Lagerkosten, Zusatzlagerkosten, Transportkosten, Versandkosten und Verkaufserlöse. Sie enthalten ganzzahlige Variablen wie sie typischerweise bei der Einbeziehung von Produktwechseln, Zuordnungen von Produkten zu Anlagen, Mindestproduktionsmengen und minimalen Transportmengen auftreten.

Neben der Behandlung dieser Probleme mit den üblichen Integer-Optimierungsverfahren werden auch andere in der Praxis gebräuchliche und relevante Lösungsverfahren wie algorithmische Verfahren, deterministische Verfahren und dynamische Optimierung besprochen.

Schließlich werden Erfahrungen und kritische Anmerkungen zum praktischen Einsatz von gemischt-ganzzahliger Programmierung angegeben.

OPTIMIERUNG ALLER MAßNAHMEN ZUR DECKUNG DES KLASSENRAUMBEDARFS AN HAMBURGER SCHULEN

S. Fleischmann, CAP debis GEI GmbH, Hamburg

In den letzten Jahren sind in der Stadt Hamburg aufgrund rückläufiger Schülerzahlen Kapazitäten an Klassenräumen abgebaut worden. Mittlerweile haben sich die Voraussetzungen geändert. Die Behörde für Schule, Jugend und Berufsbildung (BSJB) geht davon aus, daß bis zum Jahre 2005 mit einem Zuwachs in Höhe von ca. 15 Prozent zu rechnen ist. Erschwert wird das Problem dadurch, daß der Schülerzuwachs nicht homogen auf alle Stadtteile, Schulformen und Klassenstufen verteilt ist. Es ist damit zu rechnen, daß an bestimmten Schulen bis zu 30 Prozent mehr Schüler angemeldet werden.

Deckung des zusätzlichen Klassenraumbedarfs durch Neubau ist kostspielig und nicht immer möglich. Es gilt daher, die vorhandenen Kapazitäten optimal zu nutzen und Erweiterungsmaßnahmen möglichst zu vermeiden. Dabei ist eine für alle Schulen "akzeptable" Lösung zu finden.

Zur Deckung des Raumbedarfs sind folgende Maßnahmen möglich:

- Vorübergehende Bereitstellung von Räumen in Überschuß-Schulen für Mangel-Schulen (Raumverfügungen)
- Reaktivierung stillgelegter, artfremd genutzter oder vermieteter Räume
- Neubau von Pavillons oder festen Gebäuden

Für diese Maßnahmen gibt es eine ganze Reihe von einschränkenden Bedingungen. Die wichtigsten davon sind:

- Raumverfügungen sind nur zwischen bestimmten Schulen zulässig. Neben geographischer Nachbarschaft sind dabei weitere Bedingungen, die schwer in Regeln zu fassen sind, von Bedeutung.
- Im Falle einer Raumverfügung ist der Mangel-Schule eine Mindestanzahl von Klassenräumen zu überlassen.
- Um eine zu starke Zersplitterung zu verhindern, ist eine maximale Anzahl von Standorten für die einzelnen Schulen nicht zu überschreiten.

Aufgrund der Vielzahl an Handlungsalternativen und der komplexen Randbedingungen ist manuell eine optimale Lösung kaum zu finden. Die mathematische Formulierung aller Anforderungen führt aber auf ein gemischt ganzzahliges Optimierungsproblem, das für die automatisierte Lösung durch einen Rechner gut geeignet ist.

Dazu wird die Standard-Optimierungs-Software SCICONIC eingesetzt. Einerseits erleichtert SCICONIC dem OR-Fachmann die Eingabe des mathematischen Modells während der Entwicklungsphase, andererseits läßt sich SCICONIC mit seinem SQL-Interface gut in ein dialogorientiertes, anwenderfreundliches Software-System einbinden. Dies ist von besonderer Bedeutung, da die Optimierung der durchzuführenden Maßnahmen vor jedem Schuljahr durch Planer der BSJB erfolgt, die weder DV- noch OR-Erfahrung besitzen.

PROGNOSE MITTELFRISTIGER MARKTPOTENTIALE IN DER AUTOMOBILINDUSTRIE

Reinhard Hecking, Volkswagen AG Wolfsburg

Abstract:
This paper contains elements and structures of the VW mean-term forecasting-system. The mathematical model is based on multiple regression analysis.

Die mittelfristige Planung von Produktionskapazitäten, Personalbestand, Finanzbedarf etc. eines Automobilherstellers basiert wesentlich darauf, das Pkw-Marktpotential mittelfristig hinreichend gut quantifizieren zu können.

Mittelfristige Prognosen des Marktpotentials werden bei Volkswagen mit Hilfe explikativer Verfahren auf der Basis multipler Regressionsanalyse durchgeführt.

Unter Verwendung von jährlichen Neuzulassungswerten wird zunächst mit Hilfe der Kleinste-Quadrate-Methode eine Dekomposition der (historischen) Marktentwicklung in die langfristige Komponente (L_t) und die mittelfristige Komponente (M_t) erzeugt, so daß sich für das zugrundeliegende Analyse- und Prognosemodell ergibt:

$P_t = L_t \times M_t$

wobei

L_t = langfristige Entwicklung asymmetrisch logistischer Natur (Bonus-Typ)

M_t = mittelfristige Konjunktur-Komponente

$$P_t = \frac{P^*}{1+e^{a-bt}+e^{c-dy_t}} \times (f_1(V_1) + \ldots + f_n(V_n))$$

mit

P^* = Sättigungsniveau

y_t = exogene Einflußvariable

$f_i(V_i)$ = Einfluß der Konzeptvariable V_i auf die Pkw-Markt-Konjunktur (wirtschaftliche und marktspezifische Einflüsse)

Signifikanz und Stärke der Einflüsse der verschiedenen Konzeptvariablen (wie z. B. reales Masseneinkommen), die nicht-lineare und mit time-lags versehene Modellcharakteristiken besitzen, wird mit Hilfe multipler Regressionsanalyse ermittelt.

Die folgende Abbildung zeigt die Struktur des Pkw-Markt-Modells.

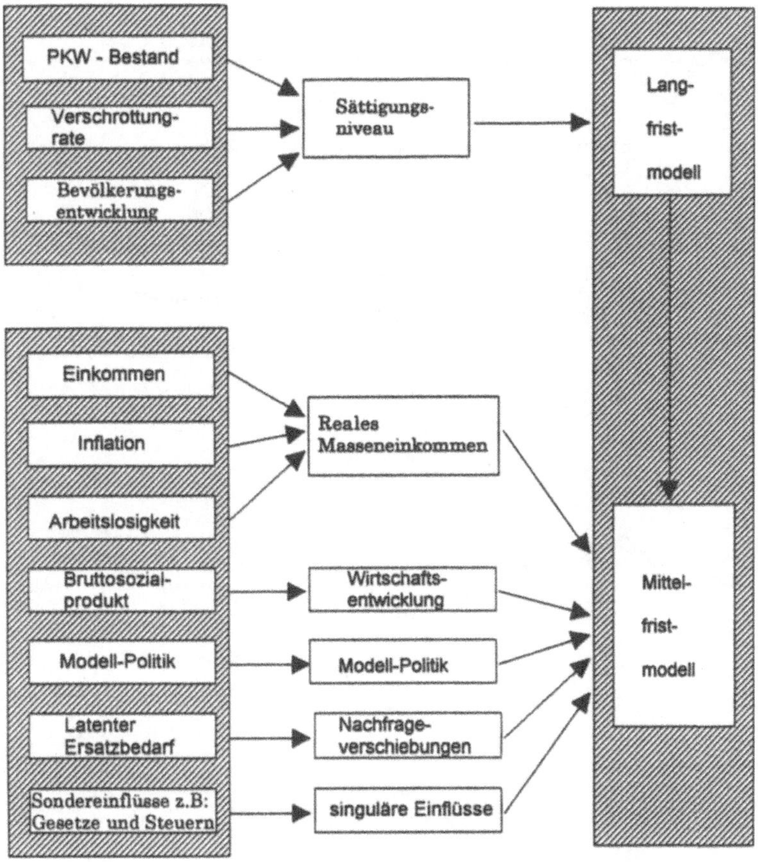

Zwei Einflußfaktoren sowie die Pkw-Gesamtmarktentwicklung und -prognose in Deutschland (West) zeigen die nachfolgenden Abbildungen.

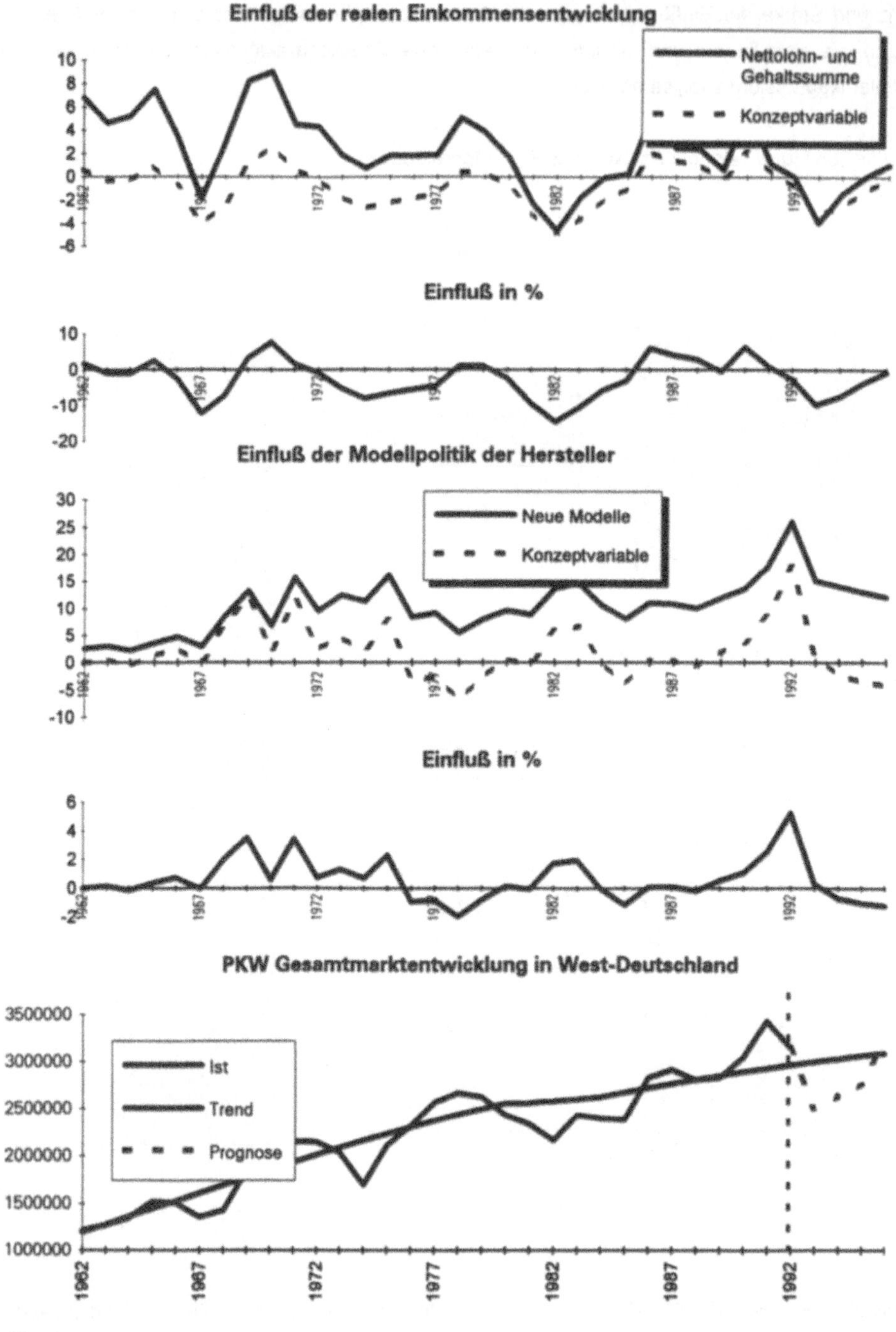

Literatur:
Lewandowski, Rudolf:
Prognose- und Informationssysteme; Band 2; Berlin; New York; W. de Gruyter (1980).

PRODUKTIONSLENKUNG UND AKQUISITION ÜBER GRAFISCHE SIMULATION

Christel Hollender, Mannesmann Datenverarbeitung GmbH, Ratingen

Diskrete Simulation ist weitestgehend anerkannt als Hilfsmittel für die Planung und Entscheidung von Investitionen und wird besonders in der Anlagen- und Fabrikplanung sowohl bei Neubauten als auch bei Modernisierungen eingesetzt. Sie ist jedoch darüber hinaus ebenfalls zur Unterstützung der Akquisition im Vorfeld großer Projekte oder nach der Inbetriebnahme zur Steuerung der Produktion geeignet. Dabei kann ein entwickeltes Simulationsmodell alle drei Phasen - Akquisition, Anlagenplanung und Produktionslenkung - oder nur eine Teilmenge dieser Phasen abdecken.
In der Akquisition ist es wesentlich, dem Kunden die angebotene Lösung anschaulich und überzeugend darzustellen. Da mit heutigen modernen Tools erstellte Simulationen immer auch eine Animation liefern, gelingt dies in sehr hohem Maß. Die Abläufe der dynamischen Prozesse werden für den Kunden sichtbar gemacht. Ebenso lassen sich verschiedene Planungsalternativen beim Kunden direkt vorführen und bewerten.
Einen wichtigen Platz nimmt eine Simulation auch in der Produktionslenkung ein. Hier gibt es zwei Aspekte, je nachdem mit welchen Daten gearbeitet wird. Stehen als Input Plandaten zur Verfügung, z.B. die Produktionsdaten eines Tages, so kann die Simulation als Planwerkzeug eingesetzt werden, um z.B. eine verbesserte Produktionsreihenfolge und entsprechende Maschineneinstellungen zu ermitteln. Sind die Eingangsdaten Istdaten, so veranschaulicht die Simulation die Produktion, um Soll-Ist-Vergleiche zu schaffen oder um aufbauend auf den erreichten Zuständen und Ergebnissen neu zu planen und die Planung durch Alternativrechnungen zu optimieren.
Diese Einsatzgebiete sollen anhand von Modellbeispielen aus der Praxis verdeutlicht werden.

MULTIFUNKTIONALER EINSATZ VON PERSONAL IN DER HOTELLERIE

Gerhard Muche, Ammersbek

Zusammenfassung: In diesem Beitrag werden die Konsequenzen auf Kosten und Qualitätsanforderungen untersucht, die aus abteilungsübergreifenden Einsätzen von Arbeitskräften in Hotelbetrieben resultieren. Es wird ein Planungsansatz präsentiert, mit dem die kostenminimale Personalstruktur in Abhängigkeit vom finanziellen Anreiz abgeleitet werden kann.

Abstract: In this paper we analyse the effects on cost and quality coming from multifunctional allocation of personnel in hotels. We present an optimisation model which allows to determine the minimal cost of personnel structur depending on financial incentive.

Der Kostendruck zwingt Dienstleistungsunternehmen, neue Wege zu gehen. Vierzig Prozent der Kosten in der Hotellerie betreffen das Personal. Das ist der größte Kostenblock. Es folgt der Wareneinsatz mit rund dreißig Prozent. Die Kosten des Wareneinsatzes unter Kontrolle zu bringen, war das Thema der letzten 10 Jahre. Derzeit werden die Personalkosten verstärkt diskutiert.

Es ist für die Hotellerie typisch, daß in einem sehr flexiblen Schichtdienst gearbeitet wird. Deshalb werden für das Personal Dienstpläne geschrieben. Im Dienstplan wird für jede Arbeitskraft eingetragen, an welchen Tagen sie frei und an welchen Tagen sie Dienst hat. Für Einsatztage wird der Beginn und das Ende des Dienstes benannt. Der Dienstplan hat eine Gültigkeitsdauer von einer Woche und sollte mindestens eine Woche im voraus erstellt und bekannt gemacht werden.

Dienstpläne werden von den Abteilungsleitern erstellt und zwar jeweils für ihren Verantwortungsbereich, das ist die Abteilung, der sie vorstehen. Die Personaleinsätze, die im Dienstplan geregelt werden sollen, müssen sich am Bedarf orientieren, wenn man betriebswirtschaftlich sinnvoll planen will. Dies ist jedoch in der Praxis bei weitem nicht immer der Fall. Dienstpläne bilden deshalb einen denkbaren Ansatzpunkt für die Steuerung der Personalkosten.

In den Abbildungen 1 und 2 sind für zwei Abteilungen die Kurven des Arbeitsanfalls für einen Tag abgebildet.

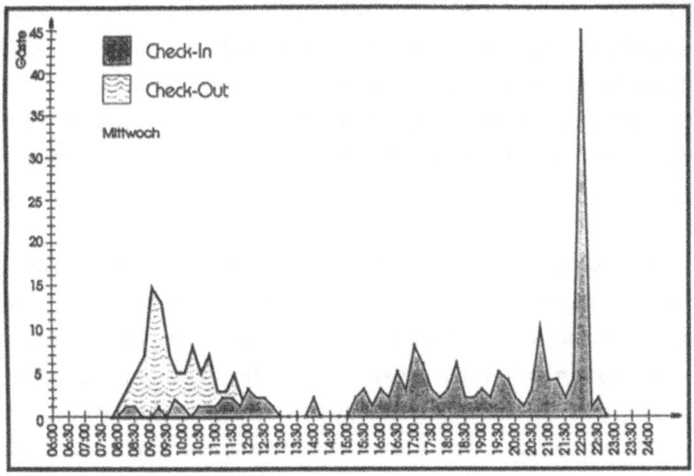

Abb. 1: Arbeitsverteilung im Bereich Front Office

Im Front Office ist morgens der typische Arbeitsanfall gekennzeichnet durch die Check-Outs. Die Abbildung zeigt, daß ab 9:00 Uhr bis 12:15 Uhr die Check-Outs stattfinden. Abends hingegen sind es vor allem die Check-Ins, die die Arbeit am Front Office darstellen.

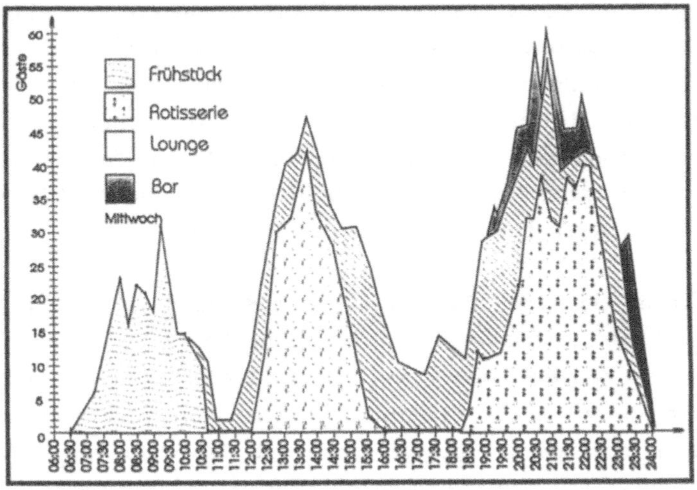

Abb. 2: Arbeitsverteilung im Bereich Restaurant

Wenn der Empfangschef den Dienstplan für seine Abteilung aufstellen soll, braucht er am meisten Leute zwischen 9 und 10 Uhr und von 16 bis 22 Uhr. Ansonsten reicht eine Mindestbesetzung.

Eine Arbeitskraft kann jedoch nur für ganze Schichten eingeteilt werden. Das heißt es werden Arbeitskräfte in der Abteilung vorgehalten, die etliche Zeit Leerlauf haben. Damit die Leerlaufzeiten nicht überhandnehmen, werden Bedarfsspitzen nicht gedeckt. Konkret bedeutet dies, daß Gäste warten müssen, bis sie bedient werden. Eine Minimierung der Leerlaufzeiten verfolgt das Ziel der **Kostenminimierung**. Eine Minimierung der Wartezeiten für die Gäste verfolgt hingegen das Ziel, daß man durch **Qualitätsstandards** setzt.

Ein großer Nutzen würde entstehen, wenn es gelingt, die Leerlaufzeiten einzuschränken. Der einzelne Abteilungsleiter kann dabei jedoch wenig ausrichten. Durch verschieden lange Schichtdauern und sogenannte geteilte Dienste kann man abteilungsweise agieren. Hinderlich hierbei ist die Prognoseunsicherheit zum einen beim Arbeitsanfall und zum anderen beim Personaleinsatz.

Vorteilhaft wäre es, wenn es einen Austausch zwischen den Abteilungen geben würde. Voraussetzung dafür ist, das in einer Abteilung wenig, während gleichzeitig in einer anderen Abteilung viel zu tun ist. Außerdem müßte man schnell reagieren können. Hier kommt der Umstand zugute, daß der Austausch zwischen zwei Abteilungen nicht eine Woche im voraus angekündigt werden muß. Diese Feineinteilung des Personaleinsatzes kann am Tag des Geschehens vorgenommen werden.

Daß der Austausch zwischen Abteilungen realistischerweise Vorteile bringen kann, macht ein Vergleich von Abb. 1 und 2 deutlich. Abb. 2 gibt die Arbeitsverteilung im Restaurantbereich desselben Tages an. Das Tal zwischen Frühstück und Mittagessen fällt gut mit den Check-Outs zusammen. Dies ist auch plausibel, wenn man bedenkt, daß ein Gast erst frühstücken wird, bevor er sich aus-checkt. Das Tal im Front-Office stimmt fast genau mit dem Bedarf für das Mittagessen überein.

Es gibt noch einen dritten Bereich, der in einen Austausch einbezogen werden könnte. Dies ist der sogenannte Housekeeping-Bereich. Jemand, der am Empfang eingesetzt werden kann, ist prinzipiell auch in der Lage, die Zimmerreinigung zu kontrollieren. D.h. in den schwachen Zeiten des Empfangs kann eine derartige Arbeitskraft die Hausdame bei der Zimmerkontrolle ergänzen. Oder ein sogenanntes Zimmermädchen, adrett gekleidet, kann beim Frühstück bedienen. Die Zimmerreinigung hat den Vorteil, daß es eine Tätigkeit mit verschiebbaren Charakter ist. Ein Zimmer kann zwar erst gereinigt werden, wenn der Gast es verlassen hat. Danach kann die Tätigkeit aber in die Bedarfstäler gelegt werden.

Zusammenfassend ist festzustellen: Wenn es gelingt, von der Abteilungsstruktur, die in den Dienstplänen abgebildet wird, fortzukommen und Personal bereichsübergreifend einzusetzen, müßte man bei gleicher Qualität mit weniger Personal auskommen; beziehungsweise bei gleichem Personalbestand könnte ein besserer Service angeboten werden.

Personalbedarfe

lfd. Nr.	Funktion / Zeit	A Restaurant	B Front Office	C Housekeep.	Summe
1	06:00	2	1	1	4
2	06:30	2	1	1	4
3	07:00	3	2	1	6
4	07:30	3	2	1	6
5	08:00	4	3	1	8
6	08:30	4	3	1	8
7	09:00	3	4	1	8
8	09:30	3	4	1	8
9	10:00	2	4	2	8
10	10:30	2	4	2	8
11	11:00	0	3	5	8
12	11:00	0	2	5	7
13	12:00	2	2	5	9
14	12:30	2	2	5	9
15	13:00	3	1	5	9
16	13:30	4	1	5	10
17	14:00	4	1	4	9
18	14:30	2	1	5	8
19	15:00	2	1	5	8
20	15:30	1	2	5	8
21	16:00	1	2	2	5
22	16:30	1	2	2	5
23	17:00	5	3	2	10
24	17:30	6	3	1	10
25	18:00	7	3	1	11
26	18:30	7	3	1	11
27	19:00	7	3	1	11
28	19:30	7	3	1	11
29	20:00	7	3	1	11
30	20:30	7	3	1	11
31	21:00	7	3	1	11
32	21:30	7	3	1	11
33	22:00	6	3	0	9
34	22:30	6	1	0	7
35	23:00	5	1	0	6
36	23:30	5	1	0	6

Abb. 3: Personalbedarfe

Im folgenden wird nicht auf die organisatorischen Aufgaben einer solchen Zusammenfassung eingegangen, sondern die Frage behandelt, welches Einsparungspotential durch Flexibilität erreichbar ist. Dies wird an einem Beispiel mit den drei genannten Bereichen A Restaurant, B Front-Office und C Housekeeping dargestellt.

Die Abb. 3 zeigt die Bedarfe an Personen im Zeitablauf. Und zwar für jede Abteilung und für jede halbe Stunde von 6 Uhr morgens bis Mitternacht. Die Schwankungen in den Abteilungen gleichen sich zum Teil aus.

Für jede Abteilung kann der Personaleinsatz separat geplant werden, also entsprechend der gegenwärtigen Situation. Die drei Abteilungen können allerdings auch unter der Berücksichtigung von Austauschbeziehungen gemeinsam geplant werden.

Wir nutzen dazu einen Modellansatz, bei dem ein strukturierter Bedarf mit einer strukturierten Personalausstattung unter Schichtalternativen abgestimmt werden kann (vgl. letzte Seite dieser Abhandlung und /1/). Der Bedarf ist bei der vorliegenden Fragestellung eine gegebene Größe. Die Anzahl von Arbeitskräften, unterschieden nach ihrer Qualifikation, ist eine gesuchte Größe. Zielkriterium ist die Minimierung der Personalkosten. Zu den Planungsvoraussetzungen des Zahlenbeispiels ist noch zu bemerken, daß die Bedarfsspitzen gekappt wurden, d.h. pro Teilperiode wurde eine Unterdeckung von 1 zugelassen. Alle Schichten haben eine Länge von 8 Stunden ohne Pausen. Zu jeder halben Stunde kann eine Schicht beginnen.

In Abb. 4 sind die separat und gemeinsam geplanten Abteilungen zusammengestellt.

Kennzahlen zum Bedarf und Einsatz des Personals

	A Restaurant	B Front Office	C Housekeep.	Summe ABC	Gesamtbereich
Personalbedarf (Std)	69,5	42,0	38,0	149,5	149,5
Personaleinsatz (Std)	88,0	48,0	40,0	176,0	160,0
Leerlauf (Std)	19,0	8,0	6,5	33,5	12,0
Unterdeckung (Std)	0,5	2,0	4,5	7,0	1,5
Abweichung (Std)	19,5	10,0	11,0	40,5	13,5
Leerl./Pers.einsatz (%)	21,6	16,7	13,8	19,0	7,5
Unterd./Pers.eins. (%)	0,6	4,2	11,3	4,0	0,9
Abw./Pers.eins. (%)	22,2	20,9	26,1	23,0	8,4

Einsparungspotential	16 Std.	9,0 %
Verringerung der Abweichung	auf ca. 1/3	

Abb. 4: Kennzahlen zum Personaleinsatz

Interessant ist der Vergleich der notwendig einzusetzenden Arbeitskräfte bei der separaten und der gemeinsamen Planung. Statt 176 werden nur noch 160 Einsatzstunden benötigt. Das entspricht zwei Arbeitskräften oder 9%. Die Unter- und Überdeckung in Summe macht die Abweichung des Einsatzes vom Bedarf deutlich. Diese verringert sich auf ca. ein Drittel. Die Unterdekkung sinkt auf ein Viertel, die "Qualität" verbessert sich entsprechend.

Man kann jedoch leider nicht davon ausgehen, daß dieses Einsparungspotential in toto realisiert werden kann, wenn dieses die vollständige qualifikatorische Flexibilität des Personals voraussetzen würde. Nicht alle Arbeitskräfte sind hinsichtlich ihres Willens und ihrer Qualifikation in der Lage, alle infrage kommenden Tätigkeiten auszuführen. Bei denen, die Willens sind und die nötigen Voraussetzungen mitbringen, muß eine Schulung durchgeführt werden. Das bedeutet Aufwand. Um Arbeitskräfte zu motivieren, muß bei einer Anreicherung der Stellen ein entsprechender Anreiz in Form eines höheren Gehaltes gegeben werden. Dies führt auf die Frage nach der notwendig bereitzustellenden Flexibilität.

Die Abb. 5 zeigt die Wirkung von Veränderungen der Kosten pro Arbeitskraft auf die Personalausstattung. Hier ist von der ganz einfachen Annahme ausgegangen worden, daß doppelt-qualifizierte Arbeitskräfte mehr kosten werden als einfach-qualifizierte Arbeitskräfte. Wenn die Einfach-Qualifizierten mit einem Kostensatz von 100 belegt werden und die Doppelt-Qualifizierten mit 100 plus x, so kann man x variieren und die Wirkung auf die Personalausstattung prüfen. Wenn man gewillt ist, bis zu 33% mehr für eine doppelt-qualifizierte Arbeitskraft auszugeben und man dafür 7 Leute motivieren kann, dann lohnt es sich noch.

Wenn höhere Kosten pro doppelt-qualifizierter Arbeitskraft erforderlich werden, ist es lohnend nur, wenn man auf 4 Leute zurückgreift. Kostet eine Doppelt-Qualifikation 100 % mehr, so lohnt ein Einsatz überhaupt nicht.

Multifunktionalität Strukturund Kosten

Kann / Mehr pro Doppelt-qual.	nur A	nur B	nur C	A und B	A und C	B und C	Summe	Kosten
0%	0	1	0	10	5	4	20	2000
1%	7	3	3	3	3	1	20	2007
33%	7	3	3	3	3	1	20	2231
34%	9	5	3	1	2	1	21	2236
99%	9	4	4	2	1	1	21	2496
101%	12	7	6	0	0	0	25	2500

Legende: Basis 100 Geldeinheiten pro einfach-qualifizierter Arbeitskraft. Bei z.B. 33% mehr pro doppelt-qualifizierter Arbeitskraft kostet diese 133 Geldeinheiten. Dann sind 7 Arbeitskräfte erforderlich, die nur A können, usw. Insgesamt sind 20 Arbeitskräfte erforderlich. Kosten entstehen in Höhe von 2231 Geldeinheiten.

Abb. 5: Multifunktionalität Struktur und Kosten

Der Ansatz läßt zu, die Kostensätze nach Art der Qualifikation zu unterscheiden, also z.B. die Situation "Es ist teurer einen Kellner als einen Empfangssekretär zum Etagendienst zu bewegen" in ihrer Konsequenz zu untersuchen. Ebenso kann man die dreifach-Qualifizierten in die Überlegungen einbeziehen und die Situation auf mehr als drei Bereiche ausweiten.

In Abb. 6 ist der verwendete Planungsansatz dargestellt. In diesem Planungsansatz ist der konkrete Personaleinsatz nicht enthalten. Dieser ist nur implizit berücksichtigt. Eine Lösung des dargestellten Ansatzes sichert jedoch auch die Existenz eines zulässigen Einsatzplanes. Die Verwendung dieses impliziten Ansatzes hat den Vorteil, daß das Modell eine Größenordnung behält, die mittels handelsüblicher Software gelöst werden kann. Der Ansatz ist lösungstechnisch gutartig, denn obwohl Ganzzahligkeit der Variablen gefordert ist, liegen die Rechnenzeiten für die angegebene Situation bei ca. 1 Minute (486er-Prozessor, 33 MHz, Hyper-Lindo, 147 Variablen, 253 Bedingungen).

$$\sum_{r=1}^{R} \sum_{p\in P_r} c_{rp}\cdot PA_{rp} \overset{!}{=} \min$$

$$\sum_{q\in\tilde{Q}} PB_{qt} \leq \sum_{r\in\bigcup_{q\in\tilde{Q}} R_q} \sum_{p\in P_r} a_{rpt}\cdot PA_{rp} \quad \text{für alle } \tilde{Q}\in\mathfrak{P}(\underline{Q}) \text{ und } t\in\underline{T}$$

$$PA_{rp} \geq 0 \text{ und ganzzahlig für alle } r\in\underline{R} \text{ und } p\in P_r$$

R	Zahl Arbeitskräftekategorien	r	Index für eine Arbeitskräftekategorie
Q	Zahl der Tätigkeitskategorien	q	Index für eine Tätigkeitsart
T	Zahl der Teilperioden	t	Index für eine Teilperiode
P	Zahl der Schichtenarten	p	Index für eine Schichtart
$\mathfrak{P}(\underline{Q})$	Potenzmenge von $\underline{Q}$	$\tilde{Q}$	Teilmenge von Q

R_q Indexmenge der r, die für Tätigkeitsart q einsetzbar sind

P_r Indexmenge der p, die für Arbeitskräfteart r zulässig sind

$\underline{N}$ gleich {1,2,...,N}, wenn N eine natürliche Zahl ist

PB_{qt} gegebener Personalbedarf für Tätigkeitsart q in Teilperiode t

PA_{rp} gesuchte Personalausstattung der Art r, die gemäß Schicht p arbeiten

c_{rp} Personalkosten für eine Arbeitskraft der Art r und Schicht p

a_{rpt} Anwesenheitskoeffizient: Arbeitskraft der Art r ist in Schicht p zum Zeitpunkt t anwesend ($a_{rpt}=1$) oder abwesend ($a_{rpt}=0$)

Abb. 6: LP-Modell zur Bestimmung von Niveau und Struktur der Personalausstattung aufgrund eines strukturierten Personalbedarfes.

Literatur

/1/ Kossbiel, H.:
Personalbereitstellungsplanung bei Arbeitszeitflexibilisierung.
ZfB Zeitschrift für Betriebswirtschaft, 62. Jg. (1992) H.2 175-198, hier Modell D, S. 181/182.

Über ein Modell der Arbeitsverteilung und Kolonnenbesetzung bei der Müllabfuhr

Jürgen Peemöller, Hamburg

Die Müllabfuhr einer großen deutschen Stadt wird sukzessive ihre Müllfahrzeuge mit einer neuen Schüttechnik ausrüsten, so daß kleine und große Müllbehälter von einem Fahrzeug (und nicht wie bisher von zwei) geleert werden können. Dies führt zu einer Veränderung der Arbeitsabläufe und insbesondere der Kolonnenbesetzung. Bisher waren die Fahrzeuge mit 4 Auflader zur Leerung der kleinen und 2 Auflader zur Leerung der großen Gefäße besetzt. Bei der Universalschüttung sind 2, 3 oder 4 Auflader zulässig (in Abhängigkeit vom Mix der Gefäße und anderer Faktoren).

Aufgabe des Projekts war die Entwicklung eines Bewertungs- und Planungsmaßstabes, der es zum einen erlaubt, unter Berücksichtigung aller relevanten Einflußfaktoren die durch einen Mitarbeiter pro Tag zu bewältigende Leistung und die Kolonnenbesetzung zu ermitteln und zum anderen praxisgeeignet ist, d.h. möglichst einfach handhabbar und für die Mitarbeiter verständlich und nachvollziehbar.

Neben der Modellentwicklung auf der Basis interner und externer Erkenntnisse und Erfahrungen waren umfangreiche empirische Erhebungen zur Überprüfung der Modellhypothese sowie zur konkreten Bestimmung der im Modell definierten Arbeitsrichtwerte erforderlich. Im Rahmen der Auswertung wurde eine kombinierte Methode von Multimoment-Technik und linearer Regression eingesetzt, wobei die Ergebnisse einer Vielzahl statistischer Prüfungen unterzogen wurden. Ergänzend wird dargestellt, wie auf Basis des Modells eine optimierende Tourenplanung mit variabler Kolonnenbesetzung durchgeführt werden kann.

EIN LINEARES OPTIMIERUNGSMODELL ZUR PRODUKTIONSPLANUNG BEI ZYKLISCHER KUPPELPRODUKTION

Christoph Pesch, Henkel KGaA, Düsseldorf

In der Fettchemie werden native Öle und Fette verarbeitet und die daraus erzeugten Produkte vermarktet. Die größten Produktgruppen sind Fettsäuren, Methylester und Fettalkohole.

Bei dem eingesetzten DV-System handelt es sich um ein Mehrperioden-Planungsmodell. Es wird zur Monats- bzw. Jahresplanung und für Alternativrechnungen (Rohstoffe, Verfahren, Verkaufsprogramme) genutzt. Es können sowohl einzelne, als auch mehrere Produktionsstandorte des Unternehmens betrachtet werden. Ziel ist es, den optimalen Einsatz von Stoffen und Kapazitäten zur Erfüllung des Verkaufsprogramms zu berechnen, so daß entweder die Gesamtkosten minimiert oder der Gewinn maximiert wird. Vorhandene Absatz-, Beschaffungs- und Kapazitätsrestriktionen werden berücksichtigt.

Die Vielzahl der verschiedenen Verfahrenswege, die Verwendung der Kuppelprodukte bei zyklischer Kuppelproduktion und der Einsatz alternativer Rohstoffe machen die Möglichkeiten der Auswahl und Zuordnungen so groß, daß eine akzeptable Lösung wegen der differenzierten Produkte auf allen Ebenen und der Kapazitätsbeschränkungen der Anlagen nur durch ein lineares Optimierungsmodell gefunden werden kann. Dazu wird die Optimierungssoftware OSL (Optimization Subroutine Library) der IBM auf einer RS6000 eingesetzt.

Alle dafür benötigten Daten werden in einer Datenbank verwaltet und gepflegt. Die Datenerfassung und die Auswertungen der Optimierungsergebnisse sind so organisiert, daß der Benutzer nur mit ihm bekannten Schlüsseln wie Artikelnummer, Stoffnummer, Rezepturnummer usw., arbeitet und die speziellen Dateien, die die Optimierungssoftware benötigt, weder sieht noch bearbeitet.

Das heutige Angebot von Hard- und Software erlaubt es, Modelle mit 14.000 Zeilen und 38.000 Spalten in vertretbarer Zeit zu lösen.

OPTIMIERUNGSAUFGABEN IN UNTERNEHMENSPROJEKTEN

Detlev Ruland, Siemens AG, München

Unternehmensprojekte werden segmentiert in die zwei Klassen:

a) Strategie-Projekte:

Ausrichtung (neuer) Geschäfte, Spezifikation der Strategie und wirtschaftliche Ziele, Ableitung von Maßnahmen

b) Restrukturierungs-Projekte:

Analyse der Ursachen für verlustträchtige Geschäfte, Identifikation von Ansatzpunkten und Spezifikation von Maßnahmen, Umsetzung und Kontrolle der Maßnahmen

In beiden Projektklassen sind zahlreiche Problemstellungen aus dem Optimierungsbereich zu lösen, die größtenteils mit traditionellen Verfahren und Algorithmen angegangen werden können. Im Rahmen des Vortrags wird daher schwerpunktmäßig auf die Problemstellungen und weniger auf die Lösungsansätze eingegangen.

Im folgenden sind einige typische Analysen diskutiert, die in Unternehmensprojekten zu Anwendungen kommen. Anhand dieser Beispiele werden die dort auftretenden Optimierungsaufgaben exemplarisch diskutiert:

1. Analyse von Losgrößen
2. Analyse der Leistungstiefe
3. Analyse von Fertigungsstandortkonzepten
4. Analyse von Vertrieb-Länderportfolios
5. Analyse von Komplexitätskosten
6. Analyse von FuE-Kosten/FuE-Programm
7. Analyse zum Target Costing

Ökonomische Projektbewertung in der Pharma-Industrie im Spannungsfeld von F&E und Marketing

Wolfgang Seyfert, Hermann-Josef Kruse, Peter-Michael Stüben
Schering AG, 13342 Berlin, Germany

Zusammenfassung: Die Entwicklung von Pharma-Produkten wird aus ökonomischer Sicht zunehmend schwieriger. Die Priorisierung von Projekten wird immer notwendiger. Für eine funktionierende Projektpriorisierung ist eine gemeinsame Zielbildung von F&E und Marketing erforderlich. Dies ist in der Pharma-Industrie wegen der enorm langen F&E-Phase von Pharma-Projekten (10-15 Jahre) und den damit verbundenen hohen Unsicherheiten bei der Marktprognose ein besonderes Problem.
Wir haben ein Bewertungssystem (EPA für *Economic Project Analysis*) entwickelt, das bei der Definition und der Analyse der Bewertungsdaten den speziellen Problemen der Pharma-Industrie Rechnung trägt. EPA verdichtet bisher vorhandene Partikulärbetrachtungen zu einer ökonomischen Gesamtbewertung von F&E-Projekten mit Hilfe von klassischen Methoden der dynamischen Investitionsrechnung sowie Periodenergebnisrechnung unter systematischer Einbindung von Hintergrundinformationen über die Spielräume in der Produktentwicklung und über Marktszenarien, wodurch eine Brücke zwischen den F&E- und Marketing-Interessen geschlagen wird.

Summary: The development of pharmaceutical projects becomes more and more difficult from the economic view-point. The necessity to priorize projects is increasing strongly. To make priorization of projects work, a common definition of targets between R&D and Marketing is needed. In the pharmaceutical industry, this is particularly difficult because of the extreme long duration of its R&D-processes (10-15 years) and the entailed unsecurities of market forecasts.
We elaborated a system for analysis and evaluation of projects (EPA for *Economic Project Analysis*), that takes into account the specific problems related to the pharmaceutical industry when defining and analyzing appraisal data. EPA condenses already existing particular analyses to a comprehensive economic evaluation of R&D-projects by applying classical methods of investment calculus as well as statements of periodic operational result-prognoses, systematically including background-information about the scope of product development and about market-scenarios, thus bridging the gap between R&D- and Marketing interests.

1. Einleitung

Die Notwendigkeit, Pharma-Entwicklungsprojekte auf ihr wirtschaftliches Erfolgspotential hin zu untersuchen, ist größer als je zuvor:

- Die Entwicklung neuer pharmazeutischer Produkte ist in den letzten 20 Jahren erheblich teurer geworden, da die Qualitätsanforderungen der Gesundheitsbehörden an die Produktentwicklung stark gestiegen sind.
- Weil dadurch Entwicklungs- und Zulassungszeiten steigen, nimmt der effektive Patentschutz ab.
- Durch die weltweit laufenden Maßnahmen zur Kostensenkung im Gesundheitswesen werden die Preisspielräume geringer.
- Wegen diagnostischer Fortschritte werden Pharma-Produkte in zunehmend spezielleren Indikationen angewendet.

Diese Tendenzen zwingen jedes Pharma-Unternehmen, seine Kräfte auf die erfolgversprechendsten Projekte zu konzentrieren, um diese dadurch schnell auf den globalen Markt zu bringen. Früher als bisher und konsequenter als bisher (d. h. auch stärker auf Abbruchentscheidungen ausgerichtet) werden deshalb Projekte ökonomisch analysiert.

Diese ökonomische Projektanalyse (im folgenden EPA für *Economic Project Analysis*) spielt sich in der Pharma-Industrie in besonderem Maße im Spannungsfeld zwischen F & E und Marketing ab:

- Der oben dargestellte Zwang zur früheren Marktorientierung ist ein relativ neuer Trend und schränkt bisher vorhandene F & E-Freiräume bezüglich der Produktprofilierung ein.
- Aufgrund der langen Entwicklung- und Zulassungszeiten müssen ökonomische Projektanalysen oft mehrere Jahre vor Markteinführung gemacht werden. Marktpotentialprognosen abzugeben ist daher schwierig und sehr unsicher, was die Absprache von Zielvereinbarungen mit Führungskräften aus Marketing und Vertrieb erschwert.
- Da zum spätesten Zeitpunkt des Einstieges in eine EPA - und zwar vor Festlegung bedeutender Kapazitäten zum Einstieg in große klinische Prüfungen - weder das Produktprofil noch die Marktanforderungen klar sind, muß eine Analyse hier auch Klarheit durch Willensbildung schaffen. Eine EPA muß dazu beitragen, daß die Energie im Spannungsfeld von F&E und Marketing produktiv wird. Die wesentliche Frage ist daher für das Folgende: Wie ist der Kommunikationsprozeß innerhalb einer EPA und wie ist der Umgang mit Kennzahlen einer EPA zu gestalten, um EPA nicht zu einem Desinformationsinstrument, sondern zu einem Zielinstrument zu machen?

Wir berichten anhand unserer Erfahrungen bei der Schering AG Berlin.

2. Der Kommunikationsprozeß der ökonomischen Analyse

Der Nutzen der Analyse von Projekten liegt zu einem guten Teil im Prozeß der Analyse selbst: "Eine klare Formulierung der erwarteten Entwicklungsergebnisse" (die im Zuge einer Bewertung zu leisten ist) "ist ein wesentlicher Erfolgsfaktor für den Unternehmenserfolg" [1]. Die ökonomische Bewertung ist Teil vielfältiger Kommunikationen. Sie trägt durch eine Fokussierung dieser Kommunikationen auf den wirtschaftlichen Erfolg per se zum wirtschaftlichen Erfolg bei. Die Auswahl der richtigen Projekte (Priorisierung) setzt voraus zu wissen, was mit einzelnen Projekten erreicht werden kann. Dazu ist ein Zielabgleich zwischen F&E und Marketing erforderlich, der mit EPA unterstützt wird.

2.1. EPA als Teil des Zielabgleichs von F&E und Marketing

Grundlage der ökonomischen Bewertung von Projekten ist nicht nur (und meistens sogar nur zum kleineren Teil) eine Analyse von Fakten, sondern ist immer auch eine Analyse von Zielen. Wie können die Ziele von F&E und Marketing gegenseitig transparent gemacht und im Sinne der Marktorientierung aufeinander bezogen werden?

Auf dem Weg der Motivation und Kommunikation kann EPA hierzu ein Hilfsmittel sein, indem es u.a. Projekte in "Profile" und "Szenarien" strukturiert.

Produktprofile und Einreichungstermin sind Zielvereinbarungen für F&E

Ein Produktprofil beschreibt das angestrebte Produkt so genau und so umfassend, daß ein operativ tätiger Produktmanager den medizinischen Wert klar erkennen und gegenüber Alternativtherapien und Konkurrenzpräparaten einordnen kann. Produktprofile und Einreichungstermin sind zwischen F&E, Marketing und Projektmanagement abgestimmt.

---

[1] Brockhoff 1988, S. 189.

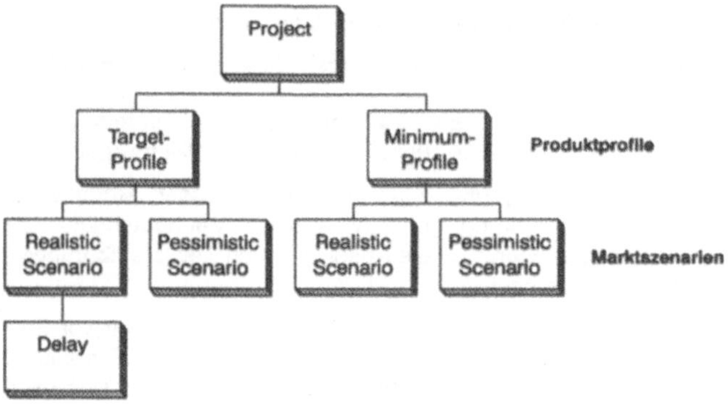

Bild 1: Projektstruktur in EPA

- *Target pharmaceutical profile*: Das Produktprofil, das F&E erreichen soll.
- *Minimum pharmaceutical profile*: Das Produktprofil, das den Mindestanforderungen des Strategischen Marketings genügt und das F&E glaubt, mit hoher Wahrscheinlichkeit erreichen zu können.

Die Profile markieren Grenzen, in denen sich das zukünftige Produkt wahrscheinlich realisieren wird (vgl. auch 2.3). Der Einreichungstermin gilt für beide Profile.

Ergebnisziele in Marktszenarien sind Ziele für das Marketing:
Mittels der Produktprofile bekommen Analytiker und Entscheidungsträger im Marketing zwei Ausprägungen eines zukünftigen Produktes zur Bewertung angeboten. Zu analysieren sind die eigenen Fähigkeiten bezüglich der Vermarktung des Produktes und die Marktentwicklung. Für letzteres sollen Szenarien gebildet werden: Definition von Einflußfaktoren und Zusammenfassung von "realistischen" Ausprägungen dieser Einflußfaktoren in ein "Realistic scenario" bzw. von ungünstigen Ausprägungen (die in der gewählten Kombination aber für denkbar gehalten werden) in ein "Pessimistic scenario". Einflußfaktoren sind z.B. Rahmenbedingungen wie wirtschaftliche Entwicklung und Gesundheitspolitik, das Verhalten der Wettbewerber, die Entwicklung des Kundenverhaltens, die Entwicklung des technologischen Umfeldes, die Entwicklung alternativer Therapiemöglichkeiten etc..

Aus dieser Analyse heraus können seitens der Entscheidungsträger im Marketing Umsatzziele und zugehörige Kostenziele für jede Produktprofil/Szenario-Variante gebildet werden. Mittels einer *Product Profitability Analysis* (PPA)[2] werden diese Ziele in EPA eingebracht. Die Szenarien sind Teil der Erwartungsbildung und werden in EPA kommuniziert. Dies ist eine Form der Risikobetrachtung in EPA (vgl. 2.3).

Vor dem Hintergrund von Profilen und Szenarien werden F & E-Kosten prognostiziert, die zusammen mit den PPA-Ergebnissen den ökonomischen Erfolg eines Projektes bestimmen. Die Analysenergeb-

---

[2] Ein Analyse- und Prognosehilfsmittel für regional und operativ tätige Produktmanager.

nisse werden vom zentralen Produktmanagement[3] zu den Töchtern und zu den Entscheidungsträgern F&E kommuniziert.

Die Kommunikation zwischen Marketing und F&E ist ein Prozeß, in dem sich über die Bewertungszyklen (Jahre) eines Projektes Änderungen des Target profile durch sich ändernde Marktanforderungen ergeben können. Für die gerade laufende Bewertung sind die Profile feststehende Zielvereinbarungen. Für die nächste Bewertung (i.d.R. in einem Jahr) können die F&E-Ziele an geänderte Markterfordernisse angepaßt werden.

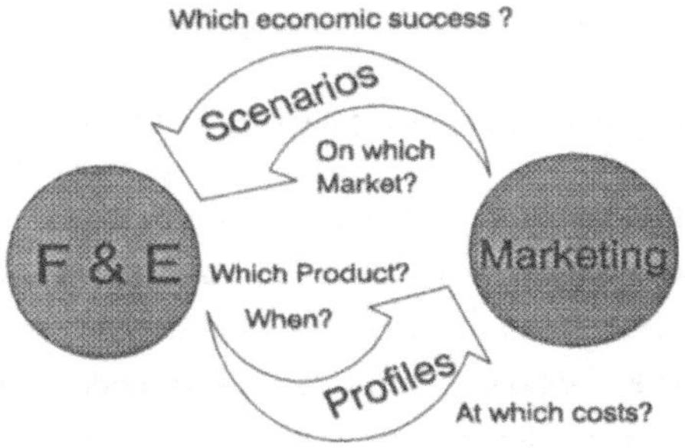

Bild 2: Informationsfluß zwischen F&E und Marketing

2.2. EPA als Hilfsmittel der Projektpriorisierung

Nach wie vor hängt die Priorisierung von Pharma-Entwicklungsprojekten nicht nur von ökonomischen Kriterien ab, und ganz sicher kann sie nicht allein mit Hilfe von Kennzahlen aus einem Analyse- und Bewertungssystem wie EPA geleistet werden. Bei SCHERING gibt es einen Strategieprozeß und einen laufenden, komplexen Projektvergleich, mit dem Projektschwerpunkte gesetzt und damit auch F&E-Budgets gelenkt werden. In diesem Projektvergleich hat EPA mit einer Methode, die die ökonomischen Wirkungen eines Projektes für SCHERING in ihrer Gesamtheit analysiert, ihren Platz:

> *"In vielen Unternehmen funktioniert diese Verteilung (von F&E-Budgets auf Projekte, d. Verf.) schlecht, weil derartige Ertragswertbetrachtungen überhaupt nicht angestellt werden und das General Management die Kapazitäten den Projektgruppen zuweist, die sich 'am lautesten' darum bemühen. Soll sich an diesem Verfahren etwas ändern, dann muß häufig mehr geschehen als bloß ein neues Instrumentarium einzuführen. Denn Verhaltensänderungen in Unternehmen sollten weit stärker auf einer 'emotionsloseren' Betrachtung gründen - und das aus der Perspektive des Marktes."* [4]

Als Waffe in einem Verteilungskampf taugt EPA nichts. Es muß der Wille zur Einigung über die für SCHERING vorteilhaftesten Projekte bei allen für den Erfolg von Entwicklungsprojekten Verantwortli-

---

[3] Bei Schering ist das zentrale Produktmanagement als "Strategisches Marketing" in strategischen Geschäftseinheiten angesiedelt. Vereinfachend sprechen wir statt von dieser firmenspezifischen Organisationsform hier im folgenden vom zentralen Produktmanagement.
[4] Duelli et al. 1991, S. 96.

chen da sein, damit EPA-Information trotz der Verschiedenheit der zu priorisierenden Projekte und trotz der mit der Bewertung verbundenen Unsicherheit eine Hilfe bei der Priorisierung sein kann. Wenn dieser Wille da ist, wird EPA einen Beitrag zur Priorisierung von Projekten über Funktionsgrenzen hinweg leisten, indem es den Entscheidungsträgern eine strukturierte Zusammenfassung vieler ökonomisch relevanter Informationen liefert, die bisher isoliert erzeugt und verwendet werden. Mit EPA wird der ökonomische Gesamterfolg analysiert und damit mehr Vergleichbarkeit hergestellt.

2.3. Umgang mit Risiko

Entwicklungsprojekte unterliegen zwei Arten von Risiko:

- dem Entwicklungsrisiko, nicht das Zielprofil für das geplante Produkt bzw. nicht die Zulassung auf bestimmten Märkten zu erhalten
- dem Marktrisiko, nicht den geplanten Deckungsbeitrag zu erreichen.

Das Entwicklungsrisiko wird weitgehend von Einflußfaktoren gesteuert, die mit dem Zustand bzw. den Aktivitäten des Unternehmens zusammenhängen. Beispiele sind: Güte der Tiermodelle, Stand der Entwicklung (Phase des Projektes), wissenschaftliches Know how, klinische Kompetenz, Qualität der externen Prüfer, technisches Know how, Qualität der Dokumentation, Image bei den Gesundheitsbehörden, Effektivität und Effizienz der Projektorganisation, Ressourceneinsatz (Menschen, Budget etc.), etc.

Das Marktrisiko wird sowohl von inneren als auch von äußere Faktoren beeinflußt. Beispiele sind: Kundenorientierung, Kundennutzen, Marktmacht/Marketingkompetenz, alternative Technologie, Aktivitäten der Wettbewerber, wirtschaftliche Entwicklung, Gesundheitspolitik, etc.

Bezüglich der ökonomischen Bewertung gilt, daß in EPA Risiko in "Situationen" abgebildet wird (vgl. 2.1). Für bestimmte Profil/Szenario-Situationen wird jeweils so getan, als ob die Situation mit Sicherheit einträte. Mit dieser Annahme werden (z.B. für das Minimumprofil im realistischen Marktszenario) Zahlungsströme geschätzt und mit diesen deterministische Kennzahlen gebildet, als ob es sich um sichere Zahlungen handeln würde. Es wird also kein Risiko in Kennzahlen "hineingerechnet", etwa über einen Risikozuschlag zum Diskontierungszins oder über die Gewichtung unsicherer Zahlungsströme mit einer wie auch immer ermittelten Eintrittswahrscheinlichkeit. Die Beurteilung des ökonomischen Gesamtrisikos eines Projektes erfolgt also nicht mit Hilfe eines Rechenmodells, sondern durch unmittelbaren Vergleich aller zu den verschiedenen Risikosituationen gehörenden Kennzahlen. Die Entscheidungsträger müssen die Risikosituation der Projekte bei Priorisierungs- bzw. Abbruchentscheidungen berücksichtigen. EPA macht das mit einem Projekt verbundene Risiko auf einfache Weise transparent.

3. Die Ergebnisse der ökonomischen Bewertung

In EPA werden Kennzahlen bestimmt, die größtenteils in der Schering AG bereits eingeführt sind. Begriffliche Vertrautheit der Entscheidungsträger mit der Bewertungsinformation ging vor eleganter

Optimierung, etwa durch simultane Entscheidungsmodelle. Der Grund hierfür ist, daß in Entscheidungssituationen, in denen die Bewertungsgrundlagen nicht wohldefiniert sind, Entscheidungen nicht algorithmischen Systemen überlassen werden können. Solange Entscheidungen auch dazu dienen, Entscheidungsgrundlagen zu festigen - nämlich durch Fokussierung der Organisation auf Ziele -, ist es sinnlos, einer Rechenprozedur eine weitgehende Vorstrukturierung der Entscheidung zu übertragen[5]. Es kommt vielmehr darauf an, mit Kennzahlen Argumente für einen interdisziplinären Willensbildungs- und Entscheidungsprozeß zu liefern. Hierzu liefern wir folgende Kennzahlen:

- Operatives Projektergebnis als Zeitreihe,
- Kapitalwert, ROI und Amortisationszeit,
- Vermögensverlust durch Verzögerung,
- Kapitalwertraten.

Die Kennzahlen müssen allerdings nicht nur interpretierbar, sie müssen auch vergleichbar sein. Für die Projektpriorisierung muß das Ergebnis von EPA deshalb **standardisiert** präsentiert werden. Daher wurden in EPA Auswertungskonventionen vereinbart. In diesen werden u.a. der Gebrauch von einheitlichen Zinssätzen und Währungsumrechnungen geregelt, die Bestimmung der relevanten Projektzeithorizonte abgesteckt sowie Form und Inhalte der Ergebnispräsentation vorgegeben. Abweichungen von den vereinbarten Standards müssen offengelegt und begründet werden.

Daneben bietet EPA allerdings auch die Möglichkeit von Zusatzauswertungen an. Hierdurch soll dem Anwender ein Playground geschaffen werden, auf dem beliebig fiktive Projektlternativen erprobt werden können. Durch implementierte Methoden (z.B. Sensitivitätsanalyse) könen anhand von What-If-Betrachtungen die Zusammenhänge zwischen Einflußgrößen und deren Auswirkungen auf die Kennzahlen analysiert werden. Eine Entscheidungsvorlage kann so vertieft analysiert werden.

4. Ausblick

Das vorgestellte Konzept wurde mit allen in den Bewertungsprozeß involvierten Bereichen im Hause kommuniziert und abgestimmt. Der erfolgreiche Einsatz des Projektbewertungstools wird aber nicht allein von der Akzeptanz der Methodik, sondern zu einem nicht unwesentlichen Teil von der Akzeptanz des zugehörigen DV-Systems abhängen. Daher werden gegenwärtig große Anstrengungen unternommen, um eine in die Anwenderwelt passende DV-Lösung zu implementieren.

Literatur:

Brockhoff, K.: *Forschung und Entwicklung: Planung und Kontrolle*. München-Wien-Oldenburg, 1988.
Duelli, J.; Walgenbach, E., Wittek, B.: *Das Dilemma der Pharma-Forschung*. In: Harvardmanager 1/1991.
Wilkinson, A.: *Developing an expert system on project evaluation*. In: R&D Management 3/1991.

[5] Ähnlich argumentiert A. Wilkinson (1991, S. 212): "If Expert Systems are to be really useful they have to be understood by practising managers, and as with other computer aids become tools helping managers to achieve their objectives more quickly and simply."

OPTIMIERENDE BETRIEBSMITTELEINSATZPLANUNG IM ÖFFENTLICHEN PERSONENNAHVERKEHR

Manfred Völker, HanseCom GmbH, Hamburg

Den wesentlichen Anteil an der Betriebsmitteleinsatzplanung in Unternehmen des öffentlichen Personennahverkehrs (ÖPNV) stellt die Einsatzplanung von Fahrzeugen und Fahrpersonal dar. Diese sogenannte Fahr- und Dienstplanung ist heute ab einer gewissen Betriebsgröße nicht mehr ohne (optimierende) DV-Unterstützung denkbar. Eine solche Unterstützung bietet zum Beispiel das Programmsystem HOT. Ein Überblick über Problemstellung und mathematische Verfahren in der Fahr-und Dienstplanung wurde in den Operations Research Proceedings 1986 gegeben. In jüngster Zeit ist das System HOT weiterentwickelt worden, sowohl die Benutzeroberfläche als auch die implementierten Algorithmen. In diesem Vortrag steht der in der Praxis erzielte Nutzen im Vordergrund.

Zunächst auf einem Großrechner realisiert, wurde HOT 1987 auf das Betriebssystem UNIX portiert. In seiner heutigen Version gliedert sich das System in die fünf Module Datenverwaltung, Sensitivitätsanalyse, Fahrzeug-Umlaufbildung, Dienstplanbildung und Dienstreihenfolgenbildung. Bis auf die Datenverwaltung basieren alle Moduln auf Optimierungsalgorithmen. Die Benutzeroberfläche ist menügesteuert, die Dateneingabe erfolgt über Eingabemasken und Grafikoberflächen.

Der bei den HOT-Anwendern beobachtete Nutzen wird einerseits anhand des Einsparpotentials beim Fahrpersonal dargestellt. So ist der HOT-Anwender in der Lage, "auf Knopfdruck" die Änderung zahlreicher, betrieblicher Vorgaben (wie z.B. maximale Dienstdauer) in ihrer Kostenwirksamkeit zu prüfen. Die Einflußmöglichkeiten der Parameter werden erläutert. Die im Vortrag dargestellte Funktionalität der Dienstreihenfolgenbildung beinhaltet neben der optimalen Verteilung der Wochenarbeitszeit eine besonders benutzerfreundliche Oberfläche.

Andererseits wird über die Optimierung beim Fahrzeugeinsatz berichtet. Mit dem Modul Sensitivitätsanalyse wurden in der Praxis Einsparungen von mehr als einer Million D-Mark erzielt. In der Umlaufbildung wurden neue Eingriffsmöglichkeiten für den Anwender zur Steuerung des Optimierungsprozesses geschaffen.

WBSGEN - Wissensbasiertes System für das Genehmigungsverfahren baulicher Anlagen

Helmut Waldorf, Essen

Die Planung und Durchführung von Bauvorhaben zu Gebäuden und Anlagen erfordert die Beachtung zahlreicher Gesetze, Verordnungen und Normen. Durch die Fülle der infrage kommenden Vorschriften und durch die Unübersichtlichkeit der darin formulierten Anforderungen wird die Erstellung von Genehmigungsanträgen für Bauvorhaben zu einem aufwendigen Problem.

WBSGEN unterstützt in der Planungsphase und bei der Formulierung des Genehmigungsantrags:
Zunächst wird das Bauvorhaben grob beschrieben. Dazu werden Angaben zum vorgesehenen Verfahren (z.B. Lagern), zum eingesetzten Stoff (z.B. Schwefelsäure), zu Volumina und Art der Lagerung gemacht. Anschließend ermittelt das System die für das vorliegende Bauvorhaben relevanten Gesetze - im derzeitigen Prototyp noch beschränkt auf das Wasserhaushaltsgesetz - und die darin enthaltenen Detailanforderungen für die bauliche Ausführung und den späteren Betrieb, ggf. mit möglichen Alternativen.

Erreicht wurde dieses über einen wissensbasierten Ansatz: Ein Gesetz wird in Einzelanforderungen zerlegt, die in einer Objekthierarchie angeordnet werden. Die Relevanzkriterien für jedes Objekt werden durch entsprechende Regeln beschrieben.

Die Implementierung erfolgte mit Hilfe der IBM Prolog Workbench, wesentliche Auswahlkriterien waren hierbei die gute Kooperation regelbasierter und objektorientierter Programmierung, der einfache Aufbau einer Benutzerschnittstelle unter dem Presentation Manager und die verfügbare Schnittstelle zum Database Manager.

ADMINISTRATIVE LOGISTIK: Wie konnten wir das vergessen?

Folkert W. Wierda, ORIGIN Information Technology Hamburg GmbH

Der Begriff *Logistik* stammt aus der Kriegskunde und beinhaltet die Problematik der Versorgung und des Nachschubes der Armee. Primär ist damit die Versorgung mit Waffen, Munition, Nahrung und eventuell Soldaten gemeint. Aber auch die Rolle der *Versorgung mit Nachrichten* über dieses Mittel ist in der Kriegskunde schon länger bekannt, wie von Clausewitz zeigt (Vom Kriege, I/6). In der heutigen wirtschaftlichen Form der Logistik bleibt unverständlicherweise das Interesse an dieser sogenannten *administrativen Logistik* im Hintergrund.

Für Nachrichten - Informationen würde man heute sagen - gelten ähnliche Aufgaben und Probleme wie für Produkte oder Teile und deren Verarbeitung. Wenn man ein Informationssystem (IS) auffaßt wie eine Abbildung des reellen Systems (RS), wird sogar bald klar, daß Engpässe im IS sofort zu Engpässen im RS führen werden. Das beste Beispiel für diese Behauptung ist der Container, der in 10 Tagen von Hamburg bis New York transportiert wird, aber erst nach 30 Tagen weiterverarbeitet werden kann, weil vorher die Transportdokumente noch nicht angekommen und verarbeitet waren.

In diesem Beitrag wird anhand einer praktischen Fallstudie bei einer Versicherungsgesellschaft erläutert, wie die administrative Logistik einer Organisation studiert und verbessert werden kann. In der Vorgehensweise werden Techniken der diskreten Ereignis-Simulation (DES) angewendet. Bekannte logistische Ziele, wie *Just-in-Time* und *Lean Production*, sind auch in der Verwaltung anzustreben und realisierbar.

Die Fallstudie zeigt eine pragmatische Anwendung einiger Methoden und Techniken aus dem Operations Research und schließt mit einem Plädoyer für dasVerringern der Distanz, die - jedenfalls im deutschsprachigen Raum - zwischen Wissenschaft und Praxis besteht, ab.

Effizienter Datenbankentwurf für den Bergbaubetrieb mit semantischer Datenmodellierung

Helmut Zander, Essen

Zusammenfassung: Effizienter Datenbankentwurf sollte Paradigmeninvarianz gegenüber Computerimplementationen garantieren. Semantische Modelle stellen nur dann eine wirkliche Hilfe dar, wenn deren Syntax durch eine mathematische Metasemantik objektivierbar ist. Die kategoriale Semantik erlaubt die Formulierung der mit semantischen Typkonstruktoren verbundenen Integritätsregeln sowie eine problemadäquate Definition der Begriffe Schema, Theorie und Modell. Am Beispiel einer Datenbank für Steinkohlenbergbaubetriebe wird die Nützlichkeit des semantischen Datenentwurfes belegt.

Summary: Efficient database design should be independent of the acutal implementations of computer paradigms. Semantic modelling, as a possible alternative, needs justification by some mathematical metasemantics, that garantees more intellectual stability. Categorical semantics allows the formulation of integrity rules, belonging to semantic type constructors, within its own context as well as a new approach in defining terms like scheme, theory and model. The usefulness of semantic DB-design is demonstrated by a special design for an underground coal mining data model.

Einleitung

Investitionen in komplexe Softwaresysteme erfordern einen immer höheren Kapitaleinsatz. Die Alterung von Software aufgrund geänderter Umgebungsbedingungen, gestiegener Anforderungen seitens der Anwender und sich ändernder Implementierungsparadigmen verkürzen die Lebenszyklen von Software und lassen komplette Anwendungssysteme obsolet werden.

Ein wesentlicher Grund für diese Situation ist die Gleichzeitigkeit der Evolution der Anwenderanforderungen und der Paradigmenwechsel der Implementierungswerkzeuge (Netzwerk-/relationale DBMS, Objektorientierte Systeme).

Als Ausweg aus dem damit verbundenen Dilemma bietet sich der verstärkte Einsatz semantischer Spezifikationswerkzeuge an, die von hoher Flexibilität gegenüber Erweiterungen und Akzentverschiebungen der Anwenderseite sind. Die heute gebräuchlichen Spezifikationssysteme unterliegen jedoch ihrerseits der paradigmatischen Spezialisierung. Dies wird belegt durch die Fülle der sich zur Zeit anbietenden Spezifikations- und Modellierungssysteme wie z.B.

- Entity Relationship Modelle
- Semantische Datenmodelle
- Objektorientierte Entwurfssysteme.

Abgesehen von den durchaus bestehenden Unvollständigkeiten dieser Systeme ergeben sich hieraus Inkompatibilitäten und terminologische Konkurrenz, die den Einsatz semantischer Entwurfswerkzeuge nur dann rechtfertigen, wenn deren begriffliche Klarstellung durch Orientierung an einer langfristig paradigmenfreien Metasemantik möglich und beweisbar ist.

Folgende Forderungen sind an eine derartige Metasemantik zu stellen:

1. *Anthropomorphie:*

In Abhängigkeit der durch die Anwendungsproblematik gegebenen Akzente sind *adäquate* Ausdrucksmittel bereitzustellen. Deren Grundtypen sind Diagrammatik, Linguistik und Logik.

2. *Integrität*

Alle anwendungsabhängigen Integritätsbedingungen müssen grundsätzlich mit den Ausdrucksmitteln dieser Metasemantik formulierbar sein.

3. *Vollständigkeit*

Alle praktisch relevanten Spezifikationssysteme sollten aus diesem System ableitbar sein.

4. *Äquivalenz*

Die unter 1. genannten Ausdrucksmittel sollten *opportunistisch*, d.h. nur unter dem Gesichtspunkt der Darstellungsoptimalität einsetzbar sein. Somit sollten innerhalb des Metasystems die Ausdrucksmittel von Diagrammatik, Linguistik und Logik *gleichmächtig*, d.h. verschiedene aber prinzipiell äquivalente Aspekte des gleichen metasemantischen Systems sein.

Im folgenden sollen die Grundelemente der kategorialen Semantik und die sie betreffenden grundlegenden Sachverhalte vorgestellt werden, die belegen, daß die Forderungen 1 bis 4 erfüllbar sind und somit Auskunft geben über die Möglichkeiten der weiteren Entwicklung praktischer semantischer Spezifikations- und Entwurfswerkzeuge.

Es wird ferner die Konformität der Konstrukte des erweiterten Entity Relationshipmodells (EERM) mit der kategorischen Semantik aufgezeigt. Am Beispiel eines für den Steinkohlenbergbau entworfenen Basistypenmodells für eine Fachübergreifende Bergwerksdatenbank (FÜDB) wird die Modellierungsmächtigkeit des EERM vorgestellt und durch die Einbeziehung des EERM in die Zusammenhänge der kategorialen Semantik belegt, daß der damit geleistete Analyseaufwand grundsätzlich geschützt ist gegenüber den unvermeidbaren Strukturbrüchen durch Evolution

- der semantischen Paradigmatik
- der Implementierungstechnologien und
- der horizontalen und vertikalen Schwerpunktverlagerung der Anwendungsanforderungen.

Die kategoriale Semantik

G sei ein gerichteter (Multi-)Graph (mit Zyklen). Einen Pfeil von G von einem Knoten A zu einem Knoten B bezeichnen wir mit $f: A \rightarrow B$ (oder auch mit $A \xrightarrow{f} B$). Eine Kategorie $\underline{C}$ ist ein gerichteter Graph mit folgenden zusätzlichen Eigenschaften:

1. Sei $f: A \rightarrow B$ und $g: B \rightarrow C$ so gibt es einen Pfeil $g \circ f: A \rightarrow C$ (Komposition)
2. Jeder Knoten X des Graphen besitzt einen Pfeil id_X auf sich selbst.
   Es gilt: $f \circ id_A = id_B \circ f = f$ für alle $f: A \rightarrow B$
3. Es gilt: $h \circ (g \circ f) = (h \circ g) \circ f$ für alle $f: A \rightarrow B$, $g: B \rightarrow C$ und $h: C \rightarrow D$.

Die Knoten heißen die *Objekte*, die Pfeile die *Morphismen* der Kategorie $\underline{C}$. Die Menge der Morphismen von einem Objekt A nach einem Objekt B bezeichnen wir mit $[A,B]_{\underline{C}}$ oder einfach $[A,B]$. Beispiele sind die Kategorie der Mengen $\underline{Set}$ mit $[A,B]_{\underline{Set}}$ gleich der Menge aller mengentheoretischen Abbildungen $f: A \rightarrow B$, ferner die Kategorie der Gruppen $\underline{G}$ mit $[A,B]_{\underline{G}}$ gleich der Menge aller Gruppenhomomorphismen der Gruppe A in die Gruppe B.

Ein *Funktor* $F: \underline{C} \rightarrow \underline{C}'$ von der Kategorie $\underline{C}$ in die Kategorie $\underline{C}'$ ist eine Abbildung, die jedem Objekt von $\underline{C}$ ein Objekt von $\underline{C}'$ und jedem Morphismus von $\underline{C}$ einen Morphismus von $\underline{C}'$ zuordnet. Für diese Zuordnung muß gelten:

1. für $f: A \rightarrow B$ in $\underline{C}$ gilt $Ff: F(A) \rightarrow F(B)$ in $\underline{C}'$
2. $F(id_A) = id_{FA}$ und
3. für $f: A \rightarrow B$ und $g: B \rightarrow C$ gilt $F(g \circ f) = Fg \circ Ff$.

Sei z.B. $\underline{G}$ die Kategorie der Gruppen, dann ist die Abbildung, die jeder Gruppe die ihr zugrundeliegende Menge zuordnet, ein Funktor $U: \underline{G} \rightarrow \underline{Set}$ (der vergeßliche Funktor). U ist "treu", d.h. für zwei Gruppen G_1, G_2 ist $U: [G_1,G_2]_{\underline{G}} \rightarrow [G_1,G_2]_{\underline{Set}}$ injektiv. Da diese Abbildung im allgemeinen nicht surjektiv ist, ist sie nicht "voll".

Aus jeder Kategorie $\underline{C}$ läßt sich eine andere, die entgegengesetzte, Kategorie $\underline{C}^{op}$ ableiten. Die Objekte von $\underline{C}^{op}$ sind die von $\underline{C}$. Die Morphismen sind gegeben durch $[A,B]_{\underline{C}^{op}} = [B,A]_{\underline{C}}$, d.h. es gilt $[A,B]_{\underline{C}} = [B,A]_{\underline{C}^{op}}$.

Ein Funktor $F: \underline{C}^{op} \rightarrow \underline{C}'$ heißt *kontravarianter Funktor* von $\underline{C}$ nach $\underline{C}'$. Ein Funktor $F: \underline{C} \rightarrow \underline{C}'$ heißt auch *kovarianter Funktor*.

Sei $\underline{C}$ eine Kategorie, A ein Objekt von $\underline{C}$. Jedem Objekt X von $\underline{C}$ können wir die Menge $F_A(X) = [A,X]_{\underline{C}}$ zuordnen. Jedem Morphismus $f: X \rightarrow Y$ können wir die Abbildung $F_A(f): F_A(X) \rightarrow F_A(Y)$ zuordnen. Diese bildet ein $g \in F_A(X) = [A,X]$ auf $f \circ g \in [A,Y] = F_A(Y)$ ab. Damit erhalten wir einen (kovarianten) Funktor $F_A: \underline{C} \rightarrow \underline{Set}$. Auf die gleiche Weise kann man jedem Objekt A aus $\underline{C}$ einen kontravarianten Funktor $G_A: \underline{C} \rightarrow \underline{Set}$ mit $G_A(X) = [X,A]$ zuordnen.

Die kontravarianten Funktoren G: $\underline{C} \to \underline{Set}$, d.h. die Funktoren G: $\underline{C}^{op} \to \underline{Set}$ heißen auch Prägarben. Ein Funktor F: $\underline{C} \to \underline{Set}$ heißt *darstellbar*, wenn $F(X) = [A,X]$ für ein bestimmtes Objekt A aus $\underline{C}$ gilt. A heißt *darstellendes Objekt* des Funktors F. Dasselbe gilt für (darstellbare) Prägarben.

Für zwei Kategorien $\underline{C}$ und $\underline{C}'$ bilden alle Funktoren F: $\underline{C} \to \underline{C}'$ wieder eine Kategorie. Dazu müssen wir definieren, was ein Morphismus T zwischen zwei Funktoren F und G ist. Eine *natürliche Transformation* T von F nach G ordnet jedem Objekt A aus $\underline{C}$ einen Morphismus $T(A)$: $F(A) \to G(A)$ in $\underline{C}'$ zu.

Dabei soll für f: $A \to B$ gelten, daß das folgende Diagramm von Morphismen in $\underline{C}'$ kommutativ ist:

$$\begin{array}{ccc} F(A) & \xrightarrow{T(A)} & G(A) \\ F f \downarrow & & \downarrow G f \\ F(B) & \xrightarrow{T(B)} & G(B) \end{array}$$

das heißt: es soll gelten $Gf \circ T(A) = T(B) \circ Ff$. Die Menge der natürlichen Transformationen von F nach G bildet dann die Morphismen [F,G] der *Funktorkategorie* Funk ($\underline{C}$,$\underline{C}'$) der Funktoren von $\underline{C}$ nach $\underline{C}'$.

Jeder Morphismus f: $A \to B$ in einer Kategorie $\underline{C}$ erzeugt eine natürliche Transformation f': $F_A \to F_B$ der den Objekten A und B zugeordneten darstellbaren kontravarianten Funktoren F_A und F_B. Da $F_A(X) = [X,A]_{\underline{C}'}$, ergibt sich die natürliche Transformation f' aus $f'(g) = f \circ g$, mit $g \in [X,A]$. Die Zuordnung f zu f' erzeugt also einen Funktor von $\underline{C}$ nach Funk ($\underline{C}^{op}$,$\underline{Set}$). Dieser (Yoneda-)Funktor stellt eine volle Einbettung der Kategorie $\underline{C}$ in die Kategorie der mengenwertigen Prägarben von $\underline{C}$ dar (Yoneda-Einbettung) (vgl. z.B. [BW]), d.h. die Morphismen f: $A \to B$ und die natürlichen Transformationen f': $F_A \to F_B$ sind durch die oben beschriebene Abbildung f --> f' bijektiv einander zugeordnet.

Damit verfügen wir über das terminologische Rüstzeug für ein Verständnis der kategorialen Semantik (vgl. auch [PAPR]). Eine solche besteht aus einer Kategorie $\underline{C}$, deren Objekte wir in diesem Zusammenhang auch als *Typen* bezeichnen. Den Begriff Morphismus können wir mit dem Programmbegriff identifizieren.

Ein *Programm* f: $A \to B$ hat demnach einen Input vom Typ A und einen Output vom Typ B. Die Mächtigkeit der kategorialen Semantik ergibt sich aus der Mächtigkeit der kategorialen Werkzeuge, insbesondere für die Formulierung von Typkonstruktoren und deren kategorial formulierten Integritätsbedingungen.

Ausgewählte Typkonstruktoren der kategorialen Semantik

Die im folgenden aufgeführten Beispiele von Typkonstruktoren verwenden den Begriff des darstellbaren Funktors. Die Existenz der darstellenden Objekte dieser Funktoren ist in jedem Fall als gesonderte Voraussetzung zu erfüllen. Für den Fall, daß ein Funktor darstellbar ist, gilt jedoch immer die Eindeutigkeit des darstellenden Objektes (bis auf Isomorphie).

Aggregation

Die *Aggregation* von Typen entspricht dem Begriff des cartesischen Produktes. Eine Kategorie $\underline{C}$ erlaubt (endliche) cartesische *Produkte*, wenn für zwei Typen (Objekte) A und B der mengenwertige Funktor $F_{A,B}(X) = [X,A] \times [X,B]$ darstellbar ist, d.h. wenn es einen Typ (ein Objekt) $A \times B$ in $\underline{C}$ gibt, für den immer gilt $[X,A \times B] \cong [X,A] \times [X,B]$. $A \times B$ heißt der Produkttyp von A und B. Die mit der Aggregation verbundene Integritätsbedingung besteht also darin, daß alle Programme mit einer Aggregation $A \times B$ als *Output* allen *Paaren* von Programmen mit den *Outputtypen* A und B entsprechen.

Array

$X^n = X \times ... \times X$, das n-fache cartesische Produkt des Typs X mit sich selbst, ist der Typ "Array vom Typ X der Länge n".

Partition

Der *Partition* entspricht das *Coprodukt*, d.h. A und B stellen eine Partition des Typs C dar, wenn C das Coprodukt A + B der Typen A und B ist, d.h. wenn C das Produkt von A und B in der Kategorie C^{op} ist. Äquivalent dazu ist die Aussage, daß A + B den kontravarianten mengenwertigen Funktor $G_{A,B}(X) = [A,X] \times [B,X]$ darstellt, d.h. wenn gilt $G_{A,B}(X) = [A + B,X]$. Die mit einer Partition verbundene Integritätsbedingung besteht also darin, daß alle Programme mit einer Partition A + B als *Inputtyp* allen *Paaren* von Programmen mit den *Inputtypen* A und B entsprechen.

Generalisierung

Der Typ B ist eine *Generalisierung* des Typs A, wenn es einen *Mono*(morphismus) i: A → B gibt. i heißt *mono*, wenn für zwei beliebige f,g: C → A mit $i \circ f = i \circ g$ immer f = g folgt, d.h. wenn die durch i (durch Komposition von links) induzierte Abbildung i′: [C,A] → [C,B] injektiv ist. A heißt auch *Spezialisierung* von B. i induziert (durch Komposition von rechts) eine Abbildung [B,C] → [A,B], d.h. alle Programme vom Inputtyp B liefern Programme vom Inputtyp A (Inklusionspolymorphismus, Vererbung).

Höhere Typen

Zu zwei Typen A und B einer Kategorie $\underline{C}$ existiert der *Exponentialtyp* B^A, wenn der mengenwertige Funktor $F_{A,B}(X) = [X \times A,B]$ darstellbar ist, d.h. wenn $F_{A,B}(X) = [X,B^A]$ ist.

$\underline{C}$ besitzt ein terminales Objekt 1, wenn für jeden Typ X gilt, daß [X,1] = {∅}, d.h. eine einelementige Menge ist. Ein Morphismus 1 → X heißt eine *Entität* vom Typ X. Da immer $X \times 1 \cong X$ ist, für jeden Typ X, gilt also $[A,B] \cong [1 \times A,B] \cong [1,B^A]$, d.h. [A,B] ist die Menge der Entitäten von B^A. Eine Kategorie $\underline{C}$ mit Exponentialtypen repräsentiert also zu je zwei Typen A und B durch B^A die Menge aller Programme mit A als Input- und B als Outputtyp. Eine solche Kategorie heißt cartesisch abgeschlossen (cartesian closed category, CCC).

Untertypen

Ein Morphismus f: A → B heißt ein *Isomorphismus*, wenn es ein g: B → A gibt, so daß $g \circ f = id_A$ und $f \circ g = id_B$ ist. Zwei Monomorphismen i_0: A_0 → A und i_1: A_1 → A heißen äquivalent, wenn ein Isomorphismus j: A_0 → A_1 existiert, derart, daß das Diagramm

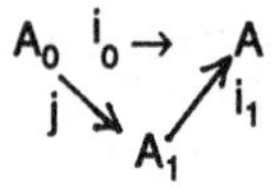

kommutativ ist, d.h. wenn gilt $i_1 \circ j = i_0$. Ein *Untertyp* von A ist eine Klasse äquivalenter Monomorphismen nach A.

Potenztypen

Eine Kategorie $\underline{C}$ besitzt *Potenztypen*, wenn für jeden Typ A der mengenwertige Funktor $F_A(X)$ = Sub(X x A) der Untertypen von X x A darstellbar ist, d.h. wenn $F_A(X) = [X,P(A)]$. P(A) heißt der Potenztyp von A. Besitzt $\underline{C}$ ein terminales Objekt 1, dann stellt somit Ω = P(1) den Funktor Sub X der Untertypen von X dar. Die mit dem Potenztyp verbundene Integritätsbedingung besteht also darin, daß die "Relationen" zwischen zwei Typen A und B den Programmen von A in den Potenztyp von B entsprechen.

Listen

$L_n(X)$ sei der Typ von Listen vom Typ X der maximalen Länge n. Dann ist $L_{n+1}(X) = L_n(X) + X^{n+1}$ der Typ von Listen vom Typ X der maximalen Länge n + 1. Demnach existiert eine Sequenz von Monomorphismen $L_0(X) = 1 \xrightarrow{i_0} L_1(X) \xrightarrow{i_1} L_2(X) \xrightarrow{i_2} \cdots$.

Ein Cokegel mit Spitze L′(X) bezüglich dieser Sequenz besteht aus einem Objekt L′(X) und einer Menge von Morphismen l'_i: $L_i(X) \to L'(X)$ mit $l'_n = l'_{n+1} \circ i_n$.

Die Zuordnung $F_X(L')$ = {Cokegel mit Spitze L′} definiert auf naheliegende Weise einen mengenwertigen Funktor auf der Kategorie $\underline{C}$ bezüglich des Typs X, da jeder Morphismus L′ → L′′ jedem Cokegel mit Spitze L′ einen solchen mit Spitze L′′ zuordnet. Wenn der Funktor $F_X(L')$ darstellbar ist, mit dem darstellenden Objekt L(X), d.h. wenn gilt $F_X(L') \cong [L(X),L']$, dann heißt L(X) der zum Typ X gehörige *Listentyp*.

Limiten, Colimiten

Der Typkonstruktor Liste stellt ein Beispiel dar für die Formulierung von *Colimiten* in der kategorialen Semantik. Der duale Begriff in $\underline{C}^{op}$ liefert den Begriff des *Limes* in $\underline{C}$.

Die Sequenz 1 → $L_1(X)$ → $L_2(X)$ → verbindet abzählbar unendlich viele Typen. Ähnliche Konstruktionen über Diagrammen mit nur endlich vielen Typen führen zum Begriff des *endlichen Limes (Colimes)*.

Die einfachsten Beispiele endlicher Limiten und Colimiten sind Produkte (Aggregationen) und direkte Summen (Partitionen). Weitere wichtige Beispiele sind Pullbacks und Pushouts.

Ein Limes L bezüglich eines Diagramms von drei Typen A, B, C der Form $A \xrightarrow{f} C \xleftarrow{g} B$ ergibt ein kommutatives Diagramm der Form

$$\begin{array}{ccc} L & \xrightarrow{P_B} & B \\ P_A\downarrow & & \downarrow g \\ A & \xrightarrow{f} & C \end{array}$$

so daß es zu jedem weiteren Typ L′ mit gleichen Eigenschaften, d.h. mit $P'_A: L' \rightarrow A$ und $P'_B: L' \rightarrow B$ sowie $P'_A \circ f = P'_B \circ g$, einen Morphismus $L' \xrightarrow{h} L$ gibt, mit $P'_A = P_A \circ h$ und $P'_B = P_B \circ h$. L heißt Pullback des Diagramms. Die gleiche Konstruktion liefert in $\underline{C}^{op}$ die Definition eines Pushouts.

Beispiel:

Sei B der Typ "Mitarbeiter" und C der Typ "Manager", sowie g ein Programm, das zu jedem Mitarbeiter den entsprechenden Vorgesetzten liefert. Sei A der Typ "Personalabteilung" und f ein Programm, das zu jeder Personalabteilung deren Leiter ermittelt. Das Pullback L liefert den Typ "Mitarbeiter von Personalabteilungen sowie ihre jeweilige Abteilung".

Die kategoriale Semantik als Metasystem für praktische Spezifikations- und Entwurfswerkzeuge

Die Mächtigkeit der kategorialen Semantik als Metasystem praktischer Modellierungssysteme ergibt sich im Lichte der eingangs formulierten Kriterien. Als prinzipiell diagrammatisches System bietet der Kategorienbegriff ein Maximum an Bildhaftigkeit durch Verzicht auf Objektatomarität und den damit verbundenen Konstruktionsvorschriften. Das primäre Anschauungsmodell des Kategorienbegriffs ist vielmehr das des Netzwerkes von Typen, deren strukturelle Granularität sich nur implizit über die Art und Mächtigkeit der Morphismusverbindungen zu anderen Typen ableiten läßt.

Mindestens auf der Grundlage cartesisch abgeschlossener Kategorien (CCC) und deren begrifflicher Verfeinerungen über die lokal cartesisch abgeschlossenen Kategorien (LCC) bis zum Begriff des Topos ergibt sich die unter 1 geforderte Anthropomorphie durch die Tatsache, daß jede CCC (mit einem natürlicher Zahlentyp) über eine *"innere Sprache"* verfügt, die einem *typisierten Lamdakalkül* entspricht und umgekehrt jedem derartigen Kalkül eindeutig eine CCC zugeordnet ist (vgl. [LS]).

Da aus logischer Sicht eine *CCC* einen *positiv intuitionistischen Aussagenkalkül mit Konjunktion* repräsentiert, sind bereits auf dieser Ebene Linguistik und Logik gleichrangig der kategorialen Semantik zugeordnet. Spezifikations- und Entwurfswerkzeuge sollten diese Vielgestaltigkeit nutzen, um unter jedem gewünschten Aspekt (Basisdaten-, Funktionenmodell) die notwendige Modellierungseffizienz zu gewährleisten.

Die Verwendung von *Typkonstruktoren* beim semantischen Entwurf impliziert die mit der *Darstellbarkeit der zugehörigen Funktoren* gegebenen *Integritätsregeln.* Diese im Fall des Pullbackkonstruktors beispielhaft wiedergegebene Formulierung verwendet dazu prädikatenlogische Aussagen über Morphismen und Objekte der Typenkategorie $\underline{C}$. Das besondere Interesse muß hierbei der Yoneda-Einbettbarkeit von $\underline{C}$ in die Kategorie der mengenwertigen Prägarben gelten, da diese immer einen *Topos* darstellt. Dabei ist ein Topos eine CCC deren Untertypfunktor Sub X durch ein Objekt Ω darstellbar ist, d.h. es gilt Sub $X \cong [X,\Omega]$ (vgl. [BW]).

Da der Toposbegriff dem der *intuitionistischen Prädikatenlogik mit höheren Typen* entspricht (Ω ist der Typ der Wahrheitswerte und nicht notwendig zweiwertig!) und die inneren Sprachen der Topoi den intuitionistischen Typtheorien (via Adjunktion) entsprechen (vgl. [LS]), bieten die Prägarbenkategorien ein Diskursuniversum für die Formulierung von Integritätsbedingungen, die die Mächtigkeit dieser linguistischen und logischen Systeme repräsentieren und weiter jedwede Metasystematik entbehrlich machen.

Diese Sachverhalte belegen somit auch die Erfüllbarkeit der Forderung nach Äquivalenz von Diagrammatik, Linguistik und Logik. Dagegen muß sich die Vollständigkeit im praktischen Einsatz ergeben. Als Beweis kann gelten, daß die kategoriale Semantik die Subsummierung von Computeranwendungen erlaubt, die auf der Ebene der Implementierungsparadigmen in keinerlei Beziehung zu stehen scheinen. Als Beispiele hierzu mögen die *linearen Logiken* von *Petrikartogien* (vgl. [MOM]) für die Implementierung paralleler Prozesse und die Interpretation von *Fuzzy Mengen* (vgl. [W]) als kategoriale (Quasi-)Topoi gelten.

Schema, Theorie und Modelle

Vor dem Hintergrund der kategorialen Metasemantik muß sich jede praktische Modellierungssemantik Notationen bedienen, welche die Erfordernisse des Metasystems berücksichtigt, jedoch dessen praktische Grenzen überwindet. Dazu gehört, daß nur die Ausschnitte der gesamten kategorialen Semantik ausformuliert werden, welche dem jeweiligen Problembereich entsprechen. Voraussetzung für eine klare Abgrenzung zwischen praktischen semantischen Entwurfsverfahren und der kategorialen Semantik ist eine genaue Klärung der Begrifflichkeit.

Grundlage praktischer Entwurfsschemata ist der Begriff des Graphen. Ein Graph mit einer ausgezeichneten Klasse von Teilgraphen (Diagramme) und von Kegeln heißt ein *(Entwurfs-)Schema*. Das einer Kategorie zugrundeliegende Schema besteht aus dem ihr zugrundeliegenden Graphen sowie der Klasse aller *kommutativen* Diagramme und *Limeskegel* ([BW]).

Ein Schemamorphismus von S nach S´ überführt als Morphismus von Graphen die Diagramme und Kegel von S in solche von S´. Ein *Modell eines Schemas* in einer Kategorie $\underline{C}$ ist ein Morphismus von S in das $\underline{C}$ zugrundeliegende Schema und überführt daher die Diagramme von S in kommutative Diagramme von $\underline{C}$ und die Kegel von S in Limeskegel von $\underline{C}$. Die Modelle von S in der Kategorie $\underline{Set}$ bezeichnen wir mit Mod(S). Mod(S) ist eine Kategorie, deren Morphismen in naheliegender Weise als "natürliche" Transformationen definiert werden.

Für große Klassen von Schemata (z.B. FP- und LE-Schemata, vgl. [BW]) existiert zu jedem Schema S eine Kategorie $\underline{S}$, so daß Mod(S) $\cong$ Mod($\underline{S}$) gilt, d.h. die Kategorie der Modelle von S ist äquivalent zu der Kategorie der limeserhaltendenen mengenwertigen Funktoren von $\underline{S}$. $\underline{S}$ heißt die von dem Schema erzeugte *Theorie*.

In der klassischen Modelltheorie besteht eine Theorie aus einer Sprache, Schlußfolgerungsregeln und Axiomen. Damit ist die Theorie ein Gegenstand der formalen Logik. Modelle sind Mengen mit einer zusätzlichen Struktur. Der neuere Ansatz behandelt Theorien als Kategorien, Modelle als Funktoren und Morphismen als natürliche Transformationen von Funktoren.

Die schemasprachlichen Ausdrucksmittel definieren also neben dem eigentlichen Schemagraphen bestimmte ausgezeichnete Untergraphen desselben (Diagramme) und bestimmte Kegel über Diagrammen, die nicht notwendig zu denen gehören, die ausgezeichnet sind. Da die Kategorie aller Graphen einen Topos bildet, bewegen wir uns daher im Bereich der Entwurfsschemata bereits auf natürliche Weise in einem kategorialen intuitionistischen (nicht booleschen) Diskursraum.

Dieser Topos entsteht als Kategorie der Prägarben (mengenwertigen kontravarianten Funktoren) auf der Kategorie $0 \rightrightarrows 1$. Ein solcher Funktor besteht aus zwei Mengen A_0 (Knoten) und A_1 (Pfeile) und zwei Abbildungen von der Menge der Pfeile in die Menge der Knoten (Zielknoten und Quellknoten)!

Die Modelle, als Entitätenmengen der Typen der Schematheorie und als die zwischen den Entitätenmengen bestehenden (Integritäts-)beziehungen, bilden als limenserhaltende Funktoren eine Kategorie, welche die Schematheorie umfaßt (alle darstellbaren Funktoren sind limeserhaltend!) und die ihrerseits wieder im Topos aller mengenwertigen Funktoren eingebettet ist.

Da jeder Update einer Datenbank mindestens eine Entitätenmenge verändert, liegt es nahe, ein Modell als momentanen Zustand einer Datenbank zu interpretieren. Mindestens unter der Voraussetzung, daß die Schematheorie cartesisch abgeschlossen ist (CCC), umfaßt dieser Datenbankbegriff neben den Basisdatentypen auch alle höheren Typen (Funktionen), da diese im Falle einer CCC darstellbar sind.

Das Typenmodell der Fachübergreifenden Bergwerksdatenbank (FÜDB)

Für die Bereiche Grubengebäude und Lagerstätte von Steinkohlenbergwerken wurde ein semantisches Datenmodell entwickelt, welches den Voraussetzungen der zuvor skizzierten Schematheorie entspricht, d.h. bis auf Notationsunterschiede erfolgte die Darstellung über gerichtete Graphen und den Einsatz der unter drei dargestellten Typkonstruktoren. Die Auswahl der zulässigen Typkonstruktoren entspricht der des Erweiterten Entity Relationship Modells (EERM).

Demgemäß wurden die folgenden Typkonstrukturen verwendet:

- Generalisierung
- Aggregation
- Partition
- Menge (= Potenztyp)
- Liste

Im Gegensatz zur Notation des Semantischen Datenmodells (SDM vgl. [H]) wurde statt der ausschließlichen Verwendung des Potenztyps noch der Beziehungstyp als Spezialisierung der Aggregation (Relation) eingesetzt. Ein Aggregationstyp heißt auch zusammengesetzter Typ, andernfalls heißt der Typ atomar.

Beispiele (Typen sind in Parenthese gesetzt):

Aggregation	"Hohlraumelement"	=	"Hohlraumbasisdaten" x "Markscheiderische Fachangaben" x "Elektrotechnische Fachangaben" x "Maschinentechnische Fachangaben" x "Fachangaben der bergmännischen Planung"
	"Gebirgskörper"	=	"Geologische Aufschlüsse" x "Scholle"
Liste	"Grubenbau"	=	Liste ("Hohlraumelement")
	"Punkt-Liste"	=	Liste ("Punkt")
	"Aufschluß"	=	Liste ("Aufschlußelement")
Menge	"Geologische Aufschlüsse"	=	P ("Geologischer Aufschluß")
Partition	"Grubenbau"	=	"Seigerer Grubenbau" + "Strecke" + "Bauhöhe" + "Bohrloch"
	"Geologischer Aufschluß"	=	"Tektonischer Aufschluß" + "Flözaufschluß" + "Sonstiger Aufschluß"

Das für die FÜDB bestehende Basisdatenmodell erlaubt an jeder Stelle Erweiterungen und Vertiefungen. So wird zur Zeit in einem weiteren Projekt der Basisdatentyp "Bohrloch" verfeinert und um die mit den Bohrdaten verbundenen Datentypen so erweitert, daß das derart entstehende Typenschema problemlos in das FÜDB-Schema integrierbar ist.

Beim derzeitigen Implementierungsstand wurde das FÜDB-Schema in eine relationale Datenbank auf UNIX-Oracle-Platformen übertragen und steht zur Zeit für die Datenübergabe in andere Fachbereiche (z.B. die Daten des offenen Grubengebäudes für die Bewetterungstechnik) und für ad-hoc-Auswertungen (SQL, SQL-Forms) zur Verfügung. Dagegen wird das Teilschema "Bohrdatensystem" in eine objektorientierte Motif/C++-Umgebung transformiert. Dieses Vorgehen belegt die Anpassungsnotwendigkeiten, die bei einem Übergang von der Modellierungsebene auf die Implementierungsebene in Abhängigkeit von den Anwendungsanforderungen (z.B. Abfrageorientierung oder Strukturkomplexität) zu beachten sind. Die Vorteile der semantischen Modellierung liegen auf der Hand, wenn z.B. in Zusammenhang mit dem objektorientierten Umbau eines bereits im Einsatz befindlichen CAE-Systems für Lagerstättenprognostik (DIGMAP) eine weitere Implementierung des FÜDB-Basisdatenmodells vorgenommen wird. In einem derartigen Fall stellt das semantische Schema die integrative Klammer dar, wenn aus technischen Gründen (Performance) auf der Implementierungsebene Datenintegrität nicht realisierbar ist und durch Datenspiegelung von objekbankbasierten CAE-Systemen mit Query-orientierten RDBMS ersetzt werden muß.

[BW] M. Barr, C.Wells [1985] Toposes, triples and theories; Springer-Verlag

[H] A. Heuer [1992] Objektorientierte Datenbanken; Addison-Wesley

[LS] J. Lambert, P.J. Scott [1986] Introduction to higher Order Categorial Logic; Cambridge University Press

[MOM] N. Marti-Oliet, J. Meseguer [1989] From Petri Nets to Linear Logic, Proceedings; Springer-Verlag

[PAPR] D. Pitt, S. Abramsky, A. Poigné, D. Rydeheard (Ed.) [1985] Category Theory and Computer Programming, Proceedings; Springer-Verlag

[W] O. Wyler [1991] Lecture Notes on Topoi and Quasitopoi; World Scientific

INVESTMENT ANALYSIS BASED ON THE COMMERCIAL SCOPE
The Production of Natural Gas

Douwe-Frits Broens, University of Groningen, The Netherlands
Willem K. Klein Haneveld, University of Groningen, The Netherlands

We give a description of results of a joint research project of our department and N.V. Nederlandse Gasunie, Gasunie for short. Gasunie buys all natural gas produced in the Netherlands and delivers all gas that is consumed on the inland market. Furthermore, Gasunie exports a great part of the inland production. After production, the gas is transported and mixed before it reaches consumers. Gasunie is enclosed between both supply-side and demand-side contracts. There is some space left between them, allowing Gasunie to find feasible production and transportation schedules. As a result of commercial developments, the space is shrinking, but it can be influenced by investments in production capacities and transportation means. Such investments mostly concern large projects, with great expenses and a lead time of several years. Because of the long lead times, Gasunie is faced with uncertainty about and within supply- and demand-side contracts, at the moment one has to decide on investment projects.
The research is focused on the role of commercial uncertainty in the investment planning. Gasunie's production planning problem is described as a system of linear (in)equalities, representing technical and contractual constraints. The goal of investments is to maintain feasiblity of the production planning problem. The central concept is the 'commercial scope': the set of future commercial scenarios yielding a feasible production planning problem. In most practical situations facing commercial uncertainty, the real objective of investments is to have a large enough commercial scope. Considering the resulting scope as a basic property of a proposed investment, both deterministic and stochastic characteristics are derived, using which one can quantify the commercial consequences of an investment proposal. In practice it is unwise to rely upon only one characteristic of the scope. We will derive new planning criteria. Well-known ones like robustness and risk are formalised in relation to the scope.
For many purposes it would be nice to know the commercial scope explicitly. Unfortunately, it turns out to be impractical to construct a complete explicit description. On the other hand it appears that it is not necessary either, for the calculation of the characteristics. What is more, one easily obtains explicit descriptions of 'induced' constraints on the uncertain variables, describing relevant parts of the boundary of the scope.
The investmet problem is elaborated in a numerical example. In the case at hand only a few induced constraints are identified. Special attention is paid to the interpretation of these constraints. They give direct insight into the effects of investments. It turns out that a sufficient characterisation of the scope can be obtained by using only these few induced constraints. Using them as an approximation of the scope gives great computational advantages. The quantitative characteristics of the scope are approximated very closely by this procedure.

TRAIN CIRCULATION PLANNING AT THE NETHERLANDS RAILWAYS

Henk van Evert, CVI Utrecht, NL

ABSTRACT

The prediction of the demand of, and the use of, passenger train types is of great importance to the Netherlands Railways. A train is composed of a certain number of train types 1 and 2. Each type has a different passenger capacity. At some stations during a train journey, the composition of the train can be changed according to the number of passengers to be transported. Up until now, deciding the mix of trains to buy of certain train types is based on a limited sensitivity analysis. This is because it is very time consuming for a planning expert to calculate manually the minimum number of train types necessary to run a train service during a day, given the planned timetable and the estimated number of passengers on each section (which varies heavily during the day). Using "conventional" Operations Research techniques it was until recently only possible to solve simplified instances of the problem on a computer (for example, a simplified instance can consist of 1200 equations with 200 integer variables and 100 binary variables). However, using a constraint satisfaction programming language, we (the Netherlands Railways, CVI, Bull and LCN) have succesfully built an operational system that suits most of the demands of the planning experts. The computer time of this model on a UNIX workstation is less than 15 minutes. This makes it possible to perform all kinds of simulations (different timetables, passenger growth, train capacity etc.). The results are equal to or sometimes even better than those of the planning experts. This means a substantial cost saving for the Netherlands Railways. Moreover, due to the nature of the CSP-language, the model is very flexible and changes can be made easily.

1. The business problem

About 4000 trains run daily on the Netherlands railroad network. There have to be enough train types at a starting point to let a train run and transport all passengers. It is fairly easy to plan these trains if the number of passengers is constant during the day. The problem is that in the morning a great number of people want to travel to e.g. Amsterdam and only few people travel the other way. In the afternoon this situation is exactly the other way round. So, a lot of train types go in the morning to Amsterdam that will be necessary only in the afternoon. If the Netherlands Railways (NS) employ train types only at the moment and the place that they are necessary, then big shunting yards will be necessary in the major cities. Besides, much more equipment would be used than necessary. Since every train type costs several million guilders, this is not acceptable. NS solves this problem by sending (as few as possible) empty train types back with the regular timetable. The organisation of this is called a train circulation.

The business problem being described in this article is concerned with the financial consequences of the purchase of train types for a medium long period, that is, for a period of 2 - 5 years. The delivery time for a train type is about 2 years. Therefore it is important to have a correct estimation of the mix of train types needed after 2 years.

NS has divided the train service into train series. A train series is the total of trains that, during a day, takes care of a connection between 2 terminal stations with a certain frequency. Example: the intercity series 2100 runs every hour between Amsterdam and Vlissingen from 6:00 until 24:00. The timetable of NS can be divided in approx. 100 of these series. For each of these series a train circulation is calculated manually to determine the mix of train types needed on a medium long period and to determine the cost of exploitation. For a train circulation the number of train types needed for each train in the timetable is calculated for a period of one day. This includes sending back empty train types. The situation at the end of the day has to be the same as the situation at the start of the day. Of course all passengers have to be transported (the number of passengers for each train is an estimation based upon the current situation). Calculating a train circulation takes 2-3 days work for one planning expert. Because of this, it is impossible to calculate different variants (different timetables, passenger growth, train capacity etc.). Therefore a continuous search has been made to solve this problem on a computer within reasonable computing time (one or two hours at the most). This article describes how this problem is solved using a constraint programming language. After a short historical review the mathematical problem will be described, followed by the solution principle. A brief note will be given on the computer architecture used and finally performance and results will be discussed.

2. History

Several publications (Cock, 1965), (Rees, 1965) prove that automation of the train circulation problem is already actual in the mid sixties with the emerging of Operations Research techniques (OR). Using integer lineair programming and network techniques, very simplified problems could be handled. But the real-life size problems could not be solved. In the mid eighties another attempt was made with the help of the University of Groningen. Again it was concluded that with the OR techniques of that moment it was impossible to solve the train circulation problem optimally and within a reasonable amount of computing time. The last few years attempts have been made to write a branch- and bound algorithm in Pascal. This program gives optimal solutions but the processing time varies heavily, which was unacceptable.

3. The mathematical problem

3.1 Introduction

As said before, the train service is divided into approx. 100 train series. On each series a number of trains is run and each train consists of one or more train types. On each series there are several stations where the composition of the train can be changed. Between two of those stations is a section. A train type consists of 1, 2, 3 or 4 carriages. On each series two train types can be employed.
The problem is: determine the minimum number of carriages and then the minimum number of train types that is needed every day. After that, determine the minimum number of carriage kilometres. Input is a (prognosis of the) timetable and a (prognosis of the) number of passengers to be transported on each section and at every moment of the day.

3.2 Constraints

The following constraints have to be taken into account when calculating a train circulation:

- For each section there is an upper limit for the number of carriages in a train due to the length of the platforms;
- At least one train type has to travel from beginning to end, allowing passengers to remain seated during the journey;
- If a train type is uncoupled it cannot be used at once: allow a redeployment time;
- At each station coupling can be done at either the front side or the back side of the train; the same goes for uncoupling;
- At each station either coupling or uncoupling can be done but not both;
- The situation at the start and the end of the day have to be the same.

3.3 Linear programming model

We will now describe the linear progamming model for a simple case: only one train type can be used, there is only one section and the redeployment time is 0.

Section AB has length l (in kilometres). Daily n runs are made.

d_i := the departure time of run i.

t := the travel time on section AB.

We number the runs 1..n such that $d_i \leq d_j$ for $i < j$.

N := $\{1, 2, ..., n\}$.

N_X := $\{ i \in N \mid \text{run } i \text{ leaves from } X\}$ $(X \in \{A, B\})$

V_X := The number of train types on station X at the beginning of the day $(X \in \{A, B\})$.

C_1 := the capacity of a train type for the first class.

C_2 := the capacity of a train type for the second class.

R_{i1} := the number of passengers for run i for the first class.

R_{i2} := the number of passengers for run i for the second class.

X_i := the number of train types that is used on run i.

M := the maximum number of train types for each train that is allowed on section AB.

Notation: $S(X_i \mid 1 \leq i \leq n)$ is the sum of all X_i with i between 1 and n.

Target:

Minimize $V_A + V_B$ (as few as possible train types) and after that:

minimize $l \times S(X_i \mid 1 \leq i \leq n)$ (as few as possible kilometres).

Constraints:

a. $C_j \times X_i \geq R_{ij}$ for $i \in N, j \in \{1, 2\}$

(all passengers have to be transported)

b. $X_i \leq V_A + S(X_j \mid j \in N_B, d_j + t \leq d_i) - S(X_j \mid j \in N_A, j < i)$ for $i \in N_A$.

$X_i \leq V_B + S(X_j \mid j \in N_A, d_j + t \leq d_i) - S(X_j \mid j \in N_B, j < i)$ for $i \in N_B$.

(the stock on a station has to be sufficient to perform a run)

c. $S(X_i \mid i \in N_A) = S(X_j \mid j \in N_B)$

(the situation at the end of the day = the situation at the start of the day)

d. $X_i \geq 1$ for $i \in N$ and $X_i \leq M$ for $i \in N$

This leads to $5 \times n + 1$ equations for n integer variables.

In the general case of 2 train types, n runs and p sections we have $12 \times n + 2 \times p$ equations with $2 \times n$ integer variables and n binary variables. For the series Amsterdam - Vlissingen with 100 runs and 3 sections this means approx. 1200 equations with 200 integer variables and 100 binary variables. Tests have shown that running this model on a PC leads to unacceptable fluctuation in performance. Besides, it is very difficult to adapt the program to new wishes of the users.

4. The solution principle

4.1 Introduction

The train circulation problem has the following general form:
"Given a finite set of objects that have to be assigned to slots, and a set of constraints (e.g. availability, connectivity, order), determine a solution satisfying the constraints and, if necessary, using some optimality criterion (e.g. time, cost, number of resources)."

These discrete constraints problems belong to the field of Operations Research. Though seeming very easy, they are extremely difficult to solve, especially due to the "combinatorial explosion": the enormous number of candidate-solutions. Conventional OR-solutions for this kind of problems, based on procedural languages, have a great disadvantage: they require a long period of development and therefore a big amount of money. Hence they lack the flexibility to solve this category of industrial problems. Besides, the performance is often not good enough.

Constraint satisfaction programming languages (CSP-languages) offer the flexibility and efficiency to solve this type of problems. Due to the declarative nature of a CSP-language the programmer only needs to describe the problem in a model using variables and constraints. Each variable has a set of possible values: the domain of the variable. The constraints limit the domains. Initially the search space consists of the cartesian product of all domains.
Finding a solution means searching the search space for a possible set of values satisfying all constraints. To do this in acceptable time is only possible by using a powerful reduction mechanism: constraint propagation.

4.2 Constraint propagation

Constraint propagation is a process that dynamically reduces the domains of the variables. We will give a short description of the principle.
First the variables with their domains are defined and after that the constraints. A variable will be present in several constraints and a constraint will include several variables. For every new constraint the reduction of the domain of each variable in that constraint is calculated. After that, these reductions are passed to all other constraints put on those variables. This can lead to new reductions which will be passed to other constraints etc. This is called constraint propagation.

Sometimes the solution is already found after defining all constraints. Mostly, a reduced search space has to be searched non-deterministically. Several techniques can be used for this. In this phase, constraint propagation remains active. If, during this process, a domain becomes empty, then a wrong assumption has been made and e.g. backtracking has to be used.

4.3 example

For solving the train circulation problem the CSP-language Charme is used (Bull, 90), (Bull, 91). We will give a simple example of a program in Charme. This is called the head-leg problem. Given are the number of heads and legs of a group of humans, horses and snakes. Determine the number of humans, horses and snakes. Written in Charme the solution reads as follows:

```
define head-leg (Head, Leg)
{
    Human ≥ 0;
    Horse ≥ 0;
    Snake ≥ 0;
    Human + Horse + Snake = Head;
    2 × Human + 4 × Horse = Leg;
    generate [Human, Horse, Snake];
    print [Human, Horse, Snake];
    fail;
}
```

If we use this procedure with a value of 5 for Head and 10 for Leg, then the following process is activated. The first 3 instructions constrain the domains of Human, Horse and Snake to the non-negative integer numbers. The fourth instruction reduces these domains to the values 0 upto and including 5. The fifth instruction further reduces these domains. This is because if there are 10 legs, the maximum number of horses is 2. They have a total of maximal 8 legs, so there must be at least 1 human. But then there can only be 4 snakes, due to the fourth constraint. So: Human = 1..5, Horse = 0..2, Snake = 0..4.

This is as far as constraint propagation can go. To find a solution, the sixth instruction is executed. The strategy pursued is: select a value from the smallest domain, so: Horse = 0. Due to constraint propagation this leads to Human = 5 (fifth constraint) and Snake = 0 (fourth constraint). So the first solution is found. To find the other solutions (if any), the instruction "fail" is executed. This means: backtrack to the last assumption and make another one. In this way the other solutions are found: Horse = 1, Human = 3, Snake = 1 and Horse = 2, Human = 1, Snake = 2.

4.4 Tackling the train circulation problem

As stated in 3.1, we want to determine for each train series a solution with a minimal number of carriages, train types and carriage kilometres (in this order). For this we need information about the series and train types used:

- At which stations can the composition of the train be changed;
- What is the length of the sections;
- Which trains are in the time table;
- What is the number of passengers on each section at every moment of the day;
- Which train types are used and what are the passenger capacities of these train types

Calculating the minimum number of carriages, train types and carriage kilometres is done in 3 phases. The variables used are Carriage, Traintype and Carriagekilometre.

First the constraints mentioned in 3.2 are set. Constraint propagation will reduce the domains of the 3 beforementioned variables.

In the second phase a strategy is followed that delivers the mimimal value for Carriage and Traintype and a first, but not necessarily minimal, solution for Carriagekilometre.

The third phase tries to minimise the value of Carriagekilometre. We use an iterative strategy that decreases the upper value of the domain of Carriagekilometre until no more solutions can be found.

Most of the time the minimum is found quickly, but sometimes the proof that a solution is the minimum takes a lot of time. Since we work with an estimated number of passengers, it is not important to find the absolute minimum: a solution which varies not more than 5% from the absolute minimum is quite acceptable. This reduces computing time.

5. Computer architecture

In most projects building the man-machine interface and processing the input data takes a lot of time. This project was no exception. The structure of the input data is complex and demands a lot of analysis and the output data has to be presented graphically. From the start development of the man-machine interface and the CSP-program has been done seperately and on different machines (IBM-PC and IBM-RS6000). This was based on several arguments:

- problems arising at both sides are quit different and demand different solving techniques;
- verification of the input can be done completely by the man-machine interface;
- to increase the performance only the CSP-program has to be analysed;
- it is a lot easier to write the interface in a suitable language than in a CSP-language.

6. Performance

As stated before, the performance of the program is the most important aspect. During development it became clear that the performance is related to the input in a not always understandable way. To increase the performance the following actions have been taken:

- Several redundant constraints have been added. They don't change the way the program works but they increase the performance. They have to be chosen carefully because not every redundant constraint leads to performance increasement.
- The requirement to find the absolute minimum of carriage kilometres is changed to the requirement to find a solution which varies not more than 5% from the absolute minimum.
- Extra constraints can be switched on and of by the user according to the problem at hand.

7. Results

The train circulation problem is very difficult to solve, even with the aid of a CSP-language. However, where conventional OR-techniques failed, CSP succeeded. The Netherlands railways now has an instrument that simulates real-life size problems and gives a (near-) optimal solution within acceptable time. With this instrument all sorts of simulations can be performed (different time tables, passenger growth, train capacity etc.) and new ideas can be tested. Moreover, changes to the model can be relatively easy made. In the near future, the model will be extended to deal with specific user needs.

8. Literature

Bull, 1990.	Introduction to CHARME. Parijs
Bull, 1991.	CHARME First Manual. Parijs
Cock, E. de, 1965.	Toepassing van Operational Research bij de bepaling van de behoefte aan en de omloop van reizigersmaterieel. In: Spoor- en tramwegen, 38e jaargang nr. 17, 16 september 1965.
Rees, J. van, 1965.	Een studie omtrent de circulatie van materieel. In: Spoor- en tramwegen, 38e jaargang nr. 18, 30 september 1965.

FREIGHT CENTERS
FUTURE NODES OF AN EUROPEAN LOGISTICS NETWORK

Dietrich Fischer, GARONOR Planungs- und Betriebs-GmbH, Taufkirchen

The heavy increase in transport, distribution and production logistics during the last decades has - along with the growing competition in the logistics market - resulted in a new concept, that is expected to increase the efficiency of the European logistics system and reduce the environmental impacts at the same time, i. e. freight centers rsp. logistic platforms, in German: Güterverkehrszentren.

These freight centers covering from 50 to 200 ha and offering 2000 - 5000 jobs are supposed to offer at least interfaces between road and rail transportation, in the ideal case also between water and air. Within the center - as part of their infra- rsp. suprastructure - they are supposed to provide all types of service of interest to a logistics company, including any meachanism to foster a voluntary cooperation between the various companies settled in the same center or in various centers to reduce long and short distance (city logistics) transportation by increasing average load factors and sharing overhead costs.

In Western Europe freight centers have developed over the last three decades, mainly as part of the GARONOR chain. In 1990 an European Organisation called EUROPLATFORM was established to ease an international exchange of experience and foster a linkage of the freight centers to a working European network.

In Germany official policy even claims combined traffic terminals incl. freight nodes conceived by the federal railway and the postal system to become an intergral part of freight centers specified in a national master plan. On the other side the development of freight centers has been delayed by the lack of terrain required and the reservations against cooperation common amongst logistics organisations.

Unfortunately also reservations against the application of OR techniques are widely spread, though freight centeres requiring an investment of up to a billion DM or even more would offer plenty of opportunities for the implementation of planning and optimization techniques.

The paper will give a survey on the freight center concept and its present implementation, incl. an outline of related OR problems.

PONTIS: AN OPTIMIZATION SYSTEM FOR IMPROVING AND MAINTAINING BRIDGE NETWORKS IN THE U.S.A.

Kamal Golabi, Optima Inc., U.S.A.

This paper presents a recently completed network optimization system to address the improvement and maintenance of bridge networks. The federally-funded bridge management system has sufficient flexibility for implementation in various states replacing current procedures which are largely based on subjective sufficiency ratings. The system would not only provide each agency with a decision tool to help in planning for MR&R (maintenance, repair and rehabilitation), improvement and replacement of its bridges, but would also provide a common basis for allocation of resources to states by federal agencies according to improvement needs and bridge conditions of each state.

The system, developed under the direction of the speaker, is named Pontis and addresses the dynamic, multiobjective bridge MR&R and improvement problem through a set of interrelated predictive, optimization and economic models. The system integrates the objectives of public safety and risk reduction, user convenience, and preservation of investment, with budgetary and program policies, and it effectively provides a vehicle for comparing preventive versus corrective maintenance policies. It gives a systematic procedure for allocation of resources to the improvement and MR&R of various bridges in the network and considers the costs and benefits of maintenance policies versus investments in improvement or replacement. Furthermore, it specifies minimum funding requirements to remove backlogs and achieve goals, and predicts future network conditions for various budget scenarios.

Pontis has been implemented in California and has undergone extensive tests in several other states. It is planned that the system would serve as a basis for bridge management decisions in all states.

RESULTS ON PREPROCESSING, POSTPROCESSING AND NODE PROCESSING IN INDUSTRIAL LP AND MIP MODELS

L. Van hamme, Beyers Innovative Software Brasschaat, Belgium
A. Van Looveren, Beyers Innovative Software Brasschaat, Belgium

OMP, a general solver for industrial LP and MIP models, is based on the simplex and branch and bound algorithms. Performance has been substantially improved through the implementation of various techniques, which reduce the model size and prune the branch and bound tree. Furthermore, the basic algorithms have been made faster, and the choice of branching variables has been sophisticated.
The model reduction techniques include variable and constraint elimination and substitution. Pruning of the branch and bound tree is obtained by sharpening variable bounds, applying coefficient reduction to the constraints, and generating cuts.
An overview of the various algorithms is presented. The modifications have been tested on a set of practical industrial LP and MIP models. Test results are evaluated on the basis of solution time. Improvements in LP models range from a factor 2 to 10. For MIP models, even more dramatic reductions in solution time have been obtained. Problems which could not be solved within a reasonable amount of time, are now solved with proven optimality within minutes.

DESIGN OF TRANSPORTATION SYSTEMS WITH A DECISION AID TOOL

Roland Hellberg, Linjegods AS, Norway
Heinz Michael Winkels, IBIES FH Dormund

This paper describes a case study of the design and evaluation of a terminal based transportation system for groupage with a high customer service level, performed by Linjegods A.S. in Norway. The project has been supported by a tailor-made decision aid tool (TERMLOC) that provides the best location of terminals under time-restrictions with respect to lowest total cost for collection, traffic-lines, distribution and handling.

In a terminal based transportation system the costs are mainly constituted by distribution/collection activities and the terminals. The quality of a groupage system mainly depends on collection and delivery time. In this sense the quality will increase with a great number of terminals and short distances to the various market areas since this implies short collecting and distribution time. On the other hand, a great number of terminals implies high cost for running these and high transportation costs for small volumes transported to each destination. That is where the multiple criteria conflict between cost minimisation and service level maximisation begins.

The problem is to find the localisation of terminals rendering the best and cheapest geographical covering. The solution includes an evaluation of the geographical covering and a comparison with the total cost for collection, terminal-handling, traffic-lines and distribution. It also contains an evaluation of alternative network structures (traffic-lines). TERMLOC is a tool based on a PC or UNIX database with additional simulation and optimisation modules. The data base is feed with all consignment flows during one year from zip code area to zip code area and the basic data of the transportation company (Linjegods) network structure.

The generation of free locations is based on an heuristic cluster analysis, where the locations are chosen in the order of the highest potential amount of collected and distributed goods. The reachable areas assigned to a specific terminal depend on the time restriction for collection and distribution. In TERMLOC the traffic-lines schedule can be manipulated interactively with respect to the use of transfer terminals and with respect to the mode of transport. After each manipulation the algorithm recalculates the container densities, the number of empty containers, and the cost of the traffic-line, etc.

Based on the existing terminal structure and with completely free locations, 21 different alternatives with connected traffic-line network and cost structure were derived. With respect to provided service level and the cost level the preferred new structure turned out. The solution will provide nearly the same service as present situation but with 12-15 MNOK lower cost. (The cost minimum is 20-25 MNOK, but then the service level in some areas will be lower.)

Liegeplatzdisposition auf einem Container Terminal - Ansätze zur Entwicklung eines entscheidungsunterstützenden Systems

Lutz Hoffarth und Stefan Voß, Darmstadt

Zusammenfassung

Für die Optimierung der Arbeitsabläufe auf dem Container Terminal Burchardkai Waltershof arbeitet die *Hamburger Hafen- und Lagerhaus-Aktiengesellschaft* zur Zeit an der Entwicklung bzw. Weiterentwicklung eines DV-gestützten Planungs- und Steuerungssystems. In dieser Arbeit befassen wir uns mit dem Teilaspekt der *Liegeplatzdisposition.* Sie hat zur Aufgabe, Schiffe so an dem Container Terminal zu disponieren, daß eine möglichst schnelle Abfertigung der Schiffe gewährleistet wird. Zur Lösung des Problems wird ein heuristischer Ansatz vorgestellt. Das Verfahren wurde anhand realer Daten sowie zufällig erzeugter Instanzen getestet.

Abstract

For optimizing the operations on the container terminal Burchardkai Waltershof, the *Hamburger Hafen- und Lagerhaus-Aktiengesellschaft* at present develops and investigates a decision support system. This paper deals with the aspects of *berth planning.* The overall objective is to guarantee a fast dispatching of the ships. For solving the problem a heuristic approach is proposed. The algorithm was tested on real-world data as well as on randomly generated instances.

1. Einführung

Am Container Terminal Burchardkai Waltershof der *Hamburger Hafen- und Lagerhaus-Aktiengesellschaft (HHLA)* werden jährlich über 3000 Schiffe abgefertigt (zum Großteil handelt es sich dabei um Containerschiffe). Für die Optimierung der Arbeitsabläufe auf diesem Terminal arbeitet die HHLA zur Zeit an der Entwicklung bzw. Weiterentwicklung eines DV-gestützten Planungs- und Steuerungssystems, in das u.a. die folgenden Komponenten integriert werden sollen:

- Yardplanung (Aufteilung der Containerstellplätze auf dem Terminal)
- Fahrwegoptimierung der Van Carrier (VC, Portalhubfahrzeuge) im Bereich Platzgang (interne Containertransporte)
- Liegeplatzdisposition
- Brückeneinsatzplanung

In dieser Arbeit befassen wir uns mit der *Liegeplatzdisposition.* Sie hat zur Aufgabe, Schiffe so an dem Terminal zu verteilen, daß eine möglichst schnelle Abfertigung der Schiffe gewährleistet wird. Hierbei sind eine Vielzahl von Prämissen und Restriktionen wie z.B. Liegezeitvorgaben, Schiffsmerkmale (Tiefgang, Länge, Breite etc.) und die Brückenverfügbarkeit zu berücksichtigen.

Eine schnelle Abfertigung der Schiffe ist u.a. von den Transportwegen der Exportcontainer von ihren Stellplätzen zu den Schiffsliegeplätzen und umgekehrt (Importcontainer) abhängig. Das Ziel der Liegeplatzdisposition ist daher die Verteilung aller angemeldeten Schiffe für einen Planungszeitraum auf die Kaianlagen des Terminals, so daß die Gesamtstrecke der anfallenden VC-Fahrwege minimiert wird.

Es wird ein heuristischer Ansatz zur Lösung der genannten Aufgabenstellung vorgestellt. Er beinhaltet ein Eröffnungsverfahren zur Bestimmung einer Ausgangslösung (Prinzip der sukzessiven Einbeziehung) sowie ein nachgeschaltetes Verbesserungsverfahren. Die Verfahren wurden anhand von praktischen Daten der HHLA sowie zufällig erzeugten Probleminstanzen getestet. Darüber hinaus wird auf die Einbindung der Verfahren in ein entscheidungsunterstützendes System (manuelle und automatisierte Disposition der Schiffe an den Kaianlagen des Terminals) sowie auf Schnittstellen zu weiteren Komponenten hingewiesen (z.B. der Brückeneinsatzplanung).

2. Problemstellung

Das Umschlagsystem des Container Terminals gliedert sich je nach Verkehrsart bzw. durchzuführenden Tätigkeiten in verschiedene Arbeitsbereiche. Die LKW- und Bahnabfertigung stellt dabei die landseitige, die Schiffsabfertigung die wasserseitige Schnittstelle dieses Systems dar.

Die *Schiffsabfertigung* bezieht sich auf Übersee-, Feeder- und Binnenschiffe. Die großen Überseeschiffe besitzen Kapazitäten von bis zu 4200 Containereinheiten (TEU, gemessen in 20 Fuß Einheiten). Ihre wasserseitige, regionale Anbindung erfolgt durch Feederschiffe mit Kapazitäten bis ca. 200 TEU, die landseitige über Brücken, die ihrerseits durch VC's bedient werden. Die angelieferten Container werden in einem Import-Export-Stock (Yard) zwischengelagert.

Die Containertransporte werden über ein an den Host der HHLA gekoppeltes Datenfunksystem gesteuert, wobei die von dem Host-Rechner generierten Fahraufträge über eine Datenfunkschnittstelle an die VC's weitergeleitet werden. Bei Beendigung eines Transportauftrages wird der neue Stellplatz per Funktelegramm vom VC an das System zurückgemeldet. Die VC-Fahrwege sind in der Regel rechtwinklig, bedingt durch die blockweise Anordnung der Container. Ihre Länge (proportional zu den anfallenden Betriebskosten der VC's) stellt unter anderem einen erheblichen wirtschaftlichen Faktor dar. Da die aktuellen Containerstellplätze zu jedem Zeitpunkt abrufbar sind, können die Transportwege einerseits über eine *Fahrwegoptimierung* der VC's (vgl. [2,3]) und andererseits bei der wasserseitigen Schnittstelle durch eine günstige Disposition der Schiffe am Kai verringert werden. Letztere fällt in den Aufgabenbereich der Liegeplatzdisposition (**LPD**).

Bei der Liegeplatzdisposition unterscheidet man zwischen der langfristigen und der kurzfristigen LPD. Die *langfristige LPD* hat zusammen mit der Yardplanung die Aufgabe, eine möglichst schnelle Abfertigung (Löschen und Beladen) der Schiffe zu gewährleisten. Hierzu wird den Schiffen frühzeitig (bei Großschiffen meist vier Wochen vor Schiffsabfahrt) ein voraussichtlicher Liegeplatz an der insgesamt ca. 2500 m umfassenden Kaianlage zugeordnet (vgl. Abb. 1 für einen Überblick der Kaianlage). Am Container Terminal Burchardkai gibt es 15 Brücken sowie zwei RoRo-Anlagen. (Von RoRo-Verkehr wird gesprochen, wenn Fahrzeuge oder durch Hilfsmittel fahrbar gemachte Ladungen über geeignete Rampen auf Schiffe rollen.)

Eine frühzeitige Zuordnung der Schiffe an die Liegeplätze ist möglich, da die Großschiffe meist einem "Fahrplan" unterliegen und daher etwaige Ankunftsdaten im voraus bekannt sind. Die

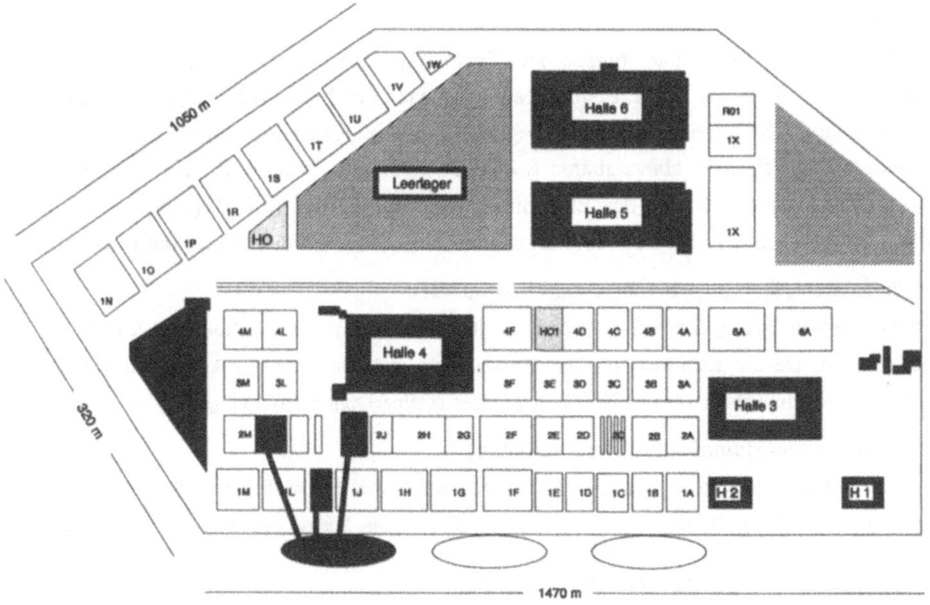

1. langfristige LPD (vier Wochen): Schiff an Liegeplatz
2. Yardplanung (anschließend): Stellplätze nahe Schiff
3. aktuelle LPD (wenige Tage): Schiff nahe Stellplätze

Abb. 1. Schematische Darstellung des Container Terminals Burchardkai / LPD und Yardplanung

anschließende Yardplanung vergibt Stellplätze für die zu erwartenden Exportcontainer dieser Schiffe, die möglichst in der Nähe ihrer geplanten Liegeplätze sein sollten. Erst wenige (ein bis vier) Tage vor der erwarteten Schiffsankunft können jedoch präzisere Daten zur Schiffsabfertigung erfaßt werden. Darüber hinaus werden weitere Schiffsankünfte (hauptsächlich kleinere Feederschiffe, welche zum Teil weitere Container für die Großschiffe anliefern bzw. abholen) angemeldet.

Diese Veränderungen erfordern eine laufende Aktualisierung der Liegeplatzverteilung. Dies ist die Aufgabe der *kurzfristigen LPD.* Ziel ist es dabei, die zur Abfertigung der Schiffe benötigten VC-Fahrwege auf dem Terminal gering zu halten. Hierbei sind eine Vielzahl von Prämissen und Restriktionen wie z.B. Liegezeitvorgaben, Schiffsmerkmale (Tiefgang, Länge, Breite etc.) und Brückenverfügbarkeit zu berücksichtigen.

Diese Aufgabe wird zur Zeit auf Grundlage eines Kaibelegungsplans mit Hilfe eines DV-gestützten Graphik-Systems durchgeführt, wobei die Schiffe den Liegeplätzen manuell zugeordnet werden. Über die Brückeneinsatzplanung werden den Schiffen dann unter Berücksichtigung baulicher und technischer Gegebenheiten Brücken oder RoRo-Anlagen zugeordnet. Die Brückeneinsatzplanung liefert als Ergebnis die Informationen, welche Brücke/RoRo-Anlage wann welche Container an welchem Schiff bewegt. Für die kurzfristige LPD, die zusammen mit der Brückeneinsatzplanung ein interaktives System bildet, soll nun ein entscheidungsunterstützendes System (weiter-) entwickelt werden. In den folgenden Abschnitten skizzieren wir dazu die dem Problem zugrundeliegenden Daten, die sich u.a. daraus ergebenden Nebenbedingungen sowie eine zu optimierende Zielsetzung.

2.1 Eingabe-Daten

Um eine LPD zu ermöglichen, müssen dem System gewisse (Input-) Daten bekannt sein. Hierzu gehören Schiffs-, Brücken- und Liegeplatzdaten, welche über Nebenbedingungen die zulässigen und damit möglichen Kaibereiche je Schiff begrenzen, sowie Kapazitätsbeschränkungen, welche die Gesamtanzahl gleichzeitig abfertigbarer Schiffe beeinflussen. Punkte am Kai werden in Meter-Marken (MM) angegeben, die der Entfernung vom Anfang der Kaianlage bis zu der jeweiligen Position entsprechen. Für die LPD sind im wesentlichen die folgenden relevanten Daten verfügbar:

- **Schiffsdaten (Stamm-, Ankunfts- und Routendaten)**
  hierzu gehören: Name, Schiff-Code, Länge, Breite, ob es sich um ein RoRo-Schiff handelt, ob Querstau möglich ist, eta/ets (estimated time of arrival/sailing, und damit die gewünschte Liegezeit), aktueller Tiefgang, Arbeitsaufwand (Anzahl der Lösch-/Lade-Container), priorisierter Liegeplatz
- **Brückendaten (Stamm- und Sperrungsdaten)**
  hierzu gehören: Brücken-Nr., Länge des Auslegers, Breite der Brücke, Fahrbereich in MM, Vorhandensein einer Drehkanzel (für Querstau), Sperrungsdaten (Dauer von/bis, Sperrungsgrund)
- **Liegeplatzdaten (Stamm- und Sperrungsdaten)**
  hierzu gehören: Soll-Wassertiefe und Lage von Untiefen, Sperrungsdaten (Dauer von/bis, Bereich von/bis)

Aus der Yardplanung sind Informationen über die Container (u.a. die Stellplätze) verfügbar. Darüber hinaus ist eine Distanzmatrix mit den folgenden Angaben gegeben: von / bis (Bezeichnung von Stellplätzen (Stellplatz-Code) oder Brücken (Brücken-Nr.)) und die Entfernung. Bei der Personal-/Gerätedecke wird für jede Schicht die Anzahl der verfügbaren Gänge (entspricht der maximalen Anzahl gleichzeitig einsetzbarer Brücken) vorgegeben.

2.2 Nebenbedingungen

Die Nebenbedingungen lassen sich in ausschließende und konkurrierende Bedingungen unterteilen. Ausschließende Bedingungen beziehen sich auf das einzelne Schiff, wohingegen die konkurrierenden Restriktionen den gegenseitigen Beschränkungen unter den Schiffen gelten. Im folgenden werden die Nebenbedingungen entsprechend dieser beiden Kriterien aufgelistet.

- **ausschließende Nebenbedingungen:**

(1) Die Brücken sind nur innerhalb bestimmter Meter-Bereiche des Kais einsetzbar. (Die Länge der Brückenfahrwege bei Brückenbewegungen sind für die Optimierung nicht relevant).

(2) Gesperrte Brücken dürfen während der Sperrungsdauer nicht mit in die Disposition einbezogen werden.

(3) Gesperrte Kaibereiche dürfen während der Sperrungsdauer nicht mit in die Disposition einbezogen werden.

(4) Der aktuelle Tiefgang eines Schiffes darf nicht größer sein als die Wassertiefe des zugeordneten Liegeplatzes.

(5) Die Breite des Schiffes muß mit der Länge des Auslegers der zugeordneten Brücke(n) korrespondieren (um ein Löschen bzw. Beladen auf voller Breite zu gewährleisten).

(6) Schiffe mit Querstau benötigen eine Brücke mit Drehkanzel (hierbei werden Container nicht nur längs im Schiff verstaut und müssen daher beim Lösch- und Ladevorgang durch die Brücke um 90^0 gedreht werden).

(7) RoRo-Schiffe benötigen entsprechende Anlagen und können daher nur bestimmten Meter-Bereichen (bzw. Liegeplätzen) zugeordnet werden.

- **konkurrierende Nebenbedingungen:**

(8) Nebeneinanderliegende Schiffe müssen zueinander einen Sicherheitsabstand von 10m (20m bei Großschiffen) haben.

(9) Die Schiffsabfertigung muß innerhalb der Liegezeit abgeschlossen werden können.

(10) Der Arbeitsbedarf gleichzeitig zu bedienender Schiffe muß mit der zur Verfügung stehenden Personal- und Gerätedecke (max. Anzahl der gleichzeitig eingesetzten Brücken) zu bewältigen sein.

2.3 Zielsetzung

Bei der kurzfristigen LPD handelt es sich um ein kombinatorisches Problem (die Schiffe sind bestimmten Meter-Bereichen der Kaianlage zuzuordnen), welches unter Zugrundelegung einer geeigneten Bewertungsfunktion zu lösen ist. Das Ziel ist die Verteilung aller für den Planungszeitraum angemeldeten Schiffe (aktuelles Datum bis zum jüngsten Ankunftsdatum; max. vier Wochen im Voraus) auf die Kaie des Terminals unter Berücksichtigung der im vorigen Abschnitt genannten Nebenbedingungen. Dabei ist die Zielsetzung

- Minimiere die Gesamtstrecke der im Planungszeitraum anfallenden VC-Fahrwege

zu verfolgen. Weitere, nachgeordnete Ziele, die zusätzlich betrachtet werden können, sind:

- Die Abweichungen von den gewünschten Liegezeiten sollen möglichst gering sein.
- Die Beachtung der Prioritäten bei der Liegeplatzvergabe (z.B. priorisierter Liegeplatz aus den Schiffsdaten) sollte nach Möglichkeit erfolgen.

3. Lösungsansatz

Ausgehend von den einem Kaibelegungsplan zugrundeliegenden Überlegungen läßt sich das Problem der kurzfristigen LPD als ein verallgemeinertes Zuordnungsproblem formulieren. Genauer handelt es sich dabei um ein dynamisches nichtlineares Semi-Zuordnungsproblem (jedes Schiff ist der Kaianlage sowie mindestens einer Brücke zuzuordnen, aber nicht jedem Meter-Bereich der Kaianlage bzw. jeder Brücke ist ein Schiff zuzuordnen; vgl. zum quadratischen Semi-Zuordnungsproblem z.B. [1,4]). Bezüglich der Komplexität des Problems sei darauf verwiesen, daß bereits das Auffinden einer zulässigen Lösung ohne Berücksichtigung der oben angegebenen Zielfunktion ein NP-schweres Problem darstellt, so daß wir uns auf die Betrachtung von Heuristiken beschränken. Im folgenden geben wir zunächst ein Eröffnungsverfahren, welches sich zur Umgehung des Zulässigkeitsproblems einer fiktiven Kaianlage bedient, und daran anschließend ein Verbesserungsverfahren an.

3.1 Eröffnungsverfahren

Mit Hilfe des Eröffnungsverfahrens werden die zur Verfügung stehenden Daten zunächst eingelesen und aufbereitet. Zur Generierung einer zulässigen Anfangslösung wird eine dem Verfahren der *sukzessiven Einbeziehung* ähnliche Vorgehensweise verwendet. Hierbei werden die Schiffe in der sortierten Reihenfolge ihrer Ist-, bzw. Soll-Ankunftszeit kostengünstigst jeweils demjenigen Kaibereich zugeordnet, der eine zulässige Lösung bezüglich aller bereits zugeordneten Schiffe ermöglicht. Alle nicht zuordenbare Schiffe werden zunächst an einem fiktiven (d.h. künstlich hinzugefügten) Kai, der keinerlei Beschränkungen unterliegt (z.B. besitzt er unendliche Kailänge), mit sehr hoher Kostenbewertung disponiert. So wird gewährleistet, daß immer eine "zulässige" Lösung existiert.

Die Zuordnung der Schiffe erfolgt in Abhängigkeit einer vom Benutzer vorzugebenden *Schrittweite*. Aufgrund dieses Parameters werden die für ein Schiff zulässigen Lösungsbereiche, jeweils vom

Anfang des Bereiches plus halber Schiffslänge bis zum Ende des Bereiches minus halber Schiffslänge mit der entsprechenden Schrittweite, auf die günstigste Einfügemöglichkeit hin überprüft. Die Berechnungen der VC-Fahrwege beziehen sich dabei auf die Schiffsmitte und werden gemäß der oben definierten Meter-Marken durchgeführt.

3.2 Verbesserungsverfahren

In einem ersten Schritt des Verbesserungsverfahrens werden zunächst gegenseitig konkurrierende Schiffe, die am fiktiven Kai untergebracht sind, daraufhin untersucht, welche kostengünstiger an die tatsächlichen Kaie umdisponiert werden können. Im zweiten Schritt des Verbesserungsverfahrens wird über ein Austauschverfahren versucht, Schiffe zu vertauschen bzw. zu verschieben, so daß sich der Zielfunktionswert verbessert, d.h. die Gesamtstrecke der VC-Fahrwege reduziert. Es handelt sich also um ein Vertauschungsverfahren, bei dem jeweils zwei Schiffe paarweise auf eine zulässige Vertauschung hin untersucht werden.

4. Implementierung und Ergebnisse eines Fallbeispiels

Die Tests zum obigen Verfahren wurden auf einem IBM-kompatiblen PC 80486DX/50 MHz mit 8 MB Speicherkapazität und 256 kByte Cache durchgeführt. Die Implementierung erfolgte in der (datenbankorientierten) Programmiersprache CLIPPER 5.01. Die im folgenden wiedergegebenen Testergebnisse beziehen sich auf den Zeitraum von drei Tagen (25.–27.05.1993), in denen 15 Schiffe (elf Großschiffe und vier Feederschiffe) in die Optimierung einbezogen waren (vgl. Tab. 1 zu Angaben über minimale, maximale und durchschnittliche Schiffs-Längen und Kosten für beide Schiffs-Typen). Weitere sechs Schiffe schränkten deren mögliche Liegeplätze ein. Diese hatten entweder schon einen früheren Ankunftstermin, und ihre Liegezeit überschnitt sich mit dem Planungszeitraum, und/oder sie waren vom Planer einem bestimmten Liegeplatz fest zugeordnet worden. Durch die nicht frei disponierbaren Schiffe entstand ein zusätzlicher konstanter Kostenwert C, der zum Zielfunktionswert addiert werden mußte, um die Gesamtkosten (Gesamtstrecke in m) innerhalb des Planungszeitraumes bestimmen zu können.

Tab. 1: Schiffs-Statistik (Angaben jeweils in Metern)

	Großschiffe			Feederschiffe		
	min	max	ϕ	min	max	ϕ
Längen	100	290	163	80	146	101
Kosten	568	465699	103794	0	42332	8815

Tab. 2: Testergebnisse zum Fallbeispiel (in Auszügen)

Schritt-	Eröffnungsverfahren (E)		Verbesserungsverfahren (V)		Verbesserungen in %		
weite	Kosten	Zeit	Kosten	Zeit	V - E	E - M	V - M
5	295842	32,52	295500	42,79	0,12	18,33	18,42
10	295849	16,91	295507	29,71	0,12	18,33	18,42
20	295851	10,43	295509	24,05	0,12	18,33	18,42
50	296049	6,98	295959	18,90	0,03	18,27	18,30
80	421121	6,38	332179	20,04	21,12	-16,26	8,30
110	407246	6,10	296174	16,59	27,27	-12,43	18,24
140	421146	6,09	332204	19,78	21,12	-16,26	8,29
170	296473	5,95	296244	17,90	0,08	18,15	18,22
200	403306	5,98	367090	19,27	8,98	-11,34	-1,34

Tab. 2 gibt die numerischen Ergebnisse für den beschriebenen Datenfall wieder. Die manuelle Disposition der Schiffe innerhalb des Planungszeitraumes (also die tatsächlich durchgeführte LPD) führte zu einer Gesamtfahrstrecke der VC's von 362237 m (bei zusätzlichem C = 36544 m). Tab. 2 zeigt zunächst die absoluten Zielfunktionswerte sowie die CPU-Zeiten des Eröffnungs- (E) und des Verbesserungsverfahrens (V) für ausgewählte Schrittweiten. Zusätzlich sind die relativen Verbesserungen des Verfahrens V zu E sowie beider gegenüber der manuellen Planung (M) gegeben.

In dem betrachteten Datenfall konnten alle Schiffe am Kai positioniert werden. Die Ergebnisse lassen sich im einzelnen wie folgt interpretieren. Mit längeren Schrittweiten nehmen die Rechenzeiten ab. Dies war zu erwarten, da höhere Schrittweiten die Anzahl der zu untersuchenden möglichen Liegeplatzpositionen verringern. Bei fast 50% der Testläufe lieferte schon das Eröffnungsverfahren einen besseren Wert als die manuelle Disposition, und in über 80% der Fälle wurde mit Hilfe des Verbesserungsverfahrens ein bis zu 19% günstigerer Kostenwert erzielt. Das Verbesserungsverfahren lieferte zum Teil über 27% bessere Ergebnisse als das Eröffnungsverfahren. Die besten Ergebnisse wurden jedoch bei denjenigen Schrittweiten erzielt, bei denen bereits das Eröffnungsverfahren sehr gute Kostenwerte (über 18% Verbesserung gegenüber der manuellen Disposition) erzielte.

5. Ausblick

In der vorliegenden Arbeit haben wir die Liegeplatzdisposition auf einem Container Terminal betrachtet. Numerische Ergebnisse eines realen Datenfalles zeigen, daß dem Planer mit Hilfe eines geeigneten Verfahrens Unterstützung bei der Entscheidung gegeben werden kann. In dem speziellen Datenfall wurden die besten Ergebnisse mit der kürzesten Schrittweite erzeugt. Weitere durchgeführte Testläufe (u.a. mit zufällig erzeugten Daten) lassen jedoch eher die Vermutung zu, daß eine datenabhängige Wahl dieses Parameters (z.B. vermöge mittlerer Schiffslänge, verschiedener Sicherheitsabstände bei Groß- und Feederschiffen) eine gewisse Rolle spielt.

Grundsätzlich bietet der vorgestellte Ansatz die Möglichkeit der Entscheidungsunterstützung im Rahmen einer interaktiven Gesamtlösung aller die VC-Fahrwege betreffenden Komponenten sowie die Brückeneinsatzplanung und die Container-Stauplanerstellung. Bessere Ergebnisse ließen sich gegebenenfalls durch die Anwendung moderner heuristischer Metastrategien (vgl. z.B. [1,4] beim quadratischen Semi-Zuordnungsproblem) sowie eine mehrfach hintereinander gekoppelte Vertauschung von Schiffen erzielen.

Literatur

[1] W. Domschke, P. Forst und S. Voß, Tabu search techniques for the quadratic semi-assignment problem, in: G. Fandel, T. Gulledge and A. Jones (Hrsg.) *New Directions for Operations Research in Manufacturing*, Springer, Berlin u.a. (1992), 389 - 405.

[2] D. Steenken, Fahrwegoptimierung am Containerterminal unter Echtzeitbedingungen, *OR Spektrum* 14 (1992), 161 - 168.

[3] D. Steenken, A. Henning, S. Freigang und S. Voß, Routing of straddle carriers at a container terminal with special emphasis on internal moves, *OR Spektrum* 15 (1993), im Druck.

[4] S. Voß, Network design formulations in schedule synchronization, in: M. Desrochers and J.-M. Rousseau (Hrsg.) *Computer-Aided Transit Scheduling*, Lecture Notes in Economics and Mathematical Systems **386**, Springer, Berlin u.a. (1992), 137 - 152.

DESIGN OF A CASSETTE-ORDERPICKING SYSTEM
A CASE STUDY

RENÉ DE KOSTER, INGENIEURSBUREAU GROENEWOUT B.V., BREDA

Summary: At Van Leeuwen Buizen, a design has been made for a cassette system, for automatic order-picking of tubes, strapping and loading of trucks. This system combines a number of known subsystems completed with new elements. The control of the system is new as well. In order to investigate the performance, the system has been simulated. It appears that the picking capacity of only one system is sufficient, but for the loading of trucks, two systems are needed.

Zusammenfassung: Bei Van Leeuwen Buizen ist ein Kassettenlagerungssystem für automatische Kommisionierung und laden von Stahlrohren entworfen worden. Dieses System kombiniert bestehende und neue Elemente. Die Steuerung ist neu entworfen worden. Um die Leistung dieses Systems zu überprüfen, ist es simuliert worden. Dabei ist deutlich geworden, daß für Kommisionierung nur ein System ausreichend ist, aber daß für das Laden zwei Systeme erförderlich sind.

1. INTRODUCTION

Van Leeuwen Buizen B.V. is a major stock-keeping distributor of steel tubes and profile tubes, with its headquarters in Zwijndrecht, the Netherlands.

After having carried out a feasibility study, Ingenieursbureau Groenewout B.V. was ordered in 1991 to develop a new logistic design, including realization. This logistic design includes a new site layout and an extension of the indoors storage with all necessary storage and material handling systems.
An important part of the design is the automatic picking of customer orders for steel tubes from a so-called cassette-orderpicking system. In the cassette system, the order picking stock of the majority of the articles is stored. About 80% of the turnover should be realized from the system.

In a cassette system, the tubes are stored in long bins (cassettes). The cassettes are open at the top and half-open at the bottom. The cassettes are stored in cantilever racks, which are grouped in aisles. Cassettes are stored and retrieved by an automatic crane.

Conventional cassette systems usually consist of one storage module for cassettes, with narrow aisles, with an overhead transfer crane. To retrieve a cassette, the crane lowers in the aisle, takes a cassette from the rack and drops the cassette on, for example, a shuttle (an automatic rail-guided vehicle). This shuttle brings the cassette to an order picking station, where an operator picks the desired quantity of tubes from the cassette.

In most cases, the transfer crane and shuttle perform double cycles; the shuttle returns a cassette to an aisle containing an empty location. The crane picks up the cassette from the shuttle and stores it. Then, the shuttle and crane travel to the aisle where the next cassette has to be retrieved, and retrieve it.

Within cassette systems, different storage strategies are observed, such as random storage, pickfrequency class-based storage and mixtures of the two.

In the team of Van Leeuwen and Groenewout, a concept has been developed in which three such systems operate together and are responsible for automatic orderpicking and loading of orderlines at four truck-loading stations. This includes automatic counting, strapping and weighing of the proper number of tubes, as well as the loading of four different trucks simultaneously from one system. On the basis of the number of daily trucks to be loaded from the system (40 trucks in 7 hours), it was estimated that two such systems were needed, with an investment of about Dfl 13 million each.
Cassette systems, combined in this way, have not been applied before. The control has to be designed newly.

In this paper, the design of the cassette system and its performance will be discussed. In order to investigate the performance, a simulation study was carried out. It appears that only one system yields sufficient picking capacity for all daily trucks to be loaded. This would imply an important saving. Unfortunately, the truck loading capacity at the system is much less.
Furthermore, the picking capacity is strongly affected by the system's control strategy employed.

2. THE CASSETTE SYSTEM

The cassette system consists of three types of coupled elements.

1. A storage part, with material handling systems. This consists of two so-called 'enkele handels-lengte' (EHL) subsystems with 7 m cassettes and one 'dubbele handels-lengte' (DHL) subsystem, with 14 m cassettes. Each subsystem has a transfer crane and a shuttle of proper dimensions.

2. Per truck-loading station, two automatic orderpick stations. One for EHL and one for DHL cassettes. At an orderpick station, the cassettes are fixed, and the desired number of tubes is automatically taken from the cassette, counted and strapped in a bundle. The strapped tubes are automatically transported on a buffer (an accumulating conveyor, with a width of 7 m or 14 m, depending on the pick station) to the loading position.

3. Four truck-loading stations. A loading station consists of a manually controlled transfer crane, that is able to load from the 7 m and the 14 m buffers and also from separate so-called route locations, opposite the truck loading location.

The EHL and DHL subsystems and the picking stations are connected via conveyors.
A number of designs have been worked out, of which the one shown in Figures 1 (perspective view) and 2 (cross section of storage part) has been chosen.

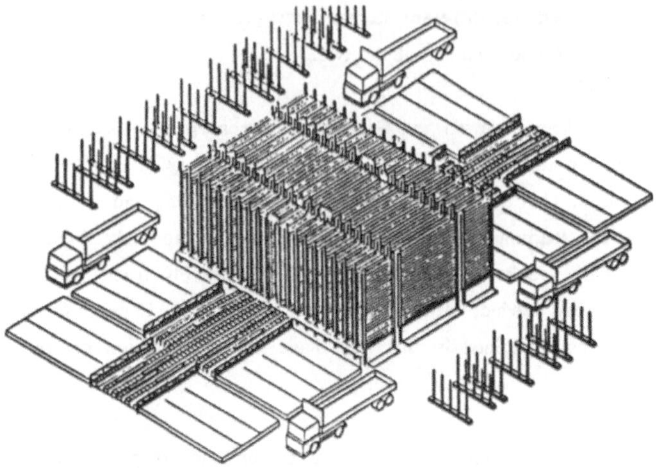

Figure 1. Three dimensional view of the cassette system.

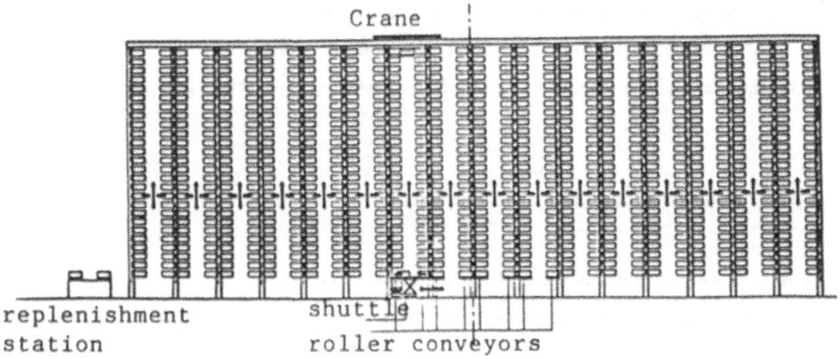

Figure 2. Cross section of a storage module.

In the DHL-subsystem, all A- and B-tubes with a length up to 14 m, that fit in the cassette, are stored. In each of the two EHL-subsystems, all A articles, but only half of the B-articles are stored. This implies, that when a truck needs B-articles of 7 m, on average half of these articles have to be retrieved from the far EHL subsystem.

At the four loading stations, four trucks can be loaded simultaneously. Each loading station has route-locations, one EHL-buffer and a DHL buffer. The route locations are for articles that have been picked from other locations than the cassette system, or for articles picked from the cassette system that have been machined. The articles in the route locations are grouped per truck, and per truck stored in LIFO loading sequence.
For large orderlines, a truck loads directly at a bulk location.

Each of the three subsystems works simultaneously for four loading stations. Since the truck-loading sequence is known in advance, each subsystem can retrieve orderlines per loading station in strict loading sequence.

3. CONTROL OF THE CASSETTE SYSTEM

Before the loading process starts, the trucks are allocated in sequence to the different loading stations. For each subsystem, the cassettes are retrieved per loading station in loading sequence of the trucks. To assure a high throughput, each of the three subsystems performs double cycles. Each of the three subsystems starts with a number (this number is a control parameter, denoted by the *workload norm per subsystem-loading station combination*) of retrievals, but then only double cycles are carried out. The higher the workload norms (there are 3x4 such norms), the higher the throughput. However there is a trade-off: the higher the workload norms, the more cassettes there will be simultaneously in transportation in the system and the longer the reponse times (this is the time span between the start of the retrieval and the arrival of the cassette at the picking station). At still higher norms, the system can become deadlocked.

The fact that each EHL subsystem should contain all A-, but only half of the B-articles, was solved by returning each cassette to the subsystem it came from. Within each subsystem a random storage strategy was chosen. This is realized, by giving the shuttle a capacity for two cassettes. In a double cycle, the shuttle first moves with the returned cassette to the next retrieval aisle. There, the crane retrieves a cassette and deposits it on the shuttle. The shuttle repositions, and the crane then takes the returned cassette from the shuttle and stores it on the same location the outbound cassette came from. In the mean time, the shuttle deposits the outbound cassette on the conveyor leading to the picking station, as soon as the drop-off position on this conveyor is free.

As can be seen in Figure 3, the conveyor that supplies one EHL or DHL pickstation, discharges an opposite EHL or DHL pickstation. Furthermore, the EHL conveyors are used simultaneuously by two EHL subsystems. This requires, with several cassettes simultaneously on each conveyor, a control of the one (DHL conveyor) or two (EHL conveyor) pick and drop (p&d) positions on the conveyor. In order to avoid deadlock it is important, that the p&d positions are kept free, until a returning cassettes is allowed to be stored.

When several returning cassettes are waiting on the conveyor near a p&d point, the oldest is stored first.

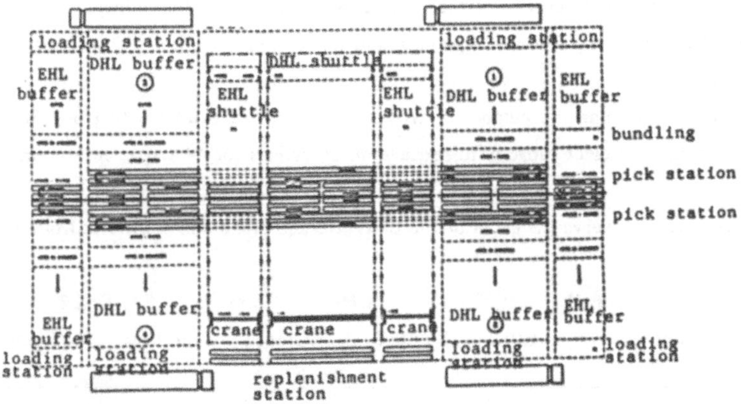

Figure 3. Top view of the system.

4. SIMULATION OF THE CASSETTE SYSTEM

In the simulation, use is made of 236 standard trucks, which have been registered during a number of weeks and analyzed. Registered are such data as loading sequence, loading times, loading location, number of pieces per orderline. In the analysis, per orderline the loading location in the new design is determined (either a bulk location, a route location, or a cassette location) and the pickfrequency class.
If loading has to take place at bulk locations, this is generally done before or after the truck has loaded at a cassette loading location.

Data for the transfer cranes, shuttles, conveyors (speed, acceleration, load pick-up and drop off times etc.) have been registered in practice as much as possible. Data for the pick stations (to be designed newly) have been estimated.

Three different simulation run types have been carried out.
First, the 'optimal' values of the workload norms have been determined. Optimal means maximum throughput capacity (number of cassettes from the system that is retrieved and stored per hour) with acceptable response times, without deadlock possibility. It appears that the throughput is very sensible to the value of the workload norms. For not too high values, the throughput is an increasing function of the different workload norms. With too high workload norms, deadlock may occur.

The optimum workload norm values appear to be:

- 2 cassettes for each of the 4 DHL subsystem-pickstation combinations.
- 3 cassettes for each of the 4 EHL subsystem-close pickstation combinations.
- 1 cassettes for each of the 4 EHL subsystem-far pickstation combinations.

Since one additional cassette is in movement within each subsystem, this means that, after startup, with sufficient work, 2x4 + 3x4 + 1x4 + 3 = 27 cassettes are in permanent movement. The response times are acceptable, about 10 minutes and deadlock can not occur.

Second, the throughput capacity of the cassette system has been determined.
This has been done, by setting all picking and truck loading times to 0, using the optimal values for the workloads.
By gradually increasing the number of trucks that have to be dealt with per 7 hours, it appears that only one system has sufficient capacity to retrieve all orderlines for 44 trucks in 7 hours.

In the next table, a survey is given of the daily average crane and shuttle utilizations (in %).

	EHL1		DHL		EHL2	
	Shuttle1	Crane1	Shuttle2	Crane2	Shuttle3	Crane3
Aver./day	75,9	81,6	35,3	35,5	75,2	79,3

Within each subsystem, the cranes are more heavily utilized than the shuttles. This is, because per storage and retrieval, the crane cycle times are longer than those of the shuttle.

Finally, the joint picking and loading capacity has been determined.

It appears, that the loading process is the bottleneck. With loading times unequal to 0, only about 18 trucks can be loaded per system in 7 hours at the four loading stations.

5. FUNCTION OF THE BUFFERS

In order to profit of the high picking capacity of the system, it is necessary to decouple the cassette retrieval and loading process. Therefore the use of EHL and DHL buffers is necessary. Since a truck at a loading station may need material from three different location types (EHL, DHL buffers and route locations), the buffers may become full. When a buffer is full, retrieval for that buffer (i.e. loading station) stops.

On the other hand, if a buffer is empty before the truck has finished loading, the truck has to wait on the cassette system. Therefore, the buffer lengths have to be so large that this rarely happens. A capacity of about 30 orderlines seems reasonable.

6. REMARKS ABOUT THE DESIGN

In spite of the high throughput obtainable by the control, there are also a number of disadvantages:

1. The fact that EHL cassettes come both from the close and the far EHL subsystem, implies that bundled orderlines are not strictly sorted in loading sequence in the EHL buffers. It is true, that each EHL subsystems retrieves strictly in loading sequence per loading station. But, a cassette from a close system may arrive sooner in the buffer than one from the far station (or the other way around).
2. The control is more complex than traditional cassette-control systems and requires careful project managament.
3. The high potential throughput can not be realized due to the slow loading process.

The disadvantage, mentioned in point 1. could partly be overcome by slightly adapting the control as follows:

a. **Orderlines have to be grouped together per truck on the buffers.**
   This can be realized with a minor adaptation of the control. Each time an EHL subsystem finishes retrieval for a far loading station, it waits until the other EHL subsystem has finished retrieval for the same truck at the loading station.

In this way, within one truck, orderlines can still be intermixed at EHL buffers.
This solution costst about 10% picking capacity.

b. **Orderlines at EHL buffers have to be sorted strictly in loading sequence.**
This requires a considerable adaptation of the control and leads to a loss in picking capacity of about 30%.

The disadvantage, mentioned in point 1. could also be overcome by using only one EHL subsystem. This subsystem is considerably larger than the old one and contains all A and B articles, in sufficient quantities. In this case, the picking capacity reduces to about 20 trucks per 7 hours per system. This is less than half of the 44 original trucks, since the EHL subsystem has become larger and hence the crane and shuttle cycle times have increased.

The disadvantage mentioned in point 3., is a consequence of the fact that there are only four loading stations per system, where simultaneous truck loading can take place. With a proper organization of the loading process (like no waiting times between truck departure and arrival at the loading station), the truck loading capacity can be increased, but the loading process will remain the bottleneck.

7. CONCLUDING REMARKS

As far as the picking capacity is concerned, one cassette system would be sufficient to pick and bundle all orderlines for at least 40 trucks in 7 hours. This would imply a potential saving of Dfl 13 million as compared with the original investment estimate. Unfortunately, the loading capacity at the four truck loading stations stays far behind. In order to profit from the high picking capacity, the loading capacity of the system has to be increased, by using more loading stations.
This requires, at least in part, a new design of the system.

On Consultancy with the help of Operations Research. Effects and Effectiveness

with examples from logistics and distribution

Bert R. Meijboom, Heineken Nederland

Summary: The application of OR in business firms can be found at different levels: in operations management as well as in tactical or even strategic decision-making. Several examples will be given, but mathematical or IT-oriented issues of the solutions developed are not the main topic.
We merely focus on the organizational and environmental conditions that influence the degree to which OR can succesfully contribute. Here 'successfully' primarily refers to satisfaction and quality as perceived by clients. The main contribution is the interpretation of OR projects as a consultancy process. As a result a list of essentially non-technical requirements (skills, behaviour) can be stated. We end up with the following paradox. Suppose one approaches an OR project according to the standard sequence of stages of traditional operational *research* (more or less like in SDM). Now the 'implementation phase' is planned almost at the end of the project. From a consultancy point of view, however, implementation should be taken care of from the very beginning of the project.

1. INTRODUCTION

The discipline *operations research* (abbreviated to OR) is of great value to practice. This would really not be gathered offhand from the term 'research'. Secondly, OR is not only used for problems on an operational level. In decision-making and in the preparation of decisions on a tactical or even a strategic level OR can make a contribution, too. Thus, the denomination of the professional discipline may give rise to wrong expectations or perceptions of OR and its practitioners.
However, the reputation and appreciation of OR may undoubtedly as well be damaged because of an 'ivory tower' approach of problems or a 'black box' use of mathematical models.
As a contribution in the desired direction, this paper addresses the *effects* of OR applications, the *effectiveness* of OR-consultants and how improvements can be achieved here.

In section 2 OR as regards content, the question 'WHAT' (*In what problem fields is OR used? For whom?*) is dealt with, and, very briefly,the methodology (*What instruments, methods does the OR discipline use ?*). The solution process shows a certain structure: certain phases, steps are gone through.
In section 3 the question 'HOW' is placed in focus (*How does an OR consultant or executive operate?*). To this end OR projects are conceived to be (management) consultancy paths. Phrased the other way around: we are considering consultancy paths in which, at least when looked upon as regards content, the problem statement leads to using, applying the operations research discipline. Within the framework of this generalization towards consultancy paths, I discuss the role and the conduct of the OR consultant. Thus we are ending up with the effects of his/her efforts, in other words the effects of consultancy with the help of OR.

It will be clear that those pursuing OR in practice are to be reckoned as belonging to the target group of the present paper. But I'd also like to have a go at potential clients of (both external and internal) OR consultants as well as those readers who are unacquainted with the OR discipline. In view of this a few examples of OR applications have been included in section 2. Last but not least the paper should be of interest to say methodology oriented OR specialist, as it may tell something of the desired degree of sophistication of tools to be developed and of 'inevitable' interdisciplinary challenges (like cooperation with EDP experts).

2. ASPECTS WITH RESPECT TO CONTENT

2.1 Target group/market and methodology

Consultancy with the help of operations research takes place in various fields. The involvement is aimed at concrete, real world problems, both on the management level and on the operational level.

In the field of distribution and logistics one may for instance have in mind:
- Location allocation issues for a producer of drinks: how many distribution depots - and in what geographical locations - are required to achieve an optimum supply of the entire country with the various products? What does it mean when points of sale are supplied direct, without the intermediary of a distribution depot?
- Selection and introduction of planning packages: the delivery of the volumes ordered at the times agreed upon with customers every day leads to a lot of planning efforts; is there a suitable software package available with which enables the planner to make a cost-efficient route planning in an accelerated way?
- Possible improvements of the goods flow control within a production plant: what do the market's wishes mean for the way in which the supply and processing of raw materials, packing material and intermediary products are organized internally?

In banking and insurance, environmental planning, public transport and defence, the application of OR is likewise met with.

Tackling problems is in principle done with the aid of mathematical and statistical techniques. However, something essential will have to precede this. Before the operational problem to be analysed is exactly known, complicated situations requiring decision-making are more or less unravelled and mapped out, structured, sometimes with the aid of existing models, but usually by means of modellike thinking. In doing this, such a complex whole has already become better controllable, although no calculating, no optimizing whatever has as yet been done (from Duyn Schouten (1990).
In view of the fact that mathematics and, if found expedient, statistics are used, parties giving assignments will be expecting 'unambiguous' and 'firm' results to begin with. Before this can ever be attained, however, the problem first must have been clearly defined: what are the preliminary considerations to go by?

2.2 Phasing of OR projects

In order to attain a correct definition of the problem it has been found in numerous practical situations that intensive and frequent consultations with those directly concerned - both management and operational executives - is an essential precondition for the success of OR application (see Tilanus et al. (1986)). The consultant takes the course 'from unstructured to structured'. His procedure is widely different from the rigidly structured methods characteristic of the classical EDP department or the software house that has been hired in. Models in OR in fact are of the nature of prototypes rather than of clearly defined, accurately specified systems.

Given the mathematical formulation of the practical problem under consideration, the choice of the problem-solving technique will be more of the 'trial and error' type.
A mode of solving the problem will sometimes only be feasible under certain assumptions. As a consequence it will subsequently have to be investigated whether these assumption actually are in line with earlier basic assumptions in the 'real world' problem statement.

In the approach to the problem a certain phasing is in fact pursued, with feedback to the client at each passing from one phase to the next. In this way the client remains in control of the consultancy process: he can apply a midcourse correction when there is a threat of 'searching for an optimimum of the wrong problem'. We'll come back to this parallel with the management consultancy work later on.

Now the first step of the investigation is the one of *stock-taking*. Many discussions and interviews are held, and a variety of data are collected. These data must be upgraded to the state of *information*. In doing this the OR consultant develops an experimenting database. This enables him to combine the collected data and to present and interpret these data from different viewpoints.

By way of the stock-taking step, structures and interdependencies are shown. These initial results with as yet a fairly qualitative bias are discussed with the assigner, who will thus obtain a better 'feel' for the problems in hand. He will consequently be in a better position to get problem statements more explicit, thus being enabled to steer the course of a follow-up investigation, if any (Step #2):

> "*Yes, but if we had taken the line that* (during the coming quarter exports to the U.S.A. would be 15% higher than estimated)".
> Or: "*If we focus our policy on* (reducing the number of transports structurally by 25% during the morning rush hour)".

Altered situations, new policy variants will be coming up for consideration. These will be sparking off '*what if*' simulations.

At this stage (Step #3) making a *decision-support* model is taking shape. These are calculation models that are outstandingly suited for a more systematic varying of certain basic assumptions like the introduction of an altered situation, a new policy, a different scenario, etc.

3. APPLICATION OF OPERATIONS RESEARCH VIEWED AS A CONSULTANCY PATH

In this section I am not considering problems with respect to content or techniques to be applied, whether computer-supported or not, anymore (For an overwiew of techniques see e.g. Harpell et al. (1989)). I merely review the method, performance, role and conduct of the OR consultant. In doing so certain parallels with management consultancy work may be appropriately employed. Viewed purely from the OR discipline, which is after all an off-shoot of the family by the name of 'Mathematics', this may be a somewhat unexpected turn in the line of argument. However, my own experience has been that the dimension coming up for discussion now has appeared to be of crucial importance to a successful practical use of OR and accordingly to achieving the effects envisioned.

In the foregoing I already referred to structure and phasing in OR studies. In OR textbooks (e.g. Daellenbach (section 1.1, 1983) many more phasings, action plans and the like are to be found. In order to structure the line of reasoning, I introduce the following (time-sequential) triple division, which by and large might be valid for every consultancy path, namely:

1. Intake
2. Analysis and formulation of recommendations
3. Implementation

For each phase of the consultancy path, we shall be addressing properties and skills that are relevant in that particular phase for optimum consultancy results.

3.1. Intake phase

When starting on a new job it may be very illuminating to ascertain whether the potential client is in fact in need of engaging the assistance of operations research. I have frequently found that, on account of the term 'research', in view of a supposed sophistication or because of the relation withmathematics, it is made out as either 'scientific' or 'impractical'.

> Just imagine: Somebody is head of a production plant, the time of delivery is under pressure, and then a 'researcher' appears saying that the stocks could well be further reduced " ... *if a little heuristics is developed with which the bottleneck in the production process is efficiently scheduled*".

Another major point deserving particular attention is making the mutual roles and responsibilities explicit.

> If, with the help of your PC and a spreadsheet or database package you are processing the first bulk data, are you then doing this in complete isolation up to the moment when all the reports have been prepared and have been duplicated, or are you going to associate the client in the analyses and is he enabled to exercise an influence on the shape and contents of the graphs and tables to be produced? And how will you react if the assigner does not want to have certain results reported?

Without entering into further details, I would yet like to mention also:

* If necessary, convincing the client of an incorrect target-setting
* Is the client going to do something with my recommendations?
* Finding out on one's own what can/will be done with my recommendations.

Accordingly, as a recapitulation: at the start of a new assignment I find it indispensable to give appropriate consideration to questions like the ones raised just now. The central theme of the intake phase in fact is winning/building up confidence, in person and approach.
The snag is that there is usually little time available for such a careful start: "*We would like to have your project approach on paper by the end of the week, and we'll have further talks next week*".

Reflecting on this it occurs to me in this connection that one might come to an unusual interpretation of the well-known Dutch saying (literally translated): "A good beginning is half the work": starting a project in the way indicated above will be requiring so much time that the 'real' work, that is to say the work in terms of content, will only be half-finished !

3.2. Analysis and formulation of recommendations

Let's now have a look at the stage at which the consultant (with the assignment, action plan and budget in his attaché case) has actually set to work. Even though the models will now be featuring largely, the OR consultant must be able to switch rapidly to and fro between model and real world. Besides model(building) skills, economic and management scientific expertise will therefore be required.
It is no uncommon thing to find that the problem initially defined as the pressure point is in fact a matter of secondary importance, or that it is caused by other pressure points that had not yet been pinpointed as such before.

But ... how do I ensure that my client is able, and willing, to go along with the tightened or readjusted identification and formulation of the problem?!
And what if this leads to a volume of work that had not been taken into account during the intake ?

The main result of working with models is that one gets a better assessment of a variety of decisions and processes. In distribution and logistics the range-of-application varies from long-term capacity planning to systems for shopfloor control. This is simply because modelling is none other than *structuring.* Matters of primary and those of secondary importance are distinguished and kept apart.

However, the question once more is: does the client meanwhile recognize the matter equally 'sharply', or did I have to leave him behind somewhere during the structuring process?

By way of example I might mention some more potentional effects as regards content, with in each case the note:

"*... and, does my client go along in this? Does he recognize the results highlighted by me?*"

Only then, and not before, I am effective to him or her! Essentially it is a matter of retaining confidence as well as, simultaneously, increasing/ improving his grasp of the content of the problem.

I have found that the following factors may be of help in this:

* Being oneself conscious of the steps made in the thinking process; to what extent can the client keep up with these, or has he not been able to keep pace?
* Assessing how deeply the client's certainty or uncertainty is rooted (or, the reverse: where does the uncertainty begin). Is it a cognitive matter?
* Am I capable of making analyses together with the client?
* Unthreateningly placing on record/reporting back of collected data

This kind of matters appear to be indispensable if one wants to make an analysis that the client recognizes as his own.

3.3. Implementation

In the case of operations research projects at this stage the following questions will be coming up:

* Can my end product be rated as an advice, a recommendation?

* Or does a model have to be programmed first?
  Can I generate 'intermediate' results 'halfway' the model building?
  What are suitable moments for feedback to the client?

* Does this model have to be integrated in the operational system?
  Or will an adapted use of the existing operational system suffice?

* Does the automation department accept me as a discussion partner?

* Do the effects of my conclusions or of the proposed system go beyond the borders of my department? And what other persons have to be involved?

There is something of a paradox in this listing. Viewed at it with respect to content it is a question of items which from the viewpoint of the phasing of the project are related to the tail end of the path. Processwise, however, they all can hardly be attended to at an early enough moment, indeed even during the 'intake'.

4. CONCLUSION

In order to focus on the effectiveness of OR we generalized OR projects towards (management) consultancy paths. Behaviour and non-technical skills of OR practitioners were classified according to the phasing of a standard consultancy path. It appeared that 'Implementation', commonly seen as the completion of an OR project, captures several activities that ought to be addressed in a much more earlier stage of the project.

Perhaps appearances are deceptive, but I for one have the feeling that the direction and therefore the effect of each consultancy job is greatly determined by everything that has and everything that has not happened during the intake. In the case of OR consultancy this implies that 'ivory-tower' perceptions among the client, 'black box' effects of OR models and other mysticisms have to be eliminated from the outset. Similarly the expected embedding of an OR software module, eventually not even developed yet, in the operational administrative system requires early involvement of the EDP experts.

Acknowledgements

The author is grateful to mr. K.F. Creutzberg (Rijntraining bv), ir. J.D. v.d.Meer (Van der Meer & Van Tilburg) and drs. J. Peters (Moret, Ernst & Young) for their critical comments and useful suggestions to earlier drafts of this paper.

REFERENCES

Daellenbach H.G., J.A. George and D.C. McNickle (1983), Introduction to Operations Research Techniques, 2nd edition, Allyn and Bacon Inc, Boston/London/Sydney/Toronto.

Duyn Schouten, F.A. v.d. (1990), Struktuur in het beslissingsproces, Kwantitatieve Methoden, 33, pp. 5-28.

Harpell, J.L., M.S. Lane and A.H. Mansour (1989), Operations Research in Practice: A Longiditunal Study, Interfaces, 19, 3, May-June 1989, pp.65-74.

Meyer zu Selhausen, H. (1989), Repositioning OR's Products in the Market, Interfaces, 19, 2, March-April 1989, pp. 79-87.

Tilanus C.B., O.B. de Gans and J.K. Lenstra (eds.) (1986), Quantitative Methods in Management: Case Studies of Failures and Successes, Chichester, Wiley.

OPTIMAL SUPPLY OF SLABS FOR A COLD ROLLING MILL

Kees van Montfort, Free University, Amsterdam
Pim Bergkamp, Hoogovens Group, The Netherlands

Summary: An algorithm for the order entry system of a cold rolling mill is developed. Given a client order for coils of steel the algorithm computes the optimal number and weights of supply slabs. Therefore supply weight intervals between the successive production operations are constructed. The influences of the seperate input variables on the computed supply of slabs are transparent.

1. Introduction

The integrated steel plant of Hoogovens Group is located in IJmuiden (The Netherlands). About 10,000 employees produce nearly 5 million tonnes of steel each year. The production contains cudgels, wires, not-rolled flat products, coated and uncoated cold rolled products, and packaging steel.

One of the cold rolling mills (KB2) rolls slabs of steel mainly used in the automotive industry and the industry of consumer goods. The total production is yearly about one and a half billion US-dollars. The rolling mill needs supply slabs of a warm rolling mill of the Hoogovens Group. The slabs will successively be stained, pre-rolled, annealed, post-rolled, worked off and transported to one of the clients.

Cold rolling is a process of great complexity. Each client specification of an order contains coils with an unique coil weight, order tonnage, flatness, steel quality, width, thick-ness, length, number of permitted coil joints, etc. Furthermore, in a rolling mill about 25 percents of the coils will totally or partly be declared unfit for sale. It is impossible to predict the moment of damage.

An algorithm for the order entry system of KB2 is developed. Given a client order for coils of steel the algorithm computes the optimal number and weights of supply slabs. Besides the client order requirements also the installation, production and logistic restrictions of the cold rolling mill are important input variables. The algorithm has a modular structure. The influences of the seperate input variables on the computed supply of slabs are clear.

In this paper the abovementioned algorithm is presented. First of all the objectives and the restrictions are discussed. Next the working-out of the algoritm is described. Finally the method is evaluated.

Only the main ideas will become attention. Many details will be omitted.

2. Objectives and restrictions

The algorithm has to compute the optimal number and weights of supply slabs for an entered order of KB2 coils. There are two objectives:
- the weights of the processed KB2 coils have to be as big as possible within the coil weight interval which is dictated by the client;
- the weights of the supply slabs have to be as big as possible.

The first objective is of importance because of client satisfaction. The second one is introduced to reduce production costs of the cold rolling mill. Among other things, both weights depend on installation, production and logistic restrictions of the slab factory and the KB2.

The structure of the algorithm has to be modular because of the desired insight in the influence of the separate relevant restrictions. Subjects like installation restrictions, client requirements, coil dividing routines, efficiency loss, joint possibilities of different coils, etc. are moduled in separate parts of the algorithm. This approach makes it also possible to fit changes of installation, production and logistic requirements into the algorithm easily.

By the production of slabs often unexpected weight losses appear. Therefore the slab factory prefers to produce slabs within a required weight-interval instead of producing slabs with required exact values for the weights.

3. The working-out of the algorithm

In figure 1, which is placed on the end of this paper, the separate production stages of the KB2 are represented: supply from the slab factory, stain, pre-roll, anneal, post-roll, finish off and transport to one of the clients. The main idea of the algorithm is to work with permitted transport (=output) and supply (=input) weight-intervals between the successive production stages. The weight-intervals between the slab factory and the stain-hal will indicate the minimal and maximal permitted weights for the supply slabs, which will be ordered.

The algorithm considers four steps:

1. Determination of the <u>technical intervals</u> between the successive production stages.
   These intervals indicate the technical possibilities of the installations, without the restrictions of the client order specifications.
2. Determination of the <u>coil weight intervals</u> between the successive production stages. These intervals depend on the <u>minimal and maximal weights for the client coils</u>, which are <u>prescribed by the clients</u>. Client coil joints and the dividing of coils are also incorporated.

2a. Determination of the <u>intersection of the technical and coil weight intervals</u>.

3. The <u>optimization</u> within the derived intervals of the previous steps.
4. If no supply slabs satisfy both the technical intervals and the client order specifications, then the resulting interval of the supply slabs is empty. In that case <u>alternative solutions</u> have to be determinated.

In the next sections the four steps wil be discussed more in detail.

3.1 Determination of the <u>technical intervals</u> between the successive production stages

The maximal and minimal possible coil weights for each separate production step, without the restrictions of the client order specifications, will be determinated. Therefore we use an <u>assistance table</u>, which we have constructed in conjunction with process engeneers. The table contains for each isolated production step the input and output <u>process-interval</u> (=minimal and maximal weights), only based on the process properties of the specific production step. These intervals are functions of the widths and the diameters of the coils.

The procedure is as follows:

1. Search in the abovementioned table the input and output process-intervals of the last operation in the cold rolling process, namely transport of the coils to one of the clients. The technical input interval of this operation will be the intersection of the input and output process-interval.
2. Search in the assistance table the input and output process-intervals of the preceding operation in the cold rolling process (finish-off). The technical output interval of this operation

will be the intersection of the output process-interval of this operation and the technical input interval of the previous operation. The technical input interval of this operation will be the intersection of the input process-interval and the technical output interval of this operation.

3,4,5,....

Repeat step 2 for each operation until the technical input interval of the operation staining is derived.

last step.

Determine the intersection of the process-interval of the slab factory and the technical input interval of the staining installation. Then the maximal and minimal technical weights for the supply slabs are determinated. These weights are functions of slab width and slab diameter.

Weight losses appear because of damages of the steel during the production process. Therefore in each of the abovementioned steps the minimal technical weights are raised using average loss-rates per operation.

Some installations have the possibility to divide coils. In that case we get under some conditions, which will not be discussed in this paper, more than one technical input interval. Suppose the case in which the installation can divide one coil into one, two or three coils. In that case we will get three technical input intervals:

- the intersection of the input process-interval and the technical output interval;
- the intersection of the input process-interval and the technical output interval multiplied by two;
- the intersection of the input process-interval and the technical output interval multiplied by three.

The annealing-hall, which consists of four furnaces, needs some additional attention. The furnaces have different input and output process-intervals and also different capacities. When an order is just entered, the planner doesn't know which furnace has enough free space. Therefore, we derive technical input and output intervals for each separate furnace. Next, the intersection of the sets of furnace intervals are computed. So, the resulting intervals of the annealing-hall are valid for all furnaces. Unfortunately, this approach leads to small intervals.

3.2 Determination of the coil weight intervals and the intersection of the technical and coil weight intervals

In the preceding section we derived the technical intervals. Using the prescribed weight intervals of the client coils we can determine the coil weight intervals for each operation.

The coil weight intervals of the coils, which will be transported to the clients, are prescribed by the clients. We call it the client coil weight intervals. These client coil weight intervals are used to determine coil weight intervals for the other operations. The main idea is as follows:

1. The client coil weight interval corresponds with the last operation in the cold rolling process. The new input coil weight interval of this operation will be the intersection of the client coil weight interval and the corresponding technical input interval.
2. In this step the weight coil interval of the preceding operation in the cold rolling process (finish-off) will be determinated. The coil weight output interval will be the intersection of the technical output interval of this operation and the coil weight input interval which was computed in the previous step. The coil weight input interval of this operation will be the intersection of the technical input interval and the coil weight output interval of this operation.

3,4,5,....

Repeat step 2 for each operation until the coil weight input interval of the operation pre-rolling is derived. Then the maximal and minimal coil weights for pre-rolling are determinated.

last step.

The stain installation is the only installation which has the possibility to joint and to divide coils. The following combinations of the number of input coils and the number of output coils are possible: (1:1), (2:1), (3:2), (4:3), (5:2), (5:3) and (5:4). Some clients don't want to have joints in their coils. The client order specifications indicate which combinations are acceptable for the clients.

A coil weight interval will be determinated for each possible combination (n:m) of the number of input and output coils:

max. input coil weight := (m/n) * max. input coil
weight of the pre-roller;

min. input coil weight := (m/n) * min input coil weight of the pre-roller.

After this we have to compute the intersections between these input coil weight intervals and the corresponding technical intervals. The resulting coil weight intervals are functions of slab width, slab diameter and joint-restrictions, which are prescribed in the client order specification during the order-entry. Furthermore each input interval corresponds with a combination of a number input coils and a number output coils.

Some installations have the possibility to divide or to joint coils. Those installations will get under some conditions, which will not be discussed in this paper, more than one coil weight interval. The approach is analogous to the method we described in the previous section (technical intervals).

3.3 The optimization within the derived intervals of the previous steps

As we mentioned in section 2 we have two objectives:
- the weights of the processed KB2 coils and
- the weights of the supply slabs.

Given a client order specification, in this stage of the algorithm several coil weight input intervals of the stain installation (=the output interval of the supply slabs) are available. Furthermore, each computed interval corresponds with a combination of a number input and a number output stain coils.

Now an interval or a part of an interval can be choosen for which a weighted combination of the scores for the two objectives is optimal. Underlimits or indifference curves for the two objectives can also be used. Using these methods the sensitivenesses of the objectives and the trade-offs between them can be detected.

Eighty percent of all orders are repeat orders. In those cases the order tonnage is of no interest. This means, the number of supply slabs, mutiplied by the slab weights, might be bigger than the order tonnage. In the other cases the number of slabs, multiplied by the slab weights, has to be in a tolarance interval around the desired order tonnage. Sometimes, given a specific client order specification, it is impossible to compute permitted numbers and weights of supply slabs.

3.4 Alternative solutions

If no supply slabs satisfy both the technical intervals and the client order specifications, then the resulting interval of the supply slabs is empty. After consulting the client, the specifications of the client order can be adapted. For instance, the minimal client coil weight, the maximal client coil weight, the number of permitted joints, the order tonnage, the width or the thickness can be changed.

4. Evaluation of the method

a. It is easy to adjust the algoritm to new installation, production or logistic requirements.

b. The sensitivenesses of the objectives and the trade-offs between these objectives can easily be seen.

c. The method is very transparent. The influences of each client order requirement can easily be detected. So, the sales department can use the algorithm during the negotiating with a client for new order specifications.

d. The choosen modelling of the annealing hall can cause very small or even empty intervals. Consequently, additional research is needed.

Reference

- Documentation of the order-entry system of Hoogovens IJmuiden, Pieterse et al. (about 5,000 pages).

Figure 1, The separate production stages of the cold rolling mill

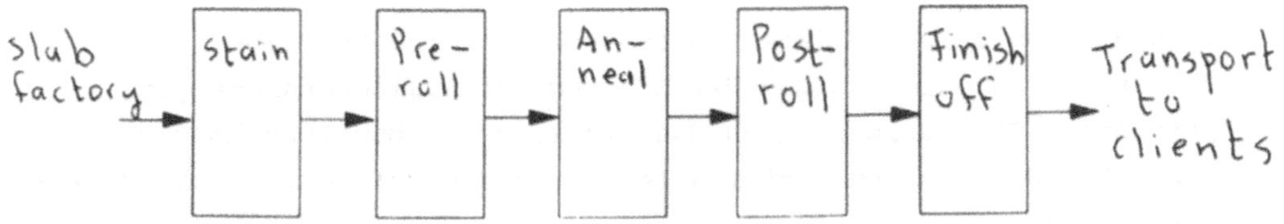

TOURENOPTIMIERUNG IM PRESSEVERTRIEB

Axel Stolze; Wolfgang Eberle
Martin-Luther-Universität Halle-Wittenberg

Die langjährige praxisorientierte Forschung zur datenseitigen Durchdringung und anschließenden Lösung und Umsetzung von Tourenproblemen im allgemeinen sowie die Beschäftigung mit spezifischen Bedingungen solcher Probleme innerhalb des Pressegroßhandels ermöglichen es, auf einige Besonderheiten eines rechnergestützten Vorgehens auf diesem Spezialgebiet einzugehen.

Die Aufgabe eines Pressevertriebes besteht darin, eine Anzahl von ca. 2000 - 2500 Artikeln (Objekten) an einen umfassenden Kundenkreis von bis zu 1000 Kunden zu verteilen.

Neben der ungewöhnlichen Größe solcher Tourenprobleme erschweren weitere Spezifika die Handhabung. Einige seien genannt:

- Vorliegen einer sehr dichten, aber ungleichmäßigen Verteilung der Kunden im Liefergebiet
- Auffinden einer geeigneten Zuordnung der Kunden im Entfernungswerk und Lösung von Reihenfolgeproblemen bei Mehrfachzuordnungen
- wochentagabhängige starke Schwankung der Auftragsmengen
- unterschiedliche Kundenzeitfenster.

Die praxisnahe Berücksichtigung der Spezifika führt in der Projektrealisierung zu folgenden inhaltlichen und methodischen Schwerpunkten:

- Gewinnung und Aufbereitung aussagefähiger Kunden(Gewichts-)daten
- empirische und operative Berücksichtigung von Kundenfluktuation und Kundenwünschen
- Umgehung von Zeitkonflikten durch unpersönliche Übergabe
- sukzessive Optimierung und Veränderung des Auslieferungsregimes
- permanente Datenpflege durch das Pressevertriebsunternehmen und wiederholte Optimierung (Intervalle: ca. 3 Monate) durch externe Auftragnehmer.

Die erzielten und zu erwartenden Einsparungseffekte betragen bezüglich der Anzahl der Touren, der Gesamtweglänge und der Kosten jeweils zwischen 10 und 15 Prozent.

AUDIT-STAFF SCHEDULING: METHODS FOR THE MEDIUM-TO-SHORT-TERM LEVEL

Thomas Bartsch, Christian-Albrechts Universität Kiel
Frank Salewski, Christian-Albrechts-Universität Kiel

The *medium-to-short-term model* is a component of an hierarchical approach with three levels, which is suitable to represent the problem of audit-staff scheduling from the user's point of view [2]. The formulation of the model is based on the results of an empirical study conducted among 200 of the biggest German audit companies. The *medium-to-short-term planning* disaggregates the results of the first level, the medium-term planning, for one week and all auditors [1]. The outcome is a schedule that specifies for each auditor - on the basis of periods of four hours - all the phases in which the auditor is involved in the considered week. The objective function minimizes the total costs of the resulting schedule.

The corresponding problem is formulated in terms of a binary optimization model which is closely related to resource-constraint project scheduling. Apart from job completion, time, resource, and precedence constraints this model comprises also setup (travel) times.

However, widely-known heuristics adopted to "classical" project scheduling are not directly applicable since they aim at minimizing project makespan. In order to minimize costs, it may be preferable to schedule jobs *not always* as early as possible. First, we present a parallel-randomized algorithm which is tailored to the structure of the problem. With this method each job gets a selection probability Second, we discuss a genetic algorithm which manipulates such probabilities.

To evaluate the efficiency of the implemented solution methods we have developed an instance generator for which the results of the empirical study served as a basis. We produced twelve types of test instances with different sizes and tractabilities. Afterwards, each heuristic was applied to these instances. Furthermore, small instances have been solved to optimality with the state-of-the-art MIP solvers OSL and LINDO, in order to provide evaluation benchmarks.

Results obtained so far indicate that the proposed methods are capable of solving instances of practical size in neglectible amounts of time.

References

[1] Salewski, F., T. Bartsch, E. Pesch (1993): Auftragsterminierung für die taktisch-operative Personaleinsatzplanung in Wirtschaftsprüfungsgesellschaften, Discussion Paper, Universität Kiel.

[2] Salewski, F., A. Drexl (1993): Personaleinsatzplanung in Wirtschaftsprüfungsgesellschaften - Bestandsaufnahme und konzeptioneller Ansatz, to appear in: Zeitschrift für Betriebswirtschaft.

AUDIT-STAFF SCHEDULING: METHODS FOR THE SHORT-TERM LEVEL

Lars Böttcher, Christian-Albrechts-Universität Kiel
Frank Salewski, Christian-Albrechts-Universität Kiel

A survey conducted among 200 of the biggest German CPA firms indicates that the process of audit-staff scheduling is usually divided into three different levels, viz. medium-term, medium-to-short-term, and short-term planning [2]. This paper addresses the *short-term planning level* of the corresponding hierarchical approach. The short-term planning is based upon the results of the medium-to-short-term level for one week and one phase [1]. It assigns the auditors involved in auditing that phase in the considered week to the corresponding audit tasks and schedules them on the basis of periods of one hour. Decision objective is the balancing of the auditors' workload.

The problem is formulated in terms of a binary optimization model which is closely related to multi-mode resource-constrained project scheduling. Thereby, different auditors are tally with different modes while sets of auditors are equivalent to sets of modes. Apart from job completion, time, resource, and precedence constraints this model comprises also so-called set-of-mode-identity constraints which guarantee that the assignment of a set of auditors to a set of tasks, once done, is never changed.

As a consequence, for solving the short-term audit-staff scheduling problem, an appropriate two-stage algorithmic scheme has been developed which can be classified as parallel-randomized. The assignment of sets of auditors to sets of tasks and of single auditors out of one corresponding set to single tasks is performed in the first stage, while the actual scheduling of the tasks takes place in the second stage.

Several resource- and critical path-based rules are proposed and evaluated. For testing purposes a set of representative instances has been generated on the basis of empirically obtained data. We produced twelve types of test instances with different sizes and tractabilities. Afterwards, several solution variants have been tested on these types of instances. In addition, some small instances have been solved to optimality with the state-of-the-art MIP solvers OSL and LINDO, in order to provide evaluation benchmarks.

The results indicate that the heuristics are able to solve instances of realistic sizes in neglectible amounts of time and still yield "acceptable" solutions.

References

[1] Salewski, F., L. Böttcher, A. Drexl (1993): Prüffeldorientierte Zuordnung von Prüfern für die operative Personaleinsatzplanung in Wirtschaftsprüfungsgesellschaften, Discussion Paper, Universität Kiel.

[2] Salewski, F., A. Drexl (1993): Personaleinsatzplanung in Wirtschaftsprüfungsgesellschaften - Bestandsaufnahme und konzeptioneller Ansatz, to appear in: Zeitschrift für Betriebswirtschaft.

Scheduling and Flexible Manufacturing

Peter Brucker, Universität Osnabrück

With new production technologies special scheduling problems have been formulated. In this survey we will discuss some specific features of scheduling in flexible manufacturing systems like general shop scheduling problems, scheduling for multi-purpose machines, combined scheduling and transportation problems, families of jobs (operations) and changeover costs, batching, and due-date scheduling.

The general shop problem is a generalization of the open-shop problem and the job-shop problem. Each job consists of a set of operations. Associated with the operations there are processing times and machines on which each operation must be processed. Between operations precedence constraints are given and we have to find a feasible schedule which minimizes some objective function.

A multi-purpose machine is a machine which is equipped with different tools. An operations can be processed on all machines equipped with the tool needed for doing the operation. The multi-purpose machine scheduling problem is to find a feasible assignment of operations to machines and to schedule the operations on the assigned machines. Special general shop problems with multi-purpose machines are discussed.

As a combined scheduling and transportation problem a general shop problem is considered in which jobs are to be transported from machine to machine. This transport is done by one or more robots and takes time . Cases with (i) only one robot, (ii) a few robots, (iii) a sufficient large number of robots are considered. Problems with families of operations and changeover times if two succeeding operations on the same machine belong to different families are equivalent to corresponding combined scheduling and transportation problems.

Another problem is the batching problem. Here the jobs to be processed on the same machine must be grouped together into batches. All jobs in the same batch are processed jointly and their finish time is the finish time of the last job in the batch (batch availability). If a batch goes onto a machine there is a set-up time. The problem is to find a partioning of all jobs into batches and a schedule for these batches such that some objective function like mean flow time is minimized.

Finally scheduling problems involving due-dates which arize in connection with just-in-time production strategies are discussed.

EINE MÖGLICHKEIT DER NUMERISCHEN BEHANDLUNG VON ZEITLICH-STOCHASTISCHEN NETZPLÄNEN

Uwe Clasen, Universität Rostock

Zusammenfassung: Gegenstand des Beitrages ist ein Verfahren zur Zeitplanung in Netzplänen mit stochastischen Vorgangsdauern. Für die Beschreibung der Termine wird dabei der Verteilungstyp der zusammengesetzten Normalverteilung genutzt. Bei der Ermittlung des Maximums und des Minimums von Terminen werden deren Abhängigkeiten von gemeinsamen Vorgängern bzw. Nachfolgern beachtet. Berechnungsergebnisse nach diesem Verfahren werden Simulationsergebnissen gegenübergestellt.

Summary: The topic of the article is a method for time planning in networks with stochastic durations of activities. The type of distribution of the compound normal distribution is used for the description of dates. To ascertain the maximum and minimum of dates you have to observe their dependence on common predecessors respectively successors. Results of calculation of this method are compared with results of simulation.

1. Zeitlich-stochastische Netzpläne

Ausgangspunkt der Zeitplanung für ein Projekt sind die Ausführungszeiten der einzelnen Vorgänge. In Abhängigkeit vom Charakter des Projektes und der jeweiligen Vorgänge können jedoch beträchtliche Unsicherheiten hinsichtlich der Dauer von Vorgängen und damit des gesamten Projektes bestehen. Ein Netzplan, in dem die Dauern der einzelnen Vorgänge Unsicherheiten aufweisen, soll als zeitlich-stochastischer Netzplan bezeichnet werden. Die Erfassung dieser Unsicherheiten und deren Beachtung bei der Zeitplanung können wichtige Informationen über das zeitliche Risiko des Projektes liefern. Auch wenn die Unsicherheit nur näherungsweise beschrieben werden kann, ist es besser, mit diesen Schätzwerten zu operieren als durch die Annahme deterministischer Vorgangsdauern den stochastischen Charakter zu verdecken.

2. Verfahren der Zeitplanung in zeitlich-stochastischen Netzplänen

Die bekannteste Methode zur Beachtung der zeitlichen Unsicherheit in Netzplänen ist die Methode PERT. Sie geht davon aus, daß jede Vorgangsdauer näherungsweise mit einer Betaverteilung beschrieben werden kann. Bei der Terminplanung verstößt PERT jedoch gegen grundlegende Gesetze der Wahrscheinlichkeitstheorie. So wird der Erwartungswert des frühesten Anfangstermins eines Vorganges als Maximum der Erwartungswerte der frühesten Endtermine seiner unmittelbaren Vorgänger bestimmt. Dabei werden die Operationen Bestimmung des Erwartungswertes und Maximierung in unzulässiger Weise vertauscht. Als Varianz des frühesten Anfangstermins eines Vorganges wird die Varianz des frühesten Endtermins des unmittelbaren Vorgängers mit dem größten Erwartungswert einfach übernommen. Hier wäre eigentlich die Varianz des Maximums von Zufallsgrößen zu bestimmen, worauf im allgemeinen alle Varianzen

der am Maximum beteiligten Zufallsgrößen Einfluß haben. Durch diese Vorgehensweise werden die Erwartungswerte systematisch unterschätzt und die Varianzen überschätzt. Kritische Wege und kritische Ereignisse werden bei PERT auf deterministische Weise festgelegt. In der Realität besitzt aber jeder Weg von der Quelle zur Senke eine positive Wahrscheinlichkeit dafür, kritisch zu sein. Unter Berufung auf den zentralen Grenzwertsatz werden alle Termine und Pufferzeiten als asymptotisch normalverteilt angenommen, obwohl die Voraussetzungen für die Anwendung des zentralen Grenzwertsatzes nicht erfüllt sind. Insgesamt führen die in der Methode PERT enthaltenen Fehler dazu, daß die Ergebnisse allenfalls als grobe Näherungslösungen angesehen werden können (siehe hierzu auch die PERT-Kritik in [1],[2],[3],[4] und [6]). Die zahlreichen Ansätze zur Weiterentwicklung von PERT, wie z.B. die Verfahren von MARTIN oder MESCHKOW (vgl.[2]), konnten zwar Verbesserungen in einzelnen Kritikpunkten erreichen, fanden aber wohl vor allem wegen der umständlichen und aufwendigen Algorithmen kaum praktische Nutzung.
Mit der Entwicklung der Computertechnik entstand die Möglichkeit, Projektabläufe zu simulieren. Dabei lassen sich unterschiedliche Verteilungstypen für die Dauern der einzelnen Vorgänge zugrundelegen und auch Abhängigkeiten zwischen den Vorgangsdauern beachten (vgl.[5]). Um zuverlässige Ergebnisse zu erreichen, muß die Anzahl der Simulationsläufe sehr hoch gewählt werden. Dadurch wird der Zeitaufwand trotz der hohen Leistungsfähigkeit heutiger Computer allerdings schnell sehr groß.
Im folgenden wird eine Methode zur numerischen Behandlung der zeitlichen Unsicherheiten in Netzplänen dargestellt, die mit vergleichsweise geringem Aufwand Aussagen liefert, deren Realitätsnähe wesentlich besser ist als bei der Methode PERT.

3. Die zusammengesetzte Normalverteilung zur Beschreibung von Terminen

Werden Vorgangsdauern als Zufallsgrößen betrachtet, dann treten folglich auch stochastische Termine auf, für deren Beschreibung jeweils ein Verteilungstyp gefunden werden muß, der die reale Verteilung des betreffenden Termins möglichst gut approximiert. Nach dem zentralen Grenzwertsatz gilt für den gemeinsamen Abschlußtermin mehrerer unmittelbar aufeinanderfolgender Vorgänge, deren Dauern unabhängig sind, daß er asymptotisch normalverteilt ist. Andererseits hat der Termin für den Abschluß mehrerer gleichzeitig beginnender Vorgänge im allgemeinen eine von Null verschiedene Schiefe. Deshalb sollte ein Verteilungstyp zur Beschreibung von Terminen bei gutem Anpassungsvermögen auch an schiefe Verteilungen die Möglichkeit bieten, den Prozeß der asymptotischen Annäherung an die Normalverteilung nachzubilden. Diese Eigenschaften besitzt die zusammengesetzte Normalverteilung.
Als *"zusammengesetzte Normalverteilung" (ZN-Verteilung)* wird dabei eine Verteilung bezeichnet, die durch "Aneinanderfügen" einer linken und einer rechten Hälfte zweier Normalverteilungen mit gleichem Modalwert entsteht, wobei die Dichtefunktionen beider Hälften mit geeigneten Faktoren gewichtet werden, damit die neue Dichtefunktion stetig ist und die Vollständigkeitsrelation erfüllt. Eine solche Dichtefunktion lautet:

$$f(x) = \begin{cases} \dfrac{2}{a+b} \cdot \phi\left(\dfrac{x-m}{a}\right) & \text{für } x < m \text{ und } a > 0 \,, \\ 0 & \text{für } x < m \text{ und } a = 0 \,, \\ \dfrac{2}{a+b} \cdot \phi(0) & \text{für } x = m \,, \\ \dfrac{2}{a+b} \cdot \phi\left(\dfrac{x-m}{b}\right) & \text{für } x > m \text{ und } b > 0 \,, \\ 0 & \text{für } x > m \text{ und } b = 0 \,. \end{cases}$$

Dabei ist ϕ die Dichtefunktion der standardisierten Normalverteilung, m der Modalwert, a der Abstand zwischen der Abszisse des linken Wendepunktes und dem Modalwert und b der Abstand zwischen dem Modalwert und der Abszisse des rechten Wendepunktes. Für a und b gelten die Bedingungen $a \geq 0$, $b \geq 0$ und $a + b > 0$.
Eine symmetrische ZN-Verteilung (a = b) ist mit einer Normalverteilung der Standardabweichung $\sigma = a = b$ identisch. Im allgemeinen beschreibt die Funktion f(x) eine "schiefe Glockenkurve", deren Schiefe Werte des Intervalls [- 0,99527; + 0,99527] annehmen kann. Die Wahrscheinlichkeit, daß eine ZN-verteilte Zufallsgröße einen Wert des Intervalls [m-3a;m+3b] annimmt, ist mit 0,9972 nahezu 1.
Die Summe zweier unabhängiger ZN-verteilter Zufallsgrößen ist zwar im allgemeinen nicht ZN-verteilt, ihre Verteilung läßt sich aber sehr gut durch eine ZN-Verteilung approximieren. Auch die Verteilung des Maximums oder des Minimums zweier unabhängiger ZN-verteilter Zufallsgrößen läßt sich mit einer ZN-Verteilung in den meisten Fällen gut beschreiben. Nur wenn beide Zufallsgrößen eine besonders starke Schiefe aufweisen oder das Verhältnis ihrer Varianzen sehr groß bzw. sehr klein (z.B. größer als 3 bzw. kleiner als 0,33) ist, weicht die Dichtefunktion des Maximums oder Minimums stärker von einer ZN-Verteilung ab. Die Berechnung von Erwartungswert, Varianz und Schiefe des Maximums bzw. Minimums bereitet keine Schwierigkeiten.
Damit bietet die ZN-Verteilung bei der Beschreibung von Terminen deutliche Vorzüge gegenüber anderen Verteilungstypen, wie z.B. Normalverteilung, Betaverteilung oder Dreiecksverteilung.
Da eine Vorgangsdauer eine nach oben und unten beschränkte, nichtnegative und meist unimodale Zufallsgröße ist, kann man davon ausgehen, daß eine ZN-Verteilung, deren Parameter die Bedingung m > 3a erfüllen, zu ihrer Beschreibung ebenso sinnvoll ist wie die häufig verwendete Betaverteilung.

4. Zeitplanung mit ZN-verteilten Terminen

In einem einfachen Vorgangsknotennetz (alle Anordnungsbeziehungen sind Normalfolgen) ergibt sich der früheste Anfangstermin FA_i eines Vorganges i als Maximum der frühesten Endtermine seiner unmittelbaren Vorgänger. Der früheste Endtermin FE_i dieses Vorganges ist dann die Summe aus dem frühesten Anfangstermin FA_i und der

Vorgangsdauer D_i . Den spätesten Endtermin SE_i erhält man als Minimum der spätesten Anfangstermine aller unmittelbaren Nachfolger und der späteste Anfangstermin SA_i ist die Differenz aus dem spätesten Endtermin SE_i und der Vorgangsdauer D_i .
Wird die paarweise Unabhängigkeit der Ausführungsdauern aller Vorgänge angenommen, so liegen bei den Operationen Addition und Subtraktion stets auch unabhängige Operanden vor. Zwei zu verknüpfende früheste Endtermine sind jedoch nur dann unabhängig, wenn die beiden Vorgänge im Netzplan keinen gemeinsamen Vorgänger haben. Ein gemeinsamer Vorgänger wirkt sich auf die Parameter beider Termine aus und erzeugt somit eine Abhängigkeit. Soll der früheste Anfangstermin eines Vorganges richtig ermittelt werden, müssen folglich die Abhängigkeiten der frühesten Endtermine seiner unmittelbaren Vorgänger in geeigneter Weise berücksichtigt werden. Analog ist bei der Bestimmung des spätesten Endtermins eines Vorganges mit mehreren unmittelbaren Nachfolgern zu prüfen, ob zwei oder mehr von ihnen gemeinsame Nachfolger haben, und eine entsprechende Abhängigkeit ihrer spätesten Anfangstermine gegebenenfalls zu berücksichtigen.
Durch Gegenüberstellung von Simulations- und Berechnungsergebnissen für den frühesten Endtermin des Netzplanes 1 wurde untersucht, ob das Maximum zweier unabhängiger ZN-verteilter Zufallsgrößen als Summe der Anteile beider Zufallsgrößen an diesem Maximum dargestellt werden kann. Ist eine solche Darstellung möglich, dann ist der früheste Anfangstermin FA_4 des Vorganges 4 die Summe der Anteile $A_{1,4}$ und $A_{2,4}$ der beiden frühesten Endtermine FE_1 und FE_2 an ihrem Maximum. Zur Berechnung des frühesten Zieltermins kann der Netzplan 1 dann durch den Netzplan 1' ersetzt werden. Dabei sind FE_3 und FE_4 die frühesten Endtermine der Vorgänge 3 bzw. 4. Da die Varianz und das zentrale Moment dritter Ordnung des Maximums zweier Zufallsgrößen invariant gegen Translation sind, kann der Erwartungswert des Anteils $A_{1,4}$ noch auf Null gesetzt werden.

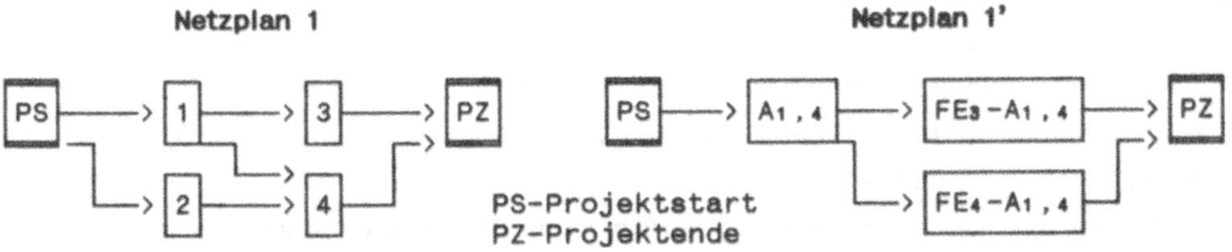

Für 100 Datenkonstellationen wurden zunächst aus je 300.000 Simulationsläufen auf der Grundlage des Netzplanes 1 die Parameter Erwartungswert, Varianz und Schiefe des frühesten Zieltermins ermittelt. Dann wurden unter Variation der Parameter Varianz und zentrales Moment dritter Ordnung des Anteils $A_{1,4}$ Berechnungen am Netzplan 1' so lange ausgeführt, bis sich für den frühesten Zieltermin näherungsweise die gleichen Parameter wie bei der Simulation ergaben. Die statistische Auswertung zeigte, daß das Maximum zweier unabhängiger Zufallsgrößen näherungsweise als Summe der Anteile dieser Zufallsgrößen dargestellt werden kann. Für die Parameter Varianz V und zentrales Moment dritter Ordnung Z3 des Anteils A_1 der Zufallsgröße X_1 am Maximum der unabhängigen Zufallsgrößen X_1 und X_2 ergaben sich mit einem Bestimmt-

heitsmaß von 0,891 bzw. 0,863 die Formeln

$$V(A_1) = \frac{V(\mathrm{MAX}\{X_1;X_2\}) - (k_1 - k_2)\cdot V(X_2)}{V(X_1) + V(X_2)} \cdot V(X_1) \quad \text{und}$$

$$Z3(A_1) = k_1 \cdot Z3(X_1) + \frac{Z3(\mathrm{MAX}\{X_1;X_2\}) - k_1 \cdot Z3(X_1) - k_2 \cdot Z3(X_2)}{V(X_1) + V(X_2)} \cdot V(X_1) \quad ,$$

wobei k_1 die Wahrscheinlichkeit $P(X_1 > X_2)$ und $k_2 = 1 - k_1$ ist.
Wenn das Maximum zweier unabhängiger ZN-verteilter Zufallsgrößen als Summe ihrer Anteile dargestellt werden kann, läßt sich jeder früheste Anfangstermin und damit auch jeder früheste Endtermin als Summe von Anteilen der Vorgänge des Netzplanes darstellen. Bei der Ermittlung des Maximums zweier frühester Endtermine FE_j und FE_k kann aus der Zerlegung dieser Termine in ihre Anteile der gemeinsame Anteil GA_i jedes Vorganges i und als deren Summe der gemeinsame Anteil GA aller Vorgänge des Netzplanes bestimmt werden. Betrachtet man nun die Zufallsgrößen X_1 und X_2 mit $X_1 = FE_j - GA$ und $X_2 = FE_k - GA$, so enthalten sie jeweils nur noch die Anteile aus den frühesten Endterminen, die der andere früheste Endtermin nicht enthält, d.h. X_1 und X_2 sind unabhängige Zufallsgrößen. Bildet man ihr Maximum und addiert dazu GA, so ist offenbar das Maximum der beiden frühesten Endtermine richtig bestimmt.

5. Vergleich von Berechnungs- und Simulationsergebnissen

Zur Beurteilung der Realitätsnähe von Berechnungsergebnissen unter Verwendung ZN-verteilter Termine wurden bei mehreren Netzplänen mit verschiedener Struktur und unterschiedlicher Komplexität Simulations- und Berechnungsergebnisse verglichen. Für jede Vorgangsdauer wurden die Parameter Erwartungswert E, Varianz V und Schiefe S zufällig gewählt. Bei den Simulationsexperimenten wurde eine ZN-Verteilung aller Vorgangsdauern zugrunde gelegt. Für jede Datenkonstellation wurden 10.000 Simulationsläufe durchgeführt und daraus die Parameter Erwartungswert, Varianz und Schiefe der Projektdauer bestimmt. Diesen Simulationsergebnissen wurden die Berechnungsergebnisse a) unter Verwendung von ZN-verteilten Terminen und b) nach der Methode PERT gegenübergestellt.
Exemplarisch soll der folgende Netzplan 2 mit der Datenkonstellation

Vorgang	1	2	3	4	5	6	7	8	9	10	11	12
E	20,0	16,6	38,4	32,3	21,1	21,3	17,7	19,6	21,6	18,5	38,3	18,6
V	2,4	7,3	4,4	12,4	3,1	2,1	4,9	5,0	4,0	6,7	4,5	2,9
S	-0,5	0,2	-0,2	-0,5	0,4	-0,3	0,4	0,3	0,0	-0,2	-0,3	-0,2

Vorgang	13	14	15	16	17	18	19	20	21	22	23	24
E	21,2	20,8	52,3	17,6	16,7	72,7	16,3	46,1	23,4	23,3	22,4	20,1
V	7,8	1,9	13,3	2,9	7,0	19,5	4,7	6,3	4,8	3,5	8,0	3,5
S	0,4	0,1	0,1	0,3	-0,2	-0,1	0,1	0,1	-0,3	0,2	0,0	-0,1

betrachtet werden. Auf einem Personalcomputer 386SX25 erforderte die Berechnung mit ZN-verteilten Terminen etwa eine Minute, während die Simulation mehr als zwei Stunden in Anspruch nahm.

Netzplan 2

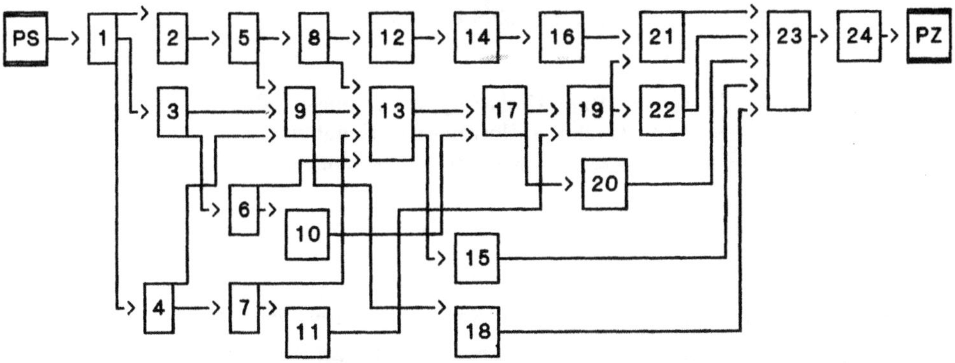

Es wurden folgende Ergebnisse bestimmt:

	Erwartungswert	Varianz	Schiefe
Simulation	209,245	35,816	0,064
Berechnung mit ZN-verteilten Terminen	209,859 *100,29%*	34,808 *97,18%*	0,112
Berechnung nach PERT	206,462 *98,67%*	43,343 *121,02%*	0

Die Prozentsätze beziehen sich jeweils auf den durch Simulation ermittelten Parameter. Unter Berücksichtigung der Ergebnisse von 7 Netzplänen mit je 70 Datenkonstellationen kann zur Berechnung mit ZN-verteilten Terminen eingeschätzt werden:

- Der für die Projektdauer bestimmte Erwartungswert weicht nur geringfügig vom Simulationsergebnis ab (maximale Abweichung unter 0,5%)
- Die ermittelte Varianz stimmt in den meisten Fällen gut mit dem Simulationsergebnis überein. In Einzelfällen treten Abweichungen bis zu 10% auf.
- Auch die berechnete Schiefe stimmt im allgemeinen gut mit dem Simulationsergebnis überein. Die maximale Abweichung lag unter 0,18.

Sind die Parameter der Vorgangsdauern hinreichend genau bestimmt, ist damit unter Verwendung ZN-verteilter Termine eine rationelle Ermittlung realitätsnaher Projekttermine möglich und so die Basis für zuverlässige Wahrscheinlichkeitsaussagen gegeben.

Literaturverzeichnis

[1] ELMAGHRABY,S.E.: *Activity Networks*. Verlag John Wiley & Sons New York/London/ Sidney/Toronto 1977

[2] GOLENKO,D.I.: *Statistische Methoden der Netzplantechnik*. Teubner-Verlag Stuttgart 1972

[3] NEUMANN,K.: *Netzplantechnik*. In: Grundlagen des Operations Research Band 2, Springer-Verlag Berlin/Heidelberg 1989

[4] SCHWARZE,J.: *Netzplantechnik*. Verlag Neue Wirtschafts-Briefe Herne/Berlin 1991

[5] SCHWARZE,J.: *Simulation von Netzplänen mit stochastischen und abhängigen Vorgangsdauern*. In: Operations Research Proceedings 1980, Springer-Verlag Berlin/Heidelberg, S.190-197

[6] UNGVARI,L.: *Beiträge zur Berücksichtigung des Zufalls in den Wirtschaftswissenschaften*. Dissertation A, Hochschule für Ökonomie Berlin 1989

POSSIBILITIES AND LIMITATIONS OF USING SCHEDULING THEORY IN THE PROCESS INDUSTRY

J.Peter van Dam, University of Groningen, NL
Gerard J.C. Gaalman, University of Groningen, NL
Gerard Sierksma, University of Groningen, NL

Scheduling in process industries is very complex, due to for example (partly) identical machines, shared resources, sequence-dependent set-up times, flow shops with limited intermediate storage, and manual operations carried out off-line. Moreover, it is not possible to define one objective function. Objectives of the scheduling are to obtain a feasible schedule, to meet due dates, to minimize the inventory, and to minimize additional costs such as costs of overtime and extra personnel.

In our view these scheduling problems contain elements of various dependent problems, as the assignment problem, the travelling salesman problem, and the flowshop scheduling problem. Although much literature can be found about these separate scheduling problems, there is not much attention for the integral problem.

In practice we observed a scheduling hierarchy of two levels. The type of decisions taken at each level, varies among the different scheduling situations. In making a schedule, the schedulers use rules of thumb. There are many other ways to make a schedule too. We think that before developing algorithms, the complex scheduling situation must be analyzed and ordered. Therefore, we define scheduling characteristics which describe the situations found in practice. The scheduling algorithms we develop, are based on these characteristics. This will by demonstrated by means of an application in a practical case.

SEQUENCING WITH NEURAL NETWORKS IN FLEXIBLE ASSEMBLING SYSTEMS
Automatic Formation of Optimal Control Strategies with TD-Method

Elgar Fleisch, Institut für Höhere Studien, Wien
Harald Hruschka, Universität Regensburg

A flexible assembling system (FAS) consisting of four partly different flexible assembling cells is to be controlled. The objective of the intelligent control system (agent) is minimization of mean flowtime, mean tardiness or any arbitrary combination of these competitive performance measures. The control problem involves sequencing and routing tasks.

The agent performs a heuristic dynamic programming approach in an adaptive heuristic critic framework to learn an optimal decision policy (strategy). The TD-Errors generated by application of the TD-method provide reinforcement signals used to adjust the parameters of the agent. The agent learns a strategic utility function and an optimal decision policy using a neural network approach. The applied design of adaptive heuristic critics works without built-in knowledge and without a model describing the controlled world.

Sequencing in FAS is a Markov decision problem and defined in terms of a discrete-time stochastic dynamical system with finite state set. At each time-step the controller observes the world's state i and selects an action k, which moves the world to state j with probability p_{ij}^k. The transition from state i to j produces a payoff r_{ij}^k, where $|r_{ij}^k| < \infty$ for each i, j and k. For any policy Π and state i one can define the expected infinite-horizon discounted return

$$E_{\Pi}\left[\sum_{t=0}^{\infty} \gamma^t r(t) \middle| i(0) = i\right], \qquad (1)$$

where r(t) denotes the payoff at time t, i(0) is the system's initial state and γ, $0 \leq \gamma \leq 1$ is a factor used to discount the future payoffs. The objective of the agent is to form a policy that maximizes the expected return defined by Equation 1 for each state i.

To provide suitable payoffs one has to measure the control-targets incrementally. The agent receives the states and the payoffs from the FAS online and performs direct learning. Applied to single-machine problems the agent converges to the well known optimal rules SPT and EDD. When applied to complex FAS problems the new control system converges to policies superior to advanced priority rules. The intricacy of tardiness-optimization strongly depends on the placed duedates. The automatically generated policies are priority rules and may be used for knowledge extraction.

References:

Barto90 Barto, Andrew G.; Sutton, Richard S.; Watkins, Christopher J.C.H; Learning and Sequential Decision Making, in: Gabriel, M. and Moore, J. (editors); Learning and Computational Neuroscience: Foundations of Adaptive Networks; The MIT Press; Cambridge, MA, 1990

Biendl84 Biendl, Peer; Ablaufsteuerung von Montagefertigungen; Haupt; Bern-Stuttgard, 1984

Tempelmeier92 Tempelmeier, Horst; Kuhn, Heinrich; Flexible Fertigungssysteme; Entscheidungsunterstützung für Konfiguration und Betrieb; Springer-Verlag; Berlin, 1992

EINE HEURISTIK ZUR OPERATIVEN PRODUKTIONSPROGRAMMPLANUNG BEI AUFTRAGSORIENTIERTER EINZELFERTIGUNG

Werner Geiger, Universität Bayreuth

Bei einer sukzessiv durchgeführten Produktionsplanung und -steuerung erfolgt für gewöhnlich zuerst die Festlegung des Absatz- und Produktionsprogramms. Speziell bei einer kundenauftragsorientierten besteht häufig das Problem, daß bei der Bestimmung des Absatzprogramms die Produktionsressourcen unberücksichtigt bleiben, was zu Schwierigkeiten bei der Termineinhaltung führen kann. Die fehlende Einbeziehung der Produktionspotentiale ist nicht zuletzt auf die Unsicherheit der zukünftigen Nachfrage zurückzuführen.

Im folgenden soll eine Heuristik vorgestellt werden, mit deren Hilfe sich bei Eingängen von potentiellen Kundenaufträgen Entscheidungen über die Auftragsannahme treffen lassen. Im Hinblick auf die unterstellte globale Unternehmenszielsetzung Gewinnmaximierung finden dabei die auftragsbezogenen Merkmale Deckungsbeitrag, Lieferzeit, Produktionsressourcenbeanspruchung und Kundenpriorität, als unternehmensbezogene Kriterien die zum Entscheidungszeitpunkt vorliegenden Kapazitätsauslastungssituationen der einzelnen Produktionsstufen sowie als marktbezogene Größen Erwartungen (Verteilungen) über die zukünftigen Kundenaufträge hinsichtlich Quantität, Deckungsbeiträge und Kapazitätsbedarfsstruktur Berücksichtigung. Durch geringfügige Modifikationen läßt sich das Modell zur Auftragsselektion in ein Verfahren zur Angebotsplanung überführen, das der Ermittlung von Auftragspreisuntergrenzen in Abhängigkeit der Kundenpriorität und der Lieferzeit dient.

Das Verfahren eignet sich speziell für Produktionssituationen, die sich wie folgt kennzeichnen lassen:

- mehrstufiger Produktionsprozeß mit mehreren potentiellen Engpässen
- kein Bevorratungsgrad an Eigenfertigungsteilen
- keine Zusammenfassung von durch unterschiedliche Kundenaufträge induzierten Fertigungsaufträgen zu gemeinsamen Fertigungslosen
- reihenfolgeunabhängige Rüstzeiten

Abschließend sollen zur Beurteilung der Leistungsfähigkeit der Heuristik die Ergebnisse (kleinerer) Studien vorgestellt werden.

Stronger Lagrangian bounds by use of slack variables with applications to machine scheduling problems

Han Hoogeveen, Princeton University, U.S.A.
Steef van de Velde, University of Twente, The Netherlands

The basic idea behind Lagrangian relaxation is to see an NP-hard problem as an 'easy-to-solve' problem complicated by a number of 'nasty' side constraints. We show that reformulating nasty inequality constraints as equalities by using slack variables leads to stronger bounds. The improvement proceeds by computing a bound on the associated so-called slack variable problem. The trick is widely applicable, but we focus on machine scheduling problems, including the single-machine problem of minimizing total weighted completion time subject to precedence constraints, the two-machine flow-shop problem of minimizing total completion time, the single-machine problem of minimizing total weighted tardiness, and the flow-shop problem of minimizing total completion time with batch machines. We derive the slack variable problem for each of these problems and point out how lower bounds can be computed for it. For each application we obtain promising computational results in the sense that a significant improvement of the Lagrangian bound can be achieved relative to the extra needed to compute a lower bound on the slack variable problem.

A Preprocessing Method for a Short-Term Capacity Assignment Problem in Job-Order Production

Heinz D. Mathes, Universität Bayreuth

Summary: This paper considers the short-term capacity assignment problem of made-to-order production in a process industry environment. The resulting binary optimisation problem can be solved by commercial software after applying a preprocessing procedure to the capacity constraints.

Zusammenfassung: Der Beitrag untersucht das Problem der kurzfristigen Kapazitätsplanung bei Auftragsfertigung mit mehrstufiger, serieller Arbeitsgangfolge. Das resultierende binäre Optimierungsproblem wird nach Anwendung einer Preprocessing Prozedur auf die Kapazitätsrestriktionen durch Standardsoftware gelöst.

1. Introduction

In this paper the multi-period, multi-stage capacity assignment (capacity planning) problem of job-order production (or made-to-order production) in process industries as for instance paper converting or metal products manufacturing is considered. A job-order or job is an accepted customer order based on a contract specifying amount, properties and delivery due date of the product to be manufactured (open shop). A job can be described by a set of processing operations interrelated by technologically determined precedence relations. When no permanently installed production lines are established, the processing operations are performed at given facilities or production stages equipped with parallel (interchangeable) machines with fixed capacity. The routing of each job (and therefore the sequence of production stages) to be pursued in the manufacturing process shall be limited to the case of serially ordered operations. Yet this assumption does not restrict all jobs to an identical routing, since it still holds if jobs start at different stages or by-pass single stages (identical routing passing). Though all machines at a production stage are capable of performing the same type of processing operation, some of them may be constructed to process only certain item sizes or materials. Due to these constructive differences between machines the assignment of a job to a machine cannot only be guided by the usual manufacturing considerations such as cost of processing, or required processing time, but also has to observe the technological compatibility of the selected machine (compatibility condition). In a multi-stage production environment with interchangeable machines the completion of a job can in general be achieved by a number of alternative machine sequences, provided a sequence contains exactly one (compatible) machine of each production stage contributing to the job´s completion and the machine order complies to the precedence relation.

When operating under a rolling horizon procedure the manufacturing unit faces at the beginning of each planning cycle the following multi-stage assignment problem: At each time period (shift, working day, week) within the planning horizon the available machine capacity has to be assigned to the accepted job-orders, so that for the whole planning intervall a feasible, least cost capacity assignment or capacity schedule is determined. A schedule is called feasible when in all periods the resulting job-machine combinations regard the precedence relations and compatibility conditions and the agreed due dates can be met considering the limited capacity per period. Such a schedule does obviously not determine the processing order of the jobs onto the different machines because neither starting nor completion times for the single operations are fixed. In the terminology of lot sizing theory this planning problem represents something like a big-bucket version of a flow-shop scheduling problem.

For the outlined planning situation a decision model shall be formulated, which meets the following requirements : (1) The model structure shall be sufficiently flexible in the sense that it can be adjusted to changes in the production environment without much effort. (2) The mathematical formulation of the model shall be tractable by commercially available software to compute optimal or near optimal schedules within reasonable time. As observed in [1] linear programming based models meet these requirements. This consideration recommends the description of the capacity assignment problem by a set of linear inequalities. However, due to properities of the underlying manufacturing process as for instance the occurrence of set-up operations a (0,1)-restriction then has to be imposed on some variables. Though the resulting structure would still comply to the flexibility requirement an obvious drawback of this approach is given by the fact, that (to the author's knowledge) all commercial software routines tackle integer programming problems by the use of branch-and-bound codes and even for comparatively small problem sizes their application usually results in prohibitive computation times. This deficiency arises mainly because the bounds generated by the LP relaxation of the integer problem deviate considerably from the (unknown) objective value of the optimal integer solution. A further objective of this study is therefore to develop a solution procedure which will speed up the numerical solution and is tractable by standard "off the shelf" codes.

For production planning problems formulated as mixed integer programms currently a number of studies are investigating preprocessing methods to tighten the generated bounds [2,3,5]. The approach put forward in this paper follows that rationale. To deal with the given multi-stage scheduling problem a preprocessing method will be presented, which applies two mathematically closely related techniques to achieve a tighter formulation of the problem at hand. First, for each capacity constraint a so-called stronger equivalent inequality with reduced coefficients and an adequately adjusted right hand side is constructed and substituted for the original constraint. These inequalities are equivalent over the (0,1)-hypercube but tighter in the continuous sense and thereby tend to strengthen the obtained bounds. A successful application of this method to a single stage problem has recently been presented in [3]. The second part of the preprocessing phase is based on certain sets of infeasible solutions of each constraint, which are required in constructing equivalent inequalities. It can be shown that these infeasible solutions can be used to generate a set of valid inequalities for the convex hull of solutions of each capacity constraint with the convenient property, that at least one of them forms a facet of the convex hull. Adding these valid inequalities completes the preprocessing. Besides leading to a considerable reduction of running times this method has the advantage, that it requires far less additional programming effort as compared to the Lagrangean relaxation or column generation methods and it can be linked easily with existing standard software.

2. Decision Model for the Capacity Assignment Problem

At the beginning of a planning cycle Z: = {1,...,T} a set R of different job-orders, $r \in R$, is given. The manufacturing process consists of n different production stages i, $i \in I: = \{1,...,n\}$. The sequence of operations required for the completion of job r is given by a particular ordered subset of I, $I(r): =\{i_k(r) \in I: i_k(r) > i_{k+1}(r)\}$ with $\min\{(i_s(r) : i_s(r) \in I(r))\}$ as the final processing operation/production stage. If job r is processed at stage i, then $s(i,r) \in I$ denotes the immediately succeeding processing operation with $s(i,r) = \{\ \}$, if the job is completed at stage i. At each stage a number of parallel machines $j \in J_i$ is available. The set $M(i,r) \subset J_i$ gives all machines at stage i, which can be employed to process job-order r (set of technologically feasible assignments of job r). The subset of orders, which at stage i can be assigned to machine $j \in J_i$, will be denoted $V(i,j) \subset R$ (set of technically feasible assignments of machine j). The capacity of each machine $j \in J_i$ is restricted to c_{ij}. For all job-orders the splitting of jobs and the aggregation of different jobs to batches will be assumed infeasible. In job-order manufacturing batching of different jobs is usually excluded by the heterogeneous set-up and processing requirements of the jobs. Due to differences between jobs in item sizes, required tolerances or materials to be processed no batches can be formed even if the routing of the jobs is identical. Job splitting tends to be avoided, because set-up operations in process industries are frequently rather time consuming and therefore could result in an unfavourable relation of the sum of set-up times to the entire processing time of a job, thus leading to a quite inefficient capacity usage. This effect could be further enhanced, because the job-size in made-to-order production can take on comparatively small values. To allow for the length of a planning period (t-1, t] to be chosen sufficiently small (i.e. according to requirements of short-term planning), each job may be processed at any period at most at one stage i.

The processing of job r by machine $j \in J_i$ incurs set-up cost K_{irj} and manufacturing cost P_{irj}. When job r is processed at stage i in period t and its immediately succeeding operation is not performed in (t+1) average inventory holding cost L_{ir} are charged per job and period of carrying inventory (for the definition of L_{ir} cf. [6]). The sum of set-up and processing times is given by b_{irj}, if at stage i job r is processed by machine $j \in J_i$. It will be assumed $b_{irj} \in Z^1_+$ and $b_{irj} \le c_{ij}$ for all feasible assignments. For each job the due date shall be given by $t_1(k,r) + 1$, if at stage $k \in I$ the final processing operation is performed. From the given routing of job r, I(r), and the assumption that a job spends one full period at each stage, its earliest possible starting date $t_0(i,r)$ and latest completion date $t_1(i,r)$ for all stages $i \in I(r)$ can be calculated in a straightforward manner. The set of periods t, at which the processing of job r at stage i is feasible, will be denoted $T(i,r) := \{t_0(i,r), t_0(i,r)+1, \ldots, t_1(i,r)\}$.

For the mathematical formulation of the planning situation a model will be used which does not contain inventory holding variables explicitly. This type of model was chosen over other representations since computational tests seemed to indicate a better running time behaviour [cf. 1]. Defining

$$x_{irjt} = \begin{cases} 1, & \text{if at stage i job r is assigned to machine j in period t} \\ 0 & \text{otherwise} \end{cases}$$

one obtains the following binary optimisation problem BOP:

$$\min \sum_{i \in I} \sum_{r \in R} \sum_{j \in J_i} \sum_{t \in T(i,r)} \left(K_{irj} + P_{irj} + L_{ir}\right) \cdot x_{irjt} \tag{2.1}$$

s.t.

$$\sum_{j \in M(i,r)} \sum_{t \in T(i,r)} x_{irjt} = 1, \qquad \textit{for all } i \in I,\ r \in R \tag{2.2}$$

$$\sum_{j \in M(i,r)} \sum_{t = t_0(i,r)}^{k} x_{irjt} - \sum_{j \in M(s(i,r),r)} \sum_{t = t_0(s(i,r),r)}^{k+1} x_{s(i,r),r,j,t} \geq 0 \tag{2.3}$$

$$\textit{for all } i \in I - \{1\},\ r \in R,\ k \in T(i,r) - \{t_1(i,r)\}$$

$$\sum_{r \in V(i,j)} b_{irj} \cdot x_{irjt} \leq c_{ij} \qquad \textit{for all } i \in I,\ j \in J_i,\ t \in T(i,r) \tag{2.4}$$

$$x_{irjt} \in \{0,1\} \qquad \textit{for all } i \in I,\ r \in R,\ j \in J_i,\ t \in T(i,r) \tag{2.5}$$

Constraints (2.2) assure that a job r is assigned to a machine exactly once at all stages required by the completion of the job. This assignment must take place at a period t belonging to the set of feasible production periods of job r, T(i,r), at all stages. If job r skips stage i, i.e. $i \notin I(r)$, because of $M(i,r) = \{\ \}$ a constraint of type (2.2) is not defined. Constraints (2.3) represent the flow conservation constraints for all jobs. They require that a job r cannot be processed at stage s(i,r) in period t if it was not completed at stage i in a previous period. (2.4) limits the capacity usage in all periods to the available capacity. Job splitting and job batching is prevented by constraints (2.5). It need not be emphasized that the addition of further constraints representing limitations on ressources other than machine capacity can easily be implemented according to (2.5).

3. A Preprocessing Method for the Binary Optimisation Problem BOP

3.1 Theoretical Background

The following considerations refer exclusively to the capacity constraints (2.4) which form the "complicating constraints" as far as the numerical solution is concerned [2]. To facilitate the analysis a constraint of type (2.4) will be represented by

$$\sum_{j \in N} b_j \cdot x_j \leq b_0 \qquad (3.1)$$

It is convenient to order the coefficients monotonically so that $b_1 \geq b_2 \geq ... \geq b_n$ for all b_j, $j \in N=\{1,...,n\}$. For obvious reasons let $\sum b_j > b_0$. (3.1) forms a typical knapsack restriction with a set of solutions defined by $S := \{x \in B^n : \sum b_j x_j \leq b_0\}$. Elements of B^n can be represented by characteristic vectors so that for $C \subset N$ with $C:=\{j_1,j_2,...,j_r\}$ and $j_1<j_2<...<j_r$, the vector x^C has components $x_j^C:=1$ if $j \in C$ and $x_j^C=0$ otherwise [4]. One obtains $x^C \notin S$ iff

$$\sum_{j \in C} b_j = b_{j1} + ... + b_{jr} = b_0 + \delta, \ \delta \geq 1.$$

For an inequality (3.1) a (stronger) equivalent inequality (3.2) with a reduced coefficient vector $a = (a_j, j \in N, a_0)$ and a set of solutions S_a can be formulated [1]:

$$\sum_{j \in N} a_j \cdot x_j \leq a_0 \qquad (3.2)$$

(3.2) is equivalent to (3.1), iff its set of solutions S_a is identical to S, and therefore (3.2) can be substituted for (3.1). Following the reduction of the coefficients (a_1, ..., a_n, a_0) the LP-relaxation of (3.2), $\left\{x \in R_+^n : \sum a_j x_j \leq a_0, x_j \leq 1, j \in N\right\}$, provides a tighter description of the convex hull of S, conv(S), than the corresponding relaxation of (3.1) thus leading in general to improved bounds when applied in branch- and bound procedures. The equivalence of inequalities is based on the concept of roofs, which was introduced in [1].

Definition (D): $C \subseteq N$ is a *roof* of (3.1), if:

(i) $x^C \notin S$

(ii) $b_{j_r} \geq \delta$

(iii) If $j_k \in C$, $(j_k + 1) \notin C$ then $b_{j_k} - b_{j_k+1} \geq \delta$, with $\delta = \sum_{j \in C} b_j - b_0$.

Sets C, which suffice at least (i) and (ii), are called minimal dependent sets [4]. For each inequality (3.1) and (3.2) a finite, nonempty set of roofs can be derived. These sets form the basis for determining the equivalence of any two inequalities, since it holds, that two inequalities are equivalent iff their corresponding roof sets are identical [1]. Besides providing a stopping rule for procedures which by an iterative process [3, 6] construct equivalent inequalities, the roof set of any inequality (3.1) possesses a further property, which can be exploited for speeding up the numerical solution of BOP. This property can be derived from a result proved in [4]. If C is a roof, the set $E(C) := C + \{k \in N - C : b_k \geq b_{j_1}\}$ is denoted the *extension* of C. From E(C) one obtains a valid inequality for S by

$$\sum_{j \in E(C)} x_j \leq |C| - 1 \qquad (3.3)$$

If C fullfills some qualifying conditions the valid inequalities (3.3) represent facets of conv(S) [4,6]. Based on these conditions one obtains the following proposition which indicates that constructing a valid inequality (3.3) for each roof in the roof set of (3.1) (or (3.2)) provides a systematic method to tighten the problem formulation even further (for a proof cf. [6]):

Proposition 1: The set of roofs of (3.1) contains at least one roof C with an extension E(C), such that (3.3) generates a facet of conv(S).

At first sight Proposition 1 appears to be not very encouraging as far as the number of existing facets is concerned. However computational tests showed that approximately 50% of all roofs generated a facet.

3.2. Preprocessing Procedure

The results of the proceeding section strongly recommend a preprocessing scheme applying two consecutive phases to transform the original problem.

Phase 1: At this stage equivalent inequalities for all capacity constraints are constructed. This can be achieved by the iterative procedure tested in [3] or its modified version (for details cf. [6]).

Phase 2: Employing the roof sets of the equivalent inequalities derived in phase 1 the corresponding extensions are formed. For all extensions valid inequalities are generated. The set of valid inequalities is then scanned for dominated inequalities and the remaining ones added to the set of restrictions. The number of facets of a constraint depends on the cardinality of its roof set, which is a function of the number of variables in the constraint.

4. Computational Results

The derived preprocessing method was applied to a randomly constructed set of 30 test problems. For the numerical computations HYPERLINDO was used running on a DX2-66. The preprocessing routine was programmed in FORTRAN. The following experimental design was selected: A manufacturing unit with four production stages was considered. At each stage 3 - 4 machines were available. For every test problem the exact number was fixed at random. The number of orders ranged from 10 to 20 different jobs. Order routing was randomly determined with the minimum number of operations set to 3. All coefficients were chosen at random and tested for feasibility to exclude S = { }. The resulting problem sizes ranged from 360 - 697 binary variables (with an average of approximately 500) and comprised 130 - 300 inequalities for unpreprocessed problems and 426 - 742 inequalities for preprocessed problems. The following measures of computational performance were chosen: (1) Percentage of problems with no feasible integer solution (NoInt). (2) Percentage of problems solved to optimality (Opt). (3) Average optimality gap (AvGap) in %. (4) Maximal optimality gap (MaxGap) in %. For problems not solved to optimality the optimality gap was given by (best integer sol. found / $\lceil$LP-optimum$\rceil$)-1. All performance measures are based on a maximal running time of 5 CPU-minutes (including preprocessing time for the preprocessed problems). The test results are summarized in the following table:

	NoInt (%)	Opt (%)	AvGap (%)	MaxGap (%)
Unpreprocessed Problems	26.67	10	3.51	15.01
Preprocessed Problems	0	50	0.664	4.46

Table 1: Summary of computational results.

Hyperlindo showed the tendency to perform better with problems of a smaller size. For preprocessed problems no significant differences in performance were found when problem size varied, so that an analysis by problem size can be omitted.

Summarizing the computational tests one can safely conclude that the presented preprocessing method arrived at rather encouraging results though a greater sample is certainly required for a more precise judgement of the method's computational performance. Currently research efforts are directed towards extending this method to other classes of problems.

References

1. Bradley, G.H., Hammer, P.L., Wolsey, L., Coefficient Reduction for Inequalities in 0-1 Variables, Mathematical Programming 7 (1974), pp. 263 - 282.
2. Eppen, G.D., Martin, R.K., Solving Multi-Item Capacitated Lot-Sizing Problems Using Variable Redefinition, Operations Research 6 (1987), pp. 832 - 848.
3. Guignard, M., Liberatore, M.J., Applying a Modified Coefficient Reduction Method to a Dynamic Production Scheduling Model, EJOR 56 (1992), pp. 119 - 130.
4. Nemhauser, G.L.,Wolsey, L., Integer and Combinatorial Optimization, New York 1988.
5. Maes, J., McClain, J.O., Van Wassenhove, L.N., Multi-Level Capacitated Lotsizing Complexity and LP-based Heuristics, EJOR 53 (1991), pp. 131 - 148.
6. Mathes, H.D., An Integer Programming Approach to the Multi-Period, Multi-Stage Capacity Assignment Problem in Made-to-Order Production (forthcoming)

Multi-resource scheduling of an FMC in discrete parts manufacturing

Geatse Meester and Henk Zijm
University of Twente, The Netherlands

In this paper we develop a scheduling procedure for a Flexible Manufacturing Cell (FMC), consisting of a number of basically identical machining centres and operating as part of a larger component manufacturing shop (basically a hybrid job shop). Each job on the FMC is subject to release and due dates; the objective of the scheduling procedure is to minimize the maximum lateness, subject to both machine and cutting tool capacity constraints.
An important aspect of the system is the presence of a tool robot which can interchange tools between a central tool store and the local tool magazines dynamically, i.e. during machine operations. However, since many cutting tools are unique (i.e. no duplicates are present in the system) the scheduling problem is multi-resource in nature: both machines and tools are scarce resources which have to be used simultaneously. A heuristic to solve this problem is proposed and test results are presented for both, theoretical tests and a test with an FMC in a company producing pneumatic and electrical actuator valves.

THE JOB SHOP SCHEDULING PROBLEM: CONVENTIONAL AND NEW SOLUTION TECHNIQUES

Erwin Pesch, University of Limburg, NL-Maastricht

A job shop consists of a set of different machines that perform operations on jobs. Each job has a specified processing order through the machines, i. e. a job is composed of an ordered list of operations each of which is determined by the machine required and the processing time on it. There are no precedence constraints among operations of different jobs. Operations can not be interrupted (non–preemption) and each machine can handle only one job at a time. Consequently, each job can be performed on only one machine at a time. The operation sequences on the machines are unknown and have to be determined in order to minimize the makespan, i. e. the time required to complete all jobs.

The minimum makespan problem of job shop scheduling is a classical combinatorial optimization problem that has received considerable attention in the operations research literature. It is well known that the job shop scheduling problem is NP–hard and belongs to the most intractable problems considered.

The history of the job shop scheduling problem, starting more than 30 years ago, is also the history of a well known benchmark problem consisting of 10 jobs and 10 machines and introduced by Fisher and Thompson in 1963. This particular instance of a 10 job 10 machine problem opposed its solution for 25 years leading to a competition among researchers for the most powerful solution procedure.

In recent years exact algorithms (Calier / Pinson (1989, 1990), Applegate / Cook (1991), Brucker / Jurisch / Sievers (1991)) and tailored approximation methods (Adams / Balas / Zawack (1988), van Laarhoven / Aarts / Lenstra (1992), Dell'Amico / Trubian (1992), Dorndorf / Pesch (1992)) have been significantly progressed.

We are going to consider the most important solution ideas and their related neighbourhood structures leading to sophisticated branching structures for the exact methods as well as powerful and generally applicable approximation techniques, e.g. local search procedures such as simulated annealing, tabu search, or genetic algorithms.

AUDIT-STAFF SCHEDULING: AN HIERARCHICAL APPROACH

Frank Salewski, Christian-Albrechts-Universität Kiel
Andreas Drexl, Christian-Albrechts-Universität Kiel

Traditional audit-staff scheduling models are single-level models which try to construct a direct assignment of auditors to audit tasks and periods. By contrast, an empirical investigation among 200 of the biggest audit companies in Germany shows that the process of audit-staff scheduling usually is divided into three different levels [3]. While the names of these levels, viz. *medium-term*, *medium-to-short-term*, and *short-term planning*, refer to their differing planning horizons, also clear distinctions can be made in terms of organizational echelon involved, length of planning periods, degree of aggregation of the audit tasks, degree of detail of the required information, and decision objectives. Furthermore, the results of a more aggregated level are used as input to a more disaggregated level. The levels can be characterized as follows:

The *medium-term planning* [4] assigns teams of auditors to audit engagements. It constructs a schedule containing all the engagements' phases, e.g. preliminary, intermediate and final audit, and determines the workload per auditor and week - on the basis of periods of forty hours - over a planning horizon of between three and twelve months.

The *medium-to-short-term planning* [1] disaggregates the results of the medium-term level for one week and all auditors. The outcome is a schedule that specifies for each auditor - on the basis of periods of four hours - all the phases in which the auditor is involved in the considered week.

The *short-term planning* [2] is based upon the results of the medium-to-short-term level for one week and one phase. It assigns the auditors involved in the auditing of that phase in the considered week to the corresponding audit tasks and schedules them on the basis of periods of one hour.

With respect to those practical requirements we present for each level an appropriate model. The models are formulated in terms of binary optimization; they are closely related to resource-constrained project scheduling. Finally, we introduce some consistency relations and assumptions about parameter values to guarantee feasibility on the more disaggregated levels.

References

[1] Salewski, F., T. Bartsch, E. Pesch (1993): Auftragsterminierung für die taktisch-operative Personaleinsatzplanung in Wirtschaftsprüfungsgesellschaften, Discussion Paper, Universität Kiel.

[2] Salewski, F., L. Böttcher, A. Drexl (1993): Prüffeldorientierte Zuordnung von Prüfern für die operative Personaleinsatzplanung in Wirtschaftsprüfungsgesellschaften, Discussion Paper, Universität Kiel.

[3] Salewski, F., A. Drexl (1993): Personaleinsatzplanung in Wirtschaftsprüfungsgesellschaften - Bestandsaufnahme und konzeptioneller Ansatz, to appear in: Zeitschrift für Betriebswirtschaft.

[4] Salewski, F., A. Schirmer, A. Drexl (1993): Auftragsorientierte Bildung von Prüferteams für die taktische Personaleinsatzplanung in Wirtschaftsprüfungsgesellschaften, Discussion Paper, Universität Kiel.

AUDIT-STAFF SCHEDULING: METHODS FOR THE MEDIUM-TERM LEVEL

Andreas Schirmer, Christian-Albrechts-Universität Kiel
Frank Salewski, Christian-Albrechts-Universität Kiel

A survey conducted among 200 of the biggest German CPA firms shows that the process of audit-staff scheduling is usually divided into three distinct levels [1]. The first of these levels, the *medium-term planning* [2], assigns teams of auditors to audit engagements; then it constructs a schedule covering all the engagements' phases, e.g. preliminary, intermediate, and final audit, and determines the workload per auditor and week - on the basis of forty-hours-periods - over a planning horizon of between three and twelve months.

The problem is formulated in terms of a binary optimization model which is closely related to multi-mode resource-constraint project scheduling. Thereby, different teams are tally with different modes. Apart from job completion, time, resource, and precedence constraints this model comprises also so-called mode-identity constraints which guarantee that the assignment of a team to an engagement, once done, is never changed.

As a consequence, for solving the medium-term audit-staff scheduling problem, a general two-stage algorithmic scheme has been developed. In the first stage, the assignment of modes to engagements is done by choosing a mode for each engagement, according to *regret values* (generalized opportunity costs) which measure the worst-case consequence of *not* selecting a considered mode. In the second stage, following the general operating scheme of *parallel scheduling*, the actual scheduling takes place, using heuristically obtained *priority values*.

Both stages use *randomization* to avoid some of the drawbacks of traditional, deterministic scheduling heuristics. Rather than selecting the candidate with the highest regret or priority value, they calculate selection probabilities proportional to these values and perform the selection process in a random way, according to the calculated probabilities. Due to this nondeterminism, repeated application will produce a *set of solutions* rather than one sole solution.

To evaluate the performance of the scheme, an experimental study was undertaken, in the course of which the scheme was applied to a sample of representative instances. The computational experience gained so far is promising. For most of the attempted instances feasible solutions were found. Further refinements on the priority rules are expected to lead to slightly better results.

References

[1] Salewski, F., A. Drexl (1993): Personaleinsatzplanung in Wirtschaftsprüfungsgesellschaften - Bestandsaufnahme und konzeptioneller Ansatz, to appear in: Zeitschrift für Betriebswirtschaft.

[2] Salewski, F., A. Schirmer, A. Drexl (1993): Auftragsorientierte Bildung von Prüferteams für die taktische Personaleinsatzplanung in Wirtschaftsprüfungsgesellschaften, Discussion Paper, Universität Kiel.

Scheduling a single machine with release dates, due dates and family setup times

Marco Schutten and Henk Zijm
University of Twente, The Netherlands

In machining environments the combined goal of efficient and effective production may lead to complex control problems. Efficient production in such an environment is translated in minimizing the loss of capacity due to setups and therefore combining jobs with similar setup characteristics. Effective production in an order-driven environment is translated in completing jobs in time, or at least minimizing tardiness. Clearly, these two objectives may be conflicting: clustering jobs with similar setup characteristics may lead to a severe delay of other "badly fitting" jobs, thereby causing a substantial job tardiness. On the other hand, an "earliest due date" schedule may lead to too many setups, hence a capacity overload and more and more late orders, contrary to its objective. Any solution to this problem should be based on a combination of batching and sequencing procedures.

We present a branch and bound algorithm for the single machine scheduling problem where job processing is subject to both technical (setup characteristics) as well as logistic constraints (release and due dates). The objective is to minimize the maximum lateness. A lower bound is presented for the lateness of a set of jobs in an optimal schedule. A tabu search algorithm in combination with a cheapest insertion algorithm provides an upper bound to be used in the branch and bound algorithm. Several side results are developed to speed up the fathoming of subsets of solutions in the algorithm. Test results are presented for up to 40 jobs.

Development of a shopfloor planning and scheduling system for hybrid job shops

Henk Zijm, University of Twente
The Netherlands

One of the research themes of the Production and Operations Management Group at the University of Twente concerns the development of a complete shopfloor planning and scheduling system for hybrid job shops. The term 'hybrid job shop' refers to a production system where jobs may have different routings and where each operation is performed in a (complex) workcell instead of just on a single machine. Such a workcell may be a set of parallel machines or even a Flexible Manufacturing Cell with additional fixture and cutting tool capacity constraints (leading to multi-resource scheduling problems). Jobs may be clustered in family structures (according to similar set-up characteristics). Parts of the job shop may operate in different shift modes. The most important objective to be met will be the timely delivery of jobs (subject to internally set or externally determined due-dates).

The system developed so far is based on experiences in several metal cutting parts manufacturing shops; prototypes of several modules have been developed in cooperation with, and are tested in various companies in the Netherlands. In this talk we present the basic structure of the scheduling system, which has a highly modular structure and is based on decomposition principles. An example of a (really existing) application site will be presented. Information and implementation aspects are briefly reviewed. In subsequent talks (Meester, Schutten and Van de Velde) details on the algorithms developed for several modules will be presented.

HEURISTIC PROCEDURES FOR THE STOCHASTIC LOT-SIZING PROBLEM

N.P. Dellaert
M.T. Melo*
Econometric Institute, Erasmus University Rotterdam

Abstract: In this paper three different production rules are proposed for a lot-sizing problem in a stochastic single item situation with production to order and fixed lead times. The performance of the rules is studied by comparing their average costs with those of the optimal production policy. One of these strategies, the so-called (x, T)-rule, is also extended to the constrained capacity case.

Zusammenfassung: Es werden drei unterschiedliche Produktionsvorschriften für ein Losgrößenbestimmungsproblem eines Produktes bei stochastischer Nachfrage und auftragsorientierter Produktion mit fester Terminvorgabe vorgeschlagen. Durch einen Vergleich der durchschnittlichen Kosten für diese Verfahren mit den Kosten der optimalen Produktionsvorgehensweise wird das Verhalten der Verfahren analysiert. Eine dieser Strategien, die sogenannte (x, T)-Regel, wird erweitert für den Fall der Kapazitätsbeschränkung.

1 Introduction

We consider a lot-sizing problem in which a single type of product is manufactured on a make-to-order basis. In many production processes with highly uncertain demand, clients are interested in having the type produced according to their wishes and request to have it delivered at short notice. Since in such situations manufacturers do not wish to keep any safety stocks, a due date resulting from an agreement between the firm and the client is set for each incoming order. Decisions about production are taken at the end of a review period where each review period has fixed length. Moreover, the firm classifies its clients in N different groups and offers each client a fixed lead time so that all clients belonging to group i, $1 \leq i \leq N$ obtain a lead time equal to i periods if they place an order. This implies that in an arbitrary period t, clients assigned to group i can order for period $t+i$. Firms with regular clients usually follow this type of policy. The distribution of the demand for each of the N periods is known, stationary and independent of the arrival period. In addition, during each period the demand of the different groups of clients is assumed to be stochastically independent. Every order placed by a client consists of one unit of product which is manufactured on one single machine and may be satisfied with backlogging.

By combining orders into production lots one hopes to save on set-up costs at the expense of some additional holding costs for the orders that are finished before their due date and penalty costs for the orders that are not ready by their due date. We are therefore interested in establishing a production scheme which minimizes the average costs per period.

This paper is organized as follows. In Section 2 the model applied to the problem is introduced. In Section 3 a method based on dynamic programming is briefly described for determining the optimal production policy. In Section 4 three different heuristic solution methods are presented:

*On leave from D.E.I.O. (Universidade de Lisboa, Portugal). This research was supported by J.N.I.C.T. (Portugal) under contract BD/2264/92-IA.

the so-called (x, T)-rule, a Silver-Meal-like approach and a cyclic production rule. Some numerical results are given in Section 5. Section 6 contains some remarks concerning the extension of one of the heuristic procedures, the (x, T)-rule, to situations with capacity constraints. Finally, some conclusions are drawn in Section 7.

2 Model Description

In order to find a good control rule for the production planning of the situation described, the problem is modelled as a Markov Decision Process with discrete state and time space and without capacity constraints.
The state space is denoted by R and consists of order state vectors $r = (r_1, r_2, \ldots, r_N)$ where each component at the end of an arbitrary period t has the following meaning:

$$r_j = \begin{cases} \text{number of orders of unit size to be delivered during} & \\ \text{period } t+1\text{, including backorders from earlier periods} & \text{if } j = 1 \\ \text{number of orders of unit size to be delivered during} & \\ \text{period } t+j & \text{if } 2 \leq j \leq N \end{cases}$$

By considering the demand distribution and the future required deliveries, one can decide at the end of a period t to delay production, to produce all known demand or to produce part of the known demand. In the first case, penalty costs are incurred for all the orders that should be delivered by the end of period $t+1$ or even earlier. It is assumed that the penalty costs are equal among all clients and p units are charged per period for each late order. Selecting one of the other two possible decisions leads to set-up costs s and holding costs of h units per order and per period for the demand that only has to be delivered during one of the periods $t+2, \ldots, t+N$. As a result, each state $r \in R$ is associated with a finite non-empty set of actions $A(r)$. Like in the deterministic version of the problem, it will always be optimal to produce the known demand for an integral number of periods. Therefore, the meaning of an action a is that all orders for the first a periods will be produced, i.e. $\sum_{i=1}^{a} r_i$. If $a = 0$ production is delayed. Hence, $A(r)$ is defined for every $r \in R$ as:

$$A(r) = \left| \begin{array}{ll} \{0\} & \text{if } r_1 = 0 \\ \{1, \ldots, N\} & \text{if } r_1 p > s \\ \{0, 1, \ldots, N\} & \text{otherwise} \end{array} \right.$$

If action a is taken on observing state r, let $Q_a(r)$ denote the state at the end of the next period, just before the new orders are added to the order state. Suppose that four groups of clients place orders and at the end of period t it is decided to delay production. This will imply that at the end of the following period $Q_0(r) = (r_1 + r_2, r_3, r_4, 0)$. Similarly, $Q_3(r) = (0, 0, r_4, 0)$.
Let $J \subseteq \mathbb{N}^N$ be the set of possible one-period demands $(j_1, j_2, \ldots, j_N)$ and let d_{ij} denote the known probability that during an arbitrary period t, j units of product are ordered for period $t+i$. The transition probability P_{rz}^a that on observing state r and choosing action $a \in A(r)$ we enter state $z = Q_a(r) + (j_1, j_2, \ldots, j_N)$ is therefore defined by $\prod_{i=1}^{N} d_{ij_i}$ for all $(j_1, j_2, \ldots, j_N) \in J$ and 0 elsewhere. Moreover, under the assumption that the holding costs for orders that are produced too early will

be paid immediately, the costs of taking action a on observing state r have the following form:

$$q_r^a = \begin{cases} p\,r_1 & \text{if } a = 0 \\ s + h\sum_{i=1}^{a-1} i\,r_{i+1} & \text{if } a > 0 \end{cases} \tag{1}$$

3 Optimal Production Policy

In case the dimension of the state space R is finite, i.e. if the one-period demand is bounded, the optimal production policy can be determined by using dynamic programming. In order to find the policy that minimizes the expected costs g per transition, we define $v_n(r)$ as the total expected costs from the next n transitions when the current state is r and an optimal policy is followed:

$$v_{n+1}(r) = \min_{a\in A(r)} \left[q_r^a + \sum_{z\in R} P_{rz}^a v_n(z)\right], \quad r \in R,\ n = 0, 1, \ldots \tag{2}$$

By [2], any production rule achieving the minima in (2) also has minimal costs per transition. Therefore, starting with $v_0(r) = 0$ for all $r \in R$, we use the above iteration scheme until a satisfactory degree of convergence is achieved. It follows that g may be estimated by $\frac{1}{2}\left[\min_{r\in R}\{v_{n+1}(r) - v_n(r)\} + \max_{r\in R}\{v_{n+1}(r) - v_n(r)\}\right]$. This estimate becomes nearly exact for large n.

4 Heuristic Procedures

Finding the optimal policy is only possible for rather small problems due to the large dimension of the state space. For this reason it is important to derive good heuristics, that is, fast heuristics with relatively low average costs.

In this section we propose three different production rules: the so-called (x, T)-rule, a Silver-Meal-like approach and a cyclic production rule.

4.1 The (x, T)-Rule

In testing the optimal production policy for a lot of examples, we noticed that in most cases the decision whether to produce or not only depends on the value of r_1 and not on the required deliveries for later periods. Moreover, for a given demand distribution, the number of periods for which the orders are produced is nearly always the same. Only in "extreme" states with a rather small probability, the optimal policy deviates from these results. Therefore, we propose a simple decision rule based exclusively on the number of orders for the first period or one of the earlier periods. If r_1 is smaller than some variable x there is no production during the first period. Otherwise (i.e. if $r_1 \geq x$), all orders with a residual lead time of T periods or less are produced during that period. Like x, T is also a decision variable. This production rule will be called the (x, T)-rule. Since a decision about production is made at the end of every period, we can produce again less than T periods after the previous one if in the meantime enough orders have arrived.

One of the main advantages of the (x, T)-rule is that the average costs per period can easily be determined for given values of x and T in the unconstrained capacity situation. In order to compute these costs we consider the order state during one production cycle. Since demand is stationary and the (x, T)-rule is a fixed production strategy, there exist time points at which the stochastic process that represents the order state, probabilistically restarts itself. One of these regenerative points takes place at the end of a period in which the required deliveries for the first T periods have been produced. Since the probabilities for each possible order state vector are the same in every regenerative point, the average costs per period are determined by considering the order state during the periods between two regenerative points.
Although $r = (r_1, r_2, \ldots, r_N)$ provides all necessary information about the order state, it is not suitable for the analysis of situations in which the dimension of the state space is very large. On the other hand, in the (x, T)-rule, a decision about production only depends on the value of r_1. However, defining a state simply by r_1 does not give enough information about the transition probabilities. The changes in the value of r_1 are influenced by the time past since the last production period and are not difficult to describe (cf. [1]). For this reason, we can use in the analysis of the (x, T)-rule a state description given by $(\textit{time since last production}, r_1)$.
If there is a probability that there are no orders during a period or if the demand can be infinitely large, the number of possible states may be infinte. However, both the decision and the costs of the decision are the same for $r_1 = x$ and $r_1 > x$. Also both the costs of an action and the transition probabilities for the component r_1 are the same for all periods except the $T-1$ first periods. Therefore, we can consider a finite recurrent Markov chain with states $(i, j) \in \{1, 2, \ldots, T-1\} \times \{0, 1, \ldots, x\}$ where i represents the number of periods since the last production period and j the value of r_1. In case $T = 1$ the time-component is limited by $\{1\}$. Figure 1 gives an illustration of this Markov chain.

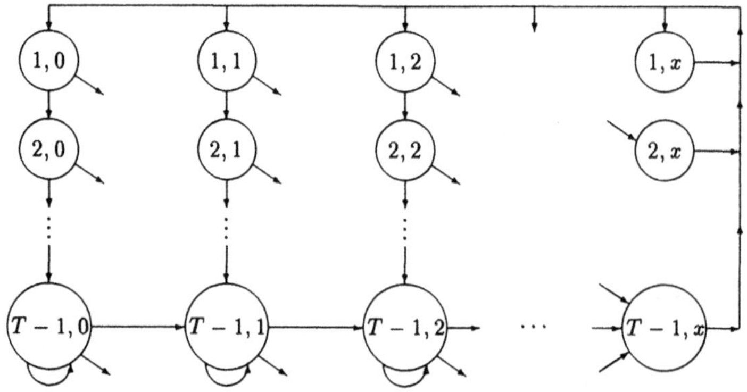

Figure 1 The states of the Markov chain for the (x, T)-rule with $T \geq 2$.

Using the above Markov chain will allow us to determine the average costs per period associated with a given pair (x, T).
Let q_{ij} denote the expected number of visits to the state (i, j) between two production periods. Due to the special structure of the Markov chain, these values are easy to obtain (cf. [1]). Moreover, the sum of the expected number of visits to all states gives the expected time between two production periods:

$$T(x, T) = \sum_{i=1}^{T-1} \sum_{j=0}^{x} q_{ij} \tag{3}$$

On the other hand, the expected costs between two production periods are determined by

$$C(x,T) = \sum_{i=1}^{T-1} \sum_{j=0}^{x} q_{ij} c_{ij} \tag{4}$$

with c_{ij} the expected costs in state (i,j). In the states (i,j) with $j < x$ we only have penalty costs. If $j = x$ we have set-up costs and also consider expected holding costs.
Hence, the average costs per period associated with the pair (x,T) are easily obtained by dividing (4) by (3).

$$g(x,T) = \frac{C(x,T)}{T(x,T)}$$

Making use of two properties of [1], the optimal pair (x,T) can be found without having to compute the value of $g(x,T)$ for all possible pairs (x,T).

4.2 The Silver-Meal Approach

In the Silver-Meal algorithm (cf. [3]), the expected costs of an action are divided by the number of periods involved in that action. The action for which this quotient is minimal is then chosen. Based on this idea, a heuristic procedure for our problem can be derived. Taking action a on observing state $r \in R$ leads to the costs q_r^a defined by (1). Under the assumption that the next production will take place a periods after the current period, also indirect costs will have to be considered. These costs can easily be estimated by taking the expected penalty costs during the first a periods (cf. [1]). Observe that if we produce during period t then the orders that arrive during periods $t, t+1, \ldots, t+a-2$ for periods $t+1, \ldots, t+a-1$ will be delivered too late and consequently penalty costs will be incurred. Denoting by $p(a)$ the penalty function associated with these costs, the Silver-Meal-like production rule takes the following form: if state $r \in R$ is observed then we choose action a' given by

$$a' := \text{argmin} \left\{ \frac{q_r^a + p(a)}{\max\{a, 1\}} : a \in A(r) \right\}$$

4.3 The Cyclic Production Rule

In the cyclic production rule a set-up occurs every T periods, unless there is nothing to produce. In addition, whenever production takes place the required deliveries for the first T periods are produced. In order to determine the average costs g_π per period for this rule, we divide by T the total expected costs during one cycle which consist of set-up costs, holding costs for the orders that are produced too early and penalty costs for the orders that will be finished after their due date. Therefore, we choose the value of $T \in \{1, 2, \ldots, N\}$ for which g_π is minimal.

5 Numerical Results

The performance of the three production rules presented in Section 4 is studied by comparing the average costs per period of these rules with those of the optimal strategy for a set of problems with $N = 4$, $p = 3$ and $d_{i1} = 1 - d_{i0} = d$, $i = 1, \ldots, 4$.
Table I contains the average costs per period obtained for different values of d, s and h for the optimal strategy (OPT), the (x, T)-rule (XT), the Silver-Meal rule (SM) and the cyclic production rule (CYC).

d	0.25	0.25	0.25	0.50	0.50	0.50	0.75	0.75	0.75
s	3.25	8.00	8.00	6.50	16.0	16.0	9.75	9.75	24.0
h	1.00	1.00	2.00	1.00	1.00	2.00	1.00	2.00	1.00
OPT	1.8895	3.7147	3.9871	4.5357	8.1705	8.8504	7.0425	8.0240	12.6002
XT	1.9074	3.7326	4.0219	4.5392	8.1965	8.9870	7.0451	8.1141	12.6125
SM	1.9002	3.7173	4.0723	4.5392	8.1793	9.0459	7.0445	8.1129	12.6054
CYC	2.2123	4.1655	4.7489	4.7373	8.4987	9.6654	7.1249	8.2499	12.7500

Table I Average costs per period of the different production strategies.

Although the SM-rule performs better in 6 out of 9 tests, it has the disadvantage of requiring considerable computational effort, just like in the determination of the optimal strategy, due to the large dimension of the state space. On the contrary, the average costs associated with the (x, T)-rule can easily be obtained which makes this rule especially useful in situations where the combination of different aspects has to be analysed.

6 The (x, T)-Rule in the Capacitated Case

Due to the good performance of the (x, T)-rule for the unconstrained lot-sizing problem, we describe in this section how to extend this production rule to situations where the available capacity is limited. When capacity limits are set upon the production size, we distinguish two situations. If $r_1 \geq x$ and the required deliveries for the first T periods do not exceed the available capacity C, we proceed like in the unconstrained case. However, in situations of capacity shortages we produce exactly C items and assume that production will not take place during periods for which the required deliveries that belong to C have been produced completely. This results in a considerable simplification of the analysis of the (x, T)-rule in these cases, since the changes in the values of $r_2, \ldots, r_T$ would be very difficult to describe if we did not proceed in this way.
Due to the fact that the values of $r_2, \ldots, r_T$ are influenced by the time elapsed since the last time the required deliveries for T periods were produced completely, we are interested in the regenerative points which take place at the end of a period in which the demand for T periods could be produced entirely. Denoting such an event by absorption, we consider for the state description the value of r_1 and also the number of periods passed since the last absorption. Therefore, a similar Markov chain to the unconstrained capacity case can be used to determine the average costs between two absorptions. The state space in the time direction is not limited but a limit can be set upon the r_1-component.

The possible transitions and their expected costs in the Markov chain are now more difficult to obtain and three different cases have to be considered. If i periods have elapsed since the last absorption we may enter a state (i, j_1) with $0 \leq j_1 \leq x-1$, a state (i, j_2) with $x \leq j_2 \leq C$ or a state $(i, C+j_3)$ with $1 \leq j_3 \leq M-C$ and M an upper bound for the value of r_1. In the first case we can move to any state $(i+1, j_1+k)$, $k \geq 0$ and only penalty costs are incurred. In the second case, absorption may occur if the demand for the next T periods does not exceed C. If we can only produce the demand for the next y periods completely ($y < T$), we make a "jump" to period $i+y$. The computation of both the expected holding costs and penalty costs associated with this jump involve many calculations. Finally, when we are in state $(i, C+j_3)$ only C items can be produced and j_3 items are left for the next period. As a result, we move to a state $(i+1, j_3+k)$, $k \geq 0$. The expected costs in this case include set-up costs and penalty costs for j_3 orders that will be delivered too late.
For a test problem with $N = 6$, $s = 50$, $h = 1$, $p = 3$ and $d_{i1} = 1 - d_{i0} = 0.4$, for $i = 1, \ldots, 6$, we present in Table II the average costs per period of the best (x, T)-rule for several capacity limits. Observe that as the available capacity becomes tighter, the average costs increase considerably and as it is expected production takes place more often.

C	18	12	9	8	7	6	5	4	3
x	6	5	5	4	3	3	2	2	1
T	4	4	4	4	4	4	4	4	3
Costs	16.6298	16.9717	18.5231	19.3550	20.8826	23.0432	26.5371	31.9002	42.5636

Table II Average costs per period for different capacity limits.

7 Conclusions

The problem considered in this paper combines several aspects which are difficult to analyse together: production to order, uncertain demand and fixed lead times. The three heuristic solution methods proposed appear to perform well especially the (x, T)-rule by offering low average costs and being easy to implement. Furthermore, the (x, T)-rule is a flexible production strategy that can be extended to more complex situations with capacity constraints.

References

[1] N.P. Dellaert. Production to order. In *Lecture Notes in Economics and Mathematical Systems 333*, Springer-Verlag, Berlin, 1989.

[2] A. Odoni. On finding the maximal gain for markov decision processes. *Operations Research*, 17:857–860, 1969.

[3] E.A. Silver and H.C. Meal. A heuristic for selecting lot size requirements for the case of a deterministic time-varying demand rate and discrete opportunities for replenishment. *Production and Inventory Management*, 14:64–74, 1973.

LÖSUNG DES DYNAMISCHEN MEHRSTUFIGEN MEHRPRODUKTLOSGRÖẞENPROBLEMS MITTELS DER LAGRANGE-RELAXATION - PARAMETEREINSTELLUNGEN BEI DER SUBGRADIENTENOPTIMIERUNG

Matthias C. Derstroff, TU Braunschweig

Es wird ein Modell des kapazitierten dynamischen mehrstufigen Losgrößenproblems betrachtet[1]. Zur Lösung dieses Modells wurde von Tempelmeier und Derstroff[2] ein iteratives Verfahren vorgeschlagen, in dessen Verlauf eine untere und eine obere Schranke für den optimalen Zielfunktionswert des Modells bestimmt werden. Die untere Schranke wird mittels der Lagrange-Relaxation des Modells gewonnen. Zur Aktualisierung der Lagrange-Multiplikatoren wird das Verfahren der Subgradientenoptimierung eingesetzt. In diesem Vortrag werden die von auf Vorschläge von Polyak[3] und Shor[4] zurückgehenden Schemata eingesetzt und die hierzu in der Literatur vorgeschlagenen Parametereinstellungen im Hinblick auf die Entwicklung der Zielfunktion untersucht.

---

1 Zur Modellierung vgl. z.B. Billington, P.J., McClain, J. O. und Thomas, L.J.: "Mathematical Programming Approaches to Capacity-Constrained MRP Systems: Review and Reduction", Management Science 29 (1983) 10, S. 1126-1141; Tempelmeier, H.: "Material-Logistik - Grundlagen der Bedarfs- und Losgrößenplanung in PPS-Systemen", Springer-Verlag 1992, S. 203 ff.

2 Vgl. Tempelmeier, H. und Derstroff, M.C.: "Mehrstufige Mehrprodukt-Losgrößenplanung bei beschränkten Ressourcen und genereller Erzeugnisstruktur", OR-Spektrum 15 (1993) 2, S. 63-73.

3 Vgl. Polyak, B. T.: "Minimization of Unsmooth Functionals", Computational Mathematics and Mathematical Physics 9 (1969), S. 14-29

4 Vgl. Shor, N. Z.: "The Rate of Convergence of the Generalized Gradient Descent Method", Cybernetics 4 (1968) 3, S. 79-80

BEDIENTHEORETISCHE MODELLIERUNG EINER BLECHTEILFERTIGUNG ALS MITTEL DES MATERIALFLUSSCONTROLLINGS

Kai Furmans, Universität Karlsruhe
Rolf Soukal, RWTH Aachen

Die Planung und Steuerung eines Fertigungsbereiches, in dem viele verschiedene Produkte in mehreren Schritten auf eine große Zahl von verschiedenen Maschinen hergestellt werden, ist eine komplexe Aufgabe. Insbesondere der Zusammenhang zwischen vorhandener Kapazität, erwartetem Durchsatz und der hierfür notwendigen Durchlaufzeit, die gleichzeitig auch den Umlaufbestand bestimmt, ist häufig Gegenstand von Diskussionen. Deshalb wird vorgeschlagen, bedientheoretische Methoden einzusetzen, um ein akzeptables Niveau für Umlaufbestand und Durchlaufzeit unter Berücksichtigung der aktuellen Kapazitätsauslastung zu bestimmen.

Um die Verwendbarkeit dieses Ansatzes zu überprüfen, wurde eine Blechteilfertigung bei einem Automobilhersteller zwei Monate beobachtet. Während dieser Zeit durchliefen etwa 800 Aufträge die 35 Maschinen und Maschinengruppen in diesem Fertigungsabschnitt. Die Bearbeitungs- und die Wartezeiten an jeder Maschine wurden für jeden Auftrag aufgezeichnet. Dazu kamen Brach- und Störzeiten der Maschinen. Auf der Basis dieser Informationen wurde ein geschlossenes und ein offenes Bediensystemnetzwerk modelliert.

Bei der Modellierung mußte einigen Besonderheiten Rechnung getragen werden. Die Maschinen wurden in verschiedenen Schichtmodellen betrieben. Dies führt zu zusätzlichen Pufferzeiten, die sich zu den Wartezeiten addieren, die aus den zugrundeliegenden stochastischen Prozessen resultieren. Gleichzeitig wird dadurch die korrekte Berechnung des Auslastungsgrades erschwert. Ein weiteres Problem stellt die gelegentliche Aufteilung von großen Aufträgen dar, die dann parallel auf mehreren Maschinen bearbeitet werden.

Die Kennwerte für die bedientheoretischen Modelle wurden mit Hilfe des Programmpaketes PEPSY berechnet, das von G. Bolch und seiner Gruppe von der Universität Erlangen-Nürnberg zur Verfügung gestellt wurde. Hiermit konnten verschiedene Näherungsverfahren verwendet und verglichen werden. Die Analyse der Ergebnisse zeigt, daß Kennwerte für das Gesamtsystem mit hinreichender Genauigkeit bestimmt werden können. Bei Werten, die sich auf einzelne Maschinen beziehen, mußten zum Teil große Abweichungen festgestellt werden. Diese konnten zum Teil auf abweichende Abarbeitungsreihenfolgen zurückgeführt werden. Es zeigte sich jedoch auch, daß es notwendig ist, die Approximationsverfahren weiter zu verbessern, um auch beispielsweise mehrgipflige Bedienzeitverteilungen besser abzubilden.

DIE BESTIMMUNG (OPTIMALER) GANZZAHLIGER LÖSUNGEN EINDIMENSIONALER ZUSCHNEIDEPROBLEME AUF DER GRUNDLAGE EINES DEKOMPOSITIONSANSATZES

Thomas Gau, Technische Universität Braunschweig

Das (relaxierte) Standardmodell eindimensionaler Zuschneideprobleme

$$\begin{aligned} \min: \quad z &= \sum_{j=1}^{n} x_j \\ \sum_{j=1}^{n} a_{ij} x_j &\geq b_i \qquad i = 1, \ldots, m \\ x_j &\geq 0 \end{aligned}$$

ist mit dem spaltenerzeugenden Verfahren von GILMORE, GOMORY im allgemeinen auch für praxisrelevante Größenordnungen effizient lösbar. Da aber mit der Lösung des relaxierten Problems nicht notwendigerweise ganzzahlige Verwendungshäufigkeiten der Schnittmuster einhergehen, ist es u.U. notwendig, die geforderte Ganzzahligkeit durch Einsatz spezieller Verfahren oder heuristischer Regeln ex post sicherzustellen.

Es ist bekannt, daß die Differenz zwischen dem Zielfunktionswert z^{int} des ganzzahligen und dem Zielfunktionswert z^{rel} des relaxierten Problems i.d.R. minimal ist, d.h. es gilt (bis auf wenige Ausnahmen) die Beziehung $z^{int} - z^{rel} \leq 1$. Diese sogenannte Integer–Round–Up (IRU) Eigenschaft liefert allerdings keinen unmittelbaren Beitrag zur Konstruktion einer ganzzahligen Lösung mit $z^{int} = \lceil z^{rel} \rceil$.

Es wird ein (heuristisches) Verfahren vorgestellt, das ausgehend von der relaxierten Optimallösung eine ganzzahlige Lösung erzeugt. Es handelt sich bei dem verwendeten Ansatz um eine sukzessive Problemdekomposition, die sowohl eine Lösung des jeweiligen Zuschnittproblems als auch des resultierenden Bin-Packing (Rest-)Problems erfordert. Auf jeder Dekompositionsstufe erfolgt ein Test auf Optimalität der gefundenen Lösung (IRU-Eigenschaft).

Für alle getesteten 111 Beispielprobleme aus Literatur und Praxis war die mit Hilfe des Dekompositionsansatzes gefundenen Lösung optimal.

SYSTEMRÜSTUNGSPLANUNG BEI DER HALBAUTOMATISCHEN BESTÜCKUNG VON LEITERPLATTEN

Manfred Gronalt, Hans-Otto Günther, Franz Piller
Universität Wien

Die halbautomatische Bestückung von Leiterplatten mit elektronischen Bauelementen läßt sich wie folgt beschreiben: aufgrund eines vorgegebenen Bestückungsprogramms wird durch einen Lichtstrahl angezeigt, an welcher Stelle der Leiterplatte ein Bauteil aufgesetzt werden soll, während gleichzeitig der Entnahmebehälter mit dem betreffenden Bauteil freigegeben wird. Der eigentliche Bestückungsvorgang erfolgt hierbei manuell, die Steuerung des Lichtstrahls und die Freigabe der Bauteilebehälter hingegen computergesteuert. Obwohl die meisten Bauteile heute vollautomatisch montiert werden, sind halbautomatische Bestückungstechniken nach wie vor weit verbreitet. Sie werden überwiegend eingesetzt, um Bedarfsspitzen abzufangen und spezielle Bauteile zu montieren.

Das untersuchte Problem der Systemrüstungsplanung besteht darin, eine möglichst günstige Zuordnung von Bauteilen zu den Bauteilebehältern eines halbautomatischen Bestückungstisches zu finden. Hierbei sind einerseits die begrenzten Magazinkapazitäten und die erheblichen Umrüstzeiten, andererseits die terminliche Dringlichkeit und die jeweiligen Prozeßpläne der Fertigungsaufträge zu beachten. Dieses Planungsproblem kann als ganzzahliges lineares Optimierungsmodell dargestellt werden, jedoch kommen aus praktischer Sicht lediglich heuristische Lösungsverfahren in Frage.

Außerdem wird die Konzeption eines Simulationsmodells erläutert, in das die Systemrüstungsplanung als zentrales Modul eingebunden ist. Die Simulation bietet hierbei Entscheidungshilfe in zweierlei Hinsicht: zum einen kann aufgrund des simulierten Fertigungsablaufs die Kommissionierung der Bauteile rechtzeitig vor Montagebeginn vorgenommen werden, so daß sich die Stillstandszeiten erheblich reduzieren; zum anderen läßt sich abschätzen, wie viele Bestückungstische in Betrieb genommen werden sollen.

A NEW PROCEDURE FOR SOLVING THE MULTI-ITEM, SINGLE-LEVEL, LIMITED CAPACITY, LOT-SIZING AND SCHEDULING PROBLEM WITH SEQUENCE-DEPENDENT SET-UP COSTS

Tore Grünert, RWTH Aachen

We consider the problem of determining lot-sizes and a production schedule for different products which are produced on the same assembly line. It is required that the demand for each product in each period and the corresponding production times are known. The objective is to minimize the sum of sequence-dependent set-up costs and inventory costs. The problem belongs to the class NP, for which no optimal solution procedures exist that solve realistic-size problems.
Many heuristic procedures solve relaxations of the described problem, for instance the Dixon/Silver and Günther heuristic for the DLSP, the Lagrange-relaxation algorithm by Fleischmann for the CLSP.
Our algorithm is based on a combination of tabu search, a meta-heuristic developed by Glover, and Dijkstra's shortest path algorithm. We demonstrate how the implemented swapping techniques and the problem-specific modelling affect the computational effectiveness. Possible extensions of the basic algorithm will also be considered.

Case Study: Lot-Sizing and Scheduling

M. Hauth and H.-J. Vaterrodt
Lehrstuhl für ABWL und Unternehmensforschung, Universität Mannheim

Starting point of the case study is the production planning for a producer of diversely coloured flat glas. The glas is basically melted in a furnace out of the raw materials sand, lime and soda and then brought into the shape in question on production lines, the so-called Feeders. The different colours of the glass are obtained by mixing the raw materials with colour components. The shift from one colour campaign to the next lasts up to 5 days. Within each colour campaign, only products with the joint characteristic basin colour can be made. Besides, a product is identified by the imprinted pattern, the thickness of the glass, the measures as well as the way of packing. Each change of product causes an individual preparation time.

The production structure is based on a two-level production process. Each of the two levels requires the solution of the lot size and sequence planning. The aim is to minimize the preparation times and the inventory levels. Practically, with 500 products and 3 colour campaigns, it is quite unrealistic to perform production planning for campaigns and products at the same time. The hierarchical approach chosen structures the problem into a medium-term campaign planning and a short-term product planning.

In this case study, the upper level of the hierarchical approach is introduced. The level below together with the product planning following afterwards is taken into account by aggregated production coefficients.

For the upper level we tested three approaches. We started with a general Mixed-Integer Program that took into account change-over times. For the time horizon of 52 weeks, the implemented Branch & Bound runs longer than 24 hours.

We than moved to an Integer Model with change-over cost by a zero switch rule, which means if we produce one product in one week we produce it during the whole week. We solved the problem by using the approach of A. Gascon and R.C [2]. Leachman which we extended to deal with change-over cost [4].

In the last approach we reduce the dynamic problem to static one. The solution method used to define the colour campaigns is built up hierarchically. In a first step, the sequences and then the lot sizes are determined. The solution method is based on the procedure by C.M. Delporte and L.J. Thomas [1] [3].

We checked the results of the three approaches against each other. All approaches provided similar solutions in spite of the different assumptions and input parameters.

[1] Delporte, C. M., Thomas, L.J.: "Lot Sizing and Sequencing for *n* Products on One Facility", *Management Science*, No. 23, S. 1090 - 1079, 1977.

[2] Gascon A., Leachman, R.C.: "A Dynamic Programming Solution to the Dynamic, Multi-Item, Single-Machine Scheduling Problem", *Operations Research*, Vol. 36, No. 1, S. 50 - 55, 1988.

[3] Hauth, M.: "Simultane Losgrößen- und Reihenfolge-Planung für den statischen Fall", Diplomarbeit, Universität Mannheim, 1992.

[4] Kempf, Th.: "Simultane Losgrößen- und Reihenfolge-Planung für den dynamischen Fall", Diplomarbeit, Universität Mannheim, 1992.

ABSTIMMUNG VON PRODUKTIONS- UND TRANSPORTLOSGRÖSSEN ZWISCHEN LIEFERANT UND ABNEHMER

Christian Hofmann, Ludwig-Maximilians-Universität München

Die Abstimmung von Produktions- und Transportlosgrößen als Bestandteil der Koordination von Unternehmungen wird mittels eines stationären Losgrößenmodells abgebildet. Betrachtet werden die Losgrößenentscheidungen der Bereiche Lieferant, Transporteur und Abnehmer, bei welchen jeweils einstufige Prozesse unterstellt werden. Die Entscheidungsgrößen dieser drei Bereiche können einerseits simultan sowie andererseits sukzessiv bestimmt werden. Schließlich werden die beiden Alternativen unterschieden, bei welchen der Transporteur einem an ihn angrenzenden Bereich zugeordnet wird und die Losgrößenplanung für diese beiden Bereiche simultan erfolgt.

Für diese vier verschiedenen Arten der Koordination der Planung werden die Gesamtkosten aller drei Bereiche bestimmt und miteinander verglichen. Dabei wird angenommen, daß die Kosten des Lieferanten und des Transporteurs auf den Abnehmer überwälzt werden. Es ergibt sich eine Systematik hinsichtlich der Zusammensetzung der Gesamtkosten in Abhängigkeit von der Art der Koordination.

Mittels des gewählten Ansatzes können Zustände angegeben werden, für welche die Art der Koordination keine Rolle bzgl. der anfallenden Kosten spielt. Ferner ist eine Abschätzung möglich, bei welchen Zuständen es kostengünstiger ist, den Transporteur dem Abnehmer zuzuordnen als dem Lieferanten.

Zur Koordination dezentraler Werkstattsteuerungsregeln

Oliver Holthaus, Hans Ziegler
Universität Passau

Zusammenfassung: In diesem Beitrag wird ein Ansatz zur Koordination dezentral arbeitender Werkstattsteuerungsregeln präsentiert und mit Hilfe der Simulation analysiert. Die Koordinationsstrategie basiert auf vorausschauenden Informationen und einem zielorientierten Auftragsangebot-/-nachfragemechanismus. Die Ergebnisse der Simulation zeigen, daß im Vergleich zu den herkömmlichen Steuerungsregeln bessere Zielerreichungsgrade realisiert werden können.

Summary: In this paper a new coordination approach for decentralized job shop scheduling rules is presented and analyzed in a simulation study. The coordination is based on look ahead informations and contains a mechanism for demanding and supplying jobs. The simulation experiment shows that the performance of conventional scheduling rules is significantly improved using the coordination mechanism.

1. Einleitung

Bei einem M-stufigen Werkstattsteuerungsproblem ist ein Fertigungssystem mit M Maschinen und eine Menge von zu bearbeitenden Aufträgen gegeben. Jeder Auftrag besteht aus einer individuellen Sequenz von Arbeitsgängen (Operationen), wobei die technologische Reihenfolge der Arbeitsgänge (Maschinenfolge) und deren Dauern auftragsspezifisch vorgegeben sind. Zu bestimmen ist für jede der M Maschinen die zeitliche Reihenfolge (Auftragsfolge), in der die Aufträge zu bearbeiten sind, um eine vorgegebene Zielsetzung (z.B. *Minimierung von Durchlaufzeiten, Einhaltung von Terminen*) bestmöglich zu erfüllen.

Zur Ablaufplanung in Werkstattfertigungssystemen werden üblicherweise Prioritätsregeln als dezentrale Steuerungsregeln eingesetzt. Hierbei werden die zwischen den Maschinen bestehenden Interdependenzen nicht oder nur pauschal berücksichtigt, so daß für jede Maschine im allgemeinen eine gute Auftragsfolge ermittelt wird, aber die Menge dieser isoliert bestimmten Auftragsfolgen nicht zu einer guten Lösung für das Fertigungssystem insgesamt führen muß. In diesem Beitrag wird ein dezentraler Koordinationsmechanismus präsentiert und analysiert, der darauf ausgerichtet ist, eine am Gesamtoptimum orientierte Steuerung zu erreichen. Die Koordinationsregel basiert auf vorausschauenden Informationen und einem zielorientierten Auftragsangebot-/-nachfragemechanismus. Hierbei finden Interaktionen zwischen den Maschinen statt, wobei eine Maschine den von einer anderen Maschine als nächsten zu bearbeitenden Auftrag direkt vorgeben kann. Zunächst wird im Abschnitt 2 das zugrundeliegende Werkstattsteuerungsproblem kurz charakterisiert. Der Abschnitt 3 präsentiert einige traditionelle Steuerungsregeln sowie einen Ansatz zur Koordination dezentraler Werkstattsteuerungssysteme. Im Abschnitt 4 wird der Koordinationsansatz mit Hilfe der Simulation analysiert. Eine zusammenfassende Betrachtung erfolgt im Abschnitt 5.

2. Charakterisierung des Werkstattsteuerungsproblems

Für das zu lösende Werkstattsteuerungsproblem wird von folgenden Prämissen ausgegangen [1, 10]:

Die Aufträge treffen zeitlich kontinuierlich in dem Fertigungssystem ein, wobei zum Ankunftszeitpunkt die Maschinenfolge und die Bearbeitungszeiten eines Auftrags bekannt sind. Die Maschinenfolge eines Auftrags muß eingehalten werden. Wenn die Bearbeitung eines Auftrags auf einer Maschine begonnen wird, dann wird die Operation ohne Unterbrechung durchgeführt. Vorzeitige Beendigungen von Arbeitsgängen sind nicht möglich. Die Bearbeitungszeiten der einzelnen Operationen sind unabhängig von der Reihenfolge, in welcher die entsprechenden Aufträge von einer Maschine bearbeitet werden. Eventuelle Rüstzeiten sind in den Bearbeitungszeiten enthalten. Jeder Auftrag wird zu jedem Zeitpunkt von höchstens einer Maschine bearbeitet. Jede Maschine bearbeitet zu jedem Zeitpunkt höchstens einen Auftrag. Wartungsbedingte Unterbrechungen bzw. stochastische Störungen der Maschinen werden nicht berücksichtigt.

Weiterhin wird folgende Notation für Auftrags- und Maschinendaten vereinbart:

N: Anzahl der Aufträge [-]
M: Anzahl der Maschinen [-]
r_i: Ankunftszeitpunkt von Auftrag i [ZE] (ZE Zeiteinheit)
k_i: Anzahl der Bearbeitungen von Auftrag i [-]
p_{ij}: Bearbeitungszeit der j-ten Operation von Auftrag i [ZE]
d_i: vorgegebener Fertigstellungstermin von Auftrag i [ZE]
d_{ij}: vorgegebener Fertigstellungstermin der j-ten Operation von Auftrag i [ZE]
C_i: Fertigstellungszeitpunkt von Auftrag i [ZE]
F_i: Durchlaufzeit von Auftrag i [ZE]; $F_i = C_i - r_i$
$\overline{F}$: Mittlere Durchlaufzeit [ZE]; $\overline{F} = 1/N \sum_{i=1}^{N} F_i$
F_{max}: Maximale Durchlaufzeit [ZE]; $F_{max} = Max\{F_i | 1 \leq i \leq N\}$
T_i: Terminüberschreitung von Auftrag i [ZE]; $T_i = Max\{0, C_i - d_i\}$
$\%T$: Prozentualer Anteil der Aufträge mit Terminüberschreitung [%];
$\%T = 100\, NT/N$, NT = Anzahl der Aufträge mit Terminüberschreitung [-]
$\overline{T}$: Mittlere Terminüberschreitung [ZE]; $\overline{T} = 1/N \sum_{i=1}^{N} T_i$
T_{max}: Maximale Terminüberschreitung [ZE]; $T_{max} = Max\{T_i | 1 \leq i \leq N\}$

Die Aufgabe der Werkstattsteuerung besteht darin, für jede der M Maschinen eine Bearbeitungsreihenfolge für die Aufträge so festzulegen, daß ein möglichst schneller und stauungsfreier Ablauf des Fertigungsprozesses und eine möglichst termingerechte Fertigstellung der Aufträge erreicht wird. Dementsprechend wird versucht, die Werte der durchlaufzeitorientierten Zielfunktionen $\overline{F}$ und F_{max} sowie die Werte der terminorientierten Zielfunktionen $\%T$, $\overline{T}$ und T_{max} näherungsweise zu optimieren [6, 8, 10].

3. Steuerung und Koordination von Werkstattfertigungssystemen

3.1. Werkstattsteuerungsregeln

Bei der dezentralen Steuerung von Werkstattfertigungssystemen wird die Auftragsfolge für jede Maschine durch eine Prioritätsregel festgelegt. Jedem wartenden Auftrag i wird ein Prioritätswert Z_i^t zugewiesen, der die Dringlichkeit der Bearbeitung ausdrückt. Der Auftrag mit dem kleinsten Wert für Z_i^t wird als nächster bearbeitet. Ein Überblick über verschiedene Prioritätsregeln wird in [6] gegeben. Innerhalb der letzten drei Jahrzehnte wurden über hundert Steuerungsregeln vorgeschlagen und mit Hilfe von Simulationsmodellen ausführlich analysiert (z.B. [2, 3, 4, 5, 9, 11, 12]). Diese Simulationsstudien belegen, daß keine allgemeingültige Aussage getroffen werden kann, welche Prioritätsregel zur Steuerung eingesetzt werden soll, da die relativen Vorteilhaftigkeiten der einzelnen Steuerungsregeln bei einer durchlaufzeit- oder terminorientierten Zielsetzung durch die Parameter des Werkstattfertigungssystems wesentlich beeinflußt werden. Diese Parameter sind die Stärke des Auftragsstromes, der die Systemauslastung beeinflußt, und die Art der Vorgabe der Fertigstellungstermine für die Aufträge [2, 9].

Einige ausgewählte *lokale* und *globale* Prioritätsregeln werden stellvertretend dargestellt [6, 9]:

FIFO (first in, first out): $Z_i^t := r_i^m$. *SPT (shortest processing-time)*: $Z_i^t := p_{ij}$.

MOD (modified operation due-date): $Z_i^t := Max\{d_{ij}, t + p_{ij}\}$.

S/OPN (slack per number of operations remaining): Falls $s_i \geq 0$ $Z_i^t := s_i/(k_i - j + 1)$, sonst $Z_i^t := s_i * (k_i - j + 1)$.

COVERT (cost over time): $Z_i^t := -c_i/p_{ij}$, bei gleichen Quotienten entscheidet SPT.

Dabei bezeichnet m die Maschine, bei der die Prioritäten berechnet werden [-]; t den Zeitpunkt, zu dem die Prioritäten berechnet werden [ZE]; j den Index der Bearbeitung, auf die Auftrag i bei Maschine m wartet [-]; r_i^m den Zeitpunkt der Ankunft von Auftrag i bei Maschine m [ZE]; s_i den Schlupf von Auftrag i [ZE] ($s_i = d_i - t - \sum_{q=j}^{k_i} p_{iq}$); c_i die geschätzten *Terminüberschreitungskosten* von Auftrag i mit $c_i := 0$ falls $s_i \geq \sum_{q=j}^{k_i} w_q$, $c_i := 1$ falls $s_i < 0$ und $c_i := (\sum_{q=j}^{k_i} w_q - s_i)/\sum_{q=j}^{k_i} w_q$ falls $0 \leq s_i < \sum_{q=j}^{k_i} w_q$, wobei w_q die geschätzte (mittlere) Wartezeit einer Operation q angibt.

In den in der Literatur beschriebenen Simulationsstudien liefert SPT in der Regel die geringsten $\overline{F}$-Werte, geringe $\%T$-Werte aber sehr große F_{max}-Werte, während sich z.B. bei FIFO der Verlauf tendenziell entgegengesetzt darstellt. Bezüglich der Zielfunktionen $\overline{T}$ und T_{max} dominieren im allgemeinen die Steuerungsregeln, die Termininformationen bei der Berechnung der Prioritäten berücksichtigen, wie z.B. MOD, S/OPN und COVERT. Die publizierten Simulationsergebnisse sind häufig sehr verschieden, da jeweils bei den Auslegungen des Werkstattfertigungssystems unterschiedliche Ausprägungen der beeinflussenden Parametern zugrunde gelegt wurden.

3.2. Koordination von Werkstattsteuerungsregeln

Wird eine Prioritätsregel als dezentraler Steuerungsmechanismus eingesetzt, dann werden die zwischen den Maschinen bestehenden Interdependenzen nicht (lokale Prioritätsregel) oder nur pauschal (globale Prioritätsregel) berücksichtigt. Durch den Einsatz einer Prioritätsregel wird zwar für jede Maschine im allgemeinen eine gute Auftragsfolge ermittelt, die Menge dieser isoliert bestimmten Auftragsfolgen muß jedoch nicht zu einem guten Ablaufplan für das Fertigungssystem insgesamt führen. Um eine am Gesamtoptimum orientierte Steuerung zu erreichen, ist es notwendig, die dezentralen Steuerungen der Maschinen zielorientiert zu koordinieren. Wird neben einer Steuerungsregel ein geeigneter Koordinationsmechanismus eingesetzt, dann kann eine Maschine m die aktuelle Prioritätsregel der Maschine $\hat{m}$ vorübergehend außer Kraft setzen und abhängig von ihrem Zustand die Prioritäten der Aufträge in der Warteschlange von Maschine $\hat{m}$ vorgeben. Eine *Koordinationsregel* beschreibt, wann und in welcher Form ein Eingriff in den Arbeitsablauf einer Maschine erfolgt. Im folgenden wird als Koordinationsregel die *vorausschauende Auftragsnachfrage (VANF)* [7] dargestellt:

Wenn der Arbeitsvorrat einer Maschine m, der sich als Summe der Bearbeitungszeiten der bei dieser Maschine wartenden Aufträge und der Restbearbeitungszeit des Auftrags, der aktuell bearbeitet wird, ergibt, einen bestimmten Wert, die als Parameter vorzugebende *Arbeitsvorratsschranke*, unterschreitet, dann leitet Maschine m eine Auftragsnachfrage ein, indem sie von allen anderen Maschinen $\hat{m}$ ($\hat{m} = 1, \ldots, M$, $\hat{m} \neq m$) Auftragsangebote anfordert. Eine jede Maschine $\hat{m}$, die von Maschine m eine Auftragsnachfrage erhält, bietet der Maschine m diejenigen Aufträge an, die nach der Bearbeitung auf Maschine $\hat{m}$ von Maschine m bearbeitet werden müssen, sofern derartige Aufträge bei Maschine $\hat{m}$ vorhanden sind und Maschine $\hat{m}$ noch keine Bearbeitungsanweisung einer anderen Maschine vorliegen hat.

Hat Maschine m von allen anderen Maschinen die Auftragsangebote vorliegen, wählt sie den Auftrag aus, der zum frühestmöglichen Zeitpunkt, jedoch nicht vor dem Zeitpunkt der Beendigung der Bearbeitung des aktuellen Arbeitsvorrats, bei ihr eintreffen kann. Falls alle angebotenen Aufträge schon vor diesem Zeitpunkt eintreffen sollten, wird der Auftrag ausgewählt, der zum spätestmöglichen Zeitpunkt eintreffen kann. Nach der Auswahl des Auftrags wird die Maschine $\hat{m}$, bei welcher dieser Auftrag auf seine Bearbeitung wartet, angewiesen, diesen Auftrag nach Abschluß der aktuellen Bearbeitung als nächsten zu bearbeiten.

Wenn die Maschine $\hat{m}$ zum selben Zeitpunkt von verschiedenen Maschinen Bearbeitungsanweisungen empfängt, wählt sie eine zufällig aus und teilt der entsprechenden Maschine mit, daß der von ihr gewünschte Auftrag vorgezogen wird. Die anderen Maschinen erhalten eine Absage ihrer Bearbeitungsanweisungen. Falls einer nachfragenden Maschine m keine Aufträge angeboten werden können oder der von ihr angeforderte Auftrag auf Maschine $\hat{m}$ nicht vorgezogen wird und nach Ablauf einer bestimmten Zeitspanne (z.B. 10% der Mindestbearbeitungszeit eines Auftrags) keine neuen Aufträge bei ihr eingetroffen sind (Aufträge, deren erste Bearbeitung auf Maschine m stattfinden muß), leitet sie eine erneute Auftragsnachfrage ein.

4. Analyse des Koordinationsmechanismus

4.1. Auslegung des Simulationsmodells

Als Grundlage für die Simulationsexperimente diente ein 5-stufiges Werkstattfertigungssystem mit einem stochastischen Auftragsstrom, wobei die Zwischenankunftszeiten der Aufträge exponential-, die Anzahl der Operationen je Auftrag gleich- (aus dem Intervall [2,10]), die Maschinenfolgen gleich- und die Bearbeitungszeiten gleichverteilt (aus dem Intervall [10,20]) erzeugt wurden. Die Systemauslastung wurde mit Hilfe der Zwischenankunftszeiten variiert. Der Fertigstellungstermin d_i eines Auftrags i wurde proportional zur Gesamtbearbeitungszeit vorgegeben [2], d.h. $d_i := r_i + c\sum_{j=1}^{k_i} p_{ij}$, wobei durch die Variation des Terminfaktors c unterschiedliche Terminspannen eingestellt wurden. Die Daten, die im folgenden angegeben werden, stellen die Mittelwerte der jeweils in 10 Simulationsexperimenten berechneten Ergebnisse dar, wobei in einem Simulationsablauf 1200 Aufträge in das System eingesteuert und eine Einschwing- und Ausschwingphase von jeweils 100 Aufträgen zugrunde gelegt wurden.

4.2. Simulationsergebnisse

Die Simulationsergebnisse zeigen, daß durch den Einsatz der Koordinationsregel $\overline{F}$ bei allen analysierten Steuerungsregeln reduziert wird. Der Umfang der Verringerung von $\overline{F}$ und der Beeinflussung von F_{max} ist abhängig vom Kapazitätsauslastungsgrad und der Arbeitsvorratsschranke *AVS*. $\%T$, $\overline{T}$ und T_{max} werden in sehr vielen Fällen reduziert, wobei der Umfang der Verbesserung zusätzlich durch den Terminfaktor c beeinflußt wird. Im folgenden wird die Vorteilhaftigkeit des Koordinationsmechanismus in Abhängigkeit von *AVS* für die Steuerungsregeln SPT und COVERT bei einem Auslastungsgrad von 90% analysiert.

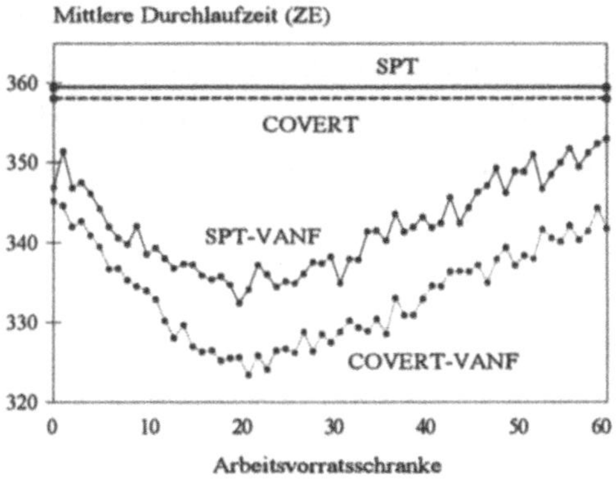

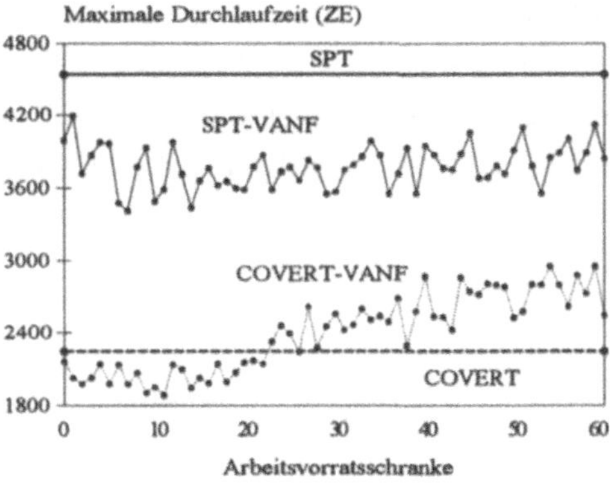

Abbildung 1: $\overline{F}$ und F_{max} in Abhängigkeit von *AVS*

In Abbildung 1 sind die Werte von $\overline{F}$ und F_{max} in Abhängigkeit von *AVS* dargestellt. *AVS* wird in Schritten von 1 ZE zwischen 0 ZE und 60 ZE variiert. Bei beiden Steuerungsregeln steigt die Verringerung von $\overline{F}$ zunächst mit zunehmender Arbeitsvorratsschranke bis zu einem Vorrausschauhorizont von ca. 10 ZE an, ist in dem Bereich zwischen 10 ZE und 30 ZE am Größten und sinkt ab 30 ZE tendenziell mit zunehmender Arbeitsvorratsschranke wieder ab. Während F_{max} bei SPT für jede *AVS* verbessert wird, tritt bei COVERT bis zu einem Vorausschauhorizont von ca. 22 ZE eine Verringerung des Zielwertes und ab diesem Zeitpunkt eine tendenziell zunehmende Vergrößerung bei steigender *AVS* ein.

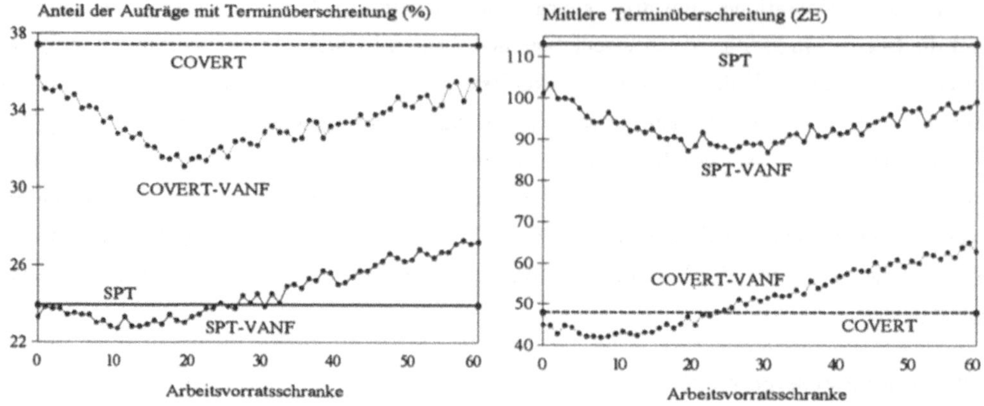

Abbildung 2: $\%T$ und $\overline{T}$ bei $c = 4$ in Abhängigkeit von AVS

In Abbildung 2 sind die Werte von $\%T$ und $\overline{T}$ bei $c = 4$ in Abhängigkeit von AVS dargestellt. Bei COVERT steigt die Verringerung von $\%T$ zunächst mit zunehmender AVS bis zu einem Vorrausschauhorizont von ca. 10 ZE an, ist in dem Bereich zwischen 10 ZE und 30 ZE am Größten und sinkt ab 30 ZE tendenziell mit zunehmender AVS wieder ab. Bei SPT sind mit einer AVS im Bereich zwischen 10 ZE und 20 ZE die stärksten Reduzierungen zu erzielen, wobei ab einer AVS von ca. 25 ZE $\%T$ mit zunehmender AVS tendenziell zunehmend vergrößert wird. Bei COVERT liegt der für die Reduzierung von $\overline{T}$ günstigste Vorausschauhorizont im Bereich zwischen 5 ZE und 15 ZE, während ab einer AVS von ca. 25 ZE $\overline{T}$ mit steigendem Vorausschauhorizont zunehmend verschlechtert wird. Im Gegensatz zu COVERT wird bei SPT $\overline{T}$ bei jeder AVS verbessert, wobei im Bereich zwischen 15 ZE und 35 ZE die stärksten Reduzierungen zu erzielen sind. Die Verläufe der Beeinflussung von T_{max} sind bei beiden Steuerungsregeln fast identisch mit den Verringerungen von F_{max}, vgl. dazu Abbildung 3.

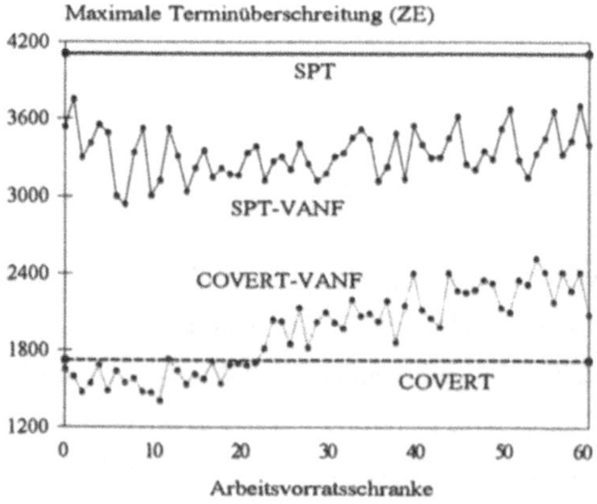

Abbildung 3: T_{max} bei $c = 4$ in Abhängigkeit von AVS

5. Zusammenfassung

In diesem Beitrag wurde ein Ansatz zur Koordination der dezentralen Steuerung mehrstufiger Werkstattfertigungssysteme präsentiert und mit Hilfe der Simulation analysiert. Der Einsatz des Koordinationsmechanismus VANF führt generell zu einer Verbesserung von $\overline{F}$ und in vielen Fällen zu einer Verringerung von F_{max}. Innerhalb eines bestimmten Bereichs für den Vorausschauhorizont, der für die Koordinationsregel zugrunde gelegt wird, können die stärksten Verbesserungen der Zielerreichung erzielt werden. Für die Reduzierung von $\%T$, $\overline{T}$ und T_{max} existieren ähnlich günstige Bereiche der Arbeitsvorratsschranke, die sich mit dem für die Verbesserung von $\overline{F}$ überschneiden. Der Vorausschauhorizont für den Koordinationsmechanismus kann also so eingestellt werden, daß alle fünf Zielfunktionswerte gleichzeitig verbessert werden.

Literatur

[1] Baker, K.R.: Introduction to Sequencing and Scheduling, John Wiley & Sons, New York 1974.

[2] Baker, K.R.: "Sequencing Rules and Due-Date Assignments in a Job Shop", *Management Science*, vol. 30, 1984, no. 9, pp. 1093-1104.

[3] Blackstone, J.H.; Phillips, D.T.; Hogg, G.L.: "A State-of-the-Art Survey of Dispatching Rules for Manufacturing Job Shop Operations", *International Journal of Production Research*, vol. 20, 1982, no. 1, pp. 27-45.

[4] Conway, R.W.: "Priority Dispatching and Work in Process Inventory in a Job Shop", *Journal of Industrial Engineering*, vol. 16, 1965, no. 2, pp. 123-130.

[5] Conway, R.W.: "Priority Dispatching and Job Lateness in a Job Shop", *Journal of Industrial Engineering*, vol. 16, 1965, no. 4, pp. 228-237.

[6] Haupt, R.: "A Survey of Priority Rule-Based Scheduling", *OR Spektrum*, 11. Jg., 1989, H. 1, S. 3-16.

[7] Holthaus, O.; Rosenberg, O.; Ziegler, H.: "Verteilte Simulation dezentraler Werkstattfertigungssysteme", *Information Management*, 8. Jg., 1993, H. 2, S. 6-12.

[8] Paulik, R.: Kostenorientierte Reihenfolgeplanung in der Werkstattfertigung, Haupt Verlag, Stuttgart 1984.

[9] Russell, R.S.; Dar-El, E.M.; Taylor, B.W.: "A comparative Analysis of the COVERT Job Sequencing Rule using various Shop Performance Measures", *International Journal of Production Research*, vol. 25, 1987, no. 10, pp. 1523-1540.

[10] Seelbach, H.: Ablaufplanung, Physica-Verlag, Würzburg 1975.

[11] Weeks, J.K.: "A Simulation Study of Predictable Due-Dates", *Management Science*, vol. 25, 1979, no. 4, pp. 363-373.

[12] Witte, T.: "Fallstudie zur Fertigungssteuerung mit Prioritätsregeln", in: Adam, D. (Hrsg.): Fertigungssteuerung II, Wiesbaden 1988, S. 107-120.

LOTSIZING AND SCHEDULING WITH SEQUENCE-DEPENDENT SETUPS VIA BATCH SEQUENCING

Carsten Jordan, Christian-Albrechts Universität Kiel
Andreas Drexl, Christian-Albrechts Universität Kiel

A single stage multi item capacitated dynamic lotsizing and scheduling model with setup-times and -costs is formulated and solved as a **batch sequencing model** (BSM). We consider the Discrete Lotsizing and Scheduling Problem with Setup Times (DLSPST, cf.[1]) which determines simultaneously the size and sequence of lots. We investigate the advantages and drawbacks of the BSM with respect to the DLSPST under modelling aspects and concerning solution algorithms.

We present a BSM (cf. [2], [3]), where lots consist of an integral number of demands, which are batched together. Demands are interpreted as jobs with a deadline, a certain processing time and earliness penalties. Analogously to the items the set of jobs is partitioned into jobclasses, and setups occur only between different jobclasses (=items). Earliness penalties of a job equal processing times multiplied with holding costs of its jobclass.

The corresponding scheduling problem now consists of scheduling the jobs between zero and their deadline on a single machine taking into account the (sequence-dependent) setup-times between jobclasses. Early completion of a job is weighted with its earliness-costs and setup-costs occur when switching from one jobclass (=item) to another. The objective is to minimize the sum of earliness- and setup-costs.

The relationship between both models is illustrated by two theorems, which are easily proven with exchange arguments:

Theorem 1 (cf. [2]): If there exists a feasible solution for the DLSPST (where a demand may be splitted into different lots), then there exists a feasible solution for the BSM (with unsplitted demand), too.

Theorem 2: In the case of identical holding costs for all items (w.l.o.g. $h_i = 1 \; \forall \; i$, so that earliness costs equal processing times) each schedule with splitted demand can be transformed into a schedule with unsplitted demand with less or equal costs. Thus there is also an optimal schedule with unsplitted demand.

The BSM is solved with an enumerative approach and compared with the dual ascent and column generation approach for the DLSPST. Therefore the DLSPST instances of [1], where holding costs are identical for all items are transformed into BSM instances. Computational results show, that the first approach outperforms the second one even for the special case of sequence-independent setups though the BSM accommodates also sequence-*dependent* setups.

References

[1] Cattrysse, D., M. Salomon, R. Kuik and L. N. van Wassenhove, 1993. A Dual Ascent and Column Generation Heuristic for the Discrete Lotsizing and Scheduling Problem with Setup Times, *Management Science*, Vol. 39, Nr. 4.

[2] Unal, A., and A. S. Kiran , 1992. Batch Sequencing, *IIE Transactions*, Vol. 24, pp.73-83.

[3] Bruno, J. and P. Downey, 1978. Complexity of task sequencing with deadlines, setup-times and changeover costs, *SIAM Journal on Computing*, Vol.7, pp. 393-404.

TOWARDS A BETTER UNDERSTANDING OF ORDER ACCEPTANCE

Hans A. ten Kate, University of Groningen, NL

This paper concerns a conceptual research on the comparison of two extreme approaches for order acceptance. It focusses on the interaction of order acceptance and production scheduling at an operational level. In the process industries, demand is often as high as or even higher than the available production capacity, while it is impossible to increase the available capacity. In combination with tendencies towards, on the one hand, an increased product variety and, on the other hand, shorter lead times, this has led to a situation in which order acceptance has more and more obtained a central position in production control.

As mentioned, the goal of this research is the comparison of two extreme approaches for order acceptance. In the first approach, a hierarchical approach, the decision whether or not to accept an order is based on aggregate characteristics of the set of already accepted orders. The production schedule is periodically updated for the orders accepted in the last period. Order acceptance and production scheduling take place on different levels of the planning hierarchy.

In the second approach, a so-called scheduling-oriented approach, order acceptance and production scheduling are integrated. The decision whether or not to accept the order is taken by determining a good production schedule which includes the new order. If this new schedule is acceptable, the order can be accepted. Obviously, in this approach order acceptance and production scheduling are on the same level in the planning hierarchy.

For a situation which is based on some of the main characteristics of process industries, both approaches for order acceptance are compared by means of a simulation model. Numerical results on a number of performance measures for both approaches will be presented for various scenarios.

PLANNED LEAD TIMES IN MRP, COST CONSIDERATIONS AND CUSTOMER SERVICE

Ton G. de Kok, Eindhoven University of Technology

In this presentation we discuss the problem of the determination of the planned lead time parameters in MRP systems. MRP systems are widely used in practice in job shop and mass assembly production systems. The MPP logic translates customer demand into time-phased demand for sub-assemblies and components. The time-phasing is based on the planned lead times of purchasing and sub-assembly. In most situations the planned lead times are estimated based on average and standard deviations of the actual lead time parameters. E.g. the planned lead time is set in such a way that the probability that the actual lead time is shorter than the planned lead time equals 90%. We present an overview of methods to arrive at the planned lead times.

In addition we present a model in which both Work-In-Progress costs and customer service criteria are taken into account to find an optimal set of planned lead times for a serial production system with stochastic lead times. Using the model we can find rules of thumb for setting planned lead times in most practical situations.

The analysis of the model turns out to allow for a full decomposition of the model. I.e. for an N-stage serial production system we have to recursively solve stage by stage an equation, which is similar to that arising in the classical Newsboy problem.

Since the equations turn out to be rather complicated expressions, we have developed an approximation scheme, which gives very accurate results in negligible time. An important problem is the determination of the actual lead time distribution. We will discuss practically useful results from queuing theory to arrive at these lead time distributions.

BEWERTUNG VON INVESTITIONSMASSNAHMEN AN EINER MONTAGELINIE - EIN ANWENDUNGSFALL

Heinrich Kuhn*, **Technische Universität Braunschweig**

Zusammenfassung: Betrachtet wird eine im Betrieb befindliche automatisierte Montagelinie. Ein Softwaresystem wurde entwickelt, welches hilft, Investitionsprojekte zur Verbesserung der Stationsverfügbarkeit zu bewerten. Die Bestimmung der Leistungsfähigkeit der Anlage erfolgte dabei anhand eines analytischen Modells.

Summary: The paper considers a real life automatic assembly line. A software system is developed which supports investment decisions of improvements in station reliability. The line efficiency is determined by an analytical model.

1. Problemstellung

Ein Unternehmen der Zulieferindustrie fertigt in großen Mengen und vielen Varianten komplexe Produkte (10.000 Stk/Tag) mit Hilfe einer elastisch verketteten, automatisierten Montagelinie. Elastisch verkettete Montagelinien zeichnen sich insbesondere dadurch aus, daß die einzelnen Stationen durch Zwischenpuffer entkoppelt sind. Folgewirkungen einer Maschinenstörung werden somit entschärft.

Während des Betriebs der betrachteten Anlage ergeben sich häufig konstruktive Verbesserungs- und Erweiterungsmöglichkeiten an den Anlagenkomponenten (z.B. an den Zuführungen für die Montageteile), die als eigenständige Investitionsprojekte betrachtet werden können. Im Rahmen der Bewertung dieser Investitionsprojekte ist es i.allg. unproblematisch, die Zunahme der technischen Komponentenverfügbarkeit zu spezifizieren und die hierfür notwendige Investitionssumme zu quantifizieren. Unklarheit existiert jedoch darüber, welcher Zusammenhang zwischen der Verfügbarkeitserhöhung an einer einzelnen Station und der resultierenden Leistungsverbesserung für die gesamte Montagelinie besteht. Erst nachdem diese Kenngröße (Zuwachs der Produktionsrate) für die Gesamtanlage bekannt ist, läßt sich der Zahlungsrückfluß der angestrebten Investition ermitteln und somit die ökonomische Vorteilhaftigkeit des Projekts verifizieren.

Zur Bestimmung der Leistungskennwerte der Gesamtanlage in Abhängigkeit von der Verfügbarkeitsverbesserung einer Einzelkomponente bieten sich zum einen die Methode der Simulation und zum anderen die Anwendung analytischer Modelle an [7]. Die praktische Untersuchung gliederte sich daher in zwei wesentliche Teilbereiche. Zum einen wurde für das oben beschriebene Montagesystem ein realitätsgenaues Simulationsmodell erstellt. Zum anderen wurden der Literatur zwei analytische Modelle entnommen und implementiert, deren Annahmen eine weitestgehende Übereinstimmung mit den realen Gegebenheiten versprachen. Das Simulationsmodell diente anschließend als Bewertungsinstrument, um die Qualität der analytischen Modelle zu überprüfen und deren Anwendbarkeit im Rahmen eines Investionsentscheidungssystems zu beurteilen.

Die weiteren Ausführungen sind wie folgt gegliedert: Zunächst (Kap. 2) wird eine explizite Beschreibung des Modells der untersuchten Montagelinie vorgenommen. Danach werden in Kapitel 3

* Der Autor dankt den Herren F. Blömer, M. Bürger, M. Geselle und S. Gstettner für die im Rahmen ihrer Studienarbeit geleistete Unterstützung zu diesem Beitrag.

die analytischen Modelle zur Leistungsbewertung erläutert. Die Analyse der Ergebnisqualität in bezug auf das untersuchte Montagesystem erfolgt in Kapitel 4. Anschließend (Kap. 5) wird ein im Rahmen des Projekts entwickeltes Softwaresystem zur Unterstützung der Investitionsentscheidung an der Montagelinie vorgestellt. Einige Schlußbemerkungen (Kap. 6) beenden den Beitrag.

2. Modellbeschreibung

Die untersuchte Montageanlage besteht aus zwei Montagekreisläufen mit insgesamt 22 automatisierten Montagestationen, in denen zusammen eine konstante Anzahl von ca. 270 Werkstückträgern zirkulieren. Die Datenaufnahme und eine erste Analyse führte zu der Erkenntnis, daß die gesamte Montageanlage als ein offenes, ununterbrochenes Fließsystem modelliert werden kann, welches einen strikt linearen Materialfluß aufweist (s. Abb. 1).

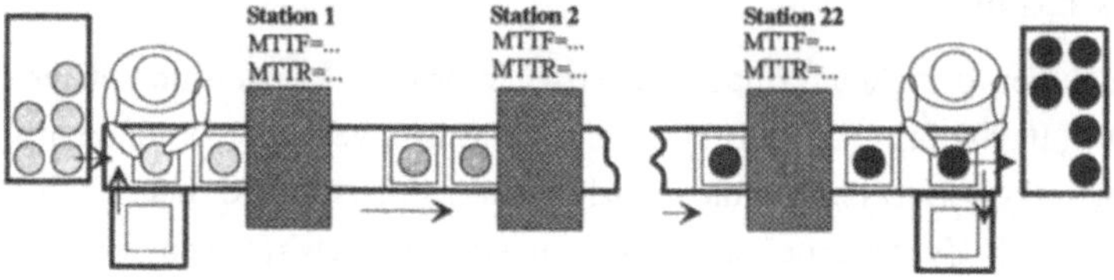

Abb. 1: Modell der untersuchten Montagelinie

Die weiteren Annahmen des Modells sind im folgenden in zusammengefaßter Form aufgelistet:

a) Die Fließlinie setzt sich aus M=22 Stationen zusammen.
b) An allen Stationen (m=1,2,...,M) besteht eine konstante Taktzeit von D=6 sek.
c) Alle Stationen produzieren nur intakte Teile (keine Berücksichtigung von fehlerhaften Teilen und Nachbearbeitungen).
d) Das Ausfallverhalten einer Station m wird anhand einer stationsspezifischen mittleren ausfallfreien Laufzeit beschrieben ($MTTF_m$, *Mean Time To Failure*). Die ausfallfreie Laufzeit wird als geometrischverteilt angenommen.
e) Unbeschäftigte Stationen können nicht ausfallen (nur *operationsbedingte* Ausfälle).
f) Ein Stationsausfall hat keinen qualitativen Einfluß auf das aktuell bearbeitete Teil (keine Zerstörung des Werkstücks bzw. Nacharbeit am Werkstück).
g) Die mittlere Zeit zur Behebung eines Stationsausfalls ($MTTR_m$, *Mean Time To Repair*) ist gegeben. Die Reparaturzeiten sind geometrischverteilt und unabhängig von dem gleichzeitigen Ausfall anderer Stationen.
h) Zwischen den Stationen besteht eine Pufferungsmöglichkeit von 7 Werkstücken (Z=7).
i) Ist der Ausgangspuffer einer Station voll belegt, dann kann ein fertiggestelltes Teil seine Station nicht verlassen (transfer bzw. production blocking). Die letzte Station (m=22) kann zu jeder Zeit verlassen werden (*never blocked*).
j) Die erste Station verfügt über einen unbegrenzten Arbeitsvorrat (*never starved*).
k) Die Transportzeiten zwischen den Stationen können vernachlässigt werden.

3. Analytische Modellierungsvarianten

Zur analytischen Leistungsanalyse einer Montagelinie unter den eben angeführten Annahmen bieten sich grundsätzlich zwei unterschiedliche Konzepte an, die im folgenden als diskreter bzw. kontinuierlicher Modellansatz bezeichnet werden. Bevor wir jedoch die beiden Konzepte genauer darstellen, wollen wir zwei Grenzwerte für das Leistungsverhalten einer Fließlinie beschreiben.

3.1 Grenzwerte der Leistungsfähigkeit einer getakteten Fließlinie

Einen wesentlichen Einfluß auf die Leistungsfähigkeit einer Fließlinie haben die *technischen Verfügbarkeiten* der einzelnen Stationen, die sich anhand der jeweiligen mittleren ausfallfreien Laufzeiten (*MTTF*) sowie der mittleren Reparaturzeiten (*MTTR*) bestimmen.

$$v_m = \frac{MTTF_m}{MTTF_m + MTTR_m} \qquad m=1,2,\ldots,M \qquad (1)$$

Neben der *technischen Verfügbarkeit* der einzelnen Stationen ist die Leistung einer Fließlinie erheblich von dem Zusammenspiel zwischen den Stationen abhängig und inwiefern die einzelnen Stationen voneinander entkoppelt sind, d.h. ob und wieviele Pufferplätze zwischen den Stationen bestehen. Hierbei gilt grundsätzlich, daß sich die Leistungsfähigkeit einer Anlage bei zunehmender Zwischenpuffergröße erhöht, da mögliche Blockiersituationen reduziert werden [2].

Eine *untere* bzw. *obere Schranke* für die *effektive Verfügbarkeit* einer *taktgebundenen* Fließlinie resultiert aus der Betrachtung eines Systems, welches keinerlei (Z=0) bzw. unendlich viele (Z=∞) Zwischenpufferplätze aufweist.

$$v(Z=0) = \frac{1}{1 + \sum_{m=1}^{M} \frac{MTTR_m}{MTTF_m}} \qquad (2)$$

$$v(Z=\infty) = \frac{1}{1 + \max_{m \in M} \frac{MTTR_m}{MTTF_m}} = \min_{m \in M} v_m \qquad (3)$$

Die Bestimmungsgleichung (2) unterstellt im Gegensatz zur Beziehung (3), daß nur ein Instandhalter zur Verfügung steht. D.h. die effektive Reparaturzeit einer Station ist von der Anzahl gleichzeitig ausgefallener Stationen abhängig. Weiterhin gilt die zuletzt genannte Bestimmungsgleichung nicht nur für *taktgebundene* Systeme, sondern ganz allgemein für alle Typen von Fließlinien, deren Stationszeiten identische Mittelwerte aufweisen. Grundsätzlich gelten beide Beziehungen für beliebige Ausfall- und Reparaturzeitverteilungen der Stationen.

Die eben angeführten Grenzwerte sind i.allg. soweit voneinander entfernt, daß sie zur Leistungsbestimmung einer Montagelinie keine ausreichende Lösungsgüte bieten. Insofern wurden in der Literatur analytische Modelle entwickelt, die explizit die Dimension der Zwischenpufferkapazität berücksichtigen. Im folgenden werden hierzu zwei Modellvarianten dargestellt, wobei sich die Betrachtungen zunächst auf Zwei-Stationen-Modelle beschränken.

3.2 Zwei-Stationen-Modelle

- **Modellierung auf der Basis diskreter Zeiten (Markov-Kette)**

Eine Zwei-Stationen-Fließlinie, die konstante Taktzeiten sowie diskrete geometrischverteilte ausfallfreie Lauf- und Reparaturzeiten aufweist, läßt sich mit Hilfe einer homogenen Markov-Kette exakt abbilden [1, 2]. Die Zustände der Markov-Kette sind (S_1, S_2, z). Sie geben an, ob sich eine Station aktuell in betriebsbereitem Zustand (S_m=W) oder in Reparatur befindet (S_m=R) und wieviele Pufferplätze gerade belegt sind (z=0,1,...,Z). Die Übergangswahrscheinlichkeiten zwischen den einzelnen Zuständen werden durch das Ausfall- und Reparaturverhalten der beiden Stationen bestimmt. Kennzeichen einer Markov-Kette ist es, daß für alle Knoten der Kette stationäre Zustandswahrscheinlichkeiten existieren, die ausschließlich anhand der vorhergehenden Zustände und der jeweiligen Übergangswahrscheinlichkeiten erklärt werden können. Insofern lassen sich, wenn die Übergangswahrscheinlichkeiten der Markov-Kette bekannt sind, alle stationären Zustandswahrscheinlichkeiten ermitteln. Zur Bestimmung der Leistungsfähigkeit (*effektive Verfügbarkeit*) der Gesamtanlage sind dann lediglich die outputrelevanten Zustände aufzusummieren.

- **Modellierung auf der Basis kontinuierlicher Zeiten (Warteschlangenmodell)**

Einen gänzlich anderen Ansatz erhält man, wenn man sich von den diskreten Zeitzuständen des Systems löst und die Operationszeit, die ausfallfreie Laufzeit sowie die Reparaturzeit als kontinuierliche Zufallsvariablen betrachtet. Unter diesen Bedingungen läßt sich für jede Station eine Belegungsdauer ermitteln, in der neben der reinen Operationszeit auch proportionale Reparaturzeiten enthalten sind. Man erhält somit anstatt der deterministischen Taktzeit eine Zufallsvariable (Belegungsdauer T_m), die in Abhängigkeit von dem Ausfall- und Reparaturverhalten eine gewisse Streuung aufweist und mit Hilfe ihres Erwartungswertes $E\{T_m\}$ und ihrer Varianz $V\{T_m\}$ beschrieben werden kann.

Die Beschreibung des Fertigungs-, Ausfall- und Reparaturprozesses einer Station mit Hilfe einer einzigen Zufallsvariable (Belegungsdauer) ermöglicht es, eine 2-Stationen-Fließlinie mit beschränkten Puffern als ein einstufiges GI/G/1/Z+2-Warteschlangensystem abzubilden [2]. Der Ankunftsprozeß bzw. Abfertigungsprozeß des Warteschlangensystems entspricht dabei dem Belegungsprozeß der ersten bzw. zweiten Station der Fließlinie. Die begrenzten Pufferplätze zwischen den Stationen werden durch den begrenzten Warte- und Bedienraum (Z+2) des Warteschlangensystems abgebildet. Im Anschluß an diesen Transformationsprozeß läßt sich dann mit Hilfe der aus der Literatur bekannten Approximationsgleichungen für die Auslastung des Bediensystems auf die *effektive Verfügbarkeit* der Zwei-Stationen-Fließlinie schließen.

3.3 M-Stationen-Modelle

Zur Bestimmung der Leistungskennwerte für eine M-Stationen-Fließlinie können die Ergebnisse der Zwei-Stationen-Systeme unmittelbar verwandt werden. Hierzu wird die Gesamtanlage (M-Stationen) in eine Menge von M-1 Zwei-Stationen-Systeme dekomponiert. Beginnend von der ersten Station werden dann die Parameter (Ausfall- und Reparaturzeiten bzw. Ankunfts- und Bedienraten) der Folgestationen derart modifiziert, daß sie das Leistungsverhalten des unmittelbar vorhergehenden 2-Stationen-Systems abbilden. Diese sukzessive Parametermodifikation wird solange fortgeführt, bis sich die abgeschätzte *effektive Verfügbarkeit* der Gesamtanlage in zwei aufeinander folgenden Iterationsschritten nur noch unwesentlich unterscheiden [2, 3, 4].

4. Ergebnisanalyse

Zur Überprüfung der Qualität der gewählten analytischen Modelle zur Bestimmung des Leistungsverhaltens der Montageanlage wurde mit Hilfe der Simulationssprache SIMAN [6] ein detailliertes Simulationsmodell implementiert. Die folgende Graphik zeigt die Ergebnisse (mittlere Produktionsrate in Stk/h), die mit Hilfe der beiden analytischen Modellansätze und der Simulation erzielt wurden, wobei die Zahl der Pufferplätze variiert worden ist.

In Abbildung 2 ist die gute Qualität des diskreten analytischen Modells deutlich erkennbar, wobei dies uneingeschränkt für alle Puffergrößen gilt. Mit dem kontinuierlichen Modell gelangt man demgegenüber, insbesondere bei wenigen Zwischenpufferplätzen, zu einer erheblichen Unterschätzung der Produktionsrate. Der schwankende Verlauf der Simulationsergebnisse gibt einen

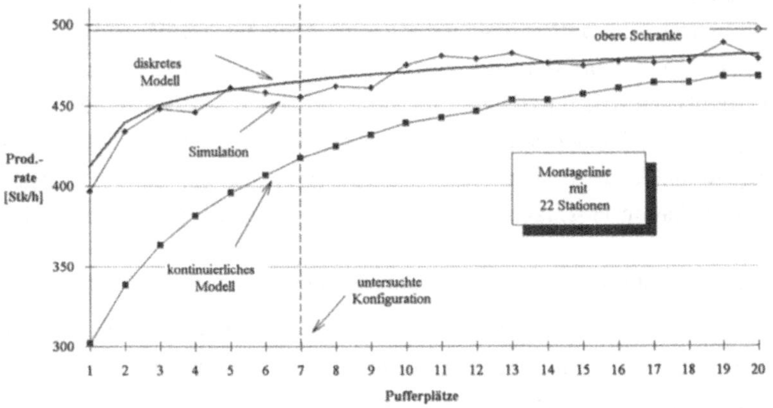

Abb. 2: Mittlere Produktionsrate in Abhängigkeit von der Anzahl der Zwischenpufferplätze

Hinweis darauf, daß die Simulationszeit (pro Konfiguration SIM-Zeit: 100.000 sek, Rechenzeit 486-33Mhz: 20 min) verlängert werden muß, um für eine einzelne Konfiguration einen statistisch abgesicherten Mittelwert zu erhalten. Insgesamt zeigen die dargestellten Ergebnisse, daß der diskrete Modellansatz bestens geeignet ist, um Leistungsveränderungen der Gesamtanlage aufzuzeigen, die aufgrund von technischen und/oder organisatorischen Maßnahmen zur Verbesserung der ausfallfreien Lauf- und Reparaturzeiten der einzelnen Stationen erzielt werden können. Im folgenden werden derartige Veränderungen der Stationsparameter vorgenommen und Leistungsveränderungen der Gesamtanlage anhand des diskreten analytischen Ansatzes bestimmt.

	Variante					
	(0 → 0-1: $MTTR_{11}/2$; 0 → 0-2: $MTTF_{22}$*10)		(0-2 → 2-1: $MTTR_{11}/2$; 0-2 → 2-3: $MTTF_{21}$*10)		(2-3 → 3-1: $MTTR_{11}/2$)	
Puffer	0 [Stk/h]	0-1 [%]	0-2 [%]	2-1 [%]	2-3 [%]	3-1 [%]
1	412	1.05	5.57	1.37	2.94	1.45
2	439	0.58	6.24	1.34	1.77	1.42
3	451	0.18	8.07	1.26	1.10	1.34
4	456	0.06	9.76	1.13	0.73	1.25
5	460	0.02	10.99	0.93	0.54	1.17
6	463	0.00	11.77	0.55	0.49	1.09
7	465	0.00	12.14	0.17	0.64	1.03
8	467	0.00	12.14	0.03	0.99	0.98
9	469	0.00	12.01	0.01	1.36	0.94
10	471	0.00	11.85	0.00	1.68	0.91

Tab. 1: Prozentuale Veränderung der Produktionsrate bei variierten mittleren ausfallfreien Lauf- bzw. Reparaturzeiten

Tabelle 1 zeigt die Wirkung von modifizierten mittleren ausfallfreien Lauf- und/oder Reparaturzeiten an den Stationen 11, 21 und 22, wobei die Zahl der Zwischenpufferplätze von 1 bis 10 variiert wurde. In der ersten Spalte (Variante 0) ist die erzielte Produktionsrate (Stk/h) der Ausgangskonfiguration angegeben. Die übrigen Varianten zeigen die prozentuale Leistungssteigerung gegenüber dieser Konfiguration bei veränderten Stationskennwerten. Die jeweilige Modifikation ist im Tabellenkopf angegeben. In Variante (0-2) ist z.B. die mittlere ausfallfreie Laufzeit der Station 22 um den Faktor 10 verlängert worden, während die Werte in Variante (2-1) die Leistungssteigerungen angeben, wenn darüberhinaus die mittlere Reparaturzeit der Station 11 halbiert wird.

Variante (0-2) zeigt einen erheblichen Leistungszuwachs der Gesamtanlage, da Station 22 den Engpaß des Montagesystems darstellt (vorher: $v_{22}=0.83$, nachher: $v_{22}=0.98$). Eine Reduktion der mittleren Reparaturzeit an Station 11 (Varianten (0-1), (2-1) u. (3-1); vorher: $MTTR_{11}=3.83$ min, nachher: $MTTR_{11}=1.94$ min) führt zu einer größeren Wirkung bei Konfigurationen mit wenigen Puffern zwischen den Stationen. Bei wenigen Zwischenpuffern und langer mittlerer Reparaturzeit besteht relativ schnell die Gefahr, daß der nachfolgenden Station das Arbeitsmaterial ausgeht (*Starving* der nächsten Station). Bei einer Taktzeit von $D=6$ sek besitzt eine Station mit sieben vorgelagerten Pufferplätzen einen maximalen Arbeitsvorrat von 42 sek. Weiterhin ist ersichtlich, daß sich der Erfolg einer verkürzten mittleren Reparaturzeit an Station 11 erhöht (s. Variante 3-1), wenn die Engpaßsituationen an den letzten beiden Stationen (St21 und St22) abgebaut werden. Verzögerungen an Station 11 schlagen dann eher bis zum Ende der Montageanlage durch.

5. Softwaresystem zur Unterstützung einer Investitionsentscheidung

Ziel des Projektes war es, dem Unternehmen ein Softwaresystem zur schnellen und hinreichend genauen Beurteilung von Verbesserungsmaßnahmen (Investitionsprojekten) an der bestehenden Montagelinie zur Verfügung zu stellen.

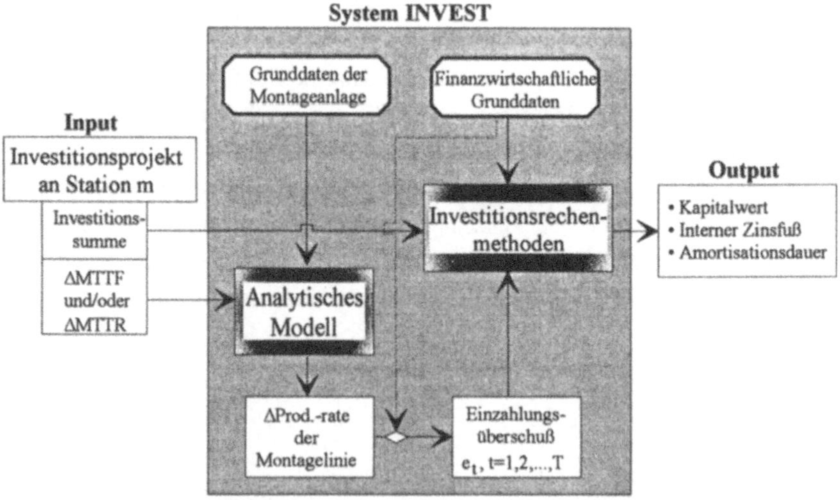

Abb. 3: Konzeption des Softwaresystems INVEST

Wesentliche Inputgröße dieses Systems sollte die Höhe der Investitionssumme sowie die mit dem Investitionsprojekt verbundene Verbesserung der technischen Komponentenverfügbarkeit sein (Verlängerung der *MTTF* und/oder Verkürzung der *MTTR* an einer Station). Diese Daten stehen i.allg. im Anschluß an einen Konstruktionsentwurf des Werkzeugbaus bzw. anhand eines Angebotes von einem Fremdlieferanten zur Verfügung. Gewünschte Outputgrößen des Systems sollten Kenngrößen sein, die einen Anhaltspunkt über die wirtschaftliche Vorteilhaftigkeit des angestrebten Investitionsprojektes geben.

Zur Lösung dieser Aufgabe unter den angesprochenen Prämissen wurde das oben als geeignet herausgestellte analytische Modell zusammen mit den Grunddaten der Montageanlage (Anzahl der

Stationen, Pufferkapazität etc.) in einem Softwaresystem zur Unterstützung von Investitionsentscheidungen (*INVEST*) eingebettet. Als Investitionsrechenmethoden wurden die Kapitalwertmethode, die Methode des internen Zinsfußes sowie die Bestimmung der Amortisationsdauer integriert [5] und unternehmensspezifische finanzwirtschaftliche Grunddaten (Abschreibungsart, Abschreibungsdauer, Kalkulationszinssatz, Steuersatz, Einzahlungsüberschuß je zusätzlicher Ausbringungseinheit etc.) in einer Datenbank abgelegt. Die Konzeption des Softwaresystems *INVEST* sowie die Beziehungen zwischen den Grundelementen des Systems zeigt Abbildung 3.

6. Schlußbetrachtungen

In dem beschriebenen praktischen Anwendungsfall konnte gezeigt werden, daß aus der Literatur bekannte, analytische Modelle geeignet sind, hinreichend genaue Abschätzungen für das Leistungsverhalten einer automatisierten Montagelinie zu ermitteln und als Bewertungsinstrument im Rahmen der angesprochenen Investitionsentscheidungen zu dienen.

Ansatzpunkte für weitere Forschungsanstrengungen bietet u.a. die Aufgabe, eine genauere Abbildung der Verteilung der Instandsetzungs- bzw. Reparaturzeiten in dem diskreten Modellansatz zu erreichen. Es ist zu vermuten, daß die Annahme von geometrischverteilten Instandsetzungszeiten deren wahre Streuung für viele Stationen der betrachteten Montagelinie erheblich überschätzt. Bei den Stationsstillständen handelt es sich i.allg. um Verklemmungen an den Zuführungen der Montageteile, so daß an der Anlage häufig eine ähnlich lange Instandsetzungszeit beobachtet werden kann.

Kontinuierliche Modellansätze, die das Ausfall- und Reparaturverhalten einer Station mit Hilfe einer modifizierten Bedienzeit (Belegungsdauer) abbilden, sind in der Lage, die Varianz der ausfallfreien Lauf- und Reparaturzeit abzubilden, jedoch führen diese Modelle aufgrund der mangelnden Abbildung der hier relevanten deterministischen Taktzeit zu erheblichen Fehlern. Diskrete Ansätze demgegenüber, die eine explizite Berücksichtigung der Varianz von Reparatur- bzw. Instandsetzungszeiten vornehmen, sind dem Autor jedoch nicht bekannt, so daß sich ein entsprechender Forschungsbedarf aufzeigt.

Ungeachtet dieser Einschränkungen leisten die bisher bekannten analytischen Modellansätze einen erheblichen Beitrag zur Leistungsanalyse von Fließsystemen, so daß vielfach auf i.allg. mit mehr Aufwand und Rechenzeit verbundene Simulationsexperimente verzichtet werden kann.

7. Literatur

[1] **Askin, Ronald G. und Standridge Charles R.**, *Modeling and Analysis of Manufacturing Systems*, New York (Wiley) 1993.

[2] **Buzacott, John A. und Shanthikumar, J. George**, *Stochastic Models of Manufacturing Systems*, Englewood Cliffs (Prentice Hall) 1992.

[3] **Dallery, Yves und Gershwin, Stanley B.**, Manufacturing flow line systems: A review of models and analytical results, in: *Queueing Systems*, 12(1992), S. 3-94.

[4] **Gershwin, Stanley B.**, An efficient decomposition method for the approximate evaluation of tandem queues with finite storage space and blocking, in: *Operations Research*, 35(1987)2, S. 291-305.

[5] **Kruschwitz, Lutz**, *Investitionsrechnung*, 4. bearb. Aufl., Berlin (de Gruyter) 1990.

[6] **Tempelmeier, Horst**, *Simulation mit SIMAN*, Berlin (Springer) 1991.

[7] **Tempelmeier, Horst und Kuhn, Heinrich**, *Flexible Manufacturing Systems: Decision Support for Design and Operation*, New York (Wiley) 1993.

NACHFRAGEUNABHÄNGIGER BANDABGLEICH BEI VARIANTENREICHER FLIESSFERTIGUNG

Claudius Link, SAP AG Walldorf
Hans-Joachim Vaterrodt, Universität Mannheim

Das Problem, alle für die Herstellung eines Produktes notwendigen Arbeitsgänge auf eine Menge von Arbeitsstationen so zu verteilen, daß die vorhandene Arbeitskapazität optimal ausgenutzt wird und die technologischen Restriktionen (Reihenfolgebedingungen zwischen den Arbeitsgängen, Zonenbeschränkungen, etc.) eingehalten werden, bezeichnet man als Fließbandabgleich (FBA). Der FBA für variantenreiche Fließfertigung (Herstellung von V Varianten eines Produktes auf derselben Fertigungslinie) wird bisher mit modifizierten Verfahren des Einproduktfalles durchgeführt. Grundlage für den FBA ist dabei der geplante Tagesbedarf D_V der Varianten V, womit eine Mischvariante bestimmt und ausgetaktet wird. Aufgrund dieser Vorgehensweise hat die Austaktung, durch schwankende Nachfragewerte $D_V(t)$, nur kurzfristige Gültigkeit, was zu Störungen im Produktionsprozeß führen kann. Es erscheint daher sinnvoll, einen mittelfristig gültigen Abgleich zu bestimmen, der unabhängig von den Nachfragewerten ist und alle Stationen für jede Variante möglichst gleichmäßig belastet. Durch eine gleichmäßige Belastung der Stationen wird die kurzfristige Reihenfolgeplanung weniger sensitiv, und die für eine gewünschte Produktionsmenge notwendige Taktzeit CT minimiert. Eine mögliche Zielfunktion, die dieses Ziel beschreibt, ist:

$$\max_{1 \leq n \leq N} \max_{1 \leq v \leq V} \left| \sum_i x_{in} * t_{iv} - w_v \right| \rightarrow \min ,$$

wobei

$x_{in} = 1$, wenn Arbeitsgang i der Station n zugeordnet wird und $x_{in} = 0$ andernfalls,
t_{iv} die Bearbeitungszeit von Vorgang i für die Variante v ist, und
$w_v = 1 / N * \sum_i t_{iv}$ das mittlere Arbeitsvolumen der Variante v.

Das oben beschriebene Problem kann mit einem Branch-and-Bound Verfahren gelöst werden. Durch Vorgabe eines Anspruchsniveaus für die größte Abweichung kann der Rechenzeitaufwand erheblich reduziert werden, so daß Probleme mit ca. 60 Arbeitsgängen auf einem 386/486-er PC innerhalb von ca. 30 Sekunden lösbar sind. Die an einem Praxisbeispiel durchgeführten Testrechnungen belegen die korrekte Auswahl der obigen Zielfunktion im Hinblick auf die reale Zielsetzung und dokumentieren die praktische Anwendbarkeit von Ansatz und Verfahren.

[1] Johnson, R.V.: *Optimally balancing large assembly lines with FABLE*,
Management Science, Vol 34 (1988), S. 240-253

[2] Rachamadugu,R., Talbot, B.: *Improving the equality of workload assignments in assembly lines*,
International Journal of Production Research Vol 29 (1991), S. 619-633

[3] Thomopoulos,N. T.: *Line balancing-sequencing of mixed-model assembly*,
Management Science, Vol 16 (1970), S.593-603

Ein Branch&Bound-Verfahren zur Abstimmung von Fließbändern bei gegebener Stationsanzahl

Armin Scholl, Darmstadt

Zusammenfassung: Im Zusammenhang mit der Einrichtung und dem Betrieb von Fließband-Fertigungssystemen ergeben sich verschiedene, miteinander verknüpfte Entscheidungsprobleme. Im Fall einer getakteten Fließfertigung handelt es sich dabei u.a. um **Probleme der Fließbandabstimmung**. Dieser Beitrag geht auf eine in der Literatur bislang wenig beachtete Variante eines Grundmodells der Fließbandabstimmung ein, die in der Minimierung der Taktzeit bei gegebener Anzahl an Arbeitsstationen besteht. Zur Lösung dieses Problems wird ein neuartiges Branch & Bound-Verfahren vorgeschlagen und mit bisher verfügbaren Lösungsansätzen verglichen.

Abstract: There are several decision problems in the context of installation and usage of assembly line production systems. In case of paced assembly **balancing problems** are of particular importance. This paper considers type 2 of the Simple Assembly Line Balancing Problem (SALBP): For a given number of stations, the cycle time has to be minimized. A new branch-and-bound algorithm is proposed and compared to existing solution procedures.

1. Problemstellung

Wir betrachten ein Grundmodell der Fließbandabstimmung (SALBP), bei dem eine zeitlich getaktete Fließfertigung unterstellt wird. Das Fertigungssystem besteht aus **m Stationen**, die gemäß der Reihenfolge von Bearbeitungsvorgängen entlang eines Fließbandes angeordnet sind. Es wird ein Produkt, dessen Produktionsprozeß in **n** elementare (unteilbare) **Arbeitsgänge** zerlegbar ist, hergestellt. Den Arbeitsgängen $j = 1,...,n$ sind deterministische, ganzzahlige **Bearbeitungszeiten** t_j zugeordnet. Zwischen ihnen bestehen zeitliche **Reihenfolgebeziehungen**, die sich in Form eines Vorranggraphen darstellen lassen. Dabei entsprechen die Knoten den Arbeitsgängen und die Pfeile den Reihenfolgebeziehungen.

Abb. 1 enthält einen Vorranggraphen, der einen Produktionsprozeß mit 11 Arbeitsgängen beschreibt, deren Bearbeitungszeiten als Knotenbewertungen dienen. Der Pfeil (1,4) drückt aus, daß Arbeitsgang 1 vor dem Beginn von Arbeitsgang 4 beendet sein muß.

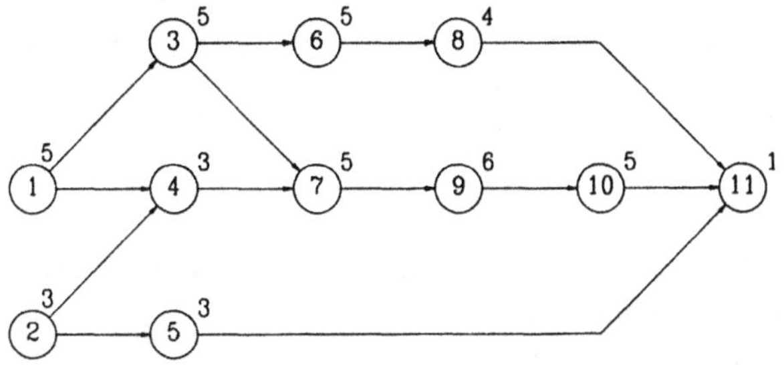

Abb. 1: Vorranggraph

Eine *Abstimmung* besteht darin, jeden Arbeitsgang unter Beachtung der Reihenfolgebeziehungen eindeutig einer Station zuzuordnen. Aufgrund des zeitlich getakteten Werkstücktransports auf dem Fließband dürfen die aus den jeweils zugeordneten Arbeitsgängen resultierenden *Belegungszeiten* der Stationen eine bestimmte **Taktzeit c** nicht überschreiten.

In der Literatur werden vorwiegend zwei Varianten dieses Grundmodells betrachtet:

Typ 1 (SALBP-1): Minimiere die Stationsanzahl m bei gegebener Taktzeit c

Typ 2 (SALBP-2): Minimiere die Taktzeit c bei gegebener Stationsanzahl m

Bei SALBP-1 ist die Produktionsrate (1/c) vorgegeben und ein Fließband-Fertigungssystem neu zu konfigurieren. SALBP-2 liegt vor, wenn vorhandene Systeme aufgrund veränderter Produktionsprozesse oder Nachfragesituationen neu abgestimmt werden müssen. Dabei wird durch Minimierung der Taktzeit (= maximale Belegungszeit einer Station) die Produktionsrate maximiert.

Zu SALBP-1 existiert in der Literatur eine Vielzahl exakter und heuristischer Lösungsverfahren, während für SALBP-2 nur wenige Lösungsansätze verfügbar sind, auf deren prinzipielle Vorgehensweisen in Abschnitt 2 eingegangen wird; vgl. Baybars (1986) sowie Domschke et al. (1993, Kap. 4).

2. Überblick über existierende Lösungsverfahren für SALBP-2

Für SALBP-2 wird in der Literatur überwiegend vorgeschlagen, SALBP-1 mit verschiedenen, jeweils vorzugebenden Taktzeiten zu lösen, um so die kleinste Taktzeit zu ermitteln, für die eine Lösung mit höchstens m Stationen existiert; vgl. z.B. Mansoor (1964), Baybars (1986) sowie Hackman et al. (1989). Dabei reduziert sich SALBP-1 jeweils auf ein *Zulässigkeitsproblem*, d.h. anstelle der Bestimmung der optimalen Stationsanzahl ist lediglich bei gegebener Taktzeit eine zulässige Lösung zu finden bzw. deren Nicht-Existenz zu beweisen. Dies läßt sich in den meisten exakten Verfahren für SALBP-1 durch zusätzliche Auslotregeln bzw. logische Tests auf einfache Weise berücksichtigen.

Derartige iterative Verfahren gehen aus von einem Intervall [LB, UB] möglicher ganzzahliger Taktzeiten. Die dabei verwendete *untere Schranke* $LB = LB_1$ ergibt sich gemäß (1) durch gleichmäßige Verteilung der Summe der Bearbeitungszeiten aller Arbeitsgänge (t_{sum}) auf die m Stationen unter Vernachlässigung ihrer Unteilbarkeit. Zusätzlich geht in LB_1 ein, daß die Taktzeit die größte Bearbeitungszeit eines Arbeitgangs (t_{max}) nicht unterschreiten darf. Weitere Möglichkeiten der Berechnung unterer Schranken werden in Abschnitt 3.3 angegeben.

$$LB_1 := \max \{ \lceil t_{sum}/m \rceil, t_{max} \} \tag{1}$$

Bei UB handelt es sich um eine *obere Schranke* für die Taktzeit, die entweder heuristisch oder theoretisch (z.B. gilt $UB \leq 2 \cdot LB$) bestimmt werden kann; vgl. Hackman et al. (1989).

In Abhängigkeit von der Suchreihenfolge innerhalb des Intervalls [LB, UB] lassen sich folgende grundlegende Vorgehensweisen unterscheiden:

LB-Methode: Beginnend mit $c = LB$, wird c so lange um 1 erhöht, bis eine zulässige Lösung für SALBP-1 mit höchstens m Stationen existiert.

UB-Methode: Beginnend mit $c = UB$, wird c so lange um 1 verringert, bis für $c-1$ keine zulässige Lösung für SALBP-1 mit höchstens m Stationen mehr existiert.

Binäre Suche: In jedem Iterationsschritt wird ein mittleres Intervallelement gewählt. Existiert für diese Taktzeit c eine zulässige Lösung mit höchstens m Stationen, so wird $UB := c$, ansonsten $LB := c+1$ gesetzt. Das Verfahren endet, wenn das Intervall einelementig ist.

Im Gegensatz zu obigen iterativen Ansätzen beschreiben Charlton und Death (1969) die Grundzüge eines (zu) allgemeinen B&B-Verfahrens, das auch für Flow und Job Shop-Probleme anwendbar ist.

3. Ein Branch & Bound - Verfahren für SALBP-2

In den folgenden Abschnitten werden die wichtigsten Komponenten eines speziellen B&B-Verfahrens für SALBP-2 kurz beschrieben; für eine ausführlichere Darstellung mit Beispielen siehe Scholl (1993).

3.1 Startschritt

Zur Bestimmung einer ersten oberen Schranke UB dient eine einfache **Eröffnungsheuristik**. Sie verwendet die in Abschnitt 2 beschriebene binäre Suche, wobei für jede zu untersuchende Taktzeit c ein *Prioritätsregelverfahren* für SALBP-1 eingesetzt wird. Diese Vorgehensweise wird mit verschiedenen Prioritätsregeln wiederholt, wobei folgende Sortierreihenfolgen betrachtet werden:

- monoton fallende Bearbeitungszeit
- monoton fallende Anzahl unmittelbarer Nachfolger
- monoton fallendes Positionsgewicht (Bearbeitungszeit des Arbeitsgangs und seiner Nachfolger)

Um zu einer möglichst verschärften oberen Schranke zu gelangen, wird anschließend ein **Verbesserungsverfahren** mit einer statischen **Tabu Search-Strategie** angewendet. Dabei werden unter Beachtung der Reihenfolgebeziehungen Arbeitsgänge aus Stationen ohne Leerzeit *(kritische Stationen)* in andere Stationen verschoben oder mit Arbeitsgängen aus anderen Stationen vertauscht. Da die Lösungsgüte des Verfahrens sehr stark von der verwendeten festen Länge der Tabuliste abhängt, wird es mit verschiedenen Listenlängen durchgeführt; vgl. Scholl und Voß (1993).

3.2 Teilprobleme und ihre Verzweigung

Teilprobleme P_μ lassen sich durch einen Vektor $z = (z_1, ..., z_n)$ beschreiben. Dabei gibt $z_j > 0$ die Nummer derjenigen Station an, der Arbeitsgang j zugeordnet ist. Wegen der Reihenfolgebeziehungen muß $z_i \leq z_j$ für alle Pfeile (i,j) gelten. Falls $z_j = 0$, besteht noch keine Zuordnung für j.

Für die Beschreibung der Verzweigung von Problemen benötigen wir einige weitere Bezeichungen:

$\mathcal{V}_j^m$ bzw. $\mathcal{N}_j^m$	Menge mittel- und unmittelbarer Vorgänger bzw. Nachfolger von Arbeitsgang j
$t(S_k) := \sum_{j \in S_k} t_j$	Belegungszeit der Station k = 1,...,m mit Belegung $S_k = \{j \mid z_j = k\}$

Die folgenden Größen beziehen sich auf eine Höchsttaktzeit (aktuelle obere Schranke) c:

$$E_j(c) := \max \{1, \lceil (t_j + \sum_{h \in \mathcal{V}_j^m} t_h)/c \rceil \} \qquad \textbf{früheste Station} \text{ für Arbeitsgang j}$$

$$L_j(c) := m - \max \{0, \lceil (t_j + \sum_{h \in \mathcal{N}_j^m} t_h)/c \rceil - 1\} \qquad \textbf{späteste Station} \text{ für Arbeitsgang j}$$

Bei der Bestimmung frühester Stationen $E_j(c)$ wird berücksichtigt, daß Arbeitsgang j frühestens dann begonnen werden kann, wenn alle seine Vorgänger in derselben oder einer vorhergehenden Station durchgeführt worden sind. Analog zu LB_1 in Formel (1) handelt es sich hierbei um eine einfache untere Schranke (von SALBP-1) für den Stationsbedarf des Arbeitsgangs j und seiner Vorgänger. Dementsprechend ergibt sich $L_j(c)$ bei Betrachtung von j und dessen Nachfolgern.

Beispiel: Für eine Höchsttaktzeit von c = 10 erhält man für das Problem aus Abb. 1 mit m = 5:

j	1	2	3	4	5	6	7	8	9	10	11
$E_j(c)$	1	1	1	2	1	2	3	2	3	4	5
$L_j(c)$	2	3	2	4	5	5	4	5	4	5	5

In der Regel lassen sich schärfere Werte für $E_j(c)$ bzw. $L_j(c)$ durch Einsatz weiterer unterer Schranken von SALBP-1 für den Stationsbedarf von j sowie Vorgängern bzw. Nachfolgern erzielen. Derartige Schranken berücksichtigen sowohl die Größe von Bearbeitungszeiten als auch die Reihenfolgebeziehungen; vgl. dazu Johnson (1988), Domschke et al. (1993, Kap. 4.3.3.2) sowie Scholl (1993).

Mit Hilfe obiger Größen kann die Menge $B_k(c)$ der bei einer Höchsttaktzeit c in Station k *einplanbaren* Arbeitsgänge definiert werden: $B_k(c) := \{ j \mid k \in [E_j(c), L_j(c)] \wedge t(S_k) + t_j \leq c \wedge z_j = 0 \}$

Eine Station heißt *geschlossen*, wenn die Menge $B_k(c)$ leer ist, ansonsten heißt sie *offen*.

Bei der Bestimmung frühester und spätester Stationen wird stets von der Höchsttaktzeit $\bar{c} := UB - 1$ einer gegenüber der aktuellen oberen Schranke UB verbesserten Lösung ausgegangen.

Im Ausgangsproblem P_0 ergibt sich damit $z_j = E_j(\bar{c})$ für alle j mit $E_j(\bar{c}) = L_j(\bar{c})$, ansonsten $z_j = 0$.

Die **Verzweigung** von Teilproblemen nach der *LIFO-Regel* (reine Tiefensuche) erfolgt durch Zuordnung *(Fixierung)* von Arbeitsgängen zu Stationen. Dabei werden zwei Strategien betrachtet:

1) Verzweigen nach Arbeitsgängen: Es wird in jedem Teilproblem ein Arbeitsgang j ausgewählt. Dabei wird eine der folgenden *Arbeitsgangregeln* oder eine Kombination angewendet:

- Bearbeitungszeitregel: Wähle ein j mit größter Bearbeitungszeit t_j.
- Vorrangregel: Wähle ein j, dessen sämtliche Vorgänger bzw. Nachfolger fixiert sind.
- Intervallregel: Wähle ein j mit kleinstem Intervall $[E_j(\bar{c}), L_j(\bar{c})]$.

Die Verzweigung in $L_j(\bar{c}) - E_j(\bar{c}) + 1$ Teilprobleme erfolgt durch Fixieren von j zu den Stationen in alternierender Folge: $k_j := \lfloor (E_j(\bar{c}) + L_j(\bar{c}))/2 \rfloor$, k_j+1, k_j-1, $k_j+2, \ldots, E_j(\bar{c}), L_j(\bar{c})$

2) Verzweigen nach Stationen: In jedem Teilproblem wird gemäß einer *bidirektionalen Strategie* entweder diejenige offene Station mit kleinstem Index kv (Vorwärtsschritt) oder diejenige mit größtem Index kr (Rückwärtsschritt) gewählt. Als Entscheidungskriterium wird eine mittlere Bearbeitungszeit $\bar{t}(k)$ von auf Station k einplanbaren oder fixierten Arbeitsgängen verwendet:

$$\bar{t}(k) := [t(S_k) + \sum_{j \in B_k(\bar{c})} (t_j / (L_j(\bar{c}) - E_j(\bar{c}) + 1))] \,/\, | S_k \cup B_k(\bar{c}) |$$

Im Fall $\bar{t}(kv) \geq \alpha \cdot \bar{t}(kr)$ wird kv, ansonsten kr gewählt. In P_0 ist $\alpha = 1$, ansonsten gilt $\alpha \geq 1$ ($\alpha \leq 1$), falls bei der vorhergehenden Verzweigung die letzte (erste) Station betrachtet wurde. Je stärker Parameter α von 1 abweicht, desto seltener ergibt sich ein Wechsel zwischen Vor- und Rückwärtsschritt.

Die Verzweigung erfolgt bei Wahl von kv durch Fixieren eines Arbeitsgangs $j \in B_{kv}(\bar{c})$, dessen sämtliche Vorgänger bereits fixiert sind. Bei Wahl von kr wird dementsprechend ein Arbeitsgang $j \in B_{kr}(\bar{c})$ mit vollständig fixierten Nachfolgern zugeordnet. Dabei können obige Arbeitsgangregeln zur Bestimmung der Betrachtungsreihenfolge derartiger Arbeitsgänge verwendet werden. Zusätzlich lassen sich Informationen aus Schrankenberechnungen einbeziehen (vgl. Abschnitt 3.3).

3.3 Untere Schranken

In diesem Abschnitt werden neben LB_1 aus (1) weitere Möglichkeiten zur Berechnung unterer Schranken $LB(P_\mu)$ für Teilprobleme P_μ angegeben. Es gilt $LB(P_\mu) \geq \max \{t(S_k) \mid k = 1,\ldots,m\}$.

- Eine Schranke $\mathbf{LB_2(P_\mu)}$ ergibt sich durch Ermittlung der kleinsten Taktzeit c, bei der eine untere Schranke für SALBP-1 die Stationsanzahl m nicht überschreitet.

- Durch Berechnung von Mindestleerzeiten für verschiedene Taktzeiten kann eine Schranke $\mathbf{LB_3(P_\mu)}$ gewonnen werden. Als *Leerzeit* einer Station in Abhängigkeit von c wird die Differenz $c-t(S_k)$ bezeichnet. Die maximale Summe von Leerzeiten ist $LZ(c) := m \cdot c - t_{sum}$.
  Eine Mindestleerzeit $ML_k(c)$ für Station k erhält man durch Berechnung maximaler Belegungszeiten. Dazu wird ein *Subset Sum-Problem* (Spezialfall des Knapsack-Problems) gelöst. Im Sinne des Knapsack-Problems werden die in Station k einplanbaren Arbeitsgänge $j \in B_k(c)$ als Gegenstände mit Gewicht t_j betrachtet. Es ist eine Kombination von Gegenständen zu wählen, deren Gewichtssumme das Maximalgewicht c möglichst gut ausnutzt. Die verbleibende Differenz entspricht einer Mindestleerzeit $ML_k(c)$. Verfahren für dieses NP-schwere, aber häufig effizient lösbare Problem beschreiben Martello und Toth (1990, Kap. 4).
  $LB_3(P_\mu)$ ergibt sich durch Ermittlung der kleinsten Taktzeit c mit $\Sigma_k\, ML_k(c) \leq LZ(c)$.
  Bei bidirektionaler Stationsauswahl (vgl. Abschnitt 3.2) ist es sinnvoll, Arbeitsgänge, die in einer maximalen Belegung für die betrachtete Station enthalten sind, bevorzugt auszuwählen.
- Bei Weglassen von Reihenfolgebeziehungen entspricht SALBP-2 einem Maschinenbelegungsproblem mit *identischen parallelen Maschinen* und der Zielsetzung der *Zykluszeitminimierung*; vgl. Domschke et al. (1993, Kap. 5.4.1). Durch Lösen dieses ebenfalls NP-schweren Problems erhält man eine Schranke $\mathbf{LB_4(P_\mu)}$. Jedoch ergibt sich zumeist ein viel zu hoher Rechenaufwand.
- Bei Unterteilung des Stationsintervalls [1,m] in Teilintervalle [1,i] und [i+1,m] mit $i = 1,\dots, m-1$ ergeben sich für eine Taktzeit c jeweils zwei aggregierte Stationen mit "Taktzeit" $i \cdot c$ bzw. $(m-i) \cdot c$. Sämtliche Arbeitsgänge j mit $L_j(c) \leq i$ können der ersten und solche mit $E_j(c) > i$ der zweiten "Station" zugeordnet werden. Für die restlichen Arbeitsgänge wird eine Aufteilung auf die beiden "Stationen" unter Beachtung der Restkapazitäten gesucht. Dies kann z.B. durch Lösen eines entsprechend modifizierten *Bin Packing-Problems* (vgl. Martello und Toth (1990, Kap. 8)) geschehen. Eine untere Schranke $\mathbf{LB_5(P_\mu)}$ ergibt sich dann aus der kleinsten Taktzeit c, bei der für alle $i = 1,\dots,m-1$ eine zulässige Aufteilung existiert.

3.4 Logische Tests

Zusätzlich zu unteren Schranken können **logische Tests** zum Ausloten (Ausschließen) von Teilproblemen P_μ oder zur Reduktion von Problemdaten dienen. Dabei lassen sich beispielsweise sämtliche $j \in B_k(\bar{c})$ in der dann geschlossenen Station k endgültig fixieren, wenn dadurch die Belegungszeit $t(S_k)$ die untere Schranke $LB(P_\mu)$ nicht überschreitet. Eine positive Differenz zwischen $LB(P_\mu)$ und $t(S_k)$ einer geschlossenen Station k ergibt eine Mindestleerzeit $ML_k(\bar{c})$. Gibt es nun eine offene Station, deren größtmögliche Belegungszeit die Mindestbelegungszeit $LB(P_\mu) - (LZ(\bar{c}) - \Sigma_k\, ML_k(\bar{c}))$ nicht überschreitet, so kann P_μ ausgelotet werden. Zu weiteren Tests vgl. Scholl (1993).

4. Test und Vergleich der Verfahren

Das in Abschnitt 3 beschriebene B&B-Verfahren wird in zwei Varianten (BB-1 und BB-2) mit zwei auf binärer Suche basierenden Verfahren verglichen. Dazu werden zwei der effizientesten B&B-Verfahren für SALBP-1, die Algorithmen FABLE von Johnson (1988) und EUREKA von Hoffmann (1992) in modifizierter Form verwendet; vgl. Abschnitt 2 sowie Hackman et al. (1989).

Typ2-FABLE: Vorwärtsgerichtetes Einplanen von Arbeitsgängen nach der Bearbeitungszeitregel in der Stationsreihenfolge 1,...,m; verschiedene Dominanztests und Schrankenberechnungen.

Typ2-EUREKA: Bildung von kompletten Stationsbelegungen, entweder vorwärts- *oder* rückwärtsgerichtet; falls dies nicht zum Erfolg kommt, wird zusätzlich eine Heuristik angewendet.

BB-1: Verzweigen nach Arbeitsgängen mit der Bearbeitungszeitregel und alternierender Stationsfolge; Ermittlung von LB_1 in allen Teilproblemen und von LB_3 in P_0; Einsatz logischer Tests sowie der Eröffnungsheuristik und des Verbesserungsverfahrens mit Tabu Search (Abbruch nach 1000 Iterationen ohne Lösungsverbesserung, Tabulistenlängen $(m+n)/3$, $(m+n)/3-3$,..., $n/4$).

BB-2: Verzweigen nach Stationen mit $\alpha = 6/5$ bzw. $5/6$; Bestimmung einer maximalen Belegung der gewählten Station k gemäß LB_3 bei zusätzlicher Einbeziehung des Vorranggraphen; bevorzugtes Fixieren darin enthaltener Arbeitsgänge nach Bearbeitungszeitregel; sonst wie BB-1.

Die Verfahren Typ2-FABLE, BB-1 und BB-2 wurden vom Autor in Pascal programmiert, während Typ2-EUREKA auf dem Fortran-Programm von Hoffmann (1992) basiert.

Zum Vergleich der vier Verfahren wird ein Datensatz mit 128 aus der Literatur zu SALBP-1 entnommenen und mit geeigneten Stationsanzahlen versehenen Problemen verwendet. Enthalten sind neben den schwierigsten Problemen aus einem Benchmark-Datensatz von Talbot und Patterson (1984) Daten aus Arbeiten von Buxey (1974), Lutz (1974) und Gunther et al. (1983). Der Test wird auf einem PC (80486-CPU, 66 MHz), bei einer Rechenzeitschranke von 500 CPU-Sekunden für jedes der 128 Probleme, durchgeführt. Für Typ 2-EUREKA wird die verfügbare Rechenzeit gleichmäßig auf vorwärts- und rückwärtsgerichtetes Vorgehen verteilt. Bei Überschreitung der Zeitschranke wird nur noch die enthaltene Heuristik zur Verringerung der *oberen* Schranke des Suchintervalls angewendet.

Die folgende Tabelle zeigt eine Übersicht über die Ergebnisse des Rechenexperimentes (für detailliertere Resultate siehe Scholl (1993)). Für jedes Problem werden die Arbeitsganganzahl n und die verwendeten Stationsanzahlen m sowie die daraus resultierende Anzahl an Probleminstanzen angegeben. Die 3 Eintragungen in den Verfahrensspalten entsprechen (1) der Anzahl optimal gelöster Probleminstanzen, (2) der Anzahl von Instanzen mit dem optimalen oder dem besten bekannten Zielfunktionswert (bei 27 Instanzen liefert kein Verfahren eine optimale Lösung) und (3) der Anzahl an Instanzen, die am schnellsten gelöst werden (evtl. sind mehrere Verfahren gleich schnell; Genauigkeit 0,05 Sek.).

Problem	n	m (Anzahl)	Typ2-FABLE	Typ2-EUREKA	BB-1	BB-2
Buxey	29	7 - 14 (8)	8 / 8 / 6	8 / 8 / 0	8 / 8 / 2	8 / 8 / 2
Sawyer	30	7 - 14 (8)	8 / 8 / 3	8 / 8 / 0	8 / 8 / 2	8 / 8 / 6
Lutz 1	32	8 - 12 (5)	5 / 5 / 5	5 / 5 / 0	5 / 5 / 1	5 / 5 / 1
Gunther	35	6 - 15 (10)	10 / 10 / 6	10 / 10 / 0	10 / 10 / 3	10 / 10 / 4
Kilbrid	45	3 - 11 (9)	9 / 9 / 2	9 / 9 / 5	9 / 9 / 8	9 / 9 / 8
Tonge	70	3 - 25 (23)	20 / 22 / 8	17 / 17 / 10	8 / 9 / 4	20 / 22 / 8
Arcus 1	83	3 - 22 (20)	4 / 4 / 0	5 / 7 / 3	2 / 4 / 3	14 / 19 / 18
Lutz 2	89	9 - 28 (20)	17 / 17 / 5	20 / 20 / 3	16 / 16 / 11	20 / 20 / 11
Arcus 2	111	3 - 27 (25)	5 / 13 / 9	4 / 5 / 1	3 / 11 / 8	6 / 17 / 16
Gesamt		(128)	86 / 96 / 44	86 / 89 / 22	69 / 80 / 42	100 / 118 / 74

Es zeigt sich, daß BB-2 für diesen Datensatz den anderen Verfahren bezüglich jeder der drei Kenngrößen überlegen ist. Es werden die meisten Probleme optimal gelöst (100 von 128) und die meisten besten Lösungen erzielt (118). In 74 Fällen benötigt BB-2 die kürzeste Rechenzeit. Etliche Instanzen (v.a. Kilbrid, Lutz 2) können bereits durch die Heuristiken optimal gelöst werden.

BB-1 schneidet, offenbar aufgrund seiner Verzweigung nach Arbeitsgängen, bezüglich der Lösungsgüte deutlich schlechter ab als die anderen, gemäß einer Stationsreihenfolge verzweigenden, Verfahren. Lösungen mit geringem Zeitbedarf beruhen überwiegend auf dem Einsatz der Heuristiken.

Die Unterschiede zwischen Typ2-FABLE und Typ2-EUREKA sind nicht sehr deutlich. Typ2-FABLE besitzt aufgrund seiner Auslotregeln Zeitvorteile und ist bei Tonge und Arcus2 überlegen. Typ2-EUREKA erzielt bei Arcus1 und Lutz2 wegen vorwärts- und rückwärtsgerichtetem Vorgehen bessere Resultate; außerdem liefert die Heuristik in einigen Fällen optimale Lösungen.

Insgesamt fällt auf, daß das Lösungsverhalten der Verfahren stark von der Problemstruktur abhängt. So kann entweder eine vorwärts- (Tonge, Arcus 2), eine rückwärtsgerichtete (Sawyer) oder eine bidirektionale Strategie (Arcus 1) günstiger sein. Daraus ergibt sich der Vorteil von BB-2, da dessen Verzweigung jede dieser Möglichkeiten bietet und flexibel mit Hilfe von Prioritätsregeln gesteuert wird.

5. Zusammenfassung und Ausblick

In diesem Beitrag wird ein neues B&B-Verfahren für SALBP-2 angegeben und mit bisherigen (indirekten) Ansätzen verglichen. Es ergibt sich, daß das Verfahren mit flexibler bidirektionaler Verzweigung nach Stationen zu den besten und mit arbeitsgangweiser Verzweigung zu deutlich schlechteren Resultaten führt. Darüber hinaus zeigt sich, daß bei vielen Probleminstanzen der Einsatz einer Tabu Search- Heuristik gute obere Schranken liefert. Demgegenüber ergeben weitere Tests, daß die Berechnung zusätzlicher unterer Schranken (vgl. Abschnitt 3.3) häufig unrentabel ist.

Daher erscheint es lohnenswert, weiteren Forschungsaufwand in die Verfeinerung von Verzweigungsstrategien und in die Weiterentwicklung von Heuristiken zu investieren.

Literatur

Baybars, I. (1986): A survey of exact algorithms for the simple assembly line balancing problem. Management Science **32**, 909 - 932.

Buxey, G.M. (1974): Assembly line balancing with multiple stations. Management Science **20**, 1010-1021.

Charlton, J.M. und C.C. Death (1969): A general method for machine scheduling. International Journal of Production Research **7**, 207 - 217.

Domschke, W.; A. Scholl und S. Voß (1993): Produktionsplanung - Ablauforganisatorische Aspekte. Springer, Berlin.

Gunther, R.E.; G.D. Johnson und R.S. Peterson (1983): Currently practiced formulations for the assembly line balance problem. Journal of Operations Management **3**, 209 - 221.

Hackman, S.T.; M.J. Magazine und T.S. Wee (1989): Fast, effective algorithms for simple assembly line balancing problems. Operations Research **37**, 916 - 924.

Hoffmann, T.R. (1992): Eureka: a hybrid system for assembly line balancing. Management Science **38**, 39 - 47.

Johnson, R.V. (1988): Optimally balancing large assembly lines with 'FABLE'. Management Science **34**, 240 - 253.

Lutz, L. (1974): Abtakten von Montagelinien. Krausskopf, Mainz.

Mansoor, E.M. (1964): Assembly line balancing - an improvement on the ranked positional weight technique. Journal of Industrial Engineering **15**, 73 - 77 und 322 - 323.

Martello, S. und P. Toth (1990): Knapsack problems. Wiley, New York.

Scholl, A. (1993): Ein B&B-Verfahren zur Abstimmung von Einprodukt-Fließbändern bei gegebener Stationsanzahl. Arbeitspapier, TH Darmstadt, Fachgebiet Operations Research.

Scholl, A. und S. Voß (1993): Maximizing the production rate in simple assembly line balancing: a tabu search approach. Arbeitspapier, TH Darmstadt, Fachgebiet Operations Research.

Talbot, F.B. und J.H. Patterson (1984): An integer programming algorithm with network cuts for solving the assembly line balancing problem. Management Science **30**, 85 - 99.

OPTIMIERUNG DES VERSCHNITTES UND DER ANZAHL VON ALUMINIUMPROFILEN

Hartmut Schwab, Universität Kaiserslautern
Dietmar Schweigert, Universität Kaiserslautern

Ein Hersteller von LKW-Aufbauten verarbeitet Aluminiumprofile. Dazu befinden sich in seinem Lager verschiedene Lagerlängen. Bei der Verarbeitung der Profile und durch die Art des Zuschnittes entsteht Verschnitt. Das Problem besteht darin, den Verschnitt an Profilen möglichst gering zu halten und gleichzeitig eine möglichst geringe Anzahl verschiedener Lagerlängen in das Lager aufzunehmen. Diese beiden Ziele widersprechen sich. Durch eine große Anzahl verschiedener Lagerlängen kann der Verschnitt gesenkt werden. Werden umgekehrt nur sehr wenige Lagerlängen aufgenommen, so wird der Verschnitt größer. Als Eingabedaten für das Problem werden die von den Kunden bestellten Längen und die Anzahl der Bestellungen benötigt.
Modelliert wird das Problem durch ein kürzestes Wegeproblem mit mehrfacher Zielsetzung in einem Graphen. Die Kanten des Graphen tragen als Gewichte zweidimensionale Vektoren. Die Anzahl der Knoten entspricht der Anzahl der von den Kunden bestellten Längen. Der Graph enthält alle gerichteten Kanten von Knoten mit höherer Nummer zu Knoten mit niedrigerer Nummer. Die speziellen Eigenschaften des Graphen können ausgenutzt werden, um einen Algorithmus zu entwickeln, der für eine bestimmte Anzahl von Lagerlängen den minimalen Verschnitt berechnet.
Am Ende des Referats wird ein Spezialfall betrachtet. Für ein speziell vorgegebenes Bestellverhalten der Kunden können mit den Hilfsmitteln der Analysis die Optimallösungen für verschiedene Anzahlen von Lagerlängen berechnet werden. Es zeigt sich dabei, daß die Optimallösungen auf einer Hyperbel liegen.
Auch an den Problemen mit echten Bestelldaten ließ sich ein ähnliches Verhalten feststellen. In diesen Fällen liegen die Optimallösungen für verschiedene Anzahl von Lagerlängen näherungsweise auf einer Hyperbel.

TABU SEARCH IN MANUFACTURING

Stefan Voß, Darmstadt

Tabu Search is a metastrategy for guiding known heuristics to overcome local optimality. Although still in its infancy, during the last few years successful applications of this kind of metaheuristic to a great variety of problems have been reported in the literature.

Several topics arising in automated manufacturing, the planning and scheduling of manufacturing systems, and group technology, just to mention some of the areas within manufacturing, may be formulated as optimization problems. In most cases these problems are too difficult to be solved exactly within reasonable time so that heuristics come into play. If one is not only content to settle for a feasible solution but for improved and hopefully near optimal solutions improvement procedures need to be investigated. In this paper we review recent developments with respect to tabu search applied to various problems arising in the area of manufacturing. This includes problems from, e.g., production planning and control and machine scheduling.

1. Introduction

Within the manufacturing domain the following problem areas are among the most versatile research fields: production planning and control, facility layout, inventory control, operations management as well as sequencing and scheduling. An immense number of the problems dealt with in these fields have been formulated as optimization problems and being dealt with exact as well as heuristic algorithms. Due to the complexity of a great variety of those optimization problems, heuristic algorithms are especially relevant when dealing with large scale instances. Among the most studied ideas are the application of priority based procedures including various dispatching rules (see e.g. Domschke, Scholl and Voß [17]).

Furthermore, a huge number of local search approaches have been developed to improve given feasible solutions. In this respect, the main drawback of algorithms such as deterministic exchange procedures is their inability to continue the search upon becoming trapped in a local optimum. This suggests consideration of recent techniques for guiding known heuristics to overcome local optimality.

Following this theme, we investigate the application of the *tabu search* metastrategy for solving optimization problems in manufacturing. In Section 2 we describe the basic ingredients of tabu search and explain it by means of a simple algorithmic evaluation. Some advanced aspects will be treated in Section 3. Most of the successful implementations of tabu search being reported in the literature are within the area of manufacturing, especially in machine scheduling. In Section 4 we shortly review the corresponding literature. Together with a large number of references this may help to explore tabu search for various problems not dealt with in this paper. Finally, some conclusions will be drawn.

2. Tabu Search: Basic Ingredients

Many solution approaches are characterized by identifying a neighbourhood of a given solution which contains other so-called *transformed* solutions that can be reached in a single iteration. A transition from a feasible solution to a transformed feasible solution is referred to as a *move* and may be described by a set of one or more *attributes*. For example, in a zero-one integer programming context these attributes may be the set of all possible value assignments or changes in such assignments for the binary variables. Then two attributes e and $\bar{e}$, which denote that a certain binary variable is set to 1 or 0, may be called *complementary* to each other. Following a steepest descent / mildest ascent approach, a move may either result in a best possible improvement or a least deterioration of the objective function value. Without additional control, however, such a process can cause a locally optimal solution to be re-visited immediately after moving to a neighbour, or in a future stage of the search process, respectively.

To prevent the search from endlessly cycling between the same solutions, tabu search may be visualized as follows. Imagine that the attributes of all moves are stored in a list, named a *running list*, representing the trajectory of solutions encountered. Then, related to a sublist of the running list a so-called *tabu list* may be defined. Based on certain restrictions, it keeps some moves, consisting of attributes complementary to those of the running list, which will be forbidden in at least one subsequent iteration because they might lead back to a previously visited solution. Thus, the tabu list restricts the search to a subset of admissible moves (consisting of admissible attributes or combinations of such attributes). This hopefully leads to "good" moves in each iteration without re-visiting solutions already encountered. A general outline of a tabu search procedure (for solving a minimization problem) may be described as follows:

Tabu Search

Given: A feasible solution x^* with objective function value z^*.
Start: Let $x := x^*$ with $z(x) = z^*$.
Iteration:
while stopping criterion is not fulfilled [1] **do begin**
(1) select best admissible move that transforms x into x' with objective function value z(x') and add its attributes to the running list
(2) perform tabu list management: compute moves to be set tabu, i.e., update the tabu list
(3) perform exchanges: $x := x'$, $z(x) = z(x')$
if $z(x) < z^*$ **then** $z^* := z(x)$, $x^* := x$ **endif**
endwhile
Result: x^* is the best of all determined solutions, with objective function value z^*.

* * *

For a background on tabu search and a number of references on successful applications of this metaheuristic see, e.g., Glover [22, 23, 24] as well as [27, 31, 32].

---

1 A possible stopping criterion can be, e.g., a prespecified time limit.

Tabu list management concerns updating the tabu list, i.e., deciding on how many and which moves have to be set tabu within any iteration of the search. We may distinguish two main approaches: static methods and dynamic methods. In the sequel we focus on a static concept, which up to now seems to be the most popular one that has been applied in the literature: the *tabu navigation method* (TNM). For dynamic concepts we refer to Section 3.1.

In TNM either complete moves or single attributes are set tabu as soon as their complements have been part of a selected move. The moves or attributes stay tabu for a distinct time, i.e. number of iterations, until the probability of causing a solution's re-visit is small. The efficiency of the algorithm depends on the choice of the tabu status duration, i.e. the length tl_size of the underlying tabu list. (In the literature often a "magic" tl_size=7 is proposed.) Although quite efficient at times, TNM seems to be a rather limited approach. Therefore, often researchers used a static approach with random variation, i.e., tl_size was randomly varied within some specific interval.

Recalling the use of the running list, the basic idea of tabu search is to keep sufficient information of the search within some memory. Considering the recent history of all moves or a subset of them, a general idea for reducing the computational effort in a tabu search algorithm is that of ***search intensification*** using a so-called *short term memory*. Its basic idea is to observe the attributes of all performed moves and to eliminate those from further consideration that have not been part of any solution generated during a given number of iterations. This results in a concentration of the search where the number of neighbourhood solutions in each iteration, and consequently the computational effort, decreases. Obviously the cost of this reduction can be a loss of accuracy.

Correspondingly, a ***search diversification*** strategy may be based on some *long term memory* to penalize often selected assignments. Then the neighbourhood search can be led into not yet explored regions where the tabu list operation is restarted (resulting in an increased computation time). The basic intention of search diversification is to explore regions of the search space that previously have not been investigated in depth, whereas search intensification tries to search within a limited space, e.g. in a promising region.

For the sake of an improved effectivity, a so-called ***aspiration level criterion*** may be considered, which permits the choice of an attribute even when it is tabu. This can be advantageous when a new best solution may be calculated, or when the tabu status of the attributes prevent any move from feasibility.

3. Advanced Issues

Most successful applications of tabu search reported in the literature so far apply TNM. Nevertheless, with this method cycling may still occur. To overcome this deficiency different concepts may apply. First of all in Section 3.1 we consider a modified tabu list management. Another idea is to incorporate some hashing function which can be implemented even in cases where moves are not represented in a proper way, i.e., they do not satisfy a sufficiency property (see [4, 70] for some references). In this respect, the choice of the neighbourhood becomes important (see Section 3.2). Finally, we shortly investigate the use of target analysis in Section 3.3.

3.1 Dynamic Tabu List Management

The static tabu list management concept, though successful in a large number of applications (compare Section 4), seems to be somehow limited. Another probably more fruitful idea is to define an attribute as being *potentially tabu* if it belongs to a chosen move and to handle it in a *candidate list* first. Via additional criteria these attributes can be definitely included in the tabu list if necessary, or excluded from the candidate list if possible. Therefore, the candidate list is an intermediate list between a running list and a tabu list. Glover [24] suggests the use of different candidate list strategies in order to avoid extensive computational effort without sacrificing solution quality. In the sequel, we describe the following dynamic strategies for managing tabu lists: the *cancellation sequence method* (CSM , in a revised version, cf. Dammeyer, Forst and Voß [12]), and the *reverse elimination method* (REM).

CSM as well as REM both use additional criteria for setting attributes tabu. The primary goal is to permit the reversion of any attribute but one between two solutions to prevent from re-visiting the older one. To find those *critical* moves, CSM needs a candidate list that contains the complements of attributes being potentially tabu. This *active tabu list* (ATL) is built like the running list where elimination of certain attributes is furthermore permitted. Whenever an attribute of the last performed move finds its complement on ATL this complement will be eliminated from ATL. All attributes between the cancelled one and its recently added complement build a *cancellation sequence* separating the actual solution from the solution that has been left by the move that contains the cancelled attribute. Any attribute but one of a cancellation sequence is allowed to be cancelled by future moves. This condition is sufficient but not necessary, as some aspects have to be taken into account so that CSM works well.

First of all we have to characterize a move in more detail. The above mentioned aspects work well for the case that a move consists of exactly one attribute, i.e., when so-called *single-attribute moves* are considered instead of *multi-attribute moves.* In addition, the corresponding parameters have to be chosen appropriately (e.g. the tabu list duration of a tabu attribute, and how to apply the aspiration level criterion). Applying CSM to multi-attribute moves needs additional criteria to prevent errors caused by uncovered special cases. For *paired-attribute moves* (moves consisting of exactly two attributes), for instance, those moves must be prohibited that may cancel a cancellation sequence consisting of exactly two attributes (because none of them is tabu when choosing a move).

In case of multi-attribute moves making a single attribute tabu prevents many moves which could lead to yet unvisited solutions. (An attribute becomes tabu if its complement is the only attribute of a cancellation sequence. It becomes tabu for one iteration if its complement is the most recent attribute of ATL. Otherwise a cancellation sequence could not be defined between these two attributes.) For building a cancellation sequence, the remaining attributes of the older and the current move are not necessarily taken into consideration. This depends on the order in which the move's attributes are added to ATL. Furthermore, those attributes of a move that did not cancel another attribute within a specific cancellation sequence are disregarded when making its last remaining attribute tabu (although they separate two solutions).

Whenever a cancellation sequence includes a smaller one the smaller sequence is said to *dominate* the larger. Then the larger cancellation sequence may be disregarded, because any of its attributes will only become tabu if they are within the smaller sequence, too.

The conditions of TNM and CSM need not be necessary to prevent from re-visiting previously encountered solutions. Necessity, however, can be achieved by REM. The idea of REM is that any solution can only be re-visited in the next iteration if it is a neighbour of the current solution. Therefore, in each iteration the running list will be traced back to determine all moves which have to be set tabu (since they would lead to an already explored solution). For this purpose, a *residual cancellation sequence* (RCS) is built up stepwise by tracing back the running list. In each step exactly one attribute is processed, from last to first. After initializing an empty RCS, only those attributes are added whose complements are not in the sequence. Otherwise, their complements in the RCS are eliminated (i.e. cancelled). Then at each tracing step it is known which attributes have to be reversed in order to turn the current solution back into one examined at an earlier iteration of the search. If the remaining attributes in the RCS can be reversed by exactly one move then this move is tabu in the next iteration. For single-attribute moves, for instance, the length of an RCS must be one to enforce a tabu move.

Obviously, the execution of REM represents a necessary and sufficient criterion to prevent re-visiting known solutions. Since the computational effort of REM increases if the number of iterations increases, ideas for reducing the number of computations have been developed (cf. [13, 24]). For an extensive example comparing TNM, CSM and REM the reader is referred to [16].

3.2 Connectivity

Very often in tabu search one of the most crucial aspects is not the method by itself but the neighbourhood under consideration. Usually any problem specific neighbourhood taking structural properties of the underlying problem into account seems to be more successful than a general purpose exchange algorithm. Consider, for example, the *quadratic assignment problem*, where the change in the objective function value of any move exchanging two objects simultaneously may be calculated very efficiently (see Section 4.5). Instead, shifting any object from a specific position to any other position looks like overkill (cf. [57]), i.e., making the evaluation of a move very time consuming.

From a theoretical point of view the following concept of *connectivity* seems to be quite obvious. It is said that a neighbourhood structure is *connected* if for any two feasible solutions π and π' solution π' is reachable from solution π by a (finite) sequence of moves. The concept of connectivity has been treated by several authors and experiments have shown that in general a "good" neighbourhood structure has to be connected (see e.g. [8, 15]).

With respect to the dynamic tabu list management strategies described above, a so-called *sufficiency property* has to be observed when defining moves or a neighbourhood structure, respectively. This property specifies that two solutions π and π' cannot be the same unless there is a sequence of moves connecting π and π' where the corresponding attributes are matching themselves such that each attribute is matched with its complementary attribute.

3.3 Target Analysis

Target analysis may be viewed as a general learning approach. Given a problem then, first of all, for a set of sample instances an extensive effort is made to obtain a solution which is optimal or close to optimality. The best solutions obtained will provide some *targets* to be sought within the next part of the approach. The tabu search algorithm under consideration will resolve the problems with the aim of finding what are the right choices to come to the already known solution (or as close to it as possible). This may give some information on how to choose the right parameters (like tl_size). For references on target analysis see, for instance, [38, 39, 42].

4. Applications

In the sequel we relate specific tabu search issues to various problems from different areas of manufacturing. In each subsection a specific problem is treated. The choice of the problems is by no means complete. Whenever possible additional references relevant to the field under consideration will be given.

4.1 Machine Scheduling and Sequencing

The remarkable success of tabu search has nowhere been more marked than in the area of machine scheduling and sequencing with an evermore growing importance in real-world manufacturing. Various methods have been employed yielding considerable improvements over previous approaches. In the sequel we survey some of the contributions in this area. For further references on some additional problems treated in the literature see [3, 42]. A general survey on scheduling and sequencing can be found e.g. in [17].

Note, that a large number of scheduling problems may be equivalently formulated as a traveling salesman problem (TSP) or a vehicle routing problem (VRP). Up to now the TSP seems to be among the most studied problems with respect to tabu search (see e.g. [20, 25, 45]). Tabu search ideas for the VRP can be found, for instance, in [21].

Flow Shop Scheduling: In a flow shop problem each of n jobs must be processed through m stages in the same order, and each job is processed exactly once at each stage. Usually, there is exactly one machine at each stage. The most common objective of flowshop scheduling used throughout the literature is that of minimizing the makespan, i.e., the maximum completion time C_{max}. Equivalently, C_{max} is the flow time of that job which is completed last and therefore the time required to complete all n jobs. Most research on this problem is concerned with finding a schedule with an identical processing order on all machines which is referred to as a *permutation schedule.*

Widmer and Hertz [67] provide the first successful tabu search implementation for $PF \mid \mid C_{max}$ (the permtation flow shop problem minimizing C_{max}). Their approach uses TNM (tl_size=7) with the neighbourhood definition being based on the swap strategy, i.e., in each iteration the best possible exchange of two jobs on their respective positions is performed. An interesting variant is used by Taillard [56]. He applys TNM with randomely varying tl_size values and defines a completely different kind of attribute, i.e., each move is represented by its objective function value instead of the jobs and their positions. Any swap move leading to an objective function value which is tabu is forbidden. Further references on tabu search implementations for PF | | C_{max} are [1, 47, 65].

Most prevalent future directions in flowshop research might be in the fields of flexible manufacturing systems as well as in applications in the chemical process industry. This leads to more general problems involving hybrid- or flexible flowshops as well as consideration of sequence-dependent setup times as well as buffer limitations between different stages of a flow line. For references on corresponding generalizations see, e.g., [14, 61, 63].

Job Shop Scheduling: More generally than for flow shop scheduling in a job shop environment each of n jobs must be processed through a subset of m stages probably in a different order. Again TNM has been applied within different settings especially concerning the definition of a good neighbourhood. The best known implementation known to date is that of Dell'Amico and Trubian [15] with additional references being [45, 55].

Again if there is more than one machine at each stage, the problem becomes more flexible. Respectively, the machines may be multi-purpose machines such that a combined routing and scheduling problem has to be solved (see [7, 34] as well as [66]).

Resource Constrained Scheduling: In resource constrained scheduling problems one attempts to find a schedule of jobs to optimize some performance criteria, while satisfying the constraints imposed by limited resources. Vaithayanathan [58] assumes a dynamic environment where the resources are subject to sudden unforeseen changes. The basic idea for a solution approach involves the decomposition of the problem into several multiconstraint zero-one knapsack problems (MCKP) each at a given instant in time. These problems are then solved by an approach combining neural networks and tabu search. For more efficient tabu search implementations for MCKP based on different versions of REM see [13, 62].

A flow shop problem (see above) with flexible resources is treated in [14].

An important subset of resource constrained scheduling problems are **Timetabling Problems.** Opposite to machine scheduling here usually a feasible schedule has to be found minimizing the number of conflicts. See [19, 28, 29, 30] for tabu search approaches for timetabling.

One Machine and Parallel Machine Scheduling Problems are treated, for instance, in [36, 37, 68, 69] and [2, 40], respectively. Scheduling of a flow-line manufacturing cell is discussed in [53, 59].

For **Packing Problems** (including bin packing) we refer to [5, 26, 33, 43, 44]. Some additional references on scheduling problems are [6, 46].

4.2 Assembly Line Balancing

An assembly line consists of a sequence of workstations performing a set of n nonpreemptive jobs with processing times t_j (j=1,...,n). These times are independent of the station on which the respective jobs are performed. Furthermore, due to technological restrictions precedence constraints between the different jobs have to be observed. In the simple assembly line balancing problem of type II (SALB-II, for short) the objective is to assign all jobs to a fixed number of workstations to maximize the production rate while observing the precedence constraints. Assume the alloted amount of time for any station being c (the cycle time) then the production rate equals 1/c, i.e., the problem is to assign all jobs to a fixed number of stations to minimize c.

Given a feasible solution we assume that all jobs are numbered topologically according to the precedence constraints. Obviously any move to a transformed solution has to be performed with respect to a *critical station*, i.e., any station of the given solution without any idle time. Then two kinds of moves may be defined. First, a job of a critical station may be shifted to a neighbouring station as long as no precedence constraints are violated. Second, two jobs of different stations may be exchanged without violating the precedence constraints, where at least one of the stations involved has to be a critical one.

Scholl and Voß [50] describe a static approach with different values of tl_size combined with a global aspiration criterion. The choice of an appropriate parameter tl_size seems to be quite sensitive with respect to SALB-II, whenever the initial feasible solution is far from optimality. If the initial feasible solution is close to optimality any value greater or equal to six seems to work well.

4.3 Lot Sizing

Lot sizing is among the most difficult tasks, especially when the production process consists of multiple levels and when there are finite production resources, i.e. capacities. Formally, we deal with the assignment of production resources to production activities such that agreed upon quantities of products are available for meeting external demand. In a multilevel capacitated lotsizing problem for assembly product structures (MLCLSP) the objective is to minimize the sum of setup and holding costs. Dependent as well as independent demands have to be fulfilled in any period either from inventory or from production in the current of T periods. Furthermore, in each period the production has to maintain within the limitations of the available capacity while backlogging is not allowed. The product structure is assumed to be assembly structure indicating that each product i is component of at most one product j.

The MLCLSP may be formulated as a mixed integer program with binary variables referring to setup decisions and continuous variables denoting the production quantities (for any product and period). The tabu search approach of Kuik et al. [35] considers problems with capacity constraints at exactly one level. It proceeds in that it involves a walk through the state space of feasible production schedules as defined by any "feasible" setup pattern. Given a setup pattern represented by a fixed number of setups, a neighbour setup pattern is any feasible pattern obtained by complementing one of the binary setup variables. From an arbitrarily chosen subset of neighbours the best one is chosen (involving the calculation of the objective function, i.e. fixing the production quantities within the capacity limitations and the setups) as long as it is not tabu. Within TNM a value tl_size=10 is assumed for a total of 500 iterations. Two versions are tested and compared to other heuristic procedures: first, the approach as sketched above and second, an approach based on an initial solution found by solving the LP-relaxation. In the latter approach the search space was further restricted due to a rounding process of some setup variables based on the LP-solution. This approach seems to give good results, however, it is slightly outperformed by a specific simulated annealing implementation of the same authors (see [35] as well as [48, 49]).

For further references in the lot sizing area see [41] and [69].

4.4 Flexible Manufacturing Systems

In flexible manufacturing systems planning problems are often decomposed into several hierarchical subproblems which are solved sequentially. In this respect, Srivastava and Chen [54] consider the part type selection problem. The problem is to determine a subset of part types having production requirements for immediate and simultaneous processing over the upcoming period of the planning horizon subject to the tool magazine and processing time limitation (compare the resource constrained scheduling problem under dynamic condition in Section 4.1). Four implementations of tabu search with different use of long term memory as well as different values tl_size were tested and compared against simulated annealing. The results show that a tl_size depending on the problem size outperformes simulated annealing as well as other tabu search implementations.

Another comparison of simulated annealing with tabu search addresses the flowpath design issue and in particular the design of unidirectional flowpaths of automated guided vehicle systems (see [11]). Within tabu search two different neighbourhood definitions are investigated.

4.5 Facility Layout

The *quadratic assignment problem* (QAP) is to find an assignment of a number of n objects to a number of n locations such that the cumulative product of flow between any two objects and distance between any two locations is minimized. More formally, we are given a distance matrix $(d_{ij})_{n\times n}$ describing the mutual distances between pairs of locations and a cost matrix $(c_{pq})_{n\times n}$ describing the pairwise connection or flow costs between pairs of objects. Then QAP is to find a permutation π of the set $\{1,\ldots,n\}$ such that $\Sigma_{i=1}^{n} \Sigma_{j=1}^{n} d_{ij} \cdot c_{\pi(i)\pi(j)}$ is minimized.

With respect to QAP nearly all different tabu search techniques have successfully been applied with the TNM algorithm of Skorin-Kapov [52] being among the first overall implementations of tabu search. Further references are [4, 9, 10, 20, 51, 57]. The currently best known results on a number of benchmark problems are reported in [64] using a clustering approach for diversification purpose together with REM.

With respect to, e.g., floor layout planning also the quadratic semi-assignment problem (QSAP) is of interest. This problem indeed is a relaxed version of QAP where one set of constraints is missing. For numerical results and comparisons for different versions of TNM, CSM and REM see [16, 60]. QSAP also arises within some scheduling problems with cost-oriented objective functions.

5. Conclusions

In this paper we have reviewed some basic principles of tabu search as well as some advanced ideas. Successful applications of this recent metastrategy within manufacturing are reported and an immense list of references is provided.

Although successful in most applications one drawback of tabu search sometimes mentioned especially when comparing it with other local search techniques is the lack of theoretical foundations. Of course, there are a few theoretical results that have to be taken into account. First of all this is the necessity and sufficiency of the reverse elimination method (see Section 3.1) as well as some modifications (cf. [24, 62]). In addition, for instance, there are some convergence results for a non-deterministic tabu search approach (cf. [18, 24]).

Finally, we should mention that a promising area of future research is in applying tabu search on parallel computers (see [62] for some basic concepts for parallel tabu search).

References

[1] B. Adenso-Diaz, Restricted neighborhood in the tabu search for the flowshop problem, *European Journal of Operational Research* **62** (1992) 27 - 37.

[2] J.W. Barnes and M. Laguna, Solving the multiple-machine weighted flow time problem using tabu search, *IIE Transactions* **25** (1993) 121 - 128.

[3] J.W. Barnes and M. Laguna, A tabu search experience in production scheduling, *Annals of Operations Research* **41** (1993) 141 - 156.

[4] A. Battiti and G. Tecchiolli, The reactive tabu search, Preprint, Univ. of Trento (1992).

[5] J. Blazewicz, P. Hawryluk and R. Walkowiak, Using a tabu search approach for solving the two-dimensional irregular cutting problem, *Annals of Operations Research* **41** (1993) 313 - 325.

[6] J. Bovet, C. Constantin and D. de Werra, A convoy scheduling problem, *Discrete Applied Mathematics* **30** (1991) 1 - 14.

[7] P. Brandimarte, Routing and scheduling in a flexible job shop by tabu search, *Annals of Operations Research* **41** (1993) 157 - 183.

[8] P. Brucker, J. Hurink and F. Werner, Improving neighborhoods for local search heuristics, Working Paper, Univ. Osnabrück (1993).

[9] J. Chakrapani and J. Skorin-Kapov, A connectionist approach to the quadratic assignment problem, *Computers & Operations Research* **19** (1992) 287 - 295.

[10] J. Chakrapani and J. Skorin-Kapov, Massively parallel tabu search for the quadratic assignment problem, *Annals of Operations Research* **41** (1993) 327 - 341.

[11] W.-C. Chiang, G.J. Gutierrez and P. Kouvelis, Simulated annealing and tabu search, in: A. Kusiak (ed.) *Intelligent Design and Manufacturing*, Wiley, Chichester (1992) 673 - 700.

[12] F. Dammeyer, P. Forst and S. Voß, On the cancellation sequence method of tabu search, *ORSA Journal on Computing* **3** (1991) 262 - 265.

[13] F. Dammeyer and S. Voß, Dynamic tabu list management using the reverse elimination method, *Annals of Operations Research* **41** (1993) 31 - 46.

[14] R.L. Daniels and J.B. Mazzola, A tabu-search heuristic for the flexible-resource flow shop scheduling problem, *Annals of Operations Research* **41** (1993) 207 - 230.

[15] M. Dell'Amico and M. Trubian, Applying tabu search to the job-shop scheduling problem, *Annals of Operations Research* **41** (1993) 231 - 252.

[16] W. Domschke, P. Forst and S. Voß, Tabu search techniques for the quadratic semi-assignment problem, in: G. Fandel, T. Gulledge and A. Jones (eds.) *New Directions for Operations Research in Manufacturing*, Springer, Berlin (1992), 389 - 405.

[17] W. Domschke, A. Scholl and S. Voß, *Produktionsplanung - Ablauforganisatorische Aspekte*, Springer, Berlin (1993).

[18] U. Faigle and W. Kern, Some convergence results for probabilistic tabu search, *ORSA Journal on Computing* **4** (1992) 32 - 37.

[19] J.A. Ferland and A. Lavoie, Exchanges procedures for timetabling problems, *Discrete Applied Mathematics* **35** (1992) 237 - 253.

[20] C.N. Fiechter, A. Rogger and D. de Werra, Basic ideas of tabu search with an application to traveling salesman and quadratic assignment, *Ricerca Operativa* **62** (1992) 5 - 28.

[21] M. Gendreau, A. Hertz and G. Laporte, A tabu search heuristic for the vehicle routing problem, Working Paper, C.R.T. Montreal (1991).

[22] F. Glover, Future paths for integer programming and links to artificial intelligence, *Computers & Operations Research* **13** (1986) 533 - 549.

[23] F. Glover, Tabu search - part I, *ORSA Journal on Computing* **1** (1989) 190 - 206.

[24] F. Glover, Tabu search - part II, *ORSA Journal on Computing* **2** (1990) 4 - 32.

[25] F. Glover, New ejection chain and alternating path methods for traveling salesman problems, in: O. Balci, R. Sharda and S.A. Zenios (eds.) *Computer Science and Operations Research*, Pergamon, Oxford (1992) 491 - 509.

[26] F. Glover and R. Hübscher, Bin packing with tabu search, Working Paper, Univ. of Colorado at Boulder (1991).

[27] F. Glover and M. Laguna, Tabu search, in: C.R. Reeves (ed.) *Modern Heuristic Techniques for Combinatorial Problems*, Blackwell, Oxford (1993) 70 - 150.

[28] F. Glover and C. McMillan, The general employee scheduling problem: an integration of management science and artificial intelligence, *Computers & Operations Research* **13** (1986) 563 - 593.

[29] A. Hertz, Tabu search for large scale timetabling problems, *European Journal of Operational Research* **54** (1991) 39 - 47.

[30] A. Hertz, Finding a feasible course schedule using tabu search, *Discrete Applied Mathematics* **35** (1992) 255 - 270.

[31] A. Hertz and D. de Werra, The tabu search metaheuristic: how we used it, *Annals of Mathematics and Artificial Intelligence* **1** (1990) 111 - 121.

[32] A. Hertz and D. de Werra, Tabu search techniques: a tutorial and an application to neural networks, *OR Spektrum* **11** (1989) 131 - 141.

[33] R. Hübscher and F. Glover, Applying tabu search with influential diversification to multiprocessor scheduling, Working Paper, Univ. of Colorado at Boulder (1992).

[34] J. Hurink, B. Jurisch and M. Thole, Tabu search for the job shop scheduling problem with multi-purpose machines, Working Paper, Univ. Osnabrück (1993).

[35] R. Kuik, M. Salomon, L.N. van Wassenhove and J. Maes, Linear programming, simulated annealing and tabu search heuristics for lotsizing in bottleneck assembly systems, *IIE Transactions* **25** (1993) 62 - 72.

[36] M. Laguna, J.W. Barnes and F.W. Glover, Tabu search methods for a single machine scheduling problem, *Journal of Intelligent Manufacturing* **2** (1991) 63 - 74.

[37] M. Laguna, J.W. Barnes and F. Glover, Intelligent scheduling with tabu search: an application to jobs with linear delay penalties and sequence dependent setup costs and times, Working Paper, Univ. of Colorado at Boulder (1992).

[38] M. Laguna and F. Glover, On target analysis and diversification in tabu search, Working Paper, Univ. of Texas in Austin (1990).

[39] M. Laguna and F. Glover, Integrating target analysis and tabu search for improved scheduling systems, *Expert Systems with Applications* **6** (1993) 287 - 297.

[40] M. Laguna and J.L. Gonzalez-Velarde, A search heuristic for just-in-time scheduling in parallel machines, *Journal of Intelligent Manufacturing* **2** (1991) 253 - 260.

[41] M. Laguna, J.P. Kelly, J.L. Gonzalez-Velarde and F. Glover, Tabu search for the multilevel generalized assignment problem, Working Paper, Univ. of Colorado (1991).

[42] M. Laguna-Castillo, The application of heuristic methods and artificial intelligence to a class of scheduling problems, Ph.D. Thesis, Univ. of Texas at Austin (1990).

[43] E.L. Mooney and R.L. Rardin, Local improvement heuristics for multiple choice vertex packing problems, Working Paper, Purdue Univ. (1991).

[44] E.L. Mooney and R.L. Rardin, Tabu search for a class of scheduling problems, *Annals of Operations Research* **41** (1993) 253 - 278.

[45] E. Pesch, Local search based learning in automated manufacturing, Habilitation thesis, TH Darmstadt (1993).

[46] A.P. Punnen and Y.P. Aneja, Categorized assignment scheduling: a tabu search approach, *Journal of the Operational Research Society* **44** (1993) 673 - 679.

[47] C.R. Reeves, Improving the efficiency of tabu search for machine sequencing problems, *Journal of the Operational Research Society* **44** (1993) 375 - 382.

[48] M. Salomon, *Deterministic lotsizing models for production planning*, Lecture Notes in Economics and Mathematical Systems **355**, Springer, Berlin (1991).

[49] M. Salomon, R. Kuik and L.N. van Wassenhove, Statistical search methods for lotsizing problems, *Annals of Operations Research* **41** (1993) 453 - 468.

[50] A. Scholl and S. Voß, Maximizing the production rate in simple assembly line balancing: a tabu search approach, Working Paper, TH Darmstadt (1993).

[51] M. Sinclair, Comparison of the performance of modern heuristics for combinatorial optimization on real data, *Computers & Operations Research* **20** (1993) 687 - 695.

[52] J. Skorin-Kapov, Tabu search applied to the quadratic assignment problem, *ORSA Journal on Computing* **2** (1990) 33 - 45.

[53] J. Skorin-Kapov and A.J. Vakharia, Scheduling a flow-line manufacturing cell: a tabu search approach, *International Journal of Production Research* **31** (1993) 1721 - 1734.

[54] B. Srivastava and W.-H. Chen, Part type selection problem in flexible manufacturing systems: tabu search algorithms, *Annals of Operations Research* **41** (1993) 279 - 297.

[55] E. Taillard, Parallel taboo search technique for the jobshop scheduling problem, Working Paper, Ecole Polytechnique Federale de Lausanne (1989).

[56] E. Taillard, Some efficient heuristic methods for the flow shop sequencing problem, *European Journal of Operational Research* **47** (1990) 65 - 74.

[57] E. Taillard, Robust taboo search for the quadratic assignment problem, *Parallel Computing* **17** (1991) 443 - 455.

[58] S. Vaithayanathan, Stochastic and deterministic neural networks for resource constrained scheduling problems under dynamic conditions, Ph.D. Thesis, Univ. of Houston (1992).

[59] A.J. Vakharia and Y.-L. Chang, A simulated annealing/tabu search approach to cell formation in cellular manufacturing, Presented at TIMS/ORSA Meeting Nashville (1991).

[60] S. Voß, Network design formulations in schedule synchronization, in: M. Desrochers and J.-M. Rousseau (eds.) *Computer-Aided Transit Scheduling*, Lecture Notes in Economics and Mathematical Systems **386**, Springer, Berlin (1992) 137 - 152.

[61] S. Voß, The two-stage hybrid-flowshop scheduling problem with sequence-dependent setup times, in: G. Fandel, T. Gulledge and A. Jones (eds.) *Operations Research in Production Planning and Control*, Springer, Berlin (1993), 336 - 352.

[62] S. Voß, Tabu search: applications and prospects, in: D.-Z. Du and P. Pardalos (eds.) *Network Optimization Problems*, World Scientific, Singapore (1993), 333 - 353.

[63] S. Voß, Fließfertigung mit Lagerkapazitätsbeschränkungen, Working Paper, TH Darmstadt (1993).

[64] S. Voß, Solving quadratic assignment problems using the reverse elimination method, Working Paper, TH Darmstadt (1993).

[65] F. Werner, On the heuristic solution of the permutation flow shop problem by path algorithms, *Computers & Operations Research* **20** (1993) 707 - 722.

[66] M. Widmer, Job shop scheduling with tooling constraints: a tabu search approach, *Journal of the Operational Research Society* **42** (1991) 75 - 82.

[67] M. Widmer and A. Hertz, A new heuristic method for the flow shop sequencing problem, *European Journal of Operational Research* **41** (1989) 186 - 193.

[68] D.L. Woodruff, Subcontracting when there are setups, deadlines, and tooling costs, in: W.T. Scherer and D.E. Brown (eds.) *Proceedings of the Intelligent Scheduling Systems Symposium*, San Francisco (1993) 337 - 353.

[69] D.L. Woodruff and M.L. Spearman, Sequencing and batching for two classes of jobs with deadlines and setup times, *Production and Operations Management* **1** (1992) 87 - 102.

[70] D.L. Woodruff and E. Zemel, Hashing vectors for tabu search, *Annals of Operations Research* **41** (1993) 123 - 137.

One-Dimensional Integer Cutting Stock Problems

Gerhard Wäscher, Technische Universität Braunschweig

Papers on cutting stock problems belong to the earliest references (Kantorovich 1939) in the field of Operations Research. Today, there is still a considerable interest in the field, which can be seen from the constantly growing number of publications.

One area which has been focused by several researchers during the last few years was the questions of how to generate integer solutions to the standard one-dimensional cutting stock problem (SCSP). Research on this subject may have been attracted by the apparent contradiction that on one hand the generation of an optimum integer solution has proven to be difficult (Gilmore 1979) while on the other hand the integer round-up property seems to hold for almost any real-world problem (Marcotte 1985).

In this paper the state-of-the-art in solving the SCSP is reported. Both exact solution methods (Branch-and-Bound) and heuristic procedures (round-off methods, repeated pattern exhaustion techniques, bin packing algorithms) are presented and compared with respect to their performance in solving a series of problems, taken from real world applications and from literature.

Our test results indicate that small and medium-sized problems can be solved to an optimum by means of standard ILP-codes when specifying simple lower bound on the objective function value. For large problems round-off methods often generate optimum solutions, while other heuristic methods only provide non-competitive solutions far from optimum.

Finally, open questions and directions for future research are pointed out.

Literature

Gilmore, P.C. (1979):
Cutting Stock, Linear Programming, Knapsacking, Dynamic Programming and Integer Programming - Some Interactions. Annals of Discrete Mathematics 4, pp. 217 - 235.

Kantorovich, L.V. (1939):
Mathematical Methods of Organizing and Planning Production. Management Science 6, 1960, pp. 366-422. (Translation of a report given on May 13, 1939, at the Leningrad State University).

Marcotte, O. (1985):
The Cutting Stock Problem and Integer Rounding. Mathematical Programming 33, pp. 82-92.

Ein Ansatz zur Anwendung genetischer Algorithmen auf das Problem der Grobterminierung

Ulrich Weingarten, Technische Universität Braunschweig

Gegenstand der Untersuchung ist die Grobterminierung. Zugrundegelegt wird dabei das Modell von Pritsker/Watters/Wolfe[1].
Aufgrund der Problemkomplexität eignen sich exakte Verfahren nur für kleinere Problemumfänge. Deterministische Heuristiken bergen das Problem, daß sie ceteris paribus in einem (sehr großen) Lösungsraum immer wieder die gleiche Lösung generieren.
Zur Überwindung dieser Nachteile wird die Anwendung des Prinzips der Genetischen Algorithmen auf das Problem der Grobterminierung vorgestellt. Gezeigt wird eine Möglichkeit der Datenrepräsentation sowie der Einfluß von drei unterschiedlichen Crossover/Reordering-Operatoren und des Operators Mutation.
Die Gene eines Chromosoms repräsentieren Arbeitsgang-Betriebsmittel-Kombinationen. Daraus resultiert, daß die Zustandsausprägung eines einzelnen Gens nicht mehr nur durch eine binäre Ziffer dargestellt werden kann[2].
Die Kombination aus Datenrepräsentation und der aus der Problemstellung resultierenden Forderung, daß jeder Arbeitsgang des zu bearbeitenden Auftragsspektrums genau einmal durchgeführt werden muß, stellt spezielle Anforderungen an den Crossover-Operator. Beispielhaft wird auf die Methodik der implementierten Crossover-Operatoren (Cycle-, Order-, Partially-Matched-Crossover) eingegangen.
Die gleiche Problematik ist bei dem Entwurf des Mutations-Operators zu berücksichtigen. Neben der Grundform für einen solchen Operator wird eine Variante vorgestellt, die besonders der Berücksichtigung alternativer Betriebsmittel Rechnung trägt.

---

1 Zur Modellierung vgl. **Domschke, W.; Drexl, A.**; Einführung in Operations Research, Springer-Verlag (1990), S. 94-95; **Pritsker, A.B.; Watters, L.J. und Wolfe P.M.**: Multiproject scheduling with limited resources: a zero-one programming approach; in: Management Science Vol. 16 (1969), No. 1, S. 93-108

2 vgl. **Goldberg, David E.**; Genetic Algorithms in Search, Optimization, and Machine Learning, Reading / Mass., Addison-Wesley (1989); **Vignaux, G. A. und Michalewicz, Z.**: A Genetic Algorithm for the Linear Transportation Problem, in: IEEE Transactions on Systems, Man and Cybernetics, 21 (1991) 2, S. 445-452

Control Strategies for Automated Storage/Retrieval Systems

Jeroen van den Berg and Noud Gademann
University of Twente, The Netherlands

Automated Storage/Retrieval Systems (AS/RSs) are widely used in manufacturing systems and distribution centers. A typical system is comprised of multiple aisles with (high-rise) storage racks alongsides and multiple storage/retrieval (S/R) machines operating on one or more aisles that transfer items from an input/output (I/O) station to the storage positions and vice versa. The throughput of an AS/RS, *i.e.*, the maximum rate at which it is possible to process storage and retrieval requests, depends on the control strategies that are incorporated.

One such strategy that is discussed, is concerned with the allocation of items to storage positions. Some well known allocation-rules are: closest open location, class-based allocation (ABC), full-turnover based allocation and the cube-per-order index rule. The latter three make use of the idea that items with a high turnover frequency should be assigned to positions close to the I/O station.

A second strategy that is discussed, is concerned with scheduling storage and retrieval requests. When the S/R machine travels from the I/O station, performs a storage or retrieval and returns to the I/O station, this is called a single command cycle. When successively a storage and a retrieval are performed within the same cycle then this is referred to as a dual command cycle. Generally, the S/R machine has a unit load capacity, so that it is not possible to perform more storages or retrievals during one command cycle. When several storage and retrieval requests are waiting to be processed, the total travel time of the S/R machine can be minimized by pairing requests to facilitate dual command cycles. Several allocation-rules as well as scheduling strategies are compared.

LOGISTICS SUPPORT OF COMPLEX SYSTEMS AND LIFE CYCLE COSTING

Bisnoedat Bhola
TNO Physics and Electronics Laboratory, NL-The Hague

Historically, the acquisition of complex systems was focused only on acquisition costs (all costs associated with research & development, design, construction, production, test and evaluation). Low acquisition costs of systems however have often resulted in high costs for operations and logistics support. This incorporates all tasks to operate the system properly, to keep the system in or to return it to the desired operational state after failures have occurred. These tasks include scenarios for operation, maintenance planning (policies & levels), personnel, training, documentation, computer resources, facilities, provisioning, spare parts and consumables, ammunition, transport and distribution.
Since 1960 especially in the USA defence community, much effort have been spent to integrate the acquisition, operation and the logistics support to acquire systems with low cost over the total life cycle. This effort resulted in a mangement philosophy called Integrated Logistics Support (ILS) which is based on the incorporation of logistics and operational requirements in the design concept using Logistics Support Analysis (LSA) and Life Cycle Cost Analysis (LCCA).
These concepts have also been studied by TNO Physics and Electronics Laboratory (FEL-TNO) for many years. This article reports the preliminary results of a research project conducted by FEL-TNO for the Ministry of Defence which aims at developing a life cycle cost methodology to integrate the acquisition, operation, logistics support and disposal of complex (military) systems.

PHYDIS - PC-GESTÜTZTE STANDORTBESTIMMUNG IN EUROPA

Karl-Heinz Brockmann, Forschungsinstitut für Rationalisierung an der RWTH Aachen

Vor dem Hintergrund der Vollendung des EG-Binnenmarktes und der damit verbundenen distributionslogistischen Konsequenzen stellt sich für viele europaweit operierende Unternehmen das Problem der Neugestaltung ihres Warenverteilungssystems.

Die *Struktur* eines Warenverteilungssystems wird dabei durch die Realisation der folgenden vier Entscheidungsvariablen determiniert:

- Zahl der Distributionsstufen
- Zahl der Läger je Stufe
- Standorte der Läger
- Liefergebiete der Läger.

In der Literatur existiert eine Vielzahl von Veröffentlichungen, die sich mit der Entwicklung von Verfahren zur Strukturierung von Warenverteilungssystemen befassen, wobei die Minimierung der Warenverteilungskosten in aller Regel die Zielfunktion bildet. Dabei existieren zwei Klassen von Lösungsansätzen: Der *diskrete* Ansatz geht von der Existenz bekannter potentieller Lagerstandorte aus, wohingegen der *homogene* Ansatz prinzipiell jeden Punkt des untersuchten Gebiets als potentiellen Standort zuläßt.

Im Rahmen dieses Vortrages wurde ein am Forschungsinstitut für Rationalisierung entwickeltes PC-gestütztes Verfahren (PHYDIS; PHYsische DIStribution) zur Optimierung von Standorten und Liefergebieten eines Warenverteilungssystems vorgestellt, das von der Modellannahme eines homogenen Territoriums ausgeht. Dabei wird zur Bestimmung optimaler Lagerstandorte auf der Basis fester Liefergebiete ein Iterationsverfahren eingesetzt, das auf der Basis eines Gradientenansatzes entwickelt wurde, während die Zuordnung von Kunden zu Lagerstandorten nach dem Kriterium der geringsten Entfernung erfolgt. Hierbei ist auch die Integration bereits bestehender Läger in ein zu strukturierendes Warenverteilungssystem - gegebenenfalls unter Berücksichtigung von Kapazitätsrestriktionen - möglich. Als Inputdaten benötigt das Verfahren die geographischen Koordinaten der Werks- und Kundenstandorte, die erwarteten bzw. in einem repräsentativen Zeitraum erhobenen Nachfragemengen der Kunden sowie die zur Kostenberechnung erforderlichen Kostensätze.

LAYOUT AND OPERATION OF DISTRIBUTION CENTRES FOR PERISHABLES

Rob A.C.M. Broekmeulen, ATO-DLO, NL-Wageningen

The efficiency of a distribution centre plays a crucial role in the marketing of perishables. Perishables like dairy products, meat, vegetables and fruits suffer from quality loss due to aging and breakdown. Quality loss can be reduced by adequate storage conditions like relative humidity and temperature, minimal handling and avoiding of product interactions like odour and hormone transmission. Most perishables from agricultural origin have strong seasonal fluctuations in production and demand. The absence of an uniform article coding system makes the operation of the distribution centre even more complex.

A distribution centre for perishables has to deal with peak stock levels and product interactions. The aim is to maximise flexibility towards the customers and to minimise quality loss during the distribution process at the same time. This has to be achieved with minimal cost using the available resources.

We have developed a multi-stage hierarchical decision approach that is capable of handling the complexity of the overall problem. The iterative decision process consists of the following steps:

1 Facility design and technology selection.

2 Item allocation.

First, product groups are assigned to storage type clusters like cold stores. This is done with multiply goal programming and quality change models. Second, the individual articles are assigned to storage locations like racks and shelves. We use a local search technique to solve a time-dependent version of the bin packing problem.

3 Order picker routing with respect to stable customer pallets.

The process is integrated in a PC-based decision support system. The system generates several good alternatives in a relative short time.

We present the application of the developed planning approach and decision models to a distribution centre of vegetables and fruits

Keywords: decision support system, distribution centres, hierarchical planning, local search techniques, logistics, operations management, perishables.

STOCKNORMS IN DISTRIBUTION SYSTEMS: THE IMPACT OF IMBALANCE

K. van Donselaar, Eindhoven University of Technology (NL),
J. Wijngaard, University of Groningen (NL)

SUMMARY

The goal of this paper is to investigate the impact of imbalance on stocknorms in distribution systems with non-identical retailers. The retailers are non-identical with respect to their lot-sizes and demand characteristics. For these distribution systems stocknorms will be determined, which are based on a pre-specified service level.

1. INTRODUCTION

In inventory control theory distribution systems with a central depot and several retailers are analogous to divergent production systems with a common component and several final products. Stocknorms for divergent systems with identical final products have been investigated already in [1]. The major conclusions for those systems were:

1) Divergent systems may face imbalance. A divergent system is out of balance if the inventory positions of the final products are not equivalent.
2) If divergent systems are controlled integrally and if the imbalance is relatively large, the integral stocknorm should take into account this imbalance.

The most straightforward way to determine stocknorms for distribution systems with non-identical retailers is to extend the results for the divergent systems with identical final products. However, if these results are to be extended to the system with non-identical retailers, several difficulties arise:

- the definition of imbalance is more complex. Simply defining imbalance as the deviation of the retailers' inventories from the average inventory no longer makes sense. In fact it will become apparent that different definitions are needed for systems with and without depot.
- the allocation is more complex. For identical retailers allocation was simple: allocate to the retailer with the lowest inventory. If mean and standard deviation of demand as well as the lot-sizes and

consequently the reorderpoints of all retailers are different, it is no wonder that the allocation has to be more complex.

- the system service level is more complex. Simply defining the system service level as the average of all retailers' service level would be unfair to the retailers with a large average demand; Raising their service level with 5% may require a larger amount of inventory than an equivalent raise of the service level of retailers with a smaller average demand. Therefore it is suggested to use as a system service level the weighted average of the retailers' service level. The weight factors are equal to the ratio's of the retailer's average demand over total average demand.

These aspects will be discussed in Section 2, where imbalance for a distribution system with non-identical retailers will be defined.
Based on this definition an estimator for the variance of imbalance is created in Section 3. This estimator can be used in the determination of stocknorms for distribution systems.

2. DEFINITION OF IMBALANCE

2.1 Implicit definition

In [1] imbalance has been studied for (in terms of the distribution system studied in this paper) a distribution system with identical retailers. There imbalance for a retailer j was defined as the difference between the actual economic inventory (the economic inventory is equal to the inventory on hand plus on order) of that retailer (I_j) and the average inventory level of all N retailers:

$$\text{imbalance}_j = I_j - \Sigma I_i/N \qquad (1),$$

When the retailers have different demand distributions and/or different lot-sizes, this definition no longer holds. Suppose for example that two retailers each have 100 products on hand. According to (1) the system would be balanced. This makes sense if both retailers have equal average demand. However if the average demand of one of these retailers is ten times as high as the average demand of the other retailer, it is crystal clear, that the system is out of balance.
A sensible way of defining imbalance seems to be

$$\text{imbalance}_j = I_j - I_j^* \qquad (2),$$

where I^*_j is the amount of inventory for retailer j after an optimal (imaginary) re-allocation of all economic inventory (ΣI_j) without lot-sizing-restrictions.

j without lotzing-restrictions.

Note that $\Sigma E[\text{imbalance}_j] = 0$ since $\Sigma I_j = \Sigma I_j^*$. The variance of imbalance for retailer j will be denoted by $\sigma^2_{imb,j}$. So $\sigma^2_{imb,j} = \text{var}(I_j - I_j^*)$.

2.2 Re-allocation

Having defined imbalance as in formula (2), the remaining problem to be solved here is finding the optimal re-allocation rule in case of a system with non-identical retailers and without lot-sizes. Once that re-allocation rule is defined, imbalance is known according to the above definition. The system with and without depot will be dealt with separately.

2.2.1 Re-allocation in a system with depot

Examples of re-allocation rules for the system with depot, which can be used to define imbalance, are:

1) Equalize the run-out-time (that is: the ratio of inventory over average demand) for all retailers. This yields:

   $I_j^* = c \cdot \mu_j$, where $c = \Sigma I_i / \Sigma \mu_i$ (since $\Sigma I_j^* = \Sigma I_j$).

   In this way all retailers will tend to run out of stock simultaneously if their leadtimes are equal.
2) Equalize the expected service-level for all retailers.

   In case of normal distributed demand this yields:

   $I_j^* = (\ell_j+1)\mu_j + k^* \sqrt{(\ell_j+1)}\sigma_j$, where $k^* = (\Sigma I_i - \Sigma(\ell_i+1)\mu_i) / \Sigma\sqrt{(\ell_i+1)}\sigma_i$.

3) Equalize the stocknorm-ratio for all retailers.

   If the stocknorm for retailer j is denoted by r_j, this results in:

   $I_j^* = c \cdot r_j$, where $c = \Sigma I_i / \Sigma r_i$.
4) Fill the inventory of the retailers up to their stocknorm and allocate the difference between total inventory available and the sum of the retailers' stocknorms according to the average demand of the retailers.

   In this way all retailers' economic inventories will tend to drop below their stocknorm simultaneously. The corresponding formula is:

   $I_j^* = r_j + c \cdot \mu_j$, where $c = \Sigma(I_i - r_i)/\Sigma\mu_i$ (3)

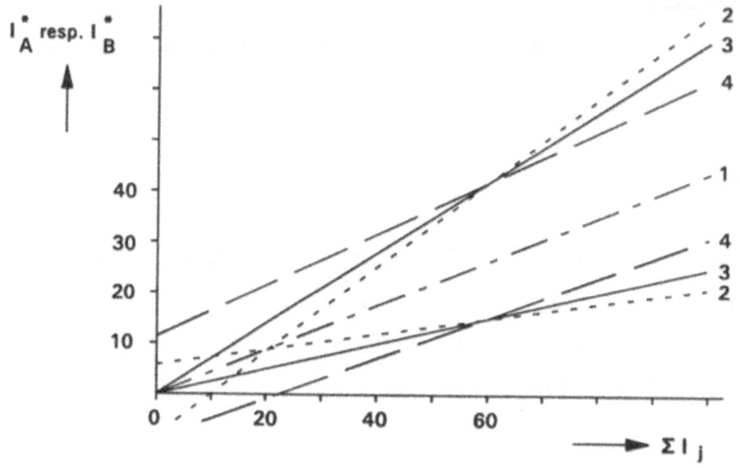

Figure 1. The proposed allocation for A using different allocation rules for a system with retailers A and B. The numbers in this figure correspond with the number of the allocation rule described above.

It appears that every rule has its weak and strong points. This will be demonstrated by considering a specific distribution system with two retailers (A and B) in more detail. Retailers A and B face normal distributed demand and have the following characteristics: $\mu(A)=\mu(B)=10$, $\sigma(A)=20$, $\sigma(B)=5$, $\ell(A)=\ell(B)=0$, $r(j)=\mu(j)+1.6\sigma(j)$ so $r(A)=42$ and $r(B)=18$.

For this example the four allocation rules will lead to the following allocations to retailer A:

1. $I_A^* = 0.5\Sigma I_i$
2. $I_A^* = -6 + 0.8\Sigma I_i$
3. $I_A^* = 0.7\Sigma I_i$
4. $I_A^* = 12 + 0.5\Sigma I_i$

In this example all rules take account of the variance for each retailer, except for the first rule. So in case $\Sigma I=\Sigma r$, the first rule will allocate $0.5\Sigma r_j$ to each retailer (see Figure 1). This implies that the expected service level of A and B will differ. In case $\Sigma I=\Sigma r$ and allocation rule 1 is used, the expected service levels will be $\Phi((I_j^*-\mu_j)/\sigma_j)$, i.e. $\Phi(1)\approx 84\%$ and $\Phi(4)\approx 100\%$ for A resp. B. With the three other allocation rules the expected service levels of A and B will be the same: they will both be equal to $\Phi(1.6)\approx 95\%$.

The second and third rule will tend to allocate too much inventory to the retailer with the largest uncertainty if total inventory far exceeds the sum of the stocknorms (e.g. due to large lot-sizes). Suppose for example that ΣI_j equals 220. According to allocation rule 2 the 'optimal' allocation is:

$$I_A^* = 10 + [(220\text{-}20)/25] \cdot 20 = 170, \quad I_B^* = 10 + [(220\text{-}20)/25] \cdot 5 = 50.$$

Both retailers have equal service levels for the next period: $\Phi(8)$. The run-out-times however for retailers A and B are resp. 17 and 5 periods. Suppose that the total inventory of 220 products was meant to last for 6 periods. The expected service level for retailers A and B for 6 periods from now will be $\Phi((170\text{-}6 \cdot 10)/20\sqrt{6}) \approx 99\%$ resp.
$\Phi((50\text{-}6.10)/5\sqrt{6}) \approx 21\%$. The reason for this imbalanced result is the greediness of allocation rule 2: It does not take into account the fact that total inventory may be needed for more than $(\ell+1)$ periods. Analogous results can be derived for allocation rule 3.

The fourth rule will tend to allocate too much inventory to the retailer with the largest uncertainty if total inventory is far below the sum of the stocknorms. Let e.g. $\Sigma I = \Sigma\mu = 20$. Then according to allocation rule 4: $I_A^* = 42 - 2 \cdot 10 = 22$ and $I_B^* = 18 - 2.10 = -2$.
The expected service level for the next period equals
$\Phi((22\text{-}10)/20) \approx 73\%$ for retailer A and $\Phi((\text{-}2\text{-}10)/5) \approx 1\%$ for retailer B.
It seems more appropriate in this case to select $I_A^* = I_B^* = 10$.

Since in this paper the attention is focussed on getting an indication of the impact of imbalance rather than on finding the over-all optimal allocation rule, a simple and robust allocation rule is selected: allocation rule number 4, i.e. formula (3).

Note that in the special case of identical retailers all these policies are the same: the optimal allocation then with each of the above policies is to allocate to each retailer the average amount of inventory.

2.2.2 Re-allocation in a system without depot

In a system without depot a different re-allocation rule has to be chosen since there are no longer stocknorms for the retailers. Federgruen and Zipkin [3] investigated the allocation of an amount of inventory among several retailers with the restriction that all inventory had to be allocated. Their distribution system without depot is replenished every now and then with a lot, which is equal to Q_{sys}. Their research revealed a.o. that a good, simple and robust allocation rule is the rule proposed by Eppen and Schrage [2]:

$$I_j' = (\ell_j+v+1)\mu_j + k'\sqrt{(\ell_j+v+1)}\sigma_j \qquad (4)$$

where $k' = [\Sigma(I_i - (\ell_i+v+1)\mu_i) + d] / \Sigma[\sqrt{(\ell_i+v+1)}\sigma_i]$

with I_j', I_j : the inverntory of retailer j after, resp. before allocation.

$v+1$: the average time between two orders $= Q_{sys}/\Sigma\mu_i$.

d : amount available to be allocated.

This is a good rule, except for the cases where the system lot-size and variances are large. For those cases Federgruen and Zipkin propose a refinement of this formula:

$$v = \min([\Sigma I_i+d - \Sigma(\ell_i+1)\mu_i]^+/\Sigma\mu_i, Q_{sys}/\Sigma\mu_i-1).$$

This refinement states that if the total inventory which is available for the retailers in the depot or downstream is small, that is: less than the lot-size of the entire system plus the average demand for the retailers times their leadtime, then the expected time up to the arrival of the next lot for the entire system should be decreased accordingly.

The allocation rule (4) is based on the fact, that allocation takes place just after an order has arrived in the imaginary depot. Since that order contains v+1 periods demand on average, the allocation is chosen such, that the expected service level after ℓ+v+1 periods is equal for all retailers. However, imbalance is measured every period and not only after an order is allocated. The allocation rule mentioned above has to be adapted for this to come up with a good re-allocation rule to measure the imbalance. A sensible adaptation seems to be:

$$I_j^* = (\ell_j+v+1)\mu_j + k\sqrt{(\ell_j+v+1)}\sigma_j \qquad (5)$$

where $k = \Sigma(I_i - (\ell_i+v+1)\mu_i) / \Sigma[\sqrt{(\ell_i+v+1)}\sigma_i]$

and $v = \min ([\Sigma I_i - \Sigma(\ell_i+1)\mu_i]^+/\Sigma\mu_i, [Q_{sys}/\Sigma\mu_i - 1\text{-}t]^+).$

with t : the number of periods after the last order arrived in the imaginary depot.

3. AN ESTIMATOR FOR THE VARIANCE OF IMBALANCE

The goal of this paper is to gain insight in the relationship between stocknorms and service level. For a distribution system with identical retailers this relationship appeared to depend on the variance of imbalance. Generalisation of that relationship to the non-identical retailers case seems to be rather straightforward. This leads to the following approximation for r_{sys}:

$$r_{sys} = \Sigma(\ell_{sys} + \ell_j + 1)\,\mu_j - \frac{Q_{sys}}{2} + k\sqrt{\left[\frac{Q_{sys}^2}{12} + \ell_{sys}\Sigma\sigma_j^2 + \{\Sigma\sqrt{[(\ell_j + 1)\sigma_j^2 + \sigma_{imb\,j}^2]}\}^2\right]} \tag{6}$$

where r_{sys} is the integral stocknorm for the entire system,
ℓ_{sys}, ℓ_j are the leadtimes for the entire system resp. retailer j,
Q_{sys} is the lot-size for the entire system and
μ_j, σ_j are the average and standard deviation of demand for retailer j.

For the relationship between the safety factor k in this formula and the expected service level, the reader is referred to [1]. According to formula (6) an estimator for the variance of imbalance is needed in order to be able to determine the stocknorm.

To derive an estimator on the variance of imbalance for the general systems considered here seems to be very complex. The only estimator at hand for the moment is the estimator for the system with identical retailers (see [1]). By assuming that the variance of imbalance for each retailer in the system with non-identical retailers is the same as in a system with all retailers being identical to that retailer, the following estimator will result:

$$\hat{\sigma}_{imb\,j}^2 = \begin{cases} 0 & \text{if } N = 1 \\ \dfrac{Q_j^2}{12} & \text{if a depot is present and } N \neq 1 \\ \dfrac{Q_j^2}{12} \cdot \dfrac{(N-1)}{N} \cdot \dfrac{(N+2)}{N} + 0{,}5\left[\dfrac{Q_{sys}}{\Sigma\mu_j} - 1\right]\dfrac{N-1}{N} \cdot \sigma_j^2 & \text{elsewhere} \end{cases} \tag{7}$$

This certainly will not be a good estimator for the variance of imbalance for retailer j. This can be easily seen from the fact that for N=2 the variance of imbalance for the two retailers should be equal (since $\text{imbalance}_1 = -\text{imbalance}_2$). If the two retailers have different lot-sizes, formula (7) however proposes unequal variances of imbalance. So formula (7) will not be correct always. Nevertheless it is possible, although the spread over all retailers is lost, that formula (7) still is a good representative of the total 'weight' of imbalance in the formula for the stocknorm.

By defining imbalance as in (2) with I_j^* defined as in (3) or (5) for a system with resp. without depot, it is possible by means of simulation to measure imbalance and to investigate its influence on the system's performance. These simulation results are reported in [4].

4. CONCLUSIONS

The determination of stocknorms for divergent systems is far more complex for a system with non-identical final products than for a system in which all final products are identical. Particularly the definition of imbalance, which is directly related to the allocation rule needed in such divergent systems, is more complex.

In this paper a definition for imbalance in divergent systems with non-identical final products is given. Based on this definition stocknorms for divergent systems have been developed.

Simulation results, reported in [4], show that the approximation suggested in this paper captures the imbalance due to lot-sizing quite well. However, at the same time the simulation results also indicate that the consequences of fully neglecting the imbalance in the determination of stocknorms are often small. In our simulations the imbalance only had a significant impact on the expected service level when the lot-size for the entire system (Q_{sys}) was large compared to the retailer's lot-size (Q_j).

LITERATURE

[1] Donselaar, K. van (1990). Integral stocknorms in divergent systems with lot-sizes. European Journal of Operational Research, 45, pp. 70-84.

[2] Federgruen, A. and P. Zipkin (1984). Allocation policies and cost approximations for multilocation inventory systems. Naval Research Logistics Quarterly, 31, page 97.

[3] Eppen, G. and L. Schrage (1981). Centralized ordering policies in a multi-warehouse system with lead times and random demand. Appeared in: Multi-level production/inventory control systems: theory and practice (ed: l.B. Schwarz). North-Holland Publ. Co., Amsterdam.

[4] Donselaar, K. van and J. Wijngaard (1988). Safety stocknorms in divergent systems with non-identical final products, internal report no. ARW-03 TUE/BDK/ORS/88/04, Eindhoven University of Technology.

QUANTITATIVE ANALYSIS OF PHYSICAL DISTRIBUTION NETWORKS WITH APPLICATIONS

Jan B.M. van Doremalen, CQM B.V., NL-Eindhoven
Hein A. Fleuren, CQM B.V., NL-Eindhoven
Matthieu C. van der Heijden, CQM B.V., NL-Eindhoven

There is a growing interest in improving the efficiency and effectiveness of international distribution networks. This is in particular true in those situations where one is considering the outsourcing of (parts of) the distribution function to service providers in the areas of warehousing and transportation. But the interest is also caused by the recent and upcoming simplifications in border-crossing traffic within the European Community which make it attractive to reconsider the country-for-country approach, followed by many multinational companies, by concentrating stocks in supra-national distribution centres. This implies that in most situations strategic and tactical decisions have to be made with respect to changing the set up and operation of the distribution function.

On the basis of earlier research done by members of our consultancy group and recent experiences in a number of consultancy projects, we are developing a quantitative analysis framework to support this decision process. In this paper we describe the framework and highlight some aspects by examples from practical situations within a multinational electronics company.

The objective of the quantitative analysis framework is to support the evaluation of quantitative performance measures in the areas of customer service levels, goods flow volumes and transportation, warehousing and stock holding costs for different distribution network scenario's. Important scenario characteristics are e.g.

(1) market demand,

(2) location of production centres and stock points,

(3) stock replenishment policies,

(4) transportation modes and associated costs and lead times,

(5) product portfolio,

(6) interest, obsolescence and price erosion factors,

(7) warehousing and stock holding costs,

(8) and product flows defined by distribution channels and customer service offers.

The analysis will yield insight in how the distribution network should look like and how it should be managed. This insight will lead to recommendations on where to locate stock points, how to replenish stocks, and how to distribute products from production centres through stock points to customers.

TASTE: ANALYSIS AND IMPROVEMENT OF DISTRIBUTION NETWORKS

Ron van Duin
TNO Physics and Electronics Laboratory, NL-The Hague

The co-ordination of goods and information flows is a challenging task in complex distribution networks where many different products and consumer groups are involved. The effects of introducing new or different control structures are difficult to predict. In many cases, a formal description of the network and a simulation of the effects of different decisions with respect to the network can gain insight into and improve the understanding of the many logistic trade-offs in such complex systems.
TASTE is a domain specific library for logistic networks, based on a (Petri-net based) formal specification language. The library contains logistic components with a generic structure and special install options. Here lies the basic assumption of TASTE that many logistic networks have simularities in the components they consist of. Besides the library, the polymorphism of the specification language and the graphical design interface (making the building process more comprehensible for policy makers), a methodology is one of the main parts of the TASTE instrument, which reduces time and costs of the simulation building process. The methodology is a result of practical know-how acquired in several business cases. The steps of the methodology will be demonstrated in one of the business cases.
Furthermore the article will show an overview of the library of logistic components. As a concluding remark the wide range of logistic applications indicates that the TASTE instrument foresees in it's needs to analyse and to improve distribution networks.

European Aspects of the Distribution Network Design. Problems and Models

Bernhard Fleischmann, Universität Augsburg

In Germany, the design of distribution networks for consumer goods has focused, in most companies in the last three years, on the extension of the current structure for supplying Eastern Germany. After the gradual changes of the eighties - concentration of the production locations, more frequent supply of smaller quantities, stagnation of rail transports - we were faced with a sudden additional increase of the volume and distances of transportation which has caused a crisis of the road traffic in Germany.

On this background, the design of distribution structures for the European market is no longer only a microeconomic problem of single industrial companies. The cost of transportation will be strongly influenced by measurements of the Governments. Logistic service companies of several countries are cooperating in building up European networks, where the distribution for many producers can be bundled. The urgent need for an increasing use of the rail has recently lead to intense planning activities for road-rail terminals, on local, regional and national levels.

We review the various planning problems and the appropriate OR models and report on some experience with planning projects in practice.

DECISION SUPPORT IN INTERNATIONAL LOGISTICS
- from operational data to decision support for management -

Lorike Hagdorn - van der Meijden, Erasmus University Rotterdam, NL
Jo A.E.E. van Nunen, Erasmus University Rotterdam, NL

Abstract
In international logistics, the topic of redesigning logistic networks of production and distribution is of high current interest. This redesign process can be supported by a Decision Support System (DSS), which is developed to assist management teams in discussing alternative logistic structures. One of the most difficult and time consuming elements in using such a DSS, is gathering the relevant data. In the past the collection of data was very time consuming. However, nowadays a lot of basic-data are electronically available. One specific element in preparing alternative international logistic network is the level of detail of data you need or want to use: a balance has to be found between the level of detail of the data gathered and the accurateness of the calculations by e.g. an LP model within the DSS. Moreover an adequate level of aggregation in the management reports has to be defined. The selection of the appropriate aggregation level is a new challenge for DSSs. In this article we present an overview of existing and new upperbounds for the error in total costs, caused by aggregation of customers to customer-groups. Experimental results on these bounds are showed.

1 Introduction

At present, many important trends and innovations in market developments, technological improvements, organizational issues, cultural changes and labour market developments, create opportunities for industrial companies to increase their competitive advantage by reconsidering the structure of their logistics network of supply, manufacturing and distribution (see figure 1).
Questions, that companies are considering in order to serve their present and future customers on a high service level on one hand and against a low cost level on the other hand, are :

- How many plants and distribution centres are needed ?
- Which locations are best ?
- What sizes of facilities are needed ?
- Which products should be produced on which plants ?
- Which products should be stored and handled in which warehouses ?
- Which customers should be delivered by which warehouses ?

etc.

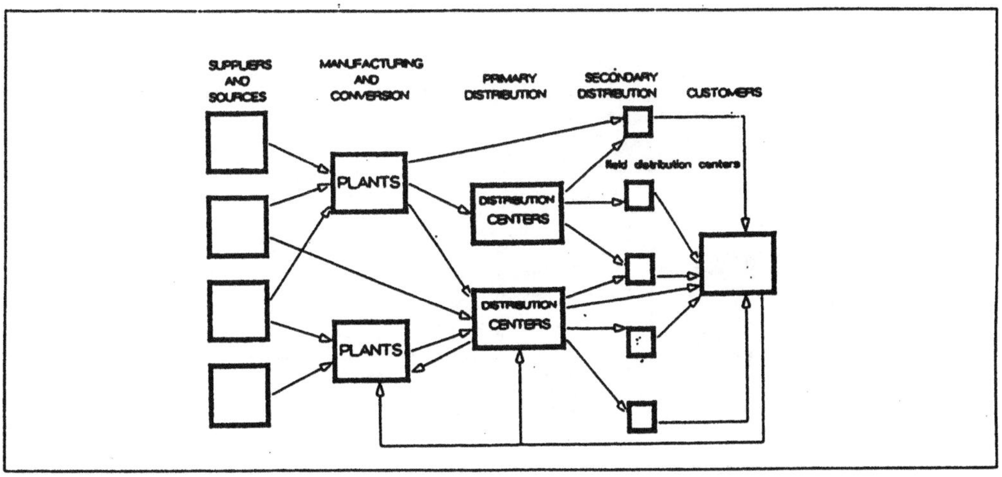

Figure 1: Schematic view of a logistics network

2 The decision process towards a logistics network

In this decision process, many parties are involved. The strategic importance of the decision clarifies the involvement of the board, which usually is supported by a task force of managers with specializations in marketing, logistics, finance, information technology and personnel & organization. The main responsibility of this task force is to support the board in deciding on an appropriate logistics network. The task force takes into account the suggestions and constraints of the board and develops many alternative logistics networks and selects the most interesting ones to discuss with the board. Often also operating companies are asked by the task force to cooperate in the design process. This is done by asking them to deliver the relevant data and by incorporating them in the design and the qualitative analysis and evaluation of different more detailed alternatives.

To assist the board and task force of a company in designing and evaluating many alternative logistics networks, we developed a Decision Support System named SLAM (Strategic Location Allocation Model) (see van Nunen (1980) and Boumans (1991)). SLAM focuses on the quantitative aspects of the logistics network design. It facilitates the task force to translate many constraints and opportunities (like internationalization, potential new locations, sales growth, new products etc.) into designs for logistics networks. Figure 2 shows the technical structure of SLAM. To illustrate the points we want to discuss in this paper we restrict the considerations to the three middle levels of figure 1 So we consider a company that produces different products p in several plants and on different production lines l. These products are transported to distribution centres (called warehouses w) and from there to the customers c. This customer echelon is a combination of the levels 'secondary distribution' and 'customers' in figure 1. To start with, the demand of each customer is supposed to be known. Now the simplified problem is to determine where to produce and store which amounts of product p and how warehouses and customers should be delivered, against minimal costs.

The model we used for this problem is the relaxation of a linear programming model :

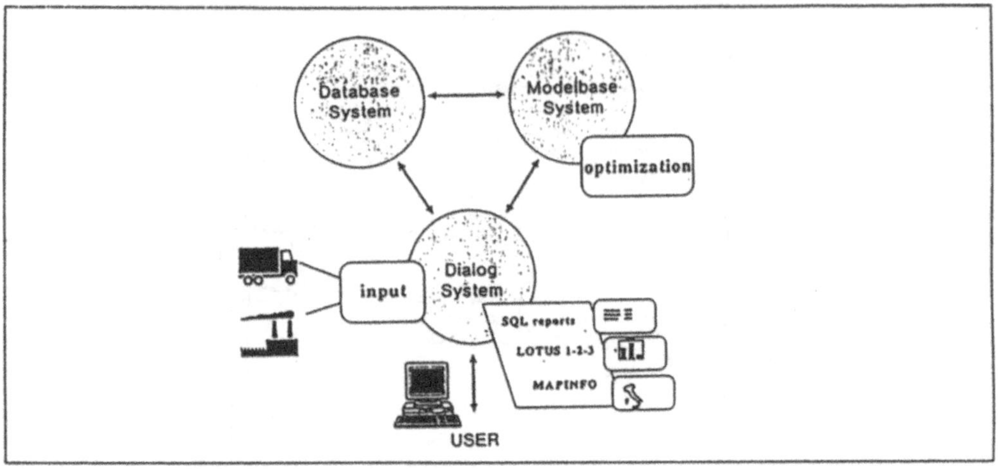

Figure 2: SLAM : a DSS to Support Logistic Network Design

Relevant data are :

c	=	index of the customers
p	=	index of the products
w	=	index of the warehouses
l	=	index of the production lines
d_{pc}	=	demand of customer c for product p
τ_{wc}	=	transportation costs from warehouse w to customer c
$upcap_w$	=	upper capacity of warehouse w
$locap_w$	=	lower capacity of warehouse w
t_{plw}	=	transportation costs for one unit of product p from line l to warehouse w
$upcap_l$	=	upper capacity on line l
$locap_l$	=	lower capacity on line l

Decision variables :

T_{plw}	=	amount of product p to be shipped from production line l to warehouse w
A_{wc}	=	fraction of the demand of customer c to be shipped to customer c from warehouse w.

Objective : Minimize Total Logistics Costs :

$$Z = \sum_p \sum_l \sum_w t_{plw} T_{plw} + \sum_p \sum_w \sum_c \tau_{wc} d_{pc} A_{wc}$$

With respect to :

Complete Delivery of Customers :

$\sum_{w} A_{wc} = 1$ for all c

Capacity Constraints :

$locap_w \leq \sum_{c}\sum_{p} d_{pc}A_{wc} \leq upcap_w$ for all w

$locap_l \leq \sum_{w}\sum_{p} T_{plw} \leq upcap_l$ for all l

Input-Output Constraints :

$\sum_{c} d_{pc}A_{wc} = \sum_{l} T_{plw}$ for all p, w

$T_{plw} \geq 0$ for all p, w

$A_{wc} \geq 0$ for all w, c

In international logistics network design processes, the size of this LP-model can grow up to about 150,000 decision variables and about 50,000 constraints.

3 Aggregation in logistics network design

Many data on sales volumes (per customer and per product) and transportation costs might be available on a very detailed level (e.g. order items, transportation invoices) and other data, concerning sales forecasts, new transportation modes etc, are available on a global level. The board and the task force only need the reports of SLAM (or another DSS, see Ballou (1991), Hagdorn (1993)) on a global level. However, for operating companies and for detailed questions, detailed data might be requested. This controversies in the level of detail of available data and the needed level of detail of the reports require some steps of aggregation and / or disaggregation in the application of a DSS like SLAM. Figure 3 gives an overview of the possibilities for aggregation and disaggregation.

In this paper we focus on the appropriate level of aggregation of basic data towards aggregated data which are input for the LP model of SLAM. The need for aggregated data as input for the LP model is threefold. By aggregating the most detailed data (e.g. aggregate products to product groups or customers to customer groups), a balance is created between the level of detail of the different data groups in the DSS. Secondly, calculating with very detailed data, suggests accuracy which might not exist. Thirdly, the calculation times are reduced by aggregating data. As there often are 10 up to 200 runs of the LP model needed in one single logistic network design project and the size of the LP model might be 150,000 decision variables and about 50,000 constraints, long calculation times will occur. This problem will decrease with the fast changes in computational power of hard-and software. Nevertheless, DSSs only can really support the creative logistic network design process in board meetings, when the calculations times of this large models are very small. At this moment, we can only reach this by reducing the problem sizes, which can be done via data aggregation. The selection of an appropriate aggregation level is often a process of trial and error. In this paper we

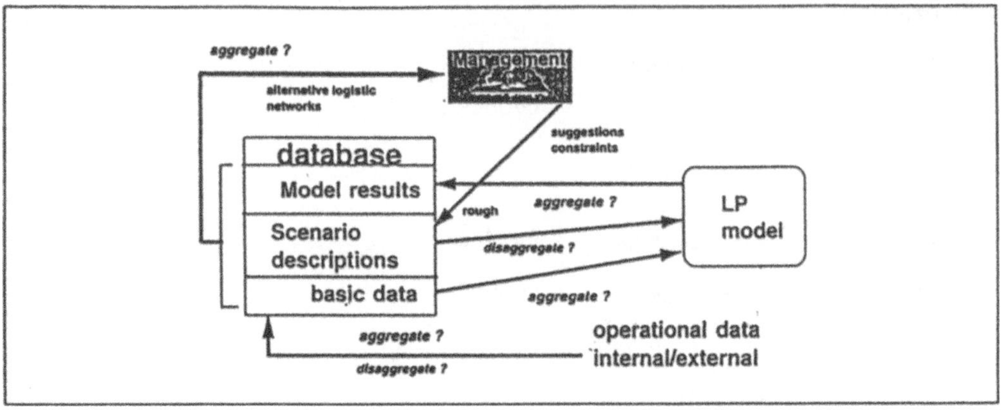

Figure 3: Aggregation in Logistics Network Design

show that several types of upperbounds on the 'aggregation error' exist and can be developed. We focus on the aggregation of customers into customer groups.

4 Customer aggregation

Customers can be aggregated into customer groups via several criteria. Some examples of aggregation rules are : combine customers with the same product-mix to customer groups or combine small customers within the same commercial region into customer groups or combine customers with the same demand frequency within the same geographical area. In this paper we apply the second set of aggregation criteria on a practical situation of the redesign of the European logistics network of a large international company with 1510 customers in Europe. Within this criteria for aggregation, we want to know to which number of customer groups we should aggregate these 1510 customers : we need insight in the error which is introduced by aggregating the customers to different numbers of customer groups.

By aggregation, several types of errors can be introduced : the total logistics costs level can differ, the costs per echelon in logistics network can differ, the product flows and/or the customer warehouse allocations can differ. We will examine the error in total logistics costs which is introduced by aggregation of customers.

Many experiments with several aggregation levels resulted in figure 4, which shows the error which is made when aggregating customers for several problem instances.

This trial and error method can be avoided, by using upper bounds on the total cost error.

5 Upper bounds on the error in total costs

Many upperbounds for linear programming models exist (see for example Geoffrion (1976), Zipkin (1980a and 1980b)). Most of these bounds focus on variable aggregation or constraint aggregation. In our problem we have a combination of these two. The existing bounds are of two types : first we have the so called a priori bound, which calculates an upper-bound of the error in logistics costs, based on the detailed data-set, the aggregation criterium and the aggregation level (so, no solutions

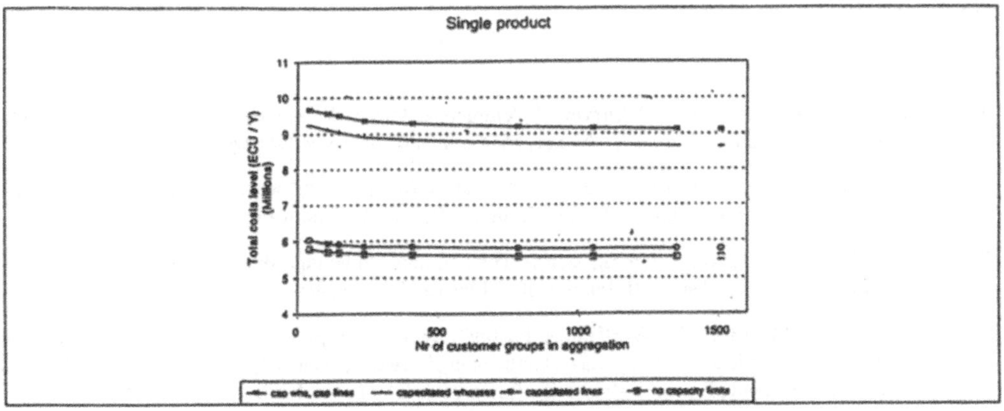

Figure 4: Effect of Customer Aggregation on Error in Total costs

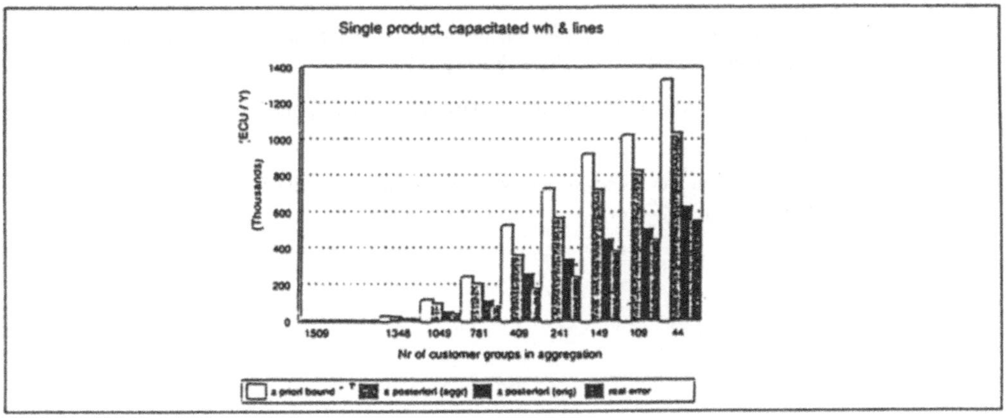

Figure 5: Experimental Results on Different Types of Upperbounds

of the detailed problem nor the aggregated problem are available). Secondly we have the so called a posteriori bound, which uses an optimal solution of the aggregated problem (so, the aggregated problem should be solved). Due to the increasing hardware and software power, nowadays we can solve the large LP models, using detailed data. However, as mentioned, this still requires (very) large computation times for realistic problems. Nevertheless, this enables us to introduce a new type of bound : the a posteriori bound, based on an optimal solution of the detailed problem. Figure 5 shows the power of this new bound.

For the theoretical description of these bounds in single product and multi product situations and for as well uncapacitated as capacitated problems, see Hagdorn (1994).

The advantage of this new type of a posteriori bounds is that, based on the solution of only the detailed problem, upperbounds for the error in total costs in the aggregated problem can be calculated in advance (without solving aggregated problems). This information can be used to determine the appropriate aggregation level for the quantitative part of the logistic network design.

For a more extensive evaluation we refer to Hagdorn (1994).

References

Ballou, R.H., Masters, J.M., 1991, A Survey of Commercial Grade Warehouse Location Models, *Annual Conference Proceedings, Volume I*, Chicago, IL: Council of Logistics Management, pp. 145-180.

Boumans, P.P.W., 1991, *A DSS for production and distribution planning based on database management and linear programming*, Doctoral Thesis, Technical University Eindhoven, The Netherlands.

Geoffrion, A.M., 1976, Customer Aggregation in Distribution Modelling, *Working Paper No. 59*, Management Science Study Center, University of California, Los Angeles.

Hagdorn - van der Meijden, E., Warffemius, P., 1993, *Decision Support Systems for Strategic Planning in Logistics -an overview-*, working paper, forthcoming in Management Report Series of the Rotterdam School of Management, Erasmus University Rotterdam, The Netherlands.

Hagdorn - van der Meijden, E., 1994, *Decision Support on Strategic Planning in Logistics - concepts, tools and applications -*, Ph.D. Thesis, Erasmus University Rotterdam, The Netherlands. (to appear)

Nunen, J.A.E.E. van, Benders, J.F, 1980, A Decision Support System for Location and Allcoation Problems within a Brewery, *Operations Research Proceedings*, pp. 96-105.

Zipkin, P.H. 1980a, Bounds for Aggregating Nodes in Network Problems, *Mathematical Programming*, Vol. 19, pp. 155-177.

Zipkin, P.H., 1980b, Bounds on the Effect of Aggregating Variables in Linear Programs, *Operations Research*, Vol. 28, No. 2, pp. 403-418.

NERVOUSNESS IN INVENTORY CONTROL

Karl Inderfurth, Universität Bielefeld

Inventory control systems usually are integrated in a rolling horizon planning environment by periodically processing new information on inventories, orders, and sales and accordingly adjusting forthcoming ordering decisions. Expecially errors in forecasting future product demands cause frequent replanning activities which give rise to an awkward syndrome of the planning system called nervousness.

Now, the degree of nervousness not only depends on external stochastics in sales but also on internal processing of demand information by the applied ordering policy.

In this paper for two frequently used control rules, the (s,nQ) and the (s,S) policy, an analytical investigation is performed which gives an answer to the question how the type of ordering policies and the fixing of the respective parameters influence the nervousness of the planning system. It is assumed that external disturbances are caused by a stationary stochastic demand process. Nervousness is measured by the replanning probability for setups in two succeeding planning cycles.

The results of the analysis are really interesting and partly surprising. In general, it turns out that a (s,nQ)-rule is connected with less nervous reactions than a (s,S)-rule. The size of the reorder point s has no impact on the planning instability whereas the fixed lot size Q and also the minimum lot size S-s to a large extent influence the degree of nervousness. Evidently for both ordering rules planning stability is very high for extremely small as well as for very large lot sizes while it is less favourable for lots of a middle size. Especially it can be shown that nervousness is considerably high if the lot size is chosen equal or nearly equal to the forecasted demand per period. This means that using a lot-for-lot ordering rule may be rather disadvantageous with respect to a performance criterion like nervousness in planning systems.

A DSS for analyzing logistic problems in a laundry

T. Robbert M. de Jong, Zeetex v.o.f., NL
Leo G. Kroon, Erasmus University Rotterdam, NL

In this presentation we describe a Decision Support System (DSS) that has been developed for a large laundry in the Netherlands.

The laundry has a number of industrial customers such as hospitals and factories. The hospitals need clean sheets for their beds and the factories need clean overalls for their employees. The industrial customers usually have contracts with the laundry. The needed goods are owned by the laundry and rented to the customers. The laundry is responsible for the permanent availability of a sufficient stock of clean goods at the customer.

The current version of the DSS can be used to estimate the required number of overalls for a new industrial customer. Aspects taken into account comprise (*i*) the customer's shift system, (*ii*) the interchange frequency at the customer, (*iii*) the transportation frequency and (*iv*) the throughput-time in the laundry. In the presentation we will give a description of the algorithm that is used for determining the required number of overalls. The algorithm uses the assumption that the customer's shift system is cyclic.

The next version of the DSS should support the management of the laundry to make proposals for contracts with new industrial customers. Contract prices include three parts: (*i*) the renting price of the goods, (*ii*) the transportation price, and (*iii*) the price of maintenance and cleaning of the goods. Also other aspects, such as the length of the contract period and the quality of the overalls will be taken into account.

Keywords: Cyclic scheduling, Decision Support Systems, inventory management.

MULTICOMMODITY NETWORK FLOW PROBLEMS WITH OVERLAPPING CAPACITIES FOR SEVERAL PRODUCTS

Andreas Kindt, Universität Augsburg

Typically the models described in the literature for multistage multicommodity distribution-systems assume, that features combining different commodities only occur in the objective. Therefore, in the case of linear or linearized distribution-system design problems the masterproblem decomposes into as many independent subproblems as there are different commodities. Therefore, each of these subproblems can be solved by known standard optimizing algorithms. Contrary to these assumptions real-life applications usually show a lot of constraints concerning two or more products. Frequently, each client has to be assigned to exactly one distribution center (single-source constraint) or there are capacitated resources at the facilities for production and throughput for all commodities or only subsets of the different commodities. Considering these constraints it's no longer possible to solve the original master-problem as a sequence of independent one-product subproblems.

For a 3-stage distribution-system design problem we will consider joint capacities in the distribution centers for not overlapping subsets of the commodities. The problem formulation used is a multicommodity network flow problem. In addition to the capacitated distribution centers we have to find unique distribution center - client assignments. Considering a real-life application we will compare different methods for solving the resulting assignment-problems. The techniques used are algorithms for transportation problems with split resolution, heuristics for the GAP as well as penalty costs methods.

The application is concerned with the evaluation of implications on the traffic due to decentralised locations for distribution centers in the surrounding of conurbations. The main objective is to avoid the heavy duty truck traffic in the innercity-regions as much as possible and to bundle today's individual traffic for delivery by means of cooperations of the interested companies. Therefore, we have to compare the performance of transportation in the long-distance traffic and the local-traffic at the present situation with the expected values in the case of decentralized locations. The present situation is characterized by a lot of different and usually central distribution centers.

EIN GENETISCHER ALGORITHMUS FÜR DAS TOURENPLANUNGSPROBLEM

Herbert Kopfer, Universität Bremen
Elmar Erkens, Universität Bremen
Giselher Pankratz, Wilnsdorf

Die Aufgaben bei der Entwicklung eines Genetischen Algorithmus (GA) bestehen in der Definition einer geeigneten Problemrepräsentation, der Konfigurierung des GA, der Implementierung und anschließender Parametrisierung.

Die hier gewählte Repräsentation von Tourenplänen durch Strings beruht auf Permutationen der Kundenorte und ist auf praktische Problemstellungen erweiterbar. In einer ersten Version des GA ordnet ein einfacher nachgeschalteter Clusterbuilder die Orte einer Permutation nach dem FIFO-Prinzip den verfügbaren Fahrzeugen zu. Die Reihung der Orte (Route) einer Tour entspricht dabei der Reihenfolge, in der die Orte in der Permutation auftreten. Außerdem wurde eine zweite Version des Clusterbuilders entwickelt, bei der zur Dekodierung der Permutation das Prinzip des Savings-Verfahrens ausgenutzt wird (intelligenter Clusterbuilder).

Bei der Konfigurierung von GAs sind Entwurfsentscheidungen bezüglich des Reproduktionsmodells, der Selektionstechnik und der Rekombinationsoperatoren zu treffen. Als Reproduktionsmodell wurde das Steady-State-Modell, zur Selektion ein ranking-basiertes Verfahren und zur Rekombination problemangepaßte permutationsbewahrende Crossover- und Mutationsoperatoren gewählt. Das Verfahren wurde in Turbo Pascal 6.0 implementiert. Zur Parametrisierung wurden geeignete Werte für die Populationsgröße, Crossover- und Muationswahrscheinlichkeit sowie weitere Parameter zur Steuerung des Selektionsmechanismus empirisch ermittelt. Eine Interpolation der Parameter wurde in die Untersuchung einbezogen.

Bei beiden obigen Repräsentationen durchläuft der GA einen Lernprozeß und setzt sich deutlich von der reinen Zufallssuche ab. Bei Verwendung des ersten (primitiven) Clusterbuilders läßt sich die Lösungsqualität verbessern, indem in Anlehnung an das Sweep-Verfahren einige gute Permutationen erzeugt und in die Anfangspopulation eingepflanzt werden (Hybridisierung durch Seeding). Der auf dem Savings-Prinzip basierende Clusterbuilder ist den beiden anderen Varianten der Repräsentation deutlich überlegen und zeigt beim Test an Hand von Literaturbeispielen ein sehr gutes Lösungsverhalten.

Im Hinblick auf die Praxisrelevanz wurde die Zielsetzung der reinen Streckenminimierung durch die Minimierung der Kosten aufgrund einer gestuften Kostenfunktion ersetzt. Die dazu erforderliche Anpassung des GA war durch eine Modifikation der Evaluierungsfunktion leicht zu realisieren. Das Lernverhalten des GA blieb davon unbeeinflußt.

Ein genetischer Ansatz für die Tourenplanung mit Kundenzeitschranken

Stefan Maziejewski, Planungsbüro Transport und Verkehr, Karlsruhe

Genetische Algorithmen (GA) haben schon oft ihre Qualitäten beim Lösen von klassischen Aufgabenstellungen des Operations Researchs gezeigt. So liegt es nahe, auch Probleme aus der Praxis mit dieser Algorithmenklasse anzugehen. In diesem Vortrag wird ein GA vorgestellt, der für Tourenplanungsprobleme mit Kundenzeitschranken gute Lösungen generiert. Die Ergebnisse sind in der Regel deutlich besser als die vom Savingsalgorithmus erzeugten Resultate. Tourenplanungsprobleme aus der Literatur von Padberg und Rinaldi konnten um bis zu 2,7% verbessert werden.

Die genetischen Algorithmen arbeiten nach folgender Vorgehensweise: zunächst wird mittels einer einfachen Heuristik eine Population aus Individuen erzeugt. Jeweils zwei Individuen aus der Population werden selektiert und mittels eines Kreuzungsoperators ein Nachkomme generiert. Dieser Nachkomme wird lokal optimiert und falls er gut genug ist, in die Population aufgenommen. Von Zeit zu Zeit mutieren einige Individuen der Population.

In der Arbeit werden verschiedene Algorithmen für die einzelnen genetischen Operatoren (Selektion, Kreuzung, Nachoptimierung und Mutation) vorgeschlagen, die speziell auf die Aufgabenstellung angepaßt sind. Diese werden mit verschiedenen Parametern (Populationsgröße, maximale Lebensdauer, Mutationsrate) getestet und verglichen. Umfangreiche Testläufe zeigen, daß eine ausgewogene Kombination von Zufall und gezielter Optimierung die besten Resultate liefert.

DISTRIBUTION OPTIMIZATION BY MEANS OF BREAK QUANTITIES: A CASE STUDY

Roel Nass, Erasmus University Rotterdam
Rommert Dekker, Erasmus University Rotterdam
Willie van Sonderen-Huisman, Erasmus University Rotterdam
The Netherlands

In this paper we present a case study on physical distribution optimization in Western Europe. Finished goods of a similar nature can be delivered either from distribution centres (DC's) or form production service centres (PSC's). The delivery time via a DC is shorter, but higher costs are involved. The selection between the two ways of delivery is based on the *stockmix* and on the *Break Quantities* (BQ's). The set of products present at a DC is called the stockmix, which for efficiency reasons is restricted. If a product present at a DC is ordered and the order quantity is below the break quantity, the order is delivered from the DC and else form a PSC. The problem is to decide on the stockmix and the break quantity for each DC.

The objective is to minimize total distribution and handling costs subject to constraints on service. The distribution costs per kilogram from a certain PSC at a certain day depend on the total volume delivered from that PSC that day. This total volume depends on the BQ's of several DC's. Therefore, it was not possible to decompose the problem to determine a BQ for each DC separately. A further complication was that service constraints were formulated for the entire network, rather than for individual DC's. Our approach was as follows. A simulation program was built to determine total costs and service performance for given stockmixes and BQ combination using all actual orders received over a four-month period. We first considered three options for the stockmixes while using historical BQ's.

The best one was: stocking the products with most orders. Next the simulation program was run for a large set of BQ combinations. We fitted a simplified function to the total costs, to allow a nonlinear optimization by a standard NLP solver. Another simplified objective function was obtained by assuming 80% truck loads for deliveries from PSC's. The methods were compared with the best generated BQ combination on performance and computational effort.

Designing European Logistic Structures

Jo A.E.E. van Nunen
Lorike Hagdorn - van der Meijden

Erasmus University Rotterdam

This presentation will give an overview of the important trends in international distribution, the challenging opportunities companies are offered, the type of decisions that have to be made and the type of distribution problems that can be solved with the assistance of Decision Support Systems.

In international distribution a lot of fast moving trends are interfering with each other, like:

- new market area's
- changing demands
- customer orientation
- cost-awareness
- new production technologies
- information technology
- specialization on logistic function
- internationalization
- reverse logistics
- environment

All these trends give companies the opportunity to develop new distribution concepts to increase their competitive advantage. One of these opportunities is the rationalization of the structure of their international distribution network. Does a company need a national or an international oriented distribution structure? Should they choose for direct or indirect deliveries? How to cope with combining the distribution of different product flows and which delivery times should be realized for which customers are some of the relevant questions. In addition, the information exchange that is required to enable an adequate service is emphasized.

The different subjects will be illustrated by some real life applications.

On (r, Q) Inventory Systems with Truncated Normal Lead Times and Poisson Demands

Ad Ridder, Vrije Universiteit, Amsterdam

Abstract: In this note we shall deal with a continuous review (r, Q) inventory control system. Demands for a single product item occur at epochs generated by a Poisson process, and the replenishment lead time has a truncated normal distribution. We shall derive expressions for the demand probabilities during lead times based on exact expressions for 'tail moments' of the standard normal distribution. The program for finding an optimal reorder point r and an optimal order quantity Q under service level constraints is solved numerically.

1 Introduction

Consider a continuous review (r, Q) inventory control system with stochastic lead times and demands. More precisely, we assume that demands occur at epochs generated by a Poisson process with rate λ and that each demand size equals one unit. The lead time L is stochastic with density

$$f_L(t) = \begin{cases} \dfrac{1}{\Phi(\mu/\sigma)} \dfrac{1}{\sigma\sqrt{2\pi}} e^{-\frac{1}{2\sigma^2}(t-\mu)^2} & t > 0 \\ 0 & t \leq 0 \end{cases} \tag{1}$$

(with $\mu > 0$ and $\sigma \neq 0$) where $\Phi(\cdot)$ stands for the standard normal cumulative distribution function. It says that L is truncated normally distributed. Mean μ_L and variance σ_L^2 of L are expressed using tail moments of the standard normal density function ϕ,

$$M_i(x) = \int_x^\infty u^i \phi(u) du \tag{2}$$

for $x \in \mathbf{R}$ and $i = 0, 1, \ldots$. In section 2 we shall derive expressions for these moments, here we use the result for $i \leq 2$:

$$M_0(x) = \Phi(-x),\ M_1(x) = \phi(x),\ M_2(x) = x\phi(x) + M_0(x)$$

Then

$$\mu_L = \mu + \sigma \frac{M_1(-\mu/\sigma)}{\Phi(\mu/\sigma)}, \quad \sigma_L^2 = \sigma^2 \left(\frac{M_2(-\mu/\sigma)}{\Phi(\mu/\sigma)} - \frac{M_1^2(-\mu/\sigma)}{\Phi^2(\mu/\sigma)} \right)$$

Let the stochastic variable W denote the total demand during lead time. Its distribution is a main issue in modelling inventory systems since it has impact on safety stocks, decision variables and performance. Commonly one approximates it by a normal distribution for getting 'easy' expressions [3]. Also weibull distributions have been used as approximations that are numerical tractable [4]. In some cases it is possible to derive the exact distributions, see [2] for an overview. The particular model of this paper yields exact expressions by using (1) and (2) and executing some algebra:

$$p_W(i) = \mathsf{P}(W = i) = \int_0^\infty e^{-\lambda t}\frac{(\lambda t)^i}{i!} f_L(t)dt = G\frac{(\lambda\sigma)^i}{i!}\sum_{j=0}^{i}\binom{i}{j}\left(\frac{\hat{\mu}}{\sigma}\right)^j M_j(-\frac{\hat{\mu}}{\sigma}) \tag{3}$$

where

$$G = \frac{e^{-\lambda\mu+\frac{1}{2}\lambda^2\sigma^2}}{\Phi(\frac{\mu}{\sigma})} \text{ and } \hat{\mu} = \mu - \lambda\sigma^2$$

Note that the mean and variance of W are

$$\mu_W = \sum_{i=0}^{\infty} i p_W(i) = \lambda\mu_L \text{ and } \sigma_W^2 = \lambda^2\sigma_L^2 + \mu_W$$

The probabilities (3) are numerically tractable only for 'slow moving' items, i.e. $\lambda \leq \mu/\sigma^2$. In [1] these probabilities are approximated by relaxing the truncation of the lead time distribution:

$$\tilde{p}_W(i) = \int_{-\infty}^{\infty} e^{-\lambda t}\frac{(\lambda t)^i}{i!}\frac{1}{\sigma\sqrt{2\pi}}e^{-\frac{1}{2\sigma^2}(t-\mu)^2}dt \tag{4}$$

For slow moving items (4) is the density of a Hermite distribution [1]. Numerical experiments show that for (absolute) small σ and for relative large λ (compared to μ) the exact and approximate values agree with 4 decimales. As for 'fast moving' items where we cannot calculate the exact probabilities or use the hermite approximation, we either use an approximation of the lead time demand (e.g. normal distribution), or we approximate the demand and/or lead time distribution in a way that a well known model arises [2].

2 Expressions for $M_i(x)$

Theorem 1 of this section expresses the tail moments $M_i(x)$ in terms of values of the standard normal distribution and density. First we introduce two collections of integers.

Definition 1 *We introduce for $n = 0, 1, \ldots$ and $k = 0, 1, \ldots, n$ the integers $a_k^{(n)}$ by*

$$\begin{aligned}
a_0^{(0)} &= 1 \\
a_k^{(n)} &= 2na_k^{(n-1)}, \quad n = 1, 2, \ldots, \quad k = 0, 1, \ldots n-1 \\
a_n^{(n)} &= 1, \quad n = 1, 2, \ldots
\end{aligned}$$

and the integers $b_k^{(n)}$ *by*

$$\begin{aligned} b_0^{(0)} &= 1 \\ b_k^{(n)} &= (2n+1)b_k^{(n-1)}, \quad n=1,2,\ldots, \ k=0,1,\ldots n-1 \\ b_n^{(n)} &= 1, \quad n=1,2,\ldots \end{aligned}$$

Then it is easy to show by induction on n the following relations of these integers.

Lemma 1

$$\begin{aligned} a_k^{(n)} &= (2k+2)a_{k+1}^{(n)}, \quad n \geq 1, \ k \leq n-1 \\ b_k^{(n)} &= (2k+3)b_{k+1}^{(n)}, \quad n \geq 1, \ k \leq n-1 \\ b_k^{(n)} &= a_k^{(n)} + \sum_{\ell=k+1}^{n} a_\ell^{(n)} b_k^{(\ell-1)}, \quad n \geq 1, \ k \leq n \end{aligned}$$

The main result of this section is the following statement.

Theorem 1 $\qquad M_i(x) = \alpha_i \Phi(-x) + \beta_i(x)\phi(x)$

with

(i) for $i = 0$

$$\alpha_0 = 1, \ \beta_0(x) = 0$$

(ii) for $i = 2n+1$ *odd* $(n \geq 0)$

$$\alpha_i = 0, \ \beta_i(x) = \sum_{k=0}^{n} a_k^{(n)} x^{2k}$$

(iii) for $i = 2n$ *even* $(n \geq 1)$

$$\alpha_i = b_0^{(n-1)}, \ \beta_i(x) = \sum_{k=0}^{n-1} b_k^{(n-1)} x^{2k+1}$$

Proof (i) is clear.

(ii) For $n = 0$ the statement says $M_1(x) = \phi(x)$ which is an easy exercise. Suppose that (ii) holds for $n \leq m$ and set $n = m+1$ $(i = 2m+3)$. Then by using a change of variable, integration by parts, the induction hypothesis and Lemma 1,

$$\begin{aligned} M_{2m+3}(x) &= \int_x^\infty u^{2m+3} \frac{1}{\sqrt{2\pi}} e^{-\frac{1}{2}u^2} du = \int_{x^2}^\infty v^{m+1} \frac{1}{2} \frac{1}{\sqrt{2\pi}} e^{-\frac{1}{2}v} dv \\ &= x^{2m+2} \frac{1}{\sqrt{2\pi}} e^{-\frac{1}{2}x^2} + (m+1) \int_{x^2}^\infty v^m \frac{1}{\sqrt{2\pi}} e^{-\frac{1}{2}v} dv \\ &= x^{2m+2} \frac{1}{\sqrt{2\pi}} e^{-\frac{1}{2}x^2} + 2(m+1) \int_x^\infty u^{2m+1} \frac{1}{\sqrt{2\pi}} e^{-\frac{1}{2}u^2} du \\ &= x^{2m+2}\phi(x) + 2(m+1)M_{2m+1}(x) = x^{2(m+1)}\phi(x) + 2(m+1) \sum_{k=0}^{m} a_k^{(m)} x^{2k} \phi(x) \\ &= \sum_{k=0}^{m+1} a_k^{(m+1)} x^{2k} \phi(x) \end{aligned}$$

(iii) First show by induction on n (and using part (ii)) that

$$M_i(x) = \sum_{k=0}^{n-1} a_k^{(n-1)} \left\{x^{2k+1}\phi(x) + M_{2k}(x)\right\}, \; i = 2n, n \geq 1 \tag{5}$$

Then when $n = 1$ ($i = 2$) (5) says

$$M_2(x) = x\phi(x) + M_0(x) = x\phi(x) + \Phi(-x)$$

This gives (iii) because $\alpha_2 = b_0^{(0)} = 1$ and $\beta_2(x) = b_0^{(0)}x = x$. Suppose the statement holds for $n \leq m$ and set $n = m+1$ ($i = 2m+2$), then by using (5), the induction hypothesis and Lemma 1,

$$\begin{aligned}
M_{2m+2}(x) &= \sum_{k=0}^{m} a_k^{(m)} \left\{x^{2k+1}\phi(x) + M_{2k}(x)\right\} \\
&= \sum_{k=0}^{m} a_k^{(m)} x^{2k+1}\phi(x) + a_0^{(m)}\Phi(-x) + \sum_{k=1}^{m} a_k^{(m)} \left\{b_0^{(k-1)}\Phi(-x) + \sum_{\ell=0}^{k-1} b_\ell^{(k-1)} x^{2\ell+1}\phi(x)\right\} \\
&= \left\{a_0^{(m)} + \sum_{k=1}^{m} a_k^{(m)} b_0^{(k-1)}\right\} \Phi(-x) + \sum_{k=0}^{m} a_k^{(m)} x^{2k+1}\phi(x) + \sum_{k=1}^{m}\sum_{\ell=0}^{k-1} a_k^{(m)} b_\ell^{(k-1)} x^{2\ell+1}\phi(x) \\
&= b_0^{(m)}\Phi(-x) + \sum_{k=0}^{m} a_k^{(m)} x^{2k+1}\phi(x) + \sum_{\ell=0}^{m-1}\sum_{k=\ell+1}^{m} a_k^{(m)} b_\ell^{(k-1)} x^{2\ell+1}\phi(x) \\
&= b_0^{(m)}\Phi(-x) + \sum_{k=0}^{m-1} \left\{a_k^{(m)} + \sum_{\ell=k+1}^{m} a_\ell^{(m)} b_k^{(\ell-1)}\right\} x^{2k+1}\phi(x) + x^{2m+1}\phi(x) \\
&= b_0^{(m)}\Phi(-x) + \sum_{k=0}^{m} b_k^{(m)} x^{2k+1}\phi(x)
\end{aligned}$$

□

3 Optimization programs

In this section we use the lead time demand probabilities (3) in the average cost function per unit time of the (r, Q) inventory system with backordering. Particularly we are interested in the effect of the lead time variability on the decision variables r and Q. Also we compare the results with those obtained applying the approximation (4).

Assuming no backordering costs we consider minimizing the average costs per unit time under service level constraint. The cost function becomes under the usual assumptions [3].

$$\mathcal{K}(r, Q) = A\frac{\lambda}{Q} + IC\left(\frac{Q+1}{2} + r - \mu w\right)$$

where A is the cost of placing an order, I the inventory carrying charge, C unit cost of the item

We shall consider two optimization programs.

(I) minimize the cost function under service level constraint on the probability of no stockout during a replenishment cycle,

$$B_1(r) = \mathsf{P}(W \le r) = \sum_{i=0}^{r} p_W(i) \ge 1 - \epsilon_1$$

(II) minimize the cost function under service level constraint on the fraction of demand satisfied directly from on-hand inventory,

$$B_2(r, Q) = 1 - \frac{1}{Q}\sum_{i=r}^{\infty}(i - r)p_W(i) \ge 1 - \epsilon_2$$

The ϵ's are typically of the order of 5%. The programs are solved by calculating the differentials

$$\Delta_r\mathcal{K}(r, Q) = \mathcal{K}(r + 1, Q) - \mathcal{K}(r, Q) \text{ and } \Delta_Q\mathcal{K}(r, Q) = \mathcal{K}(r, Q + 1) - \mathcal{K}(r, Q)$$

Then we use structural properties of these differentials and apply standard numerical procedures to find the optimal reorder point r^* and optimal order size Q^*. The following tables summarize some numerical experiments for slow moving items, keeping λ and μ fixed and varying the lead time variance via σ. The cost factors are $A = 500, IC = 25$, the ϵ's are 5%.

	Program I		Program II	
σ	r^*	Q^*	r^*	Q^*
0.05	8	6	5	9
0.5	8	6	5	9
0.75	8	6	6	9
1.25	8	6	6	9
1.5	9	6	6	9
1.75	9	6	6	9
1.95	9	6	7	9

$\lambda = 1, \mu = 4, 0 < \sigma < 2$

	Program I		Program II	
σ	r^*	Q^*	r^*	Q^*
0.05	5	6	3	8
0.25	5	6	3	8
0.5	5	6	3	8
1.0	5	6	3	8
1.25	6	6	3	9
1.4	6	6	4	9

$\lambda = 1, \mu = 2, 0 < \sigma < 1.414$

	Program I		Program II	
σ	r^*	Q^*	r^*	Q^*
0.05	6	8	4	10
0.5	6	8	4	10
0.75	7	8	4	10
1.0	7	8	4	11
1.15	8	8	5	10

$\lambda = 1.5, \mu = 2, 0 < \sigma < 1.155$

	Program I		Program II	
σ	r^*	Q^*	r^*	Q^*
0.05	8	9	5	11
0.6	8	9	5	11
0.7	8	9	5	12
0.95	9	9	6	12

$\lambda = 2, \mu = 2, 0 < \sigma < 1.0$

From these results we conclude that the optimal reorder point and optimal order size are (almost) insensitive for σ (and hence for variability in lead time). We performed the same experiments for the hermites approximation of the lead time demand. It turns out that we find (in most cases) the same numbers. The approximation works so well because we need rounding to integers.

References

[1] U. Bagchi, J.C. Hayya and J.K. Ord. The Hermite Distribution as a Model of Demand during Lead Time for Slow-Moving Items. *Decision Sciences*, 14:447 - 466, 1983.

[2] U. Bagchi, J.C. Hayya and J.K. Ord. Modeling Demand during Lead Time. *Decision Sciences*, 15:157 - 176, 1984.

[3] E.A. Silver and R. Peterson. *Decision Systems for Inventory Management and Production Planning*. John Wiley, New York, second edition, 1985.

[4] P.R. Tadikamalla. Applications of the Weibull Distribution in Inventory Control. *Journal of the Operational Research Society*, 29:77 – 83, 1978.

Modelling and Optimization of a Dynamic Supply, Storage, and Distribution Problem

Günther Ruhe
Hans-Martin Wallmeier

IBM Science Center Heidelberg

Summary

We provide an optimal strategy which determines the purchase, transport and storage of natural gas in a multi-month planning period. The problem is modelled using minimum-cost network flows with additional side constraints. These constraints are due to the objective of minimizing the peak requirements in the purchase of gas. Additionally, confidence constraints are introduced which provide a safety stock against changes of the uncertain consumers' demand.

The underlying network is essentially composed by copies of a fundamental graph describing the physical dependencies between customers, storage devices and delivery nodes. The number of copies of the fundamental graph coincides with the number of periods taken into account. Dynamical aspects are realized by links between the storage devices of consecutive periods.

Computational results for different scenarios with networks of varying size and structure are presented.

Zusammenfassung

Wir betrachten das Problem der Bestimmung einer optimalen Strategie für den Erwerb, den Transport und die Speicherung von Erdgas. Der Planungszeitraum beträgt ca. 2 Jahre, wobei die Festlegung des Verhaltens monatsbezogen erfolgt. Die Fragestellung wird vermittels kostenminimaler Netzwerkflüsse unter zusätzlichen linearen Restriktionen modelliert. Die Zusatzbedingungen resultieren aus der Forderung, die maximale Abnahmemenge - bezogen auf das Planjahr - zu minimieren und ferner aus der geforderten Berücksichtigung von Konsistenzbedingungen zur Absicherung gegen auftretende Änderungen in den unsicheren Bedarfswerten der Verbraucher.

Das zugrundeliegende Netzwerk setzt sich im wesentlichen aus Kopien eines Basisgraphen zusammen, der die physischen Verbindungen zwischen Anbietern, Verbrauchern, sowie Speicher- und Transportknoten beschreibt. Die dynamischen Aspekte werden durch Verbindungen zwischen den Speicherknoten aufeinanderfolgender Perioden realisiert.

Es werden numerische Ergebnisse zu verschiedenen Szenarien mit Netzwerken unterschiedlicher Größe vorgestellt.

Problem Statement

Natural gas is to be transported within a given pipeline system, which is operated by a trading company. The pipeline system interconnects a set of suppliers/producers, some storage devices and a large number of customers. The quantities being procured are immediately delivered to the customers or transferred to the storage devices. The demand for natural gas comes from two distinct groups of customers, industrial enterprises and public energy supply companies. While the demands of the first group is comparatively stable, the requirements of the second group strongly depend on weather conditions.

The natural gas can be delivered out of the trading companies' storage devices, or directly from the producers. There exists a (usually) small number of producers of natural gas. These are located at different sites of the pipeline system. The quantities that they are able to deliver differ as well as the contractual terms do. Due to the connectivity of the pipeline system it is possible to transport natural gas from any producer and storage device to any customer. Thus it must be ensured that the total demand for natural gas is less than or equal the supplies delivered plus the volume of gas available from the storage devices.

We present a planning tool based on a network flow optimization algorithm. The application program determines the quantities of natural gas coming into the pipeline system as well as the quantities to be stored. The objective of the planning tool consists of fulfilling the demands of the customers at minimum costs for the trading company. The costs depend non-linearly on the quantities being purchased. The tool is designed to cover a planning period of about two years on a monthly basis. It can be also used for short-time planning on daily base. Due to the length of the basic period the problem of controlling the gas transport within the pipeline system is neglected. The complexity of the planning problem on one hand results from the fact that a large number of time periods has to be considered. On the other hand it comes from the structure of the purchase price of the natural gas. Its value depends on the maximum quantity being purchased in the basic intervals belonging to the gas year. The purchase price consists of a fixed and a variable component.

The Model: Network Flows with Side Constraints

Optimization algorithms that exploit the network structure can be awaited to be faster than general LP-solvers. Furthermore, the possibility of displaying the graph-theoretical structure in two-dimensional drawings greatly simplifies the insight into the problem and the interpretation of the results. The practical problem described above is modelled using minimum-cost network flows with additional side constraints. We use the standard notation of network flow theory as described in Ruhe [8]. The flow conservation rules read

$$\sum_{(j,k)\in A} x(j,k) - \sum_{(i,j)\in A} x(i,j) = \begin{cases} \text{val(x) if } j = 1 \\ -\text{ val(x) if } j = n \\ 0 \text{ otherwise} \end{cases} \tag{1}$$

where val(x) denotes the value of the flow. The flows satisfy the capacity constrains, thus

$$0 \le x(i,j) \le cap(i,j) \quad \text{for all } (i,j) \in A \tag{2}$$

However, a number of additional constraints are needed in order to obtain a sufficiently exact model of reality. These constraints are stated in the form

$$S \times x + P \times y \le r \tag{3}$$

In (3), y denotes a non-network related variable, and S and P are (hopefully sparse) general matrices of corresponding dimension. With the right-hand side vector r and upper bounds *ub* for the side columns

$$0 \leq y \leq ub \tag{4}$$

the feasible area is summarized in the polyhedron (X,Y).

Let c and d denote cost vectors for the variables x and y, respectively. Then the network with side constraints can be represented in the form

$$\min\{c^T x + d^T y / (x,y) \in (X,Y)\} \tag{5}$$

The Transformation

Within the present section we describe how to formulate the original problem in terms of (5). The underlying graph $G = (V,A)$ is composed by copies of a basic subgraph which reflects the topology of the pipeline system.
The typical structure of such a subgraph is shown in figure 1.

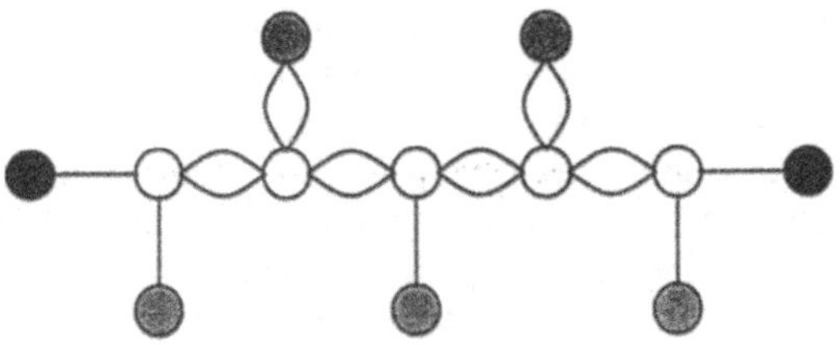

Figure 1: Example of a basic subgraph with

two delivery nodes (dark)
two storage devices (medium shaded)
three customers (lightly shaded)
five transition nodes (bright)

The purchase of natural gas has to be planned for consecutive time intervals. Thus the planning problem exhibits a dynamic structure. For dynamic network problems there has been developed a number of algorithms, each of them well suited to solve the optimization problem under specific assumptions. The survey article of Aronson [2] provides an overview. In the present context we are able to map the dynamic problem structure into a static one as was done in Grossmann et al. [5] for a dynamic assignment problem. The method behind is to copy the basic subgraph where the number of copies corresponds to the number T of time intervals. These subgraphs are connected by a source vertex for each supplier and arcs connecting vertices which represent storage devices of consecutive time intervals. Additionally, to describe the volumes of gas in the storage devices at the beginning of the first period we introduce additional vertices connected with the corresponding vertices of the devices in the first period.

The composition of G for the basic subgraph of figure 1 is illustrated in figure 2.

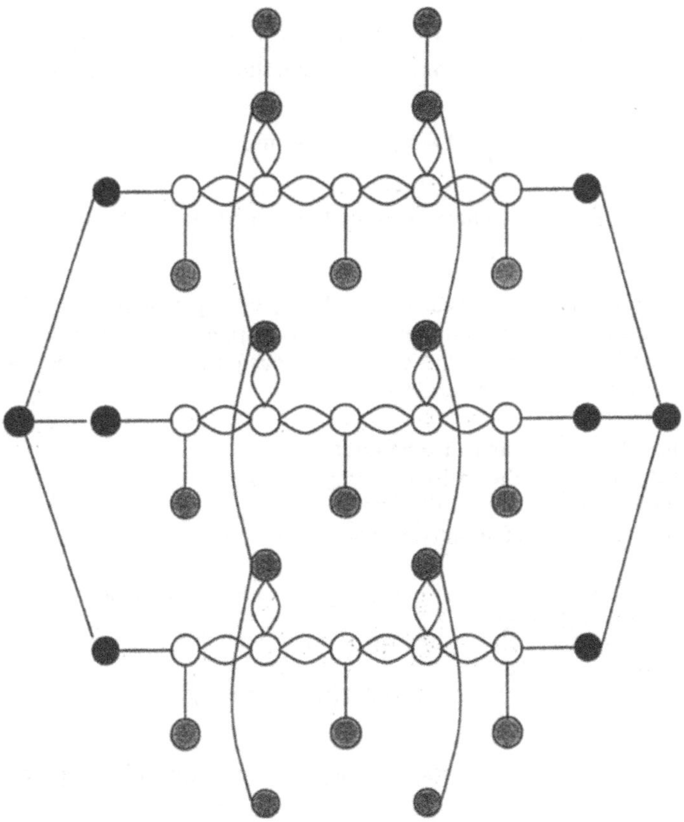

Figure 2: Composition of G by three copies of the basic subgraph.

The purchase price of natural gas is typically a piece-wise linear function. It is well known (cf. Ahuja et al. [1]) that convex cost flows can be transformed to the linear case by introducing parallel arcs. For each of these arcs there is a upper and lower bound of the flow in correspondence to the stability interval with respect to the marginal costs. In the following we assume a convex and piece-wise linear cost function.

Ericson et al. [4] consider a network flow problem with additive and concave flow costs. They show that, in the general case, exponential time is required to find the optimum flow. Guisewite & Pardalos [6] prove that the single-source uncapacitated version of the problem, where all arcs except one have linear cost, is solvable in polynomial time.

The purchase price of natural gas strongly depends upon the maximum quantity received within all time intervals. The trading company's objective is to minimize the peak requirements by an appropriate storage policy. This is realized by a penalty component within the objective function and additional constraints of the form

$$supply(i,t) \leq \lambda(i) \quad \text{for all producers i and } t = 1, \ldots, T \tag{6}$$

In (6), $supply(i,t)$ denotes the supply of producer i for time interval t. With corresponding penalties $p(i)$ and

$$f_1 = \sum_i p(i) \times \lambda(i) \to \min \tag{7}$$

we ensure that the peak requirements are minimized. The second component f_2 of the objective function reflects the variable component of the purchase costs and is defined by

$$f_2 = \sum_i \sum_t^T C(i) \times supply(i,t) \to \min \tag{8}$$

where C(i) is the (piecewise linear) purchase price of the i-th producer. Now, the objective function becomes $f = f_1 + f_2$.

We remark that there is no difficulty to incorporate also operating expenses into the model by defining corresponding cost parameters on arcs. There is a trade-off between the purchase costs and the level of confidence to meet the demand of customers. We use the following notation to formulate these confidence constraints:

$\alpha(t)$	- security factor for the t-th time interval
β	- security factor for the entire planning period
dem(j,t)	- demand of customer j within time interval t
store(i,t)	- volume of gas stored in device i within time interval t.

For each time interval t the total volume of gas being available must be a pre-determined multiple of the expected sum of demands:

$$\sum_i store(i,t) + \sum_i supply(i,t) \geq \alpha(t) \times \sum_j dem(j,t) \tag{9}$$

for all time intervals t = 1, . . . , T. Correspondingly, for the entire planning period the inequality

$$\sum_t (\sum_i store(i,t) + \sum_i supply(i,t)) \geq \beta \times \sum_t \sum_j dem(j,t). \tag{10}$$

has to be satisfied.

The Solution Algorithm

The most popular method to solve the network flows with side constraints problem (5) is the primal partitioning procedure (for a description see Ruhe [8]). The main idea is to apply the simplex method where the basis and its inverse are maintained in partitioned form. This allows to handle a part of the basis in correspondence to a spanning tree which is computationally efficient. Barr et al. [3] combine this approach with an LU factorization of the basis inverse as known from modern linear programming systems. Kennington & Whisman [7] presented a Fortran code called NETSIDE for solving (5) which is mainly based on the results of Barr et al. [3]

Considering (5) we remark that it contains two important special cases:
The minimum-cost flow problem (if there are no additional constraints) and linear programming (if there are no network-related constraints). Due to the authors, NETSIDE is designed for problems in which the number of rows in (1) (i.e., the number of vertices of the

graph) is about ten times the number of rows in (3). This is also the approximate bound for which this specialized code should perform better than general linear programming systems.

To test the computational behaviour of the algorithm we randomly generated a series of test examples. As input parameters for the generation procedure we used

- the number of time periods
- the number of customers
- the number of storage devices
- the number of deliverers
- levels of confidence
- intervals for cost coefficients (equally distributed)
- variation of demands

The computations have been performed on an IBM 3090/300E VF under VM. In the table, CPU-time and number of iterations until reaching optimality are reported. We considered a problem with three storage devices and two producers. The arc capacities as well as the level of confidence were assumed to be constant. The piecewise linear cost function given in (8) is composed of three intervals.

Table:

	consumers a)	periods b)	nodes c)	arcs d)	network restr. e)	non-network restr. f)	CPU g)	iter. h)
1.	20	6	294	464	1222	19	2.6	623
2.	30	6	414	644	1702	19	4.4	837
3.	50	6	654	1004	2662	19	7.3	947
4.	100	6	1254	1904	5062	19	21.0	1504
5.	20	12	576	926	2428	37	11.4	1487
6.	30	12	816	1286	3388	37	17.5	1760
7.	50	12	1296	2006	5308	37	31.1	2090
8.	100	12	2496	3806	10108	37	97.6	3629
9.	20	24	1140	1850	4840	73	46	3139
10.	30	24	1620	2570	6760	73	74	3725
11.	50	24	2580	4010	10600	73	145	4872
12.	100	24	4980	7610	20200	73	508	9103

The table gives the numerical results obtained from 12 randomly generated test examples. The columns read

a) number of consumers
b) number of time periods
c) number of nodes
d) number of arc
e) number of network related constraints
f) number of non-network related constraints
g) CPU-time in seconds
h) number of iterations to reach optimality

The main conclusion that we obtained from these numerical investigations is that we can solve problems of realistic size up to optimality in a reasonable time.

Acknowledgement
We greatly appreciate the support of Jeff Kennington who provided the program NETSIDE.

References

[1] **AHUJA,R.K.; MAGNANTI,J.B.; ORLIN,J.B.** *Network Flows: Theory, Algorithms, Applications.* Prentice Hall 1993.

[2] **ARONSON,J.E.** *A Survey of Dynamic Network Flows.* Annals of Operations Research 20, 1-66, 1989.

[3] **BARR,R.S.; FARHANGIAN,K.; KENNINGTON, J.L.** *Networks with Side Constraints: An LU Factorization Update.* The Annals of the Society of Logistics Engineers 1, 66-85, 1986.

[4] **ERICSON,R.E.; MONMA,C.L.; VEINOTT,A.F.** *Send-and-split Method for Minimum-concave-cost Network Flows.* Mathematics of Operations Research 12, 634-664,1987.

[5] **GROSSMAN,G.; GUARISO,G.; HITZ,H.; WERTHNER,H.** *A Min-cost Flow Solution for Dynamic Assignment Problems in Networks with Storage Devices.* To appear in Management Science, 1993.

[6] **GUISEWITE,G.M.; PARDALOS,P.M.** *A Polynomial Time Solvable Concave Network Flow Problem.* Networks 23, 143-147, 1993.

[7] **KENNINGTON,J.L.; WHISMAN,A.** *NETSIDE user's guide.* Technical Report 86-OR-01 (Revised). Southern Methodist University, Department of Operations Research 1988.

[8] **RUHE,G.** *Algorithmic Aspects of Flows in Networks.* Kluwer Academic Publishers, 1991.

VARIABLE RECIPES: TO USE OR NOT TO USE

W.G.M.M. Rutten, Eindhoven University of Technology, NL
J.W.M. Bertrand, Eindhoven University of Technology, NL

Abstract: Process industries often obtain their raw materials from mining or agricultural industries. These raw materials usually have variations in quality which often lead to variations in the recipes used for manufacturing a product. Another reason for using variable recipes is to minimize production costs by using the cheapest materials which still lead to a satisfactory quality in the product. A third reason for using variable recipes is in case that not all materials for the standard recipe are available. This last use of variable recipes can be avoided by keeping sufficient safety stock. This means that keeping safety stock can be balanced with using variable recipes. In this paper we study this issue by means of a small scale model of the situation. We construct a simple, quantitative model of the situation under investigation and for this simple situation we derive a decision procedure that balances safety stock costs and increased production costs.

1. Introduction

During the last decade, various articles have been published on production control in process industries. Most of these articles focus on the typical characteristics of process industry as compared to the discrete manufacturing situations. In this body of literature two extreme types of process industry can be distinguished; the process/flow industry and the batch/mix industry (Fransoo and Rutten 1993). Process/flow is defined as: a manufacturer who produces with minimal interruptions in any one production run or between production runs of products which exhibit process characteristics such as liquids, fibres, powders, gases etc. Batch/mix is defined as a process business which primarily schedules short production runs of products (Connor 1986). In this paper we concentrate on the batch/mix process industry.

Process industries often obtain their raw materials from mining or agricultural industries. These raw materials have natural variations in quality. For example, crude oils from different oil fields have different sulphur contents and different proportions of naphta, destillates and fuel oils. Oil refinery designs, production plans and operating schedules must account for this variability in crude oil qualities (Taylor et al 1981). May (1984) mentioned that material variability implies that the real characteristics of the material are usually not known until the production process is started. The variability in raw materials sometimes even determines which product will be produced (Rice and Norback 1987).

Variations in raw material quality often lead to variations in bills of material (recipes[1]) (May 1984, Cokins 1988). For example, variation in the moisture contents, acidity, viscosity or concentration of active ingredients in different raw materials may cause variations in raw material proportions required to make finished product quality specifications (Taylor et al 1981). This variation in raw material quality is one reason for using variable recipes.

[1] A recipe is the specification of the amount of each raw material which is required per unit of measure of the finished product.

A second reason for using recipes that are dependent on the quality of available materials is to minimize the total materials costs to produce the finished product. Then for each production order the recipe is determined such that the finished product quality specification is met with a combination of available raw materials which produces least costs. For example, a pet food may have specifications for the minimum amount of protein, carbohydrates and fat per pound of pet food; however, the proportions of various materials may be varied depending on their current price, quality and availability. In this way recipe variability is used to minimize product costs.

A third reason for varying the recipe of a product is that not all the materials which are required for the standard recipe are available at the time of production. This option is available if the finished product can be made with different recipes, whereby a specific cost is associated with each recipe. Of course, the recipe with the minimum costs is the standard recipe, but this standard recipe can only be applied if all raw materials required are available. If not all required raw materials are at hand, an alternative recipe may be applied which uses available materials, but ends up with higher costs. From this it follows that in this situation the use of variable recipes should be avoided. Two ways to avoid the use of alternative, more expensive recipes are:

1. to postpone production of the finished product until all raw materials in the standard recipe are available,
2. to always have sufficient raw materials available to cope with the demand for raw materials which follows from the demand for finished products.

The first condition can be satisfied if the market accepts variable leadtimes which depend on materials availability. Generally however, the finished products are commodities which are delivered to many customers and for which a short, standard leadtime must be used in the market place to maintain competitiveness. Thus postponing production often is not a realistic option. The second condition can be satisfied if either the demand for finished products can be predicted with certainty over a horizon which covers the production throughput time plus the raw materials replenishment throughput time (the stacked leadtime) or, sufficient raw materials safety stock is kept to cope with the cumulated demand uncertainty over the stacked leadtime. In a commodity market, the customers usually demand short delivery leadtimes. Therefore, unless production and replenishment leadtimes are very short, the only means to avoid the use of alternative recipes is to have sufficient safety stock.

An implicit assumption in the previous discussion is that all raw materials received are exactly according to their quality specification. Often however, raw materials have to be rejected or graded differently than ordered. This supply uncertainty in the replenishment orders of specific raw materials must be added to the requirements uncertainty to get the total uncertainty that the materials planner has to cope with.

Now the question arises under what conditions, or to what extent, raw materials safety stocks should be used to cope with uncertainty, and under what conditions, or to what extent, alternative recipes should be used. It will be clear that the answer to this question depends on:

- the uncertainty to cope with
- the costs of keeping safety stock
- the increase in production costs due to the alternative recipe.

In this paper we study this issue by means of a small scale model. The purpose of our study is to construct a simple, quantitative model of the situation under investigation and to derive a decision procedure for this simple situation that balances safety stock costs and increased production costs. The remainder of this paper is as follows. First, we will consider a simple model without variation in raw materials quality and derive an equation for balancing safety stock costs and increased production costs. Next, we will extend the model to situations with variable qualities in raw materials and show that the value of the balance equation increases, which means that using variable recipes becomes more attractive. In the last section, some conclusions are given.

2. The simple model

We consider the situation with two finished products denoted x and y, and two raw materials, denoted a and b. Product x uses per unit demand one unit of raw material a, and product y uses per unit demand one unit of raw material b. However, product y can also be made out of raw material a (one unit per unit demand); but then the production costs increase. Raw materials are replenished using an $(S-1, S)$ replenishment system, with for both materials a replenishment leadtime of four periods. The order-up-to level S for each material is based on the average and variance in demand for its main product. The demand process is modelled as follows: Each period for each product one order is generated with order size according to an (n, λ) Erlang distribution function. Orders are processed in the same period in which they are generated; there are no capacity limitations, only materials availability limitations. We measure the α service level, which is defined as the probability of no stockout in a period. The S-level has been set such that 95% ($=\alpha$) of all orders can be produced in the ordering period, assuming use of alternative recipes. For the time being we only consider demand uncertainty and neglect supply uncertainty. If due to lack of raw materials, an order for product x cannot be completely produced, the rest of the order is backlogged and has priority in the next period when competing for materials. If due to lacking materials an order for product y cannot be completely produced, the (rest of the) order is produced out of material a. If there is not enough material a, the rest of the order for product y is also backordered.

Now it will be clear that when using the alternative recipe for product y two things will happen. First, in 5% of the cases for product y an out of stock situation occurs and the alternative recipe will be tried for. This will be succesfull in at most 95% of the cases (service level of material a is 95%, we neglect the interactions). This will result in an upperbound for the service level for product y of $0.95+0.05\times0.95=0.9975$. Also this will result in an increase in the costs of product y. Suppose that producing product y out of material a costs p times the costs of producing it out of material b, denoted by c_p. Then under the conditions indicated the average costs for product y would be: $0.95\times c_p+0.05\times p\times c_p$. The increase in cost per unit then is: $(p-1)\times 0.05\times c_p$. Second, in approximately 5% of the cases, material *a* will be used for product *y* and this will decrease its availability for product *x*. This will lead to a reduction in service level for product *x*. The magnitude of the reduction depends on the level of safety stock.

From the reasoning above it follows that when using recipe flexibility it is impossible to control the service levels of *x* and *y* only by manipulating the *S*-levels of *a* and *b*. For instance, we could decrease the S_b-level such that the actual service level for product *y*, using recipe flexibility, is equal to 95%. This however will increase the number of occasions where *a* will be used for *y* and will decrease the availability of *a* for product *x*. Thus the service level for product *x* will decrease. Suppose we try to compensate that by increasing the S_a-level to the extent that the service level of *x* is again 95%. Then the service level for *y* will increase also, because the probability of succes when trying the alternative recipe increases.

In short, due to the single-direction recipe flexibility, the service level of product *y* will always be larger than the service level of product *x*. This phenomenon is illustrated in Table 1 which shows the actual service levels α_x and α_y, the (relative) number of cases where the alternative recipe is used, $\#a_y$, the mean fraction of material *a* in product *y*, *fa*, and the average material stock levels Inv_a and Inv_b, as a function of the *S*-level for material *b*, for two combinations of the variation coefficients in demand for *x* and *y*.

Table 1 Simulation results when using variable recipes. The S-level$_a$ is 3661, which is the 95% service level for product x with a leadtime of four periods. S-level$_b$ is changed in order to decrease the α_y. The results given are the means of 15 runs of 4.000 periods each.
The number of cases where the alternative recipe is used, $\#a_y$, is lower than expected since we use an order-up-to level which is based on a backorder situation. The demand for raw material b, however, can be considered as a lost sales situation (if no b is available, a is used and thus demand for b is lost). Using a 'backorder order-up-to level' in a lost sales situation results in a higher α (Rutten et al. 1992), and therefore the resulting $\#a_y$ will be lower.

σ_x/μ_x	σ_y/μ_y	S-level$_b$	α_x (%)	α_y (%)	fa (%)	$\#a_y$ (%)	Mean Inv$_a$	Mean Inv$_b$
400/400	400/400	3661 (95%)	94.33	99.38	2.88	3.09	1832.43	1908.82
		3197 (90%)	93.67	98.67	5.46	5.82	1789.28	1495.53
		2907 (85%)	93.03	97.90	7.91	8.47	1748.69	1254.10
		2688 (80%)	92.35	97.00	10.27	10.95	1709.44	1082.83
		2510 (75%)	91.65	96.20	12.58	13.36	1671.41	952.17
		2356 (70%)	90.95	95.25	14.91	15.65	1633.42	846.25
400/400	200/400	2788 (95%)	94.76	99.68	1.34	3.09	1858.09	1018.75
		2590 (90%)	94.57	99.37	2.56	5.81	1837.05	843.57
		2462 (85%)	94.38	99.02	3.76	8.32	1816.52	737.96
		2363 (80%)	94.20	98.64	4.96	11.02	1796.01	661.44
		2281 (75%)	93.97	98.30	6.18	13.48	1775.32	602.19
		2208 (70%)	93.79	97.89	7.43	15.91	1754.03	552.66

One way to tackle this control problem is to let the use of the alternative recipe depend on the cumulative measure of the service level, which is defined as the actual service level measured in the current subrun (2.000 periods). If the cumulative service is below the target (95%) then the use of the alternative recipe is allowed for the production order considered. If not, then the alternative recipe is not used, the order is backlogged to the next period. Using this decision procedure the control of the service levels improves a lot, as is shown in Table 2.

Table 2 Simulation results with using variable recipes and tuning α_y. The target α for controlling the service level of product y is 95% in all situations. The situations are identical to the ones in Table 1. The results given are the means of 15 runs of 4.000 periods each.

σ_x/μ_x	σ_y/μ_y	S-level$_b$	α_x (%)	α_y (%)	fa (%)	#a$_y$ (%)	Mean Inv$_a$	Mean Inv$_b$
400/400	400/400	3661 (95%)	94.87	95.65	0.51	0.48	1872.14	1889.24
		3197 (90%)	94.09	94.98	3.42	3.49	1823.84	1483.63
		2907 (85%)	93.36	94.91	6.46	6.62	1773.17	1248.64
		2688 (80%)	92.56	94.84	9.27	9.64	1726.47	1081.40
		2510 (75%)	91.81	94.75	11.92	12.54	1682.55	951.89
		2356 (70%)	91.03	94.45	14.54	15.21	1639.78	846.77
400/400	200/400	2788 (95%)	94.95	95.72	0.20	0.43	1877.34	1008.18
		2590 (90%)	94.71	95.04	1.51	3.04	1854.95	836.23
		2462 (85%)	94.48	94.99	2.90	5.75	1831.35	732.91
		2363 (80%)	94.26	94.97	4.26	8.75	1808.29	658.83
		2281 (75%)	94.03	94.96	5.58	11.40	1785.78	600.56
		2208 (70%)	93.83	94.93	6.93	14.06	1762.60	552.45

There is still interaction between the service levels of the two products, but since the service level of y is under control for each order-up-to level of S_b, the order-up-to level of S_a can be determined experimentally (for instance via simulations) such that the service level α_x is equal to 95%, as is shown in Table 3.

Table 3 Simulation results with using variable recipes and tuning α_y. The S-level$_a$ is increased until α_x = 95%. The situations are identical to the ones in Table 1. The results given are the means of 15 runs of 4.000 periods each.

σ_x/μ_x	σ_y/μ_y	S-level$_b$	increase S-level$_a$	α_x (%)	α_y (%)	fa (%)	#a$_y$ (%)	Mean Inv$_a$	Mean Inv$_b$
400/400	400/400	3661 (95%)	10	94.97	95.65	0.51	0.48	1881.63	1889.24
		3197 (90%)	100	94.96	94.98	3.41	3.50	1918.72	1483.46
		2907 (85%)	195	95.01	94.92	6.43	6.50	1958.38	1248.81
		2688 (80%)	265	94.97	94.90	9.32	9.53	1975.99	1082.65
		2510 (75%)	330	94.93	94.88	11.92	12.40	1993.80	954.52
		2356 (70%)	415	94.98	94.87	14.60	14.97	2029.72	849.54
400/400	200/400	2788 (95%)	0	94.95	95.72	0.20	0.43	1877.34	1008.18
		2590 (90%)	30	94.97	95.04	1.52	3.04	1883.35	836.31
		2462 (85%)	57	94.95	94.99	2.91	5.75	1885.26	733.04
		2363 (80%)	90	94.98	94.97	4.27	8.66	1893.43	659.19
		2281 (75%)	110	95.00	94.97	5.58	11.34	1889.94	600.68
		2208 (70%)	130	94.95	94.94	6.94	14.00	1885.61	552.69

Now suppose that we know all combinations of S'_a and S'_b which produce $\alpha_x=\alpha_y=95\%$, and compare these with the levels S_a and S_b which lead to $\alpha_x=\alpha_y=95\%$ in the situation without use of alternative recipes. Then it will be clear that: $S'_a-S_a< S_b-S'_b$, since with the use of the alternative recipe, some of the demand uncertainty for material a and b is puzzled. As a result the total of inventory decreases with an amount: $S_b-S'_b+S_a-S'_a$.

This decrease in inventory must be valued and balanced against the increase in production costs, which can be modelled as: $fa \times (p-1) \times c_p$ per unit production y, where fa is the mean fraction of product y which is made of alternative material a and $(p-1)$ is the increase in production costs per unit. To obtain the annual increase in production cost we multiply this by D_y, the annual demand for product y. Since with each combination (S'_a, S'_b) a fraction of alternative recipe use of fa is involved, denoted by fa', the entire range of combinations can be valued as follows:

Determine from among all combinations (S'_a, S'_b) the combination such that $\alpha_x = \alpha_y = 95\%$ for which the following expression takes a maximum value:

$$(S_b - S'_b + S_a - S'_a) \times c_i - fa' \times (p-1) \times c_p \times D_y \tag{1}$$

where the costs c_i consist of capital investment cost related to the product, valued at the interest rate, plus the costs of the capital invested in inventory related infrastructure (silos etc.)

If this maximum value is negative than it does not pay to use the alternative recipe. Then the increase in production costs is never offset by the decrease in inventory holding costs. However if this maximum value is positive then the use of alternative recipes can lead to a decrease in total costs, provided of course that the right combination of S_a and S_b is used. Figure 1 illustrates equation (1), using the numbers from Table 3. As can be seen in Figure 1, the maximum p for which it still pays to use alternative recipes under the given conditions, equals 1.0159 and therefore the maximum per unit increase in product costs $((p-1) \times c_p)$ in case of alternative recipe use equals 1.59%. Since the profit margins are very small in the commodity market, this value is of significant importance.

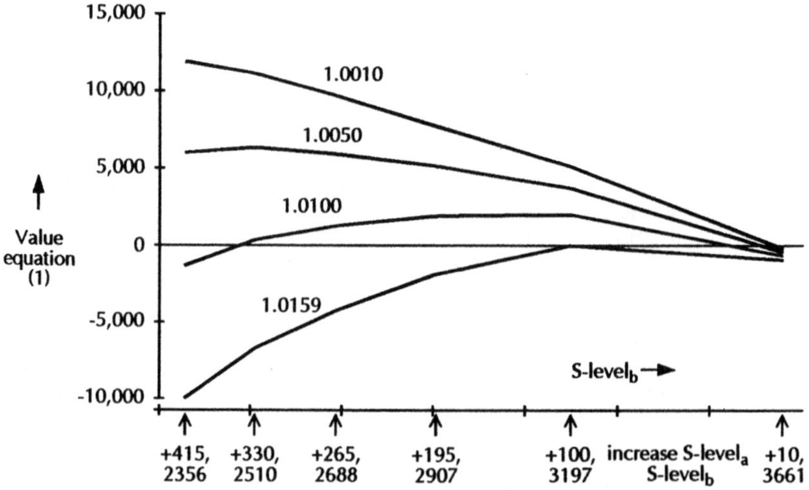

Figure 1 Equation (1) for the situation with $\sigma_x/\mu_x = \sigma_y/\mu_y = 400/400$, for several factors p. The different (S'_a, S'_b) combinations are marked on the x-axis, the y-axis gives the value of equation (1). The factor c_i is calculated as the capital investment cost related to the product, valued at an interest rate of 15%. Factor c_p equals 100 and annual demand for product y is taken 100,000 ($250 \times \mu$) units.

3. Variation in raw materials quality

In many process industries, raw materials are graded at arrival. For instance in animal food industry, soya is used in which the fraction of protein is the main ingredient to classify soya. If 'soya class II' is ordered, the arriving replenishment lot can be in class I, II or III, where a higher class stands for more protein. Thus the grading process can yield a lower, but also a higher quality than ordered. We still consider the case with two products and two raw materials, which means that grading raw materials is worked out as follows: raw material *a* is reordered and at delivery we detect (with probability *q)* if the raw material indeed is raw material *a*, or if it can be classified as raw material *b* instead. The same procedure is followed for raw material *b* turning out to be *a*.

Due to this supply uncertainty, the S_a and S_b levels have to be increased, even if we do not use variable recipes. So first we have to find the inventory levels which are necessary to achieve the service level of 95% without using variable recipes (the (S_a, S_b) values). Second, we will use variable recipes, and an equal procedure as shown in the previous section can be followed; we decrease the S_b-level and control the service level of product *y* and we increase the S_a-level to achieve the situation where $\alpha_x=\alpha_y=95\%$. In Table 4, the results of this procedure are shown for one combination of variation coefficients in demand for *x* and *y* and for two values of the probability that the material received equals the material ordered *(q)*.

Table 4 Simulation results with grading of raw materials. We use two values for the probability that the material received equals the material ordered (q). In the situation without using variable recipes, first the results are shown if no correction on the S-levels is made and second the corrected situation is shown. The S-levels for the corrected situation are the (S_a, S_b) combination for keeping safety stock. In the situation with variable recipes, several (S'_a, S'_b) combinations are given for which $\alpha_x = \alpha_y = 95\%$. The results given are the means of 15 runs of 4,000 periods each.

σ_x/μ_x	σ_y/μ_y	q	S-level$_a$	S-level$_b$	α_x (%)	α_y (%)	fa (%)	#a_y (%)	Mean Inv$_a$	Mean Inv$_b$
without using variable recipes, keeping safety stock										
400/400	400/400	0.90	3661	3661	93.10	93.10	0.00	0.00	1943.50	1951.98
			3911	3911	95.03	95.01	0.00	0.00	2178.80	2187.22
		0.70	3661	3661	87.79	88.05	0.00	0.00	2244.55	2282.32
			4561	4561	94.94	95.03	0.00	0.00	3071.15	3110.19
with using variable recipes										
400/400	400/400	0.90	3911	3911	95.02	95.78	0.45	0.45	2164.21	2194.29
			3911	3661	94.95	95.21	1.37	1.32	2135.43	1970.84
			4025	3197	94.93	94.98	4.75	4.61	2164.06	1593.56
			4131	2907	95.02	94.94	7.81	7.54	2192.85	1389.33
			4186	2688	94.91	94.91	10.09	9.96	2189.73	1219.51
			4271	2510	94.99	94.93	12.74	12.62	2210.78	1104.86
			4326	2356	94.93	94.91	15.20	15.26	2208.62	1004.28
		0.70	4561	4561	95.30	96.18	0.67	0.55	2983.81	3091.48
			4696	3661	95.02	95.06	3.31	3.17	2840.51	2454.84
			4761	3197	95.06	95.02	6.42	5.97	2764.65	2086.44
			4826	2907	95.17	94.95	8.68	8.17	2720.66	1925.60
			4946	2688	94.94	94.93	10.00	9.71	2735.41	1870.97
			4971	2510	94.90	94.94	11.74	11.36	2701.85	1721.13
			5071	2356	94.90	94.94	12.88	12.43	2703.66	1679.70

Finally, equation (1) can be used again to find the maximum per unit increase in production costs $((p-1)\times c_p)$ in case of alternative recipe use, for which it still pays to use alternative recipes. Under the given conditions, the maximum equals 3.44%, as is shown in Figure 2. For q=0.90, the maximum equals 2.73%.

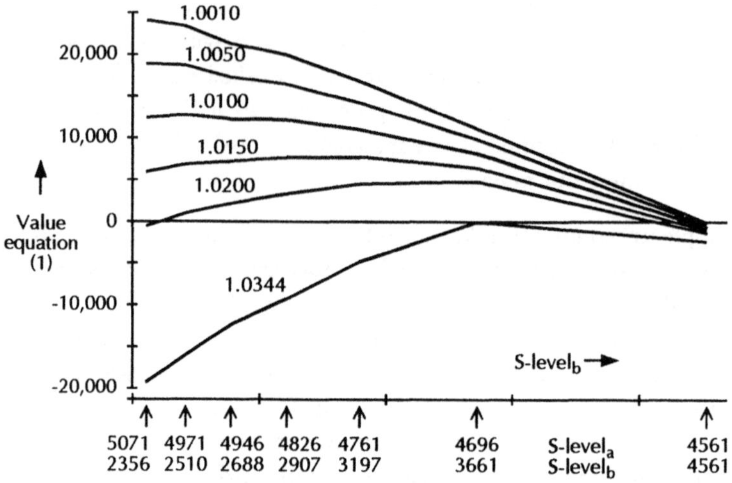

Figure 2 Equation (1) for the situation with $\sigma_x/\mu_x = \sigma_y/\mu_y$ = 400/400 and variation in raw materials (q=0.70), for several factors p. The different (S'_a, S'_b) combinations are marked on the x-axis, the y-axis gives the value of equation (1). The factor c_i is calculated as the capital investment cost related to the product, valued at an interest rate of 15%. Factor c_p equals 100 and annual demand for product y is taken 100,000 (250×μ) units.

4. The analysis of the model

It will be clear that the outcome of the balancing process heavily depends on the parameters of the model used, and thus depends on the characteristics of the production process. The purpose of this paper is to determine for the simple situation considered the set of parameter values for which the use of the alternative recipe is economically justified. For that purpose we first have determined (for a range of values of combined uncertainty in demand and supply (grading of an arriving replenishment lot) in both products) the (S'_a, S'_b) values that are required to realise a service level of 95% in both products with alternative recipe use, and the (S_a, S_b) values without alternative recipe use.

Next we have determined the value of the factor p for which equation (1) is equal to zero. This yields the maximum per unit increase in product costs $((p-1)\times c_p)$ in case of alternative recipe use, for which it still pays to use alternative recipes. For higher per unit cost increase values, the variable recipes should never be used under the given conditions, and all uncertainty should be compensated by holding safety stock.

In conclusion, if the variability in raw materials increases (lower q), the use of variable recipes becomes more attractive. Even if there is no variability in raw materials, there still are situations in which it is favorable to use variable recipes instead of holding safety stock. We developed a decision procedure to find the balance in keeping safety stock and using variable recipes. The conclusions are of course limited to the simple model we used. Nevertheless, considering the results for this simple model, examining more complex and realistic situations can be argued.

References

COKINS G.M., 'Control systems for process industries', *Manufacturing Systems*, vol. 6, no. 5, 1988, pp. 79-85.

CONNOR S.J., *'Process Industry Thesaurus'*, American Production and Inventory Control Society, 1986.

FRANSOO J.C. and W.G.M.M. RUTTEN, *'A typology of production control situations in process industries'*, Research report TUE/BDK/LBS/93-02, Eindhoven University of Technology, Graduate School of Industrial Engineering and Management Science, 1993.

MAY N.P., 'The functions needed for process-oriented systems', *APICS 1984 Conference proceedings*, 1984, pp. 29-32.

RICE J.W. and J.P. NORBACK, 'Process industries production planning using matrix data structures', *Production and Inventory Management Journal*, second quarter, 1987, pp. 15-23.

RUTTEN W.G.M.M., K. VAN DONSELAAR, A.G. DE KOK and G.J.K. REGTERSCHOT, *'The lost sales inventory model'*, Research report TUE/BDK/LBS/92-04, Eindhoven University of Technology, Graduate School of Industrial Engineering and Management Science, 1992.

TAYLOR S.G., S.M. SEWART and S.F. BOLANDER, 'Why the process industries are different', *Production and Inventory Management Journal*, fourth quarter, 1981, pp. 9-24.

CAPACITATED WAREHOUSE LOCATION PROBLEMS WITH ECONOMIES OF SCALE IN OPERATING COSTS

Birgit Schildt, TH Darmstadt

In most industries distribution costs represent a major expense, so that a careful design of a company's distribution system appears to be of vital importance. Relevant costs to be considered concern the location of the distributing facilities as well as the allocation of customers to those locations chosen. Minimization of distribution costs focuses on a tradeoff between transportation costs favouring decentralisation and operating costs associated with the facilities, where a potential loss of economies favours centralization. Realistic assumptions about the underlying cost structure are an important precondition for obtaining a reasonable solution.

Though economies of scale are very common in real life production-transportation planning situations, most publications concerning nonlinear discrete location problems consider some type of convex objective function. This may be encouraged by the fact, that in convex minimization any local optimal solution is known to be a global optimum, which makes things a lot easier. Nevertheless, in most practical applications it is more realistic to assume decreasing marginal operating costs associated with each potential site. Respective models lead to concave minimization problems, that so far have only been discussed by a few authors.

Considering the problem of finding a minimal total costs warehouse design wherein individual warehouses have limited capacity and exhibit continuous economies of scale in variable warehousing costs, we investigate a number of different exact and heuristic solution procedures. Heuristic methods include a cross decomposition approach for deriving good upper and lower bounds on the optimal objective function value as well as the application of recent improvement metastrategies. Besides, a branch and bound procedure is introduced, where lower bounds are obtained via Lagrangean relaxation; the relaxed subproblems may be solved to optimality by inspection.

Presentation includes a comparison of respective numerical results. Test data were generated reflecting different real world cost situations.

MULTI-RETAILER SYSTEMS: PULL OR PUSH STRATEGY?

Erik van der Sluis, University of Amsterdam, The Netherlands

We consider stochastic multi-retailer systems under periodic review. Retailers are replenished by a central depot which in turn places orders at an outside supplier. Both retailers and depot incur fixed cost for placing order and holding cost for each unit stocked at the end of a period. Two rolling horizon procedures based on a pull or a push approach are examined. In the former retailers replenish independently and the decisions at the depot are based on the expected order schedule of the retailers. In the latter all replenishment decisions are taken simultaneously with the depot ordering either at constant intervals or at variable intervals. The objective of this paper is to point out which approach performs best in a stochastic system.

Simulation experiments showed that a straightforward implemented rolling horizon procedure can be improved by adjusting the order decision of the retailers and the depot. This resulted in time-dependent (s,S_t)-strategies for retailers in a push system. Cost and demand structure of the retailers may indicate which approach gives the best results.

EIN STEUERUNGSKONZEPT FÜR AUTOMATISCHE BLOCKLAGER

Hartmut Stadtler, TH Darmstadt
Gert Schade, TH Darmstadt

Zunehmend werden automatische Blocklager erstellt, die sich durch einen höheren Raumnutzungsgrad als die bekannten Hochregallager auszeichnen. Automatische Blocklager werden je nach Anbieter und eingesetzter Technik auch Satelliten-, Kanal- oder Dynamische Blocklager genannt. Kennzeichnend für diese neuen Lagersysteme ist die Lagerung der Ware in einseitig zugänglichen Kanälen mit Kanaltiefen von bis zu 21 (Roll-) Paletten. Für die Auslagerung gilt das LIFO-Prinzip(!).

Ein Problem entsteht, wenn bei nicht-artikelreiner Kanalbelegung auf eine bestimmte Ware zugegriffen werden muß, die nur hintere Stellplätze im Kanal belegt. Hierzu müssen dann zunächst (Roll-) Paletten freigeräumt werden, was zu einer Verringerung der Ein-/Auslagerungsleistung führen kann.

Für einen derartigen Lagertyp wird ein Steuerungskonzept vorgestellt, das mit Hilfe der bekannten oder erwarteten Auslagerungstermine Kanalbelegungen bestimmt, die eine hohe Ein-/Auslagerungsleistung auch bei hohem Füllgrad des Lagers ermöglichen. Zur Bestimmung dieser Auslagerungstermine wird auf die üblicherweise in der 'Materialwirtschaft' oder im 'Vertrieb' verfügbaren Informationen - z.B. Bedarfsprognosen - zurückgegriffen. Das Steuerungskonzept besteht i.w. aus den Teilen

- Einlagerung,
- Auslagerung und
- Reorganisation.

Die Einlagerung bestimmt für jede Einlagerungscharge die Ziel-Gasse(n) und die Ziel-Kanäle, hierbei kommen einfache heuristische Regeln sowie ein Rucksackmodell zum Einsatz. Die Auswahl der auszulagernden (Roll-) Paletten läßt sich als binäres Entscheidungsmodell auffassen, zu dessen Lösung ein Tabu-Search Verfahren vorgeschlagen wird. Um die Kanalbelegungen an die aktualisierten Auslagerungstermine anzupassen, wird in beschäftigungsarmen Zeiten eine Reorganisation des Lagers vorgenommen.

Der Vortrag endet mit ersten Ergebnissen der Simulation eines Dynamischen Blocklagers mit ca. 35000 Stellplätzen unter Verwendung praxisnaher Produktions- und Absatzpläne.

DISKOVER - Ein Verfahren zur Optimierung von Sicherheitsbeständen

Bertram Stausberg, Forschungsinstitut für Rationalisierung an der RWTH Aachen

Im Zuge der für die meisten Unternehmen seit Jahren anhaltenden Bestandsproblematik wurde am Forschungsinstitut für Rationalisierung das stochastische Dispositionsverfahren DISKOVER (**DIS**position mit Hilfe von **KO**nfidenzbereichen unter Berücksichtigung der tatsächlichen Lagerabgangs**VER**teilung) entwickelt, welches die Bestimmung von Grund- und Sicherheitsbeständen bei vorzugebender Lieferbereitschaft ermöglicht.

Eine im Vorfeld der Verfahrensentwicklung in mehreren Unternehmen durchgeführte Datenerfassung hatte gezeigt, daß in der Praxis zahlreiche, verschiedene theoretische Verteilungstypen sowohl der Lagerabgänge als auch der Wiederbeschaffungszeiten auftreten. Des weiteren konnte ein großer Teil der beobachteten Verteilungen keinem theoretischen Verteilungstyp zugeordnet werden. Dieser Tatbestand wird jedoch im Rahmen der bestehenden stochastischen Lagerhaltungsmodelle nicht berücksichtigt. Ziel war es daher, mit DISKOVER ein verteilungstypunabhängiges Verfahren zur Bestimmung notwendiger Lagerbestände zu entwickeln. Dies erfolgte in zwei Teilschritten.

Das erste Verfahrensmodul (DISKOVER I) ermöglicht eine auf der tatsächlichen Lagerabgangsverteilung basierende Berechnung der Menge zu lagernder Artikel, die notwendig ist, den für einen bestimmten Zeitraum zu erwartenden Bedarf zu befriedigen. Da dieser durch den Lagerbestand zu überbrückende Zeitraum für jeden Artikel beliebig vorgegeben werden kann, ermöglicht das Verfahren unter anderem, die Auswirkungen unterschiedlicher Bestellstrategien sowie durchlaufzeitverkürzender Maßnahmen auf die Höhe der erforderlichen Bestände zu ermitteln. Darauf aufbauend ermöglicht die Verfahrenserweiterung DISKOVER II die Berechnung zusätzlicher, zum Ausgleich real auftretender Wiederbeschaffungsschwankungen zu lagernder Sicherheitsbestände. Diese Berechnung basiert auf den beobachteten Fehlmengen pro Nachfrage, die sich aufgrund von Lagerabgangs- und Wiederbeschaffungsschwankungen einstellen. Die Grundlage beider Berechnungsmodule bilden die Lagerbewegungsdaten der Vergangenheit, die mit Hilfe nichtparametrischer statistischer Berechnungsmethoden ausgewertet werden. Dabei wird davon ausgegangen, daß der Bedarf keinen planmäßigen Einflüssen unterliegt und somit nicht prognostiziert werden kann.

Im Rahmen dieses Vortrages wurden die Vorteile des Verfahrens DISKOVER gegenüber konventionellen Verfahren der Bestandsberechnung sowie die Berechnungsgrundlagen und die beiden Module des als PC-Version realisierten Verfahrens dargestellt.

TOURENPLANUNG IN DER STADT MIT ZEITABHÄNGIGEN GESCHWINDIGKEITEN: DATENMODELLIERUNG UND VERFAHREN

Richard Vahrenkamp, Universität GH Kassel

Sammel- und Verteilverkehre in städtischen Zonen und dem Umland zu modellieren, erfordert, die durchschnittlichen Fahrtzeiten von der Tageszeit abhänig zu machen. Die Spitzen im Berufsverkehr, die in Mittelstädten besonders ausgeprägt sind, können damit modelliert werden. Bisher gibt es nur wenig Veröffentlichungen zum Thema der Zeitabhänigkeit der Geschwindigkeiten in der Tourenplanung. Unser Ansatz bildet die Staus im Berufsverkehr mit unsymmetrischen Distanzen und unsymmetrischen Kürzeste-Wege-Matrizen ab, während bisher mit symmetrischen Größen in der Literatur gearbeitet wurde. Morgens der Weg in die Stadt und abends aus der Stadt wird mit "Staufaktoren" verlängert. Vier verschiedene Zeitzustände bilden die Befahrbarkeit der Straßen über die 24 Stunden des Tages ab. Erforderlich werden damit vier Matrizen mit kürzesten Wegen. Das Stadtgebiet wird in 10 Lieferzonen aufgeteilt, die in das vereinfachte Straßennetz der Stadt aufgehängt werden. Das Straßennetz des Umlandes wird detailliert mit drei verschiedenen Straßentypen abgebildet und mit Übergängen zum Stadtnetz versehen.
Die verschiedenen Verfahren der Tourenplanung, wie paralleles Saving, sequentielles Saving, Einfügen und 2-Opt, werden auf ihre Eignung für die zeitabhängigen Daten untersucht. Zur Modellierung von Zeitfenstern bei den Kunden werden Warte- und Umwegzeiten vorgesehen. Werden die Zeitfensterrestriktionen aufgehoben, ergibt die Tourenplanung eine betriebswirtschaftliche Bewertung der Zusatzkosten von Zeitfenstern in Abhängigkeit von deren Größe. Das Modell wird mit den Straßendaten der Stadt Kassel und Umland für ein Containerdispositionssystem in der Abfallwirtschaft erprobt.

EIN MODELL ZUR PROGNOSE DES SCHADENSPOTENTIALS VON ARBITRAGEHANDEL FÜR EINEN GEBRAUCHSGÜTERHERSTELLER

Imaan Bukhari, Universität Erlangen-Nürnberg

Zusammenfassung: Aufgrund der enormen Preisdifferenzen für identische Produkte innerhalb Europas existiert ein starker Anreiz für Absatzmittler zur Durchführung von Arbitragehandel. Mit der zunehmenden Internationalisierung des Einzelhandels steigt die Gefahr von Parallelimporten, die für die Konsumgüterindustrie ein beachtliches Schadenspotential beinhalten. In diesem Beitrag wird ein Modell zur Ermittlung des maximalen quantitativen Schadenspotential am Beispiel eines international tätigen Sportartikelherstellers entwickelt. Der Zweck des Modells liegt in der Entscheidungsunterstützung des internationalen Preismanagements.

Abstract: The large price differences that exist within Europe for identical products represent a strong incentive for wholesalers and retailers to perform arbitrage. Due to the internationalization of the retail industry, there is an increasing danger of parallel imports, which entail considerable damage potential for consumer goods firms. A model for estimation of the maximum quantitative damage potential resulting from cross-border arbitrage is developed here using the case of an international sporting goods producer. The aim is to provide decision support for international price management.

1. Problemstellung und Ziel der Untersuchung

Die Preispolitik stellt für viele in Europa tätige Kosumgüterhersteller zur Zeit eine große Herausforderung dar. Innerhalb der EG-Länder bestehen noch enorme Preisunterschiede, die bei Konsumgütern durchschnittlich 20% betragen. In einigen Fällen sind Abweichungen von 100% keine Seltenheit [1,12]. Die bisherige Praxis der internationalen Preisdifferenzierung wird jedoch insbesondere durch die Europäisierung des Einzelhandels [6] zunehmend in Frage gestellt. Aufgrund dieser Entwicklung steigt die Markttransparenz und somit die Gefahr von Querimporten sowie das Potential eines zentralen Einkaufs zum europaweit niedrigsten Preis [1,4]. Simon/Wiese [13] schlagen daher fünf Schritte eines europäischen Preismanagements vor. Nach der Festlegung der europäischen Preisdeterminanten (1. Schritt), der Bestimmung des länderspezifischen Optimalpreises (2. Schritt) und der Analyse der sich im Zuge des Binnenmarktes ergebenden Veränderungen der Einflußfaktoren (3. Schritt), ist die Frage zu beantworten, ob bei bestehenden Presidifferenzen Arbitragehandel stattfindet, bzw. damit zu rechnen ist (4. Schritt). Ist dies der Fall, sollte ein europäischer Preiskorridor zur Einschränkung von Arbitrage implementiert werden (5. Schritt).

Das Ziel dieses Beitags ist es, ein Modell zu entwickeln, welches einem Hersteller eine Schätzung des Umsatzverlustpotentials durch Arbitragehandel liefert und als Entscheidungsgrundlage für die Gestaltung einer internationalen Preispolitik dient. Dies erfolgt am Beispiel eines international tätigen Herstellers für Sportartikel. Des weiteren soll eine einfache Schätzung der Umsatzverluste durch Einführung eines europäischen Preiskorridors durchgeführt und mit dem Schadenspotential von Arbitragehandel verglichen werden.

2. Arbitragehandel: Möglichkeiten, Motive und Konsequenzen

Unter Arbitragehandel versteht man die "Ausnutzung von Preisdifferenzen zwischen gleichen Produkten in verschiedenen Ländern, wobei die grenzüberschreitende Warenbewegung nicht über die geplanten Distributionskanäle vollzogen wird" [14]. Synonym hierfür werden auch die Begriffe Parallelimporte, Querimporte und Reimporte verwendet [13]. Arbitragehandel kann grundsätzlich auf der Absatzmittlerebene und der Endverbraucherebene stattfinden. Arbitrage durch Konsumenten ist in der Regel mengenmäßig nicht signifikant [7]. Beim Arbitragehandel zwischen Absatzmittlern sind mehrere Fälle der internationale Warenbewegung denkbar [3]. Erstens kann Arbitrage zwischen Händlern derselben Handelsorganisation stattfinden. Zweitens kann ein nicht-autorisierter Händler der Querimporteur sein. Drittens besteht die Möglichkeit, daß ein Importeur mit Großhandelsfunktion die Ware sowohl an autorisierte als auch nicht-autorisierte Händler in einem anderen Land veräußert. Schließlich ist es auch denkbar, daß sich eine eigene Tochtergesellschaft als Parallelexporteur betätigt. Dieser Fall ist aber vernachlässigbar, da die Kontroll- und Sanktionsmöglichkeiten des Herstellers dies i.d.R. verhindern, auch wenn einige derartige Fälle bekannt sind [9].

Das Motiv eines Parallelimporteurs zum Querbezug von Ware liegt i.d.R. in einem niedrigeren Bezugspreis. Darüber hinaus kann bei selektiven Vertrieb das Motiv zum Querimport auch in der zeitversetzten Belieferung der Länder oder im Ausschluß des Händlers vom Bezug der Ware liegen. Weiterhin ist ein Parallelimport aufgrund eines kurzfristigen Mehrbedarfs des Händlers denkbar. Die Beweggründe des Parallelexporteurs können zum einen im Abbau von Lagerbeständen liegen. Zum anderen kann der Arbitragehandel das Resultat einer gezielten internationalen Einkaufspolitik sein, wenn Querimporteur und -exporteur derselben Handelsorganisation angehören. Hierdurch werden nicht nur die internationalen Preisdifferenzen ausgenutzt, es können auch aufgrund der höheren Bestellmengen im Einkaufsland u.U. bessere Konditionen erzielt werden [3].

Internationaler Arbitragehandel führt zu einem direkten Konzernumsatzverlust einer internationalen Unternehmung. Da der Hersteller i.d.R. keine Kosteneinsparungen durch den Arbitragehandel erzielt, schlägt sich dieser Umsatzverlust auch direkt in einer Verminderung des Deckungsbeitrags bzw. des Gewinns in gleicher Höhe nieder [11]. Insofern kann ein zunächst gering erscheinender Umsatzverlust durch Parallelimporte gravierende Auswirkungen auf die Gewinnmarge einer Unternehmung haben. Neben diesen unmittelbaren kurzfristigen Folgen von Arbitragehandel sind aber auch mittelbare, z.T. langfristige Auswirkungen zu beachten. Parallelimporte führen letztendlich zu einem Kontrollverlust über die eigenen Marken auf nationaler und internationaler Ebene. Wenn Arbitrage in erheblichem Umfang stattfindet, kann eine Unternehmung nicht mehr gezielt die Distribution seiner Produkte steuern, wenn diese auch über nicht-vorgesehene Absatzkanäle an den Endverbraucher gelangen. Dies kann zu Imageverlusten der Produkte bzw. der Marke führen, v.a. wenn die reimportierte Ware in preisaggressiven oder beratungsinkompetenten Absatzkanälen auftaucht [7]. Darüber hinaus kann ein durch Arbitrage induziertes Absinken der Preise auf das international niedrigste Niveau die Marke schwächen [2]. Weiterhin können Parallelimporte zu erheblichen Störungen und Vertrauensverlusten im Distributionssystem eines Herstellers führen. Durch Arbitrage geraten authorisierte Händler u.U. in Konkurrenz zu Händlern, die die gleiche Ware billiger bezogen haben und daher preisgünstiger anbieten können [8]. Schließlich kann der Fall eintreten, daß sich die Konsumenten an die niedrigeren Preise von Querimporteuren gewöhnen und die höherpreisigen authorisierten Händler ihre Kundschaft verlieren [2].

3. Datenbasis und Modellspezifikation

Der Vertrieb des Sportartikelherstellers findet in den betrachteten Ländern (Deutschland, Frankreich, Großbritannien, Belgien und die Niederlande) über Tochtergesellschaften statt, wobei diese von einer zentralen Produktionsgesellschaft beliefert werden. Grundlage der Schätzung ist der Umsatz, den dieser Hersteller mit vier internationalen Schlüsselkunden in fünf europäischen Ländern mit drei Warengruppen erzielt. Diese vier Euro-Key-Accounts sind sowohl von ihrer Umsatzbedeutung in den untersuchten Ländern (ca. 23% des Sportartikelumsatzes des Herstellers) als auch von ihrer internationalen Präsenz und dem daraus resultierenden Querimportpotential von herausragender Bedeutung für den Hersteller. Datenbasis der Schätzung sind die getrennt ausgewiesenen Verkaufsmengen und

Nettoabgabepreise von über 100 Artikeln der Warengruppen in den jeweiligen Ländern für die Jahre 1991 und 1992. Bei den Artikeln bestehen Preisunterschiede zwischen den Ländern bis zu 40 DM, so daß hier ein erheblicher Anreiz zur Arbitrage besteht.

3.1 Modellannahmen

Bei der Spezifikation des Modells und der daraus resultierenden Schätzung des Schadenspotentials müssen einige vereinfachende Annahmen getroffen werden. Erstens kann hier nur das unmittelbar quantifizierbare Schadenspotential geschätzt werden. Eventuell langfristige Imageverluste der Marke finden keine Berücksichtigung. Zweitens wird nur der Umsatzverlust berücksichtigt, der durch Querimporte zwischen autorisierten Händlern einer Handelsorganisation, d.h. der 4 Euro-Key-Accounts, stattfindet. Weiterhin wird angenommen, daß ein Artikel dann querimportfähig ist, wenn dieser Artikel in mindestens zwei Ländern vertrieben wird und eine Netto-Einstandspreisdifferenz, die höher als die Arbitragekosten ist, zwischen den Ländern besteht. Als Arbitragekosten werden nur die Transportkosten berücksichtigt. Diese wurden aufgrund von Expertenaussagen auf durchschnittlich DM 1 pro Stück geschätzt. Darüber hinaus wird angenommen, daß ein querimportfähiger Artikel immer aus dem Land mit dem niedrigsten Netto-Einstandspreis des Key-Accounts bezogen wird, da das maximale Schadenspotential ermittelt werden soll. Schließlich ist noch anzumerken, daß über die genaue Bestellmenge einzelner Schuhmodelle der jeweiligen Key-Accounts keine Daten verfügbar waren, sondern nur über deren Gesamtbestellmenge. Dagegen liegt diese Einzelartikelinformation auf der Ebene der Tochtergesellschaften vor. Daher wurde angenommen, daß die Sortimentsstruktur der Key-Accounts mit derjenigen der Tochtergesellschaften in den jeweiligen Ländern identisch ist. Diese Annahmen werden in Tab. 1 zusammengefaßt.

(1)	Auswirkungen der Reimporte in der Nachperiode werden nicht berücksichtigt.
(2)	Das Motiv für Parallelimporte liegt nur in bestehenden Preisdifferenzen, d.h. es findet keine Berücksichtigung von Absatzschwankungen (konstanter Endverbraucherabsatz) und des Lieferungszeitpunktes statt, und der Key-Account im Niedrigpreisland (Exporteur) verkauft an den Parallelimporteur nicht unter Einstandspreis.
(3)	Der Key-Account im Niedrigpreisland kann die Nachfrage nach Waren aus allen Hochpreisländern zu 100% befriedigen.
(4)	Die Key-Accounts der jeweiligen Länder führen alle Modelle im Sortiment bzw. können alle Modelle bestellen, die von der Vertriebsgesellschaft ihres Landes angeboten werden.
(5)	Der Beschaffungsanteil pro Artikel der Key-Accounts ist identisch mit dem Beschaffungsanteil der Vertriebsgesellschaften im jeweiligen Land.
(6)	Ein Artikel ist dann "querimportfähig", wenn dieser Artikel in mindestens zwei der untersuchten Länder vertrieben wird und eine Netto-Einstandspreisdifferenz, die höher als die Transportkosten ist, zwischen den Ländern besteht.
(7)	Wenn ein Artikel querimportfähig ist, dann bezieht der jeweilige Key-Account ihn aus dem Land, in dem dieser Artikel am billigsten ist.

Tab. 1: Modellannahmen

3.2 Modellbeschreibung

Im folgenden soll nun das der Schätzung des Schadenspotentials zugrundeliegende Modell dargestellt werden. Hierbei handelt es sich um eine ex-post Prognose, d.h. es wird geschätzt, welches Arbitragepotential in den Jahren 1991 und 1992 vorlag. Ausgangspunkt der Betrachtung ist der einzelne reimportfähige Artikel. Hierdurch wird ermöglicht, das maximale Umsatzverlustpotential zu erfassen, da bei einem Artikel z.B. Frankreich das Niedrigpreisland sein kann, während bei einem anderen Artikel Frankreich das Hochpreisland ist. Darüberhinaus bietet die artikelbezogene Betrachtungsweise die Möglichkeit zu analysieren, welche Artikel für den Großteil des Schadenspotentials verantwortlich sind. Die verwendeten Variablen und die Ermittlung des Umsatzverlustpotentials werden in

Tab. 2 dargelegt. Zur besseren Übersicht wird zunächst nur der Fall für einen Euro-Key-Account dargestellt. Das gesamte Umsatzverlustpotential ergibt sich dann aus der Aufsummierung der Schadenspotentiale pro Key-Account.

i= A,B,C,D und N	Länderindex, wobei N = Niedrigstpreisland für den jeweiligen Artikel
l = 1,...,L	Index der querimportfähigen Artikel
X_i	gesamte (geplante) Beschaffungsmenge aller Artikel der Tochtergesellschaft in Land i
k_{il}	Beschaffungsmenge des Artikels l des Key-Accounts von der Tochtergesellschaft in Land i
R_{il}	Anteil des Artikels l an der Beschaffungsmenge der Tochtergesellschaft in Land i
r_{il}	Anteil des Artikels l an der Beschaffungsmenge des Key-Accounts von der Tochtergesellschaft in Land i; $R_{il} = r_{il}$ (Annahme 5)
$r_{il}k_{il} := q_{il}$	Beschaffungsmenge des Key-Accounts des querimportfähigen Artikels l von der Tochtergesellschaft in Land i = potentielle Querimportmenge des Artikels l in Land i
$\Sigma_{(l=1..L)}\,[\,q_{il}\,]$	potentielle Gesamtquerimportmenge aller Artikel im Land i
p_{il}	Nettoeinstandspreis des Artikels l im Land i
$[p_{il}(i=A..D) - p_{Nl}(i=N)] :=$	$\Delta p'_{il}$ Netto-Einstandspreisdifferenz für den Artikel l in den Ländern i=A,B,C,D (Höherpreisländer) im Vergleich zum Land i=N (Niedrigstpreisland)
$q_{il}\,\Delta p'_{il} := Q_{jl}$	Umsatzverlust des Herstellers durch Querimport des Artikels l
$\Sigma_{(l=1..L)}\,[\,q_{il}\,\Delta p'_{il}\,] = Q_j$	Umsatzverlust des Herstellers durch Querimport aller querimportfähigen Artikel in Land i
$\Sigma_{(i=A..D)}\,Q_j$	maximaler Umsatzverlust bzw. Schadenspotential des Herstellers durch Arbitragehandel

Tab. 2: Variablendefinition und Ermittlung des Schadenspotentials pro Key-Account

Zunächst wird die Beschaffungsmenge des querimportfähigen Artikels für die einzelnen Höherpreisländer bestimmt. Daraufhin wird die Differenz der Netto-Einstandspreise in DM von den jeweiligen Höherpreisländern zum Niedrigstpreisland ermittelt, wobei die Leitkurse des Europäischen Währungssystems (EWS) als Umrechnungsfaktor dienen. Die Multiplikation der Netto-Einstandspreisdifferenz mit der Beschaffungsmenge des querimportfähigen Artikels ergibt dann den Umsatzverlust, den der Hersteller erleiden müßte, wenn der Key-Account in einem Höherpreisland diesen Artikel vollständig aus dem Niedrigstpreisland bezieht. Diese Größe kann dann beliebig nach Ländern, Key-Accounts oder Artikel aufsummiert werden, um Umsatzverlustpotentiale eben nach diesen Größen zu erhalten.

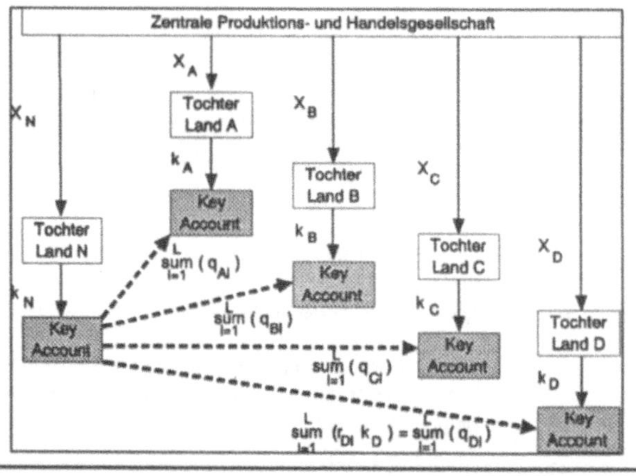

Abb. 1: Ein Modell zur Prognose der potentiellen Querimportmenge

Hier wurde nun zunächst nach Artikeln und dann nach Ländern aufsummiert, um ein Schadenspotential pro Euro-Key-Account zu erhalten. Das zugrundeliegende Modell ist in Abb. 1 dargestellt.

4. Ergebnisse der Schätzung

Die Anwendung des Modells ergab ein Umsatzverlustpotential von ca. 2,6 Mio. DM für das Jahr 1991 bzw. 3,3 Mio. DM in 1992. Für 1992 entspricht dies ca. 3,5% des Umsatzes, den der Hersteller mit diesen Euro-Key-Accounts in den 5 Ländern tätigt. Die Zunahme des Schadenspotentials im Jahr 1992 könnte bei Betrachtung der

Einzeldaten zunächst verwundern, da im Durchschnitt die Preisdifferenzen der einzelnen Artikel abgenommen haben. Bei genauerer Analyse zeigt sich jedoch, daß der Zuwachs am Umsatzverlustpotential aus vom Hersteller vorgenommenen internationalen Sortimentsstandardisierung, v.a. bei höherpreisigen Artikeln, resultiert Hierdurch wird der Konflikt zwischen internationaler Produkt(programm)standardisierung und Preisdifferenzierung deutlich.

Jahr	Key-Acc.1	KA 2	KA 3	KA 4	Summe
1991	2.371	152	74	26	2.623
1992	2.955	277	51	29	3.312

Tab. 3: Schadenspotential des Arbitragehandels in 5 Ländern nach Euro-Key-Accounts - Angaben in Tausend DM

Bei der Analyse des Schadenspotentials nach Key-Accounts (siehe Tab. 3) zeigt sich, daß der Euro-Key-Account 1 für den Großteil des Schadenspotentials verantwortlich ist. Dies liegt darin begründet, daß dieser Euro-Key-Account der einzige ist, der in allen fünf Ländern vertreten ist. Die anderen sind bis heute jeweils nur in zwei Ländern aktiv. Hieraus wird ersichtlich, welche Gefahr eine weitere Internationalisierung der anderen Key-Accounts für den Hersteller in sich birgt.

Jahr	D	F	UK	NL	B	Summe
1991	1.310	401	384	377	151	2.623
1992	1975	353	266	507	210	3.312

Tab. 4: Schadenspotential des Arbitragehandels durch 4 Euro-Key-Accounts nach Ländern - Angaben in Tausend DM

Bei der Betrachtung der Schadenspotentiale pro Land (siehe Tab. 4) wird deutlich, daß das größte Umsatzverlustpotential für Deutschland besteht, obwohl Belgien und die Niederlande im Durchschnitt über alle Artikel die Hochpreisländer sind. Der Grund hierfür liegt in der insgesamt wesentlich höheren Umsatzbedeutung des deutschen Marktes. Hier zeigt sich nochmals der Vorteil einer artikelbezogen Schätzung, da eine zu frühe Aggragation der Daten, z.B. bei einer Schätzung des Umsatzverlustpotentials mittels Warengruppendurchschnittspreise, u.U. eine Verfälschung der länderbezogenen Schadenspotentiale zum Resultat haben kann.

5. Sensitivitätsanalyse mit Wechselkursen

Die Attraktivität des Querimports von Ware kann sich durch Wechselkursschwankungen verändern. Zwar bestehen zwischen den im EWS eingebunden Ländern relativ feste Wechselkurse, dennoch können diese auch bis zu 2,25% (bei Italien bis zu 6%) innerhalb der festgelegten Grenzwerte um den Leitkurs schwanken [5]. Außerdem gilt es zu berücksichtigen, daß sich das EWS seit dem Austritt von Großbritannien und Italien im September 1992 in einer Krise befindet [10]. Insofern erscheint es durchaus sinnvoll, eine Sensitivitätsanalyse bezüglich der Wechselkurse vorzunehmen.

Hierbei wurden zwei Wechselkursvarianten näher analysiert. Bei der ersten Alternative wurde der Niedrigstkurs (unterer Grenzwert im EWS) für Frankreich und der Höchstkurs (oberer Grenzwert) für die anderen Länder (Wechselkursvariante 1) angenommen. Diese Alternative wurde gewählt, da Frankreich im Durchschnitt das Niedrigpreisland darstellt und daher die größten Effekte auf das Schadenspotential durch Kursschwankungen innerhalb der EWS Bandbreiten bei einer Abwertung des französischen Francs vermutet

Wechselkursvariante 1:							
Jahr	D	F	UK	NL	B	Summe	Δ Tab. 3
1991	1.490	223	683	393	179	2.968	+345
1992	1.956	187	506	489	249	3.387	+75
Wechselkursvariante 2:							
Jahr	D	F	UK	NL	B	Summe	Δ Tab. 3
1991	2.941	968	3	658	194	4.764	+2.141
1992	4.789	1.376	11	903	276	7.358	+4.046

Tab. 5: Schadenspotential durch Arbitragehandel der 4 Euro-Key-Accounts in 5 Ländern bei zwei Wechselkursvarianten - Angaben in Tausend DM

wurden. Als zweite Alternative wurde die Auswirkung des Austritts des britischen Pfunds aus dem EWS untersucht, wobei der amtliche Kurs vom 16.3.93 in Höhe von 2,3980 DM zugrundegelegt wurde (Wechselkursvariante 2). Die Ergebnisse dieser Analyse sind in Tab. 5 wiedergegeben. Die größte Auswirkung ergibt sich durch den EWS-Austritt Großbritanniens, das dadurch bei fast allen Artikeln zum Niedrigstpreisland wird. Als Folge davon erhöht sich das Schadenspotential in allen anderen Ländern. Hierdurch wird die Bedeutung der Berücksichtigung von Wechselkursschwankungen bei der Analyse des Schadenspotentials deutlich.

6. Schätzung des Umsatzverlustes bei Einführung eines Preiskorridors

Wie bereits im einleitenden Abschnitt dargelegt, stellt die Verringerung der bestehenden internationalen Preisdifferenzen durch Einführung eines Preiskorridors ein Mittel dar, um Parallelimporte einzuschränken [13]. Die Bildung eines Preiskorridors, den Simon als "Anteil in Prozent, um den der höhere Preis unterschritten werden darf" [11] definiert, stellt einen Kompromiß zwischen den länderspezifischen Optimalpreisen und einem europaweit einheitlichen Preis dar [13]. In diesem Abschnitt soll nun am Beispiel des Sportartikelherstellers untersucht werden, welche Umsatzwirkung die Einführung eines Preiskorridors von 10% bzw. 5% hat. Die für eine Prognose des Umsatzverlustes notwendigen Preiselastizitäten wurden anhand von Expertenschätzungen ermittelt. Dies stellt in der Praxis eine durchaus gebräuchliche und bedeutende Methode zur Ermittlung von Preis-Absatz-Funktionen dar [11]. Im vorliegenden Fall wurden die Preiselastizitäten getrennt für die drei untersuchten Warengruppen in jedem Land geschätzt, da zwischen diesen Unterschiede vermutet wurden. Zur Schätzung des Umsatzverlusts wurde wie folgt vorgegangen. Bei allen Artikeln wurden die Preise so nach oben angepaßt, daß diese maximal um 10% (bzw. 5%) unter dem internationalen Höchstpreis der jeweiligen Artikel liegen. Daraufhin wurde mittels der geschätzen Preiselastizitäten ermittelt, welchen Umsatzverlust der Hersteller aufgrund der vorgenommenen Preiserhöhungen bei den 4 Euro-Key-Accounts erfährt. Diese Ergebnisse sind in Tab. 7 dargestellt.

	D	F	UK	NL	B
Warengruppe 1	-2,5	-3,3	-2,0	-6,1	-6,1
Warengruppe 2	-3,8	-4,7	-5,9	-6,1	-6,1
Warengruppe 3	-3,0	-3,7	-6,4	-7,1	-7,1

Tab. 6: Geschätzte Preiselastizitäten für einen repräsentativen Artikel der 3 Warengruppen in %

Korridorbreite:	5 %	10 %
1991	23.227	14.081
1992	19.085	10.097

Tab. 7: Geschätzer Umsatzverlust durch Vorgabe eines Preiskorridors (Tsd. DM)

Einschränkungen dieser Vorgehensweise sind darin zu sehen, daß erstens nur der Umsatzverlust bei den 4 Euro-Key-Accounts geschätzt wird. Eine Preiserhöhung hätte jedoch sicherlich auch geringere Umsätze bei anderen Abnehmern zur Folge. Hierbei gilt es jedoch zu bedenken, daß auch das Umsatzverlustpotential durch Arbitrage nur diese 4 Abnehmer berücksichtigt. Zweitens beruht die Umsatzverlustprognose auf einer sehr einfachen Schätzung der Preiselastizitäten. Drittens wurde der Preiskorridor auf der Basis des jeweils höchsten Preis gesetzt. Ein "optimaler" Preiskorridor kann jedoch durchaus von einen niedrigeren Preis ausgehen. Schließlich können auch bei einem Preiskorridor von 10% bzw. 5% immer noch Querimporte, allerdings in wesentlich geringerem Umfang, stattfinden. Andererseits liefert diese einfache und auch sehr kostengünstige Schätzung der Umsatzverluste dennoch Anhaltspunkte für die Gestaltung einer internationalen Preispolitik, insb. wenn diese in Beziehung zum Schadenspotential von Arbitrage gesetzt wird (siehe Abb. 2).

Der Vergleich zeigt, daß der Umsatzverlust bei Einführung eines Preiskorridors, insb. eines 5%igen, wesentlich höher liegt als das Umsatzverlustpotential von Querimporten. Im Jahr 1992 gleichen sich jedoch diese Größen an. Der Umsatzverlust des 10%igen Preiskorridors ist nur noch ca. 2,6 Mio. DM höher als das Arbitrage-Schadenspotential der zweiten Wechselkursvariante, die zumindest z.Zt. die realistischste darstellt. So kann bei akuter Gefahr der

Realisierung des Schadenspotentials durch einen international aktiven Einzelhandel die Einführung eines Preiskorridors von 10% u.U. die bessere Alternative im Vergleich zur Inkaufnahme von Querimporten darstellen, v.a. wenn zum einen die längerfristigen Auswirkungen des Arbitragehandels berücksichtigt und zum anderen die Preisanpassungen nicht am jeweils höchsten Preis ausgerichtet werden und eine Unterstützung durch den Einsatz anderer Marketinginstrumente erfolgt.

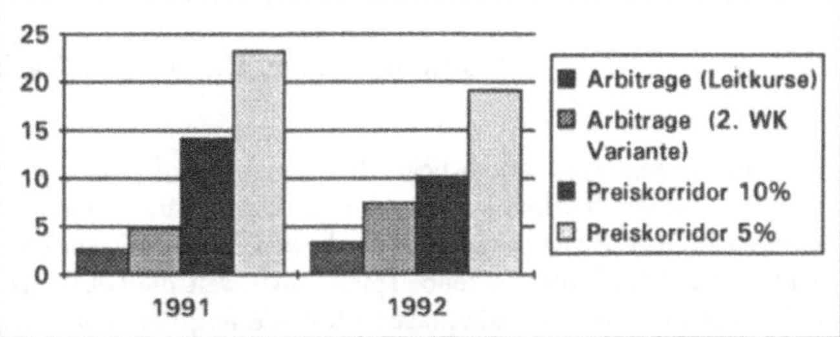

Abb. 2: Arbitrage-Umatzverlustpotential und geschätzter Umsatzverlust durch Vorgabe eines Preiskorridors - Angaben in Mio. DM

7. Zusammenfassung

Das hier vorgestellte Modell zur Prognose des Schadenspotentials von Parallelimporten ist auf den spezifischen Fall eines bestimmten Sportartikelherstellers zugeschnitten. Dennoch ist die Grundstuktur des Modells ohne weiteres auf andere Fälle übertragbar. Im Rahmen der Schätzung mußten, im wesentlichen aufgrund von Datenlimitationen, Annahmen getroffen werden, welche die Genauigkeit der Prognose beeinflussen können. Dabei ist jedoch zu beachten, daß der Zweck des Modells darin liegt, ein Potential zu schätzen und hierbei v.a. die Größenordnung eines maximalen Umsatzverlusts aufgrund von Arbitragehandel zu ermitteln ist. Diese Information ist für Konsumgüterhersteller, die sich mit der Gefahr von Querimporten konfrontiert sehen, bereits von erheblicher Bedeutung für das internationale Preismanagement. Neben dem quantifizierten, kurzfristigen Schadenspotential gilt es jedoch einerseits, die weiteren negativen, nicht direkt quantifizierbaren Auswirkungen von Querimporten zu berücksichtigen und andererseits, die Umsatzwirkungen einer Preisangleichung, z.B. durch Implementierung eines Preiskorridors, mit dem Schadenspotential von Arbitragehandel abzuwägen.

Literaturverzeichnis

[1] Bukhari, I., Pricing Conditions in the European Common Market: An Empirical Cross-Country Analysis, Arbeitspapier Nr. 17 des Lehrstuhls für Marketing, Universität Erlangen-Nürnberg, Nürnberg 1992.

[2] Cavusgil, S.T.; Sikora, E., How Multinationals can Cope with Grey Market Imports, Report No. 87-109, The Marketing Science Institute, Cambridge Mass. 1987.

[3] Cordes, C., Preispolitische Rahmenbedingungen für den Absatz von Sportartikeln im EG-Binnenmarkt, unveröffentlichte Diplomarbeit am Lehrstuhl für Marketing, Universität Erlangen-Nürnberg, 1993.

[4] Diller, H., Preispolitik der internationalen Unternehmung, in: Wirtschaftswissenschaftliches Studium, 16. Jg., H.6 1987, S. 269-275.

[5] Dudley, J.W., 1992 - Strategien für den gemeinsamen Markt, München 1989.

[6] George, G.; Diller, H., Internationalisierung als Wachstumsstrategie des Einzelhandels, in: Trommsdorf, V. (Hrsg.), Handelsforschung 1992, Wiesbaden 1993, S. 165-186.

[7] Euler, G.R.; Berqvist, S., Pan-European Pricing of Consumer Goods, Booz, Allen & Hamilton (Hrsg.), Consumer Goods Viewpoint, New York 1991.

[8] Kreutzer, R., Global Marketing - Konzeption eines länderübergreifenden Marketing, Wiesbaden 1989.

[9] o.V., Kleinkrieg unter Reebok-Töchtern, in: markt intern, XI: Jg., 1.10.1992, S. 1.

[10] o.V., Dann ist Maastricht erledigt, in: Der Spiegel, 15/1993, S. 114-117.

[11] Simon, H., Preismanagement, 2. vollst. überarb. u. erw. Aufl., Wiesbaden 1992.

[12] Simon, H., Preisbombe, in: manager magazin, Februar 1992, S. 86-89.

[13] Simon, H.; Wiese, C., Europäisches Preismanagement, in: Marketing ZFP, Heft 4 1992, S. 246-256.

[14] Tietz, B., Euro-Marketing - Unternehmensstrategien für den Binnenmarkt, 2. völlig überarb. und erw. Aufl., Landsberg/Lech 1990.

MODELLGESTÜTZTE OPTIMIERUNG DES MARKETING-MIX AUF DER BASIS VON KAUFVERHALTENSDATEN

Reinhold Decker, Universität Karlsruhe (TH)

Zusammenfassung: Im vorliegenden Beitrag werden ökonometrische Marktanteilsmodelle hinsichtlich ihrer Verwendbarkeit im Rahmen von Gewinnmaximierungsproblemen auf Konsumgütermärkten untersucht. Die Arbeit setzt sich zu diesem Zweck mit einigen formalen Aspekten auseinander und zeigt anhand konkreter Datenbeispiele, inwieweit die betrachteten Ansätze in der Lage sind, potentielle Anwender bei der adäquaten Modellierung gegebener Marktsituationen und der Bestimmung gewinnsteigernder Marketing-Mix-Konstellationen zu unterstützen.

Summary: In the present paper econometric market share models are investigated with respect to their suitability for profit maximization problems on consumer goods markets. Therefore, after the discussion of some theoretical aspects, selected empirical results will be presented to demonstrate how far those approaches are able to support potential users both in the adequate modeling of given market situations and in the determination of profit increasing marketing mix constellations.

1. Problemstellung

Im Rahmen der modellgestützten Kaufverhaltensforschung kommt der Analyse und Prognose der Einflüsse des Marketing-Mix-Instrumentariums auf die gegenwärtige und zukünftige Absatzentwicklung eines Produkts zentrale Bedeutung zu. Besonderen Stellenwert besitzt dabei die Bestimmung jener Marketing-Mix-Konstellationen, die – aus heutiger Sicht – eine potentielle Gewinnsteigerung für zukünftige Zeiträume ermöglichen.

Schwerpunkte der Arbeit bilden deshalb die Darstellung formaler Rahmenbedingungen für die Anwendung modellbasierter Gewinnfunktionen im Rahmen von Optimierungsfragestellungen, die Bestimmung geeigneter Ansätze zur Modellierung von Marktanteils- und Gewinnentwicklungen und deren empirische Überprüfung anhand konkreter Datenbeispiele.

Die im vorliegenden Fall zum Einsatz kommenden ökonometrischen Marktanteilsmodelle MUL (MULtiplikativer Ansatz), MCI ('Multiplicative Competitive Interaction'-Modell) und MNL (MultiNomiales Logit-Modell) zeichnen sich u.a. durch ihre einfache Anwendbarkeit auf die im Bereich der quantitativen Kaufverhaltensforschung zentrale Bedeutung besitzenden Scanner- und Paneldaten aus. Da diese Ansätze bereits ausführlich in verschiedenen einschlägigen Publikationen beschrieben und diskutiert wurden, beschränken wir uns an dieser Stelle auf einen kurzen Verweis auf entsprechende literarische Quellen [4,5,11].

Auch speziell zum Thema "Modellgestützte Gewinn- bzw. Marketing-Mix-Optimierung" existieren bereits einige Veröffentlichungen [6,9,10,12]. Die genannten Arbeiten haben in erster Linie die auf modelltheoretischer Ebene angesiedelte Ableitung von Theoremen und Handlungsempfehlungen zum Inhalt, während z.B. in [1,3] auch Ergebnisse diesbezüglicher empirischer Untersuchungen beschrieben werden. Da aber gerade die theorieorientierten Abhandlungen zum vorliegenden Themenbereich oftmals auf umfassenden Annahmengeflechten aufbauen, die sich ohne empirische Überprüfung teilweise nur schwer motivieren lassen, wird in diesem Beitrag beiden Gesichtspunkten gleichermaßen Aufmerksamkeit geschenkt.

2. Methodische Aspekte des Modellierungsansatzes

Ausgangspunkt der weiteren Betrachtungen ist die für Marke i und Periode t definierte und zu maximierende Gewinnfunktion

$$G_{it} := V \cdot MA_{it}\,(p_{it} - k_i) - \sum_{l=1}^{L} m_{ilt}.$$

Dabei beschreibt MA_{it} den Marktanteil, p_{it} den Stückpreis und m_{ilt} die l-te Marketing-Anstrengung (z.B. Werbeaufwand) der Marke i in Periode t. Weiterhin wird angenommen, daß MA_{it} selbst wiederum als Funktion der genannten Marketing-Instrumente dargestellt werden kann. Die markenspezifischen Stückkosten k_i und das Marktvolumen V seien dabei zeitinvariant.
Da sich die Standardformulierungen der in dieser Arbeit verwendeten Marktanteilsmodelle (vgl. Tabelle 1) nach unseren bisherigen Erfahrungen nur in eingeschränktem Maße für Marketing-Mix-Optimierungen eignen, wird zusätzlich ein alternativer Modellierungsansatz vorgeschlagen. Ausgangspunkt hierfür sind die folgenden Annahmen: Neben dem für Konsumgüter des täglichen Bedarfs typischen negativen Einfluß von Preissteigerungen auf die Absatzentwicklung des betreffenden Produkts und der umgekehrten Wirkung verstärkter Marketing-Anstrengungen wird unterstellt, daß für die betrachteten Marken gewisse – maximal am Markt durchsetzbare – Preisobergrenzen p_i^{max} existieren und z.B. die Wirkung zusätzlicher Werbeaufwendungen mit zunehmendem (bereits realisiertem) Werbevolumen immer geringer wird. Eine mögliche funktionale Umsetzung dieser Aspekte besteht darin (vgl. Tabelle 1), die urspüngliche Preisvariable p_{it} (im Standardansatz) durch die Beziehung $(p_i^{max} - p_{it})^2$ und die Variable m_{ilt} durch deren Wurzel zu ersetzen.

Modell	Basisgleichungen
MUL	$MA_{it} := \exp(\alpha_i)\tilde{p}_{it}^{\beta_i} \prod_{l=1}^{L} \tilde{m}_{ilt}^{\gamma_{il}}$
MCI	$MA_{it} := \dfrac{\exp(\alpha_i)\tilde{p}_{it}^{\beta_i} \prod_{l=1}^{L} \tilde{m}_{ilt}^{\gamma_{il}}}{\sum_{j=1}^{I} \exp(\alpha_j)\tilde{p}_{jt}^{\beta_j} \prod_{l=1}^{L} \tilde{m}_{jlt}^{\gamma_{jl}}}$
MNL	$MA_{it} := \dfrac{\exp\left(\alpha_i + \beta_i \tilde{p}_{it} + \sum_{l=1}^{L} \gamma_{il}\tilde{m}_{ilt}\right)}{\sum_{j=1}^{I} \exp\left(\alpha_j + \beta_j \tilde{p}_{jt} + \sum_{l=1}^{L} \gamma_{jl}\tilde{m}_{jlt}\right)}$

Standardansatz
$\tilde{p}_{it} := p_{it}$ $\tilde{m}_{ilt} := m_{ilt}$
Modifizierter Ansatz
$\tilde{p}_{it} := (p_i^{max} - p_{it})^2$ $\tilde{m}_{ilt} := \sqrt{m_{ilt}}$

Tabelle 1: Gleichungen der verwendeten Marktanteilsmodelle

Die aus dem (in die Analyse mit einbezogenen) Datenmaterial zu schätzenden Parameter β_i bzw. γ_{il} beschreiben die Richtung und das Ausmaß des Einflusses des Preises bzw. der l-ten Marketing-Anstrengung der Marke i auf die betreffenden Marktanteile. Das Intercept α_i hingegen korrespondiert mit dem Grundnutzen bzw. der Grundattraktivität der einzelnen Marken. Gegenüber den restlichen Modellen besitzen die – mathematisch einfach zu handhabenden – MUL-Modelle den Nachteil, daß sie nicht das Kriterium der logischen Konsistenz erfüllen, d.h., daß die mit ihrer Hilfe geschätzten Marktanteile nicht zwangsläufig zwischen 0 und 1 liegen müssen und sich auch nicht in jedem Fall zu 1 addieren.

3. Bemerkungen zur Maximierungsproblematik

Die Überprüfung des durch obige Gewinnfunktion spezifizierten Optimierungsproblems hinsichtlich der Existenz von lokalen Maxima erfolgt i.d.R. über den Nachweis der positiven Definitheit der mit -1 multiplizierten Hessematrix. Da dies aber aufgrund der vorgenommenen Reparametrisierung der ursprünglichen Gewinnfunktion mittels der soeben skizzierten Marktanteilsmodelle u.U. mit beträchtlichem rechentechnischem Aufwand verbunden sein kann und damit auch interpretatorische Auswertungen dieses Optimalitätskriteriums auf modelltheoretischer Ebene kaum mehr möglich sind, wurde im vorliegenden Fall eine andere Vorgehensweise gewählt [9].
Die im folgenden abgeleitete Funktionalbeziehung erleichtert den Nachweis der strengen Quasikonkavität der Gewinnfunktion durch die ihrerseits wiederum eine effiziente Anwendung heuristischer Optimierungsverfahren – z.B. eines 'Pattern Search'-Ansatzes – ermöglicht wird. Weiterhin wird ein Satz angegeben, der die Eindeutigkeit des Maximums bei streng quasikonkaven (Gewinn-)Funktionen gewährleistet. Auf diese Weise läßt sich die Berechnung der Hessematrix umgehen.
Ausgehend von dem Satz ([14], S. 427):

Alle Hauptdiagonalelemente einer positiv semidefiniten Matrix sind größer oder gleich 0.

und unter Berücksichtigung des nachfolgend dargestellten Zusammenhangs ([13], S. 112) zwischen der Existenz eines lokalen Minimums einer Funktion f und der positiven Semidefinitheit der zugehörigen Hessematrix H_f:

Es sei f eine auf der offenen Menge $M \subset \mathbf{R}^n$ definierte und zweimal differenzierbare Funktion.
Ist x^* ein lokales Minimum von f auf M, so gilt:
a) $\operatorname{grad} f(x^*) = 0$,
b) $H_f(x^*)$ *ist positiv semidefinit.*

läßt sich eine hinreichende Bedingung für die Nichtexistenz eines lokalen Minimums formulieren ([10], S. 481):

Eine zweimal differenzierbare Funktion f nimmt in x kein lokales Minimum an, wenn an dieser Stelle gilt: $\frac{\partial^2 f(x)}{\partial x_i \partial x_i} < 0$, für mindestens ein x_i.

Weiterhin findet sich in ([2], S. 78) nachstehende notwendige und hinreichende Bedingung für die strenge Quasikonkavität einer Funktion:

Es sei f eine auf der offenen konvexen Menge $M \subset \mathbf{R}^n$ definierte differenzierbare Funktion. In diesem Fall ist f genau dann streng quasikonkav, wenn für jedes $x \in M$ und $v \in \mathbf{R}^n$ (mit $v'v = 1$ und $v' \operatorname{grad} f(x) = 0$) die Funktion $g(t) := f(x + tv)$ kein lokales Minimum an der Stelle $t = 0$ annimmt.

Aus den zitierten Sätzen läßt sich nun eine einfach zu überprüfende hinreichende Bedingung für die strenge Quasikonkavität ableiten ([8], S. 23):

Eine auf der offenen Menge $M \subset \mathbf{R}^n$ definierte und zweimal differenzierbare Funktion f ist streng quasikonkav, wenn für jedes $x \in M$ gilt: $\frac{\partial^2 f(x)}{\partial x_i \partial x_i} < 0$, für mindestens ein x_i.

Darüber hinaus gilt ([2], S. 80 und [7], S. 108):

Eine auf der konvexen Menge $M \subset \mathbf{R}^n$ definierte streng quasikonkave Funktion nimmt ihr Maximum bezüglich M höchstens in einem Punkt an, d.h. jedes lokale Maximum von f ist zugleich auch globales Maximum.

Unter Berücksichtigung der soeben dargestellten mathematischen Zusammenhänge lassen sich über die zweiten partiellen Ableitungen der reparametrisierten Gewinnfunktion G_{it} nach p_{it} (mit $k_i < p_{it} < p_i^{max}$) bzw. m_{ilt} (Ungleichheits-)Beziehungen für die Überprüfung der strengen Quasikonkavität der betrachteten Funktionsvarianten (vgl. Tabelle 2) extrahieren. Eine Funktionsvariante erfüllt die genannte Eigenschaft, wenn mindestens eine der im zugehörigen Tabellenfeld aufgeführten Bedingungen erfüllt ist. In diesem Fall kann unter der jeweils vorliegenden Variablen- und Parameterkonstellation eine Zielfunktionsmaximierung mit Hilfe eines geeigneten Suchverfahrens durchgeführt werden.

Modell	Bezug	Standardansätze	Modifizierte Ansätze
MUL	p_{it}	$\frac{1}{\beta_i} < \frac{k_i - p_{it}}{p_{it} + k_i}$	$\frac{1}{\beta_i} > \frac{k_i - p_{it}}{\frac{1}{2}(p_{it} + k_i) - p_i^{max}}$
	m_{ilt}	$\gamma_{il}(\gamma_{il} - 1) < 0$	$\frac{1}{\gamma_{il}} > \frac{1}{2}$
MCI	p_{it}	$MA_{it} > \frac{1}{2}\left(1 + \frac{p_{it} + k_i}{\beta_i(p_{it} - k_i)}\right)$	$MA_{it} > \frac{1}{2}\left(1 + \frac{p_{it} + k_i - 2p_i^{max}}{2\beta_i(p_{it} - k_i)}\right)$
	m_{ilt}	$MA_{it} > \frac{1}{2}\left(1 - \frac{1}{\gamma_{il}}\right)$	$MA_{it} > \frac{1}{2}\left(1 - \frac{2}{\gamma_{il}}\right)$
MNL	p_{it}	$MA_{it} > \frac{1}{2}\left(1 + \frac{2}{\beta_i(p_{it} - k_i)}\right)$	$MA_{it} > \frac{1}{2}\left(1 + \frac{3p_{it} - k_i - 2p_i^{max}}{2\beta_i(p_{it} - k_i)(p_i^{max} - p_{it})^2}\right)$
	m_{ilt}	$MA_{it} > \frac{1}{2}$	$MA_{it} > \frac{1}{2}\left(1 - \frac{1}{\gamma_{il}\sqrt{m_{ilt}}}\right)$

Tabelle 2: Bedingungen für die strenge Quasikonkavität der Gewinnfunktion

Die in Tabelle 2 aufgeführten Beziehungen geben bereits erste (theoretische) Hinweise auf die Eignung der einzelnen Funktionsvarianten für die modellgestützte Gewinnoptimierung. So sind z.B. die Bedingungen für strenge Quasikonkavität beim Standard-MNL-Modell – falls nur die Marketing-Anstrengungen Gegenstand der Optimierung sind – lediglich für diejenige Marke j einer Produktklasse erfüllbar, die eine dominierende Marktposition ($MA_{jt} > \frac{1}{2}$) innehat. Für die beiden anderen nicht-modifizierten Ansätze fallen die entsprechenden Bedingungen weniger restriktiv aus. Bei praktischen Anwendungen der einzelnen Ansätze hat sich im übrigen gezeigt, daß die vorgeschlagene Modellmodifikation in einigen Fällen zu einer Lockerung der bestehenden Restriktionen führen kann. Am deutlichsten wird dies wieder beim MNL-Modell, das in der modifizierten Form – unter bestimmten Variablen- und Parameterkonstellationen – auch für Gewinnoptimierungen (der oben beschriebenen Art) bei nicht-dominierende Marken anwendbar ist.

4. Anwendungsbeispiele

Ausgangspunkt der in diesem Abschnitt skizzierten empirischen Anwendungen sind zwei Datensätze, von denen der eine synthetische Daten enthält, die gezielt "generierte" Wirkungszusammenhänge zwischen zwei Marketing-Mix-Variablen (Preis, Displayaufwendungen) und den Marktanteilen dreier Marken beschreiben, während der zweite Datensatz die im Rahmen einer Panelerhebung erfaßten realen Marktanteile und Abverkaufspreise von 6 Kaffeemarken wiedergibt. Weiterhin wurden die markenspezifischen Stückkosten in beiden Fällen – in Ermangelung der genauen Werte und in Anlehnung an entsprechende Vorschläge aus der Literatur – als um 25% unter den durchschnittlich erzielten

Preisen und die Maximalpreise als um 25% über den Durchschnittswerten liegend angenommen. Das Marktvolumen wurde beidemal als zeitinvariant mit 200000 Mengeneinheiten festgesetzt.
Auf Basis dieser Daten erfolgte dann die Modellkalibrierung, d.h. die Schätzung der unbekannten Parameter α_i, β_i und γ_{i1} (für die Displayvariable). Dabei konnte mit dem modifizierten MCI-Modell für beide Datensätze die beste Anpassung erzielt werden. Tabelle 3 gibt die den anschließend beschriebenen Variablenoptimierungen zugrunde gelegten Parameterkonstellationen, die mittleren relativen Abweichungen (MRA_i) zwischen den beobachteten und den geschätzten Marktanteilen und die entsprechenden Theilschen Ungleichheitskoeffizienten ($Theil_i$) für die in den beiden Datenbeispielen betrachteten Marken wieder. Letztere sind auf das Intervall [0; 1] beschränkt, wobei der optimale Wert bei 0 liegt. Darüber hinaus zeichnet sich die auf dem modifizierten MCI-Modell aufbauende Gewinnfunktion noch dadurch aus, daß bei ihr in beiden Datenbeispielen die für die strenge Quasikonkavität erforderlichen Bedingungen (siehe letzte Zeile der Tabelle) erfüllt sind. Dies gilt im übrigen auch für den Fall des etwas schlechter angepaßten Standard-MCI-Ansatzes, während bei den MNL-Anwendungen und – in geringerem Maße – auch bei Anwendungen der MUL-Modelle die relevanten Bedingungen nicht immer (d.h. für jede Marke) erfüllt waren.

	Fiktiver Markt			**Kaffee-Markt**					
	Marke (i)			**Marke (i)**					
Bezeichner	1	2	3	1	2	3	4	5	6
α_i	0.076	0.237	0.714	1.056	-0.123	-0.080	0.032	0.716	1.166
β_i	0.842	0.693	0.634	0.814	0.962	0.875	0.848	0.643	0.969
γ_{i1}	0.447	0.411	0.350	–	–	–	–	–	–
MRA_i	4.24	4.95	3.45	5.46	14.72	10.66	12.48	9.85	5.72
$Theil_i$	0.028	0.031	0.021	0.032	0.075	0.071	0.061	0.064	0.037
q.-konkav	ja, ja	ja, ja	ja, ja	ja	ja	ja	ja	ja	ja

Tabelle 3: Kalibrierungsergebnisse des modifizierten MCI-Modells

Zu den Parameterschätzern in Tabelle 3 ist zu sagen, daß sowohl β_i als auch γ_{i1} in allen Fällen die bei Konsumgütern des täglichen Bedarfs zu erwartenden Vorzeichen aufweisen. Dabei ist zu berücksichtigen, daß das – korrekte – positive Vorzeichen bei den Preisgewichten β_i aus der bei der Modellmodifikation vorgenommenen Differenzenbildung $p_i^{max} - p_{it}$ resultiert. Beim Standard-MCI-Modell beispielsweise ergaben sich hingegen durchweg negative Preisgewichte.
Kommen wir nun zu den Ergebnissen der auf Basis des modifizierten MCI-Modells durchgeführten Variablen-Optimierung.
In Tabelle 4 sind die erzielten Resultate auszugsweise dargestellt. Die für Periode $T+1$ angegebenen Werte sind – abgesehen von der letzten Tabellenzeile – das Ergebnis einer simultanen Optimierung, bei der unterstellt wurde, daß alle am Markt vertretenen Anbieter unter Berücksichtigung der jeweiligen Konkurrenzaktivitäten versuchen, ihren Marketing-Mix zu optimieren. Aufgeführt sind jeweils die in der letzten Kalibrierungssperiode T erzielten Marktanteile und Gewinne, die zugehörigen Variablenkonstellationen sowie die Werte, die sich nach der Optimierung ergeben. Als Vergleichskriterien wurden zusätzlich die markenspezifischen Zuwachsraten beim Gewinn (ΔG_i^{sim}) und die über den Kalibrierungszeitraum gemittelten Marketing-Mix-Größen angegeben.
Die tabellierten Werte zeigen, daß in beiden Datenbeispielen für jede Marke eine Verbesserung der Gewinnsituation erzielt werden konnte. Allerdings fallen die Ergebnisse teilweise recht unterschiedlich

aus, was eine differenzierte Interpretation erforderlich macht.
Bei den synthetischen Daten ist z.B. der im Vergleich zu den Konkurrenten starke Gewinnzuwachs bei Marke 1 bemerkenswert. Ein Grund hierfür ist die Tatsache, daß bei dieser Marke – bei ähnlicher Displaypolitik aller drei Marken – im Gegensatz zu den Konkurrenten eine Preissenkung (mit entsprechendem Marktanteilsgewinn) vorgeschlagen wird.

	Fiktiver Markt			**Kaffee-Markt**					
	Marke (i)			**Marke (i)**					
Bezeichner	1	2	3	1	2	3	4	5	6
MA_{iT}	0.190	0.390	0.420	0.146	0.157	0.116	0.120	0.158	0.303
$MA_{i,T+1}$	0.232	0.326	0.442	0.179	0.143	0.129	0.105	0.148	0.296
G_{iT}	10908	21258	31754	68578	42177	35456	35685	49203	99195
$G_{i,T+1}$	17091	27882	38577	82728	48666	41457	39621	57541	107984
ΔG_i^{sim} (in %)	56.68	31.16	21.49	20.63	15.39	16.93	11.03	16.95	8.86
p_{iT}	2.90	2.60	2.70	7.99	7.22	7.30	7.14	7.13	6.87
$p_{i,T+1}$	2.86	2.74	2.71	7.96	7.58	7.39	7.53	7.52	7.05
$\bar{p}_i$	2.75	2.71	2.73	7.53	7.84	7.70	7.54	7.43	6.97
m_{i1T}	21000	23000	23000	–	–	–	–	–	–
$m_{i1,T+1}$	19973	18291	19897	–	–	–	–	–	–
$\bar{m}_{i1}$	21900	21800	22700	–	–	–	–	–	–
ΔG_i^{seq} (in %)	100.94	42.55	29.94	18.81	17.02	28.51	9.06	18.90	8.04

Tabelle 4: Ausgewählte Resultate der Variablenoptimierung

Bei den Resultaten für den Kaffee-Markt fällt u.a. auf, daß die markenspezifischen Preise – sieht man einmal von Marke 1 ab – in Periode T zum Teil deutlich unter den jeweiligen Durchschnittswerten liegen. Demgemäß erscheint auch die durch die simultane Optimierung nahegelegte "allgemeine Preiserhöhung" durchaus realistisch. Dies gilt insbesondere vor dem Hintergrund, daß es sich bei den hier betrachteten Kaffeemarken um hochgradig substitutive Produkte handelt, d.h., daß größere Preiserhöhungen nur dann am Markt durchzusetzen sind, wenn alle, oder doch zumindest ein Großteil der Anbieter ihre Preise anheben. Die dazu konträre Strategie einer leichten Preissenkung bei Marke 1 resultiert aus deren auffallend hohem Preis in Periode T.
Ein zumindest von der Tendenz her sehr ähnliches Bild erhält man, wenn die Variablenoptimierung (in Periode $T + 1$) für jede Alternative getrennt vorgenommen wird. Bei dieser als sequentielle Optimierung bezeichneten Vorgehensweise bleiben – im Gegensatz zum simultanen Ansatz – mögliche (Gegen-)Maßnahmen der Konkurrenz außer acht. Auch in diesem Fall wurden erwartungsgemäß für beide Datenbeispiele Variablenkonstellationen gefunden, die eine Verbesserung der gegenwärtigen Gewinnsituation (vgl. Tabelle 4, letzte Zeile) ermöglichen.
Vergleicht man die Gewinnzuwächse (ΔG_i^{sim} und ΔG_i^{seq}) der beiden Optimierungsstrategien miteinander, so fällt auf, daß die durch die sequentielle Optimierung erzielten Zuwachsraten bei den Marken 1, 4 und 6 des Kaffeemarktes geringer ausfallen als beim simultanen Ansatz. Einer der Gründe hierfür ist der bereits erwähnte Umstand, daß eine Preissteigerungsstrategie wie sie auch – allerdings in deutlich geringerem Maße – durch die sequentielle Optimierung für die Mehrzahl der Marken nahegelegt wird, idealerweise vom überwiegenden Teil der Anbieter getragen werden sollte, da die preispolitischen Spielräume bei "Alleingängen" merklich geringer ausfallen.

5. Schlußbemerkungen

Die vorausgegangenen Ausführungen haben gezeigt, daß man bereits mit einfachen ökonometrischen Ansätzen in der Lage ist, realitätsnahe Marketing-Mix-Optimierungen vorzunehmen, und daß die auf diese Weise zu erzielenden Resultate nützliche Hinweise auf mögliche Strategieverbesserungen liefern können. Die vorgestellte empirische Untersuchung, bei der mit einem modifizierten MCI-Ansatz die beste Anpassung an die vorliegenden Daten erreicht wurde, legt die Vermutung nahe, daß die Ergebnisse einer sequentiellen mit denen einer simultanen Optimierung zumindest von der Tendenz her übereinstimmen, was auch durch eine in [3] dokumentierte Studie bestätigt wird.
Zukünftige Aktivitäten im Rahmen der vorliegenden Themenstellung könnten z.B. die Ausdehnung der Simulationsstudien auf längerlebige Gebrauchsgüter und die Überprüfung der Zweckmäßigkeit einer zeit- bzw. variablenabhängigen Definition des Marktvolumens zum Gegenstand haben.

Literatur

(1) Allenby, G. M.; Rossi, P. E.: Quality Perceptions and Asymmetric Switching Between Brands, Marketing Science, Vol. 10, No. 3 (1991), 185 - 204.

(2) Avriel, M.; Diewert, W. E.; Schaible, S.; Zang, I.: Generalized Concavity, New York: Plenum Press, 1988.

(3) Carpenter, G. S.; Cooper, L. G.; Hanssens, D. M.; Midgley, D. F.: Modeling Asymmetric Competition, Marketing Science, Vol. 7, No. 4 (1988), 393 - 412.

(4) Cooper, L. G.; Nakanishi, M.: Market Share Analysis: Evaluating Competitive Marketing Effectiveness, Boston: Kluwer Academic Publishers, 1988.

(5) Decker, R.: Analyse und Simulation des Kaufverhaltens auf Konsumgütermärkten - Konzeption eines modell- und wissensorientierten Systems zur Auswertung von Paneldaten, Dissertation, Universität Karlsruhe (TH), 1993.

(6) Eliashberg, J.; Chatterjee, R.: Analytical Models of Competition with Implications for Marketing: Issues, Findings, and Outlook, Journal of Marketing Research, Vol. 22, August (1985), 237 - 261.

(7) Elster, K.-H.; Reinhardt, R.; Schäuble, M.; Donath, G.: Einführung in die nichtlineare Optimierung, Leipzig: Teubner, 1977.

(8) Gruca, T. S.; Kumar, K. R.; Sudharshan, D.: A Technical Note to Accompany 'An Equilibrium Analysis of Defensive Response to Entry Using a Coupled Response Function Model', Working Paper, No. 91-19, University of Iowa, Iowa City, 1991.

(9) Gruca, T. S.; Kumar, K. R.; Sudharshan, D.: An Equilibrium Analysis of Defensive Response to Entry Using a Coupled Response Function Model, Marketing Science, Vol. 11, No. 4 (1992), 348 - 358.

(10) Gruca, T. S.; Sudharshan, D.: Equilibrium Characteristics of Multinomial Logit Market Share Models, Journal of Marketing Research, Vol. 28, November (1991), 480 - 482.

(11) Hruschka, H.: Marktreaktionsfunktionen mit Interaktionen zwischen Marketing-Instrumenten, Zeitschrift für Betriebswirtschaft, Jg. 61, Heft 3 (1991), 339 - 355.

(12) Karnani, A.: Strategic Implications of Market Share Attraction Models, Management Science, Vol. 31, No. 5 (1985), 536 - 547.

(13) Luenberger, D. G.: Introduction to Linear and Nonlinear Programming, Reading (Mass.): Addison-Wesley, 1973.

(14) Noble, B.; Daniel, J. W.: Applied Linear Algebra, Englewood Cliffs: Prentice-Hall, 1977.

MARKETING DECISIONS & OPERATIONS RESEARCH METHODOLOGY

Wolfgang Gaul, Universität Karlsruhe (TH)

Against the background that interconnections between "marketing – market research – data analysis – statistics" seem to be better established than those between "marketing – operations research – optimization" we start with discussing reasons and giving examples with respect to this phenomenon.

Within a framework in which computer-assisted decision support (for marketing and market research) is in the centre of interest we explain why mutual fructification – especially between marketing and operations research – is asymmetric and give hints how within the well-known decision calculus discussion one has tried to make marketing managers use models already years ago. We take MARK²MAN (MARKet research & MARKeting MANagement), a software tool developed at our institute (see, e.g., Gaul, Baier (1993)), to demonstrate how typical situations for decision support in marketing can be tackled without going back to the mathematical modeling level.

After this introductory part, a list of applications is presented where operations research methodology has been used to formulate and solve marketing problems. As examples sales force management and product line optimization are selected for a more thorough mathematical description, a demonstration of arising difficulties, and a discussion of solution possibilities.

Sales force management is a typical field where marketing and OR can overlap. Determination of sales force size, time-effort allocation, and sales territory assignment are important problems in this field, for which decision support has to be made available.

In product line optimization one has to deal with the situation, that new products of a special product category should be introduced into a market, where already established products (own ones and, normally, such of competitors) are available and successful.

Finally, we report on recent (own) research from the area of new-product-introduction and give an outlook with respect to marketing problems for which application of OR methodology should be of interest.

References

Gaul, W.; Baier, D.: MARKtforschung und MARKeting MANagement (Buch mit Diskette): München, Oldenbourg Verlag, 1993.

CLUSTERING-BASED MARKET SEGMENTATION USING NEURAL NETWORK MODELS

Harald Hruschka, Universität Regensburg

Martin Natter, Institut für Höhere Studien, Wien

Most segmentation studies proceed in two steps, determining segments in the first step and looking for discriminating characteristics (segment descriptors) in the second step. Data-analytic methods usually applied in these steps are cluster analysis and discriminant analysis techniques, respectively. Artificial neural networks represent alternatives to better known statistical techniques. Certain types of artificial neural networks are closely related to well-known statistical methods. Principal components analysis, for example, is a special case of certain artificial neural network models. Therefore use of neural network models in market segmentation seems to be justified considering their greater generality.

In this paper two types of feedforward neural networks with logistic activation functions are specified. Models of the first type determine segments on the basis of segmentation criteria. These classification models are feedforward neural networks using segmentation criteria both as input variables and output variables. Between input and output variables we put a layer of hidden units. The task of the classification and discrimination models is more encompassing. Models of this type simultaneously form segments and discriminate between these segments on the basis of additional segment descriptors. Parameters of all models are estimated by an extended version of backpropagation.

The main marketing-related objective of the study presented here is to gain a general understanding of usage patterns of the main brands of a non-durable product class. Therefore usages of brands in different situations are chosen as segmentation criteria. The respondents are a representative random sample of 1007 persons. After deletion of incorrect data and limitation to the more frequent brands and situations, the data base consists of 854 respondents.

Using nonlinear neural networks for clustering-based segmentation lead to lower error measures compared to their linear counterparts. Interpretation of model results is emphasized in view of its great importance in Marketing applications.

MAKE-OR-BUY DECISIONS OF THE MARKETING FUNCTIONS (ADVERTISING, SALES PROMOTION, PUBLIC RELATIONS)

Heinz-Theo Peters, Universität zu Köln

The decision as to whether operational functions are to be delegated or performed oneself is not systematically dealt with in practice, nor in the literature relating to organisational and decision-making theory. On the one hand the majority of existing decision aids (portfolio analyses, checklists etc.) are concerned with the delegation of production segments and not so much with operational functions. On the other they only partly reflect the overall decision-making situation. Thus, the extreme alternatives of "make" or "buy" are often the only ones which are investigated, the sole objective of "efficiency" is taken as a basis or assumptions are based on evasively defined decision contexts.

The aim of this paper is to suggest a solution to this predicament by presenting a systematic and complete decision-making technique of how this make-or-buy decision can be approached. A model will be presented which fully covers the three components of objectives, alternatives and environmental conditions and which examines their interrelations. This approach will be explained taking the delegation of the marketing functions advertising, sales promotion and public relations as an example.

The model is elaborated in three stages:

- In the first stage an objective network is construed. This objective network represents all relevant fundamental objectives for the respective function (e.g. sales promotion) which have to be taken into account in the make-or-buy decision. The objectives network contains not only the objectives but their influencing parameters too and places them in a relationship to each other.
  The objective network developed from literary studies will then be checked to see whether it contains typical and realistic decision contexts. Typical decision contexts reflect the company's decision-making situation and consist of a variety of certain environmental factors (context factors). These differ significantly with respect to the identification of objectives. The identification of typical decision contexts facilitates the decision-maker's classification of their own corporate location and simplifies their choice of objectives.
- In a second stage the objectives and alternatives will be juxtaposed (e.g. full-service agency or a-la-carte agency), i.e. which alternatives are best suited to implement the various objectives. The influence of uncertain environmental factors on the objective development is to be ignored as far as possible.
- In a final stage the correlation between different decision contexts and certain alternatives will be disclosed and concrete suggestions made for the choice of alternatives as a function of these decision contexts.

With the aid of this model it is on the one hand possible to present the decision-making situation transparently and ultimately to derive recommendations for action with practical relevance from this.

EIN MODELL ZUR MEHRPERIODIGEN WERTPAPIERPLANUNG

Klaus Hellwig, Universität Ulm

Im Gegensatz zu den herkömmlichen, auf einer Anwendung des Bernoulli-Prinzips basierenden Modellen der Portfolioplanung wird ein Ansatz formuliert, der ohne explizite Vorgabe einer Risikonutzenfunktion auskommt. Statt dessen wird eine Planung auf der Grundlage zweier Grundprinzipien - Arbitragefreiheit und Erhaltung des Ertragswertes - vorgeschlagen. Neben dem Verzicht auf die Ermittlung und Aggregation geeigneter Nutzenfunktionen bietet der Ansatz ein Lösungskonzept, das insbesondere für Investmentgesellschaften geeignet erscheint, die im Interesse ihrer Anleger die finanzielle Leistungsfähigkeit der von ihnen verwalteten Fonds erhalten wollen.

Zum Nachweis der Existenz einer den Grundprinzipien genügenden Lösung wird das Modell in ein Fixpunktproblem transformiert. Dabei wird unterstellt, daß das Entscheidungsproblem durch einen endlichen Zustandsbaum charakterisiert werden kann. Ferner wird angenommen, daß in jedem Zustand endlich viele beliebig teilbare Anlagealternativen bestehen, wobei Leerverkäufe zugelassen sind. Mit Hilfe des Fixpunktsatzes von Kakutani wird eine Lösung des Problems unter sehr allgemeinen Voraussetzungen nachgewiesen. Für die Lösung werden einige Eigenschaften abgeleitet und diskutiert. Insbesondere kann gezeigt werden, daß das Modell für den Fall einer einperiodigen Planung mit dem Modell der renditeorientierten Portfolioplanung übereinstimmt (vgl. Hellwig, K., Renditeorientierte Portfolioplanung, Zeitschrift für Betriebswirtschaft, Januar 1993). Für diesen Fall können damit die dort gewonnenen Ergebnisse, wie die Kompatibilität mit Bernoulli - Prinzip bei logarithmischer Nutzenfunktion, übertragen werden.

Für den Fall, daß nicht nur die Erhaltung, sondern das Wachstum des Ertragswertes mit a priori gegebenen (u. U. zustandsabhängigen) Wachstumsraten gewünscht wird, lassen sich entsprechende Ergebnisse ableiten.

THE ASYMPTOTIC EFFICIENCY OF THE ECOMOR TREATY

ERHARD KREMER; HAMBURG

Summary: The excess-of-loss and ECOMOR treaties are compared with respect to a certain asymptotic efficiency measure. The excess-of-loss cover turns out to be better than the ECOMOR cover under general conditions and to be practically equivalent to the ECOMOR-cover under the Pareto-model.

Zusammenfassung: Die Schadenexzedenten- und ECOMOR-Verträge werden verglichen mittels eines bestimmten Effizienz-Maßes. Der Schadenexzedenten-Vertrag erweist sich unter allgemeinen Bedingungen als besser als der ECOMOR-Vertrag und als praktisch äquivalent zum ECOMOR-Vertrag unter dem Pareto-Modell.

1. The reinsurance treaties

Consider a collective of insurance risks producing claims with (positive) sizes $X_1, X_2, X_3, \ldots$ each year. Let N denote the random variable describing the number of claims per year. The classical excess-of-loss cover is defined by the first insurer's retained claims amount:

$$R_N = \sum_{i=1}^{N} \min(X_i, P),$$

where the nonnegative number P is the so-called priority. For defining the classical ECOMOR-treaty (s. Thépaut (1950)) take the claims ordered in increasing amounts:

$$X_{N:1} \leq X_{N:2} \leq \ldots \leq X_{N:N} \quad .$$

The claims amounts retained by the first insurer in case of the ECOMOR cover is then nothing else but:

$$R_N^* = \sum_{i=1}^{N} \min(X_i, X_{N:(N-p+1)})$$

where p is a given nonnegative integer. Obviously R_N can be reformulated according:

$$R_N = \sum_{i=1}^{N} \min(X_{N:i}, P)$$

and R_N^* according:

$$R_N^* = \sum_{i=1}^{N-p+2} X_{N:i} + p \cdot X_{N:(N-p+1)} \quad .$$

Both claims amounts are special cases of the following more general one:

$$S_N = \sum_{i=1}^{N} c_{Ni} \cdot h(X_{N:i}) + d \cdot h(X_{N:(N-p+1)}) \tag{1.1}$$

where $d, c_{n1}, c_{n2}, \ldots$ are certain coefficients and h is a measurable function on $(0,\infty)$. A treaty with the first insurer's retained total claims amount of type (1.1) can be called reinsurance treaty based on ordered claims (see Kremer (1984)).

2. The asymptotic efficiency

For the comparison of two different reinsurance treaties based on ordered claims the author introduced in 1990 an asymptotic efficiency. The concept was based on investigating a sequence of growing collectives indexed with integers $k=1,2,3,\ldots$. For the claims number N_k of collective no. k it was assumed that:

$$\lim_{k\to\infty} E(N_k) = \infty \ , \ \lim_{k\to\infty} Var(N_k)/E(N_k) = c \ ,$$

with a given constant c. The claims sizes were taken to be the $X_1, X_2, \ldots,$ independently of k. They are assumed to be identically distributed with distribution function F, mutually independent and independent of the N_k. Then the retained claims amount S_N depends on the collective, say $S_N = S_{N_k}^{(k)}$, since $N=N_k$ depends on the collective and $d=d^{(k)}$, $c_{ni}=c_{ni}^{(k)}$ are allowed to depend also on k. Under certain regularity conditions one can prove that there exist certain numbers μ_F and σ_F^2 depending on the special treaty and on F, such that one has:

$$\lim_{k\to\infty} E(S_{N_k}^{(k)})/E(N_k) = \mu_F \ , \tag{2.1}$$

$$\lim_{k\to\infty} \mathrm{Var}(S_{N_k}^{(k)})/E(N_k) = \sigma_F^2 + c\cdot\mu_F^2 \ . \tag{2.2}$$

For the general formulas for μ_F, σ_F^2 the reader is referred to Kremer (1984). Taking two reinsurance treaties based on ordered claims with corresponding numbers $\mu_F(i)$, $\sigma_F^2(i)$ $(i=1,2)$ of (2.1),(2.2) such that:

$$\mu_F(1) = \mu_F(2)$$

the value:

$$\mathrm{Eff}_F(1:2) = \sigma_F^2(2)/\sigma_F^2(1)$$

is called asymptotic efficiency of treaty no. 1 relative to treaty no. 2 (see Kremer (1990)). In case that $\mathrm{Eff}_F(1:2)$ is equal to or greater than one, one says that the treaty no. 1 is equivalent or better than the treaty no. 2.

3. Efficiency of the ECOMOR-treaty

In Kremer (1990) the above described efficiency measure was used for comparing the classical excess-of-loss with the largest claims reinsurance treaty. The ECOMOR treaty was not considered since only general treaties with $d=0$ were treated. In the following the excess-of-loss and ECOMOR treaties shall be compared in the context of growing collectives. Take the priority P of the excess-of-loss cover independently of the collective number k, but the integer p dependent of the number k, say $p=p_k$. Suppose one has an arbitrary, but fixed value s between zero,one and such that:

$$\lim_{k\to\infty} \left(\frac{p_k}{E(N_k)}\right) = s.$$

One assumes in addition that:

$$F(P_s-\varepsilon) < F(P_s) < F(P_s+\varepsilon) \quad \text{for all positive } \varepsilon, \tag{3.1}$$

F^{-1} is differentiable at $(1-s)$,

where:

$$P_s = F^{-1}(1-s)$$

and F^{-1} is the pseudo-inverse:

$$F^{-1}(u) = \inf\{x: F(x) \geq u\}.$$

One has the main result:

Theorem:
Under the given conditions one has $\mathrm{Eff}_F(1{:}2) < 1$, where the first treaty is the ECOMOR-treaty and the second one the excess-of-loss treaty (with $P=P_s$). This means that the excess-of-loss cover is better than the ECOMOR-cover.

Proof:
According to the results in Kremer (1984) one has:

$$\mu_F(1) = \int_0^{P_s} x\, F(dx) + s\cdot P_s \quad ,$$

and for $P=P_s$:

$$\mu_F(2) = \int_0^{\infty} \min(x,P_s)\, F(dx) \quad ,$$

giving by a routine argument:

$$\mu_F(1) = \mu_F(2) \quad .$$

Furthermore one has (because of theorem 1 in Kremer (1984) part (b) for $l=1$):

$$\sigma_F^2(1) = \int_0^{P_s}\int_0^{P_s} (\min(F(r),F(t))-F(r)\cdot F(t))\, dr\, dt \; +$$

$$+\; 2\cdot s\cdot (F^{-1})'(1-s)\cdot \int_0^{P_s} (\min(F(t),1-s)-F(t)\cdot(1-s))\, dt$$

$$+\; s^3\cdot(1-s)\cdot((F^{-1})'(1-s))^2.$$

Here:

$$\int_0^{P_s}\int_0^{P_s} (\min(F(r),F(t))-F(r)\cdot F(t))\, dr\, dt =$$

$$= 2\cdot \int_0^{P_s} F(r)\cdot \int_r^{P_s} (1-F(t))\, dt\, dr$$

$$\int_0^{P_s} (\min(F(t),1-s)-F(t)\cdot(1-s))\, dt = (\frac{s\cdot(1-s)^2}{2}).$$

Finally (also because of theorem 1 in Kremer (1984) for I=0):

$$\sigma_F^2(2) = 2\cdot\int_0^{P_s} F(r)\cdot\int_r^{P_s} (1-F(t))\,dt\,dr. \tag{3.2}$$

Consequently:

$$\sigma_F^2(1) = \sigma_F^2(2) + \rho_F(s), \tag{3.3}$$

with:

$$\rho_F(s) = s^2\cdot(1-s)^2\cdot(F^{-1})'(1-s) + s^3\cdot(1-s)\cdot((F^{-1})'(1-s))^2. \tag{3.4}$$

Because of assumption (3.1) one has:

$$(F^{-1})'(1-s) > 0,$$

giving:

$$\rho_F(s) > 0 \ . \ \Delta$$

From (3.3),(3.4) one can easily derive the asymptotic efficiency for special claims size model assumptions. For example one has the:

<u>Corollary:</u>
Suppose that F is a Pareto-distribution function, i.e.:

$$F(x) = 1 - x^{-a} \ , \ x\geq 1 \ .$$

with a larger than 2. Then:

$$\mathrm{Eff}_F(1:2) = (1 + \rho_a(s)/\gamma_a(s))^{-1} \ ,$$

with:

$$\rho_a(s) = (\tfrac{1}{a})^2\cdot(1-s)\cdot s^{-2/a+1} + (\tfrac{1}{a})\cdot(1-s)^2\cdot s^{-1/a+1}$$

$$\gamma_a(s) = (\tfrac{1}{a-1})^2\cdot s^{2(1-a)} + (\tfrac{2}{a-2})\cdot s^{2-a} -$$

$$- 2\cdot\frac{a}{(a-1)^2}\cdot s^{1-a} - \frac{a}{2\cdot(a-2)(a-1)^2} \ .$$

In the situation of the Corollary one gets the following values for $\mathrm{Eff}_F(1:2)$:

	a = 2.5	a = 3.0	a = 4.5	a = 6.0
s = 0.1	0.9995	0.9999	1.0000	1.0000
0.2	0.9951	0.9990	0.9999	0.9999
0.3	0.9803	0.9939	0.9998	0.9999

Obviously for practical conditions (of motor insurance) the efficiency of the ECOMOR-cover relative to the excess-of-loss cover is near one (in practice s will not exceed 0.2), meaning that for the Pareto-model both treaties are practically equivalent.

REFERENCES:

Kremer, E. (1984): Rating of nonproportional reinsurance treaties based on ordered claims. A paper in the book: Premium calculation in Insurance, edited by Goovaerts, M., De Vylder, F. and Haezendonck, J., Nato ASI series.

Kremer, E. (1990): The asymptotic efficiency of largest claims reinsurance treaties. ASTIN Bulletin.

Thépaut, A. (1950): Une nouvelle forme de réassurance. Le traité d'excédent du cout moyen relatif (ECOMOR). Bulletin Trimestriel, Vol. 49, no. 192.

ZUR LÖSUNG DES NUTZUNGSDAUERPROBLEMS MIT DYNAMISCHER OPTIMIERUNG

Karl Luhn, Technische Universität Ilmenau

Die Bestimmung optimaler Nutzungsdauern bei endlichem Planungshorizont und nichtidentischen Investitionsketten ist prinzipiell mittels dynamischer Optimierung möglich. Als Ziel ist bei Unterstellung eines vollkommenen Kapitalmarktes der maximale Kapitalwert der Investitionskette zu definieren. Die optimalen Nutzungsdauern der Einzelinvestitionen lassen sich als bedingt-optimale Entscheidungen interpretieren. Dadurch entsteht die Möglichkeit, die Kapitalwerte der Einzelinvestitionen so miteinander zu kombinieren, daß aus der Vielzahl der Varianten (bei einem Planungshorizont T=11 beispielsweise 28646) jeweils nur die bedingt-optimale Verkettung der Einzelinvestitionen berücksichtigt werden muß.

Für die Vorwärtsrechnung kann folgender Algorithmus angewendet werden (Ausschnitt):

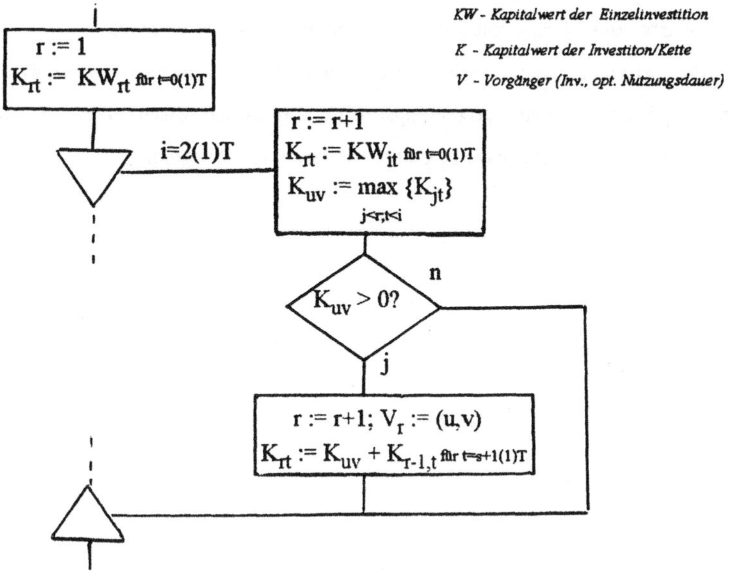

Ist das Ende des Planungshorizontes erreicht, so läßt sich die Investitionskette rückwärts aus den zum jeweiligen maximalen Kapitalwert gehörenden Indexpaaren (u,v) ermitteln. Im Vortrag wird die Vorgehensweise an einem Beispiel mit T=10 sowie 10 verschiedenen Einzelinvestitionen erläutert.

STOCHASTISCHE METHODEN BEI DER BEWERTUNG VON OPTIONSKONTRAKTEN

Peter Martens, Westdeutsche Landesbank Düsseldorf

Ziel des Vortrags ist es, die bei der Bewertung von Optionen angewendeten stochastischen Modelle detailliert darzustellen.

Der auch als "Irrfahrtenmodell" bekannte einfache Random Walk wird als stochastischer Prozeß definiert und der Wiener Prozeß als Grenzprozeß einer Familie von einfachen Random Walks hergeleitet. Sodann geben wir die Dichte der Erstpassierzeit eines Wiener Prozesses an und zeigen, wie ihre Bestimmung auf die Lösung der bekannten prospektiven Diffusionsgleichung mit einer zusätzlichen Randbedingung zurückgeführt werden kann.

Den Bezug zur Optionspreistheorie stellen wir her, indem wir die üblicherweise angenommene geometrische Brownsche Bewegung für das Basisgut als Wiener Prozeß mit Berücksichtigung von Zinseinflüssen darstellen. Auf der Grundlage dieser Kursmodellierung können dann Optionsrechte risikoneutral als diskontierte Erwartungswerte zukünftiger Zahlungen bewertet werden.

Das genannte mathematische Instrumentarium wenden wir dann am Beispiel von Barrier Options auf die Bewertung von Optionen mit pfadabhängigem Auszahlungsprofil an. In diesem Zusammenhang erläutern wir die in der Theorie der stochastischen Prozesse bekannte Methode der Spiegelung.

VERHALTENSMUSTER PRIVATER ANLEGER

Andreas Oehler, FernUniversität Hagen

Zusammenfassung: Der Beitrag beschäftigt sich mit Anlagestrategien (Verhaltensmustern) privater Anleger. Zunächst werden wesentliche in der Literatur genannte Verhaltensmuster kurz beschrieben, von denen sich die jeweiligen Anleger eine höhere Gewinnchance versprechen, die jedoch der Markteffizienzhypothese widersprechen, weil sie nicht allein an zukünftigen fundamentalen Entwicklungen ausgerichtet sind. Für die Untersuchung wird ein experimentelles Design entwickelt, welches eine differenziertere Analyse privaten Anlageverhaltens zuläßt, als dies mit Daten realer Finanzmärkte möglich wäre. Die Ergebnisse des tracking (Verkaufsseite) und chasing (Kaufseite) sprechen deutlich für nicht-theoriekonforme Verhaltensmuster, die u.a. auf der Nutzung historischer Preise basieren. Als mögliche Erklärungen können u.a. sowohl Fehleinschätzungen von Eintrittswahrscheinlichkeiten zukünftiger Preisänderungen als auch eine asymmetrische Bewertung von Gewinnen und Verlusten ("loss aversion") herangezogen werden.

Summary: This paper deals with patterns in private investor behavior. Based on a short literature survey the author describes basic behavior patterns contrary to the efficient market hypothesis: investors trading on the basis of price, because they believe that they can improve their chances of making a profit by doing so. An experimental design is run to analyse the private investor behavior in a more detailed way than it would be possible in real financial markets. We find much evidence for behavior patterns: subjects using historical prices to create their investment decisions. One part of the investors might think that winning stocks will fall and losing stocks will rise again (the behavior caused thereby called tracking); another group of subjects might think that winning stocks will rise further on and losing stocks will fall further on (the behavior caused thereby called chasing). The results can be explained by subjects' misperception of probabilities of future price changes or by an asymmetric valuation of gains and losses ("loss aversion").

1. Problemstellung

Das Verhalten privater Anleger an Finanzmärkten, insbesondere an Aktienmärkten, läßt sich in der Realität nur sehr eingeschränkt beobachten und analysieren. Üblicherweise werden hierzu Informationen durch Befragungen gewonnen oder auf aggregiertem Niveau (Marktebene) untersucht, ohne daß individuelle Entscheidungen sichtbar werden oder eine Trennung der Verhaltensweisen privater und institutioneller Anleger möglich wäre.

Im bank- und finanzwirtschaftlichen Schrifttum wird über Anlagestrategien berichtet, denen die Erwartung zugrundeliegt, mit der Nutzung vergangener Kursinformationen die Chance für Gewinne zu erhöhen. Solche Überlegungen stehen im Widerspruch zur Markteffizienzhypothese, da Anleger mit solchen Strategien dann auch nicht-fundamentale Informationen in ihr Kalkül einbeziehen.

Überlegungen zu solchen Verhaltensweisen werden vor allem durch empirische und deduktive (verhaltenslogische) Argumente begründet (für einen Überblick vgl. Shleifer/Summers 1990, De Long et al. 1990, Camerer 1987). Es wird u.a. hervorgehoben, daß bei der Informationsaufnahme fundamentale Entwicklungen eine geringere Rolle spielen als Informationen über vergangene Kursbewegungen oder über das Verhalten anderer Marktteilnehmer.

Eine Untersuchung, inwieweit private Anleger im Sinne der Kapitalmarkttheorie nicht-rationale Verhaltensweisen erkennen lassen, läßt sich in realen Finanzmärkten kaum durchführen, da die Erwartungen und die individuellen Entscheidungen der Anleger nicht kontrolliert und beobachtet werden können. Daher wird ein experimentelles Design entwickelt, das die gewünschten Analysen für private Anleger als (via Finanzintermediäre) indirekte Marktteilnehmer ermöglichen soll (vgl. hierzu z.B. Weber/Camerer 1991, Tietz 1990, Copeland/Friedman 1987, 1991 und Oehler 1991).

2. Verhaltensmuster, verwandte Untersuchungen und Hypothesen

Es lassen sich zwei grundlegende Verhaltensmuster, die eher auf der Kurshistorie als auf der Erwartung zukünftiger fundamentaler Entwicklungen basieren, differenzieren:

- Die Anlagestrategie des tracking oder negative-feedback trading läßt sich durch das Kaufen (Verkaufen) von Aktien, deren Kurse gefallen (gestiegen) sind, beschreiben.
- Für das Verhaltensmuster des chasing oder positive-feedback trading ist dagegen ein Kaufen (Verkaufen) nach steigenden (fallenden) Kursen charakteristisch.

Diese Verhaltensweisen werden in der Literatur u.a. wie folgt zu erklären versucht:

- chasing:
  - Erwartung, daß sich ein Trend fortsetzt (Lakonishok et al. 1991)
  - Repräsentativitäts-Heuristik, angewendet auf die letzte Information ("probability matching"; Andreassen 1988)
- tracking:
  - Erwartung einer Trendumkehr
  - Repräsentativitäts-Heuristik, angewendet auf die durchschnittliche Länge einer Zeitreihe ("gambler's fallacy"; Andreassen 1988)
  - Im Ergebnis identische Verhaltensweise auf der Verkaufsseite wie der auf einer "loss aversion" beruhende "disposition effect" (vgl. Kahneman et al. 1990, Shefrin/Statman 1985, Weber/Camerer 1991), der jedoch das Resultat einer asymmetrischen Bewertung von Gewinnen und Verlusten und nicht das Ergebnis eines Erwartungbildungsprozesses darstellt.

		KAUFSEITE		
		tracking	kein Muster	chasing
VERKAUFSSEITE	tracking	1	2	3
	kein Muster	4	5	6
	chasing	7	8	9

Abb. 1: Mögliche Verhaltensmuster

Differenziert man das trendkonträre (tracking) und das trendkonforme (chasing) Verhalten in solches der Kauf- und der Verkaufsseite, so lassen sich grundsätzlich vier

Verhaltensmuster identifizieren. Bezieht man schließlich noch ein Anlageverhalten mit ein, daß auf Kauf- oder Verkaufsseite keines der Muster erkennen läßt, dann existieren insgesamt neun denkbare Verhaltensmuster (vgl. Abb. 1).

Empirische und experimentelle Untersuchungen beziehen sich zumeist auf das Verhalten institutioneller Investoren und dabei besonders auf Pensionskassen und Investmentfonds. So kommen Lakonishok et al. (1991) zu dem Ergebnis einer schwachen empirischen Evidenz für das Verhaltensmuster des chasing. Eine umfassende experimentelle Analyse führte Andreassen (1987,1988) mit studentischen Versuchspersonen durch. Das Phänomen des "disposition effect" fanden z.B. Weber/Camerer 1991, Tietz 1990 und Kahneman et al. 1990.

Aufgrund der experimentellen Befunde von Andreassen, die am ehesten auf das Verhalten privater Anleger übertragbar scheinen, wird für die folgende eigene experimentelle Untersuchung angenommen, daß private Investoren das Verhaltensmuster des tracking zeigen. Daraus folgen die beiden Hypothesen:

- H1: Die Probanden kaufen bei fallenden Kursen mehr als bei steigenden Preisen.
- H2: Die Teilnehmer verkaufen bei steigenden Kursen mehr als bei fallenden Preisen.

Beide Hypothesen sollen in summarischer Analyse über alle Probanden und in einer Einzelfallanalyse untersucht werden.

3. Design der experimentellen Untersuchung

Die Experimentteilnehmer konnten in jeder der insgesamt 11 Anlageperioden in fünf Aktien und eine niedrigverzinsliche sichere Anlage investieren. Die Kauf- und Verkaufsorders wurden abgegeben, nachdem die aktuellen Kurse zu Beginn jeder Periode per Post zugegangen waren. Die Kurse der fünf Aktien waren unabhängig von den Orders der Anleger, Leerverkäufe und Kreditaufnahme waren nicht zulässig. Die Anfangsausstattung betrug 50.000 Geldeinheiten. Das Experiment wurde als schriftliches Fragebogenexperiment durchgeführt. Eine Periode umfaßte eine Woche; aufgrund der Postlaufzeiten betrug die effektive Entscheidungszeit der Probanden ca. 2 Tage.

Die Kursinformationen zu Beginn jeder Periode stellten neue Informationen über den Preistrend dar. Den Teilnehmern war durch die Teilnehmeranleitung, in der u.a. kurz Entscheidungssituationen an einer Aktienbörse beschrieben wurden, bekannt, daß allen Kurse reale Daten zugrundeliegen, die für das Experiment u.a. in Namen und Zeitraum angepaßt worden waren. Zugleich erhielten alle Probanden zu Beginn des Experimentes eine Kursprognose für einen Zeitpunkt nach dem Ende der Untersuchung (3 Monate vom Beginn entfernt) für jede Aktie in Form einer Kursspanne. Diese Spanne wurde dadurch ermittelt (den Probanden nicht bekannt), daß jeweils zu dem letzten Preis der Kurshistorie die Standardabweichung der gesamten Kursreihe der Vergangenheit und des Experimentverlaufes addiert (oberer Wert der Spanne) bzw. von diesem vergangenen Kurs subtrahiert wurde (unterer Wert der Spanne). Um eine Gleichheit des Erwartungswertes der Prognose mit dem letzten historischen Kurs zu vermeiden, wurden die Grenzen der Prognosespanne jeweils um 3 Prozent erhöht. Den Probanden kam in der Teilnehmeranleitung zur Kenntnis, daß die realisierten Kurse der Prognoseperiode zu ca. 70 % in der jeweiligen Spanne liegen würden.

Das Design wurde in zwei Varianten durchgeführt, die sich darin unterschieden, daß in Typ I Korrelationsinformationen zu den Aktien erst in der Mitte des Experimentes und in Typ II bereits zu Beginn zur Verfügung gestellt wurden. Diese Differenzierung erfolgte für eine andere Fragestellung, die hier nicht weiter verfolgt wird.

Als Probanden nahmen an beiden Experiment-Typen Arbeitnehmer, Selbständige und Freiberufler sowie Nicht-Berufstätige und Hausfrauen/-männer teil, die als Voll- oder Teilzeitstudenten im Erst-, Zweit- oder Gasthörer-Status an der FernUniversität Hagen im Hauptstudium des Lehrgebietes Bank- und Finanzwirtschaft immatrikuliert waren. Aufgrund des hohen Anteils an berufstätigen Probanden (oft in Mehrpersonen-Haushalten) mit eigenem Einkommen und einem überdurchschnittlichen Geldvermögen (Vergleich: alte Länder der Bundesrepublik) weisen die Teilnehmer günstige Eigenschaften in Bezug auf das Untersuchungsziel auf (vgl. Oehler 1992, Bitz/Oehler 1993).

4. Ergebnisse

Auf der Basis des Experimentdesigns bestanden zwei grundsätzliche Möglichkeiten, die Kauf- und Verkaufsorders zu stellen:

- Ein theoriekonformer Anleger würde zukunftsorientiert die Kursprognosen und im Falle von Risikoneutralität die Erwartungswerte der Prognosespannen verwenden. Entsprechend wäre zu erwarten, daß solche Anleger bei aktuellen Kursen, die unter dem jeweiligen Erwartungswert der Prognose liegen (Unterbewertung), kaufen und bei aktuellen Kursen, die über dem Erwartungswert liegen (Überbewertung), verkaufen.
- Anleger, die sich an der Kurshistorie orientieren, würden die aktuellen und die vergangenen Preisänderungen analysieren (Preistrend feststellen) und im Falle des tracking (chasing) trendkonträr (trendkonform) ihre Orders stellen.

In Hypothese 1 war angenommen worden, daß die Probanden bei fallenden Kursen mehr Kauforders stellen als bei steigenden Kursen. In Hypothese 2 wurde unterstellt, daß die Anleger bei steigenden Kursen mehr Verkaufsorders stellen als bei fallenden.

Zur Operationalisierung der Hypothesen werden alle Entscheidungssituationen (je Aktie eine Entscheidungssituation pro Periode) in die vier möglichen Preistrend-Kategorien "++", "-+", "+-" und "--" unterteilt und getrennt für diese Kategorien die Anzahl und die Stückvolumina der abgegebenen Kauf- und Verkaufsorders ermittelt (z.B. bedeutet ++ einen Preisanstieg in zwei aufeinanderfolgenden Perioden). Die errechneten Werte wurden im Sinne einer Standardisierung durch die Anzahl der zugrundeliegenden Entscheidungssituationen dividiert, da nicht alle Trendkategorien gleich häufig besetzt sind. Alle im folgenden dargestellten Ergebnisse basieren auf der Anzahl der abgegebenen Orders. Auf die Präsentation der Stückvolumina der Orders wird aus Platzgründen verzichtet. Dies scheint vertretbar, da die Resultate unter Verwendung der Stückvolumina fast identisch zu denen mit der Anzahl der Orders ausfallen. Abb. 2 gibt die Daten der beiden Experiment-Typen über alle Teilnehmer wieder:

Preistrend	Experiment I* Anzahl Kauforders	%	Anzahl Verkaufsorders	%	Experiment II* Anzahl Kauforders	%	Anzahl Verkaufsorders	%
- -	19,3	14	11,7	13	20,4	16	10,2	11
+ -	30,2	22	19,7	22	35,4	26	23,5	24
Σ -	**23,6**	**36**	**14,9**	**35**	**26,3**	**42**	**15,4**	**35**
+ +	46,5	38	30,9	38	50,3	48	31,3	37
- +	30,5	26	22,5	27	9,3	10	23,2	28
Σ +	**41,2**	**64**	**28,1**	**65**	**36,6**	**58**	**28,6**	**65**

* Jeweils ca. 300 Probanden.

Abb. 2: Anzahl Kauf- und Verkaufsorders in Abhängigkeit vom Preistrend

In beiden Experiment-Typen zeigt sich auf der Verkaufsseite ein deutliches tracking (und ein deutlicher "disposition effect"), da ca. zwei Drittel der Verkaufsorders in Entscheidungssituationen mit "+" - Trend gestellt werden. Für die Kaufseite läßt sich konstatieren, daß in ähnlicher Deutlichkeit das Verhaltensmuster des chasing und nicht das erwartete tracking zu beobachten ist.

Gerade die experimentelle Untersuchung ermöglicht die Analyse des Verhaltens jedes einzelnen privaten Anlegers. Für die Einzelfallanalyse müssen zunächst geeignete Kennzahlen definiert werden, um mögliche Verhaltensmuster aufdecken zu können.

Für die Kaufseite wird $\epsilon_i = (K-_i - K+_i) / (K-_i + K+_i)$

mit $K-_i$ = Anzahl der Kauforders des Teilnehmers i, wenn der Kurs gefallen ist,

$K+_i$ = Anzahl der Kauforders des Teilnehmers i, wenn der Kurs gestiegen ist

und damit $1 \geq \epsilon \geq -1$ definiert.

Für die Verkaufsseite wird analog $\alpha_i = (V+_i - V-_i) / (V+_i + V-_i)$

mit $V+_i$ = Anzahl der Verkaufsorders des Teilnehmers i, wenn der Kurs gestiegen ist,

$V-_i$ = Anzahl der Verkaufsorders des Teilnehmers i, wenn der Kurs gefallen ist

und damit $1 \geq \alpha \geq -1$ definiert (vgl. hierzu (α_i) auch Weber/Camerer 1991).

Für $\epsilon_i > 0$ und $\alpha_i > 0$ liegt tracking, für $\epsilon_i < 0$ und $\alpha_i < 0$ liegt chasing vor.

Ausprägung Kennzahlen	Experiment I Wert ε	Anteil TN in %	Wert α	Anteil TN in %	Experiment II Wert ε	Anteil TN in %	Wert α	Anteil TN in %
> 0	0,60	34,9	0,70	54,8	0,67	37,8	0,77	54,5
= 0	0	9,0	0	11,0	0	12,4	0	9,1
< 0	-0,66	56,1	-0,60	34,2	-0,63	49,8	-0,77	36,4
Durchschn.	-0,16*		0,22*		-0,06		0,16*	

* Signifikant von null verschieden auf $\alpha = 0,01$. TN=Teilnehmer

Abb. 3: Einzelfallanalysen zu den Verhaltensmustern

Die Daten der vorstehenden Abb. 3 bestätigen den Befund der summarischen Analyse auch in der Einzelfallanalyse mit ϵ und α. Ca. 55 % aller Teilnehmer unterliegen dem tracking auf der Verkaufsseite und nur ca. 35 % (Typ I; 38 % Typ II) tracking auf der Kaufseite. Hypothese 2 (tracking Verkaufsseite) ist also nicht zu verwerfen, während Hypothese 1 (tracking Kaufseite) in dieser allgemeinen Form zu verwerfen ist.

5. Diskussion

Die Ergebnisse der Untersuchungen zeigen, daß bestimmte Verhaltensmuster der Probanden beobachtet werden können, die eine Nutzung vergangener Preise vermuten lassen. Ein postexperimenteller Fragebogen, der auch Items zur Nutzung von Informationen durch die Teilnehmer enthält, ist derzeit noch nicht ausgewertet.

Es soll jedoch trotz der Befunde ergänzend geprüft werden, ob sich Anhaltspunkte für die eingangs beschriebenen theoriekonformen Verhaltensweisen (Kauforders bei Unterbewertung, Verkaufsorders bei Überbewertung) finden lassen. Die Auswertungen für beide Experiment-Typen zeigen, daß sowohl Kauf- als auch Verkaufsorders in Entscheidungssituationen mit Überbewertung im Vergleich zu solchen mit Unterbewertung überwiegen. Der Befund fällt für die Verkaufsorders allerdings deutlicher aus. Kauforders dominieren Verkaufsorders in der Überbertung und in der Unterbewertung, allerdings ist diese Dominanz in Entscheidungssituationen der Unterbewertung ausgeprägter. Insgesamt zeigt die summarische Analyse keine deutlichen Hinweise für die beschriebene theoriekonforme Verhaltensweise. Dies schließt jedoch nicht aus, daß sich einzelne Teilnehmer in diesem Sinne rational verhalten haben. Eine Einzelfallanalyse steht noch aus.

Zu den beobachteten Verhaltensmustern des tracking (Verkaufsseite) und chasing (Kaufseite) lassen sich einige interessante Resultate ergänzen, die aus Platzgründen ohne Zahlenmaterial aufgeführt werden:

- Ein erneuter Blick in Abb. 2 zeigt, daß bei nachhaltigem Preistrend (++, --) im Vergleich zu wechselnden Preistrends (-+, +-) die beobachteten Verhaltensmuster deutlicher auftreten.
- Mit stärkeren Preisänderungen geht das Verhaltensmuster des chasing (Kaufseite) deutlich zurück, während das tracking (Verkaufsseite) deutlich zunimmt. Die Abschwächung des chasing (Kaufseite) und die Verstärkung des tracking (Verkaufsseite) lassen sich mit einer häufiger auftretenden Trendumkehrerwartung oder einer stärker auftretenden Verkaufshemmung ("loss aversion") bei den Probanden erklären.
- Gleichzeitig führen stärkere Preisänderungen zu niedrigeren Ordervolumina (Stück und Anzahl) als schwächere Preisänderungen.
- Die Ordervolumina liegen bei positiven Preisänderungen deutlich höher als bei negativen Preisänderungen.

Die Ergebnisse der experimentellen Untersuchung verdeutlichen, daß das individuelle Anlageverhalten von Privatpersonen vor einem Aktienmarkthintergrund (indirekte Marktteilnehmer) u.a. von der historischen Kursentwicklung beeinflußt wird. Die Resultate legen es nahe, in weiter differenzierenden Analysen zu prüfen, inwieweit Gruppen von privaten Anlegern unterschiedliche Verhaltensmuster aufweisen (vgl. Abb. 1). Erste Ergebnisse deuten darauf hin, daß drei ähnlich große Fraktionen von privaten Anlegern existieren: Solche die tracking, solche die chasing und solche die die gemischte Strategie des tracking (Verkaufsseite) und des chasing (Kaufseite) betreiben (jeweils ca. ein Viertel der Probanden).

Insgesamt zeigen die Resultate der experimentellen Vorgehensweise, daß Arbeiten der experimentellen Kapitalmarktforschung zur Erklärung realer Phänomene an Finanzmärkten beitragen können.

Dr. Andreas Oehler, Lehrgebiet Bank- und Finanzwirtschaft, FernUniversität Hagen.

Die Arbeit wurde von der Deutschen Forschungsgemeinschaft im Rahmen des Schwerpunktprogramms "Empirische Kapitalmarktforschung" und vom Forschungspool zur Förderung des wissenschaftlichen Nachwuchses der FernUniversität Hagen finanziell unterstützt.

Zitierte Literatur:

Andreassen, P.B. (1987), On the Social Psychology of Stock Market; in: Journal of Personality and Social Psychology 1987, S. 490 - 496

Andreassen, P.B. (1988), Explaining the Price-Volume Relationship: The Difference between Price Changes and Changing Prices; in: Organizational Behavior and Human Decision Processes 1988, S. 371 - 389

Bitz, M. / Oehler, A. (1993), Überlegungen zu einer verhaltenswissenschaftlich fundierten Kapitalmarktforschung - ein Erklärungsansatz zum Anlegerverhalten; erscheint in Kredit und Kapital, Heft 2 und 3/1993.

Camerer, C.F. (1987), Do Biases in Probability Judgment Matter in Markets? Experimental Evidence; in: American Economic Review 1987, S. 981 - 997

Copeland, T.E. / Friedman, D. (1987a), The Effect of Sequential Information Arrival on Asset Prices: An Experimental Study; in: Journal of Finance 1987, S. 763 - 797

Copeland, T.E. / Friedman, D. (1991), Partial Revelation of Information in Experimental Asset Markets; in: Journal of Finance 1991, S. 265 - 295

De Long, J.B. / Shleifer, A. / Summers, L.H. / Waldmann, R.J. (1990), Positive Feedback Investment Strategies and Destabilizing Rational Speculation; in: Journal of Finance 1990, S. 379 - 395

Kahneman, D. / Knetsch, J.L. / Thaler, R.H. (1990), Experimental Tests of the Endowment Effect and the Coase Theorem, Working Paper, University of California, Berkeley 1990

Lakonishok, J. / Shleifer, A. / Vishny, R.W. (1991), Do Institutional Investors Destabilize Stock Prices? Evidence on Herding and Feedback Trading, Working Paper No. 336, Center for Research in Security Prices, University of Chicago, 1991

Oehler, A. (1991), Methodische Basis einer verhaltenswissenschaftlich fundierten Kapitalmarktforschung, Diskussionspapier Nr. 173, Diskussionsbeiträge, Fachbereich Wirtschaftswissenschaft der FernUniversität Hagen, Hagen 1991

Oehler, A. (1992), Empirisch-experimentelle Analyse des Verhaltens privater Anleger im Kapitalmarktzusammenhang - erste Ergebnisse, Diskussionspapier Nr. 194, Diskussionsbeiträge des Fachbereich Wirtschaftswissenschaft der FernUniversität Hagen, Hagen 1992

Shefrin, H.M. / Statman, M. (1985), The Disposition to Sell Winners too Early and Ride Losers too Long: Theory and Evidence; in: Journal of Finance 1985, S. 777 - 792

Shleifer, A. / Summers, L.H. (1990), The Noise Trader Approach to Finance; in: Journal of Economic Perspectives, Spring 1990, S. 19 - 33

Tietz, R. (1990), An Endowment Effect in Market Experiments?, Frankfurter Arbeiten zur experimentellen Wirtschaftsforschung, Nr. A 29, Frankfurt, Oktober 1990

Weber, M. / Camerer, C. (1991), The Disposition Effect in Securities Trading: An Experimental Analysis, Manuskripte aus den Instituten für Betriebswirtschaftslehre der Universität Kiel, Kiel 1991

Weber, M. / Camerer, C. (1992), Ein Experiment zum Anlegerverhalten; in: Zeitschrift für betriebswirtschaftliche Forschung 1992, S. 131 - 148

Rebalancing Strategies

C.A.G.M. Reniers
Institute for Research and Investment Services
Coolsingel 120, 3011 AG Rotterdam, The Netherlands

1 Introduction

Rebalancing the weights of stocks in a portfolio means *selling* (the increase of the proportion invested in each) stock after an increase in price and *buying* after a decrease. The attractiveness of such a procedure is that it explores the "buying low, selling high" principle. The added value is supposed to be an extra return on top of the "buy-and-hold" strategies' return. The simplest way to rebalance a portfolio is to keep the weights of the stocks constant within the portfolio. We call this "strict rebalancing". In this paper the implementation of a specific rebalancing strategy is presented, covering both strict rebalancing and buy-and-hold as special cases. The appealing quality of this strategy, as presented by Cover [1], is that it asymptotically outperforms the best stock in the market. In order to explain the differences between the various methods we shall also consider a more general rebalancing model, based on mean-variance optimisation, as stated in Merton [2].

The implementation of the method presented by Cover is amplified, because the method is based on a number of rather theoretical assumptions that are not met in practice. The most important assumption is about the selection of stocks. As a criterion stocks should have a positive weight in the best strict rebalanced portfolio. Only those stocks with a positive weight in this particular (not achievable) portfolio are portfolio candidates. Cover assumes that at any time every single stock has a positive weight in the best rebalanced portfolio, whereas, in practice, this is generally not true.

We will see the results of a straightforward implementation of Cover's methodology as well as the other three strategies'. The straightforwardness lies in the fact that any change in the set of active stocks has an immediate effect on the portfolio. New active stocks are bought and inactive stocks are sold. It is probably this modification why the results do not live up to expectation. It is easy to see that at every change in the active set there is an increasing discrepancy between the theoretical and practical wealth of the portfolio.

Though the portfolio, as dictated by Cover's method, depends on the returns of all previous periods, it is explained that it is also very natural to drop the oldest returns when new ones are added. From this viewpoint it is no longer obvious that the discrepancy between the theoretical and practical wealth of the portfolio should always increase. A first suggestion therefore, is to use a moving formation-period of fixed length to calculate the weights of the portfolio. This will improve results.

From the new methodology it only takes a small step to make another change. Though it may seem that one is moving further and further away from the original method, this is not the case at all. The final attachment to the methodology concerns a signalling function. This function signals whether or not to move the formation-period and change the active set at the same time. No signal means no moving formation-period and no change in the active set (while the best rebalanced portfolio might change). Between two signals, the methodology is just the same as the original methodology proposed by Cover. The important differences are that the process will start again at specific times and will choose a promising formation-period.

Cover's method as well as an overview of the differences concerning the weight schemes among all four methods are presented in section 2. Section 3 shows the results achieved with the various methods. The two succeeding dynamic methods we developed are presented in section 4. Concluding remarks and suggestions for further research are stated in section 5.

2 Cover's method

2.1 The basic ideas behind Cover's method

The mathematical representation of the method can be found in [1][*]. A very simple example illustrates what the model does. Suppose there are two stocks labelled A and B. The investment process starts at time T=0, by investing in $60 equally divided over three subportfolios labelled AA, BB and AB. The three subportfolios will be rebalanced in each period. After some time, at $T=t_1$, both A and Bs' stock price increased by 50%, as is shown in table 2.1.

Table 2.1

	T = 0	$T = t_1$	$T = t_1+1$
Price of stock A	66,7	100	110
Price of stock B	66,7	100	90
Money invested in subportfolios			
AA = 100% A	20	30	33
BB = 100% B	20	30	27
AB = 50% A + 50% B	20	40	40
Proportions invested in stocks A and B			
A	50%	50%	55 → 53%
B	50%	50%	45 → 47%

Of course, only subportfolio AB is able to profit from rebalancing and it did. At time $T=t_1$ we have $100. The very next period shows an increase of $10 for stock A and a decrease of $10 for stock B. Of course, at the beginning of this period the proportion invested in each stock was 50% and has been changed into 55% and 45% respectively, by means of the stock price movements. After realising returns, however, subportfolio AB is rebalanced. The necessary trade involved is a switch from A to B for 5% of the value of portfolio AB. This equals 2% of the total portfolio value.

Total wealth at any time is the summation of all subportfolios' wealth. The return on the overall portfolio equals therefore the average of each subportfolios' return. As stated in appendix A, the method works because the average of exponentials has the same asymptotic growth rate as the maximum. The idea is therefore that each subportfolio functions as an "exponential."

There is one more important aspect to mention, which is about stock selection. Only stocks with a positive weight in the best rebalanced portfolio are considered. If for example the investment process was started at time t_1, the best rebalanced portfolio over the period from t_1 till t_1+1 is a 100% investment in stock A. At time t_1+1 the 45% invested in B would be sold to buy more of A. Stocks with a positive weight in the best rebalanced portfolio are called "active".

The intuitive justification for the type of rebalancing proposed above, comes from the fact that strict rebalancing is not by definition better than buy-and-hold. Rebalancing is more attractive when stocks are risky and correlations are small. When geometric returns differ much, buy-and-hold will become favourable in time [*]. The method proposed by Cover contains elements of both rebalancing as well as buy-and-hold.

2.2 The weight schemes of all four models

A fourth model closely related to the subject discussed in this paper concerns the mean-variance optimisation technique. The goal of this method is to construct a portfolio with the highest expected return for a given level risk level. For a proper comparison it is fair to use historical return and risk parameters and not their expected equivalents as input for this model. Furthermore, the model needs a risk aversion parameter. It can be shown [*] that a risk-aversion parameter of ½ will generate a portfolio that is well approximated by the best rebalanced portfolio. Since the best rebalanced portfolio is the criterion for the selection of active stocks, the only difference between a mean-variance and a Cover optimal portfolio consists of the difference in weights for the stocks. The mean-variance optimal portfolio is therefore suited for explaining some results.

In order to understand the differences between the various methods, it is best to have a look at the weight schemes they dictate. An overview is given in table 2.2, which is based on the example as used in table 2.1. At time t_1 each method is equally invested in stocks A and B.

The weights of the stocks after the restructure are obvious for both buy-and-hold and strict rebalancing. The resulting weights according to Cover's method are explained above. In order to understand the result of restructuring according to the mean-variance method, it is noted that the new optimal portfolio equals the new best strict rebalanced portfolio. At the time of restructuring this portfolio (at $T=t_1+1$) it is experienced that those stocks having gained weight by means of the price movements, will be bought for an additional amount. The weight of stock A becomes more than 55%. According to intuition this is correct, because the best rebalanced portfolio will be consisting of stock A entirely if both stocks' prices keep moving in different directions.

Table 2.2

	Period t_1	Period t_1+1
Price of stock A	100	110
Price of stock B	100	90
C = 50% A plus 50% B	100	100
Composition of portfolio C after restructure in period 2		
	Proportion invested in A	Proportion invested in B
Buy-and-hold	55%	45%
Strict Rebalancing	50%	50%
Cover	53%	47%
Mean-variance optimal	> 55%	< 45%

3 Results

In order to analyse the differences between the four methods, we decided to use actual rather than simulated data. This is because we amplify the applicability in practice. Simulated data, therefore, might not show some specific properties of the return generating process. We used daily data from 28 French stocks, being CAC-40 index constituents from the beginning of March 1983 until the end of February 1993.

From the findings of Arnott e.a. [3]: "*more frequent rebalancing is beneficial and a wide tolerance for departures from policy is not*", we adopt the principle of calculating weights daily and rebalancing when the departures exceed a one percentage point tolerance. Transaction costs are assumed to be 2% for a switch. We abstract from the fact that dividends, contributions and withdrawals can be used to adjust the weights. Trading costs, therefore, should be much less in practice.

Furthermore, one assumption is made concerning the number of active stocks. We want to avoid that just one or two stocks are active, since a portfolio of two stocks does not comply with the general portfolio managers' intention to diversify risk. Therefore, the best rebalanced portfolio consisting of at least five stocks is used. When the set of active stocks changes, the implied switches between stocks are taken into account.

Table 3.1 shows the results, which are achieved after a warming-up period of one year. A warming-up period is necessary to avoid very frequent changes in the set of active stocks in the beginning. Not only return and risk, but also the distributions of the returns are analysed. For this purpose the weekly returns of the strict rebalancing strategy are sorted into five groups. The first group contains weeks with a return less than -2.1%. The weeks in the second group show a return between -2.1%

and -0.3%, and so on. The fifth group contains weeks with returns above 3.1%. For each strategy and each group the number of weeks that did better than rebalancing is expressed as a percentage. For example, if the weekly return of rebalancing is within the range of 1.2% to 3.1%, Cover did better in 58 per cent of those weeks.

Table 3.1

	Return intervals and strategy performance					% per year	
	-2,1%	-0,3%	1,2%	3,1%		Return	Risk
Rebalance						20.2%	27%
Buy-and-hold	44%	46%	40%	54%	53%	17.6%	26%
Cover	45%	45%	50%	58%	51%	15.0%	29%
MV portfolio	51%	44%	41%	43%	40%	5.2%	36%

Some very remarkable conclusions follow. First, though strict rebalancing and buy-and-hold are invested in all stocks all the time, their small differences in stock weights lead to a difference in return of 2.6% per annum in favour of rebalancing. In view of the mean-variance optimal portfolio it seems that this method is acting systematically wrong (except in case of market drops). Note that this and Cover's portfolio is invested in exactly the same stocks all the time. Conditional on the mean-variance stock selection procedure, Cover's method is doing very well. The difference that the mean-variance portfolio buys more of stocks that have already gained weight, as opposed to Cover's method that sells some weigth of those stocks, is the explanation for the results in table 3.1.

Compared to rebalancing and buy-and-hold, however, the result is not what was being expected. It follows that Cover's method outperformances rebalancing in rising markets but shows under-performances in falling markets. From the fact that Cover's method is outperforming rebalancing more often than buy-and-hold does and in view of their differences in average return, it follows that occurring underperformances are rather large. This means that, in the case of a drop of the market, Cover's method generally shows a larger drop. This is also of special interest for many investors.

The most important conclusion here concerning Cover's method is that renewing the active set of stocks (possibly every day) was not a good choice. The decision to do so was not dictated by the model, but was forced by the fact that the model did not foresee any changes in the set of active stocks. The chosen way of renewal has two disadvantages. First of all, it is a bad selection criterion. Secondly, at every change the dichotomy between the theoretic wealth of the portfolio and the actual wealth will increase. In order to understand this, let us assume that the entire portfolio is sold after the end of the day and bought again the very next day before trading starts. Remember that the portfolio you hold is a summation of rebalanced portfolios that are invested in the active stocks. Buying all portfolios consisting of active stocks could be done perfectly if those stocks were already active the day before. In that case the portfolio to buy is the portfolio you have. Typically, stocks become inactive after a bad performance. In such a case you have two groups of portfolios, those with and those without the stock that has become inactive. If it is assumed that the new active set was active before (and no other stocks were active), it can also be assumed that the latest return equalled the return of the latter group. Your performance, however, was a mixture of both groups' performances. While the first group will show a worse performance, because of the bad performance of the stock, your wealth will not equal the theoretical wealth. The difference is a sort of "reinvestment deficit" as will be the name later on. Another setback is faced when a new stock becomes active. The more changes in the set of active stocks the more the wealth stays behind the theoretical wealth.

4 More dynamic methods

As mentioned before, Cover's theoretical method yields an exponential growth factor that asymptotically equals the growth factor of the best rebalanced portfolio. Denote T_1 the current day, T_0 hundred days ago and T_2 the day hundred days from now. From appendix A something can be said about the difference between both factors over any period of time. Suppose your calculation of the maximum difference for the period [T_0, T_2] is 0.1 per cent. Since the current day is T_1 you are able to calculate the realised difference over [T_0, T_1]. Let say this is 0.15 per cent. If the average difference is (maximum) 0.1 per cent over two hundred days, and you already realised 0.15 per cent in the first hundred days, the (maximum) difference over the remaining hundred days is 0.05 per cent. This maximum underperformance equals 1.0005^100 - 1 = 5.1 per cent in hundred days.

This means that building your portfolio at current time T_1 according to Cover's method by using historical information about the preceding hundred days, and applying the method from now till T_2, will yield an underperformance of 5.1 per cent at most. Since this is relative to the best rebalanced portfolio (possibly better than each stock in the market) this might well be quite attractive. Of course you can also calculate a maximum underperformance going back 101 days, 102 days, etc. Eventually you will use every day in your historical database in order to choose the best starting portfolio.
Unfortunately, the suggested procedure has the disadvantage that a massive amount of calculations is involved. The idea though, contains an interesting element. Before, all information available was used to build a portfolio today. The number of days involved can be defined as the formation-period. Now it is no longer obvious that the formation-period must increase every day.

A simple suggestion is to use a certain formation-period the very first time you are building a portfolio and to move the formation-period while time passes. The results of the implementation are added to table 3.1, as table 4.1 shows. As one day is added to the formation-period the oldest day is removed. At any time the information used for calculating the portfolio weights, consists of the prior 250 days. These 250 trading days approximately equal one year.

Table 4.1

	Return intervals and strategy performance					% per year	
	-2,1%	-0,3%	1,2%	3,1%		Return	Risk
Rebalance						20.2%	27%
Buy-and-hold	44%	46%	40%	54%	53%	17.6%	26%
Cover	45%	45%	50%	58%	51%	15.0%	29%
MV portfolio	51%	44%	41%	43%	40%	5.2%	36%
Cover format.	81%	55%	47%	40%	22%	21.3%	24%

The improvements are spectacular in return as well as risk. There is an outperformance compared to strict rebalancing, of 0.9% per year. Somewhat strange is the fact that the new methodology works very well when the market as a whole falls (score of 81%), but does worse when the market rises (score of 22%). An explanation, worth further research, is a possible influence by some process in the return data. The behaviour, however, is the reason for the outperformance. In order to understand this, it should be noted that the performance within the different intervals is in accordance with the low risk (the weekly returns are closer to the average). From appendix B we know that risk functions as a penalty on return. The penalty, that equals half the variance, is 3.6% for rebalancing and 2.4% for the moving formation-period approach. This difference of 1.2% exceeds the outperformance of 0.9%.

An intuitive explanation for the far better result is as follows. Suppose in a given period stock A outperforms all other stocks. The best rebalanced portfolio consists of this particular stock. In a second period stock B is the outperformer. The best rebalanced portfolio contains B, however, not until the overall performance of B reaches the performance of A. This is the case when using all historical information available. Deleting old days from the database, those days when A was outperforming, means that stock B will be selected much earlier. Secondly, by adding one day the set of active stocks may change, implying a reinvestment deficit. When old days are dropped, the wealth of the portfolio to buy does not only depend on the returns of the most recent day, but also on the oldest day. Now there might be a deficit or a surplus.

Although the introduction of a formation-period of fixed length seems to be a good idea, the method now has become dependent on this parameter. A formation-period of two hundred days may work for a while but might not be as good as a formation-period of three hundred days some time later. Moving the formation-period by one day can be seen as a two-step process: the addition of a new day and the deletion of the oldest. Of course the first action is unavoidable. The execution of the latter, however, maybe dependent on some criterion. The criterion wished for is the one that says whether or not the formation-period is explored enough.

A criterion can be found within Cover's initial idea. Remember that, if there are no changes in the active set, the average growth factor tends to move to the growth factor of the best rebalanced portfolio. From this paragraph follows that next periods' maximum underperformance is relatively small when the underperformance in the formation-period is large. The idea is to start with a formation-period that contains a large difference between the maximum and average growth rate of portfolios consisting of active stocks, and to rebalance the portfolio according to Cover's method (without deleting or buying any other stock) for some time in order to yield an average growth factor roughly equal to the maximum. We calculate the following:

degree of satisfaction	=	(gfp - gfwp) / (gfbp - gfwp)
gfp	=	growth factor of the portfolio in the formation-period
gfwp	=	growth factor worst rebalanced portfolio
gfbp	=	growth factor best rebalanced portfolio

The degree of satisfaction always lies between zero and one hundred per cent. Each next trading day, one day is added and the degree of satisfaction is calculated. If the degree of satisfaction is higher than 50 per cent the oldest day is deleted and a new degree of satisfaction is calculated. If one day is deleted, more days are deleted concurrently until the degree of satisfaction is below 30 per cent or until a minimum of 200 remaining days is reached. For practical (programming) reasons a maximum of 450 days is applied. The initial length of the formation-period is 250 days. The results of this far more dynamic method is given in table 4.2.

Table 4.2

	Return intervals and strategy performance					% per year	
	-2,1%	-0,3%	1,2%	3,1%		Return	Risk
Cover format.	81%	55%	47%	40%	22%	21.3%	24%
Cover dynamic	54%	50%	44%	49%	45%	23.6%	27%

The risk of the dynamic method equals the risk of a strict rebalanced portfolio. As a result there is nothing to be gained from a more attractive return pattern, as was the case when a formation-period of fixed length was used. Therefore, the outperformance must come from a higher average return. As this does not show from the performance on several intervals, it must be that the less frequent outperformances are relatively larger than the more frequent underperformances. This is exactly the intention of the dynamic method. It selects a portfolio with a relatively bad performance (though invested in the best stocks) and drops it after a relative good performance. The fact that outper-formance is higher when returns are lower is also an appreciable quality, not shown by the theoretical method.

The degree of satisfaction works as a signalling function. As long as the upper limit is not reached, days are added to the formation-period. The method works just the same as Cover intended. There are no reinvestment deficits. When the upper limit is reached, two things may happen. One of them fulfils your initial goal to profit from a return that tends to approximate the maximum. This is true because the average growth rate has become closer to the maximum. It also may happen that the higher degree of satisfaction is forced by an underperformance of the worst rebalanced portfolio (consisting of active stocks). This is a positive contribution as well, because it implies a new calculation for the active set, which is a good thing to do at that time. It does not mean that the difference between the average and maximum growth rate has increased for certain, nor does it automatically imply a reinvestment deficit.

5 Conclusions

Rebalancing portfolio weights can improve the return of the portfolio. Even when the weights of stocks in a strict rebalanced portfolio do not differ much from the weights in a buy-and-hold portfolio, the difference can annually be a few percentage points of return. The rebalancing method proposed by Cover, however, targets higher than a few per cents per annum. According to Cover, this method will outperform the best stock in the market. However, the assumption that all stocks are active all the time is often not met, when considering actual data. Changing the active set not only generates turnover, but also implies a so called reinvestment deficit.

Fixing the problem of the reinvestment deficits can be effected by using a moving formation-period. Attractive results are achieved when a signalling function tells whether or not to move the formation-period. This final method is actually the same as succeedingly applying the original method by Cover. The intuitive correctness of the signalling function is clear because a signal occurs after a relative good performance of the portfolio or at times when a change in the active set is plausible.

As a result, the method by Cover is implemented with some practical additions. Still, there is no assumption concerning the distributions of stock returns. Maybe this is for the best, as Cover states in his article. Though, if daily returns are not independent in the cause of time, the dynamic method might improve some more if one is able to implement a successful decision rule that adapts a specific kind of return generating process. Specifically future research might find a better alternative for stock selection, while the method still depends on the stock selection procedure implied by Cover.

REFERENCES

[1] Cover, T.M. (1991): "Universal Portfolios," *Mathematical Finance*, vol. 1, no. 1, p. 1-29.

[2] Merton, R.C. (1990): "Continuous-Time Finance," Oxford: *Basil Blackwell Ltd*, p. 131-136.

[3] Arnott, R.D. and R.M. Lovell (1993): "Rebalancing: Why? When? How Often?," *The Journal of Investing*, vol. 2, p. 5-10.

[*] This is explained in the full article, which can be obtained from the author.

THE DISTRIBUTION AND WASTE DISPOSAL PROBLEM

Jacqueline M. Bloemhof-Ruwaard , Landbouw Universiteit Wageningen, NL
Marc Salomon, Erasmus Universiteit Rotterdam, NL
Luk N. Van Wassenhove, INSEAD Fontainebleau Cedex, France

We study the problem of the simultaneous design of a distribution network with plants and waste disposal units, and the coordination of product flows and waste flows within this network. Problem objective is to minimize total relevant costs, consiting of fixed costs for opening plants or waste disposal units, and variable costs, related to product and waste flows. The problem is complicated by the presence of

(i) service requirements (i.e. production must cover total demand) and

(ii) waste - that arises from production - must be disposed at waste disposal units.

We consider the uncapacitated version of the problem, as well as its capacitated version, in which capacity constraints exist on plants and waste disposal units.

We discuss alternative mathematical model formulations for the two-level distribution and waste disposal problem with and without capacity constraints. Also, lower bounding and upper bounding procedures are analysed with respect to their quality. Finally, the results of a computational study are reported.

Keywords:
capacitated facility location, mixed integer programming, relaxations, heuristics

HEURISTISCHE KONFIGURATION GUTER AUFBEREITUNGSANLAGEN FÜR BAURESTMASSEN

Marcus Nicolai, TH Karlsruhe
Thomas Spengler, TH Karlsruhe
Otto Rentz, TH Karlsruhe

Die Förderung einer rückstandsarmen Kreislaufwirtschaft im Sinne geschlossener Stoffkreisläufe ist gemäß dem Entwurf des Kreislaufwirtschafts- und Abfallgesetzes der Bundesregierung primäres Ziel zukünftiger Abfallgesetzgebung. Die Realisierung erfordert eine Verwertung von Reststoffen auf hoher Qualitätsstufe. Im Jahre 1989 fielen in der Bunderepublik Deutschland 22,6 Mio t Bauschutt an, lediglich 16% hiervon wurden stofflich verwertet. Dabei wurden 52% als Tragschichten im Straßenbau, der Rest in Lärmschutzwällen, Dämmen und im Landschaftsbau eingesetzt. Auf gleicher Qualitätsstufe, etwa als Betonzuschlag, fand keine Verwertung statt.

Der Stand der Technik von Aufbereitungsanlagen für Baurestmassen ist dementsprechend durch niedrige Produktqualitäten bei geringen spezifischen Aufbereitungskosten gekennzeichnet, obwohl hohe Produktqualitäten prinzipiell erreichbar sind. Der ökonomische Rahmen der Anlagen wird durch erzielbare Annahmepreise und das Preisniveau konkurrierender konventioneller Baustoffe eingeschränkt. Zunehmend restriktive Nutzungsmöglichkeiten von Deponiekapazitäten sowie restriktivere Genehmigung des Abbaus von Naturbaustoffen führen zu steigenden Deponiekosten und Baustoffpreisen, wodurch der wirtschaftliche Spielraum für die Entwicklung und Implementierung aufwendigerer Aufbereitungstechniken gegeben ist. Dabei bestimmen die eingesetzten Aggregate einerseits und deren Konfiguration andererseits entscheidend die erzielbaren Produktqualitäten, Verwertungsquoten und die Aufbereitungskosten.

Ziel des Vortrags ist es, einen heuristischen Ansatz zur Konfiguration guter Aufbereitungsanlagen für Bauschutt vorzustellen. Ausgehend von einer Analyse von 30 bestehenden Anlagen in der Region Oberrhein und einer mathematischen Modellierung der technischen Aggregate wird zunächst ein auf dem Flow-Sheeting chemischer Anlagen beruhendes Simulationsmodell zur Ermittlung von Stoff- und Energiebilanzen sowie Aufbereitungskosten beliebig konfigurierter Anlagen entwickelt. Mit Hilfe eines heuristisch modifizierten Branch and Bound Verfahrens werden verschiedene Anlagenkonfigurationen erzeugt, die zugehörigen kostenminimalen Aggregatelisten erstellt und so das zugrundeliegende kombinatorische Optimierungsproblem gelöst. Die Sensivität der entwickelten "optimalen" Anlagen bezüglich Aufgabegut, Annahmepreise und Produktpreise wird untersucht. Mindestverwertungsquoten, -produktqualitäten sowie einzuhaltende Grenzwerte für Schadstoffe in den Produkten können exogen vorgegeben werden.

SIMULATION DER DYNAMIK DES RESSOURCENABBAUS UNTER BERÜCKSICHTIGUNG VON RECYCLINGPROZESSEN

Carsten Oder, Universität Karlsruhe
Hans-Dietrich Haasis, Universität Karlsruhe
Otto Rentz, Universität Karlsruhe

Der Abbau von Ressourcen, ihr Verbrauch und der Zuwachs der Abfalldeponierung werden in zunehmendem Maß zu einem Problem industrieller Staaten. Zur Lösung des Problems wird versucht, durch steigende Recyclingquoten den Ressourcenabbau zu verlangsamen. Ziel dieses Papers ist, das dynamische Verhalten dieses Prozesses zu erfassen und unter verschiedenen Annahmen zu simulieren. Zu diesem Zweck wird aus dem Wirkungsgefüge der Ressourcendynamik unter Berücksichtigung von Recycling das dynamische Verhalten in Form von Differentialgleichungen quantifiziert, in einem Strukturbild dargestellt und unter Variation der Recyclingquote simuliert. Das System ist dabei in zweifacher Weise negativ über den Preis, der für die Ressource zu zahlen ist, rückgekoppelt. Zum einen werden die externen Folgekosten der Abfalldeponierung internalisiert, zum anderen steigt mit zunehmendem Preis die Rentabilität von Recyclingverfahren, die jedoch mit steigender Recyclingquote energetisch aufwendiger und damit teurer werden. Die Recyclingquote wird so eine nichtlineare Funktion des Preises. Das dynamische Verhalten des Systems, insbesondere die zeitliche Entwicklung von Abfallmengen und Ressourcenvorrat, werden simuliert, graphisch dargestellt und interpretiert. Von besonderem Interesse sind dabei hohe Recyclingquoten (zwischen 90 und 99%), wobei sich schon für Recyclingquoten von 95% ein drastischer Abfall des Ressourcenbestandes ergibt, der sich aufgrund der Rückkopplung der internalisierten Deponierungskosten und damit zurückgehender Nachfrage einem von der Recyclingquote abhängigen stationären Wert annähert.

Als Erweiterung des Modells werden andere mögliche Internalisierungsstrategien untersucht, die beispielsweise die nichtlineare Problematik bei Kumulation eines bestimmten Abfallstoffes wiederspiegeln. Als Ergebnis ergibt sich hierbei, daß eine stärkere Berücksichtigung der Abfallproblematik durch nichtlineare Kosteninternalisierung zu einer früheren Stabilisierung des Ressourcenvorrats auf höherem Niveau führt.

THE ECOLOGICAL PERFORMANCE OF JIT LOGISTICS

Joachim Reese, Universität Lüneburg

JIT logistics is widely accepted through most large national and international industries. As distances and volumes of JIT transportation grow more and more on the one hand and the vehicle load decreases on the other hand, the ecological defects of JIT are broadly discussed in the public. Nevertheless, it is shown in this contribution that the primary concern of this debate is a lack of optimization.

On the first stage a strategic implementation of a JIT concept is necessary. It is first of all an organizational problem, whether to utilize a regional forwarding agent, a JIT inventory, a good traffic center, or a pick-a-back system. Consequently, there result layout problems, loading problems, and travelling salesman problems. On the second stage it should belong to the firms' main interests to minimize ecological effects by loading and routing the vehicles to their best.

The different problem structures are modelled and discussed. Numerical examples are presented which stress that economy and ecology are not only contradictory with respect to the JIT principle. The dominance of a JIT concept depends on the way, social costs are considered. Therefore, a tax solution must be separated from solutions, when fixed costs arise. Postoptimal analysis is applied to demonstrate the stability of a JIT concept with regard to federal ecological policy.

Optimal Intertemporal Allocation of Emission Reduction Measures

Peter Ruß, Otto Rentz

French–German Institute for Environmental Research (DFIU/IFARE),
University of Karlsruhe

Summary: Global warming caused by the so–called greenhouse effect and measures to mitigate global warming currently are in the center of the attention. Especially possible strategies for reducing carbon dioxide emissions are widely discussed. CO_2, being the most important greenhouse gas, primarily is emitted through combustion of fossil fuels. In contrast to other pollutants CO_2 is not toxic and temporary high emission levels are not harmful. The negative environmental effect of CO_2 emissions is caused by the concentration of CO_2 in the atmosphere. Higher CO_2 concentrations will lead to an increase of the global temperature.

Studies dealing with CO_2 reduction strategies, so far were based on exogenous, predefined emission reduction pathways. However, compared with an optimal intertemporal allocation of emission reduction measures as proposed in this paper, a predefined emission pathway leads to significantly higher reduction costs. This is shown using the energy and environmental model EFOM–ENV. EFOM–ENV is a multi-period linear optimization model, which covers the whole chain of energy conversion starting from the primary energy supply and ending in the demand sectors. By using the model as an analytic tool, cost-effective mixtures of technological emission control measures, including fuel switching, substitution of energy conversion technologies, and application of emission control technologies are being identified to comply with different emission control targets.

In this paper, the methodological approach and selected results are presented.

Zusammenfassung: Die Erwärmung der Erde aufgrund des Treibhauseffektes sowie Maßnahmen zur Begrenzung dieser Erwärmung stehen derzeit im Mittelpunkt der Diskussion von Umweltfragestellungen. Breiten Raum nehmen dabei Strategien zur Minderung der Kohlendioxidemissionen ein. Dieses wichtigste Treibhausgas gelangt im wesentlichen durch die Verbrennung fossiler Brennstoffe in die Atmosphäre. Im Gegensatz zu anderen Schadstoffen ist CO_2 nicht toxisch und vorübergehende erhöhte CO_2-Emissionen sind unschädlich. Entscheidend ist lediglich die CO_2-Konzentration in der Atmosphäre. Eine höhere Konzentration führt zu einem Anstieg der durchschnittlichen globalen Temperatur.

Bisherige Studien, die sich mit Strategien zur CO_2-Minderung beschäftigten, gingen von vorgegebenen, fixen Verläufen der Emissionsminderung aus. Dies führt gegenüber einer optimierten intertemporalen Allokation der Emissionsminderungsmaßnahmen, wie sie an dieser Stelle vorgeschlagen wird, zu signifikant höheren Minderungskosten. Gezeigt wird dies mit Hilfe des Energie- und Umwelt–Modells EFOM–ENV. EFOM-ENV ist ein mehr-periodisches, lineares Optimierungsmodell, in dem das gesamte Energiesystem von der Resourcenbereitstellung bis zur Energienachfrage in unterschiedlichen Sektoren abgebildet ist. Das Modell erlaubt es, kosten–effiziente Maßnahmenbündel zu identifizieren, mit deren Hilfe verschiedene Emissionsreduktionsziele erreicht werden können. Zu den in Frage kommenden Maßnahmen gehören Brennstoffsubstitution, alternative Energieumwandlungstechnologien, Energiesparmaßnahmen, sowie die Anwendung von Emissionsminderungstechnologien.

In diesem Artikel wird die angewandte Methodik beschrieben sowie ausgewählte Ergebnisse vorgestellt.

1. Introduction

Nowadays, there is no doubt that the "enhanced" greenhouse effect is real. The Intergovernmental Panel on Climate Change (IPCC) stated that "it is certain, that man–made emissions are substantially increasing the atmospheric concentrations of the greenhouse gases and that these increases will enhance the greenhouse effect, resulting in an additional warming of the Earth's surface" /4/.

It is important to note that, in the case of CO_2 and the enhanced greenhouse effect, peaks in emissions do not matter. What is important is the concentration of CO_2 in the atmosphere and, hence, the total, cumulated amount of CO_2 released into the atmosphere: A strategy aiming at the mitigation of global warming therefore has to limit the overall cumulated emissions. This is in contrast to other pollutants like SO_2 and NO_x. Here peaks, i.e. high emissions of these pollutants in a particular year, might cause acid depositions beyond critical levels and, hence, lasting damage to the ecosystems.
A number of studies have been carried out dealing with greenhouse gas emission reduction (for a review of the studies and approaches see e.g. /12/). The lesser share of these studies applied the bottom-up approach (e.g /3/). However, the set of reduction options considered in these studies was limited. Especially energy conservation measures have hardly been taken into account and the future option of recovery of CO_2 from power plants has not been considered at all. Studies dealing with CO_2 reduction strategies, so far were based on exogenous, predefined emission reduction pathways. In contrast to this approach, in this paper the optimal intertemporal allocation of emission reduction is investigated. However, this study is not trying to derive optimal emission levels (like e.g. in /5/). Rather optimal strategies for achieving certain predefined emission ceilings are developed.
The aim of this study is the development of national, cost-effective strategies for CO_2 emission reduction. The study focuses on the techno-economic aspect of emission control, i.e it deals with technological strategies. Therefore, a strategy in the sense of this study is characterized by a specific combination of energy conversion and emission reduction technologies as well as energy conservation measures, quantitatively defined by capacities and energy flows, and, which is especially important for CO_2 reduction strategies, its development over time. Consequently, a bottom-up or engineering approach has been chosen. The approach is a normative one, i.e. the results are not meant to be a realistic forecast of the future, but they should show the optimal future energy system for certain exogeneously imposed assumptions.
The EFOM-ENV model of the Commission of the European Communities has been used as a tool. For the specific purpose, the model has been extended by a number of additional CO_2 reduction options, such as energy conservation measures, high efficiency energy conversion techniques, and energy conversion on the basis of renewable energy sources. Moreover options for the recovery of CO_2 from power plants have been investigated. The most promising techniques have been chosen and investigated with regard to their cost-effectiveness. The time horizon of the study is 2010.

2. Methodology

The energy-environmental model EFOM-ENV is a multi-period, linear optimization model. It is driven by an exogenous demand for useful or final energy. The whole energy chain, starting from the primary energy supply, passing through the intermediate sectors (e.g. electricity generation) and ending in the demand sectors (industry, households, transportation, etc.) is represented by linear equations. The model is organized in a modular way with each module or sub-system containing a set of alternative energy conversion techniques i. Besides already existing energy conversion techniques possible future high-efficient, low-emission techniques, e.g. fluidized bed combustion or combined cycle processes with integrated coal gasification for electricity production, which could possibly be used by the end of this century, are also included. The techniques are represented and described by conversion efficiency, installed capacity, investment, cost, etc.
The optimization criterion in use is the minimization of the annual costs C (present value) for the time span from the starting year T_o to the horizon year T_P. The annual costs for the processes on links i are summed up and discounted from the 31st December of each year to the 1st January of the base year using the present worth factor α. The total cost C discounted at 1st January of the starting year is composed of the variable cost coefficient $V_{i,t}$ (proportional to the energy or material flow x), the fixed cost coefficient $F_{i,t}$ and the capital cost coefficient $I_{i,t}$ (both proportional to the new invested capacities y). The totaling applies for all years between $T_0 + 1$ and T_P and for the complete

set of links N, respectively.

$$C = \sum_{t=T_0+1}^{T_P} \left(\alpha \sum_{i=1}^{N} \left(V_{i,t} x_{i,t} + y_{i,t} \left(F_{i,t} + I_{i,t} \right) \right) \right) \tag{1}$$

The present worth factor is α, which discounts a given quantity from 31st December of year t to the 1st January of the base year T_0, r being the discount rate.

$$\alpha = \frac{1}{(1+r)^{t-(T_0+1)}} \tag{2}$$

In order to reduce the size of the problem, the whole period (T_0–T_P)) is divided into P sub–periods: (T_0–T_1),(T_1–T_2),...(T_{P-1}–T_P). Thus, the cost function becomes:

$$C = \sum_{p=1}^{P} \left(\sum_{i=1}^{N} \left(V_{i,p} x_{i,p} + y_{i,p} \left(F_{i,p} + I_{i,p} \right) \right) \right) \tag{3}$$

The computation of the modified coefficients is made by equivalencing the corresponding terms of Equation 1 and 3. A complete deduction of the economic coefficients from the model data base by interpolation rules, formation of annuities, etc. can be found in /13/.

More than 30 categories of restrictions can be distinguished: economic, technical and political constraints, e.g. input/output restrictions at nodes, upper and lower bounds on flows or capacities, demand equations, exogenous constraints. A detailed description of all constraints can be found in /13/.

The originally energy oriented model has been extended by environmental modules /7/. For these various environmental modules additional (in-)equations have to be modeled. The main constraint is the limitation of the total amount L of pollutant k (CO_2, SO_2, NO_x), which can be introduced for each year t or sub–period p by the following set of inequations:

$$\sum_{i \in J} E_{i,k,p} x_{i,p} \leq L_{k,p} \tag{4}$$

where $E_{i,k,p}$ represents the emission factor for pollutant k of the technique on link i in the sub–period p. The total applies for all links i of the set of links J representing emitting techniques and emission reduction measures. J and the emission limit may refer to a specific technology or to the country's or sector's total emissions. These limits may be fixed for an individual period or else for all periods (see Equation 4). It has already been emphasized that, in the case of climate change mitigation strategies, the total cumulated CO_2 emissions are decisive. Since the EFOM–ENV model so far did not provide the opportunity to limit the total, integrated amount of emissions, an additional constraint has been added. In this study the emissions are integrated numerically using the Newton–Cotes method. Disregarding the remainder, the total cumulated emissions $\tilde{L}$ of pollutant k are

$$\frac{T_P - T_0}{K_m} \sum_{p=0}^{P} \left(\sum_{i \in J} E_{i,p} x_{i,p} \right) k_{m,p}, \leq \tilde{L}_k; \tag{5}$$

with

$$K_m = \sum_{p=0}^{P} k_{m,p}.$$

The value of $k_{m,p}$ depends on the number P of supporting points p, i.e. here on the number of periods considered.

Despite the simple structure of the equations, the model is a complex tool due to its comprehensive representation of the entire energy system. Up to about 12,000 rows and columns have been generated, the density (share of non-zero elements) of the matrix is about 0.06 %.

The model allows the investigation of different reduction measures as well as the evaluation of their effectiveness. The reduction measures, which can be assessed with the EFOM–ENV model can be grouped together according to the following characteristics:

- fuel switching and improvement of fuel quality,
- emission reduction techniques (CO_2 removal),
- technology substitution, i.e. use of less-emitting techniques,
- energy conservation and efficiency improvements.

Moreover, these measures can be combined to form numerous mixed strategies in order to achieve pre-defined standards or general emission levels.

3. Selected Results for CO_2 Emission Reduction

Throughout the last couple of years the EFOM–ENV model has been used to derive emission reduction strategies for the major part of European countries (see e.g. /1, 3, 8/).
Most of the evaluations focused on NO_x and SO_2 emissions /1, 8/, but the model recently also has been used to derive CO_2 emission reduction strategies /3/. For this purpose the model was extended to be able to take into account energy conservation measures and, recently, also CO_2 removal options have been introduced /10/.
For the definition of the scenarios, general assumptions have to be made concerning the economic development, e.g. the gross domestic product, the net industrial production and the market prices, and the final/useful energy demand. To obtain a consistent economic framework, the "Energy 2010"-study of the Commission of the European Communities (CEC) was used /2/.

A set of various "cost-effectiveness" cases has been defined. As already outlined, in this study strategies for the mitigation of global warming are obtained. For this purpose the total, i.e. the cumulated carbon dioxide emissions have to be reduced. Consequently, cost-effectiveness here is interpreted as the achievement of a predefined level of total national carbon dioxide emissions at the lowest possible cost, obtained by an optimal intertemporal allocation of energy conversion technologies and emission reduction measures. This intertemporal allocation of CO_2 reduction is shown in Figure 1 for selected reduction cases. For the case of Germany, at least for low reduction levels, the reduction is evenly distributed over time. However, for a 15% reduction of CO_2 emissions, the cost–effective reduction strategy is to "wait" until 2005, instead of reducing much in the year 2000. This is due to the necessary replacement of power plants in 2005, which offers a cost–effective option to reduce CO_2 emissions. Since restructering measures (technology and fuel substitution) are especially cost–effective, when old equipment has to be replaced, the emission reduction levels (compared with the reference case) tend to be higher towards the end of the period of time under consideration. Nevertheless, in all cases the cost–effective strategy is to start with the reductions immediately. A detailed description of the measures to be taken is beyond the scope of this short article (see e.g. /10/).
The optimization can be used for trade–off analyses, since cost versus emission reduction functions are obtained showing the additional energy system cost for different levels of emission reduction ("cost function"). The cost function is derived by comparing the total discounted system costs for a specific scenario with the respective system costs for the reference scenario. Thus, for every reduction level the additional system costs caused by reduction measures is obtained. Figure 3 shows the cost curve for the reduction of the total CO_2 emissions in the period 1988 to 2010.

The scenario results presented above all started from the basic idea, that an effective strategy to curb global warming has to aim at a reduction of the total, i.e. cumulated, emissions of carbon dioxide into the atmosphere. However, this approach is not the one commonly used for the derivation of CO_2 reduction strategies with optimizing energy–emission models. Rather the basis of the strategies has been exogenous reduction pathways. These pathways were either taken from IPCC scenarios /4/, e.g. in /6/, or they reflect policy targets. The latter was the case in /3/, where the emission reduction

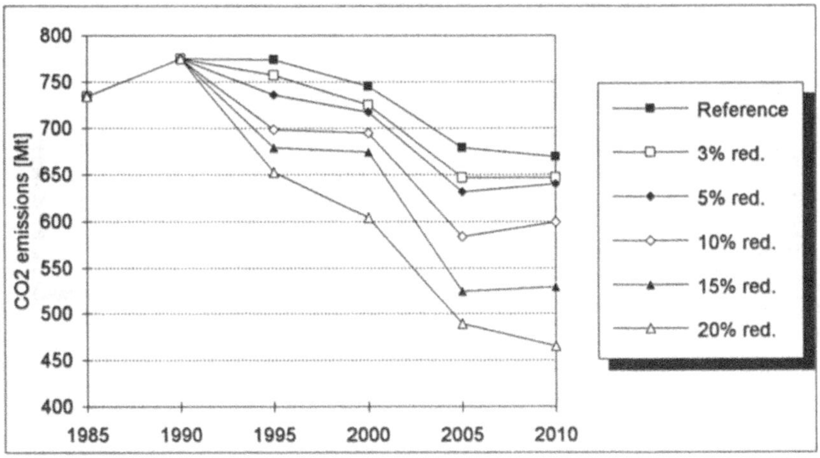

Figure 1: CO_2 emission reduction with optimal intertemporal allocation of the reduction measures

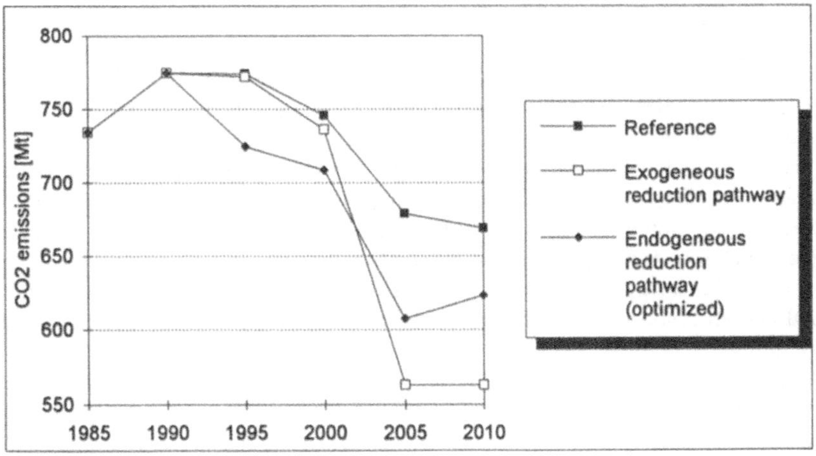

Figure 2: Comparison of of optimized (endogenous) and exogenous emission reduction pathways

pathways were defined based on the target of the European Commission to stabilize CO_2 emissions by 2000 and to reduce them thereafter. There a step function was used, implying a stabilization of CO_2 emissions at 1988 levels by 1995 and 2000, and the target reduction by 2005 and 2010.

The exogenous definition of emission reduction pathways does not consider individual features of the different energy systems. The intertemporal allocation of reduction measures is given exogeneously, rather than being optimized by the model. Hence, the results obtained by such an approach do not represent optimal reduction strategies for carbon dioxide.

To show the differences between the exogenous and endogenous determination of emission pathways, a scenario was defined, in which the emission reduction pathways was implied exogenously to the model. For the scenario emission reduction targets announced by the Commission of the European Community (stabilization of CO_2 by 2000 at 1988 levels) as well as the commitment of the German Government (25% reduction of 1987 levels by the year 2005) were implied to the model as upper emission limits. In addition, the maximum emissions in 2010 were also limited to be 25% lower than CO_2 emissions in 1988. The results of this scenario will be compared with the scenarios with

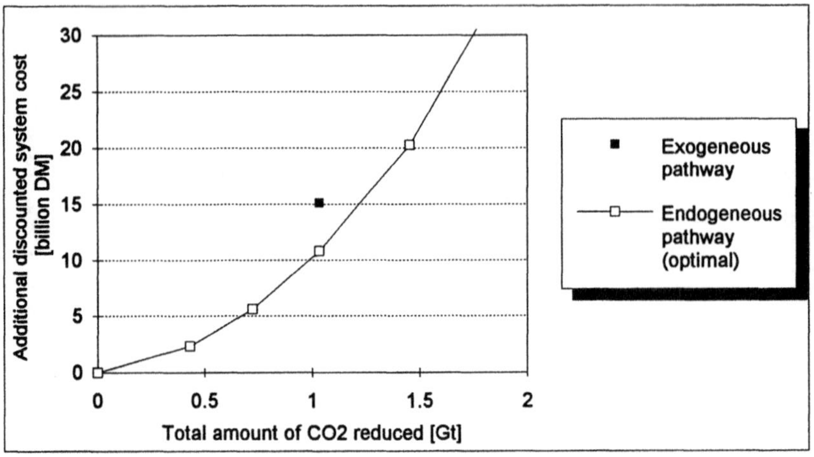

Figure 3: CO_2 reduction cost for exogenous and endogenous (optimal) reduction pathways

an endogenous, i.e. optimized, emission reduction pathway. Figure 2 shows the evolution of CO_2 emissions for this case (exogenous reduction pathway). The emission pathway is compared with the evolution of CO_2 for the reference case, i.e. without any CO_2 reduction implied to the model. The third curve shows the evolution of CO_2 emissions for a scenario in which the total CO_2 emissions for the period 1988 to 2010 were limited (endogenous reduction pathway). The maximum total emissions in this case were set at the very level, which results for the exogeneously defined emission pathway. This means that the effect of both reduction cases on global warming is the same, i.e. they have the same effectiveness. However, the target reduction of the total CO_2 emissions can be achieved following different strategies and pathways at different costs.

The different strategies for achieving a reduction of CO_2 emissions can be compared with regard to the additional system costs (i.e. the differences of the values of the objective function), caused by the respective reduction strategy.

In Figure 3 besides the cost–curve for the optimal reduction also the costs for the scenario with exogenously defined emission pathway are depicted (black square). The curve at approximately 1 Gt CO_2 reduction has to be compared with this particular point. Obviously the reduction cost for this scenario are much higher. The additional total discounted system cost amount to 5 bill. DM, which is 48% more than in the optimal case.

The cost curve in Figure 3 shows the limitations of the area of possible reduction strategies. All possible strategies lie over, or, if they are optimal/efficient, on this cost curve. A reduction at lower additional system costs is not possible for the underlying assumptions (cost data, conversion efficiencies, evolution of the energy demand etc.). The definition of exogeneously defined emission reduction pathways will usually lead to suboptimal results, i.e. higher reduction costs. The reduction pathway should be determined, taking into account the individual conditions of the energy system (e.g. capacity structure of power plants, necessary capacity expansion).

4. Conclusions

Apart from economic objectives, the reduction of emissions has become a major objective, when developing energy supply strategies resulting in the need for cost-effective emission reduction strategies. For this purpose, energy-environmental models can be used as tools. In this paper, selected results of the application of the model EFOM–ENV for the elaboration of CO_2 emission reduction strategies for the Federal Republic of Germany have been presented.

The results indicate that for cost-effective reduction of CO_2 emissions the optimal intertemporal allocation of the emission reductions plays a decisive role. The potential savings of an optimal strategy in comparison with a strategy which is not considering the individual conditions of the energy system amounts to almost 50% (for the example given in this paper). Future studies have to consider the evolution of the emissions over time. Further research is underway to to derive global warming mitigation strategies taking into account greenhouse gases other than CO_2 /11/.

The applied decision support model EFOM-ENV has been implemented for the major part of the European countries, based on data bases on a comparable aggregation level. Hence, a tool is available for decision makers, which is able to quantify relationships between energy supply and emission reduction on a national as well as on an international level.

References

/1/ Commission of the European Communities (ed.): *Energy and Environment - Methodology for Acid Air Pollution Assessment in Europe*, Brussels, CEC (1990)

/2/ Commission of the European Communities: *Energy for a New Century: The European Perspective*; Energy in Europe, Special Issue, June 1990, Brussels.

/3/ Commission of the European Community (ed.): *Cost-Effectiveness Analysis of CO_2 Reduction Options*, Brussels, 1991.

/4/ Houghton, J.T.; Jenkins, G.J.; Ephraums, J.J.: *Climate Change - The IPCC Scientific Assessment*; Cambridge University Press for the Intergovernmental Panel on Climate Change (IPCC), Cambridge, 1990.

/5/ Nordhaus, W.D.: *An optimal Transition Path for Controlling Greenhouse Gases*; paper presented at the International Energy Workshop, International Institute for Applied Systems Analysis (IIASA), Laxenburg, June 18-20, 1991.

/6/ Okken, P.A.; Lako, P; Gerbers, D.; Kram, T.; Ybema, J.R.: *CO_2 removal in competition with other options for reducing CO_2 emissions*, Energy Conversion and Management 33(1992)5–8, pp. 737–746.

/7/ Remmers, J.; Morgenstern, Th.; Schons, G.; Haasis, H.-D.; Rentz, O.: *Integration of Air Pollution Control Technologies in Linear Energy-Environmental Models*; European Journal of Operational Research, 47(1990)3, 306-316.

/8/ Rentz, O.; Haasis, H.–D.; Jattke, A.; Ruß, P.; Wietschel, M.: *Impacts of Energy Supply Structure on National Emission Reduction Potentials and Emission Reduction Costs*; Research Report for the Federal Ministry for the Environment/Federal Environmental Agency (Umweltbundesamt), Karlsruhe 1992.

/9/ Ruß, P.; Haasis, H.-D.; Morgenstern, Th.; Perelló-Aracena, F.; Plinke, E.; Remmers, J.; Schons, G.; Rentz, O.: *Development of Strategies for Reducing Air Pollutant Emissions in the European Community*. International Journal of Energy Research, 14(1990) 833-847.

/10/ Ruß, P., Rentz, O.: *Cost-Effectiveness of CO_2 Removal Techniques*; First International Conference on Carbon Dioxide Removal (ICCR), Amsterdam, March 1992.

/11/ Ruß, P.; Rentz, O.: *Cost-effective Strategies for Global Warming Mitigation Using an Integrated Approach*; International Symposium on Non-CO_2 Greeenhouse Gases: Why and how to control?; Maastricht, 1993.

/12/ United Nations Environment Programme (UNEP): *UNEP Greenhouse Gas Abatement Costing Studies - Phase One Report: Analysis of Abatement Costing Issues and Preparation of a Methodology to Undertake National Greenhouse Gas Abatement Costing Studies;* UNEP Collaborating Centre on Energy and Environment Risø National Laboratory, Denmark, 1992.

/13/ Van der Voort, E.; Donni, E.; Thonet, C.; Bois D'Enghien, E.; Dechamps, C.; Guilmot, J.F.: *Energy Supply Modelling Package EFOM 12C Mark I, Mathematical Description*, Louvain-la-Neuve, CEC, 1984.

Planning Models for the Optimal Dismantling and Recycling of Domestic Buildings with Regard to Environmental Aspects

Th. Spengler, M. Nicolai, O. Rentz, University of Karlsruhe

Summary: Demolition waste ranges with an amount of about 23 million tons per year in Germany in the same order as municipal waste. Recycled demolition waste can be reused not only in landfilling but also in road construction, in the production of paving blocks, in the manufacturing of wall blocks as well as a concrete aggregate. These secondary raw materials mainly should be used in order to replace natural raw materials. The total dismantling costs of buildings and recycling costs of the resulting materials are a function of the considered dismantling techniques, the sorting techniques, the recycling techniques and the transportation distances. Furthermore, the recycling capacities of the different reuse options may be limited in the considered geographical region. In this paper the integrated dismantling and recycling planning problem will be described and a sophisticated mixed integer linear programming model will be formulated. Different solution procedures will be analysed and a Benders' Decomposition based heuristic algorithm will be developed. In order to obtain integrated dismantling and recycling strategies for the Upper Rhine Valley (Alsace and Baden), the model will be implemented on a Personal Computer and evaluated with specific regional database.

Zusammenfassung: Das gesamte jährliche Bauschuttaufkommen in der Bundesrepublik Deutschland liegt mit 23 Millionen Tonnen in derselben Größenordnung wie Hausmüll. Aufbereiteter Bauschutt wird momentan überwiegend im Straßenbau oder zum Bau von Lärmschutzwällen eingesetzt. Zur Einsparung natürlicher Ressourcen sollte dieses aus abfallwirtschaftlicher Sicht unbefriedigende "Downcycling" in Zukunft durch ein Recycling auf möglichst gleicher Qualitätsstufe ersetzt werden. Ziel des diesem Beitrag zugrundeliegenden Forschungsprojekts ist die Konzeption eines EDV-gestützten Planungssystems zur umweltgerechten Demontage und Verwertung von Wohngebäuden und dessen Anwendung auf die Region Oberrhein (Baden/Elsaß). Ausgehend von einer detaillierten Analyse der in der Region vertretenen Gebäudetypen werden detaillierte Gebäudestücklisten ermittelt und zugehörige Demontagetechniken und -kosten analysiert. Das in der Produktionsplanung bekannte Konzept der Montage-Vorrang-Graphen wird auf die Demontageplanung übertragen und für jeden Gebäudetyp wird der zugehörige Demontage-Vorrang-Graph entwickelt. Unter Berücksichtigung materialspezifischer Aufbereitungskosten und Verwertungskapazitäten wird ein gemischt-ganzzahliges Optimierungsproblem zur Bestimmung kosteneffizienter umweltgerechter Demontage- und Recyclingstrategien formuliert und mit Hilfe einer auf der Benders-Dekomposition basierenden Heuristik gelöst.

1 Introduction

Although recycling and reuse is generally possible, in Germany only 16% of the building materials were recycled in 1989. Construction waste was almost completely disposed. The objectives of the German legislation are to achieve a recycling rate of 60% for building rubble and 40% for construction waste for the year 1995. Initiatives in France aim at the direct reuse of materials from the dismantling of buildings.

In reality, reuse options are often limited to road construction or soundproof barriers which represent a downcycling and are therefore insufficient from an economical point of view. Only recycling-products of a high standardized quality level can compete with traditional materials. Therefore, environmental legislation aims at selective dismantling and recycling of buildings. Major instruments are higher disposal prices for demolition waste, waste duties, taxes on natural construction materials or the integration of dismantling schemes into demolition permits The objective of the presented research project is the development and implementation of an integrated dismantling and recycling planning system for domestic buildings in the region of the Upper Rhine Valley. Due to the high complexity of the considered planning problem, the system will be based on a mixed integer linear programming model. With the help of this model, the costs and interactions of dismantling and recycling are evaluated and cost-efficient combinations of dismantling procedures, recycling techniques and reuse options will be determined (cf. Figure 1).

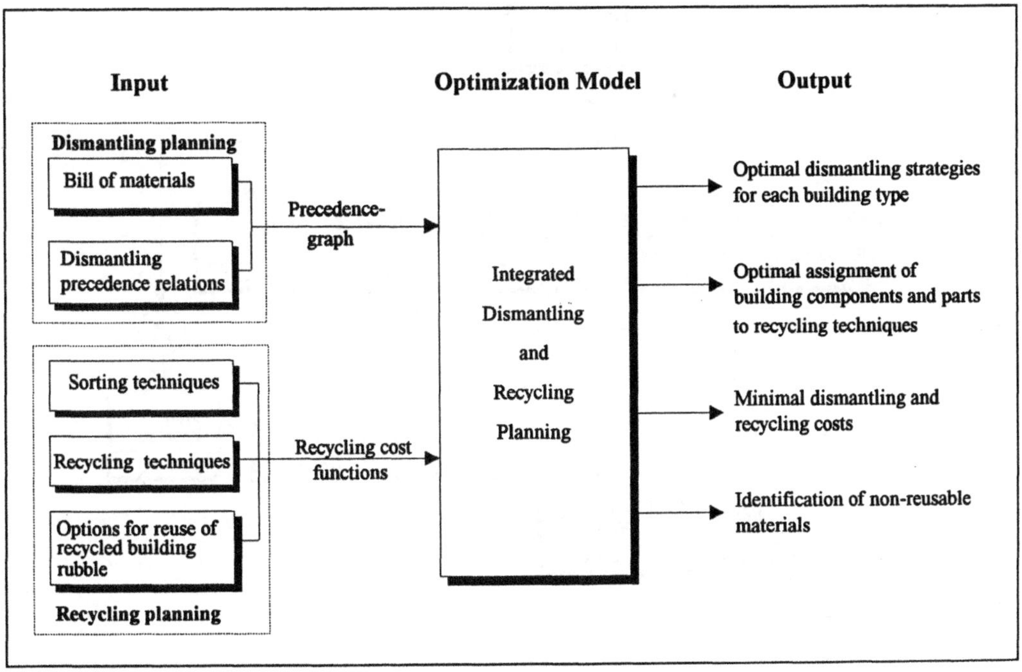

Figure 1: Structure of the integrated dismantling and recycling planning system

2 Development of dismantling-precedence-graphs

As the amount and composition of the materials from demolition show important variation with the type of building, the buildings in the region of the Upper Rhine Valley have to be classified into different types. For each of the eleven predominant types a representative building will be chosen and a detailed bill of materials will be prepared. On the basis of these bills of materials the different dismantling techniques will be analysed and aggregated into so called dismantling activities. For these dismantling activities the associated dismantling costs and the precedence-relations will be determined, so that for each identified class of buildings the following dismantling-precedence-graph can be developed (cf. Figure 2).

Following the engineering industry, where the concept of the precedence-graphs is used in the Materials-Requirement-Planning, the precedence-graph is the starting point of the dismantling planning process. By the application of a set of dismantling activities $v_1,...,v_5$ the corresponding building is disassembled into building components and parts. The decision whether a certain dismantling activity will be applied or not depends on the comparison of the alternative recycling costs of the corresponding building component with the dismantling costs and the minimal recycling costs of the resulting building components and parts.

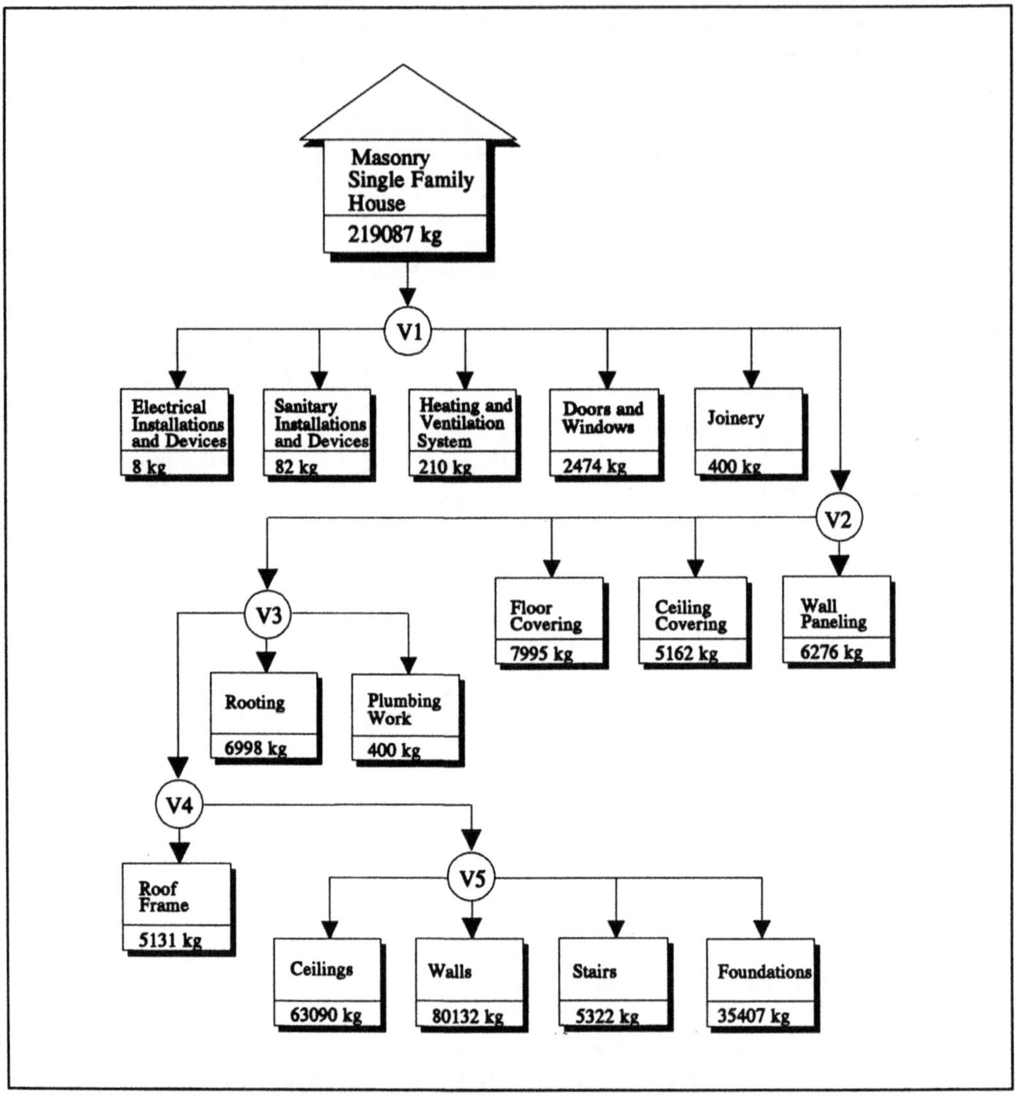

Figure 2: Dismantling-precedence-graph for a single family masonry house

The objective of recycling planning is the design of optimal recycling techniques for processing dismantled materials and building components into reusable materials. Depending on the stage of dismantling, the feed can be either a single material or a mix of all the building materials. For some materials such as glass and metals, classic recycling techniques already exist and recycling planning is a simple coordination or pretreatment. For other materials, techniques are not yet developed or have to be modified according to the composition of the specific feed, e.g. for plastic and composite materials. Toxics like asbestos require specific treatments for disposal.

3 Integrated dismantling and recycling planning

The dismantling planning process only can be carried out if the recycling costs for all building components and building materials are already determined. This means the recycling planning process has to precede the dismantling planning process. On the other hand the recycling planning process requires information about the amount and composition of the dismantled building components and materials. Due to the fact that the recycling capacities of the different reuse options are limited in the considered geographical region, an optimal allocation of the dismantled components and materials to the available reuse options is necessary in order to determine the minimal recycling costs. Following this idea the dismantling planning process has to precede the recycling planning process. Because of these interdependances the dismantling and recycling planning process, can't be carried out independently; furthermore an integrated approach is desirable (c.p. Figure 3).

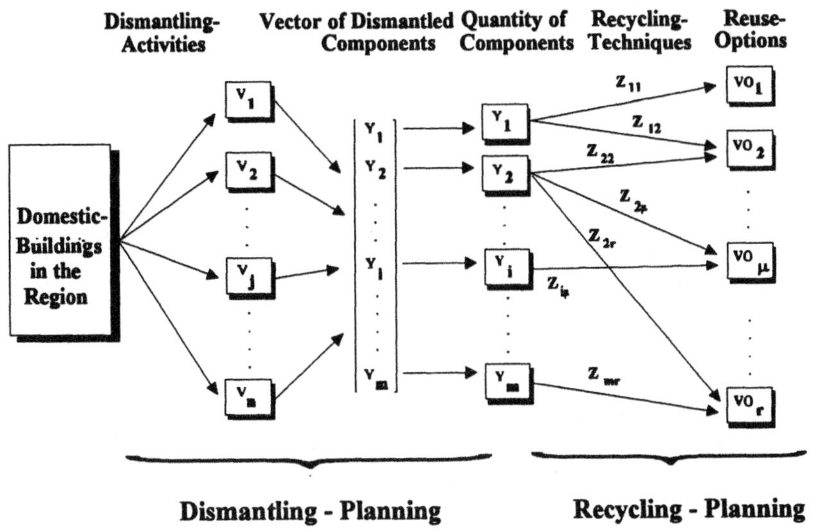

Figure 3: Structure of the integrated dismantling and recycling planning problem

The dismantling-precedence-graph of the whole region of the Upper Rhine Valley is composed of the eleven dismantling-precedence-graphs of the predominant types of buildings. These buildings can be dismantled by the application of n different dismantling activities v_j such that the quantities of the dismantled components and materials can be represented by the vector $y^T = (y_1,...,y_m)$. Taking into consideration the feasible reuse options, the achievable reuse proceeds and the capacity constraints, an optimal allocation has to be determined.

The integrated dismantling and recycling planning problem can be formulated as a mixed-integer linear programming model (MIP):

$$MAX \quad \sum_{i=1}^{m} \sum_{\mu \in T_i} c_{i\mu}^{v} \cdot z_{i\mu} - \sum_{j=1}^{n} c_j^{z} \cdot x_j \qquad (4.1)$$

Subject to

$$y_i = y_i^a + \sum_{j=1}^{n} x_j \cdot v_{ij} \qquad i = 1,\dots,m \qquad (4.2)$$

$$\sum_{\mu \in T_i} z_{i\mu} - \gamma_i \cdot y_i = 0 \qquad i = 1,\dots,m \qquad (4.3)$$

$$\sum_{i \in I_\mu} z_{i\mu} - Q_\mu \le 0 \qquad \mu = 1,\dots,r \qquad (4.4)$$

$$x_j \in IN_0 \qquad j = 1,\dots,n \qquad (4.5)$$

$$z_{i\mu} \ge 0 \qquad i = 1,\dots,m \ ; \ \mu \in T_i \qquad (4.6)$$

$$y_i \ge 0 \qquad i = 1,\dots,m \qquad (4.7)$$

where

- i : index for the different components of the product ($i = 1,\dots,m$),
- j : index for the different dismantling activities ($j = 1,\dots,n$),
- μ : index for the different reuse options ($\mu = 1,\dots,r$),
- $c_{i\mu}^{v}$: proceeds per ton of component i which is recycled by reuse option μ,
- c_j^{z} : dismantling cost of dismantling activity j,
- y_i : number of dismantled components i,
- y_i^a : initial number of dismantled components i,
- v_{ij} : number of dismantled components i resulting from one application of dismantling activity v_j,
- γ_i : tons per unit of component i,
- Q_μ : capacity limit of reuse option m,
- T_i : $\{\mu \,/\, \text{component } i \text{ can be recycled by reuse option } \mu\}$,
- I_μ : $\{i \,/\, \text{component } i \text{ can be recycled by reuse option } \mu\}$,
- $z_{i\mu}$: decision variable for the amount [t] of component i to be recycled by reuse option μ,
- x_j : decision variable for the number of applications of dismantling activity j.

The subject of the optimization problem is the maximization of the total achievable marginal income (reuse proceeds - dismantling costs) (5.1) subject to

- dismantling constraints (4.2),
- complete recycling of the dismantled components and materials (4.3),
- capacity constraints of the reuse options (4.4),
- integer and non-negativity constraints (4.5), (4.6), (4.7).

4. Solution procedures of the planning problem

The problem can be solved by the application of various kinds of solution procedures for combinatorial optimization problems, such as

- Cutting-Plane-Algorithms
- Branch-and-Bound-Algorithms
- Dynamic-Programming-Algorithms
- Decomposition Algorithms.

General mixed - integer problems are NP-complete so that the required solution time will grow exponentially with the complexity of the considered planning problem. Standard optimization software packages (e.g. LINDO, [Schrage 91]) generally use sophisticated **Branch-and-Bound-Techniques** based on the Algorithm of Dakin [Dakin 65]. Upper bounds are generated by an **LP-Relaxation** of the mixed-integer problem. The upper bounds computed in this way are not often strong enough to reduce the required solution time drastically. In this case it can be advantageous to introduce so called upper-bound-functions such as the **Lagrangean function** [Geoffrion 74].

Both kinds of Branch-and-Bound-Algorithms don't take any advantage of the special structure of the considered dismantling and recycling planning problem, which can be partitioned into an all-integer linear dismantling problem and a linear recycling problem. The **decomposition algorithm of Benders** [Benders 62] uses this special feature in order to find an optimal solution by the iterative solution of the all-integer and the linear problem [Salkin 89].

For this the mixed-integer problem (MIP) will be transformed as follows:

$$\underset{x_j \in IN_0}{MAX}\left(-\sum_{j=1}^{n} c_j^z \cdot x_j + MAX\left\{\begin{array}{l}\sum_{i=1}^{m}\sum_{\mu\in T_i} c_{i\mu}^v \cdot z_{i\mu} \,/\, \sum_{\mu\in T_i} z_{i\mu} = \gamma_i \cdot \left(y_i^a + \sum_{j=1}^{n} x_j \cdot v_{ij}\right); \\ \sum_{i\in I_\mu} z_{i\mu} \le Q_\mu ; z_{i\mu} \ge 0 \; \forall i = 1,\dots,m; \mu \in T_i\end{array}\right\}\right) \quad (5.7)$$

The interior linear optimization problem can be dualised for given integer variables $\hat{x}_j \;\; (j = 1,\dots,n)$ (DL):

$$MIN \quad \sum_{i=1}^{m} \gamma_i \cdot \left(y_i^a + \sum_{j=1}^{n} \hat{x}_j \cdot v_{ij} \right) \cdot u_i + \sum_{\mu=1}^{r} Q_\mu \cdot \pi_\mu \tag{5.8}$$

Subject to

$$u_i + \pi_\mu \geq c_{i\mu}^v \qquad i = 1,\dots,m \; ; \; \mu \in T_i \tag{5.9}$$

$$\pi_\mu \geq 0 \qquad \mu = 1,\dots,r \tag{5.10}$$

$$u_i \in IR \qquad i = 1,\dots,m \tag{5.11}$$

where

π_μ: dual variable of the capacity constraints $(\mu = 1,\dots,r)$
u_i: dual variable of the dismantling constraints $(i = 1,\dots,m)$

With the help of the extreme points $u_i^P, \pi_\mu^P \quad (p = 1,\dots,P)$ of the convex polyhedron (5.9),...,(5.11), the following all-integer problem (IP), which is equivalent to the mixed-integer problem (MIP), can be formulated:

$$MAX \quad DB \tag{5.12}$$

Subject to

$$DB \leq -\sum_{j=1}^{n} c_j^z \cdot x_j + \sum_{i=1}^{m} \gamma_i \cdot \left(y_i^a + \sum_{j=1}^{n} x_j \cdot v_{ij} \right) \cdot u_i^P + \sum_{\mu=1}^{r} Q_\mu \cdot \pi_\mu^P$$
$$p = 1,\dots,P \tag{5.13}$$

$$x_j \in IN_0 \qquad j = 1,\dots,n \tag{5.14}$$

where

$$DB := \left(-\sum_{j=1}^{n} c_j^z \cdot x_j + \underset{p=1,\dots,P}{MIN} \left\{ \sum_{i=1}^{m} \gamma_i \cdot \left(y_i^a + \sum_{j=1}^{n} x_j \cdot v_{ij} \right) \cdot u_i^P + \sum_{\mu=1}^{r} Q_\mu \cdot \pi_\mu^P \right\} \right) \tag{5.15}$$

In order to find the optimal solution of (MIP), it is often not necessary to generate all restrictions (5.13) of (IP). The idea of Benders'Decomposition is an iterative solution of (DL) and (IP) whereby each optimal solution of (DL) leads to a new restriction (5.13) of (IP). A new lower bound is generated by each solution of (DL) and a new upper bound by each solution of (IP). The iteration will stop with the optimal solution if the lower bound equals to the upper bound. In each iteration an optimal solution of the all - integer problem (IP) has to be computed. The required solution time can be reduced drastically, if only feasible solutions of (IP) are computed [Hummeltenberg 85], [Geoffrion 74]. This heuristic modification of

Benders'Decomposition Algorithm will compute "good" solutions of the considered dismantling and recycling problem in a reasonable time.

5. Application of the planning model and first results

A prototype of the above presented dismantling planning model has been implemented on a Personal Computer and evaluated with specific regional database [Spengler 93], [Nicolai 93]. The database of the implemented prototype consists of four different types of single family houses which are predominant in the region of the Upper Rhine Valley and around 35 recycling installations for demolition waste. The mixed integer linear programming problem will be formulated automatically by the planning system and solved with the help of the commercial software package LINDO and a simple heuristic for the generation of a "good" first solution which deals as a lower bound.

Taking into account disposal prices of at least 74.- DM/t for mixed demolition waste and acceptance fees at recycling installations of 24.- DM/t for demolition waste after the application of dismantling activty v_1, the application of dismantling activity v_1 is already advantageous from an economic point of view. Today, these conditions are already realized in Baden (Germany) but not in Alsace (France), where disposal prices are still too low, if they exist. By the introduction of duties on demolition waste resulting from the pull-down of the various types of domestic buildings (56.- DM/t - 72.- DM/t), the complete dismantling of the domestic buildings can be achieved in the region of the Upper Rhine Valley.

Literature

[Benders 62] Benders, J.F.: Partitioning Procedures for Solving Mixed-Variables Programming Problems, in: Numerische Mathematik 4, 1962, S. 238 - 252

[Dakin 65] Dakin, R.: A Tree Search Algorithm for Mixed Integer Programming Problems, in: Computer Journal 8(3) , 1965, S. 250 -255

[Geoffrion 74] Geoffrion, A.: Lagrangean Relaxation for Integer Programming, in: Mathematical Programming Study 2, S. 82 - 114

[Hummeltenberg 85] Hummeltenberg, W.: Capital Budgeting with Benders Decomposition, in: EJOR, 1985, S. 318 - 329

[Nicolai 93] Nicolai, M.; Spengler, Th.; Ruch, M.; Valdivia, S.; Hamidovic, J.; Rentz, O.: Minimization, Recycling and Reuse of Demolition Waste - A Dismantling and Recycling Planning System for the Region of the Upper Rhine Valley, in: Proc. of International Recycling Congress Re'93, Genf, 1993

[Salkin 89] Salkin, H.; Mathur, K.: Foundations of Integer Programming, New York, Amsterdam, London, 1989

[Spengler 93] Spengler, Th.; Hamidovic, J.; Nicolai, M.; Ruch, M.; Valdivia, S.; Rentz, O.: Konzeption eines EDV gestützten Planungssystems zur optimalen Demontage- und Verwertungsplanung von Wohngebäuden im Oberrheingraben, in: Proc. Informatik im Umweltschutz, 7. Symposium, Ulm 1993, Springer Verlag, 1993

ON THE IMPACT OF PRODUCT REMANUFACTURING AND RECYCLING ON OPERATIONS MANGEMENT

M.C. Thierry, M. Salomon, Erasmus Universiteit Rotterdam, NL
A.A.N. Ridder, Vrije Universiteit Amsterdam, NL
L.N. Van Wassenhove, INSEAD Fontainebleau Cedex, France

Many organizations are looking into the concepts of remanufacturing and recycling of used products in order to reduce the environmental impact of their business operations, and to make money. The principle of remanufacturing is to manufacture 'as new' products that fulfill predeterminde quality standards. The remanufacturing process involves extensively testing, repair, and removal of used components on all product levels, and the procurement and assembly of new components, when necessary.

Recycling refers to the conversion of used products into their basic materials, which can eventually be reused in the production of new components. This presentation reports on continuing research in product remanufacturing and recycling, and their impact on operations management. It will

(i) contain an overview of the potential advantages and disadvantages of remanufacturing and recycling
(ii) give a brief literature review of remanufacturing, recycling and operations management
(iii) discuss an adaptation of MRP in order to deal with remanufacturing and
(iv) propose some statistical inventory control models that account for remanufacturing and recycling.

Keywords:
remanufacturing, recycling, logistics, MRP, statistical inventory control models

Development of Emission Oriented Production Control Strategies Based on Fuzzy-Expert-Systems, Neural Networks and Neuro-Fuzzy-Approaches

A. Tuma, H.-D. Haasis, O. Rentz
Institute for Industrial Production (IIP)
University of Karlsruhe

Summary: In industrial production processes, materials and different forms of energy are provided, transformed respectively converted, stored and transported. In common with this process joint products in different states of aggregation are emitted. Environmental impacts can be identified at any stage of the energy and material flow process. Due to the fact that production units and processes are interconnected with energy and material flows, it is of special interest to develop production control mechanisms, which control the energy and material streams in a way that utilise available resources most efficiently and reduce emissions and by-products caused by the production process. These production control strategies have to consider variations in the input and output flows of succeeding and preceding production units.

The development of production control strategies depends especially on the structure of integrated production systems. If it is possible to influence the energy and material flows by the selection of special production processes and an adequate allocation of jobs and aggregates, the construction of production control strategies can be reduced to a combined scheduling and technology selection problem.

Methodical production control strategies can be based on optimal algorithms (e.g. dynamic programming) heuristics (e.g. rule based approaches) and methods of machine learning (e.g. Neural Networks). Due to the complexity of real production systems, it is advisable to use rule based approaches or Neural Networks depending on the structure of the available production knowledge.

Zusammenfassung: In industriellen Produktionsprozessen werden zur Herstellung von Marktprodukten Stoffe und Energiearten bereitgestellt, umgeformt bzw. umgewandelt, gelagert und transportiert. Gemeinsam mit diesem Prozeß der Leistungserstellung werden als Kuppelprodukte Stoffe in flüssigen, gasförmigen und festen Aggregatzuständen emittiert. Da auf betrieblicher und überbetrieblicher Ebene Produktionsprozesse durch Stoff- und Energieströme miteinander verbunden sind, ist die Entwicklung von Produktionsabstimmungsmechanismen von besonderer Bedeutung, welche diese so steuern, daß die zur Verfügung stehenden Ressourcen möglichst effizient ausgenutzt und durch den Produktionsprozeß enstehende Emissionen und Abfallstoffe, soweit dies technisch möglich ist, vermindert werden. Hierbei sind insbesondere die Rahmenbedingungen vor- und nachgeschalteter Produktionsstufen zu berücksichtigen.

Entwicklung und Auswahl geeigneter Produktionsabstimmungsmechanismen hängen in erster Linie von der Art integrierter Produktionsprozesse ab. Insoweit als es unter Berücksichtigung produktionstechnischer Restriktionen möglich ist, Energie- und Stoffströme durch geeignete Auswahl von Produktionsverfahren bzw. Maschinen-/Aggregatebelegungsplänen signifikant zu beeinflussen, kann deren Entwicklung auf ein kombiniertes Scheduling- und Verfahrensauswahlproblem zurückgeführt werden.

Methodisch können Produktionsabstimmungsmechanismen sowohl auf optimierenden Verfahren (z. B. Dynamische Optimierung, Branch&Bound-Verfahren) oder Heuristiken (z. B. Prioritätsregelverfahren) sowie Methoden des Maschinellen Lernens basieren. Je nach Struktur des verfügbaren Planungswissens empfiehlt sich eine Verwendung von regelorientierten Ansätzen oder Neuronalen Netzen.

Analysis of the production system under investigation

To analyse the behaviour of different production oriented control mechanisms e.g. based on Fuzzy Expert Systems, Neural Networks and Neuro-Fuzzy approaches, the described methods are verified by an exemplary production system from the textile industry (figure 1). The production system under investigation consists of a dye house, a boiler house, a hydro power plant and a flue gas neutralisation facility. The dye house covers two stages of production, the dyeing process and the drying of the dyed yarns. The two stages of production require steam/hot water and electric power, which are supplied by the preceding power plants. The flue gas of the boiler house is used to neutralise the mainly alkaline waste water of the dye house at the flue gas neutralisation facility. The storage of steam/hot water as well as the capacity of the waste water reservoir are limited. The capacity of all preceding and succeeding production units are variable. Variations can be caused by external factors (e.g. smog, variations of the water level of the inlet of the hydro power plant).

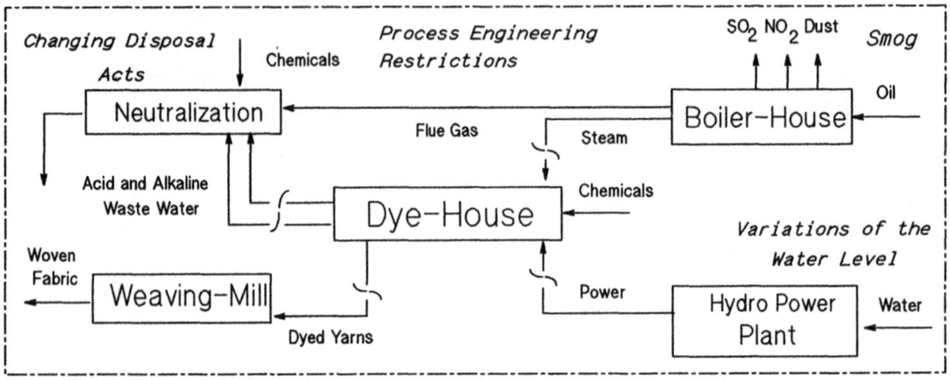

Figure 1: Structure of an interconnected production system

For the investigation of different production scenarios (e.g. different operating modi of the power plants, smog events, machinery disturbances) the production system is modelled with a simulation tool (SLAM). The physical structure of the production system (e.g. aggregates, potential energy and material flows between the aggregates), available resources and system functions (e.g. queues, the allocation of aggregates) are modelled graphically [1]. Process and job specific data (e.g. process parameters, recipe formulations, energy demand functions) are modelled in a C-database. Certain production rules (e.g. for the resetting of the equipment) and interfaces to intelligent systems (e.g. Fuzzy Expert Systems and Neural Networks) are programmed in FORTRAN and C.

A system analysis of the investigated production system shows that

- emission oriented goals, such as an increase of the efficiency of the flue gas neutralisation facility, reduction of CO_2 emissions, reduction of supplementary chemicals (HCl, H_2SO_4) for the neutralisation process in cases of a waste water excess and a reduction of waste heat losses as well as
- economical goals, such as an increase of the utilisation of the equipment, a shortage of the average waiting time

correlate with the harmonising of energy and material flows, which can be influenced by the selection of certain dyeing processes and an adequate allocation of dye batches and dye vats.

Emission oriented production control strategies based on Fuzzy Expert Systems

Owing to the structure of the decision problem (number of serial and parallel production processes, multi-criteria goal function, dynamic behaviour of the energy and material flows, fuzziness of the production knowledge) Fuzzy Expert Systems are implemented to perform the planning decisions described. In any planning situation the corresponding Fuzzy Expert System is evoked and calculates a priority number for every potential combination of a dye batch and applicable dyeing process. This number is relatively high, if the energy demand (steam/hot water, electric power) and the characteristics of the waste water implied by a certain job correlate with the current state of the system (pH-value in waste water reservoir, energy supply). Figure 2 shows a typical structure of a Fuzzy Expert Controller with rule blocks for the harmonisation of the demand and supply of energy, and material streams (e.g., NaOH, CH_3OOH, CO_2, SO_2) for the neutralisation process. The rule blocks consist of 27 to 81 individual rules [2, 3, 4].

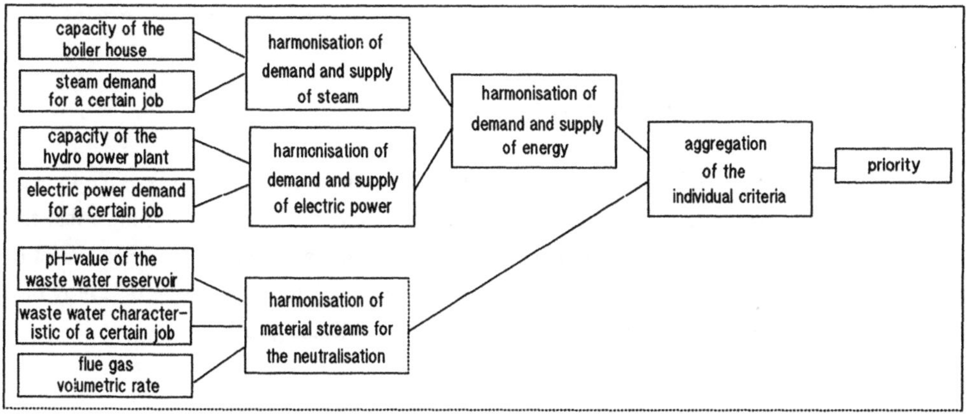

Figure 2: Structure of a Fuzzy Expert System for controlling energy and material flows

The development of Fuzzy Expert Systems requires the definition of membership functions, selection of aggregation operators, assignment of a degree of sensibleness for each rule and the selection of a defuzzification method [5]. Figure 3 shows exemplary membership functions and rules of a Fuzzy Expert Controller. An investigation of different membership functions, aggregation operators, controller architectures and strategies for the assignment of the degrees of sensibleness shows that the most critical point seems to be the assignment of a degree of sensibleness for each rule (figure 4). This determines the influence of individual rules and represents the inference structure of the Fuzzy Expert Controller. To adjust the degrees of sensibleness, it is important to have a consistent theory regarding a proper model of the controlling task. The system represented by bar 10 in figure 4 e.g. is based on the idea that the pH-value of the waste water reservoir is the key parameter for controlling the material flows for the neutralisation process. The adjustment of the degrees of sensibleness of the system represented by bar 9 is more oriented at the flue gas volumetric rate.

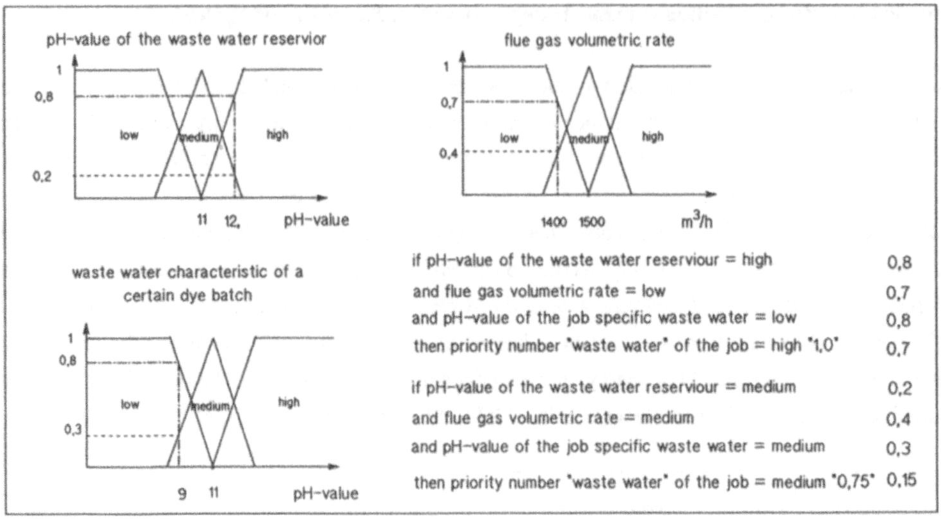

Figure 3: Exemplary membership functions and rules

In contrast to the efficiency of the neutralisation facility, it has up to now not been possible to set up a Fuzzy Expert System which fulfils the economical goals. Therefore, it is important to notice that the dependencies of the parameters and the corresponding inference structure concerning the achievement of the mentioned economic goals, which are correlated in a certain way with the harmonising of energy demand and supply, is much more complicated compared to the emission oriented goals.

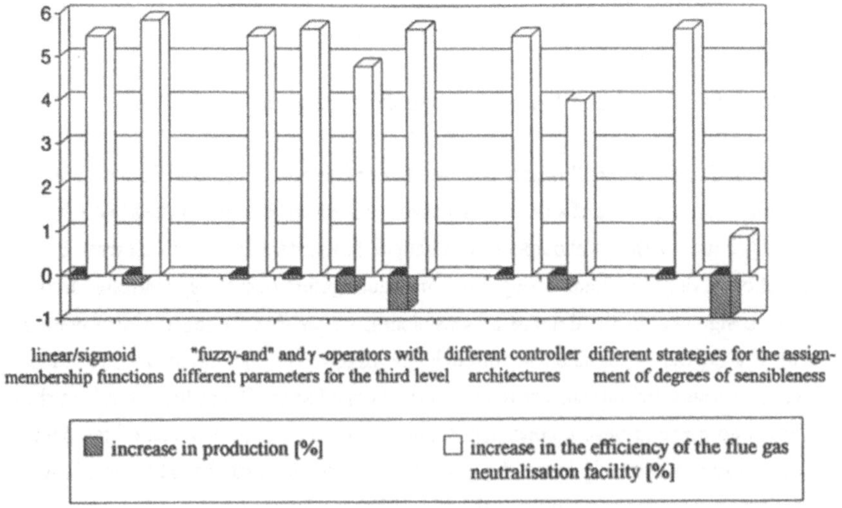

Figure 4: A comparison of different Fuzzy Expert Systems for the controlling of material and energy flows

Emission oriented production control strategies based on Neural Networks

If it is not possible to construct a consistent model i.e. to formulate explicit rules, implicit knowledge can be used. Implicit planning knowledge is e.g. included in representative production examples. One way to operationalise implicit knowledge is to use Neural Networks. The construction of production control strategies based on Neural Networks requires the formulation of the controlling task in a manner which can be processed by an adequate network architecture, acquisition of representative training examples, selection, teaching and testing of adequate network architectures.

The described scheduling and technology selection problem can be formulated as a forecasting problem. At any time, when a job has to be scheduled, the corresponding Neural Networks are evoked. For every possible combination of a dye batch and an applicable dyeing process, the expected processing and waiting time and the expected variation of the pH-value of the waste water reservoir are predicted [2, 3, 4].

The acquisition of representative production examples is based on the analysis of different simulation scenarios. 200 scenarios (different operating modi of the power plants, disturbances of preceding and succeeding production units) are chosen at random from a set of 6912 possible scenarios. For each scenario 8 to 12 break points, representing certain states (pH-value of the waste water reservoir, flue gas volumetric rate, available power of the power plants), are chosen at random. At these break points, different planning alternatives are scheduled. The most critical point in this context is the selection of evaluation parameters and the determination of the time, when the influence of the different alternatives should be evaluated. If for example the chosen evaluation time for the single alternatives is too late, the influence of a certain decision could be covered by succeeding decisions.

Due to the requirements of the forecasting problem, a backpropagation network with three layers is selected (figure 4). The input function is the weighted summation, the transfer function is the sigmoid function or the tangens hyperbolicus, the output function is the direct output [6].

As a X^2-test shows, it is possible to control the mentioned economic goals using Neural Networks. On the other hand, it was up to now not possible to control the efficiency of the flue gas neutralisation facility as successfully as with Fuzzy Expert systems. This implies that in fields, where a consistent theory can be constructed, it is advisable to use rule based systems such as Fuzzy Expert Systems. If however, this not possible due to the complexity of the controlling task, Neural Networks should be applied.

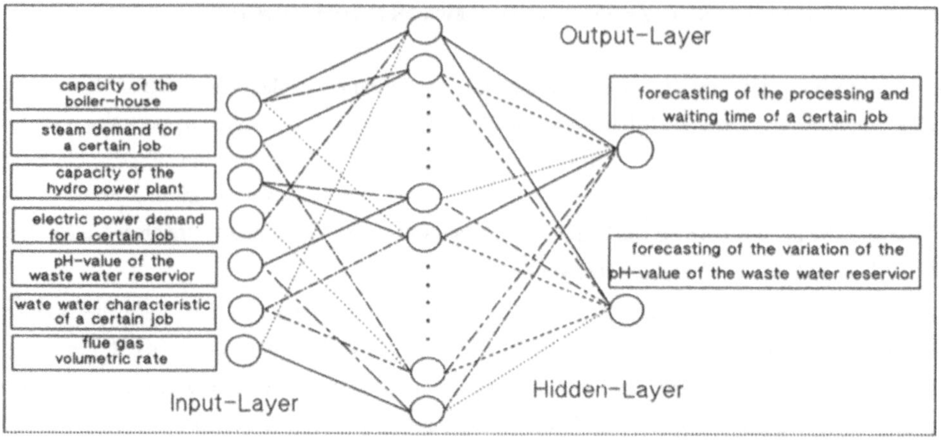

Figure 5: Structure of a Neural Network for controlling energy and material flows

Emission oriented production control strategies based on Neuro-Fuzzy Approaches

In order to combine the advantages of Fuzzy Expert Systems dealing with explicit knowledge and Neural Networks dealing with implicit knowledge, a Neuro-Fuzzy approach is developed to control energy and material flows. In principle, the rule structure of the Fuzzy Expert System is applied. The assignment of the degrees of sensibleness for certain rule blocks is achieved by machine learning algorithms. Under certain conditions (e.g. the weighted summation as input function of the Neural Network and the average operator as composition operator of the corresponding Fuzzy Expert System), Fuzzy Expert Systems can be interpreted as backpropagation networks and vice versa [2, 3, 4, 7, 8].

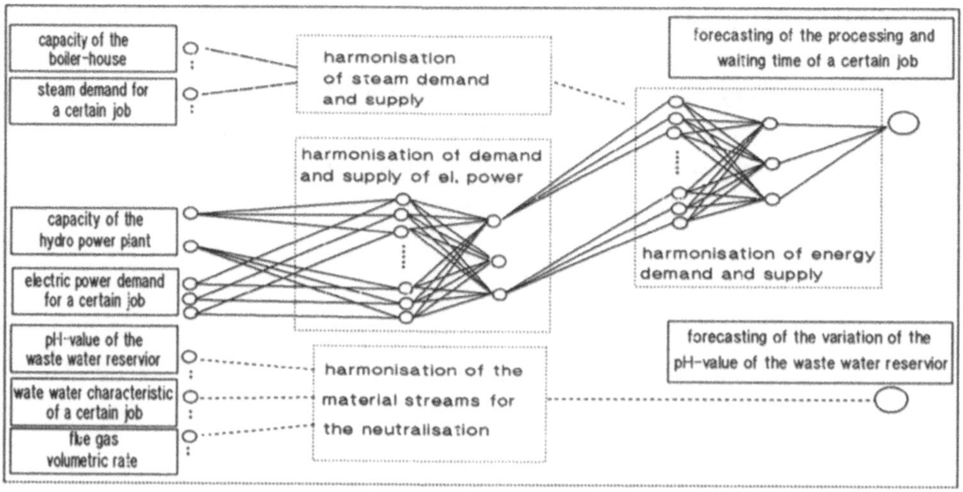

Figure 6: Structure of a Neuro-Fuzzy System to control energy and material flows

The adjustment of the system covers two steps. In a first step the degrees of sensibleness are set from the production examples (figure 7, bar 3). In a second step the interpretable weights (degrees of sensibleness) of these rule blocks, for which a consistent theory exists (e.g. the controlling of the flue gas neutralisation facility) are adjusted manually (figure 7, bar 4). This procedure combines the capabilities of machine learning, evaluating implicit knowledge, and the human capabilities for constructing a consistent theory of a closed problem with respect to the advantages of Fuzzy Expert Systems and Neural Networks. This is of special interest in fields of ambiguous knowledge, such as the controlling of energy and material flows, taking into consideration emission oriented and economic goals.

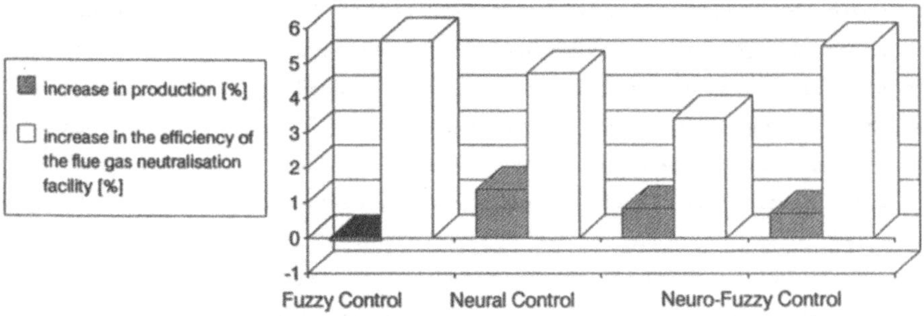

Figure 7: Comparison of different controlling methods

References:

[1] PRITSKER, A.: Introduction to Simulation and SLAM, New York, Chichester, Brisbane, 1986.

[2] TUMA, A.; HAASIS, H.-D.; RENTZ, O.: A Comparison of Fuzzy Expert Systems, Neural Networks and Neuro-Fuzzy-Approaches Controlling Energy and Material Flows. In: Ecological Modelling, Elsevier, Amsterdam, 1994.

[3] TUMA, A.; HAASIS, H.-D.; RENTZ, O.: Emission Oriented Production Control Strategies Based on Fuzzy Expert Systems, Neural Networks and Neuro-Fuzzy-Approaches. In: International Joint Conference on Neural Networks 93, Nagoya, Japan, 1993.

[4] TUMA, A.; HAASIS, H.-D.; RENTZ, O.: Entwicklung emissionsorientierter Produktionsabstimmungsmechanismen auf der Basis von Fuzzy-Expert-Controllern, Neuronalen Netzen und Neuro-Fuzzy-Ansätzen. In: Information Management, erscheint 1994.

[5] ZIMMERMANN, H.-J.: Fuzzy Set Theory and its Applications, Boston, Dordrecht, London 1991.

[6] RUMELHART, D. E.; MCCLELLAND J. A. et al.: Parallel Distributed Processing, Cambridge Mass., 1986.

[7] BERENJI, H.: Neural Networks and Fuzzy Logic in Intelligent Control. In: 5th IEEE International Sysmposium on Intelligent Control, Vol. 2, 1990, 916-920.

[8] ICHIHASHI, H.: Fuzzy Systems - Optimisation by Neuro-like Fuzzy Model, Proceedings of the Joint Japanese-European Symposium on Fuzzy Systems, Berlin, 1992.

ENTWICKLUNG EINES INTEGRIERTEN PLANUNGSMODELLS ZUR EVALUIERUNG VON EMISSIONSMINDERUNGSSTRATEGIEN

Martin Wietschel, Universität Karlsruhe
Hans-Dietrich Haasis, Universität Karlsruhe
Otto Rentz, Universität Karlsruhe

Bei der Entwicklung von Emissionsminderungsstrategien für SO_2 und eingeschränkt auch für NO_x steht der Energieumwandlungssektor im Mittelpunkt der Betrachtung. Dies spiegelt sich in den bisher entwickelten Energieflußmodellen, die eine ökonomische und umweltpolitische Ausrichtung haben und mit deren Hilfe Emissionsminderungsstrategien entwickelt und bewertet werden, wider. Der Einbezug von CO_2 erfordert nun jedoch die Erweiterung der Energienachfragesektoren in den bestehenden Modellansätzen, um Optionen der Rationellen Energieverwendung und des Lastmanagements, welche ein bedeutendes Potential zur Senkung von CO_2-Emissionen in den kommenden Jahren darstellen, abbilden und bewerten zu können.

Die Zielsetzung ist es, einen technisch-wirtschaftlichen Bewertungsansatzes zu entwickeln, der konsistent Emissionsminderungsoptionen auf Endverbraucherebene untereinander vergleicht und diese Minderungsoptionen anderen Minderungsoptionen auf der Energieversorgungsseite gegenüberstellt.

Als methodisches Hilfsmittel wird ein integriertes Planungsmodell entwickelt. Hierzu wird das mehrperiodige Energieflußmodell EFOM-ENV, welches Minderungsoptionen auf der Versorgungsseite bereits adäquat berücksichtigt und zum Teil Energieeinsparpotentiale in den Sektoren Haushalte/Kleinverbraucher und Verkehr integriert hat, weiterentwickelt und modifiziert. Die Weiterentwicklung und Modifikation des Modells ist notwendig, um alle Energieeinsparoptionen (incl. vom Lastmanagement, aufgeteilt nach technischen und tariflichen Maßnahmen, sowie der Rationellen Energienutzung in der Industrie) entsprechend berücksichtigen zu können.

Zur Abbildung von tariflichen Maßnahmen wird ein Gleichgewichtsalgorithmus in das Modell implementiert. Die exogen vorgegebene Nachfrage wird durch eine preisabhängige Nachfragefunktion ersetzt, die anhand von empirischen Analysen bestimmt wird. Unter Verwendung eines Cobweb-Algorithmus, der bei der Nachfrageberechnung die Konvexkombination der beiden letzten Lösungen einbezieht, wird das partielle Gleichgewicht berechnet. Da preisabhängige Nachfragefunktionen in der Regel einen nichtlinearen Verlauf haben, muß das bisher auf lineare Gleichungen ausgerichtete Modell um nichtlineare Gleichungen erweitert werden.

Um die Auswirkungen der tariflichen und technischen Optionen des Lastmanagements auf die Energieumwandlungssektoren abbilden zu können, müssen die Jahreszeit- und Lastzeitparameter modifiziert werden.

Parameter Estimation in Differential Equations, using Random Time Transformations

Bruno Bassan, *Dipartimento di Matematica, Politecnico di Milano*
Ruth Marcus, Hovav Talpaz, *The Volcani Agricultural Research Center, Israel*
Isaac Meilijson, *School of Mathematical Sciences, Tel Aviv University*

Consider the differential equation $dX(t) = A(X(t))dt$, $X(t_0) =$ given, where the function A depends on some unknown (vector) parameter θ to be estimated from a realization $\tilde{X}(t_0) = x_0, \tilde{X}(t_1) = x_1, \ldots, \tilde{X}(t_n) = x_n$ of a process $\tilde{X}$ related to X but which incorporates measurement errors. As an example, consider the logistic equation for population growth in a finite environment and its solution

$$A(x) = rx(1 - x/k) \ ; \ X(t) = k/(1 + (k/X(t_0) - 1)\exp\{-r(t - t_0)\}) \ .$$

Two of the most common approaches to the estimation of the unknown parameters in the presence of errors are Regression and Diffusion. We will follow an approach which models the randomness as pertaining inherently to the time lag between apparent time $X^{-1}(x_j)$ and chronological time t_j. The observed process $\tilde{X}(t)$ is viewed as $\tilde{X}(t) = X(T(t))$.

The rationale behind the random time transformation is illustrated by the following examples. The "age" of a machine is better modelled by the hours of operation than by the time elapsed since the date of purchase. If a plant disease is transmitted by insects, incidence of infection at time t within a closed field is better modelled by the total number of insect-hours than by t itself. If environmental or climatic local conditions increase temporarily the insect population, then "time" runs faster.

Brownian Motion time transformations correspond to a White Noise *rate of change* of time. A smoother Gaussian rate of change would lead to a non-Markovian time change, the Integrated Ornstein-Uhlenbeck time transformation.

We model multivariate random time transformations for the analysis of multiple-field data, consisting of

$$\tilde{X}_i(t_{i,j}) = X_i(T_i(t_{i,j})) = x_{i,j} \ ; \ 1 \leq i \leq p \ ; \ 0 \leq j \leq n_i$$

coming from p differential equations subject to a dependent noise structure. The Brownian Motion Time Transformation model $\mathbf{T}(t) = (T_1(t), T_2(t), \ldots, T_p(t))$ assumes that the process $\mathbf{T}$ has stationary, independent increments, and that such an increment, say $(T_1(t + \Delta) - T_1(t), T_2(t + \Delta) - T_2(t), \ldots, T_p(t + \Delta) - T_p(t))$, is jointly normally distributed with mean vector $(\Delta, \Delta, \ldots, \Delta)$ and covariance matrix Σ. As we shall see, the maximization of the likelihood over Σ can be brought to the standard form of finding a (constrained or unconstrained) ML estimate of a covariance matrix based on an i.i.d. sample from a $N(0, \Sigma)$ distribution. Any existing package for such a purpose (such as LISREL) can serve as a "black box" yielding an objective function - the *profile likelihood* - in terms of the parameters θ exclusively.

SIMULATION ALS INTERPRETATIONSINSTRUMENT FÜR ERGEBNISSE STATISTISCHER ANALYSEN

Jochen Benz, Fachhochschule Konstanz

Die Interpretation von Ergebnissen statistischer Erhebungen sowie deren Umsetzung in konkrete Maßnahmen bereiten in der Praxis oft erhebliche Schwierigkeiten. Dies ist insbesondere dann zu beobachten, wenn Multivariate Methoden für die Datenanalyse zum Einsatz kommen.

Eine Hilfestellung können beispielsweise Wissensbasierte Systeme liefern. Insbesondere für die Interpretation und Demonstration von Ergebnissen linearer Strukturmodelle bietet sich jedoch die Simulation als Instrumentarium an. Hierbei ist es unerheblich, ob diese Kausalanalysen mittels einer Kombination "traditioneller Verfahren" oder simultaner Methoden durchgeführt wurden.

Hybride Expertensystem-Shells - wie zum Beispiel Kappa-PC - erlauben sowohl die Gestaltung von Wissensbasierten Systemen als auch von Simulationsmodellen mit relativ geringem Programmieraufwand.

In diesem Beitrag soll die Realisierung eines Simulationsmodells zur Interpretation von Kausalanalyseergebnissen vorgestellt werden. Zielsetzung dieses Modells ist es, die Erarbeitung konkreter Maßnahmen zur Verbesserung von Schwachstellen, aufgedeckt mittels statistischer Daten und Analysen, zu unterstützen.

Der Anwender hat die Möglichkeit, an das System "was-wäre-wenn-Fragen" zu stellen. Damit erhält er einen Eindruck davon, wie sich einzelne Maßnahmen auswirken. Zugleich hat er aber auch die Möglichkeit, komplette Szenarien in kurzer Zeit durchzuspielen.

Da das vorgestellte System mittels einer hybriden Expertensystem-Shell entwickelt wurde, demonstriert es gleichzeitig die Mächtigkeit dieser Werkzeuge.

ROBUST AND ADAPTIVE TESTS FOR THE TWO-SAMPLE LOCATION PROBLEM

Herbert Büning, FU Berlin

In the important two-sample location problem the application of the t-test depends on very restrictive assumptions such as normality and equal variances of the two random variables X and Y and it is well-known that the t-test is not robust with respect to level α or power β in the case of departures from normality and for unequal variances. Thus if the assumptions of the t-test are not satisfied it is more appropriate to apply a robust version of the t-test, like the trimmed t-test, or a nonparametric test, like the Wilcoxon. But the choice of a suitable robust or nonparametric test also depends on the underlying distribution, which may be symmetric or asymmetric and may have light, middle or heavy tails. Since in most cases the practising statistician has no clear idea about the underlying distribution an adaptive test should be applied which takes the given data set into account. Such real data sets are presented in order to demonstrate departures from the model of the t-test and to suggest the application of a robust or adaptive test. Some of these tests are discussed and compared with each other and with the classical t-test under different models of non-normality, like heavy-tailed or asymmetric distributions. It is shown that an adaptive test behaves well over a broad class of distribution functions. Although it is not always the best one, in most cases it is second best for each of the distributions considered. But this is just the philosophy of an adaptive test. It works in the sense of the "safety first" principle.

Hogg, R. V. (1976). A new dimension to nonparametric tests. Commun. Stat., Ser. A, Vol.5, 1313 - 1325

Büning, H, (1991). Robuste und adaptive Tests. De Gruyter, Berlin.

SCHÄTZVERFAHREN FÜR DIE DISPOSITION VON PASSUNGSTEILEN

Thomas Christ, TH Karlsruhe

Passungsteile (Ausgleichsscheiben, Lamellen, Abstandsringe, ...) werden insbesondere zum Ausgleich von produktionsbedingten Toleranzen benötigt. Die Disposition dieser Teile bereitet Schwierigkeiten, da ihr Bedarf selbst bei determinierten Losgrößen von den Unregelmäßigkeiten des Produktionsprozesses abhängt, also stochastischer Natur ist. Das Problem tritt z.B. in der Motoren- und Getriebefertigung oder bei der Montage von Servolenkungen auf.

Anstatt den zukünftigen Bedarf mit Hilfe konventioneller Prognosemethoden (gleitende Mittelwerte, exponentielles Glätten, ...) aus den Vergangenheitsdaten zu ermitteln, wird hier versucht, aus den Verbrauchsdaten der Passungsteile die Funktionalparameter der Verteilung der zugrundeliegenden Störgröße zu schätzen. Dies erlaubt nicht nur eine Prognose des zukünftigen Bedarfs, sondern ermöglicht auch einen tieferen Einblick in die Stochastik des Fertigungsprozesses (Verteilung der Störgröße, zeitliche Veränderung, ...).

Um mit möglichst wenig verschiedenen Passungsteilen auszukommen, wird meist eine Kombination mehrerer Passungsteile gemäß einer definierten Einbaustrategie verwendet, wobei eine gewisse Resttoleranz verbleiben darf. Geht man davon aus, daß die Störgröße nicht direkt gemessen werden kann, sondern aus den Verbrauchsdaten der Passungsteile geschätzt werden muß, dann erhält man ein Schätzproblem mit gruppierten Daten (aufgrund der Resttoleranz), bei dem die einzelnen Klassen — im Gegensatz zur herkömmlichen Problemstellung — nicht zusammenhängend und teilweise sogar überlappend sind (aufgrund der Kombinierbarkeit). Konventionelle Schätzverfahren können also nicht ohne weiteres verwendet werden.

Mittels Variationen der Maximum-Likelihood- und der χ^2-Minimum-Methode sowie numerischer Approximationsverfahren (scoring system) wird eine Schätzung möglich. Ergebnisse aus einer Simulationsstudie mit verschiedenen Einbaustragien und Losgrößen zeigen, daß beide Verfahren bei großen Stichproben gute Resultate liefern.

EVALUATION OF FORECASTS IN AR MODELS WITH OUTLIERS

Regina Elsebach, Universität der Bundeswehr Hamburg

An important problem in time series analysis is forecasting when outliers are present. Outliers are modelled as "additive" or "innovational" with respect to a basic time series process. Since it is a main characteristic of outliers that their occurence is not predictable, the aim of any forecast is the prediction of how the series will develop under the basic process without outliers. Ex-post analysis of forecasts should pay attention to the fact that extreme forecast errors can be caused by outliers, not by the basic process itself. Such extreme errors should not dominate the ex-post evaluation.

Based on the well-known AR model of order p,

$$X_t - \mu = \sum_{i=1}^{p} \phi_i (X_{t-i} - \mu) + \varepsilon_t, \qquad \varepsilon_t \sim \mathbf{N}(0,\sigma^2) \quad iid,$$

the effect of additive and innovational outliers on the standard MSE-optimal linear τ-step-ahead forecast (with $(p,\tau-1) := \min\{p,\tau-1\}$)

$$\hat{x}_t(\tau) = \sum_{i=1}^{(p,\tau-1)} \phi_i \hat{x}_t(\tau-i) + \sum_{i=(p,\tau-1)+1}^{p} \phi_i x_{t+\tau-i} + \mu(1-\sum_{i=1}^{p} \phi_i)$$

and the resulting forecast errors will be examined for $\tau=1$. Two alternative methods to evaluate forecasts based on the realized errors - robust generalizations of the standard mean squared (MSE) and mean absolute error (MAE) called MRSE resp. MRAE - are presented, and their behavior is visualized by a simulation study.

Especially in the case of additive outliers, these methods compare favourably to the standard MSE and MAE. Given some competing forecasts based on robust and non-robust estimates of the parameters of the underlying AR model, the MSE (and to a lesser extent also the MAE) very often identifies a forecast based on very poor estimates as best. This is considered to be inadequate. MRSE and MRAE are much better suited to detect those forecasts which are based on the true underlying AR parameters.

AGGREGATION AND DISAGGREGATION IN DEMAND FORECASTING

Paul Giesberts, Eindhoven University of Technology, NL
Jacob Wijngaard, University of Groningen, NL

An important element in production planning and control, especially in make-to-stock and assemble-to-order situations, is demand forecasting. Two, related, questions with respect to aggregation and disaggregation are important here. The first question is how to forecast an aggregate (capacity) variable. It is interesting to compare here direct aggregate forecasting and aggregation of detailed forecasts. The second question is how to forecast detailed (item) variables. In this case it is interesting to compare direct detailed forecasting and disaggregation of aggregate forecasts.

There are two important performance elements in forecasting. The first is that the forecast reacts well on a systematic change, the second is that it does not overreact on disturbances. The following simple (auto regressive) time series model appears to be useful in getting insight in these two sides of the performance of direct aggregate forecasting versus aggregating item forecasts.

$$x(t) = x(t-1) + \epsilon(t)$$
$$d(t) = x(t) + \eta(t)$$

With $x(.)$, $d(.)$, $\epsilon(.)$, $\eta(.)$ vectors and $\epsilon_i(t)$ and $\eta_i(t)$ normally distributed.

To compare direct item forecasts with disaggregating aggregate forecasts it is more appropriate to use a two phase demand model: In the first phase total demand is determined, in the second phase the distribution over the various categories is determined.

The results give insight in the strengths and weaknesses of aggregate versus detailed forecasting.

THE QUALITY OF NON-UNIFORM RANDOM NUMBERS

Wolfgang Hörmann, WU Wien

Summary: The quality of non-uniform random numbers is not only influenced by the quality of the uniform generator that is used but also by the transformation method applied to the uniform random numbers. This differences in quality between "exact" methods were almost entirely neglected in literature. So we compare the behaviour of four different transformation methods when combined with a linear congruential uniform generator (LCG). Heuristic considerations, the computation of two measures of approximation and a statistical test show that the inversion method performs best. Among the others rejection, when combined with a LCG with small multiplier, and ratio of uniforms perform worse. Their use could slightly change the results of some simulation studies.

Zusammenfassung: Die Qualität von nicht gleichverteilten Zufallszahlen wird nicht nur vom benützten Gleichverteilungsgenerator sondern auch von der verwendeten Transformationsmethode beeinflußt. Dieser Unterschied in der Qualität verschiedener "exakter" Transformationsmethoden wurde bisher in der Literatur fast völlig vernachlässigt. Darum vergleichen wir das Verhalten vier verschiedener Transformationsmethoden in Verbindung mit linearen Kongruenzgeneratoren (LKG). Heuristische Überlegungen, die Berechnung von zwei Maßen für die Approximation und ein statistischer Test zeigen, daß die Inversionsmethode am besten abschneidet. Unter den anderen Verfahren schneiden die Verwerfung – in Kombination mit einem LKG mit kleinem Multiplikator – und die Quotientenmethode am schlechtesten ab. Es ist nicht auszuschließen, daß ihre Verwendung die Resultate von Simulationsstudien verfälscht.

1. Introduction

The literature of random number generation falls into two parts. There are a lot of papers dealing with the quality of uniform pseudo-random number generators. On the other hand many papers discuss the generation of non-uniform variates by transforming an independent sequence of uniform random numbers and are mainly concerned with the speed or simplicity of the proposed algorithms. Concerning quality it is only stated that the method is exact which means that perfect uniform random numbers (which are not available) would be transformed into independent random numbers of the correct distribution. Investigations dealing with the effects that can occur when such an exact transformation method is combined with a pseudo-random sequence (this is done in every simulation which needs non-uniform random numbers) are very rare (see [1], [8], [7] and references there). But there are important differences in quality between exact transformation methods when combined with the same linear congruential pseudo-random number generator (LCG). This old but still very popular uniform generator is based on the simple recursion $x_i = (a \cdot x_{i-1} + c) \bmod m$; to get random numbers uniformly distributed on (0,1) this sequence is divided through m. It is well known that n-tuples produced by successive calls to a LCG form a lattice or grid and that the properties of this lattice depend on the choice of the multiplier a (see e.g. [9]).

In order to assess the quality of a uniform random number generator it is generally accepted that it is sensible (and more powerful than statistical tests) to investigate the distribution of all n-tuples that can be returned by that generator. For $n = 1$ this is necessary to guarantee the uniform distribution of the numbers, whereas the higher values are important for the order of the sequence and thus for the independence of the variates. A good uniform generator should have a regular distribution of all generated n-tuples for n between 1 and (at least) 20 as for a single simulation only a small portion of these points is taken as a sample. (In [10] it is demonstrated that not more than the 2/3 power of the period should be taken.)

2. The one-dimensional distribution

As we want to discuss the quality of non-uniform random numbers we will compare the empirical distribution function G_n of all possible points for several methods of generating non-uniform variates when a LCG is used as uniform generator. One of our measures for the quality of the approximation of the theoretical cumulative distribution function F by G_n will be the classical concept of discrepancy:

$$D_n(F) := \sup_x (G_n(x) - F(x)) + \sup_x (F(x) - G_n(x))$$

In [1] numerical comparisons of the discrepancy for various normal generators combined with LCG's are presented. But the main disadvantage of discrepancy is that it is sensitive only to the greatest difference. So we agree with Monahan in [12] that measuring the L_1-distance between G_n and F should give more information about the quality of the approximation.

$$E_n(F) := \int_R |F(x) - G_n(x)|\, dF(x)$$

In [12] accuracy of random variate generation is discussed under the assumption that truly random bits are available. In this paper we use distance criteria D and E to compare the quality of different transformation methods when combined with LCG's. For a LCG with maximal period m we have for the discrepancy $D = 1/m$. The value of E depends on how the integers x_i are transformed into floating point numbers u_i between 0 and 1. If $u_i = x_i/m$ we have $E = 1/(2m)$, for $u_i = (x_i + 0.5)/m$ (which was used for all computations of this paper) $E = 1/(4m)$. The only generation method for non-uniform deviates that preserves this good approximation is inversion. As long as we neglect numerical inaccuracies when computing the inverse distribution function, non-uniform random deviates generated by inversion have the same values for D and E as the LCG itself, all other methods are much worse. Before we present the empirical results we want to summarize the heuristic and theoretical considerations contained in [1], [8] and [7]. The accuracy of a LCG is better in low dimensions than in higher ones. Therefore a method that transforms one random number of the LCG into one non-uniform random number (like inversion) should be better than a method that transforms two or more uniform random numbers in one non-uniform deviate. But this simple rule does not explain all observations concerning the quality of non-uniform variate generation, mainly because most methods do not transform a fixed number of uniform deviates. The expected number of uniforms required by a certain transformation method and the rule, less uniforms imply better quality, give only a very crude rule of thumb for the quality of a transformation method when combined with a LCG. More insight is up to now only available for two of the most important transformation methods: For ratio of uniforms (see for example [3]) it is easy to see that all pairs (u, v) lying on one line through the origin are transformed into the same random number. Due to the lattice structure of the pairs generated by a LCG (see for example [9]) there must always be relatively large intervals without a random number when the ratio of uniforms method is combined with a LCG. (For details and bounds for the discrepancy see [8].) For the rejection method the simple observation that a LCG with a small multiplier (smaller than or close to $\sqrt{m}$) forms a two dimensional lattice with the shortest vector almost parallel to the y-axis has important consequences: Rejection algorithms combined with such LCG's (which were often recommended in literature because of their ease of implementation) have a bad approximation of the desired distribution. (For details see [7].)

In order to compare the quality of the approximation of the one-dimensional distribution for various transformation methods we computed criteria D and E for three different exponential generators: Transformed rejection (TR), a rejection technique that uses inversion to sample from the dominating distribution (see [6] and [5]), algorithm TRD that uses transformed rejection combined with decomposition to reduce the expected number of uniforms required (see [6] and [5]), and ratio of uniforms (RoU) (see eg. [3]). We restricted our attention to the exponential distribution only as it seems sensible to assume (and was checked by several numerical experiments) that the shape of a continuous distribution influences the quality of approximation of the different methods only slightly. To study the influence of different LCGs we combined each of the transformation methods with the following four LCGs. All of them have a good lattice structure up to 20 dimensions.

LCG 1: $m = 2^{32}$, $a = 663608941$, $c = 0$, Period=2^{30} suggested in [2]
LCG 2: $m = 2^{32}$, $a = 1099087573$, $c = 0$, Period=2^{30} suggested in [4]
LCG 3: $m = 2^{32}$, $a = 69069$, $c = 1$, Period=2^{32} suggested in [11]
LCG 4: $m = 2^{32}$, $a = 1589013525$, $c = 1$, Period=2^{32} suggested in [14]

To determine the exact value of D and E it is necessary to sort all random numbers that can be generated by a certain method. This was done for algorithm TR, for which it is easy to compute all generated random numbers in ascending order without storing them. For the remaining two algorithms we computed upper and lower bounds by counting the number of random variates, which fall into $2 \cdot 10^7$ intervals of equal probability. Table 1 contains the results of our computations for the distance criteria D and E divided through the value of D and E of the used LCG.

Table 1

	Distance E			Distance D		
	TRD	TR	RoU	TRD	TR	RoU
LCG1	(231|432)	48	(213|424)	(2671|2719)	126	(50088|50196)
LCG2	(318|522)	77	(26|184)	(3150|3258)	68	(29366|29474)
LCG3	(43248|44962)	4545	(48|638)	(29159|29589)	18559	(117233|117663)
LCG4	(266|999)	143	(212|947)	(1013|1443)	131	(109322|109753)

The results of Table 1 show that there are large differences between the quality of the approximation of the desired distribution. Inversion has the best properties (the values for inversion in Table 1 would all be one), TR combined with LCGs 1, 2 and 4 ranks second. Both criteria show the very

bad approximation of the rejection methods (TRD and TR) when combined with a LCG with small multiplier whereas the problems of RoU can only be seen in the very high values of D.
We want to emphasize again that E and D measure the quality of the approximation of the desired distribution by all numbers that can be returned by a generator. The question how these properties influence the behaviour of small samples can be tackled in two ways: The first and theoretical more sound possibility would be to study the properties of n-tuples of successive points for several values of n. With the exception of inversion, where the results for LCGs can be applied directly, it seems impossible to assess the quality of the approximation of the distribution in higher dimensions theoretically as there is no lattice any longer. The empirical investigation by computing bounds for the discrepancy or other measures of approximation is not possible as computing- and memory-requirements increase with the power of n. So we can only get some insight into the structure of the two-dimensional set of points generated by inversion, rejection and the ratio of uniforms method, which is done in Section 3. In Section 4 we present a statistical test that seems well suited to measure the degree of randomness of small samples.

3. The two-dimensional distribution

A fist and very helpful step to get some insight into the structure of the set of all pairs generated by a transformation method combined with a LCG is to plot pictures of these pairs for small uniform generators. We used the LCG with $m = 2^{11}$, $a = 605$ and $c = 1$, which has good lattice properties. Figure 1 shows all pairs generated with inversion together with that LCG and demonstrates that the lines of the lattice of the LCG are just transformed into curves. Strictly speaking a simple computation shows that by inversion the line $y = kx + d$ of the two-dimensional grid of the uniform generator is transformed into the curve $y = F^{-1}(k \cdot F(x) + d)$, which is $y = -log(1 - k(1 - exp(-x)) - d)$ for the exponential distribution. Thus we see that a good two-dimensional distribution of the LCG is preserved by inversion.

Figure 1

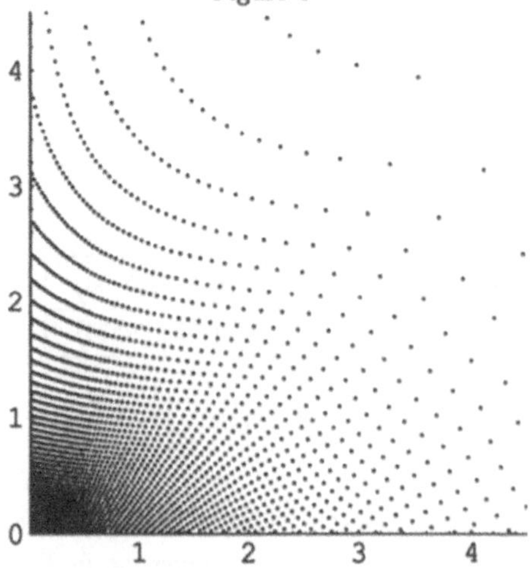

Figure 2

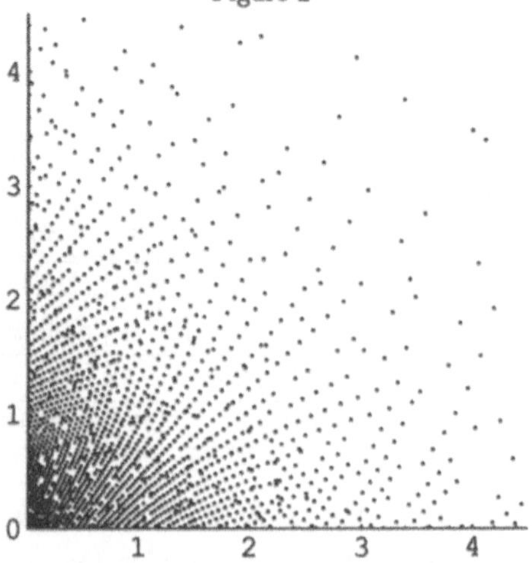

Figure 3

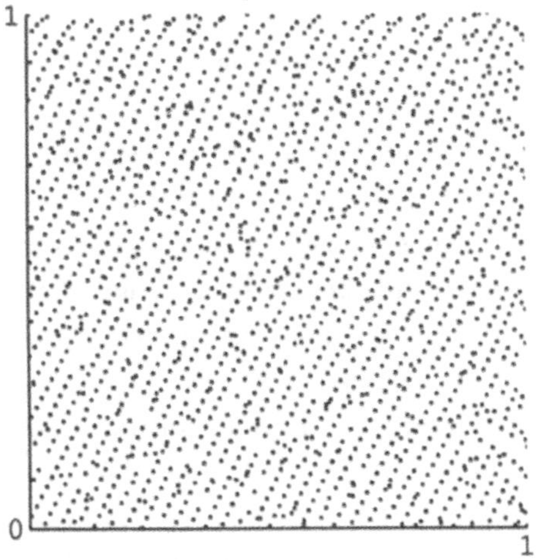

Figure 4

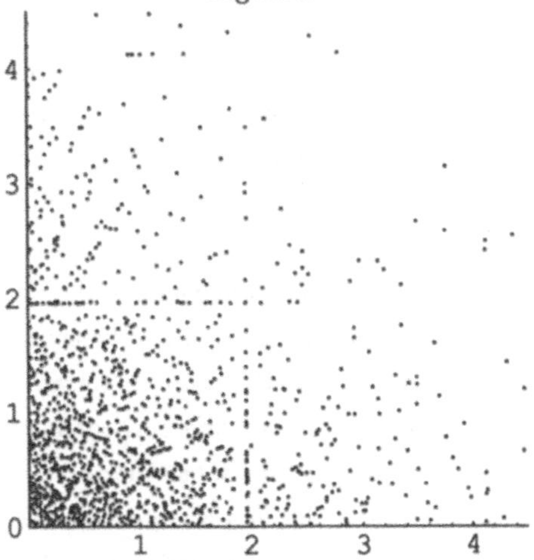

Figure 2 shows the set of all pairs generated with algorithm TR and the same LCG. At a first glance Figure 1 and Figure 2 look quite similar. This is not astonishing as the dominating distribution of TR is quite close to the exponential distribution and is generated by inversion. But it is not the 2-dimensional grid of all consecutive pairs of the LCG that is transformed. Due to the rejection algorithm a second uniform random number is necessary and so we see the transformed grid of all pairs of the LCG with one random number left out. Algorithm TR transforms at least 4 uniform random numbers into two exponential variates. So Figure 2 shows most points of the grid of all 4-tuples of the LCG transformed by the mapping $(x_1, x_2, x_3, x_4) \rightarrow (G(x_1), G(x_3))$ with G close to F^{-1} of the exponential distribution. Figure 3 shows the same points as Figure 2 but transformed by F of the exponential distribution. The grid is even more evident, but the lines are a bit curved due to the difference between the dominating distribution of TR and the exponential distribution. It is evident, especially when looking at Figure 3, that not all points lie in the regular grid and that some points of the regular grid are missing. This is due to the rejection logic that throws away some points of the main grid (about 16%) and replaces it by points of the transformed grid of all pairs of the LCG with three, five or more random numbers left out.

For TRD the two-dimensional distribution is quite similar. But for RoU (Figure 4) the set of all pairs looks entirely different. The grid structure of the LCG seems to have entirely disappeared. On the other hand RoU is quite similar to rejection as pairs of uniform random numbers are accepted or rejected. Thus the set of all points generated by RoU again consists mainly of points generated by transforming 4-tuples of the LCG. But this transformation $((x_1, x_2, x_3, x_4) \rightarrow (x_2/x_1, x_4/x_3)$, is not so simple as in the case of TR. One-dimensional affine subspaces of R^4 are transformed into hyperbolas of the form $y = (k_1 x + d_1)/(k_2 x + d_2)$ and any two-dimensional subspaces of R^4 containing the origin is transformed into one single point. Therefore the lattice structure of the LCG disappears entirely.

The above pictures and considerations seem to indicate that the quality of the approximation of the two-dimensional theoretical distribution in combination with LCGs is best for the inversion method, among the others ratio of uniforms seems to be worst. But as stated above it is very hard to compute measures for the quality of the approximation.

4. Properties of small samples

In [10] Maclaren suggests a test to detect excess of uniformity for random number generators. It is based on a result of Moran (see [13]), who shows that the sum of squares of the spacings of a true random U(0,1) sample of size N is asymptotically normal with mean $2/N$ and variance $4/N^3$. Maclaren uses this test to show that any uniform pseudorandom number generator that is based on sampling without replacement (like LCG's for example) returns too uniform samples (the sum of squares of the spacings is too small) if the sample size exceeds the 2/3 power of the period of the generator. Of course the test can also be used to detect too little uniformity if the sum of squares of the spacings is too large. By transforming variates with the cumulative distribution function the test can also be applied to non-uniform random numbers. We tested all of the transformation methods and all LCG's contained in Table 1 for samples of the size $N = 10^5$, 10^6 and $3 \cdot 10^6$. Table 2 gives the number of cases the sum of squares of the spaces was larger than the median (the sample was less uniform than an average random sampling) out of the hundred replications we made for each combination. For a real random sample the value lies with probability 0.95 between 40 and 60.

Table 2

	$N = 10^5$				$N = 10^6$				$N = 3 \cdot 10^6$			
	INV	TRD	TR	RoU	INV	TRD	TR	RoU	INV	TRD	TR	RoU
LCG1	48	43	44	46	17	9	0	96	0	0	0	100
LCG2	44	42	42	49	13	10	1	71	0	0	0	100
LCG3	52	48	66	55	38	100	100	57	23	100	100	93
LCG4	48	40	51	52	35	46	46	52	26	11	13	92

The results of Table 2 support the consideration in [10]: The uniformity of the samples of INV (i.e the results of the pure LCG) grows with the sample size. For $N = 3 \cdot 10^6$ they are definitely too uniform. Apparent are also the differences between LCGs 1 and 2 with period 2^{30} and LCG's 3 and 4 with period 2^{32}. For the other transformation methods it is evident that the whole period considerations of Section 2 and 3 have an influence on the properties of shorter samples. RoU and rejection (TR and partly TRD) combined with a LCG with a small multiplier (LCG 3) are less uniform than the used LCG and also less uniform than a truly random sample. For RoU the deviation is smaller but it occurs for all LCGs. The results for TRD and TR with LCGs 1, 2 and 4 are even more uniform than the results of INV. An explanation could be the fact that the period of TRD and TR is shorter than the period of the LCG itself.

5. Conclusions

We also did some simulation of an M/M/1-queue to find out, if the differences of the transformation method would influence the simulation results. As long as we used LCGs with good lattice properties all of the above transformation methods lead to correct results when compared with the theoretical solution. On the other hand a test with a LCG with bad lattice structure in three dimensions showed that the results for RoU were much worse than for the other transformation methods. Therefore even the simulation of this simple queue supports the claim that not only the choice of the uniform generator but also the choice of the transformation method can influence the results of a simulation. We are convinced that the theoretical and empirical considerations of this paper justify the recommendation: In combination with LCGs use inversion or, for the case that this is very slow, rejection with a large multiplier for the LCG. The use of ratio of uniforms combined with any LCG or rejection combined with a LCG with small multiplier could possibly influence the results of simulation studies.

References

[1] Afflerbach, L. and Hörmann, W.: Nonuniform random numbers: a sensitivity analysis for transformation methods. in: International Workshop on Computationally Intensive Methods in Simulation and Optimization, U. Dieter and G. Ch. Pflug, eds., Lecture Notes in Econom. Math. Systems 374, Springer-Verlag, New York (1992)

[2] Ahrens, J. H., Dieter, U. and Grube, A.: Pseudo-random numbers a new proposal for the choice of multiplicators. Computing 6, 121-138 (1970)

[3] Devroye, L.: Non-Uniform Random Variate Generation. Springer-Verlag, New York (1986)

[4] Fishmann G. S.: Multiplicative congruential random number generators with modulus 2^β: An exhaustive analysis for $\beta = 32$ and a partial analysis for $\beta = 48$. Math. Comp. 54, 331-344 (1990)

[5] Hörmann, W. and Derflinger, G.: The transformed rejection method for generating random variables, an alternative to the ratio of uniforms method. Manuskript, Institut f. Statistik, Wirtschaftsuniversität Wien, (1991)

[6] Hörmann, W.: New generators of normal and poisson deviates based on the transformed rejection method. in: Operations Research Proceedings 1992, ed. K.-W. Hansmann et al., Springer Verlag, Heidelberg (1993)

[7] Hörmann, W. and Derflinger, G.: A portable uniform random number generator well suited for the rejection method. ACM Transact. on Math. Softw., (to appear)

[8] Hörmann, W.: A note on the quality of random variates generated by the ratio of uniforms method. ACM Transactions on Modeling and Computer Simulation, (to appear)

[9] Knuth, D. E.: The Art of Computer Programming. Vol. II. Addison-Wesley, Menlo Park, London, Amsterdam, Don Mills, Sydney (1981)

[10] Maclaren, N.M.: A limit on the usable length of a pseudorandom sequence. J. Statist. Comput. Simul. 42, 47-54 (1992)

[11] Marsaglia G.: The structure of linear congruential sequences. in: Applications of Number Theory to Numerical Analysis, S. K. Zaremba, ed., Academic Press, New York (1972)

[12] Monahan, F.: Accuracy in random number generation. Math. Comp. 45, 559-568 (1985)

[13] Moran, P.A.P.: The random division of an interval. J. Royal Statist. Soc. (Supplement) 9, 92-98 (1947)

[14] Niederreiter, H.: Quasi-monte carlo methods and pseudo-random numbers. Bulletin of the American Math. Soc. 84, 6, 957-1041 (1978)

ENTSCHEIDUNGEN MIT SEMI- UND NICHTPARAMETRISCHER VORINFORMATION

Stefan Huschens, Universität Heidelberg

Zusammenfassung: Die folgende Entscheidungsstruktur $D := (\mathbf{A}, \mathbf{Z}, u, \mathbf{F}_\Pi, U)$ wird behandelt: **A** und $\mathbf{Z} = \mathbb{R}$ sind die Aktionen- und Zustandsmenge, u: $\mathbf{A} \times \mathbf{Z} \to \mathbb{R}$ ist eine Nutzenfunktion, $\mathbf{F}_\Pi$ charakterisiert die partielle Kenntnis einer A-priori-Verteilung F auf **Z** und U: $\mathbf{A} \times \mathbf{F}_\Pi \to \mathbb{R}$. $U(a,F) = \mathbf{E}[u(a,Z)]$ mit $Z \sim F$ ist die Erwartungsnutzenfunktion.
Eine nicht- bzw. semiparametrische Verteilungsfamilie $\mathbf{F}_\Pi \subseteq \mathbf{F}$, wobei **F** die Menge aller Verteilungsfunktionen bezeichnet, liegt vor, wenn die Wahrscheinlichkeiten $\mathbf{P}\{c_{j-1} < Z \leq c_j\} = \pi_j$ bekannt oder durch $\boldsymbol{\pi} \in \boldsymbol{\Pi}$ mit $\boldsymbol{\Pi} \subseteq \{\mathbf{x} \in \mathbb{R}^m \mid x_j \geq 0, \sum_j x_j = 1\}$ restringiert sind. Dabei ist $-\infty = c_0 < c_1 < \ldots < c_{m-1} < c_m = +\infty$ eine vorgegebene Zerlegung von **Z**.
Verschiedene Fälle führen zur Struktur D: (a) subjektive Wahrscheinlichkeitsschätzung (elicitation); (b) Empirical-Bayes-Schätzungen mit unvollständigen Daten; (c) die statistische Inferenz für F führt zu $\mathbf{F}_\Pi$ und ist von der Entscheidung, der Wahl einer Aktion aus **A**, getrennt.
Es werden Entscheidungskriterien für D, die resultierenden Optimierungsprobleme und für den Fall (b) die Auswirkung von Schätzfehlern diskutiert.

Summary: The following decision structure $D := (\mathbf{A}, \mathbf{Z}, u, \mathbf{F}_\Pi, U)$ is discussed: **A** and $\mathbf{Z} = \mathbb{R}$ are sets of actions and states, u: $\mathbf{A} \times \mathbf{Z} \to \mathbb{R}$ is a utility function, $\mathbf{F}_\Pi$ characterizes the partial knowledge of the prior distribution F on **Z**, and U: $\mathbf{A} \times \mathbf{F}_\Pi \to \mathbb{R}$, $U(a,F) = \mathbf{E}[u(a,Z)]$ with $Z \sim F$ is the expected utility function. A non- or semiparametric distribution family $\mathbf{F}_\Pi \subseteq \mathbf{F}$ with **F** denoting the set of all distribution functions is given, if the probabilities $\mathbf{P}\{c_{j-1} < Z \leq c_j\} = \pi_j$ are known or are restricted by $\boldsymbol{\pi} \in \boldsymbol{\Pi}$ with $\boldsymbol{\Pi} \subseteq \{\mathbf{x} \in \mathbb{R}^m \mid x_j \geq 0, \sum_j x_j = 1\}$. $-\infty = c_0 < c_1 < \ldots < c_{m-1} < c_m = +\infty$ is a given partition of **Z**.
Several cases lead to the structure D: (a) subjective probability estimation (elicitation); (b) empirical Bayes estimation with incomplete data; (c) the statistical inference for F resulting in $\mathbf{F}_\Pi$ is separated from the decision, the choice of an action from **A**.
Decision criteria for D, the resulting optimization problems and for the case (b) the effects of estimation errors are discussed.

1. Semi- und nichtparametrische Vorinformation

A-priori-Restriktionen $\mathbf{P}\{c_{j-1} < Z \leq c_j\} = \pi_j$ für die Verteilung einer reellwertigen Zufallsgröße Z, wobei $-\infty = c_0 < c_1 < \ldots < c_{m-1} < c_m = +\infty$ eine Zerlegung von $\mathbb{R}$ und $\boldsymbol{\pi} = (\pi_1, \pi_2, \ldots, \pi_m)$ ein Element aus

$$S := \{\mathbf{x} \in \mathbb{R}^m \mid x_1 \geq 0,\ldots,x_m \geq 0, \textstyle\sum_j x_j = 1\} \tag{1.1}$$

ist, definieren die Verteilungsfamilie

$$\mathbf{F}_{\boldsymbol{\pi}} := \{F \in \mathbf{F} \mid F(c_j) = s_j,\ j = 1,\ldots,m-1\} \tag{1.2}$$

mit $s_1 = \pi_1$, $s_2 = \pi_1 + \pi_2$, ..., $s_{m-1} = \pi_1 + \pi_2 + \ldots + \pi_{m-1}$. $\mathbf{F}$ ist die Menge aller Verteilungsfunktionen. Im folgenden ist $m \geq 1$ fixiert und $J := \{1,\ldots,m\}$. Bildet man die Intervalle $I_j := (c_{j-1},c_j]$, so ist die Vorinformation $Z \sim F \in \mathbf{F}_{\boldsymbol{\pi}}$ äquivalent zur Kenntnis der Wahrscheinlichkeiten $\mathbf{P}\{Z \in I_j\}$ $(j \in J)$. $\mathbf{F}_{\boldsymbol{\pi}}$ ist - ebenso wie $\mathbf{F}$ - eine nichtparametrische Verteilungsfamilie, da sie nicht durch eine Teilmenge von $\mathbb{R}^d$, sondern nur durch Teilmengen von $\mathbb{R}^\infty$ oder allgemeinere Funktionenräume parametrisiert werden kann.

Vagere Formen des Vorwissens über die Verteilung von Z können durch eine allgemeinere Verteilungsfamilie

$$\mathbf{F}_{\boldsymbol{\Pi}} = \cup_{\boldsymbol{\pi}\in\boldsymbol{\Pi}} \mathbf{F}_{\boldsymbol{\pi}} \quad \text{mit} \quad \boldsymbol{\Pi} \subseteq S \tag{1.3}$$

beschrieben werden. Beispielsweise kann $\boldsymbol{\Pi}$ durch lineare Ungleichungen $\mathbf{B}\boldsymbol{\pi}^{\mathrm{T}} \geq \mathbf{b}^{\mathrm{T}}$ definiert sein,

$$\boldsymbol{\Pi} = \{\boldsymbol{\pi} \in S \mid \mathbf{B}\boldsymbol{\pi}^{\mathrm{T}} \geq \mathbf{b}^{\mathrm{T}}\}. \tag{1.4}$$

$\mathbf{F}_{\boldsymbol{\Pi}}$ ist eine semiparametrische Verteilungsfamilie (vgl. [6]), wobei $\boldsymbol{\pi}$ ein Parametervektor mit Werten in $\boldsymbol{\Pi}$ ist und den parametrischen Teil der Verteilungsfamilie charakterisiert, während $\mathbf{F}_{\boldsymbol{\pi}}$ für jeweils festes $\boldsymbol{\pi}$ eine nichtparametrische Verteilungsfamilie ist. In [5] sind die $\mathbf{F}_{\boldsymbol{\Pi}}$-Verteilungsfamilien im Rahmen der Maximum-Likelihood-Inferenz behandelt.

2. Die Entscheidungsstruktur

Folgende Anwendung der $\mathbf{F}_{\boldsymbol{\Pi}}$-Verteilungsfamilien wird behandelt: $\mathbf{Z} = \mathbb{R}$ ist die *Zustandsmenge* einer Entscheidungsstruktur mit der *Aktionenmenge* $\mathbf{A}$, der *Nutzenfunktion*

$$u\colon \mathbf{A} \times \mathbf{Z} \to \mathbb{R} \tag{2.1}$$

und der *Erwartungsnutzenfunktion*

$$U\colon \mathbf{A} \times \mathbf{F}_{\boldsymbol{\Pi}} \to \mathbb{R},\ U(a,F) = \mathbf{E}[u(a,Z)] = \textstyle\int_{\mathbb{R}} u(a,z)dF(z) \tag{2.2}$$

mit $Z \sim F \in \mathbf{F}_{\boldsymbol{\Pi}}$. $\mathbf{F}_{\boldsymbol{\Pi}}$ (mit dem Spezialfall $\mathbf{F}_{\boldsymbol{\pi}}$, wenn $\boldsymbol{\Pi}$ einelementig ist) charakterisiert die partiell bekannte Verteilung F, d. h. es liegt das Vorwissen $F \in \mathbf{F}_{\boldsymbol{\Pi}}$ vor. Somit ergibt sich die Entscheidungsstruktur

$$D := (\mathbf{A},\mathbf{Z},u,\mathbf{F}_{\boldsymbol{\Pi}},U). \tag{2.3}$$

Damit $U(a,F)$ für alle $(a,F) \in \mathbf{A} \times \mathbf{F}_{\boldsymbol{\Pi}}$ definiert ist, müssen u und $\mathbf{F}_{\boldsymbol{\Pi}}$ bestimmte Anforderungen erfüllen. Hinreichend für die Existenz von U ist z. B., daß die Funktionen $u(a,\cdot)$ stetig und beschränkt sind oder daß die Funktionen $u(a,\cdot)$ stetig sind und $\pi_1 = \pi_m = 0$ gilt.

Verschiedene Fälle führen zu einer Entscheidungsstruktur D:

Fall a: Wenn Wahrscheinlichkeiten auf Teilmengen von $\mathbf{Z}$ durch subjektive Wahrscheinlichkeitsschätzung (elicitation, assessment) festgelegt werden, wobei anzugeben ist, welche Wahrscheinlichkeiten bestimmten Intervallen zukommen, oder welche Relationen zwischen diesen Wahrscheinlichkeiten bestehen, so resultiert als Ergebnis des Schätzverfahrens partielle Wahrscheinlichkeitsinformation der Art $\mathbf{F}_{\boldsymbol{\pi}}$ oder $\mathbf{F}_{\boldsymbol{\Pi}}$.

Fall b: Statistische Beobachtungen X können mit einer Entscheidungsstruktur $(\mathbf{A}, \mathbf{Z}, u, F, U)$ auf verschiedene Arten verknüpft sein. Das Modell der statistischen Entscheidungsfunktionen untersucht Aktionen als Funktionen der Beobachtungen. In Empirical-Bayes-Modellen werden Beobachtungen auch zur Inferenz über die A-priori-Verteilung F verwendet. Ein spezieller Empirical-Bayes-Ansatz liegt vor, wenn aus unabhängigen und identisch verteilten Beobachtungen $Z_i \sim F$ $(i = 1, \ldots, n)$ eine Schätzung für F bestimmt wird und diese als Bayessche A-priori-Verteilung auf $\mathbf{Z}$ verwendet wird. Hieraus entsteht eine Entscheidungsstruktur mit partieller Wahrscheinlichkeitsinformation über die A-priori-Verteilung, wenn folgende Situation unvollständiger Daten vorliegt: anstelle der Z_i kann jeweils nur beobachtet werden, in welchem Intervall I_j die Beobachtungen Z_i liegen. Dann ist die gesamte Stichprobeninformation in einer Realisation $\mathbf{y}$ von $\mathbf{Y} = (Y_1, \ldots, Y_m) \sim \mathrm{MULT}_m(n, \boldsymbol{\pi})$ enthalten. Die relativen Häufigkeiten $\mathbf{p} = n^{-1}\mathbf{y} \in S$ als Punktschätzung für $\boldsymbol{\pi}$ führen dann zur Verteilungsklasse $\mathbf{F}_{\mathbf{p}}$ als partieller Wahrscheinlichkeitsinformation für eine Bayessche Verteilung auf $\mathbf{Z}$. Bereichsschätzungen für $\boldsymbol{\pi}$ führen zu partieller Wahrscheinlichkeitsinformation der Art $\mathbf{F}_{\boldsymbol{\Pi}}$.

Fall c: Ist die eigentliche ökonomische Entscheidung (Wahl einer Aktion) von der statistischen Inferenzentscheidung institutionell getrennt, so resultiert die Entscheidungsstruktur D, wenn in einer ersten Phase F aufgrund von Beobachtungen X geschätzt wird und in die zweite Phase eine Bereichsschätzung der Form $\mathbf{F}_{\boldsymbol{\pi}}$ oder $\mathbf{F}_{\boldsymbol{\Pi}}$ für F eingeht.

3. Entscheidungskriterien und Optimierungsprobleme

Bei bekannter Bayesscher A-priori-Verteilung F ist $(\mathbf{A}, \mathbf{Z}, u, F, U)$ eine Entscheidungsstruktur bei Risiko. Entsprechend kann man $(\mathbf{A}, \mathbf{Z}, u, \mathbf{F}_{\boldsymbol{\Pi}}, U)$ als eine Entscheidungsstruktur mit unscharf spezifizierter A-priori-Verteilung (partieller Wahrscheinlichkeitsinformation) interpretieren. In [1] wird ein Überblick über Entscheidungskriterien für Entscheidungsstrukturen mit einer partiell bekannten Wahrscheinlichkeitsverteilung auf einer endlichen Zustandsmenge gegeben. Diese Entscheidungskriterien lassen sich teilweise direkt und teilweise analog auf die hier untersuchte Entscheidungsstruktur mit unendlicher Zustandsmenge übertragen. Die meisten Entscheidungskriterien führen zu dem Problem, für jede Aktion a den minimalen und/oder den maximalen Erwartungsnutzen

oder den Bereich möglicher Erwartungsnutzen zu bestimmen. Hier seien nur das MaxEmin-Kriterium und das Dominanzkriterium erwähnt.
Das MaxEmin-Kriterium (in der statistischen Entscheidungstheorie auch als Γ-Minimax-Kriterium bekannt) bezeichnet bei gegebener Entscheidungsstruktur D eine Aktion als optimal, die den minimalen Erwartungsnutzen maximiert; dies entspricht der Bestimmung einer Maximin-Aktion (analog einer Minimax-Aktion bei Verlustfunktion) in der Entscheidungsstruktur bei Ungewißheit

$$D^* = (\mathbf{A}, \mathbf{Z}^*, U), \tag{3.1}$$

die man erhält, wenn man $\mathbf{Z}^* = \mathbf{F}_{\mathbf{\Pi}}$ als neue Zustandsmenge interpretiert. Für jede vorgegebene Aktion a bezeichne

$$U(a, \mathbf{F}_{\mathbf{\Pi}}) := \{U(a, F) \mid F \in \mathbf{F}_{\mathbf{\Pi}}\} \tag{3.2}$$

die Menge aller mit $\mathbf{F}_{\mathbf{\Pi}}$ verträglichen Erwartungsnutzenwerte,

$$U_a := \inf U(a, \mathbf{F}_{\mathbf{\Pi}}) \quad \text{und} \quad U^a := \sup U(a, \mathbf{F}_{\mathbf{\Pi}}) \tag{3.3}$$

bezeichnen den minimalen und maximalen Erwartungsnutzen. Die Bestimmung von U_a und U^a kann als zweistufiger Minimierungs- bzw. Maximierungsprozeß durchgeführt werden, indem man zunächst die minimalen und maximalen Nutzenwerte innerhalb der Intervalle I_j bestimmt,

$$u_{aj} := \inf \{u(a,x) \mid x \in I_j\}, \quad u^a{}_j := \sup \{u(a,x) \mid x \in I_j\},$$

und dann U_a und U^a mit Hilfe der Vektoren

$$\mathbf{u}_a := (u_{a1}, \ldots, u_{am}), \quad \mathbf{u}^a := (u^a{}_1, \ldots, u^a{}_m)$$

aus

$$U_a = \inf \{\mathbf{u}_a \boldsymbol{\pi}^{\mathrm{T}} \mid \boldsymbol{\pi} \in \mathbf{\Pi}\}, \quad U^a = \sup \{\mathbf{u}^a \boldsymbol{\pi}^{\mathrm{T}} \mid \boldsymbol{\pi} \in \mathbf{\Pi}\}$$

bestimmt. Wenn $\mathbf{\Pi}$ kompakt ist, gilt

$$U_a = \min \{\mathbf{u}_a \boldsymbol{\pi}^{\mathrm{T}} \mid \boldsymbol{\pi} \in \mathbf{\Pi}\}, \quad U^a = \max \{\mathbf{u}^a \boldsymbol{\pi}^{\mathrm{T}} \mid \boldsymbol{\pi} \in \mathbf{\Pi}\}.$$

Das Dominanzkriterium beruht darauf, daß $b \in \mathbf{A}$ durch $a \in \mathbf{A}$ dominiert wird, falls

$$\Delta(a,b) := \inf \{U(a,F) - U(b,F) \mid F \in \mathbf{F}_{\mathbf{\Pi}}\}$$

nichtnegativ ist. Bei Dominanzuntersuchungen ist daher

$$\Delta(a,b) = \inf \{\mathbf{d}\boldsymbol{\pi}^{\mathrm{T}} \mid \boldsymbol{\pi} \in \mathbf{\Pi}\}$$

mit $\mathbf{d} = (d_1, \ldots, d_m) = \mathbf{d}(a,b)$,

$$d_j := \inf \{u(a,x) - u(b,x) \mid x \in I_j\}$$

für je zwei Aktionen zu bestimmen.
Wenn $\mathbf{\Pi}$ durch lineare Ungleichungen charakterisiert ist, führt daher die Bestimmung von U_a, U^a oder $\Delta(a,b)$ zu linearen Optimierungsproblemen mit der Zielfunktion $\mathbf{c}\boldsymbol{\pi}^{\mathrm{T}}$ ($\mathbf{c} \in \mathbb{R}^m$) und der Restriktionsmenge (1.4):

$$\max/\min \{\mathbf{c}\boldsymbol{\pi}^{\mathrm{T}} \mid \boldsymbol{\pi} \in \mathbf{\Pi}\} \quad \text{mit} \quad \mathbf{\Pi} = \{\boldsymbol{\pi} \in S \mid \mathbf{B}\boldsymbol{\pi}^{\mathrm{T}} \geq \mathbf{b}^{\mathrm{T}}\}. \tag{3.4}$$

Allgemein gilt $U(a, \mathbf{F}_{\mathbf{\Pi}}) \subseteq [U_a, U^a]$. Falls $\mathbf{\Pi}$ konvex ist, wenn z. B. der Fall (3.4) vorliegt, ist $U(a, \mathbf{F}_{\mathbf{\Pi}})$ ein Intervall.

4. Extreme Erwartungsnutzen bei geordneten Wahrscheinlichkeiten

Im folgenden wird unterstellt, daß $\mathbf{\Pi}$ durch mehrere Ungleichungen der Form $\pi_i \geq \pi_j$ charakterisiert ist, und es wird die Bestimmung der maximalen und minimalen Erwartungsnutzen U_a und U^a diskutiert.

$\mathbf{\Pi}$ ist in diesem Fall darstellbar als Durchschnitt endlich vieler Mengen

$$\mathbf{\Pi}(i,j) := \{\boldsymbol{\pi} \in S \mid \pi_i \geq \pi_j\} \tag{4.1}$$

mit $(i,j) \in J^2 := J \times J$. Jede Kombination von Ungleichungen ist eindeutig durch eine nichtleere Teilmenge $\mathbf{M}$ von J^2 charakterisiert, wenn die Vereinbarung

$$\mathbf{\Pi}(\mathbf{M}) := \cap \{\mathbf{\Pi}(i,j) \mid (i,j) \in \mathbf{M}\} \tag{4.2}$$

getroffen wird. Es gilt dann z. B. $\mathbf{\Pi}(j,j) = S$ für alle $j \in J$, da $\pi_j \geq \pi_j$ immer erfüllt ist; die Gleichheit von zwei Komponenten π_i und π_j ist durch $\mathbf{M}_{ij} := \{(i,j),(j,i)\}$ charakterisiert,

$$\mathbf{\Pi}(\mathbf{M}_{ij}) = \{\boldsymbol{\pi} \in S \mid \pi_i = \pi_j\}, \tag{4.3}$$

und $\mathbf{\Pi}(J^2)$ enthält als einziges Element die Gleichverteilung $(1/m, 1/m, \ldots, 1/m)$. Durch die Menge

$$\mathbf{O} := \{(1,2),(2,3),\ldots,(m-1,m)\} \tag{4.4}$$

ist eine vollständige Ordnung der Wahrscheinlichkeiten charakterisiert,

$$\mathbf{\Pi}(\mathbf{O}) = \{\boldsymbol{\pi} \in S \mid \pi_1 \geq \pi_2 \geq \ldots \geq \pi_m\}. \tag{4.5}$$

Für jede nichtleere Teilmenge $\mathbf{M}$ von J^2 ist $\mathbf{\Pi}(\mathbf{M})$ eine kompakte Teilmenge von S und bildet ein konvexes Polytop (beschränktes konvexes Polyeder) mit endlicher Eckpunktemenge $\mathbf{\Pi}^*$. Eine lineare Zielfunktion $\mathbf{c}\boldsymbol{\pi}^T$ nimmt ihr Maximum (analog das Minimum) in einem Eckpunkt an,

$$\max \{\mathbf{c}\boldsymbol{\pi}^T \mid \boldsymbol{\pi} \in \mathbf{\Pi}\} = \max \{\mathbf{c}\boldsymbol{\pi}^T \mid \boldsymbol{\pi} \in \mathbf{\Pi}^*\}. \tag{4.6}$$

Im Fall $\mathbf{\Pi}(\mathbf{M})$ ist es einfacher, $\mathbf{\Pi}^*$ zu bestimmen und dann (4.6) zu verwenden, als für jede Aktion (3.1) zu lösen. Zusammen mit den Überlegungen aus dem dritten Abschnitt folgt unmittelbar der folgende Satz, der die Bestimmung von U_a und U^a auf die Bestimmung von $\mathbf{u}_a$ und $\mathbf{u}^a$, d. h. die Lösung von $2m$ gewöhnlichen Optimierungsproblemen, sowie auf das Problem der Bestimmung von $\mathbf{\Pi}^*$ zurückführt.

Satz 1: Es sei $\mathbf{M} \subseteq J^2$, $\mathbf{M} \neq \emptyset$. $\mathbf{\Pi}^*$ sei die Eckpunktemenge von $\mathbf{\Pi} = \mathbf{\Pi}(\mathbf{M})$. Dann gilt

$$U_a = \min \{\mathbf{u}_a\boldsymbol{\pi}^T \mid \boldsymbol{\pi} \in \mathbf{\Pi}^*\}, \quad U^a = \max \{\mathbf{u}^a\boldsymbol{\pi}^T \mid \boldsymbol{\pi} \in \mathbf{\Pi}^*\}.$$

Für $\mathbf{\Pi}(\mathbf{O})$ ist wohlbekannt (siehe z.B. [3] und die Quellen in [2]), daß die Eckpunkte gerade die Verteilungen $(1,0,\ldots,0)$, $(1/2,1/2,0,\ldots,0)$, $(1/3,1/3,1/3,0,\ldots,0)$, ..., $(1/m,1/m,\ldots,1/m)$ sind. Auch für den allgemeinen Fall $\mathbf{M} \subseteq J^2$ ist eine relativ einfache Ermittlung aller Eckpunkte von $\mathbf{\Pi}(\mathbf{M})$ möglich. Wenn $\mathbf{M}$ als gerichteter Graph interpretiert wird, läßt sich zeigen (vgl. z. B. [2]), daß ein Eckpunkt gerade dadurch charakterisiert ist, daß der zugehörige gerichtete Teilgraph zusammenhängend ist und zu jedem Element die Vorgänger enthält. Z. B. lassen sich für $m = 5$ und $\mathbf{M} = \{(1,3),(2,3),(5,4)\}$ die Eckpunkte $(1,0,0,0,0)$, $(0,1,0,0,0)$, $(1/3,1/3,1/3,0,0)$, $(0,0,0,0,1)$ und $(0,0,0,1/2,1/2)$ von $\mathbf{\Pi}(\mathbf{M})$ unmittelbar angeben.

5. Zur Fortpflanzung des statistischen Schätzfehlers

Es wird der *Fall b* von Abschnitt 2 mit $\mathbf{F_\Pi} = \mathbf{F_p}$ vorausgesetzt. Wird eine optimale Aktion für $(\mathbf{A},\mathbf{Z},u,\mathbf{F_p},U)$ bestimmt, so werden implizit Funktionen der Form $\mathbf{c}\boldsymbol{\pi}^{\mathrm{T}}$ (mit $\mathbf{c} \in \mathbb{R}^m$) durch $\mathbf{c}\mathbf{p}^{\mathrm{T}} = n^{-1}\mathbf{c}\mathbf{y}^{\mathrm{T}}$ geschätzt.
Beispielsweise ist $U_a(\mathbf{p}) = \mathbf{u}_a\mathbf{p}^{\mathrm{T}} = n^{-1}\mathbf{u}_a\mathbf{y}^{\mathrm{T}}$ eine Realisation der reellwertigen Zufallsvariablen $U_a(n^{-1}\mathbf{Y}) = n^{-1}\mathbf{u}_a\mathbf{Y}^{\mathrm{T}}$ und ein Schätzwert für $U_a = \mathbf{u}_a\boldsymbol{\pi}^{\mathrm{T}}$; analog ist $U^a(n^{-1}\mathbf{Y}) = n^{-1}\mathbf{u}^a\mathbf{Y}^{\mathrm{T}}$ ein Schätzer für U^a; $U_a(\mathbf{p}) - U_b(\mathbf{p}) = (\mathbf{u}_a - \mathbf{u}_b)\mathbf{p}^{\mathrm{T}}$ ist ein Schätzwert für $U_a - U_b$, der Differenz der Erwartungsnutzenminima von zwei Aktionen a und b.
Um die Auswirkung des Schätzfehlers zu untersuchen, kann die Punktschätzung $\mathbf{p}$ durch eine Bereichsschätzung ersetzt werden. Das Konfidenzellipsoid

$$\mathbf{\Pi} = \mathbf{\Pi}(\mathbf{p},n,r) = \{\mathbf{x} \in S \mid n(\mathbf{x} - \mathbf{p})\mathbf{D_p}^{-1}(\mathbf{x} - \mathbf{p})^{\mathrm{T}} \le r^2\} \tag{5.1}$$

mit

$$\mathbf{D_p} = \mathrm{diag}(\mathbf{p})$$

basiert auf einem Vektor beobachteter relativer Häufigkeiten $\mathbf{p} > 0$ und der Neymanschen Chiquadrat-Statistik (oder Wald-Statistik)

$$W := n(\boldsymbol{\pi} - \mathbf{p})\mathbf{D_p}^{-1}(\boldsymbol{\pi} - \mathbf{p})^{\mathrm{T}}.$$

n ist der Stichprobenumfang. (Mit $\mathbf{D}_{\boldsymbol{\pi}}$ anstatt $\mathbf{D_p}$ erhält man die Pearsonsche Chiquadrat-Statistik X^2.) Der Konfidenzradius $r = r(\alpha) > 0$ legt die asymptotische Überdeckungswahrscheinlichkeit

$$\alpha = \mathbf{P}\{\boldsymbol{\pi} \in \mathbf{\Pi}\}$$

fest; für ein vorgegebenes α ist r^2 als α-Quantil einer Chiquadratverteilung mit $m - 1$ Freiheitsgraden, der asymptotischen Verteilung von W, zu wählen. Ein Konfidenzintervall zum asymptotischen Konfidenzniveau α für eine Linearkombination $\mathbf{c}\boldsymbol{\pi}^{\mathrm{T}}$ kann man konstruieren, indem man die Extrema der linearen Zielfunktion $\mathbf{c}\mathbf{x}^{\mathrm{T}}$ auf $\mathbf{\Pi}$ bestimmt (siehe [4]):

$$\mathbf{c}\boldsymbol{\pi}^{\mathrm{T}} \in \mathbf{c}\mathbf{p}^{\mathrm{T}} \pm r(n^{-1}\mathbf{c}\mathbf{V_p}\mathbf{c}^{\mathrm{T}})^{1/2} \tag{5.2}$$

mit

$$\mathbf{V_p} := \mathrm{diag}(\mathbf{p}) - \mathbf{p}^{\mathrm{T}}\mathbf{p}.$$

Zu vorgegebenem α erhält man mit einem gleich großen Koeffizienten $r(\alpha)$ zu einer beliebigen Menge $\mathbf{C} \subseteq \mathbb{R}^m$ simultane Konfidenzaussagen der Form (5.2) für alle Linearkombinationen $\mathbf{c}\boldsymbol{\pi}^{\mathrm{T}}$ mit $\mathbf{c} \in \mathbf{C}$.
Alternativ kann man $T := n^{-1}\mathbf{c}\mathbf{Y}^{\mathrm{T}}$ auch als arithmetisches Mittel von Beobachtungen unabhängiger und identisch verteilter Zufallsvariablen X_i ($i = 1,\ldots,n$) mit $\mathbf{P}\{X_i = c_j\} = \pi_j$ interpretieren. Dann ist die Stati-

stik T als Schätzer für $\mathbf{c}\boldsymbol{\pi}^{\mathrm{T}} = \mathbf{E}[X_i]$ erwartungstreu, konsistent, effizient und asymptotisch normalverteilt. Es gilt

$$\mathbf{E}[T] = \mathbf{c}\boldsymbol{\pi}^{\mathrm{T}} \quad \text{und} \quad \mathbf{Var}(T) = n^{-1}\mathbf{c}\mathbf{V}_{\boldsymbol{\pi}}\mathbf{c}^{\mathrm{T}} \quad \text{mit} \quad \mathbf{V}_{\boldsymbol{\pi}} := \mathrm{diag}(\boldsymbol{\pi}) - \boldsymbol{\pi}^{\mathrm{T}}\boldsymbol{\pi},$$

wobei $\mathbf{V}_{\boldsymbol{\pi}}$ die Kovarianzmatrix eines $\mathrm{MULT}_m(1,\boldsymbol{\pi})$-verteilten Zufallsvektors ist. $\mathbf{V}_{\boldsymbol{\pi}}$ kann durch $\mathbf{V}_{\mathbf{p}}$ und $\mathbf{Var}(T)$ durch $n^{-1}\mathbf{c}\mathbf{V}_{\mathbf{p}}\mathbf{c}^{\mathrm{T}}$ geschätzt werden. Daher ist durch

$$\mathbf{c}\boldsymbol{\pi}^{\mathrm{T}} \in \mathbf{c}\mathbf{p}^{\mathrm{T}} \pm z(n^{-1}\mathbf{c}\mathbf{V}_{\mathbf{p}}\mathbf{c}^{\mathrm{T}})^{1/2} \tag{5.3}$$

ein asymptotisches Konfidenzintervall für $\mathbf{c}\boldsymbol{\pi}^{\mathrm{T}}$ gegeben, wobei z^2 das α-Quantil einer Chiquadratverteilung mit einem Freiheitsgrad ist, da $(T - \mathbf{c}\boldsymbol{\pi}^{\mathrm{T}})(n^{-1}\mathbf{c}\mathbf{V}_{\mathbf{p}}\mathbf{c}^{\mathrm{T}})^{-1/2}$ asymptotisch standardnormalverteilt ist.

Die Intervalle aus (5.2) und (5.3) weisen die gleiche formale Struktur auf. (5.3) ergibt für eine einzelne Konfidenzaussage im Vergleich zu (5.2) ein kleineres Konfidenzintervall. Werden simultane Konfidenzintervalle mit (5.3) und der Bonferroni-Ungleichung konstruiert, so nimmt z mit wachsender Anzahl der Linearkombinationen zu, während r konstant bleibt, so daß für eine große Anzahl simultan zu vergleichender Linearkombinationen (5.2) vorteilhaft ist.

Literatur:

[1] Brachinger, H. W.
Entscheidungskriterien bei partieller Wahrscheinlichkeitsinformation - Eine klassifikatorische Übersicht.
Operations Research Proceedings 1992, 407-413 (1993)

[2] Bühler, W.
Characterizing extreme points of polyhedra.
Zeitschrift für Operations Research 19, 131-137 (1975)

[3] Fishburn, P. C.
Analysis of decisions with incomplete knowledge of probabilities.
Operations Research 13, 217-237 (1965)

[4] Huschens, S.
Entscheidungen mit partieller Information über die Wahrscheinlichkeiten einer Zerlegung der Zustandsmenge.
Diskussionsschrift Nr. 197 der Wirtschaftswissenschaftlichen Fakultät der Universität Heidelberg (1993)

[5] Huschens, S.; Stahl G.
Nichtparametrische Maximum-Likelihood-Inferenz mit A-priori-Restriktionen.
Operations Research Proceedings 1992, 374-381 (1993)

[6] Wellner, J. A.
Semiparametric models: progress and problems.
Bulletin of the International Statistical Institute 51 (4), 23.1.1.-23.1.20 (1985)

EIN NEUES PRINZIP ZUR BILDUNG VON VALIDATIONSKRITERIEN FÜR DIE SELEKTION VON MODELLSTRUKTUREN BEI UNGENÜGENDER DATENQUALITÄT

Tatjana Lange, Fachhochschule für Technik und Wirtschaft Berlin

Zusammenfassung: Im Beitrag werden Probleme betrachtet, die bei der Bildung von Strukturselektionskriterien für neuronale Algorithmen der Methode der gruppenweisen Erfassung der Argumente auftreten können. Es werden die Mehrkriterialität der Aufgaben der Parameterschätzung und der Strukturselektion diskutiert, speziell das Problem der Wechselbeziehungen zwischen der Dispersion und der Verschiebung der Modelle bei Datenunsicherheit. Es wird eine neue Strategie zur Bildung von Strukturselektionskriterien bei ungenügender Datenqualität vorgeschlagen.

Summary: This contribution deals with problems which may occur at the construction of structure selection criteria of neuronal algorithms of Group Method of Data Handling. The multicriteriality of tasks of parametric estimation and of structure selection is discussed, in particular the problem of correlation between dispersion and bias of models in case of data uncertainty. A new strategy to construct structure selection criteria in case of insufficient data quality is proposed.

1. Probleme bei der Bildung mathematischer Modelle mit neuronalen GMDH-Algorithmen

(a) Neuronale Algorithmen zur Modellierung komplizierter stochastischer Prozesse und Systeme, die auf dem Prinzip der Selbstorganisation basieren und zu denen auch die mehrstufigen Algorithmen der Methode der gruppenweisen Erfassung der Argumente (Group Method of Data Handling - GMDH) [1, 2, 3] gehören, lösen zwei Aufgaben:

- die Parameterschätzung und
- die Selektion der Modellstruktur.

Die Struktur des Modells wird in Form der Regression (bzw. Autoregression) anhand der gemessenen Daten der Ein- und Ausgangsvariablen gesucht, wobei auf den Ausgang des Objektes stochastische Störungen wirken.

Dabei ist der Entscheidungsbaum für die Strukturselektion so gestaltet, daß die Modellstruktur von Stufe zu Stufe komplizierter wird, bis eine sogenannte optimale Kompliziertheit erreicht wird, bei der das Strukturselektionskriterium sein Minimum besitzt. Offensichtlich sind bei einer solchen Art der Modellbildung die zu einfachen Modelle auf den unteren Stufen der Selektion **nicht erwartungstreu schätzbar**, was mathematisch leicht gezeigt werden kann [4]. Dadurch wird das Problem der Strukturauswahl besonders kompliziert.

(b) Die Aufgabe der Parameterschätzung und ebenso die Aufgabe der Strukturauswahl sind im allgemeinen Fall **zweikriterielle** Aufgaben. Dabei ist jede dieser zweikriteriellen Aufgaben im gewissen Sinne **spiegelsymmetrisch** zur anderen, so daß **unterschiedliche Strategien zur Bildung der Modellierungskriterien** erforderlich sind.

(c) Des weiteren führt auch die **Existenz von Unsicherheit** bei schlechter Datenqualität [5, 6] zur Änderung der Strategie für die Bildung der Strukturselektionskriterien. Dabei ist die Bewertung des Grades der Unsicherheit der Daten selbst auch keine triviale Aufgabe.

2. Mehrkriterialität der Aufgaben der Parameterschätzung und der Strukturauswahl. Die Wechselwirkung dieser Aufgaben

Es soll kurz gezeigt werden, daß die Aufgaben der Parameterschätzung und der Strukturauswahl im statistischen Sinne gegensätzliche Ziele haben.

Die Aufgabe der Strukturauswahl ist im allgemeinen immer eine zweikriterielle Aufgabe, da bei der Auswahl des Modells sowohl die Minimierung der Dispersion des Modells als auch die Minimierung der Verschiebung (Bias, Verzerrung) des Modells wünschenswert sind, was sowohl für die Interpolation und noch mehr für die Extrapolation wichtig ist. Der Zusammenhang zwischen der Verschiebung, der Dispersion und einer fehlerhaften Modellstruktur wird in [4] näher erläutert.

In [4] wurde am Beispiel eines Knotens im Entscheidungsbaum des neuronalen Algorithmus der Strukturselektion gezeigt, wie "untermodellierte" (zu einfache) Modellstrukturen zur Verschiebung führen, was bei einer hinreichend großen Datenstichprobe ein schwerwiegenderer Mangel ist als die Vergrößerung der Dispersion, die mit der Überkompliziertheit der Modelle korrespondiert.

Die klassische Methode des statistischen Hypothesentests wie die Neyman-Pearson-Aufgabe führt die zweikriterielle Aufgabe auf eine einkriterielle Aufgabe zurück, indem durch die Vorgabe einer festen Irrtumswahrscheinlichkeit η (Fehler 1.Art) nur der Fehler 2.Art minimiert wird.

Verwendet man nun die Begriffe "untermodellierte" (zu einfache), "adäquate" (bzw. "wahre", "echte") und "übermodellierte" (zu komplizierte) Strukturen, so zeigt sich, daß bei der klassischen Herangehensweise der Grad des Kompromisses zwischen der Minimierung der Dispersion und der Minimierung der Verschiebung des Modells festgeschrieben wird, da die Wahrscheinlichkeit der Auswahl einer untermodellierten Struktur minimiert wird.

Das kann durch die Minimierung des Funktionals

$$P = P_1 + P_2 \rightarrow \min.$$

ausgedrückt werden, wobei hier

$$P_1 = R^2, \quad P_2 = \gamma D$$

mit

R - Verschiebung des Modells,

D - Dispersion des Modells,

γ - eine mit der Irrtumswahrscheinlichkeit η eindeutig zusammenhängende Größe

ist.

Bei der Schätzung der Koeffizienten wird gewöhnlich nur die Dispersion der Schätzung minimiert. Die Annahme, daß die Struktur der Regression adäquat und die Störung eine Gauß-Markov-Störung ist, führt dazu, daß die Aufgabe zu einer einkriteriellen Aufgabe wird.

Im Fall solcher Mängel bezüglich der Daten wie z.B. zu kurze Beobachtungsintervalle, schlechte Determiniertheit der Fehlermatrix $(X^T X)$ u.a. (vgl. [5]) wird ein zweites Kriterium in Form des Lagrange-Multiplikators eingeführt (Ridge-Schätzungen bzw. Ridge-ähnliche Schätzungen) [7]. Aber in diesen Fall ist im Funktional

$$P = P_1 + P_2 \rightarrow \min$$
$$P_1 = D$$
$$P_2 = \varphi(R)$$

und im Unterschied zum statistischen Hypothesentest wird die Dispersion minimiert, während die Funktion $\varphi(R)$, die mit der Verschiebung des Modells verbunden ist, fixiert ist.

Im Fall robuster Schätzungen bei Wirken von Outlier-Störungen wird die Verschiebung nur durch Veränderung des ersten Funktionals P_1 ohne Einführung des zweiten Funktionals P_2 berücksichtigt.

Die Zweikriterialität der Aufgaben der Koeffizientenschätzung und der Strukturauswahl kann graphisch in Form der Summe der Abstände zweier Modelle, gemessen, einerseits, anhand ihrer Ausgänge und, andererseits, anhand ihrer Koeffizienten, dargestellt werden.

In Abb.1 stellen Q_I und Q_{II} zwei MKQ-Funktionale (MKQ - Methode der kleinsten Quadrate) im vollständigen Koeffizientenraum a_0, a_1, ... a_{M+1} dar, die zwei unterschiedlichen Tabellen (Stichproben) der gemessenen Ein- und Ausgangsdaten des zu modellierenden Objektes entsprechen.

Der Abstand der Modelle, die aus verschiedenen Stichproben bestimmt wurden, drückt sich nicht nur im Abstand zwischen den geschätzten Modellausgängen $\rho_l = (\mathbf{y}_I, \mathbf{y}_{II})$ aus, was dem Unterschied der Dispersionen $\rho = (Q^*_I, Q^*_{II})$ entspricht, sondern auch im Abstand der Koeffizientenvektoren $\rho_{M+1}(\hat{\mathbf{a}}^*_I, \hat{\mathbf{a}}^*_{II})$, welcher die unterschiedliche Verschiebung jedes der Modelle widerspiegelt.

Der Abstand der Modelle ist: $P = P_I + P_{II} = \rho\ (Q^*_I, Q^*_{II}) + \rho_{M+1}(\hat{\mathbf{a}}^*_I, \hat{\mathbf{a}}^*_{II})$

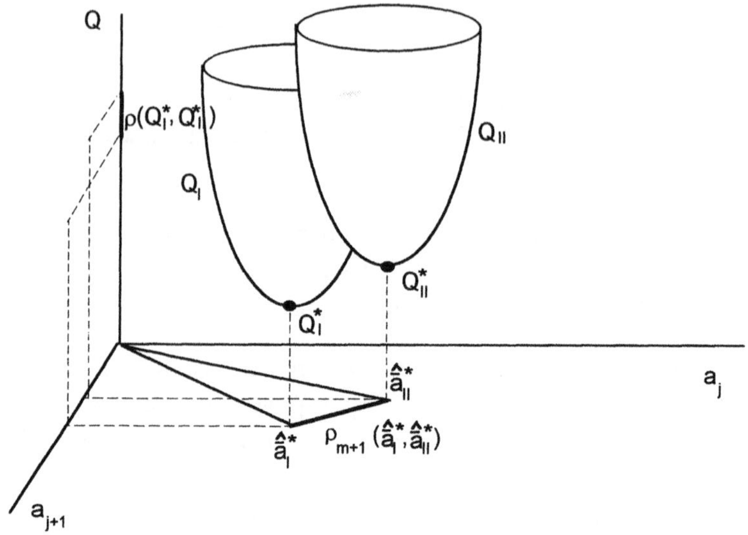

Abb. 1: Abstand zwischen zwei aus unterschiedlichen Stichproben gebildeten Modellen

3. Das Problem der Wechselwirkung der Dispersion und der Verschiebung bei Existenz von Unsicherheit

Es ist bekannt, daß beim statistischen Hypothesentest die Wahl einer zu kleinen Irrtumswahrscheinlichkeit η zum Verlust wichtiger Regressionssummanden führen kann, d.h. zur Simplifizierung des Modells. In der Sprache der Polyoptimierung kann man das dadurch erklären, daß sich mit der Verschärfung von Einschränkungen bezüglich der Minimierung des Fehlers 1.Art der Wert des Fehlers 2.Art (entspricht der Verschiebung) von seinem Minimum weiter wegbewegt.

Andererseits zeigen die Experimente, daß das klassische Schema der Minimierung des Fehler 2.Art bei kurzen Stichproben zu überkomplizierten Modellen führt. Folglich kann bei kleinem Aussteuerungsbereich eine Regression höheren Grades ausgewählt werden, die stark den stochastischen Störungen folgt und nur gering das reale Objekt widerspiegelt (Abb. 2).

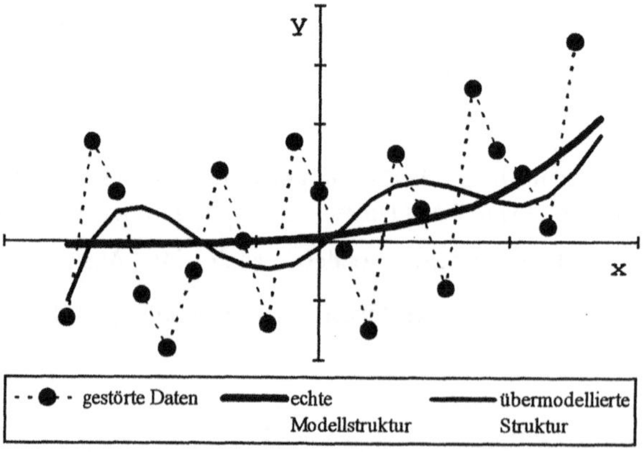

Abb. 2: Vergleich einer echten und einer übermodellierten Struktur

Es drängt sich die Schlußfolgerung auf, daß die Strategien der Strukturselektion mit neuronalen Algorithmen für lange Stichproben einerseits und für kurze Stichproben andererseits unterschiedlichen Charakter haben müssen. Des weiteren hängt die Unsicherheit nicht nur von der Länge der Stichprobe und der Größe des Aussteuerungsbereiches ab, sondern auch von anderen Bedingungen, wie z.B. der Art und der Intensität der Störung, der konkreten Form der Matrix $(X^T X)$ und anderen [5].

Es stellen sich folgende zwei Fragen:

1. Ist es möglich, die Unsicherheit anhand der gemessenen Daten zu bewerten ?
2. Ist es möglich, in Abhängigkeit vom Grad der Unsicherheit den Wert der Irrtumswahrscheinlichkeit η einzustellen, oder, in der Sprache der Polyoptimierung formuliert, das Verhältnis zwischen den zwei Strukturselektionskriterien $P_1 = \rho(\mathbf{y}_I, \mathbf{y}_{II})$ und $P_2 = \rho(\mathbf{a}_I, \mathbf{a}_{II})$ zu regulieren ?

4. Bildung von Strukturselektionskriterien bei ungenügender Datenqualität

Der Grad der Unsicherheit (bzw. die Datenqualität) läßt sich z.B. mit Hilfe des "**Criterion of Data Reliability**" (**CDR**) bestimmen, das in [6] definiert wurde.

Die Bestimmung des quantitativen Zusammenhangs zwischen der Qualität der Daten und der Größe η ist jedoch kompliziert, da die klassische Herangehensweise als Minimierung der Dispersion und der Verschiebung des gesuchten Modells gegenüber dem s.g. idealen Modell interpretiert werden kann, wobei man von einem idealen Modell nur bei Vorliegen einer hinreichenden Menge von Daten sprechen kann. Das läßt sich am Beispiel der Werte eines der Koeffizienten eines Modells mit adäquater Struktur bei unterschiedlicher Anzahl der gemessenen Daten zeigen (s. Abb.3).

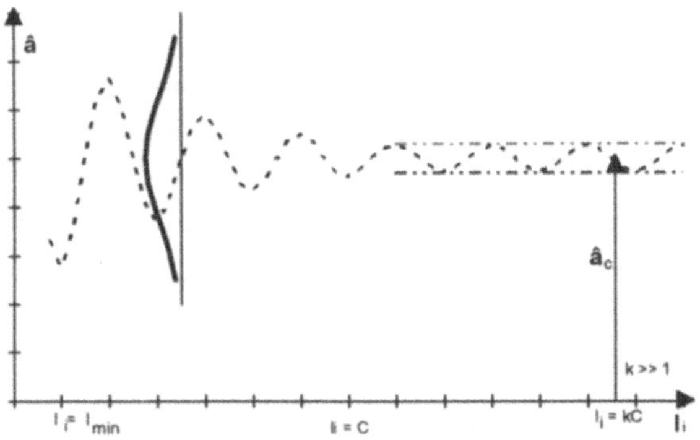

Abb. 3: Änderung der Koeffizientenverteilung in Abhängigkeit von der Stichprobenlängen

Ab einer Stichprobenlänge von $l_i = kC$ bleibt die experimentelle Verteilung der Koeffizienten unverändert und das klassische Schema ist anwendbar.
Bei einer Stichprobenlänge $l_i < C$ ändert die Verteilungsdichte ihre Parameter, was nicht nur bei einer kleinen Anzahl von Daten zu beobachten ist, sondern auch bei Wirken von Störungen, die nicht den Gauß-Markov-Bedingungen gehorchen.

Deshalb ist es sinnvoll, immer dann nicht die klassischen Herangehensweisen anzuwenden (z.B. die Strukturkriterien von Fisher, Akaike), wenn das CDR eine Datenunsicherheit anzeigt. Statt dessen sollten Validationskriterien verwendet werden. Validationskriterien lassen sich als Kriterien interpretieren, die Modelle mit der geringsten Empfindlichkeit gegenüber einer Datenänderung auswählen.

Ausgehend von den Ausführungen über die Zweikriterialität der Aufgaben der Parameterschätzung und der Strukturauswahl erhebt sich hier auch die Frage nach einer geeigneten Kombination der Kriterien. Die einfache Summierung mit Gewichtskoeffizienten, wie sie im "Kriterium der minimalen Verschiebung" und im Regularitätskriterium verwendet wurden [1, 7], erbrachte keine entscheidende Verbesserung.

Für neuronale Algorithmen der Strukturselektion vom GMDH-Typ, bei denen von Stufe zu Stufe des Selektionsbaums partielle Zwischenmodelle geschätzt werden, die äußerlich die gleiche Form besitzen[1], soll nun folgendes Prinzip für die Bildung des Selektionskriteriums vorgeschlagen werden:

$$P = P_1 + P_2 = Sp\left[\rho_l(\mathbf{y}_B, \hat{\mathbf{y}}_{B(A)}) \cdot (X_B^T X_B + \alpha \cdot I)^{-1}\right] + \rho_h(\hat{\mathbf{a}}_A, \hat{\mathbf{a}}_B)$$

Hier sind

A, B	-	Anzahl der Datenpunkte in zwei Stichproben A und B,
$\hat{\mathbf{a}}_A, \hat{\mathbf{a}}_B$	-	Koeffizientenvektoren,, die aus den verschiedenen Datenstichproben berechnet wurden,
$\mathbf{y}_B$	-	gemessener Ausgang aus der Stichprobe B,
$\hat{\mathbf{y}}_{B(A)}$	-	geschätzter Ausgang mit den Eingangsdaten der Stichprobe B, wobei die Koeffizienten des Teilmodells aus der Stichprobe A berechnet wurden,
ρ_l, ρ_h	-	metrische Abstände im l- und h-Raum,
l	-	Anzahl der Datenpunkte in der jeweiligen Stichprobe,
h	-	Anzahl der Koeffizienten,
α	-	Ridge-Parameter.

In Abhängigkeit von der Information, die bezüglich der Daten vorhanden ist, kann ρ_l euklidisch oder absolut sein oder die Form einer robusten Schätzung besitzen, die nicht unbedingt dem absoluten Abstand entspricht. Die Koeffizienten des Modells für die Schätzung des Vektors $\mathbf{y}_{B(A)}$ müssen in diesem Fall durch eine Art der robusten Schätzung bestimmt worden sein. Das soll am nachfolgendem Beispiel gezeigt werden:

---

1 Ein partielles Modell kann z.B. die Form $v=Az_i+Bz_iz_j$ besitzen, wobei z_i und z_j Zwischen-Eingangsvariable und v die Zwischen-Ausgangsvariable auf der laufenden Stufe sind.

$$P = SP\left[\frac{\sum_{i=1}^{B}\left|y_{B_i} - x_i^T \cdot \mathbf{\hat{a}}_A\right|^r}{l-h} \cdot (X_B{}^T X_B + \alpha \cdot I)^{-1}\right] + \sum_{i=1}^{h}(\hat{a}_{Ai} - \hat{a}_{Bi})^2$$

mit $1 \leq r < 2$,

$\hat{a}_A, \hat{a}_B$ - robust geschätzte Koeffizienten.

Der Parameter α ist nur dann ungleich Null, wenn (X^TX) eine schlecht definierte Matrix ist.

Der Sonderfall eines euklidischen Abstands ρ_l tritt bei Anwendung des Selektionskriteriums LDUC (Local Data Uncertainty Criterion) [2,6] auf. In [8] sind Untersuchungen zu diesem Kriterium dargestellt, die zeigen, daß es im Fall von Nicht-Gauß'schen Störungen akzeptable Ergebnisse liefert.

Referenzen:

[1] Madala, H.R.; Ivakhnenko, A.:
Inductive Learning Algorithms for Complex Systems Modeling.
CRC Press, Inc., 2000 Corporate Blcd., N.W., Boca Raton, Florida, 1993

[2] Lange, T.:
Automatische Selektion optimaler Modellstrukturen für die Identifikation nichtlinearer Systeme bei Unsicherheit.
Operation Research Proceedings, Papers of the 21st Annual Meeting of DGOR in Cooperation with ÖGOR, Springer-Verlag, 383-390 (1992)

[3] Rosenblueht, A.:
Analytische Methoden für die Untersuchung von Neuronennetzen. Ausländische Funkelektronik. (russ.)
Verlag "Sovetskoye Radio", Moskau (1965)

[4] Lange, T.:
Strukturselektionskriterien für die Modellbildung bei Existenz von Unsicherheit.
Schrift zum Erwerb des Dr.-Ing. habil. (in Vorbereitung)

[5] Lange, T.,; Gnatowski, K.;Umbreit, S.:
Zur Wahl geeigneter Kriterien für die Struktursuche bei Modellierungsaufgaben.
Wiss. Z. TH Ilmenau 33(1987), Heft 2, S. 67 - 78

[6] Lange, T.:
New Structure Criteria in GMDH.
The First US/Japan Conference on Frontiers of Statistical Modeling, Tennessee / USA, May 24-29, 1992

[7] Lange, T.:
Ein Beitrag zur strukturellen Modellierung bei kleinen Datenmengen unter Anwendung der Methode der gruppenweisen Erfassung der Argumente.
Diss., TH Ilmenau, 1983

[8] Voigt, J.:
Vergleichende Untersuchung von Strukturselektionskriterien für mehrstufige MGEA-Algorithmen.
Diplomarbeit, Fachhochschule für Technik und Wirtschaft Berlin, 1993

A Conceptual Framework for the Design and Validation of Discrete-Event Simulations

Franz Liebl, Ludwig-Maximilians-Universität München

For every project anew, simulation practitioners have to specify the design features of a simulation model. At issue are: which type of performance measures to use, which way to validate the model, and, of course, the question if a steady-state simulation is appropriate in order to generate useful results. In resolving these issues, however, practitioners are left alone: with the exception of some casuistic approaches (Bratley/Fox/ Schrage 1987, Hoover/Perry 1989, Law/Kelton 1991), there is no encompassing framework available which supports the classification of a system to be simulated.

Our argument is backed by two case studies. We present two real-life systems—a clinical laboratory and a data communication network—that show rather similar characteristics as to the life cycles of entities and the daily workloads. However, these two systems require rather different modelling approaches. While a steady-state approach lends itself to simulate the clinical laboratory, it does not in the case of the data network. Thus, a closer scrutiny shows that also practical considerations have to be incorporated into a classification scheme that establishes the link between the real-world system and the design issues.

Drawing on these experiences we propose a structured process that helps to identify the appropriate type of simulation model to be implemented. This structured process can be reduced to a three-dimensional framework that takes into account the following criteria:

- the stationarity properties of the simulated system,
- the time horizon of the real-life system compared to the time horizon of the simulation runs, and
- the research purpose.

This framework not only allows to meaningfully categorize simulation problems but also to assign the appropriate methods for tackling the start-up problem, collecting output data, implementing variance-reduction techniques, and validating the simulation model (cf. Liebl 1992).

References:

Bratley, P.; Fox, B. L.; Schrage, L. E.: A Guide to Simulation (Second Edition); New York 1987
Hoover, S. V.; Perry, R. F.: Simulation – A Problem-Solving Approach; Reading, MA 1989
Law, A. M.; Kelton, W. D.: Simulation Modeling and Analysis; New York 1991
Liebl, F.: Simulation – Problemorientierte Einführung; München 1992

VERKEHRSRICHTUNGSPROGNOSE FÜR TELEKOMMUNIKATIONSNETZE

Norbert Matthes, Bonn

Zusammenfassung: Der vorliegende Beitrag befaßt sich mit der Prognose von gerichtetem Telekommunikationsverkehr. Als Prognosemethode dient die Minimierung des Informationsgewinns, ein aus der Raumwirtschaftslehre bekanntes und für räumliche Interaktionen in den unterschiedlichsten Bereichen genutztes Verfahren. Ausgehend von dem Standardansatz wird die Problemformulierung modifiziert, um den speziellen Gegebenheiten bei neu zu errichtenden Telekommunikationsnetzen, insbesondere der sich ständig verbessernden Datenlage, gerecht zu werden.

Abstract: This paper deals with the forecasting of directed telecommunications traffic. The forecasting method to be used is the minimization of information gain, which is well known from regional science and which is implemented for spatial interactions in various fields of research. Assuming the standard approach, the formulation of the problem is modified to take account of the special conditions existing in new telecommunications networks, in particular the continually improving supply of data.

1 Prognoseschritte bei der Netzplanung

Der Betreiber eines Telekommunikationsnetzes paßt seine Netzkapazitäten im allgemeinen in diskreten Ausbaustufen an die Nachfrage nach Kommunikationsdienstleistungen an. Bereits bei der Planung einer Ausbaustufe, also mit teilweise beträchtlichem zeitlichen Vorlauf, müssen konkrete Vorstellungen über das erwartete Nachfrageverhalten vorliegen. Die Prognose des von den Nachfragern gewünschten Telekommunikationsverkehrs läßt sich in drei Schritte einteilen:

1. Prognose des gesamten Verkehrsaufkommens im Versorgungsgebiet
2. Prognose der Aufteilung des gesamten Verkehrs auf Quell- und Senkeregionen
3. Prognose einer Verkehrsmatrix der räumlichen Interaktionen (Verkehrsrichtungsprognose)

Die folgende Abbildung 1 beschreibt schematisch die Zusammenhänge zwischen den räumlichen Interaktionen A_{ij} (Richtungsverkehre), den in den Zeilen- und Spaltensummen $Q_i = \sum_{j=1}^{N} A_{ij}$ und $S_j = \sum_{i=1}^{N} A_{ij}$ zum Ausdruck kommenden räumlichen Verteilungen auf die einzelnen Quellregionen $i = 1, \ldots, N$ beziehungsweise Senkeregionen $j = 1, \ldots, N$ sowie dem gesamten Verkehrsaufkommen $V = \sum_{i=1}^{N} Q_i = \sum_{j=1}^{N} S_j$.

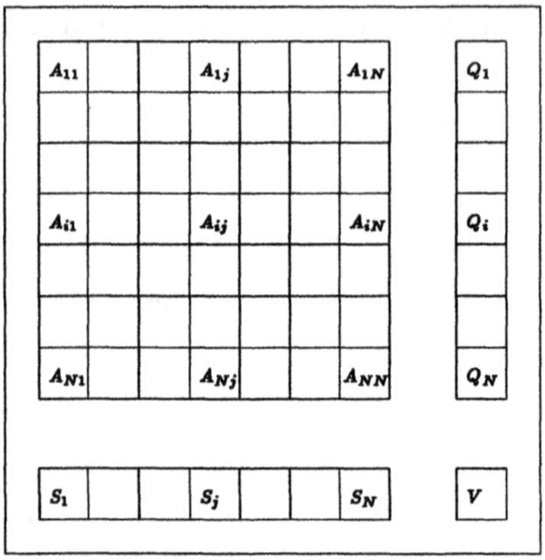

Abb. 1: Verkehrsmatrix mit Randsummen

Den Schwerpunkt des vorliegenden Beitrags bildet der dritte Prognoseschritt; annahmegemäß seien V, Q_i und S_j bereits im Rahmen der Prognoseschritte 1 und 2 ermittelt.

2 Grundlagen des Prognoseverfahrens

Als Ausgangspunkt der Entwicklung eines geeigneten Verfahrens zur Verkehrsrichtungsprognose für neue Telekommunikationsnetze kann der Standardansatz zur Minimierung des sogenannten Informationsgewinns unter Nebenbedingungen gewählt werden (vgl. /7/, S. 143 i.V.m. S. 152). Das informationstheoretische Maß des Informationsgewinns I und der Spezialfall der Entropie wird in /8/, S. 5ff., aus Axiomen hergeleitet.

Angewendet auf den Telekommunikationbereich wird diejenige Verteilung $\hat{p}$ ($p = (p_{ij})$ mit $i, j = 1, \ldots, N$, wobei $p_{ij} := A_{ij}/V$, so daß $\sum_{i=1}^{N}\sum_{j=1}^{N} p_{ij} = 1$) der Gespräche, der Gesprächsminuten oder des Telekommunikationsverkehrs für einen zukünftigen Planungszeitpunkt gesucht, die eine bekannte Verteilung möglichst gut reproduziert. Diese bekannte Verteilung, die im folgenden als a-priori-Verteilung $q = (q_{ij})$ mit $\sum_{i=1}^{N}\sum_{j=1}^{N} q_{ij} = 1$ bezeichnet wird, kann auf Verkehrsmessungen oder sonstigen Erfahrungen und Vorgaben basieren. Man kann zeigen, daß der Ansatz der Minimierung des Informationsgewinns äquivalent zu der Maximierung bestimmter Wahrscheinlichkeitsverteilungen ist (vgl. /2/, S. 36ff. und S. 87).

Mit gegebenen Randsummen Q_i, S_j und Preisen c_{ij} für Gespräche (Gesprächsminuten etc.) aus Region i nach Region j sowie der im System aufgewendeten Ausgabensumme für Kommunikationszwecke C läßt sich das Problem wie folgt formal schreiben:

$$\min_{p} I(p;q) := \sum_{i=1}^{N}\sum_{j=1}^{N} p_{ij} ln(p_{ij}/q_{ij}) \qquad (1)$$

unter den Nebenbedingungen

$$\sum_{j=1}^{N} p_{ij} = Q_i/V, \qquad i = 1,\ldots,N \qquad (2)$$

$$\sum_{i=1}^{N} p_{ij} = S_j/V, \qquad j = 1,\ldots,N \qquad (3)$$

$$\sum_{i=1}^{N}\sum_{j=1}^{N} c_{ij}p_{ij} = C/V \qquad (4)$$

Die Nebenbedingungen (2) und (3) stellen sicher, daß die prognostizierten A_{ij} konsistent im Hinblick auf die einzelnen vorgegebenen Randsummen sind, das heißt: $\sum_{j=1}^{N} p_{ij}V = \sum_{j=1}^{N} A_{ij} = Q_i$ und $\sum_{i=1}^{N} p_{ij}V = \sum_{i=1}^{N} A_{ij} = S_j$. Neben diese Randsummenbeschränkungen tritt die systemweite Ausgabenrestriktion $\sum_{i=1}^{N}\sum_{j=1}^{N} c_{ij}A_{ij} = C$; nach Division durch V in der Form (4).

Spezialfälle des Standardansatzes (1)-(4) sind die biproportionale Matrixanpassung und die Entropiemaximierung. Ein Ansatz der biproportionalen Matrixanpassung (vgl. /4/, S. 687ff.), der im Telekommunikationsbereich als Kruithof-Methode bekannt ist (vgl. z. B. /3/, S. 266ff., oder /6/, S. 5.3A.3.1ff.) entsteht, wenn auf die Ausgabenrestiktion (4) verzichtet wird. Mit $q_{ij} = 1/N^2$, gleichbedeutend mit dem Fehlen jeglicher a-priori-Informationen, geht (1)-(4) in das Problem der doppelt beschänkten Entropiemaximierung über (vgl. /9/, S. 4ff.).

Neben den genannten Spezialfällen läßt sich der Standardansatz recht einfach modifizieren, um weitere, leicht abweichende Problemstellungen betrachten zu können. Insbesondere ist das Verfahren sehr flexibel in bezug auf die Einführung weiterer Restriktionen (vgl. auch /2/, S. 43) sowie in bezug auf die Berücksichtigung von nur teilweise vorliegenden a-priori-Informationen, wie im nächsten Abschnitt gezeigt wird.

3 Übertragung auf neue Telekommunikationsnetze

Die Möglichkeit der praktischen Anwendung des Standardansatzes (1)-(4) auf ein neu zu errichtendes Telekommunikationsnetz soll am Beispiel eines modernen Mobilkommunikationsnetzes diskutiert werden. Zunächst ist zu beachten, daß deutsche Mobilkommunikationsnetze derzeit keine nationalen Preiszonen aufweisen. Wenn also im Unterschied zu der obigen Formulierung identische Preise c für alle Interaktionen gelten, entfällt die betreffende Restriktion (4): $\sum_{i=1}^{N}\sum_{j=1}^{N} p_{ij}c_{ij} = c\sum_{i=1}^{N}\sum_{j=1}^{N} p_{ij} = c = C/V$, so daß ein Problem der biproportionalen Matrixanpassung verbleibt.

Hinsichtlich der Problematik von a-priori-Daten ist anzumerken, daß Informationen aus anderen Kommunikationsnetzen vermutlich nur sehr begrenzt auf die Mobilkommunikation übertragbar sind. Selbst wenn dem Betreiber des betrachteten Mobilkommunikationsnetzes derartige Informationen über Richtungsverkehre in anderen Netzen vorliegen sollten, ist demnach zu empfehlen, vor allem auf die - sich im Verlaufe des räumlichen Ausbaus ständig verbessernden (vgl. Abbildung 2) - Verkehrsmeßdaten aus dem eigenen Mobilkommunikationsnetz zurückzugreifen.

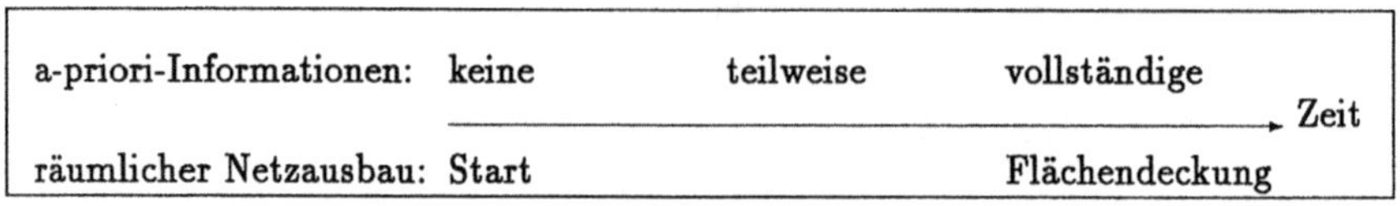

Abb. 2: Entwicklung der Datenlage

Die im folgenden betrachtete Situation ist typisch für Netze kurz nach der Einführungsphase. Es wird angenommen, daß der derzeitige Netzausbau eine unvollständige Flächendeckung erlaubt und daß für die Zukunft eine Ausdehnung der Versorgungsfläche geplant ist. Liegen nun aktuelle Verkehrsmeßwerte aus dem heutigen teilausgebauten Netz vor, ist die für die weitere Netzausbauplanung erforderliche a-priori-Matrix unvollständig. Damit sind der Standardansatz (1)-(4) beziehungsweise die Fassung (1)-(3) so nicht anwendbar. Über die Methode zur Behebung des Problems unvollständiger a-priori-Information muß im Einzelfall entschieden werden (vgl. auch /1/, S. 69). Ein denkbarer Weg in der beschriebenen Situation wäre die Setzung von $q_{ij} := 1/N^2$ wie im Grenzfall der Entropiemaximierung. Nachteilig an einem solchen Vorgehen ist jedoch die Vernachlässigung vorhandener Teilinformationen. Hier wird daher vorgeschlagen, eine vollständige a-priori-Matrix q auf der Grundlage aller verfügbaren Daten zu erzeugen (vgl. Abbildung 3).

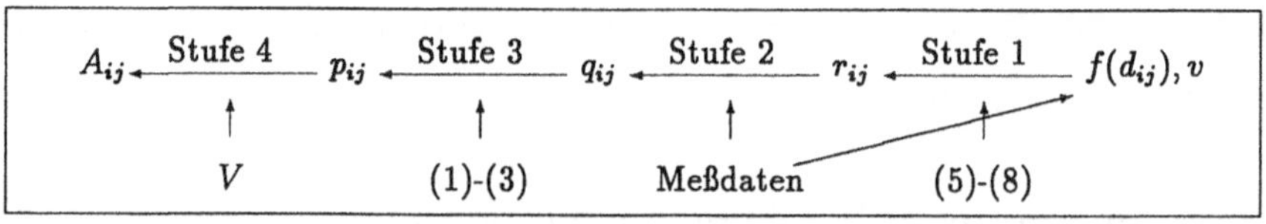

Abb. 3: Stufenweises Prognoseverfahren

Mit Hilfe des aus einem (modifizierten) Ansatz der Minimierung des Informationsgewinns herleitbaren Interaktionsmodells wird in Stufe 1 eine Matrix $r = (r_{ij})$ mit $i, j = 1, \ldots, N$ berechnet. Diese Matrix r kann in Stufe 2 für diejenigen Elemente korrigiert werden, für die Meßwerte aus dem bereits ausgebauten Teilnetz vorliegen. Als Ergebnis entsteht die Matrix q, wobei die Elemente mit $i, j = 1, \ldots, M < N$ aktuelle Meßwerte darstellen; für $i, j = M, \ldots, N$ dagegen soll $q_{ij} := r_{ij}$ gelten. Die Meßwerte werden dabei so normiert, daß ihre Summe dem Ausdruck $\sum_{i=1}^{M} \sum_{j=1}^{M} r_{ij}$ entspricht. Die eigentliche Prognose schließt sich in den Stufen 3 und 4 an.

Der Rest dieses Beitrags beschränkt sich auf die Ableitung eines modifizierten Interaktionsmodells zur Bestimmung (Prognose) der Matrix r, also auf die erste Stufe aus Abbildung 3. Als Werte für die a-priori-Matrix dieses modifizierten Interaktionsmodells werden Funktionswerte einer Funktion $f(d_{ij})$ für die Nichtdiagonalelemente und die Konstante v für die Diagonalelemente benutzt. Die unabhängige Variable d_{ij} bezeichnet die geographische Distanz zwischen dem Zentrum der Quellregion i und dem Zentrum der Senkeregion j. Gestalt und Parameter von f können aus den für das versorgte Teilgebiet vorliegenden Meßdaten $(i, j = 1, \ldots, M < N; i \neq j)$ ökonometrisch geschätzt werden. Auf diese Weise werden die in einer gemessenen Verkehrsmatrix zum Ausdruck kommenden Zusammenhänge auf die fehlenden Elemente übertragen.

Die Diagonalelemente $(i = j)$ werden nicht in die ökonometrische Schätzung einbezogen, da keine angemessenen Distanzdaten d_{ii} vorliegen; eine unterstellte Entfernung von Null oder nahe Null für alle Gespräche etc. innerhalb einer Region führt vermutlich zu Schätzfehlern. Somit sind die a-priori-Diagonalelemente auf andere Art aus den Meßdaten abzuleiten: Nimmt man an, daß alle Diagonalelemente a priori gleich groß anzusetzen sind, ist nur noch ihre Größe v offen.

Eine mögliche Alternative besteht darin, die a-priori-Diagonalelemente mit dem Durchschnitt der auf der Hauptdiagonale gemessenen Verkehrswerte zu belegen. Will man aber über die Einflußnahme auf die a-priori-Matrix hinausgehen und die sich in den Meßdaten widerspiegelnden Verhältnisse für die Diagonalelemente der Prognosematrix r sichern, kann man eine zusätzliche Restriktion für die Hauptdiagonale von r einführen. Wie noch darzulegen ist, braucht v bei dieser im folgenden beschiebenen Alternative vorab nicht bekannt zu sein. Da man auch zeigen kann, daß eine Normierung der Elemente der a-priori-Matrix mit dem Normierungsfaktor $(\sum_{i \neq j} \sum_j f(d_{ij}) + \sum_i v)^{-1}$ unerheblich für das Optimum ist, genügt es, den um diesen Normierungsfaktor bereinigten Informationsgewinn I' anstelle von I zu minimieren:

$$\min_r I'(r; f, v) = \sum_{i=1; i \neq j}^{N} \sum_{j=1}^{N} r_{ij} ln(r_{ij}/f(d_{ij})) + \sum_{i=1}^{N} r_{ii} ln(r_{ii}/v) \tag{5}$$

unter den Nebenbedingungen

$$\sum_{j=1}^{N} r_{ij} = Q_i/V, \qquad i = 1, \ldots, N \tag{6}$$

$$\sum_{i=1}^{N} r_{ij} = S_j/V, \qquad j = 1, \ldots, N \tag{7}$$

$$\sum_{i=1}^{N} r_{ii} = W/V \tag{8}$$

wobei $W = \sum_i A_{ii}$ den vorzugebenden gesamten Verkehr auf der Hauptdiagonale der zu berechnenden Matrix r bezeichnet. Aus der Lagrangefunktion dieses modifizierten Problems

$$\mathcal{L} = I'(r; f, v) + \sum_{i=1}^{N} \lambda_i (Q_i/V - \sum_{j=1}^{N} r_{ij}) + \sum_{j=1}^{N} \mu_j (S_j/V - \sum_{i=1}^{N} r_{ij}) + \tilde{\nu}(W/V - \sum_{i=1}^{N} r_{ii}) \tag{9}$$

können die notwendigen Bedingungen für ein Optimum, die nach /2/, S. 37ff., im allgemeinen auch hinreichend sind, gebildet werden: $ln\hat{r}_{ij} + 1 - lnf(d_{ij}) - \lambda_i - \mu_j = 0$ für $i = 1,\ldots,N; i \neq j; j = 1,\ldots,N$ und $ln\hat{r}_{ii} + 1 - lnv - \lambda_i - \mu_i - \tilde{\nu} = 0$ für $i = 1,\ldots,N$ zuzüglich (6)-(8). Anhand der Definition $ln\nu := \tilde{\nu} + lnv$ sieht man, daß die vorzugebende Höhe von v unerheblich für die Prognosematrix $\hat{r}$ ist. Wählt man zum Beispiel $v := 1$, ergibt sich $\nu = e^{\tilde{\nu}}$. Bei zusätzlicher Verwendung von $B_i := e^{(\lambda_i - 1)}/Q_i$, $D_j := e^{\mu_j}/S_j$ besteht das Gleichungssystem für $\hat{r}_{ij}$, B_i, D_j und ν schließlich aus (6)-(8) sowie:

$$\hat{r}_{ij} = B_i Q_i D_j S_j f(d_{ij}), \qquad i = 1,\ldots,N; i \neq j; j = 1,\ldots,N \qquad (10)$$

$$\hat{r}_{ii} = B_i Q_i D_i S_i \nu, \qquad i = 1,\ldots,N \qquad (11)$$

Die Lösung dieses nicht-linearen Gleichungssystems, das das modifizierte Interaktionsmodell beschreibt, kann analog zur Lösung des Standardansatzes mit iterativen Methoden erfolgen (zu den Lösungseigenschaften von Interaktionsmodellen vgl. /2/, S. 97ff., und /4/, S. 690ff.).

4 Ökonomische Interpretation

Nach dem Standardansatz (1)-(4) sowie nach der modifizierten Fassung (5)-(8) ergeben sich die geschätzten Elemente der Verkehrsmatrix aus statistischen Gesetzmäßigkeiten. Aus Sicht der Ökonomik ist das Fehlen von Verhaltensfunktionen sehr unbefriedigend. Daher sind zahlreiche Versuche unternommen worden, die resultierenden Interaktionsmodelle im Sinne der mikroökonomischen Nachfrage- und Gleichgewichtstheorie zu fundieren (vgl. die kritischen Anmerkungen in /2/, S. 40ff., und /5/, S. 102ff.). Auf diese Diskussion kann hier nicht umfassend eingegangen werden. Vielmehr wird im folgenden unterstellt, daß die Funktionen (10) und (11) die Nachfrage nach Telekommunikationsverkehr von i nach j ausdrücken. Die ökonomische Interpretation soll auf die Rolle der a-priori-Information, das heißt auf $f(d_{ij})$ in (10) und ν in (11), beschränkt bleiben.

Man beachte zunächst, daß es sich bei (10) um typische Gravitationsgleichungen handelt, die auch im traditionellen Entropiemaximierungsansatz entstehen, wenn man in der systemweiten Ausgabenrestriktion (4): $c_{ij} := f(d_{ij})$ spezifiziert. Während d_{ij} demnach als Indikator für eine auf der Distanz beruhenden Preisgestaltung angesetzt ist, fehlt eine systemweite Ausgabenrestriktion im Problem (5)-(8); in den aus dem modifizierten Interaktionsmodell resultierenden Gleichungen (10) steht $f(d_{ij})$ für räumliche Präferenzen der Nachfrager. Damit wird das an anderer Stelle dargelegte Ergebnis (vgl. /5/, S. 228ff.), daß Preisvorteile und Präferenzvorteile analoge Rollen in der Nachfragetheorie spielen, bestätigt. Die zugrundeliegende Hypothese für den in (10) beschriebenen Zusammenhang könnte folgendermaßen formuliert werden: Mit entscheidend für die zukünftige

Nachfrage $\hat{r}_{ij}$ nach gerichtetem Telekommunikationsverkehr sind die in den a-priori-Werten zum Ausdruck kommenden heutigen (oder auch in der Vergangenheit angelegten) Kommunikationsgewohnheiten. Eine direkte Modellierung dieser Kommunikationsgewohnheiten ist kaum möglich, so daß als Indikator die geographische Distanz benutzt wird. Da sich private und geschäftliche Kontakte verstärkt in der näheren Umgebung des Wohnortes beziehungsweise des Arbeitsplatzes von Wirtschaftssubjekten ausprägen, ist $\partial f / \partial d_{ij} < 0$ und damit $\partial r_{ij} / \partial d_{ij} < 0$ zu vermuten. Diese Überlegung läßt sich anhand der oben erwähnten Schätzung von f empirisch bestätigen.

Genau wie $f(d_{ij})$ in (10) kann ν in (11) als Indikator für die Präferenzen der Nachfrager interpretiert werden. Angenommen ν wäre modellexogen bekannt, könnte auf die Restriktion für die Hauptdiagonale (8) verzichtet werden. Anders als die exogenen Funktionswerte von $f(d_{ij})$ wird der Präferenzindikator ν jedoch endogen bestimmt. Im Verlauf des verwendeten Algorithmus zur Lösung des obigen Gleichungssystems wird das Modell so kalibriert, daß (8) erfüllt ist, daß also ein vorgegebener Anteil des Gesamtverkehrs auf die Hauptdiagonale der Matrix $\hat{r}$ entfällt.

Literatur:

/1/ Batten, D. F.
Spatial analysis of interacting economies.
Boston: Kluwer-Nijhoff (1983)

/2/ Bröcker, J.
Interregionaler Handel und ökonomische Integration.
München: Florentz (1984)

/3/ CCITT (Hrsg.)
GAS 10 Handbook, Planning data and forecasting methods, Vol. 1.
Geneva (1987)

/4/ Macgill, S. M.
Theoretical properties of biproportional matrix adjustments.
in: Environment and Planning A 9, 687-701 (1977)

/5/ Matthes, N.
Handelsströme.
München: Florentz (1991)

/6/ Stordahl, K.
Methods for traffic matrix forecasting.
ITC 12, Torino 1988, 5.3A.3.1-3.7

/7/ Snickars, F., Weibull, J. W.
A minimum information principle.
in: Regional Science and Urban Economics 7, 137-168 (1977)

/8/ Theil, H.
Economics and information theory.
Amsterdam: Elsevier (1967)

/9/ Wilson, A. G.
Entropy in urban and regional modelling.
London: Pion (1970)

SELBSTORGANISATION VON VORHERSAGEMODELLEN

Johann-Adolf Müller, Hochschule für Technik und Wirtschaft Dresden

Erfahrungen bei der Analyse und Vorhersage komplizierter Prozesse u.a. in Ökonomie, Ökologie und Meteorologie haben gezeigt, daß viele der üblichen Voraussetzungen auf der Grundlage der tatsächlich vorhandenen A-priori-Information kaum oder gar nicht überprüft werden können. Eine praktikable Modellbildung muß deshalb Elemente des interaktiven Modellentwurfs auf der Grundlage der theoretischen und experimentellen Systemanalyse mit Elementen der automatischen Modellerstellung verbinden, wobei die vorhandene Unbestimmtheit bei der Wahl der geeigneten Modelle und ihrer Struktur durch parametrische bzw. nichtparametrische Auswahlalgorithmen vermindert werden kann.
Die zur Selbstorganisation mathematischer Modelle entwickelten Algorithmen beruhen auf Prinzipien (Aufteilung der Beobachtungen in Lern- und Prüffolge, Generierung von Hypothesen (Mutationen), Selektion), die auch von Neuronalen Netzen genutzt werden. Im Unterschied zu Neuronalen Netzen, die durch assoziativen Wissenserwerb in der Trainingsphase ein internes Modell aufbauen, ermöglichen die entwickelten Algorithmen zur Selbstorganisation die Einbeziehung vorhandener A-priori-Information in den Auswahlprozeß sowie die explizite Ermittlung von Modellen optimaler Kompliziertheit. Bei geeigneter Wahl der Basisfunktionen können diese Modelle analytisch angegeben werden. Bei der Anwendung der Algorithmen auf unterschiedliche Datensätze in Ökonomie, Ökologie und Meteorologie hat sich ein zweistufiges Herangehen als zweckmäßig erwiesen:

- Nucleus-Bildung mit Hilfe nichtparametrischer Auswahlalgorithmen, d.h. Aufteilung der Systemgrößen in disjunkte Teilmengen ähnlicher (bezüglich Zeitreihenmuster oder Dynamik) Systemgrößen;
- Anwendung parametrischer Auswahlalgorithmen zur Ermittlung dynamischer Modelle (z.B. lineare oder nichtlineare Differenzengleichungssysteme) für die Repräsentanten dieser Teilmengen.

Große Aufmerksamkeit muß der Validierung der erhaltenen Modelle gewidmet werden, d.h. der Überprüfung, ob das ermittelte Modell optimaler Kompliziertheit in der Lage ist, kausale Zusammenhänge zwischen Eingangsgrößen und Ausgangsgrößen zu erfassen. Entsprechend geeignete Testgrößen wurden ausgehend vom Prinzip der Randomisierung begründet und mit Monte-Carlo-Simulation ermittelt. Alternativ werden nichtparametrische Auswahlalgorithmen zur Vorhersage angewendet, die auch für kombinatorische Vorhersagen genutzt werden können.

MOMENTS OF MIXING DISTRIBUTIONS: APPLICATIONS AND ESTIMATION IN MIXTURE MODELS

Wilfried Seidel, Universität der Bundeswehr Hamburg

Summary: Knowledge about generalized moments of mixing distributions is often useful. For example, in decision making, partial prior information can be modelled in this way. Applications in Quality Control are discussed. Estimators from observations of a mixture may be constructed in various situations by a general approach.

1. The Mixture Model

Let $\mathcal{G} = \{G(\cdot \mid \theta) : \theta \in \Theta\}$ be a family of cumulative distribution functions (cdf). A cdf will also be used to denote the associated distribution, for example, we may write $G(\cdot \mid \theta) = N(\mu, \sigma^2)$, the normal distribution with parameter $\theta = (\mu, \sigma^2)$. The *mixture* H of $\mathcal{G}$ with mixing distribution F on the parameter set Θ is defined as

$$H(x) = \int_{\Theta} G(x \mid \theta) dF(\theta) \ .$$

A mixture is called *finite*, if the mixing distribution F has finite support. In this case, H may be written as

$$H(x) = \sum_{i=1}^{k} \pi_i G(x \mid \theta_i)$$

for some $k \in \mathbb{N}$, the number of components, with weights $0 < \pi_i < 1$, the *mixing proportions*. Let us assume that $\mathcal{G}$ is known, whereas F is unknown, and that we observe a random sample $X_1, \ldots, X_n$ from H. Our aim is to get information about F.

Much work has been done on estimating mixing distributions from mixtures, see for example [6]. Here we shall concentrate on estimating generalized moments of F: Given a function $\gamma: \Theta \to \mathbb{R}$, the *generalized moment* m_γ of F with respect to γ is defined as

$$m_\gamma := \int_\Theta \gamma(\theta) dF(\theta) \; .$$

In particular, if $\gamma(\theta) = \theta^\ell$, then $m_\ell := m_\gamma$ is the ℓth moment of F about the origin. Many other parameters of F can be described as generalized moments. The next example plays some role in modelling partial prior information:

Example 1.1 For $a \in \Theta \subset I\!R$, let $A := \{\theta : \theta \leq a\}$, and I_A be the indicator function of A, i.e., $I_A(\theta) = 1$, if $\theta \in A$, and $I_A(\theta) = 0$ otherwise. For $\gamma(\theta) := I_A(\theta)$, we obtain $m_\gamma = F(a)$.

In Section 2, we shall discuss some applications, with special emphasis on modelling partial prior information in Statistical Quality Control. Estimators have been derived for a variety of models. They may be difficult to survey, but fortunately, most of them can be obtained from a simple construction principle. Section 3 considers its application in situations relevant especially in Quality Control; in Section 4, the hypergeometric distribution is dealt with in more detail.

2. Applications

Finite mixtures have many applications [12]. In order to estimate the mixing proportions and the parameters of the component distributions, the likelihood approach is usually more efficient than the method of moments. However, solving the likelihood equations as well as testing the number of components of the mixture may be difficult here [6]. In these cases, estimated values of moments of the mixing distribution can be used to derive good starting values for numerical solution of the likelihood equation and to construct simple, but powerful tests for the number of components [1], [5].

Arbitrary (not necessarily finite) mixtures frequently occur in decision making, for example in Statistical Quality Control: Let us consider a sequence of independent decision problems with a common structure, for instance decisions about the incoming lots from one supplier received in succession. In the ith problem, we observe a random variable X_i with distribution $G(\cdot \mid \theta_i)$, in acceptance sampling this being usually a test statistic calculated from a sample from the i th lot. Here, θ_i describes the quality of the lot and $G(\cdot \mid \theta)$ may be a hypergeometric, binomial, normal or noncentral t distribution; the latter in the case of a normally distributed quality characteristic with unknown expectation and variance and a one-sided specification limit.

The loss depends on the unknown parameter θ_i, which is assumed to vary independently from problem to problem according to a distribution F. Taking into account the randomness of θ_i, X_i is an observation of the mixture of $\{G(\cdot \mid \theta)\}$ with mixing distribution F. In each stage, we have information about F from the past observations X_i, which should be taken into account for the actual decision. If a Bayesian decision rule is used with prior $\hat{F}$, an estimator of F, we speak of an empirical Bayes strategy. An alternative is to consider only *partial prior information*, which is frequently assumed to be represented by an inequality

$$\mathbb{P}(\theta \leq a) \geq \gamma \,, \tag{2.1}$$

in our notation, $F(a) \geq \gamma$. For applications in Quality Control see [4]. The problem is, that it may be difficult to estimate the parameters a and γ from the past, if one does not observe θ directly. Now $F(a)$ is a generalized moment of F (see Example 1.1), and in [10] it is shown how decision rules based on inequalities on arbitrary generalized moments can be calculated numerically. Therefore partial prior information can be modelled by generalized moments which can be estimated more easily.

3. Estimation of Generalized Moments

Let f be a real-valued function on the sample space, integrable with respect to H, such that

$$\gamma(\theta) = \int f(x) dG(x \mid \theta) \tag{3.1}$$

holds for each $\theta \in \Theta$. If X is a r.v. with distribution H, then

$$E(f(X)) = \int_{\Theta} \Big(\int f(x) dG(x \mid \theta) \Big) dF(\theta) = \int_{\Theta} \gamma(\theta) dF(\theta) = m_\gamma \ ,$$

therefore $\hat{m}_\gamma := \frac{1}{n}(f(X_1) + \cdots + f(X_n))$ is an unbiased and consistent estimator of m_γ.

Equation (3.1) may be read in two directions. Firstly, let us start from f. Then (3.1) allows us to construct a parameter function γ such that $E(f(X))$ (which is easy to estimate) is the generalized moment of F with respect to γ.

From here on, we shall concentrate on univariate observations X and one-dimensional parameters θ.

Example 3.1 If prior information (2.1) seems meaningful but is hard to obtain, one may consider the following modification: Let $c \in \mathbb{R}$ and $f(x) := I_{(-\infty,c]}(x)$, then $E(f(X)) = \mathbb{P}(X \leq c) = H(c)$, and the estimator $\hat{m}$ is the proportion of observations falling below c (for instance, the proportion of accepted lots in Quality Control up to this time), possibly with a correction that protects against underestimating. The corresponding parameter function $\gamma(\theta)$ is $G(c \mid \theta)$, which can easily be used to construct decision rules based on the information $m_\gamma = H(c) \geq \beta$.

Now let us start from γ and look for an estimator of m_γ. For this purpose, we seek a function $P_\gamma(x) := f(x)$ that satisfies (3.1). This means, that $P_\gamma(X)$ is an unbiased estimator for the parameter function $\gamma(\theta)$ in the model $\{G(\cdot \mid \theta) : \theta \in \Theta\}$, which in many cases can be found using standard statistical theory: In natural exponential families with quadratic variance functions, analytical functions $\gamma(\mu)$ of the mean value parameter μ can be estimated, [5],[7],[8]. In particular, there exists an estimator $P_\ell(X)$ for $\gamma(\theta) = \theta^\ell$, which yields an estimator for the ℓth moment m_ℓ of F about the origin. The normal, Poisson, gamma, binomial and negative binomial family can be treated in this way, estimators for additional models are given in [12].

Frequently, the estimator P_ℓ can be constructed as follows:

Example 3.2 Let $X \sim G(\cdot \mid \theta)$ and assume that moments or factorial moments of X can be represented as polynomials in θ:

$$E(X^\ell \mid \theta) = a_{\ell 0} + a_{\ell 1}\theta + \cdots + a_{\ell\ell}\theta^\ell \ , \ \ell = 1,2,\ldots \tag{3.2}$$

or

$$E(X(X-1)\cdots(X-\ell+1) \mid \theta) = a_{\ell 0} + a_{\ell 1}\theta + \cdots + a_{\ell\ell}\theta^\ell \ , \ \ell = 1,2,\ldots . \tag{3.3}$$

If $a_{\ell\ell} \neq 0$ for each ℓ, then (3.2) or (3.3) can be solved recursively for powers of θ. We have, then, for $\ell = 1,2,\ldots$

$$\theta^\ell = c_{\ell 0} + c_{\ell 1}E(X \mid \theta) + \cdots + c_{\ell\ell}E(X^\ell \mid \theta) = E(P_\ell(X) \mid \theta),$$

where $P_\ell(X) := c_{\ell 0} + c_{\ell 1}X + \cdots + c_{\ell\ell}X^\ell$ is the desired estimator. Let us give two examples:

a) If $t(k,\theta)$ denotes the noncentral t-distribution with k degrees of freedom and noncentrality parameter θ, then for $X \sim t(k,\theta)$ the following holds ([3], p. 203) :

$$E(X^\ell \mid \theta) = (k/2)^{\ell/2}\frac{\Gamma((k-\ell)/2)}{\Gamma(k/2)} \sum_{0\le j\le \ell/2} \binom{\ell}{2j}\frac{(2j)!}{2^j j!}\theta^{\ell-2j} \ .$$

In particular, $a_{\ell\ell} \neq 0$ holds for $k \geq 2$. The coefficients $a_{\ell i}$, $i,\ell \leq 4$, are tabulated in [2] for various degrees of freedom. The solution of the corresponding system of equations is facilitated by the fact that many coefficients vanish.

b) If $G(x \mid \theta) = G(x-\theta)$ is a location family generated by a distribution G on $\mathbb{R}$, then $X \sim G(\cdot \mid \theta) \Leftrightarrow X = \theta + Y$, $Y \sim G$. Let us assume that $m_i(G) := E(Y^i)$ exists for sufficiently many i; these moments are of course known. From

$$E(X^\ell) = E((\theta+Y)^\ell) = \theta^\ell + \binom{\ell}{0}\theta^0 m_\ell(G) + \cdots + \binom{\ell}{\ell-1}\theta^{\ell-1}m_1(G)$$

one may construct $P_\ell(X)$ recursively as $P_0(X) = 1$ and

$$P_\ell(X) = X^\ell - \binom{\ell}{0}P_0(X)m_\ell(G) - \cdots - \binom{\ell}{\ell-1}P_{\ell-1}(X)m_1(G).$$

4. The Hypergeometric Distribution

For $k \leq N \in \mathbb{N}$ let $\Theta := \{0,1,\ldots,N\}$ and

$$h(x \mid \theta) = \binom{\theta}{x}\binom{N-\theta}{k-x}\Big/\binom{N}{x}, \quad x = 0,1,\ldots,k$$

be the probability function of the hypergeometric distribution with parameters N, k (considered as fixed here) and $\theta \in \Theta$. In Quality Control, θ is the number of defective items in a lot of size N and the observation x is the number of defective items in a sample of size k drawn from the lot without replacement.

Each function f on Θ has an orthogonal expansion

$$f(\theta) = a_0 t_0(\theta) + \cdots + a_N t_N(\theta)$$

in terms of the *Chebyshev polynomials* $t_0(\theta), \ldots, t_N(\theta)$, see [13]. In [13], polynomials $\lambda_j(\theta),\ 0 \le j \le k$, are constructed that satisfy

$$\lambda_j(0)h(0 \mid \theta) + \cdots + \lambda_j(k)h(k \mid \theta) = t_j(\theta) \quad \forall \theta \in \Theta,\ j = 0, \ldots, k\ .$$

Here the polynomials λ_j are used to estimate the first $k+1$ coefficients in the Chebyshev-expansion of the mixing distribution in a mixture model with $\mathcal{G} = \{\ h(\cdot \mid \theta) : \theta \in \Theta\}$. In our terminology, λ_j is an unbiased estimator of $t_j(\theta)$. Let $\gamma(\theta)$ be a polynomial in θ of degree $\le k$, then $\gamma(\theta)$ has an expansion

$$\gamma(\theta) = b_0 t_0(\theta) + \cdots + b_k t_k(\theta)$$

with known coefficients $b_0, \ldots, b_k$. For $X \sim h(\cdot \mid \theta)$, it is easy to see that

$$P_\gamma(X) = b_0\lambda_0(X) + \cdots + b_k\lambda_k(X)$$

is an unbiased estimator for $\gamma(\theta)$, so any polynomial of degree $\le k$ can be estimated.

With a slight modification of the model, *any* function of the parameter can be estimated: Let us assume that lots are inspected by a sampling plan (k, c) with $c < k < N$, i.e., a lot is accepted iff $X \le c$. Let us further assume that in the case of rejection, a complete check of all N articles in the lot is carried out. Let $Y := X$ if $X \le c$ and $Y := \theta$ if $X > c$ be the number of the defective items found in the whole procedure. Schmetterer [9], p. 272, proves that for any function γ on I there exists an unbiased estimator $P_\gamma(Y)$ for $\gamma(\theta)$. The estimator can often be easily obtained, for further details see [11].

References

[1] Furman, W.D. and Lindsay, B.G. (1993); Testing for the Number of Components in a Mixture of Normal Distributions Using Moment Estimators, Preprint.

[2] Hogben, D., Pinkham, R.S. and Wilk, M.B. (1961); The Moments of the Noncentral t Distribution, Biometrika, 48, 465–468.

[3] Johnson, N.L. and Kotz, S. (1970); Continuous Univariate Distributions - 2, Boston, Houghton Mifflin Company.

[4] Krumbholz, W. (1982); Die Bestimmung einfacher Attributprüfpläne unter Berücksichtigung von unvollständiger Vorinformation, Allgemeines Statistisches Archiv, 66, 240–253.

[5] Lindsay, B.G. (1989); Moment Matrices: Applications in Mixtures. Annals of Statistics, 17, 722–740.

[6] McLachlan, G.J. and Basford, K.E. (1988); Mixture Models. Inference and Applications to Clustering, New York, Marcel Dekker.

[7] Morris, C.N. (1982); Natural Exponential Families with Quadratic Variance Functions, Annals of Statistics, 10, 65–80.

[8] Morris, C.N. (1983); Natural Exponential Families with Quadratic Variance Functions: Statistical Theory, Annals of Statistics, 11, 515–529.

[9] Schmetterer, L. (1974); Introduction to Mathematical Statistics, Berlin, Heidelberg, New York, Springer.

[10] Seidel, W. (1992); Minimax Regret Sampling Plans Based on Generalized Moments of the Prior Distribution, in Lenz et al. (eds.), Frontiers in Statistical Quality Control, 4, Heidelberg, Physica.

[11] Seidel, W. (1993); A Note on Estimating Parameters of Partial Prior Information, accepted by Statistical Papers.

[12] Titterington, D.M., Smith, A.F.M. and Makov, U.E. (1985); Statistical Analysis of Finite Mixture Distributions, New York, Wiley.

[13] Walter, G.G. and Hamedani, G.G. (1988); Empiric Bayes Estimation of Hypergeometric Probability, Metrika, 35, 127–143.

BOOTSTRAP BEI VAGER VORINFORMATION

Gerhard Stahl, Universität Heidelberg

Vages Vorwissen der Form $P(X \in I_j) = \pi_j$ für die Verteilungsfunktion einer reellwertigen Zufallsvariablen X, wobei $I = (I_1, ..., I_m)$ eine Partition von $\mathbb{R}$ und $\pi = (\pi_1, ..., \pi_m)$ ein Element aus

$$\Pi \subseteq S_m = \{ y \in \mathbb{R}^m \mid y_j \geq 0, \Sigma_j y_j = 1\}$$

ist, definiert eine nichtparametrische Teilklasse

$$\Theta_\Pi = \{ F \in \mathbb{F} \mid \int 1_{I_j} dF = \pi_j, \pi \in \Pi \}$$

der Menge $\mathbb{F}$ aller Verteilungsfunktionen.

In solchen, durch Θ_Π–Klassen konstituierten, statistischen Modellen steht in der nichtparametrischen Maximum–Likelihood–Methode ein Inferenzprinzip zur Schätzung von $F^\circ \in \Theta_\Pi$ aus unabhängig identisch verteilten Beobachtungen $X_i \sim F^\circ$ zur Verfügung.

Gegenstand der Betrachtung ist die Anwendung der Bootstrap–Methode, unter Verwendung des durch Θ_Π– Klassen spezifizierten Vorwissens, zur Approximation von Fehlerverteilungen von Mittelwertstatistiken $T_n = \frac{1}{n} \sum_{i=1}^{n} X_i^k$.

Eine Modifikation der naiven Bootstrap–Methode auf Θ_Π–Modelle ermöglicht die Approximation der Fehlerverteilung von T_n bezüglich F°, $\mathcal{L}_{F^\circ}(T_n - \mu(F^\circ))$, durch $\mathcal{L}_{F_{ML}}(T_n - \mu(F_{ML}))$ im Gegensatz zu $\mathcal{L}_{F_n}(T_n - \mu(F_n))$, der Approximation von $\mathcal{L}_{F^\circ}(T_n - \mu(F^\circ))$ durch die naive Bootstrap–Methode.

Die Vorgehensweise,die Bootstrap–Methode in Θ_Π–Modellen zur Approximation von Verteilungen oder Kenngrößen von Mittelwertstatistiken zu verwenden, wird durch Bereitstellung von Konsistenzsätzen asymptotisch gerechtfertigt.

Da F_{ML} und F_n im allgemeinen, bei gleichen Beobachtungen, verschiedene Schätzungen für F° ergeben, wird am Beispiel des arithmetischen Mittels die Auswirkung der Verwendung von Vorinformation auf die Bootstrap–Schätzung von $\mathcal{L}_{F^\circ}(\bar{X} - \mu(F^\circ))$ in Θ_Π–Modellen durch Simulationen mit der naiven Bootstrap–Methode für endliche Stichprobenumfänge untersucht und verglichen.

A Heuristic Block-Rolling-Forward Principle for Sequencing Problems

Roland Fahrion, Heidelberg

Summary: In the class of job-shop scheduling problems a combination between a resource constrained project scheduling technique and the use of heuristics for sequencing job-machine assignments will be proposed. A job-blocking and block-rolling-forward principle is considered which principally allows sequence planning over large time horizons. An appropriate model for actual production scheduling would require a significant relaxation of the basic assumptions.
Thus, in a reassignment procedure the jobs of the first block are loaded into main storage. Then the jobs are reassigned in such a way that for each machine the operations are shifted as far as possible to the left margin of the block. The same procedure will be executed with the jobs previously assigned to the second block. Before the third block is loaded into memory the assignment of the first block will be written back to disk storage. This reassignment of jobs is repeated iteratively until the prior and problem-adequate measure for the jobs' tardiness with respect to due dates is attained and overlapping assignments are removed.
For the production of a software prototype Prolog points out as a suitable instrument because of the advantages in using the depth search of the Prolog interpreter. Furthermore, due to the convenient inclusion and variation of additional restrictions, and the proximity of Prolog predicates to relations in databases, Prolog might be preferred. Finally, an example is presented.

Zusammenfassung: Zur Bestimmung bestmöglicher Auftrag-Maschine Zuordnungen in Jop-Shop Scheduling Problemen (JSSP) wird die Kombination zwischen einer aus der Literatur bekannten Vorgehensweise bei der Projektplanung unter Berücksichtigung gegebener Ressourcenbeschränkungen, und einer aufwandsreduzierenden Selektionsheuristik empfohlen. Hierzu werden, nach einer topologischen Vorsortierung der Aufträge, Blöcke von Aufträgen betrachtet, von denen jeweils der aktuelle und der Vorgängerblock im Hauptspeicher gehalten werden müssen. Damit läßt sich im Grundsatz die Reihenfolgeplanung für große Planungshorizonte durchführen.
In einer Mehrfachschleife werden die Auftrag-Maschine Zuordnungen des i-ten Blocks aufgrund der Ergebnisrechnung im Rahmen einer mehrstufigen Bewertungsfunktion 'geometrisch' möglichst dicht plaziert. Danach wird ein weiterer Block i+1 mit Aufträgen geladen und unter Berücksichtigung des Vorgängerblocks ebenfalls möglichst dicht plaziert. Vor dem Laden des i+2-ten Blocks wird der i-te Block zurückgeschrieben und der heuristische Zuweisungsprozeß für Block i+1 und i+2 durchgeführt ($i \geq 1$).
Für den Umordnungsprozeß ist die logische Progammiersprache Prolog mit der dort implementierten Tiefensuche (zumindest in der Prototypphase) gut geeignet, auch wegen der Anpassungsnähe von Prolog-Prädikaten an Datenbankrelationen und der problemlosen Einbeziehung von Zusatzrestriktionen. Exemplarisch wird ein Fallbeispiel durchgeführt.

1. Basic Principle and Denotations

Let M be the number of machines, N the number of jobs, and m (j) the number of operations or tasks of job j, j = 1,...,N. Basically, the problem of finding a solution of the job-shop scheduling problem consists in determining a permutation matrix (p_{ijm}) in the three-dimensional space of operations, jobs and machines, with the property $p_{ijm} = 1$, if operation i of job j is assigned to machine m (i = 1,..., m(j); j = 1,...,N; m = 1,...,M). Further denotations concern processing times t_{ijm} , buffer times b_{ijm}, start and due dates s_{ijm} and f_{ijm} respectively. The due dates can be determined by a functional specification $f_{ijm} = g(s_{ijm}, t_{ijm})$, e.g. $f_{ijm} = s_{ijm} + t_{ijm}$ when machine m is working without preemption. Furthermore, it is assumed

(a) that for all jobs j there exist earliest and latest start and termination dates S_j and F_j such that

$$\sum p_{1jm}\, s_{1jm} \geq S_j\,, j=1,\ldots,N,$$

i. e. the first operation of a job j starts at the earliest with S_j (j = 1,...,N).

(b) The buffer time $b_{m(j)jm}$ separates the last operation of a job j from the latest due date of job j, i. e.

$$\sum_{m=1}^{M} p_{m(j)jm}\left(f_{m(j)jm} + b_{m(j)jm}\right) < F_j, \qquad j=1,\ldots,N.$$

(c) Two subsequent operations are separated at least by the buffer time of the previous operation,

$$\sum p_{i-1,jm}\left(f_{i-1,jm} + b_{i-1,jm}\right) \leq \sum_{m=1}^{M} p_{ijm}\, s_{ijm}$$

(d) Two different operations must not utilize one machine simultaneously.

As to the definition of an evaluation function and to priority rules the job shop scheduling problem is a too restrictive formulation that provides results which are anything more than suggestive for actual sequencing planning. Thus, an appropriate model would require a significant relaxation of the basic assumptions of the classical JSSP formulation. A survey of priority rule-based scheduling has been given by Haupt (1989).

The priority rules in the following section will take into account due dates, processing times, and number of operations. A scalar value is assigned to each waiting job, the best of which will determine the job to be selected in the actual planning step. This seems to be a reasonable preliminary approach for tailoring priority rules to a particular problem and to planning data at hand.

2. First Stage: Preliminary Job-To-Block Assignments

The algorithmic procedure for the heuristic evaluation of the operation-job-machine assignments is described by the following steps:

(1) Estimate the expected total duration T (days) of a production plan by orienting at the most utilized machine.

(2) Divide the time span [0,T] into approximately time-equidistant block periods of length B = [N/M].

(3) Sort the jobs topologically in descending order with respect to their operation depth.

(4) Start with the first job of the sorted sequence and assign it to the first block such that a simple additive priority rule defined by

$$\mathrm{Pr} = \sum_{i=1}^{3} c_i\, PR_i \quad , \; c_i \text{ prior given },$$

will achieve its highest value, In doing so, the partial priority rule PR_1 evaluates the best possible fitting of a job into a block frame, where the job's span has to be taken into account. PR_2 evaluates the utilization of the machines which are assigned to the corresponding operations of the job, and pursues the objective of assigning higher priorities to those capacities with less utilization in the block frame. PR_3 gives some priority to those block frames with machines already utilized, but which are not required by the actual job operation. Altogether, the three components of PR are defined as follows:

$$PR_1 = \begin{cases} -a+(d/B)^2, & \text{if there is a delay period between the begin time of the block frame and the one of the operation under consideration. Hereby, a is a free parameter which must be chosen adequately (e. g. a = 10 in the definition of the evaluation function in section 4),} \\ -a, & \text{if the start time precedes the start time of the operation under consideration ,} \\ 0 & \text{if the block frame is suitable for the operation actually checked .} \end{cases}$$

For the definition of PR_2 we consider all machines which must be assigned to the operations of a job. For each operation i of job j we determine

$$PR_{i,2} := \begin{cases} k, & \text{if k percent of the required machines for executing all operations of a job are available,} \\ 0, & \text{if the covering of the jobs coincides with the a priori given covering ,} \\ -k & \text{if the covering of the jobs supercedes the ideally possible covering by k / 10 percent, by k / 10 percent,} \end{cases}$$

and compute PR_2 as arithmetic mean

$$PR_2 := \frac{1}{m(j)} \sum_{i=1}^{m(j)} PR_{i,2} .$$

For the definition of PR_3 we consider all machines, not only those which are required by the operations of an actually considered job j. For each machine m = 1,...,M we determine

$$PR_{m,3} = \begin{cases} K, & \text{if machine m is required for executing the operations of job j ,} \\ -K, & \text{if machine m is not required for the operations of job j .} \end{cases}$$

and compute PR_3 as mean

$$PR_3 := \frac{1}{M} \sum_{m=1}^{M} PR_{m,3}$$

(5) Define a clause priority (job_id, block_id) which computes the priority function PR for the actual job and block identifiers.

(6) Define a clause assign-job-to-block (job_id, block_id) which holds the assignment job-to-block and computes the actual utilization of the machines. Furthermore, using the given job file data, for each job j being assigned during this procedure, Sj and Tj will receive the start and finishing times of the actual block. These will be used in the reassigning process (see section 3) as criteria for the tardiness of the job under consideration.

(7 The job-to-block assignment is executed in a nested loop, where assign-job-to-block(.) is run for all valid combinations 'operation-job-block'.

3. Second Stage: Iterative Reassignment of Job Operations and Block-Rolling-Forward Principle

After having achieved a first preliminary assignment of jobs to blocks by the procedure of section 2, the jobs of the first block are loaded into main storage. Then these jobs are reassigned within a multiple loop described subsequently. Geometrically spoken, the operations on each machine are rearranged by shifting them as far as possible to the left margin of the block. Then, the same procedure will be executed with those jobs which have been assigned to the second block in the preliminary assignment stage. Furthermore, the free places in a neighborhood of the right boundary of the first block and the left one of the subsequent second block are attempted to be filled in order to achieve a 'geometrically' best assignment of the machines.

Before the third block is loaded into memory, the first block together with its last assignments will be written back to a disk file. Thus, by rolling forward this process, we receive a sequence of blocks b = 2,...,B-1, where block i-1 is fixed and blocks i and i+1 are active (i = 2,...,B-1). In each step of this process only two blocks are in main storage simultaneously .

The criterion for terminating the reassignment process of job operations within a block is the prior defined assignment density degree. Principally, overlapping between two active blocks is admitted. But gaps which could not be assigned within block i may be filled with adequate job operations of block i+1.

The reassignment of jobs within a block is repeated iteratively until the prior and problem-dependent measure for a job's tardiness has come up to the due date of the job, and overlapping assignments are removed. This is done by successively selecting those jobs whose values of the evaluation function are worst. The following specification of the evaluation function represents a priority rule which has been found by numerous empirical tests: With the assumptions and denotations

- $\tau =$ Integer parameter in a given interval $[0,\bar{\tau}]$, $\tau \in \mathbb{N}$, weighting the priority either to meet the start and due dates S_j and F_j of each job j or to attain non-overlapping jobs. This preference factor will get various values during the reassignment process, and has a complementary effect in the specification of the evaluation function,
- $\sigma =$ Normalized measure for the deviation of an operation duration from the mean operation duration.,
- $\pi =$ Priority for an actual operation,
- $\eta =$ Sum of operation priorities attached to overlappings,
- $\mu =$ Number of job displacements up to the actual step,
- $\nu =$ Number of displacements of overlapping job operations,
- $\delta =$ Number of overlappings of an actually considered job,

the empirically found evaluation function for a job operation is

$$f := \begin{cases} -\left[\frac{1}{2}\delta\tau + a(\eta-\pi) + 5(\nu-\mu) + \frac{1}{a}\sigma^2(E-\tau)\right], \\ \qquad \text{if there is a delay with respect to } \tau, \\ -\left[\frac{1}{2}\delta\tau + a(\eta-\pi) + 5(\nu-\mu) + \varepsilon(\sigma+a)^2(E-\tau)\right], \\ \qquad \text{otherwise, } \varepsilon \text{ small .} \end{cases}$$

In a nested loop over blocks, $\bar{\tau}$, and jobs, the reassignment process of job operations, given ε and $\bar{\tau}$, is executed.

4. Example

For the problem characteristics

Number of jobs (N)	: 60	Maximum of operations m (j)	: 11
Number of Machines (M)	12	Sum of operations	: 304
Planning Period (days)	: 60	Earliest Start	: 0
Block size (# jobs)	: 7	Earliest Due date	: 60
Number of blocks	: 10	Latest Due date	: 66

Assignment Layout : Text mode, jobs are represented as ASCII characters in the range of ASCII (33) and ASCII (255). The sequence of operations of a job is represented by the sequence of ASCII-characters which identifies the job. Machines are represented as rows in the machine-job layout screen.

the following steps illustrate the block-rolling-forward principle, where subsequently fixed block patterns are shadowed:

Block 1 fixed : D b A G F H V
Block 2 active: a L j I K m Q
Block 3 active: M N J P Y X W

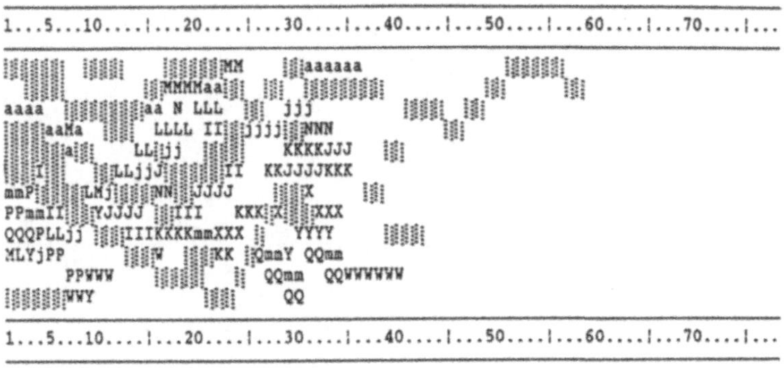

Block 2 fixed : a L j I K m Q
Block 3 active: M N J P Y X W
Block 4 active: 1 ` R s S Z !

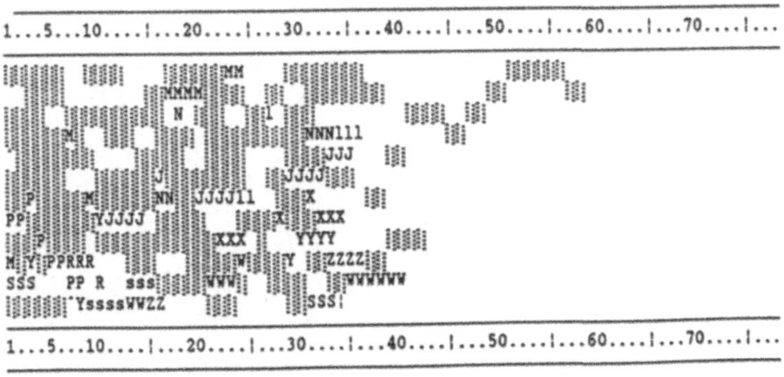

Block 3 fixed : M N J P Y X W
Block 4 active: 1 ` R s S Z !
Block 5 active: B d _ ' r [T

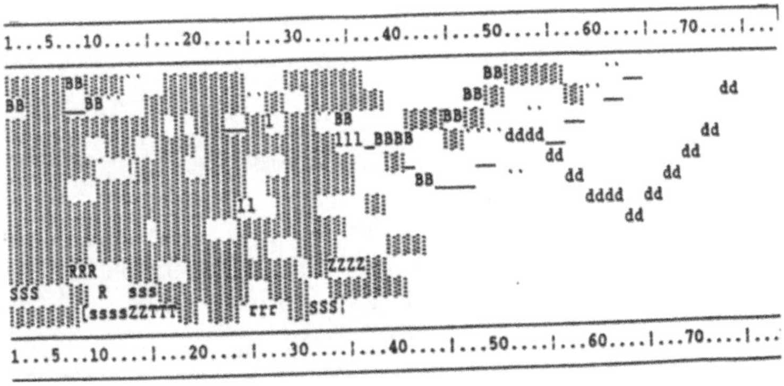

Block 4 fixed : 1 ˆ R s S Z ¦
Block 5 active: B d _ ` r [T
Block 6 active: E c C e f t o

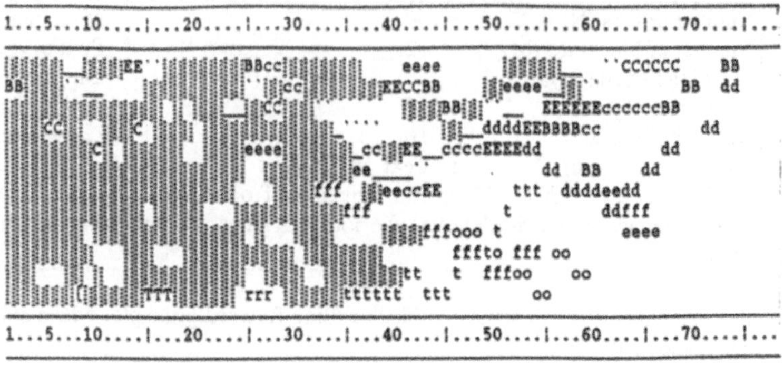

Block 5 fixed : B d _ ` r [T
Block 6 active: E c C e f t o
Block 7 active: k g i O w n h

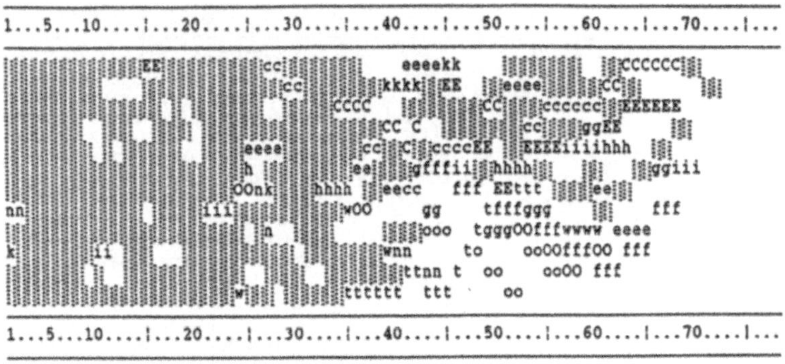

Block 6 fixed : E c C e f t o
Block 7 active: k g i O w n h
Block 8 active: v p x u U q \

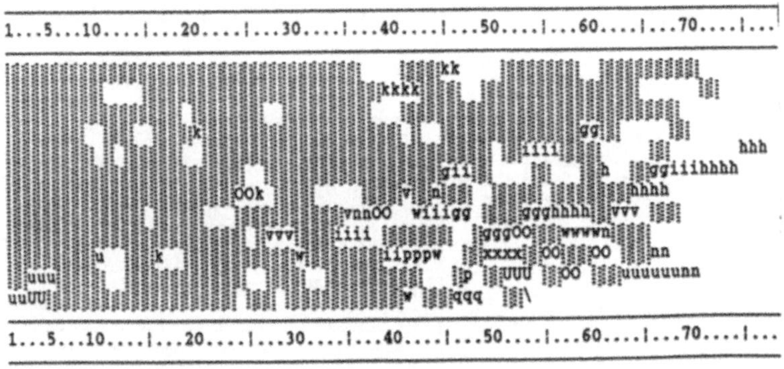

Block 7 fixed : k g i O w n h
Block 8 active: v p x u U q \
Block 9 active: ¦ y] z

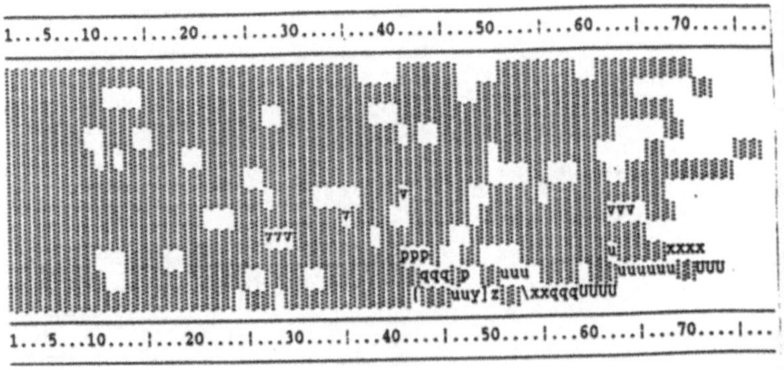

Block 8 fixed : v p x u U q \
Block 9 active: { y] z
Block 10 active: - - - - - - -

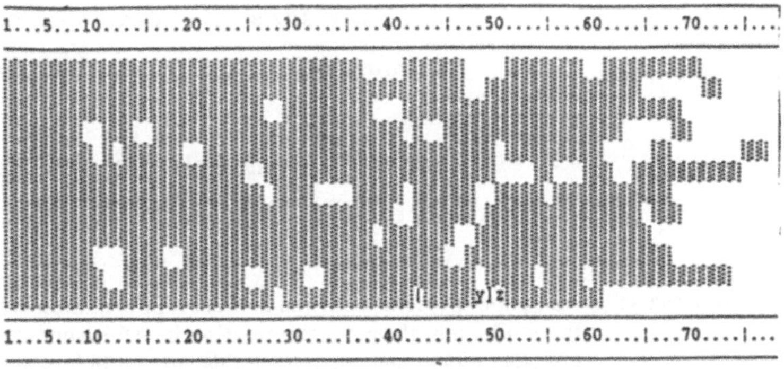

Block 9 fixed : { y] z
Block 10 active: - - - - - - -
Block 11 active: - - - - - - -

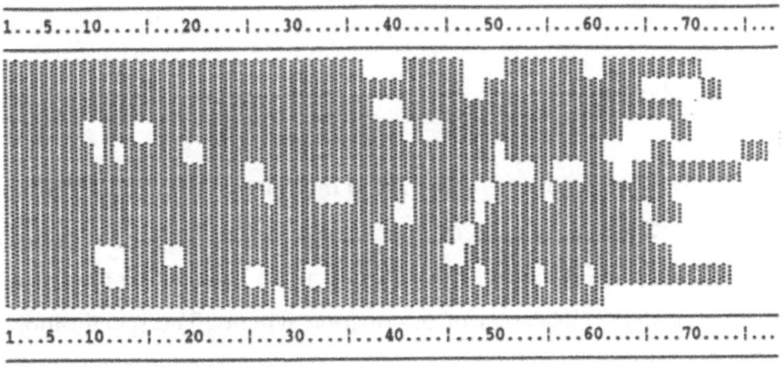

The final assignments are summmarized in the following table:

step i	block i-1	S_{i-1}	block i	S_i	block i+1	S_{i+1}
2	D b A G F H V	a	- - - - - - -		- - - - - - -	
3	D b A G F H V	a	a L j I K m Q	a	- - - - - - -	
4	D b A G F H V	f	a L j I K m Q	a	M N J P Y X W	a
5	a L j I K m Q	f	M N J P Y X W	a	1 ^ R s S Z |	a
6	M N J P Y X W	f	1 ^ R s S Z |	a	B d _ ` r [T	a
7	1 ^ R s S Z |	f	B d _ ` r [T	a	E c C e f t o	a
8	B d _ ` r [T	f	E c C e f t o	a	k g i O w n h	a
9	E c C e f t o	f	k g i O w n h	a	v p x u U q \	a
10	k g i O w n h	f	v p x u U q \	a	{ y] z	a
11	v p x u U q \	f	{ y] z	a	- - - -	
12	{ y] z	f	- - - - - - -		- - - - - - -	

S_i = block-status in step i;   a = active, f = fixed

5. References

Fahrion R. (1986): A Parallelization Approach for Sequencing Problems. In: Methods of Operations Research (Domschke W. et al. (eds.), vol. 57, 607-615).

Haupt R. (1989): A Survey of Priority Rule-Based Scheduling, OR-Spektrum 11, 3-16.

A Generalized Classification Model for Information Architecture Specification

Derk Jan Kiewiet, University of Groningen

Summary: This paper presents a formal model for information architecture specification. Because the model has also an interpretation in cell manufacturing formation, and reasonable effective techniques for solving this latter problem are available, these techniques can also be applied to information architecture specification.

Introduction

Many organizations pay attention to Information Planning. Information Planning can be defined as: the decision making process that leads to the definition of an information system plan for the organization (Boersma 1989). This plan (the Information Plan) can be used by the organization for the analysis and design of information systems. The specification of all the information systems together in such an Information Plan is the information system architecture of the organization. The specification of an information system architecture is a key element in Information Planning (Mantz et. al. 1990).

Information Architecture Specification

Information systems are designed to replace or support information processing activities in an organization. So, the starting point for the specification of an information architecture are the information processing activities in the organization, the so-called business processes. These business processes operate on entity types: aggregates of data in the organization. Usually, there are three possibilities for business processes to operate on entity types:

- a business process only uses data of a certain entity type
- a business process creates or changes data of a certain entity type
- a business process does nothing with the data of a certain entity type

Business processes, entity types and operations are represented in an information matrix, with the rows representing the business processes, the columns representing the entity types, and the cells representing the type of operation (IBM 1984).

	gegevensklasse 1	gegevensklasse 2	gegevensklasse 3	gegevensklasse 4	gegevensklasse 5	gegevensklasse 6	gegevensklasse 7	gegevensklasse 8
bedrijfsproces 1	U	U	C			C	U	
bedrijfsproces 2	C		C			U		
bedrijfsproces 3	C		C				U	
bedrijfsproces 4			U		U		C	U
bedrijfsproces 5	U				C		C	U
bedrijfsproces 6					C		U	
bedrijfsproces 7	U	C	U					U
bedrijfsproces 8		C			U	U	U	C
bedrijfsproces 9		U						C
bedrijfsproces 10				C				C

Figure 1 Information matrix for 10 business processes and 8 entity types. Relations (0, U, C) between business processes and entity types are in the cells of the matrix.

Usually, an information planner does not design one integral information system for all information processing activities together, but several information systems are designed, each for only a limited number of business processes and entity types. In an information architecture, business processes and entity types in the information matrix are grouped in several information areas, which are the bases for information systems. One commonly used family of methods for information architecture specification transforms the information matrix in such a way that the columns of the information matrix are transposed so that the 'C-cells' of the matrix are on the topleft-bottomright diagonal of the matrix. Around this diagonal rectangles are distinguished, each rectangle identifying an information area. In these processes of transforming and distinguishing, three objectives are considered (Martin 1982, Theeuwes 1986):

- localization of data-creation
- minimization of data-exchangement
- maintainability of (potential) information systems in an information area

One major criticism of these information architecture specification methods is that these objectives are *not* operationalized. Because of this, information architecture specification has substantive subjective component. Research carried out by our organization has shown that experienced information planners were not able to rank information system architectures in a consistent and consequent manner (Ebels 1991). More objective methods and techniques for information architecture specification have to be looked for.

A Generalized Classification Model

One possible area for such techniques stems from the recognition that information architecture specification can be considered as an interpretation of a classification model:

- there exist two sets B and D:
  - $B = \{b_1,...,b_M\}$
  - $D = \{d_1,...,d_N\}$
- there is a mapping u: $B \times D \rightarrow Z$, Z is a set with $|Z| \geq 2$ and in Z is a void-element. The mapping u has the following constraints:

$$\begin{cases} (\exists\, Z' \subseteq Z\backslash\{void\})\ (\forall\, d \in D)\ (\exists\, b \in B)\ u(b,d) \in Z' \\ (\exists\, Z'' \subseteq Z\backslash\{void\})\ (\forall\, b \in B)\ (\exists\, d \in D)\ u(b,d) \in Z'' \end{cases}$$

- there exists a mapping p_1: $Z \rightarrow N$
- define the matrix $X = [x_{ij}]$, $i = 1,...,M$, $j = 1,...,N$:
  - row i corresponds to b_i, $i = 1,...,M$
  - column j corresponds to d_j, $j = 1,...,N$
  - $x_{ij} = p \circ u(b_i,d_j)$, $i = 1,...,M$, $j = 1,...,N$
- find those permutations $\pi_1 = (i_1,...,i_M)$ of the rows of X, and $\pi_2 = (j_1,...,j_N)$ of the columns of X such that ME $(\pi_1 \circ \pi_2 (X))$ is maximal. ME is defined for a matrix $A = [a_{ij}]$, $i = 1,...,M$, $j = 1,...,N$, as follows:

$$ME(A) = \frac{1}{2}\sum_{i=1}^{M}\sum_{j=1}^{N}(a_{ij})^k \cdot \left[(a_{i,j+1})^k + (a_{i,j-1})^k + (a_{i+1,j})^k + (a_{i-1,j})^k\right]$$

  with $a_{0,j} = a_{M+1,j} = a_{i,0} = a_{i,N+1} = 0$, $i = 1,...,M$, $j = 1,...,N$
- there exists a mapping p_2: $Z \rightarrow \{0,1\}$
- in the matrix X, x_{ij} becomes: $x_{ij} = p_2 \circ u(b_i,d_j)$, $i = 1,...,M$, $j = 1,...,N$
- find (for an intuitive meaning of order- and block-partition see figure 2):
  - an order-partition $P(B_{\pi1})$ of $B_{\pi1} = \{b_{i1},...,b_{iM}\}$
  - an order-partition $P(D_{\pi2})$ of $D_{\pi2} = \{d_{j1},...,d_{jN}\}$
  - a block-partition $P(B_{\pi1} \times D_{\pi2})$ of $\pi_1 \circ \pi_2 (X)$ such that

    is maximal; with:
    - S_1:

$$F = \frac{\frac{1}{W}\sum_{w=1}^{W} S_2(B_w,B_w)}{1 + \frac{1}{W(W-1)}\sum_{\substack{w,v=1\\ w\neq v}}^{W} S_2(B_w,B_v)}$$

- S_1: $(B \times D) \times (B \times D) \rightarrow N$
- $$S_1((b_i,d_j),(b_k,d_l)) = p_2 \circ u(b_i,d_l) + p_2 \circ u(b_k,d_j)$$

- S_2:
  - S_2: $P(B \times D) \times P(B \times D) \rightarrow N$
  - $$S_2(B_{w_1},B_{w_2}) = \frac{1}{|B_{w_1}|\cdot|B_{w_2}|} \sum_{(b_i,d_j) \in B_{w_1}} \sum_{(b_k,d_l) \in B_{w_2}} S_1((b_i,d_j),(b_k,d_l))$$

Information architecture specification is an interpretation of this classification model:

- B is the set of business processes, and D the set of entity types
- u: $B \times D \rightarrow \{void, z_1, z_2\}$
  u (b,d) = 0 = void, business process b does nothing with data of entity type d,
  u (b,d) = C = z_1, business process b creates and modifies data of entity type d,
  u (b,d) = U = z_2, business process b only uses data of entity type d.
  Both constraints for u hold because data of every entity type has to be created, and every business has to process data of an entity type
- p_1: $\{0, U, C\} \rightarrow \{0, 1, 10\}$; $p_1(0) = 0$, $p_1(U) = 1$, $p_1(C) = 10$
- finding permutations π_1 and π_2 such that ME is maximal is an operationalization of localization of data-creation
- p_2: $\{0, U, C\} \rightarrow \{0, 1\}$; $p_2(0) = 0$, $p_2(U) = 1$, $p_2(C) = 1$
- the numerator of the objective function F is an operationalization of maintainability, the denominator of minimization of data-exchange

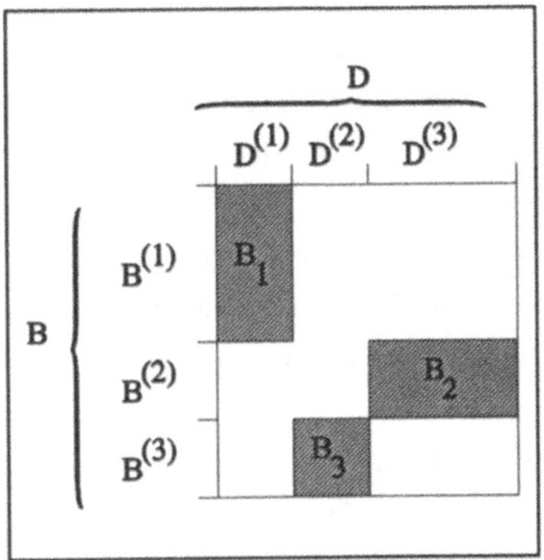

Figure 2 Order-partitions $P(B) = \{B^{(1)}, B^{(2)}, B^{(3)}\}$ and $P(D) = \{D^{(1)}, D^{(2)}, D^{(3)}\}$, and block-partition $P(B \times D) = \{B_1, B_2, B_3\}$

So information architecture specification is a classification problem. It can be shown that this classification problem (i.e. the specification of the partitions P(B) and P(D)) is NP-complete. Therefore, it is highly unlikely to find an effective algorithm for determining such partitions.

Cell Manufacturing Formation

In solving information architecture specification as a classification problem we have to rely on heuristics: classification techniques. Although the application of classification techniques for information systems architecture specification is described in the literature, applying such techniques is not widespread. Therefore, little research has been done on this topic, and the strengths and weaknesses of using classification techniques are not ascertained. However, there exists a problem that is an interpretation of the same classification model of which information architecture specification is an interpretation. This problem is cell manufacturing formation in group technology. In this problem, machines and parts (both needed for producing a certain product) are arranged into cells, making use of routing information. This routing information describes which parts are processed on which machines. The cells are so constructed that machines that process the same parts, and parts that are processed by the same machines, are grouped into one machine-part cell. Objectives for cell formation are:

- minimization of inter-cell movements
- maximization of cell usage

The claim is that cell manufacturing formation is also an interpretation of the proposed classification model:

- B is the set of machines, and D the set of parts
- $u: B \times D \rightarrow \{0,z_1\}$

  $u(b,d) = 0$: part d is not processed on machine b

  $u(b,d) = z_1 = 1$: part d is processed on machine b

  Both constraints for u hold because every machine has to process at least one part (otherwise it is useless in the production process), and every part has to be processed by at least one machine (parts are not finished products)
- the mapping p_1 is trivial, as is the matrix X
- finding permutations π_1 and π_2 such that ME is maximal ensures is an operationalization of localization of parts-processing
- the mapping p_2 is trivial
- the numerator of the objective function F represents maximization of cell usage, the denominator represents minimization of inter-cell movements

So, information architecture specification and cell manufacturing formation resemble each other: they are interpretations of the same classification model. The benefit of this similarity is that several classification techniques, used often in the area of cell manufacturing formation, can be applied in the area of information systems architecture specification.

Transfer of techniques from CMF to IAS

One family of transferable techniques are cluster techniques (Jain and Dubes 1988), the application of these techniques is called cluster analysis. Cluster analysis can be defined as: the dividing of a set of objects into several subsets with the use of similarity measurements. Each subset is called a cluster, and all clusters together form a clustering. The similarity measurements are the result of the comparison of pairs of objects. Given similarity measurements, a cluster analysis can result in different clusterings. These differences originate in the exact order in which objects and already existing clusters are joined to form new clusters.
In general, one can say that a clustering of objects depends on:

- how the similarity measurements are defined
- which cluster technique is used: how the objects and already existing clusters are joined to form new clusters

The similarity of objects can be established in several ways. For example, the objects can be compared directly, by asking people their opinion on the similarity of objects. However, this direct determination of similarity measurements has several methodological difficulties. It is also possible to compare the objects indirectly by means of the scores on a number of attributes. These scores are then combined into one single measurement. The combining of attribute scores into one measurement can be done in a lot of different ways. This results in different similarity measurements.
The definition of similarity measurements depends on:

- the attributes that are used and the scores on these attributes
- the similarity measure that is used: in what way the several attribute scores are combined into one single measurement

The use of cluster techniques for the specification of an information architecture have shown promising results (Stegwee 1991).

An other established family of techniques are neural networks. Neural networks can be described as a network of connected processors. There are three types of processors: input, output and intermediate. The connections between the processors are weighted. These weights can be adapted in a so-called learning phase of the neural network. The output of a processor is calculated from local data and inputs received via the connections. Although neural networks are used in several settings, they are very suited for classification tasks.

Discussion

Because the process of information architecture specification is open for subjective interpretation, a more formal approach was sought. This approach, information architecture specification as an interpretation of a generalized classification model, gives the information planner a deeper understanding which choices he has to make, and when. Also, by identification of cell manufacturing formation as an other interpretation of the generalized classification model, the information planner has the availability of new techniques for information architecture specification. Although the research hasn't finished yet, results are promising.

References.

Boersma, S.K.Th. *Beslissingsondersteunende Systemen: een praktijkgerichte benadering (Decision Support Systems: a practice base approach).* Schoonhoven, The Netherlands: Academic Service, 1989.

Ebels, E.J. *A prototyping cycle in the development of a DSS for IA specification*. Groningen, The Netherlands: Masters Thesis for Department of mathematics and computing science, 1991.

International Buseniss Machines. *Business Systems Planning - Information Systems Planning guide.* GE20-0527, Fourth edition. June 1984.

Jain, A.K., and Dubes, R.C. *Algorithms for Clustering Data*. Englewood Cliffs, New Jersey: Prentice Hall, 1988.

Mantz, E.A., Lieshout J.T.H.C. van, and Zijden F.A.P. van der. Omgaan met informatiebeleid en informatieplanning; bevindingen van een vijfde praktijkonderzoek. *Informatie*, Nr. 1, p. 23-31, 1990.

Martin, J. *Strategic Data-Planning Methodologies.* Englewood Cliffs, New Jersey: Prentice Hall, 1982.

Stegwee R.A. *Division for Conquest. Decision support for Information Architecture Specification.* Groningen,The Netherlands: Thesis, 1992.

Theeuwes, J.A.M. *Voorzien van informatie. Methoden voor informatiebeleidsvorming en informatieplanning.* Eindhoven, The Netherlands: Thesis, 1986.

UNTERNEHMUNGSWEITE ALLOKATION UND REPLIKATION VON DATEN IN EINER MULTI-DATENBANK-SERVER-ARCHITEKTUR

Hans-Jürgen König, Universität Gießen

Das Erzielen von Wettbewerbsvorteilen erfordert von einer Unternehmung eine schnelle und flexible informationslogistische Versorgung ihrer Aufgaben- und Entscheidungsträger mit entscheidungsrelevanter Information. Die Gestaltung des informationstechnischen Systems einer Unternehmung hat sich an dieser informationslogistischen Zielsetzung zu orientieren. Insbesondere sind dabei die sich im Zeitablauf verändernden ökonomischen Ziele und Rahmenbedingungen zu berücksichtigen. Die hierzu erforderliche flexible Nutzung von informationellen Ressourcen der Unternehmung wird erstmalig durch dezentrale interoperable Systemarchitekturen eröffnet. Derartige Systemarchitekturen ermöglichen hierbei eine evolutionäre Integration historisch gewachsener zentralistischer oder isolierter IS-Strukturen der Unternehmung. Aus einer datenorientierten Sichtweise heraus stellen vernetzte Multi-Datenbank-Server-Architekturen eine adäquate Möglichkeit dar, der genannten Zielsetzung zu genügen. Im Beitrag wird in diesem Kontext ein Modell zur Bestimmung der optimalen Allokation und Replikation von Daten vorgestellt, welches auf einem ganzzahligen nichtlinearen Optimierungsansatz basiert. Im Modell wird eine Datenbankumgebung auf Basis des relationalen Datenmodells unterstellt. Berücksichtigt wird die Topologie der lesenden und schreibenden Informationsnachfrage, die Topologie der Multi-Datenbank-Server-Architektur sowie das in Kauf zu nehmende Ausmaß an Redundanz. Das Modell wird im Beitrag letztlich für betriebswirtschaftlich relevante Szenarien eingesetzt, um die Gestaltung einer durchgängigen dezentralen Datenarchitektur in der Unternehmung zu unterstützen.

DER EINSATZ EINES WISSENSBASIERTEN, PROBABILISTISCHEN SYSTEMS BEI DER ANALYSE MAKROÖKONOMISCHER ZUSAMMENHÄNGE

Friedhelm Kulmann, FernUniversität Hagen
Heinz Peter Reidmacher, FernUniversität Hagen

Kurzfassung: Bei der Analyse statistischer Daten und dem Entwurf von Modellen werden oft Annahmen über die Unabhängigkeit einzelner Variabler getroffen. Diese Annahmen beruhen entweder auf der Einschätzung von Experten, insbesondere auf deren Erfahrungswissen oder sind Ergebnis aufwendiger Berechnungen. An einem Beispiel aus dem Gebiet der Makroökonomie wird gezeigt, daß es möglich ist, Abhängigkeitsstrukturen mit Hilfe des wissensbasierten Systems ***SPIRIT*** zu analysieren, die fundierte Basis weiterer Untersuchungen sein können.

Abstract: Given statistical data it is often necessary to have informations about the dependencies between the observed variables. To find out you have to make complex analysis. The knowledge based system ***SPIRIT*** informs you about local dependencies which are usefull to build concret economical models.

1 Probabilistische Systeme zur Entscheidungsunterstützung

Wissensbasierte Systeme sind nach einer anfänglichen Phase der rein deterministischen Regelauswertung immer mehr in der Lage, auch vages, unsicheres und subjektives Wissen zu verarbeiten. Geht man davon aus, daß Experten ein Wissensgebiet analysieren und als Regelbasis formulieren, können einem Entscheidungsfäller mit einem Expertensystem wertvolle Hilfen für die Entscheidungsfindung an die Hand gegeben werden. Wichtig für die problemadäquate Modellierung ist jedoch die Kenntnis über bestehende Zusammenhänge der Problemvariablen.

Ein Forschungszweig richtet sein Interesse auf die probabilistische Erfassung und Auswertung von Daten, da Wahrscheinlichkeiten im frequentistischen Sinne besonders geeignet sind, die durch eine Vielzahl von einzelnen Beobachtungen ableitbaren Zusammenhänge darzustellen.

Duda verwendet in dem Ende der 70er Jahre entwickelten Expertensystem PROSPECTOR auf Wahrscheinlichkeiten basierende Schlußstärkefaktoren, um Abhängigkeiten darzustellen [2, 3]. In anderen Repräsentationen, wie beispielweise Bayes- oder Markov-Netze werden Abhängigkeiten visualisiert und unter Angabe von Wahrscheinlichkeitsverteilungen quantifiziert. Ausführliche Darstellungen dieser Methoden findet man bei Pearl [5] und Neapolitan [4].

Obwohl theoretisch anspruchsvoll ist den genannten Konzepten gemeinsam, daß ein Experte vorab das Wissen strukturieren muß. Nur wenn das Abhängigkeitsmodell nahe genug an der Realität liegt, kann ein solches System eine echte Entscheidungshilfe sein.
Das probabilistische System ***SPIRIT*** benutzt Beobachtungsdaten, um eine Wissensbasis zu strukturieren, die Abhängigkeiten zwischen den beobachteten Variablen auf der Basis der Wahrscheinlichkeitstheorie zu quantifizieren und in der Lage ist, diese Informationen für Prognosen zu nutzen.

2 *SPIRIT*

Die wichtigsten Komponenten von ***SPIRIT*** sind Wissenbasis, Inferenz- und Lernkomponente.
Die in der Wissensbasis abgelegten Aussagen sind, repräsentiert durch Zufallsvariable, in Gruppen als sogenannte LEGs (local event groups) organisiert, zu denen jeweils eine Wahrscheinlichkeitsverteilung gespeichert ist. Durch das Mehrfachauftreten bestimmter Variabler in verschiedenen LEGs werden Verbindungen hergestellt, die für die Wissensverarbeitung von zentraler Bedeutung sind. Man spricht von LEG-Netzen, die auf ihren jeweiligen Schnittmengen dieselben Randverteilungen besitzen.
Der Inferenzmechanismus führt einen Update des gesamten LEG-Netzes aufgrund veränderter Randverteilungen durch. Diese Propagation von Informationen durch das Netz geschieht entropieoptimal, d.h. ohne Verwendung systemimmanenter Funktionen oder Parameter. Sie wird einerseits dazu benutzt, Prognosewahrscheinlichkeiten für eine frei wählbare Menge von Hypothesen zu berechnen, andererseits ist sie der zentrale Algorithmus bei der Adaption neuen Wissens. Die Wissensbasis wird zunächst mit der Gleichverteilung, die eine völlige Unkenntnis über jegliche Zusammenhänge repräsentiert, initialisiert. Die eingegebenen Beobachtungen führen zu einer Instantiierung der zugeordneten LEGs und mittels der oben angedeuteten Propagation zu einer Aktualisierung des gesamten Netzwerks. (Eine ausführliche Darstellung findet der Leser in [6].)
Das Wissen über Abhängigkeiten wird dann am besten adaptiert, wenn in der Wissensbasis die gemeinsame Verteilung abgelegt ist. Aus speicherplatztechnischen Gründen ist dies jedoch nicht immer möglich und aus rechenzeittechnischen Gründen auch nicht immer sinnvoll. An einem einfachen Beispiel aus der Makroökonomie wird aufgezeigt, wie sich eine Reduzierung der LEG-Struktur auf die Güte der Prognoseergebnisse auswirken kann.
Bevor in Kapitel 4 beschrieben wird, wie mit ***SPIRIT*** statistische Daten ausgewertet und strukturiert werden können, erfolgt in Kapitel 3 zunächst eine Beschreibung der verwendeten makroökonomischen Größen.

3 Makroökonomische Größen

Bei unseren Untersuchungen beschränken wir uns auf das Zusammenwirken der 5 Variablen Ersparnis, Volkseinkommen, Geldmenge, Nominalzins und Investition, die in der Makroökonomie den Güter- und Geldmarkt bestimmen.

Mittels komplexer mathematischer Analysen wird in der Makroökonomie ein funktioneller Zusammenhang zwischen den genannten Größen abgeleitet. Unterstellt man eine geschlossene Volkswirtschaft ohne staatliche Aktivitäten, so ergibt sich zum Beispiel, daß die Ersparnis S der privaten Haushalte bei steigendem Einkommen Y anwächst und die Investitionen I bei steigendem Nominalzins i abnehmen.

In makroökonomischen Modellen werden in der Regel Annahmen der obigen Art gemacht sowie konkrete funktionelle Zusammenhänge zwischen den betrachteten Größen unterstellt. Diese Einschränkung bewirkt, daß davon abweichende, in den vorgegebenen Daten implizit enthaltene Zusammenhänge nicht abgebildet werden können.

Eine allgemeine Möglichkeit, Abhängigkeiten zu repräsentieren, erlaubt das wissensbasierte System ***SPIRIT***. Im nächsten Kapitel wird gezeigt, wie mittels ***SPIRIT*** Zusammenhänge zwischen den 5 oben genannten Größen auf der Basis statistischen Datenmaterials der Bundesrepublik Deutschland der Jahre 1962-76 [1] abgebildet und schließlich analysiert werden können.

4 Modellierung mittels *SPIRIT* und erste Resultate

Da in ***SPIRIT*** Aussagen repräsentiert werden, ist die direkte Darstellung von stetigen Größen nicht möglich. Darüberhinaus ist es sinnvoll, relative Werte zu betrachten, da eine Veränderung makroökonomischer Variabler prognostiziert werden soll. Aus diesen Gründen betrachten wir für jede Größe die Zuwachsraten in % vom Vorjahr und unterscheiden jeweils vier gleich große Klassen, die die Bandbreite der möglichen Werte abdecken (siehe hierzu Tabelle 1). Im folgenden werden sowohl der Lernvorgang als auch die Prognose auf der Basis dieser Klassen durchgeführt.

Variable	S		Y		M		i		I	
Klasse	von	bis	von	bis	von	bis	von	bis	von	bis
1	-14	-5.1	0.5	2.7	-1.1	1.3	-25.5	-14.1	-14	-6
2	-5	3.9	2.8	5.0	1.4	3.8	-14	-2.6	-5.9	2.1
3	4	12.9	5.1	7.3	3.9	6.3	-2.5	8.9	2.2	10.2
4	13	21.9	7.4	9.6	6.4	8.8	9	20.4	10.3	18.3

Tabelle 1: Klassenbildung (Angaben in % Abweichung vom Vorjahr)

Nach der Aufbereitung der Daten werden sukzessive alle Realisierungen mittels ***SPIRIT*** erlernt. Dabei wird als Wissensbasis eine gemeinsame Verteilung über alle Variablen zugrundegelegt, so daß alle denkbaren Abhängigkeiten erfaßt werden können. Aufgrund des dafür erforderlichen hohen Speicheraufwands ist es wünschenswert, die Struktur zu reduzieren und damit auch die Berechnungen zu beschleunigen. Es wurde bereits angedeutet, daß durch eine Aufspaltung der gemeinsamen Verteilung in LEGs der Speicheraufwand reduziert werden kann. Ein möglicherweise dadurch verursachter Informationsverlust wirkt sich jedoch negativ auf die Güte von Prognosen aus.

Ohne Berücksichtigung des zugrundeliegenden Problems werden im folgenden 5 verschiedene LEG-Netze untersucht. In der ersten Struktur werden die Knoten isoliert (1) abgelegt; die Struktur (2) enthält über alle Paare von Variablen, die Struktur (3) über alle Dreiergruppen und die Struktur (4) über alle Vierergruppen jeweils ein LEG. Die gemeinsame Verteilung über alle 5 Variable bezeichnen wir mit Struktur (5).

Zur Beurteilung dieser ausgewählten Strukturen wurden die in der Wissensbasis gelernten Zusammenhänge zwischen den makroökonomischen Größen anhand des vorliegenden Datenmaterials überprüft. Dazu werden unter der Kenntnis einer Teilmenge der verwendeten Größen die restlichen Variablen prognostiziert. Auf der Grundlage der Klasseneinteilung wird dann der Fehler einer Prognose ermittelt. Dieser sei gleich 0, falls die geschätzte mit der realen Klasse übereinstimmt und gibt sonst den Abstand zwischen diesen Klassen an. Die durchschnittlichen Fehler in Abhängigkeit der gewählten Struktur sind in Abbildung 1 dargestellt.

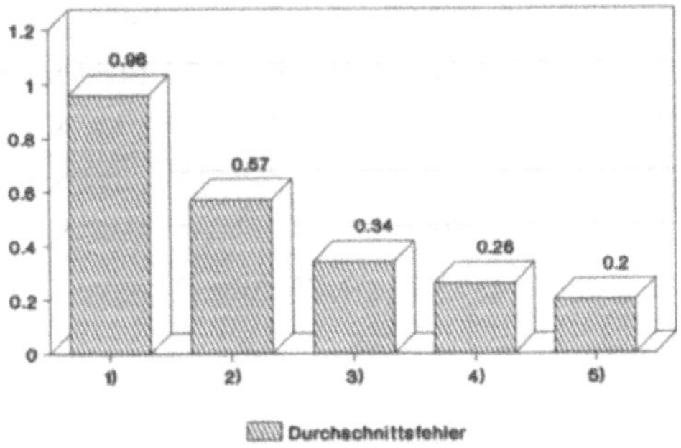

Abbildung 1: Durchschnittsfehler der Prognosen bei verschiedenen Wissenstrukturen

Wie erwartet, verringert sich der Fehler bei Anreicherung der Struktur und besitzt sein Minimum bei der vollständigen Vernetzung der 5 betrachteten Variablen. Da bei der Vergrößerung der Wissensstruktur nicht nur die Güte von Prognosen zunimmt, sondern auch die erforderliche Speicherkapazität wächst, muß in Abhängigkeit des konkreten Problems und der zur Verfügung stehenden Hardware eine Kompromißlösung gefunden werden.

In den symmetrischen Strukturen (1)-(5) werden die eigentlichen Abhängigkeiten, die sich aus der konkreten Problemstellung ergeben, nicht berücksichtigt. Zweckmäßiger ist es, unter Zugrundelegung der in der Wissensbasis erkennbaren Abhängigkeiten zwischen den abgespeicherten Variablen eine sinnvolle Struktur zu bestimmen. Variable, die stark abhängig sind, bleiben in einem LEG zusammengefaßt, während zwischen unabhängigen Größen die Verbindungen aufgespalten werden. Bei dieser Untersuchung wird als Reduktionskriterium der Fehler der Prognosen zwischen Teilmengen der makroökonomischen Größen zugrundegelegt.

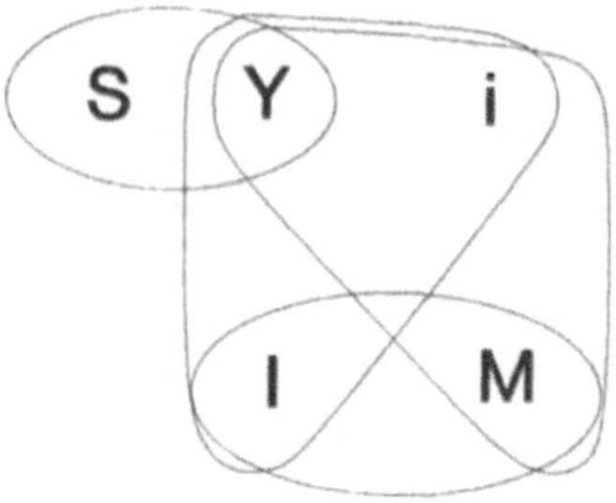

Abbildung 2: Suboptimale Wissensstruktur (0)

Eine resultierende, geeignete Struktur ist in Abbildung 2 dargestellt. Abbildung 3 zeigt, daß das Verhältnis von erforderlichem Speicheraufwand und Durchschnittsfehler für die Struktur (0) im Vergleich zu anderen Strukturen besonders günstig ist. Dieses Ergebnis steht auch mit den Erkenntnissen der Makroökonomie im Einklang.

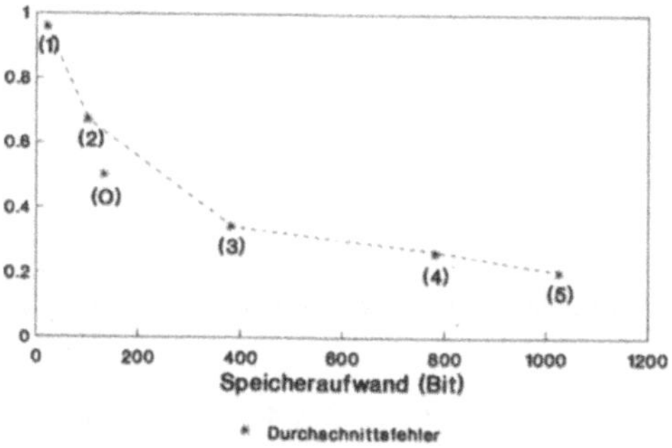

Abbildung 3: Der durchschnittliche Fehler und die Speicheraufwand der Strukturen (0)-(5)

5 Zusammenfassung und Ausblick

Es wurde deutlich, daß die Wissensstruktur einen wesentlichen Einfluß auf die Abbildung der in realen Daten enthaltenen Zusammenhänge besitzt. Aus der Bewertung durchgeführter Prognosen lassen sich Hinweise für eine Verbesserung der Wissensstruktur bezüglich der Relation des erforderlichen Speicheraufwands zur Prognosegüte ableiten.

Bei Automatisierung der oben beschriebenen Vorgehensweise ist ***SPIRIT*** in der Lage, zu einem konkreten Problembereich eine günstige Wissensstruktur zu ermitteln. Aus der darin festgelegten Gruppierung der betrachteten Variablen lassen sich Hinweise für eine makroökonomische Modellbildung ableiten.

6 Literaturverzeichnis

[1] CEZANNE, W.: Grundzüge der Makroökonomik, (Oldenbourg Verlag, München, 1982).

[2] DUDA, R.O.; GASCHNIG, J.; HART, P. : Model Design in the PROSPECTOR system for mineral exploration, in: Michie, D.: Expert Systems in the microelectronic age, (Edinburgh Press, Edinburgh, 1979), 153-167.

[3] DUDA, R.O.; HART, P.; KONOLIGE, K.; REBOH, R. : A Computer-based Consultant for Mineral Exploration , Artificial Intelligence Center, SRI International, Menlo Park, CA, September (1979).

[4] NEAPOLITAN, R.E. : Probabilistic Reasoning in Expert Systems (John Wiley & Sons, New York, 1989).

[5] PEARL, J. : Probabilistic Reasoning in Intelligent Systems (Morgan Kaufmann, Palo Alto, CA, 1988).

[6] RÖDDER, W.: Symmetrical Probabilistic Intensional Reasoning in Inference Networks in Transition, Proceedings of The World Congress on Expert Systems, Orlando, Florida, 530-540, 1991.

DAS PROBABILISTISCHE, WISSENSBASIERTE SYSTEM SPIRIT

Wilhelm Rödder, FernUniversität Hagen
Heinz Peter Reidmacher, FernUniversität Hagen

SPIRIT ist als wissensbasiertes System im weitesten Sinne den Expertensystemen zuzuordnen, obwohl es in seiner Konzeption und den verwendeten Methoden besondere Charakteristika aufweist, die es von herkömmlichen Entwicklungen unterscheidet. In diesem Artikel werden Konzepte, die insbesondere die Wissensrepräsentation, -akquisition, -verarbeitung und die Präsentation des Wissens in *SPIRIT* betreffen, vorgestellt.

Informationen, die in Form von Aussagen über eine beliebige Wissensdomäne vorliegen, werden probabilistisch durch eine Menge von Randverteilungen über Aussagengruppen repräsentiert. Die Wissensakquisition erfolgt in Abhängigkeit von statistischen Daten oder durch die Eingabe von regelbasiertem, subjektivem Expertenwissen; realisiert wird sie durch eine Bayessche Aktualisierung mit exponentieller Glättung von Wahrscheinlichkeiten der beobachteten Ereignisse. Die Strukturierung der Wissensbasis in Clustern von Aussagen erfolgt dynamisch über eingegebene Beobachtungen. Durch diesen induktiven Lernprozeß werden auch Zusammenhänge zwischen den repräsentierten Aussagen adaptiert, die Grundlage für die Strukturierung der Wissensbasis, d.h. insbesondere die Bildung der Gruppen, sind. Dadurch wird genau zwischen den Aussagen, über die starke Abhängigkeiten beobachtet wurden, auch eine reiche Informationsstruktur zur Verfügung gestellt. Da die Wissensverarbeitung entropieoptimal durchgeführt wird, sind Prognosen garantiert, die nicht durch methodisch bedingte Annahmen beeinflußt werden. Prognoseergebnisse und erlernte Informationen, die zunächst als Randverteilungen über Hypothesenmengen bzw. Aussagengruppen vorliegen, werden analysiert und als aussagekräftige wenn-dann-Regeln dargestellt. Dazu wird die Menge aller signifikanten Abhängigkeiten zwischen Aussagenmengen, die implizit in einer Wahrscheinlichkeitsverteilung enthalten sind, solange reduziert, bis mit einer möglichst kleinen Regelbasis die wichtigsten Informationen dargestellt werden können.

Abschließend werden typische Anwendungsbereiche skizziert, in denen die Vorteile von *SPIRIT*, insbesondere Abhängigkeiten selbständig zu erkennen, die erlernten Zusammenhänge entropieoptimal zu verarbeiten und die erlernten und prognostizierten Resultate verständlich darstellen zu können, zum tragen kommen.

OPTIMIERUNG VON REGELMENGEN IN EXPERTENSYSTEMEN DURCH ANALYSE ÄHNLICHER UND PLAUSIBLER FÄLLE

Heribert Schimpe, RWTH Aachen

Viele Expertensystem-Projekte werden nach der Erstellung eines Prototypen abgebrochen. Oftmals liegt der Grund hierfür in der Schwierigkeit, auch die "komplizierteren" Fälle und Ausnahmesituationen zu benennen und in die Regelmenge zu integrieren. Aufgrund der Komplexität dieses Problems besteht zunehmend Bedarf nach vollautomatischen Verfahren zur Optimierung einer Regelmenge in Expertensystemen.
Dabei ist insbesondere die Verwendung von "Fällen", d.h. von Beispielsituationen, bekannt. Hierzu existieren bereits Verfahren aus dem Bereich des maschinellen Lernens, die entweder einzelne Fälle analysieren und das Ergebnis zu verallgemeinern versuchen oder eine Vielzahl von Fällen auf Gemeinsamkeiten untersuchen. Ergebnis ist in jedem Fall die Generierung neuer Regeln.
Im vorliegenden Beitrag wird aufgezeigt, daß zum einen "ähnliche" Fälle bei der Entscheidungsfindung eines Expertensystems hilfreich sein können und zum anderen eine "plausible Ähnlichkeit" definiert werden kann, die insbesondere bei der Suche nach falschem oder unvollständigem Wissen von Nutzen ist.
Wie auch bei den in der Literatur weit verbreiteten Verfahren des *Erklärungsbasierten Lernens* geschieht die Analyse eines Falles über den Aufbau eines sogenannten *Erklärungsbaumes*. Zur Nutzung von als "ähnlich" klassifizierten Fällen in neuen Anwendungssituationen ist eine spezielle Darstellung des Erklärungsbaumes notwendig, um auf alle Elemente des Baumes gezielt zugreifen und eine vorwärtsgerichtete Inferenz simulieren zu können.
Ähnlichkeit von Fällen wird hier in einer deduktiven Form definiert, d.h. als "beweisbare" Ähnlichkeit, die anders als induktive Ähnlichkeit auf Vermutungen verzichten kann. Dabei liegt die Philosophie zugrunde, daß zwei Fälle nur dann ähnlich sein können, wenn sie wenigstens in Teilaspekten identisch sind.
Plausible Erklärungen eines Falles sind nicht deduktiv herleitbar, sondern haben einen induktiven Charakter: es handelt sich lediglich um eine Hypothese, daß die Erklärung Gültigkeit besitzt; im allgemeinen allerdings eine Vermutung mit hoher Wahrscheinlichkeit. Zum Aufbau einer plausiblen Erklärung wird die Verwendung (induktiver) Default-Regeln vorgeschlagen. Plausible Erklärungen sind einerseits eine "Rettung" in Situationen, in denen das Hintergrundwissen nicht ausreicht, eine exakte (deduktive) Erklärung zu finden, vor allem aber sind plausible Erklärungen Indizien und Ansatzpunkte für falsches oder unvollständiges Wissen in Expertensystemen und können somit eine tiefliegende Wissensrevidierung oder -erweiterung leiten.

EIN EXPERTENSYSTEM ZUR UNTERSTÜTZUNG DER SUCHE IN LITERATURDATENBANKEN

Jürgen Siemer, Rheinische Friedrich-Wilhelms-Universität Bonn

Schon seit einiger Zeit liegen eine Fülle von Literaturdatenbanken auf elektronischen Medien vor, die als Referenzdatenbanken Hinweise geben über Autor, Titel, Erscheinungsort und -jahr sowie oft über Kurzzusammenfassungen des Inhalts verfügen. In der Regel wird auf diese online unter Zuhilfenahme eines Dienstleisters, eines sogenannten Informationsvermittlers, zugegriffen.

Das Fachwissen dieses Informationsvermittlers liegt einerseits in expliziter Form vor, was vor allem für das Wissen über die Existenz und den Inhalt der verschiedenen Datenbanken sowie der verschiedenen Retrievalsprachen gilt, und ist andererseits in impliziter Form im Verhalten des Experten beim konkreten Problemlösen "versteckt". Als implizit ist das Wissen zu charakterisieren, das zur Identifizierung und Übersetzung des Informationsbedürfnisses in eine Datenbankabfrage nötig ist. Diese Datenbankabfrage wird mit Hilfe der Booleschen Algebra formuliert und verfolgt das Ziel, bei Minimierung der Kosten und Vermeidung unnötiger Information die relevanten Dokumente möglichst vollständig zu finden. Dieses implizite Wissen wurde nun durch konkurrente Protokollanalysen der Arbeit der Experten mit Tonband und Papier und anschließende Diskussion akquiriert. Die sich auf diese Weise bildende Sammlung von Fallbeispielen wurde dann auf Verhaltensmuster der Experten untersucht. Aus der Diskussion dieser Verhaltensmuster, die sich einer begrenzten Anzahl von Problemstellungen zuordnen lassen, ergaben sich schließlich Lösungsvorschläge für das Expertensystem.

Der mit Nexpert Object und Toolbook programmierte Prototyp, der sich auf einige landwirtschaftliche Datenbanken beschränkt, wendet Expertenwissen an, das es dem Nutzer ermöglicht, die Suchstrategie zu verbessern und übliche Fehler zu vermeiden. Er könnte damit ein wichtiger Schritt hin auf das Ziel des verbesserten Zugangs zu den Literaturdatenbanken werden.

PROZESSMODELLIERUNG MITTELS FUZZY PATTERN-KLASSIFIKATION

Norman Bitterlich, Technische Universität Chemnitz-Zwickau

Abstract

The fuzzy pattern classification turns out to be an efficient method to model complicated and complex processes. Doing this the interpretation of vague measuring data and incomplete knowledge inserts to the classification model. The classes are described by fuzzy sets in the feature space. The problem adequate shapes of the corresponding membership functions are based on the computer aided analysis of prototypical examples - the classificator design and optimization during the learning phase. The method is applicable to decision making and process monitoring if a pattern comparison is carried out by experts usually. The fuzzy pattern classification was tested in a lot of applications for instance in quality assurance in industriel fields successfully.

Zusammenfassung

Die Fuzzy Pattern Klassifikation ist eine leistungsfähige Methode, um komplizierte und komplexe Prozesse zu modellieren. Hierbei fließt die Auswertung von vagen Meßdaten und unvollständigen Prozeßwissen in das Klassifikationsmodell ein. Die Klassen werden durch unscharfe Mengen im Merkmalsraum beschrieben. Die problemangepaßte Form der entsprechenden Zugehörigkeitsfunktionen wird computergestützt aus prototypischen Beispielen ermittelt - der Klassifikatorentwurf und die -optimierung während der Lernphase. Die Methode ist für Entscheidungsfindung und Prozeßüberwachung anwendbar, wenn der Experte üblicherweise einen Mustervergleich auführen würde. Die Fuzzy Pattern Klassifikation wurde erfolgreich in zahlreichen Anwendungen, beispielsweise der industriellen Qualitätssicherung, erprobt.

1. Einführung

In vielen Bereichen beherrscht der Mensch - gestützt auf seine Erfahrungen und sein Wissen - selbst komplizierte und komplexe Prozesse. Er ist in der Lage, sich effizient und zielorientiert für eine Handlungsstrategie zu entscheiden, indem er zahlreiche Phänomene überblickt und ihre Einflüsse bewertet, ohne die funktionellen Zusammenhänge genau kennen zu müssen. Vielmehr nutzt er die unvollständigen und ungenauen Prozeßinformationen, um die momentane Situation auf ihre Ähnlichkeit zu bekannten Prozeßzuständen einzuschätzen. Dabei realisiert der Experte oft eine enorme Informationsreduktion auf wenige Vergleichszustände.

Klassifikationsprinzipien können offensichtlich einen Beitrag leisten, diese Vorgehensweise zu objektivieren und zu automatisieren. Für eine adäquate Prozeßmodellierung ist jedoch die in der Entscheidungsfindung innewohnende Unschärfe nicht nur zu berücksichtigen, sondern bewußt auszunutzen. Hierfür bieten die Methoden, die auf der Theorie der Fuzzy Sets basieren, einen leistungsfähigen Lösungsansatz, nämlich die vagen Eingangsinformationen der unsicheren Prozeßkenntnisse mathematisch zu beschreiben und zu verarbeiten.

2. Das Klassifikationsmodell

Diese Verknüpfung von Klassifikationsprinzipien mit den Grundideen der Fuzzy Set Theorie wurde in Chemnitz anfangs der 70er Jahre begonnen [1]. Das daraus entwickelte Verfahren der Fuzzy Pattern-Klassifikation [2] eröffnete durch die Nutzung experimenteller und wissensbasierter Formen der Modellbildung neue Wege der Prozeßmodellierung, um in Ergänzung oder anstelle aufwendiger theoretischer Prozeßanalysen die praktischen Erfahrungen von Experten umzusetzen.

Das Fuzzy Klassifikationsmodell wird in einem - möglicherweise hochdimensionalen - Informationsraum durch einen Satz von Klassen in Form von Fuzzy Sets gebildet. Dieser Raum wird durch Merkmale aufgespannt, d.h. durch Prozeßinformationen, die zur Klassifikation einen Beitrag zu leisten scheinen. Diese Merkmale können unmittelbar Meß- oder Prozeßkenngrößen sowie mittels signalvorverarbeitenden Algorithmen abgeleitete Werte sein. Hierbei tritt Vagheit vorwiegend als Meßfehler oder Störgrößeneinfluß in Erscheinung. Aber auch verbale Beschreibungen mit all ihren subjektiven oder sprachlichen Unbestimmtheiten können als Merkmale in die Modellierung eingehen.

Die Klassen besitzen eine inhaltliche Bedeutung, beispielsweise Qualitäts-, Fehler- oder Steuerklassen. Sie sind damit eng mit Entscheidungen verbunden, etwa mit Qualitätseinstufungen, Nacharbeitsempfehlungen oder Handlungsstrategien.

3. Die parametrischen Zugehörigkeitsfunktionen

Entsprechend der Fuzzy Set Theorie wird jede Klassen durch eine Zugehörigkeitsfunktion charakterisiert.
Einheitlich werden als Funktionstyp verallgemeinerte Aizermansche Potentialfunktionen gewählt, die im eindimensionalen Fall (für reelle x) die folgende Form haben:

$$\mu(x) = \frac{1}{1+(b^{-1}-1)\left|\frac{x-s}{c}\right|^{d}} .$$

Dabei sind b, c, d und s gut interpretierbare Parameter, nämlich

s ... der Klassenschwerpunkt, für den stets $\mu(s) = 1$ gilt,
c ... die Klassengrenze und
b ... die Zugehörigkeit an dieser Klassengrenze mit $\mu(s+c) = \mu(s-c) = b$ sowie
d ... als Formparameter für die Unschärfe der Klasse.

Die n-dimensionale Verknüpfung ($(x_1, ..., x_n) \in R^n$) erfolgt nun durch den Ansatz

$$\mu(x_1, ..., x_n) = \frac{1}{1+\sum_{m=1}^{n}(b_m^{-1}-1)\left|\frac{x_m-s_m}{c_m}\right|^{d_m}} .$$

Mit diesen Zugehörigkeitsfunktionen, deren Parameter (auch links- und rechtsseitig vom Klassenschwerpunkt verschieden wählbar) an die konkrete Problemstellung anzupassen sind, können bereits unscharfe Einzelobjekte beschrieben werden. Die Unschärfen sind in diesem Fall als Meßungenauigkeiten interpretierbar. Sie könnten aber auch deutlich unterschiedlich auftreten, wie es bei Berücksichtigung der Aussagekraft von Meßverfahren für das Klassifikationsziel möglich wäre.

4. Klassifikatorentwurf und -einsatz

Die Ermittlung problemangepaßter Parameter der Klassen-Zugehörigkeitsfunktionen erfolgt in einer Lernphase. Um zu geeigneten Merkmalsdaten zu gelangen, ist eine intensive Prozeßanalyse notwendig. Für "learning by doing" sind aktive oder passive Experimente zu planen. Experten werden nach ihren Prozeßkenntnissen und -erfahrungen befragt, um aufbauend auf diesem Basiswissen nach geeigneten Merkmalen zu suchen. Mit meßtechnischen und (auch spezifischen) signalverarbeitenden Methoden erhält man schließlich die Datenbasis der Lernobjekte.

Ein wesentlicher Schritt zum Klassifikatorentwurf besteht in der Strukturbildung. Der Experte legt Klassen fest und ordnet die Lernobjekte in diese ein, wobei er seine mannigfaltigen

Informationsmöglichkeiten nutzen sollte. So können bei aktiven Experimenten die Einstellbedingungen Grundlage der Bewertung sein, auch visuelle und akustische Eindrücke werden die Einschätzung anstelle eher technischer Prozeßgrößen stärker prägen. Bei einigen Problemstellungen werden aber auch mathematische Clusterverfahren wertvolle Unterstützung zur Strukturbildung leisten können..

Es schließt sich die eigentliche unscharfe Beschreibung der Klassen an, indem aus den Merkmalsdaten der jeweils zu einer Klasse gehörenden Objekten die Parameter der Klassen-Zugehörigkeitsfunktionen erechnet werden. Abbildung 1 demonstriert ein 4-Klassen-Modell.

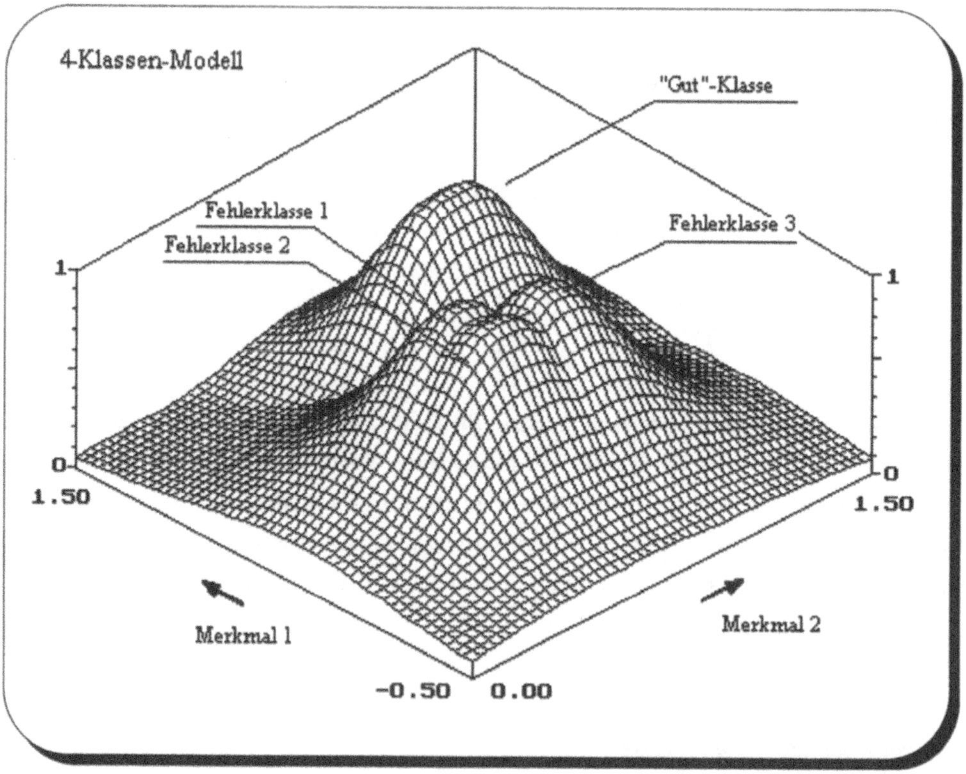

Abb. 1: Unscharfe Beschreibung von Klassen

Aufgrund der vielfätigen Eingriffsmöglichkeiten beim Klassifikatorentwurf - von der Merkmalsauswahl über subjektive Manipulation der Parameter bis zur wissensgestützten Konstruktion von Klassen, die im Experiment nicht zugänglich sind -, entstehen alternative Klassifikationsmodelle. Ihre Leistungsfähigkeit und Robustheit sind anhand erprobter Gütewerte einzuschätzen, um eine den Klassifikationszielen adäquate Auswahl zu treffen. Mit der Realisierung eines Einsatzsystemes mit prozeßnahem oder -integriertem Klassifikator findet die Lernphase ihren Abschluß.

Während des Klassifikatoreinsatzes sowohl im off-line-Betrieb als Beratungssystem als auch unter on-line-Bedingungen zur Überwachung und Steuerung laufender Prozesse werden Merkmalsdaten von Objekten unbekannter Klassenzuordnung dem Klassifikator vorgeführt. Entsprechend der berechneten Zugehörigkeitswerte zu jeder Klasse "feuern" die Klassen unterschiedlich. Als typische Klassifikationsergebnisse können folgende Situationen auftreten:

- die Zugehörigkeit zu einer Klasse ist im Vergleich zu denen der anderen Klassen deutlich höher (risikoarme Entscheidung),
- zwei (oder gar mehr) Klassen besitzen hohe Zugehörigkeitswerte, aber die (quantitativen) Unterschiede sind gering (risikobehaftete Entscheidung) oder
- alle Zugehörigkeitswerte übertreffen eine (festzulegende) Identifikationsschwelle nicht, d.h., das Objekt kann keiner Klasse zugeordnet werden (Abweisung).

Doch nicht allein die momentane Zustandsbewertung trägt zur Prozeßüberwachung bei. Von besonderen Interesse sind die Trendbeobachtungen der einzelnen Klassenzugehörigkeiten. So lassen sich Zustandsveränderungen und mögliche Klassenübergänge frühzeitig erkennen.

Abbildung 2 veranschaulicht schematisch die Lern- und Einsatzphase mit ihren Arbeitschritten und Zusammenwirken [3].

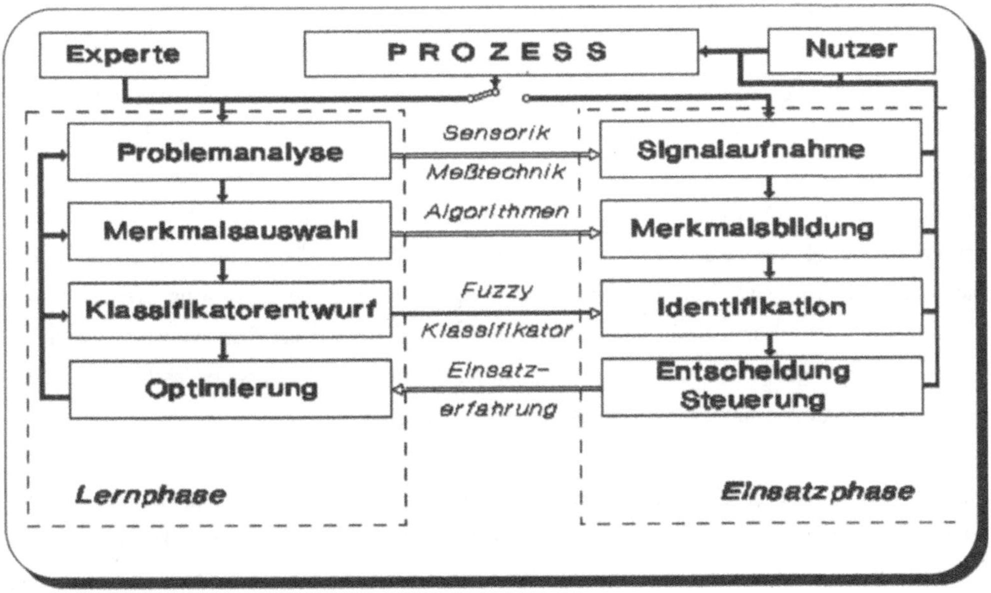

Abb. 2: Arbeitsschritte von Lern- und Einsatzphase

5. Applikationsbereiche

Die Einsatzmöglichkeiten sind aufgrund der allgemeinen Modellierungsmethode sehr breit, so beispielsweise in

- der explorativen Datenanalyse, um umfangreiches Datenmaterial vorzustrukturieren,
- der Mustererkennung für die Bild-, Zeitreihen- und Frequenzanalyse,
- der technischen und nichttechnischen Diagnostik,
- der Überwachung von Prozessen zur Verlaufsvorhersage,
- der klassenorientierten Steuerung.

Der allgemeine Lösungsansatz zur Fuzzy Pattern Klassifikation und die Entwurfsstrategie für Fuzzy Klassifikatoren ist für alle Anwendungsbereiche prinzipiell gültig. Dennoch können die Schwerpunkte sehr unterschiedlich liegen. So ist zur Datenanalyse die "intelligente" Vorverarbeitung großer Datenmengen erforderlich, wie sie in multisensoriellen Systemen der Mikrosystemtechnik zunehmend anfallen. Die Auswertung von Zeitreihen mit Echtzeit-Anforderungen bei technischen Überwachungsprozessen oder mit sehr hoher Komplexität wie in der Umwelttechnologie setzt neue Maßstäbe für die Leistungsfähigkeit des Verfahrens, die auch hardwareseitig zu unterstützen ist [4].

6. Fuzzy Pattern-Klassifikation zur Qualitätssicherung

Seit Mitte der 80er Jahre wurde die Leistungsfähigkeit des Verfahrens - gestützt auf ein umfangreiches Softwarepaket "FUCS - Fuzzy Classification System" - in industriellen und nichttechnischen Anwendungen vielfach erfolgreich nachgewiesen. In der Qualitätskontrolle und -sicherung sind Anwendungen besonders dann interessant, wenn bislang der Mensch in die Bewertung eingreift, zum Beispiel bei subjektiven akustischen oder visuellen Prüfverfahren [5,6]. Zudem erscheint in der Qualitätsbestimmung eine Absolutaussage schwieriger als ein relativer Mustervergleich mit repräsentativen Beispielen. Eine wesentliche Schwierigkeit beim Klassifikatorentwurf besteht in derFestlegung von geeigneten Merkmalen, um entweder das menschliche Vorgehen nachzuahmen oder technische Kenngrößen mit vergleichbarer Aussagekraft aufzufinden.

Die Problematik kann am Beispiel der Schweißprozeßautomatisierung [7] verdeutlicht werden. Der Schweißprozeß ist durch die Wechselwirkungen seiner Prozeßgrößen hochgradig komplex. Starke Nichtlinearitäten treten auf, die Prozeßstruktur bleibt im Detail unbekannt und Modelle mittels rein theoretischer mathematischer Methoden sind zur Zeit nicht verfügbar. Der erfahrene Schweißer hat durch Ausbildung und Praxis gelernt, diesen komplizierten Prozeß sehr genau einzuschätzen. Er stützt sich dabei nur auf seine visuellen und akustischen Eindrücke. Offenkundig ist er aber in der Lage, aus diesen komplexen Signalmustern klare Bewertungen vorzunehmen.

Aus diesen Beobachtungen erscheint eine Bild- und Schallanalyse als naheliegender Lösungsansatz. Der damit verbundene technische Aufwand wäre jedoch unangemessen hoch. Folglich müssen leichter verfügbare Größen genutzt werden: Der Verlauf von Lichtbogenstrom und -spannung erweist sich als charakteristisch für die momentan ablaufende Prozeßsituation. Eine Vielzahl von Merkmalen läßt sich nun aus diesen Signalen extrahieren. Eine objektivierte Einschätzung der Relevanz dieser Merkmale in Hinblick auf interessierende Fehlerklassen konnte die Informationen auf 8 Merkmale reduzieren, darunter beispielsweise die Tropfenfrequenz, das Verhältnis Kurzschluß-Brennphase im Spannungsverlauf oder der Mittelwert des Stroms.

In der Lernphase konnten typische Fehlerzustände - neben der "Gut"-Klasse insgesamt 5 Störungsklassen, u.a. Verunreinigungen der Werkstückoberfläche, fehlendes Schutzgas oder behinderter Drahttransport - in aktiven Experimenten durchlaufen werden. Damit lag die Zuordnung der Merkmalsdaten aus der Lernphase zu diesen Klassen vor und die Bestimmung und Optimierung der Parameter des Fuzzy Klassifikators konnte erfolgen. Die erreichte Klassifikationsgüte, gemessen an der Treffsicherheit der Identifikation in Testläufen, war überzeugend.

Literatur

[1] Peschel, M.; Bocklisch, S.F. (Hrsg.): Unscharfe Modellbildung und Steuerung. Fortschrittsberichte der GDR-Working Group for Fuzzy Sets and Systems, Band I bis VI, TH Karl-Marx-Stadt (Chemnitz), 1977-1983.
[2] Bocklisch, S.F.: Prozeßanalyse mit unscharfen Verfahren. Verlag Technik, Berlin, 1987.
[3] Bocklisch, S.F.; Burmeister, J.; Priber, U.: FUCS: Diagnose, Überwachung und Steuerung von Prozessen mit Hilfe der unscharfen Klassifikation rechnergestützt. In: Der Maschinenschaden (65) 1992, Nr. 1, S. 14-18.
[4] Bocklisch, S.F.; Müller, D. u.a.: Ein Akzeleratorsystem zur unscharfen Klassifikation - Theorie und Realisierung. In: VDI-Fachtagung "Technische Anwendungen von Fuzzy-Systemen", Dortmund 1992.
[5] Bitterlich, N.; Totzauer, R.: Geräuschbewertung mittels Fuzzy Klassifikation. In: 6. Kongreßmesse für industrielle Meßtechnik "MessComp'93", Wiesbaden, 1992, S. 140-145.
[6] Priber, U.: Oberflächenprüfung mittels Beugungslichtsensor und unscharfer Klassifikation. In: 7. Tagung "Meßinformationssysteme, Chemnitz 1991.
[7] Burmeister, J.: Fuzzy Logic - nicht nur ein Modetrend. In: Schweißen und Schneiden (43) 1992, Heft 9, S. 533-537.

ANWENDUNGEN DER FUZZY DATENANLYSE ZUR QUALITÄTSSICHERUNG

Manfred Fochem, MIT GmbH Aachen
Richard Weber, MIT GmbH Aachen

Die Qualitätssicherung stellt eine wesentliche Aufgabe in Unternehmen des produzierenden Gewerbes dar. Ziel der entsprechenden Tätigkeiten ist es, eine gleichbleibend hohe Qualität der hergestellten Produkte bei möglichst wirtschaftlicher Vorgehensweise sicherzustellen. Aus diesem Grund werden in zunehmendem Maße CAQ-Konzepte zur Automatisierung der Qualitätssicherung eingesetzt [2]. Insbesondere die Auswertung von Prüfdaten kann durch entsprechende Methoden und Konzepte unterstützt werden.
In diesem Beitrag wird ein Ansatz zur akustischen Qualitätskontrolle vorgestellt, wobei die Prüfung von Glas-, Porzellan- und Keramikprodukten im Vordergrund steht.
Bei der akustischen Prüfung von Keramikfliesen werden diese angeschlagen und das dabei entstandene Geräusch beurteilt. Damit lassen sich bestimmte Fehler aufdecken.
Eine Automatisierung dieses Vorgangs scheiterte bislang oft daran, daß die verschiedenen Fehlerklassen sich nicht eindeutig voneinander trennen lassen und daher oft falsche Zuordnungen vorgenommen wurden. Mit Methoden der Fuzzy Datenanalyse läßt sich eine dem Problem angemessenere Beurteilung von Fliesen vornehmen [1][3].
Es wird ein mögliches System zur Automatisierung der akustischen Qualitätskontrolle vorgestellt. Die von einer angeschlagenen Fliese ausgehenden Schallmuster werden über einen A/D-Wandler als Zahlenwerte im Rechner erfaßt. Dieses Zeitsignal wird durch eine Fourier-Transformation in das entsprechende Frequenzspektrum umgewandelt. Daraus werden charakteristische Merkmale zur Beurteilung der Fliesen abgeleitet.
Die an einem Versuchsaufbau erzielten Ergebnisse deuten auf die Leistungsfähigkeit der verwendeten Fuzzy Methoden hin.

Literatur

[1] Angstenberger, J., Lieven, K., Weber, R., Zimmermann, H.-J.: Fuzzy Technologien - Prinzipien, Werkzeuge, Potentiale. Düsseldorf, 1993.
[2] Bläsing, J.P.: CAQ - Qualitätssicherung unter CIM-Zielen. Braunschweig, Wiesbaden, 1990.
[3] Zimmermann, H.-J.: Fuzzy Set Theory - And Its Applications, 2nd, rev. ed., Boston, 1991.

Qualitätsmessung im betrieblichen Nichtfertigungsbereich als unscharfes Multi-Criteria-Problem

Markus Kahn, Niedersächsische Forstliche Versuchsanstalt, D-37079 Göttingen

Zusammenfassung: Subjektiv verbal charakterisierte Prozesse und Produkte aus dem betrieblichen Nichtfertigungsbereich sind einer quantitativen Analyse, etwa durch eine Statistische Qualitätsregelung, nur schwer zugänglich. Zum Zweck einer solchen quantitativen Analyse wird 'Qualität' als eine linguistische Variable im Sinne der Fuzzy Set-Theorie definiert, wobei wegen der mehrkriteriellen Merkmalsstruktur von 'Qualtität' eine vektorwertige Basisvariable angenommen wird. Das sich daraus entwickelnde multikriterielle Qualitätsmeßproblem besteht darin, die mehrdimensionalen Merkmalsvektoren auf die reellen Zahlen abzubilden, um eine Statistische Qualitätsregelung zu ermöglichen. Zunächst werden die Koeffizienten der Merkmalsvektoren, die in Form von sprachlichen Ausdrücken vorliegen, auf unscharfe Zahlen transformiert und dann auf die reellen Zahlen abgebildet. Auch die primär metrisch skalierten Koeffizienten werden fuzzifiziert. Die Aggregation der einzelnen Merkmalsvektoren zu jeweils einer reellen Zahl wird anschließend unter Verwendung unscharfer Aggregationsoperatoren durchgeführt.

Summary: In the non-manufacturing area products and processes are often characterized subjectively and verbaly and they are therefore hard to quantify and analyze for example by a statistical process control. For such a use 'quality' will be described as a linguistic variable with fuzzy set-theoretic mathematics, and because of the multidimensional structure of 'quality' a vectorvalued basevariable has to be applied. To make a statistical process control possible the vectors have to be mapped onto the real numbers, which is conducted using methods from multi-criteria analysis. At first the verbal coefficients of the feature vectors are characterized as fuzzy sets and then approximately reduced to real numbers. The primarily numerical coefficients of the feature vector have to be fuzzified too. Thus the vectors are aggregated to real numbers by using fuzzy aggregation operators.

1. Einleitung

Die Messung der Qualität von Prozessen und Produkten im betrieblichen Nichtfertigungsbereich ist ein noch weitgehend ungelöstes Problem aus dem Umfeld des unternehmensweiten Qualitätsmanagements (vgl. [7]). Die damit eng verknüpfte statistische Regelung von Nichtfertigungsprozessen ist mit standardisierten Methoden (z.B. Statistical Process Control SPC bzw. Statistische Qualitätsregelung) relativ einfach möglich, wenn das Meßproblem erst einmal gelöst ist. Dieses ist darauf zurückzuführen, daß Produkt- und Prozeßqualitäten im Nichtfertigungsbereich häufig aufgrund von Wert- und Schätzurteilen subjektiv verbal charakterisiert werden und zusätzlich oftmals mehrkriterielle Merkmalsstrukturen aufweisen. Die Motivation zur Untersuchung dieses Qualitätsmeßproblems ergibt sich daraus, daß die statistische Regelung aller relevanten betrieblichen Prozesse die Grundlage für den erfolgreichen Einsatz eines unternehmensweiten Qualitätsmanagements ist ([5]). Weil 80-85% der Qualitätsprobleme der Unternehmen i.d.R. ihren Ursprung in solchen Bereichen haben, die nicht der Fertigung zuzuordnen sind ([1], S. 37), muß den 'non-manufacturing areas' bei der Qualitätssicherung besondere Aufmerksamkeit gewidmet werden. Hierzu ergänzt CROSBY, daß "die Kosten für unsachgemäße Arbeit in (...) 'Dienstleistungsbetrieben' zweimal so hoch wie in der 'industriellen Fertigung' " ([4], S. 126) sein können.

Nach JURAN hat der Qualitätsbegriff eine Vielzahl von Bedeutungen ([8], S. 13). Eine der Bedeutungen ist der Produktnutzen, den ein Kunde (vgl. dazu auch das Drei-Rollen-Konzept nach JURAN), der Nutzer des Produkts, diesem zuordnet. Der Produktnutzen ergibt sich aus den Produkteigenschaften, und diese können zwar beispielsweise durch subjektive, unscharfe Begriffe wie 'schnell' und 'gut' verbal beschrieben, aber quantitativ nur unzureichend erfaßt werden. Unmittelbar nicht erleichtert wird die Quantifizierung von Qualität durch ihre definitorische Festlegung in der DIN 55350 als die "Beschaffenheit einer Einheit bezüglich ihrer Eignung, festgelegte und vorausgesetzte Erfordernisse zu erfüllen" ([6]; vgl. [4], S. 69). Die im betrieblichen Nichtfertigungsbereich verbreitete begriffliche, umgangs- oder fachsprachlich bedingte und durch mehrkriterielle Merkmalsstrukuren geprägte Unschärfe bei der Bewertung von Produktqualitäten kann jedoch geeignet mit Hilfe des Instrumentariums der Fuzzy Set-Theorie ([9], [14], [15]) und Ansätzen der Multi-Criteria-Analyse ([16]) einer quantitativen Analyse, etwa im Zusammenhang mit einer SPC, zugänglich gemacht werden.

2. Qualitätsmessung und Fuzzy Sets

Beispielhaft wird im folgenden gezeigt, wie unter Verwendung von Verfahren der Fuzzy Set-Theorie die sowohl verbal subjektiv als auch multiattributiv charakterisierte Produktqualität eines 'Geschäftsbriefes', als einem Kommunikationsmedium zwischen dem Unternehmen und seiner Umwelt, quantifiziert, d.h. auf die reellen Zahlen abgebildet werden kann. Das Vorgehen zur Quantifizierung eines Gesamtqualitätsurteils gliedert sich dabei in fünf Schritte, woran sich eine statistische Qualitätsregelung anschließen kann:

Schritt 1: Bestimmung des Merkmals- und Merkmalsausprägungsvektors
Schritt 2: Fuzzifizierung primär metrisch skalierter Koeffizienten
Schritt 3: Konkretisierung von linguistischen Variablen
Schritt 4: Reduktion unscharf repräsentierter sprachlicher Ausdrücke zu reellen Zahlen
Schritt 5: Abbildung des Merkmalsausprägungsvektors auf die reellen Zahlen

2.1. Die Bestimmung des Merkmals- und Merkmalsausprägungsvektors

Nach ZIMMERMANN und GUTSCHE ist "Messung (...) die Zuordnung von Zahlen zu beobachteten Objekten oder zu deren Eigenschaften nach bestimmten Regeln" ([16], S. 11). Ein wesentlicher Aspekt der Messung ist die Festlegung von Meßmerkmalen, womit hier also solche Merkmale gesucht sind, welche die Qualität des Produktes 'Geschäftsbrief' determinieren. Nach WERNERS wird zunächst "zwischen im engeren Sinne qualitativen und quantitativen Merkmalen differenziert" ([11], S. 3 f.). Für quantitative Merkmale wird demnach vorausgesetzt, daß die Merkmalsausprägungen über reelle Zahlen ausgedrückt werden können. Für qualitative Merkmale gilt dies nicht, weshalb häufig verbale Charakterisierungen getroffen werden. Ursprünglich kann die Anzahl der Qualitätsmerkmale sehr groß sein, doch ist damit zu rechnen, daß ein Teil davon redundant ist. Dazu ist eine geeignete Merkmalsreduktion durchzuführen (z.B. [2], S. 124 ff.), die etwa im Rahmen einer Faktoren- oder Diskriminanzanalyse erfolgen kann.

Exemplarisch sollen hier vier Qualitätsmerkmale als relevant erkannt worden sein und durch die folgenden Fragen beschrieben werden:

1. Wurde der Geschäftsbrief zügig bearbeitet und fertiggestellt?
2. Ist der Geschäftsbrief inhaltlich vollständig und dem Zweck (z.B. Mahnschreiben) angemessen?
3. Wieviele Tippfehler pro Seite enthält der Geschäftsbrief im Durchschnitt?
4. Ist der Geschäftsbrief formal in Ordnung, etwa in bezug auf Anschrift, Anrede, Absender, Unterschrift?

Diese 4 Merkmale bilden den Merkmalsvektor der Produktqualität eines Geschäftsbriefes. Ein Merkmal dient der charakteristischen Beschreibung einer Produktqualität hinsichtlich einer Zielgröße. Die Zielgröße für Merkmal 3 ist die Minimierung der durchschnittlichen Anzahl von Tippfehlern pro Seite in dem Geschäftsbrief. Die Merkmalsausprägung ist eine gegebene Realisierung, z.B. durchschnittlich 1,7 Tippfehler pro Seite, und bildet gemeinsam mit den Ausprägungen der 3 anderen Qualitätsmerkmale den Merkmalsausprägungsvektor. Hier liegt nur die Ausprägung des dritten Merkmals von vorneherein auf metrischem Skalenniveau vor, ist also ein primär metrisch skalierter Koeffizient des Merkmalsausprägungsvektors. Alle diese Merkmalsausprägungen sind hinsichtlich ihrer spezifischen Zielerreichung derart zu bewerten, daß letzlich alle Koeffizienten dieses 'zielbewerteten' Vektors auf demselben Skalenniveau und in derselben Einheit bzw. dimensionslos vorliegen, weil dann eine Aggregation und damit Abbildung des mehrdimensionalen Vektors auf die reellen Zahlen erfolgen kann (Schritt 5). Dies leitet zunächst über zu Schritt 2, der Fuzzifizierung primär metrisch skalierter Koeffizienten.

2.2. Die Fuzzifizierung primär metrisch skalierter Koeffizienten

Zur Bewertung der Ausprägung des Merkmals 3 hinsichtlich seiner Zielgröße wird eine an dieser orientierte unscharfe Menge $\tilde{T}$ 'tippfehlerfreie Geschäftsbriefe' gebildet. Die Definition als unscharfe Menge bietet den Vorteil, daß nicht nur zwischen 'fehlerfrei' und 'nicht fehlerfrei' unterschieden werden kann, sondern kontinuierliche Zielerreichungsgrade ausgewiesen werden können (s. Abb. 1) und

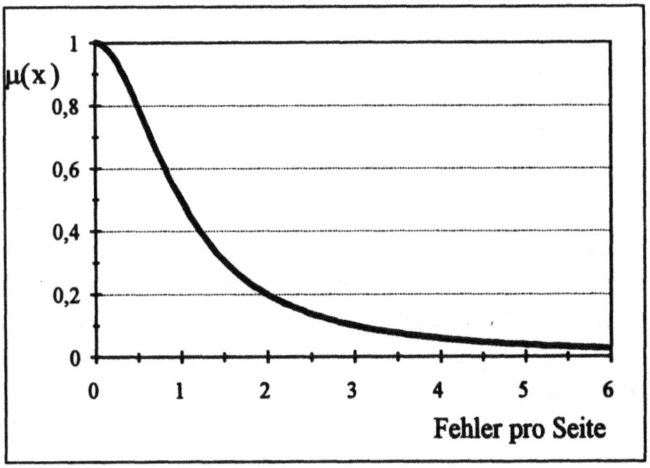

Abb. 1: Die unscharfe Menge 'tippfehlerfreie Geschäftsbriefe'

somit eine genauere Differenzierung bzgl. der Zielerreichung möglich ist. Dazu wird eine Zugehörigkeitsfunktion $\mu(x)$ definiert als $\mu(x) = \frac{1}{x^2 + 1}$. Damit erhält ein Geschäftsbrief mit durchschnittlich 1,7 Tippfehlern pro Seite eine dimensionslose graduelle Zugehörigkeit zur Menge der tippfehlerfreien Geschäftsbriefe von $\mu(1,7) = 0,26$. Die Ausprägungen der Merkmale 1, 2 und 4 liegen nicht primär auf metrischem, sondern auf ordinalem Skalenniveau vor, und die diesbezüglichen Koeffizienten des Merkmalsausprägungsvektors sind verbal. Ein Geschäftsbrief kann 'sehr zügig' beantwortet worden sein, inhaltlich 'unvollständig' und formal 'vollkommen korrekt' sein. Die Quantifizierung dieser Begriffe soll unter Verwendung linguistischer Variablen erfolgen.

2.3. Die Konkretisierung von linguistischen Variablen

Das Konzept einer linguistischen Variablen auf Basis der Theorie unscharfer Mengen wurde erstmals von ZADEH vorgestellt ([12],[13]). Eine linguistische Variable wird (nach [15], S. 132; vgl. [9], S. 64f.; [10]) charakterisiert durch das Quintupel $(x,T(x),U,G,\mathcal{M})$, wobei x den Namen der Variablen bezeichnet. T(x) beschreibt die 'Term-Menge' von x, also die Menge der Namen als linguistische Ausprägungen (sprachliche Ausdrücke) von x. Jeder sprachliche Ausdruck aus T(x) kann durch eine unscharfe Menge dargestellt werden. Die sprachlichen Ausdrücke decken vollständig einen Wertebereich U (Definitionsbereich) mit einer Basisvariablen u ab. Die Basisvariable u kann vektorwertig sein. Die Grammatik G ist eine Vorschrift, wie die Elemente der Termmenge ausgehend von einem Grundbegriff gebildet werden können. $\mathcal{M}$ schließlich bezeichnet eine Semantik, die durch die Zugehörigkeitsfunktionen repräsentiert wird.

Hier werden für die Merkmale 1, 2 und 4 insgesamt drei linguistische Variable benötigt, die 'termingerechte Fertigstellung' (Merkmal 1), 'inhaltliche Vollständigkeit' (Merkmal 2) und 'formale Richtigkeit' benannt werden. Die Termmengen dieser Variablen seien dann T(termingerechte Fertigstellung) = {gering, gering-mittel, mittel, mittel-hoch, hoch}, T(inhaltliche Vollständigkeit) = {sehr gering, gering, mittel, hoch, sehr hoch} und T(formale Richtigkeit) = {gering, mittel, hoch}. Problematisch sind nun sowohl die Zuweisung von Zugehörigkeitsfunktionen zu den einzelnen Termen der jeweiligen Termmengen als auch die Festlegung der Basisvariablen. Beide Probleme lassen sich relativ einfach lösen, wenn vordefinierte linguistische Variablen verwendet werden mit dimensionslosen, auf das Intervall [0,1] normierten Basisvariablen und mit von der Kardinalität und ordinalen Spreitung der Termmenge abhängigen Zugehörigkeitsfunktionen. So haben hier die linguistischen Variablen 'termingerechte Fertigstellung' und 'formale Richtigkeit' unterschiedliche Kardinalitäten ihrer Termmengen, wohingegen 'termingerechte Fertigstellung' und 'inhaltliche Vollständigkeit' in ihren Termmengen eine unterschiedlich große ordinale Spreitung aufweisen. CHEN und HWANG haben 1992 ein solches System von vordefinierten linguistischen Variablen vorgestellt ([3]), auf die hier zurückgegriffen werden soll.

Neben den Merkmalen 1, 2 und 4 ist auch die Gesamtqualität des Geschäftsbriefes als linguistische Variable 'Qualität' zu konkretisieren, allerdings mit einer vektorwertigen Basisvariablen. Die Koeffizienten dieser Vektoren sind die Merkmalsausprägungen der Qualitätskriterien. Die Termmenge wird festgelegt mit T(Qualität) = {sehr gering, gering, mittel, hoch, sehr hoch}, entspricht also in ihrer Kardinalität und ordinalen Spreitung der linguistischen Variablen 'inhaltliche Vollständigkeit' (s. Abb. 2).

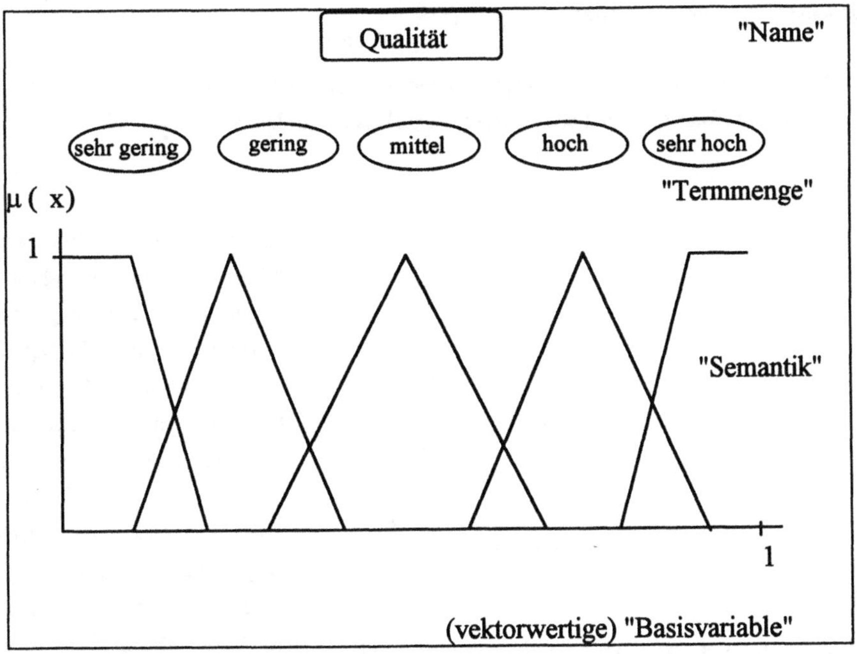

Abb. 2: 'Qualität' als linguistische Variable

2.4. Die Reduktion unscharf repräsentierter sprachlicher Ausdrücke

Die sprachlichen Ausdrücke der Termmengen liegen in quantitativer Form als unscharfe Mengen vor, die aufgrund ihrer gemäß den Normvariablen (s. [3]) speziellen Zugehörigkeitsfunktionen als unscharfe Zahlen bezeichnet werden können (vgl. [15]). Diese unscharfen Zahlen sind rechentechnisch nur unbequem zu handhaben und sollen deshalb zu jeweils einer reellen Zahl reduziert werden. Aus der Multi-Criteria-Analyse sind dazu zahlreiche Verfahren (vgl. [3]; [14]) bekannt, die u.a. für die Rangordnung durch unscharfe Zahlen repräsentierte Entscheidungsalternativen entwickelt worden sind. Hier soll einem Verfahren von CHEN und HWANG gefolgt werden ([3], S. 474), das einer unscharfen Menge $\tilde{M}$ den reellen Zahlenwert

$$\mu_T(\tilde{M}) = \frac{\mu_R(\tilde{M}) + 1 - \mu_L(\tilde{M})}{2} \text{ zuordnet,}$$

wobei $\mu_R(\tilde{M}) = \sup_x \min \{\mu_{\tilde{M}}(x), \mu_{max}(x)\}$ und $\mu_L(\tilde{M}) = \sup_x \min \{\mu_{\tilde{M}}(x), \mu_{min}(x)\}$ gelten.

$\mu_{max}(x)$ ist die sogenannte maximierende Menge $\mu_{max}(x) = \begin{cases} x, & 0 \leq x \leq 1 \\ 0, & \text{sonst} \end{cases}$ und

$\mu_{min}(x)$ die sogenannte minimierende Menge $\mu_{min}(x) = \begin{cases} 1-x, & 0 \leq x \leq 1 \\ 0, & \text{sonst} \end{cases}$

Für den Begriff 'hoch' aus der Termmenge der linguistischen Variablen 'inhaltliche Vollständigkeit' führt die Reduktion von einer unscharfen zu einer reellen Zahl beispielsweise zu dem Wert μ_T('hoch') = 0,72 (s. Abb. 3).

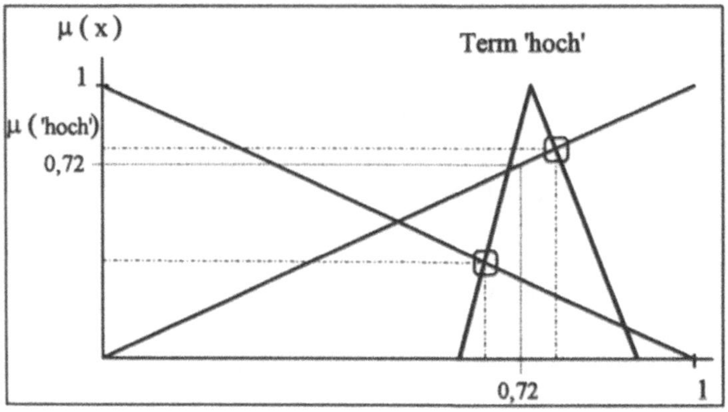

Abb. 3: Die Reduktion einer unscharfen zu einer reellen Zahl

Es sei nun ein konkreter Merkmalsausprägungsvektor V gegeben, der bzgl. der termingerechten Fertigstellungsdauer einen Geschäftsbrief als 'mittel', bzgl. der inhaltlichen Vollständigkeit als 'hoch' und der formalen Richtigkeit als 'sehr hoch' eingeschätzt wurde und durchschnittlich 1,7 Tippfehler pro Seite aufwies. Dann folgt daraus nach Fuzzifizierung des primär metrisch skalierten Koeffizienten (Schritt 2) und der Reduktion der unscharfen zu reellen Zahlen der modifizierte, zielbewertete Vektor V^* :

$$V = \begin{pmatrix} \text{mittel} \\ \text{hoch} \\ 1{,}7 \\ \text{sehr hoch} \end{pmatrix} \xrightarrow[\text{Fuzzifizierung}]{\text{numerische Approximation}} V^* = \begin{pmatrix} 0{,}5 \\ 0{,}72 \\ 0{,}26 \\ 0{,}83 \end{pmatrix}$$

2.5. Die Abbildung des Merkmalsausprägungsvektors auf die reellen Zahlen

Die Koeffizienten des Vektors V^* sind dimensionslos und im Zugehörigkeitsraum, das ist hier das abgeschlossene Intervall [0,1], metrisch geordnet. Jeder der Koeffizienten gibt an, inwieweit die entsprechende Zielgröße erfüllt ist. Durch Aggregation soll der Vektor V^* auf die reellen Zahlen abgebildet und ein Gesamturteil über die Qualität des Geschäftsbriefes ermöglicht werden (vgl. [10]). Intuitiv könnte über die Gesamtqualität von V^* vorab angenommen werden, daß sie mindestens 0,26 und höchstens 0,83 betragen wird. Der Mindestwert (Minimum) bzw. Höchstwert (Maximum) wird genau dann erreicht, wenn zwischen den Zielerreichungen keine bzw. vollständige Kompensation zugelassen wird. Unter Verwendung bestimmter Operatoren sind allerdings auch aggregierte Werte möglich, die niedriger als das Minimum bzw. höher als das Maximum sind. Wird als unscharfer genereller Aggregationsoperator das linear-konvexe kompensatorische UND verwendet ([17]), so folgt als Gesamtqualität ein Wert, der zwischen dem Minimum und Maximum liegt. Bei einem Kompensationsgrad γ mit γ=0,5 ergibt sich demnach für $\mu(V^*)$:

$$\mu(V^*) = (1-\gamma)*\min\{0,5;0,72;0,26;0,83\}+\gamma*\max\{0,5;0,72;0,26;0,83\} = 0,55.$$

Semantisch ist dieser $\mu(V^*)$-Wert als eine mittlere Qualität zu interpretieren (vgl. Abb. 1). Mit den Werten $\mu(V^*)$ liegen die ursprünglich verbal und numerisch in Vektoren zusammengefaßten Ausprägungen der Qualitätsmerkmale als reelle Zahlen vor, womit die vektorwertige Basisvariable der linguistischen Variablen 'Qualität' für die Statistische Qualitätsregelung eine geeignete Transformation erfahren hat. Sprachliche Unschärfen lassen sich somit mathematisch formal sowie rechentechnisch effizient quantifizieren, und Merkmalsvektoren sind zumindest theoretisch plausibel derart zu aggregieren, daß sie als Eingabegrößen für die SPC dienen können.

3. Literatur

[1] Atkinson, P.E.: Creating culture change: The key to successful total quality management, IFS Ltd, 1990

[2] Bocklisch, S.F.: Prozeßanalyse mit unscharfen Verfahren, Verlag Technik, Berlin, 1987

[3] Chen, S.J., C.-H. Hwang: Fuzzy Multiple Attribute Decision Making - Methods and Applications, Springer Verlag, Berlin u.a., 1992

[4] Crosby, P.B.: Qualität ist machbar, Verlag McGraw-Hill, Hamburg, 1990

[5] Deming, W.E.: Out of the crisis, MIT, Massachusetts,1982

[6] DIN 55 350: Begriffe der Qualitätssicherung und Statistik - Teil 11: Grundbegriffe der Qualitätssicherung, Beuth Verlag, Berlin, 1987

[7] Fischer, K., G. Hertel: Dienstleistungsqualität versus Produktqualität, Qualität und Zuverlässigkeit (37) 5, S. 255-258, 1992

[8] Juran, J.M.: Handbuch der Qualitätsplanung, 2. Aufl., Verlag moderne industrie, Landsberg, 1990

[9] Rommelfanger, H.: Entscheiden bei Unschärfe - Fuzzy Decision Support-Systeme, Springer-Verlag, 1988

[10] Rommelfanger, H.: Fuzzy-Logik basierte Verarbeitung von Expertenregeln, OR-Spektrum 15:31-42, 1993

[11] Werners, B.: Modellierung und Aggregation linguistischer Terme, Arbeitsbericht 90/03, Lehrstuhl für Unternehmensforschung, Institut für Wirtschaftswissenschaften, RWTH Aachen, 1990

[12] Zadeh, L.A.: The concept of a linguistic variable and its application to approximate reasoning, Teil 1 und 2, Information Sciences (8), S. 199-249, 1975

[13] Zadeh, L.A. : The concept of a linguistic variable and its application to approximate reasoning, Teil 3, Information Sciences (9), S. 43-80, 1976

[14] Zimmermann, H.-J.: Fuzzy Sets, Decision Making, and Expert Systems, Kluwer Academic Publishers, Boston u.a., 1987

[15] Zimmermann, H.-J.: Fuzzy set theory and its applications, 2. Aufl., Kluwer Academic Publishers, 1991

[16] Zimmermann, H.-J., L. Gutsche: Multi-Criteria-Analyse, Einführung in die Theorie der Entscheidungen bei Mehrfachzielsetzungen, Springer Verlag, Berlin u.a., 1991

[17] Zimmermann, H.J., P. Zysno: Latent connectives in human decision making, Fuzzy Sets and Systems (4), S. 37-51, 1980

FUZZY CLUSTERING METHODS FOR OBJECT LOCALIZATION

Raghu Krishnapuram, European Laboratory for Intelligent Techniques Engineering, Aachen

Object localization (i. e., determination of the location and orientation of objects) plays an important role in many computer vision applications such as automated assembly and target tracking. In this paper, we consider the problem of object localization when the object model is known. It is assumed that the object boundary (in the case of two-dimensional objects and intensity images) or the object surface (in the case of three-dimensional objects and range images) can be represented by a polynomial of arbitrary degree. However, the number of objects present in the image may or may not be known, and some of the objects may be occluded.

Fuzzy objective function based clustering methods have been used extensively for pattern recognition and computer vision. The Fuzzy C Means (FCM) algorithm due to Bezdek, and its derivatives have been used to detect clusters of various shapes. More recently, this approach has been extended to the case of shell clusters, i. e., clusters with no interior points. The boundary or the surface of the clusters is assumed to be quadric (second-degree) in shape. The prototypes of the clusters are represented by a parameter vector, which consists of the coefficients of the quadric. In shell clustering algorithms, the parameter vector is assumed to be unknown. However, in the present application, we assume that the curve or surface parameters are known, but only the translation and rotation parameters are to be determined. To achieve this, we construct a suitable fuzzy objective function, and design a fuzzy clustering algorithm based on the objective function to determine the rotation matrices and translation vectors for all the objects in the image.

* On leave from: The Department of Electrical and Computer Engineering, University of Missouri, Columbia, MO 65211, USA.

PROJEKTPLANUNG BEI UNSCHARFEN DATEN. INFORMATIONSVERARBEITUNG UND EINBEZIEHUNG VON BESCHLEUNIGUNGSKOSTEN

Heinrich J. Rommelfanger, J.W. Goethe-Universität Frankfurt a.M.

Netzpläne stellen ein geeignetes Instrumentarium zur Projektplanung dar, wie die umfangreiche Literatur zu diesem Themenkomplex und zahlreiche Realisierungen in der Praxis zeigen. Dies gilt aber nur unter der Annahme, daß die Ablaufstruktur bekannt ist und die Zeitdaten der einzelnen Teilprojekte exakt angegeben werden können. Da insbesondere die Dauer vieler Vorgänge als zukunftsorientierte Größe zum Planungszeitpunkt nur größenordnungsmäßig festgelegt werden kann, wird in neueren Arbeiten vorgeschlagen, Netzpläne mit Fuzzy-Daten als Planungsgrundlage zu verwenden. In /1/ wird gezeigt, daß sich die Konzepte der MPM-Netzplantechnik auch auf unscharfe Netzpläne übertragen lassen, wenn man beachtet, daß im Gegensatz zum reellen Zahlenraum bei Fuzzy Zahlen die erweiterte Subtraktion i. a. nicht die Umkehr der erweiterten Addition darstellt.

In diesem Beitrag wird untersucht, ob durch Verwendung unscharfer Netzpläne die Qualität der Projektplanung gesteigert werden kann und ob die damit verbundene zielgerichtete Informationsbeschaffung zu einer Senkung der Planungskosten und der Planungszeit führt. Weiterhin wird analysiert, wie Beschleunigungskosten in Netzpläne mit Fuzzy-Daten integriert werden können. Ein Modell zur Berechnung des kostenminimalen Maximalflusses wird vorgestellt und an einem Zahlenbeispiel erläutert.

/1/ Rommelfanger, H.: Netzplantechnik mit Fuzzy-Daten. Operations Research Proceedings 1992, Springer-Verlag Berlin Heidelberg 1993

Anwendung neuronaler Lernverfahren im Gebiet der Fuzzy Set Theorie

Jan Schaefer, RWTH Aachen

Zusammenfassung: Seit der ersten Veröffentlichung von Lotfi Zadeh im Jahre 1965 hat sich die Fuzzy Set Theorie zu einem sehr aktiven und fruchtbaren Forschungsgebiet entwickelt [Zadeh, 1965]. Seit kurzem sind auch einige neuro-fuzzy Systeme in der Literatur beschrieben worden [Kosko, 1992; Kong und Kosko, 1992; Kawamura et al., 1992]. Mittels neuro-fuzzy Systemen sollen die Vorteile von neuronalen Netzen und fuzzy Systemen miteinander verknüpft werden, ohne deren Nachteile in Kauf nehmen zu müssen.
In Kapitel zwei wird ein allgemeiner Vergleich zwischen neuronalen Netzen und fuzzy Systemen durchgeführt, wobei anschließend drei verschiedene Vorgehensweisen beschrieben werden, um die beiden Systeme zu verknüpfen. Ein konkretes neuro-fuzzy System, welches von einem verallgemeinerten Mamdani Controller abgeleitet worden ist, wird in Kapitel drei beschrieben.

Summary: Since the first publication from Lotfi Zadeh in the year 1965 fuzzy set theory has developed to a very active and fruitful area of research [Zadeh, 1965]. Recently many neuro-fuzzy systems have been described in literature [Kosko, 1992; Kong und Kosko, 1992; Kawamura et al., 1992]. Neuro-fuzzy systems combine the advantages and avoid disadvantages of neural networks and fuzzy technologies, respectively.
Section two contains a general comparison of neural networks and fuzzy technologies and there are three main approaches for the combination of both areas described. One concrete neuro-fuzzy system based on a generalized Mamdani controller is explained in section three.

1. Einführung

Ein Hauptanwendungsgebiet der Fuzzy Set Theorie liegt in der regelbasierten Wissensverarbeitung. Dies gilt sowohl für die Forschung als auch für die Anwendung.
Mit Hilfe der Fuzzy Set Theorie kann die Unsicherheit, Unschärfe und Unvollständigkeit der realen Welt adäquat in ein regelbasiertes System abgebildet werden. Dabei muß sich die Wahl der Zugehörigkeitsfunktionen, der Operatoren, des Inferenzmechanismus usw. an dem entsprechenden Anwendungsfall orientieren. In dieser Wahl liegt aber auch gleichzeitig die Problematik dieses Ansatzes, denn es ist bei einer gegebenen Problemstellung nicht klar, welche Operatoren oder welche Zugehörigkeitsfunktionen passend sind.
In diesem Beitrag wird gezeigt, wie neuronale Lernverfahren (supervised und unsupervised Learning) auf fuzzy Systeme übertragen werden können. Weiterhin wird demonstriert, wie durch die Anwendung solcher Lernverfahren Parameter für Regeln, Operatoren, Zugehörigkeitsfunktionen usw. gelernt werden können.

2. Verbindungen von neuronalen Netzen und der Fuzzy Set Theorie

Sowohl neuronale Netze als auch fuzzy Systeme werden eingesetzt, um die Leistungsfähigkeit des Gehirns nachzubilden. Im Falle der neuronalen Netze wird dies auf eine indirekte Weise versucht, indem künstliche Neuronen entworfen werden. Durch geschicktes Zusammenschalten dieser Neuronen soll also die Hardware des Gehirns reproduziert werden. Im Gegensatz dazu steht der Ansatz, welcher die Fuzzy Set Theorie benutzt. In diesem Fall wird auf direkte Weise versucht, menschliches Entscheidungsverhalten nachzubilden, indem man linguistische Variablen, Operatoren zur Aggegration, Regeln usw. benutzt. Es drängt sich natürlicherweise die Frage auf, wann welcher Ansatz geeignet ist.
Welche Voraussetzungen benötigt man bei der Benutzung von neuronalen Netzen und fuzzy Systemen und welche Vorteile haben die beiden Verfahren?
Neuronale Netze können benutzt werden, wenn das Wissen entweder unstrukturiert oder ungenau ist oder kein exaktes mathematisches Modell zur Verfügung steht. Im Falle des überwachten Lernens müssen genügend Beispieldaten, welche das gesamte Wissen abbilden, vorhanden sein. Im Falle des nicht überwachten Lernens ist eine Bewertungsfunktion, welche die Leistungfähigkeit des neuronalen Netzes beurteilt, erforderlich [Wassermann, 1989; Schaefer, 1992].
Bei der Benutzung von fuzzy Systemen müssen ähnliche Voraussetzungen gefordert werden, wobei aber ein Experte benötigt wird, der entsprechende Regeln auf linguistische Weise formulieren kann [Buckley et al., 1993]. Im mathematischen Sinne ist dieses Wissen natürlich nicht sehr genau, aber man benötigt auch kein exaktes mathematisches Modell, um ein fuzzy System zu formulieren.
Aus diesen Gründen beschreibt Kosko beide Systeme als "model free estimators" [Kosko, 1992].

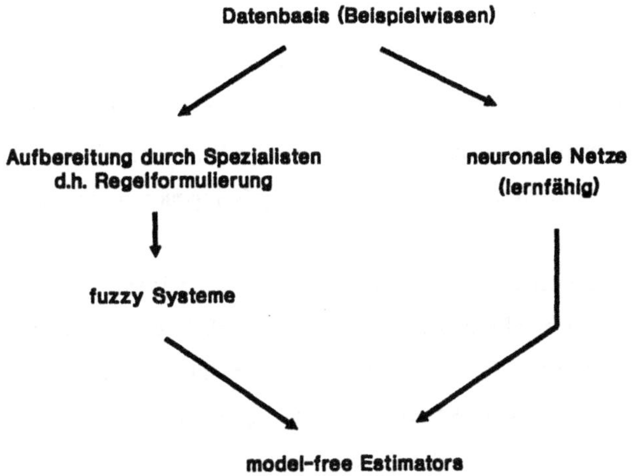

Abb. 1: Model free estimators.

In Tabelle 1 werden die wichtigsten Charakteristika von neuronalen Netzen und fuzzy Systemen miteinander verglichen.

Fuzzy Systeme	Neuronale Netze
Strukturierte Darstellung	Black Box
Einfach zu modifizieren	Schwierig zu modifizieren
Viele Variationsmöglichkeiten	Viele Variationsmöglichkeiten
Experte erforderlich	Beispiel Daten oder Bewertungsfunktion ist erforderlich
Nicht lernfähig	Lernfähig
Rechenzeit ist problemabhängig	Großer Rechenzeitbedarf in der Trainingsphase
Hohe Fehlertoleranz	Hohe Fehlertoleranz

Tab. 1: Eigenschaften von neuronalen Netzen und fuzzy Systemen.

Wie man der Tabelle entnehmen kann liegen die Vorteile der fuzzy Systeme im wesentlichen in der strukturierten Darstellung; während die Vorteile der neuronalen Netze in ihrer Lernfähigkeit liegen. Wie können diese beiden Ansätze miteinander verknüpft werden, so daß die Lernfähigkeit einerseits und die strukturierte Darstellung andererseits erhalten bleiben?
Im allgemeinen kann man drei verschiedene Möglichkeiten unterscheiden, wie neuronale Netze und fuzzy Systeme gekoppelt werden können:

1. Verwendung von neuronalen Lernmethoden in fuzzy Systemen.
2. Verbesserung von neuronalen Netzen mit der Fuzzy Set Theorie. Die neuronalen Netze werden so modifiziert, daß sie neben binären und numerischen Eingaben auch unscharfe Inputs verarbeiten können.
3. Hybride Systeme, welche sowohl neuronale als auch fuzzy Komponenten enthalten.

Im folgenden wird hier nur der erste Ansatz weiterverfolgt und beschrieben.
Betrachtet man zunächst nur den Informationsverarbeitungsprozeß in einem Neuron im Vergleich zu einem regelbasierten fuzzy System (Abb. 2), ist eine große Ähnlichkeit beider festzustellen. Mathematisch betrachtet findet in beiden Systemen zunächst eine Aggegration der Eingangsvariablen statt. Anschließend wird dieser aggregierte Wert auf den Ausgangsraum abgebildet und mit Hilfe einer Ausgaberegel weiterverarbeitet. Auch wenn man geschichtet neuronale Netze und hierarchische Regelsysteme betrachtet, bleibt die prinzipielle Ähnlichkeit in der Informationsverarbeitung erhalten.
Verfolgt man diesen Gedanken konsequent weiter, so kann ein fuzzy Controller zum Beispiel wie folgt

als neuronales Netz dargestellt werden (Abb. 3).

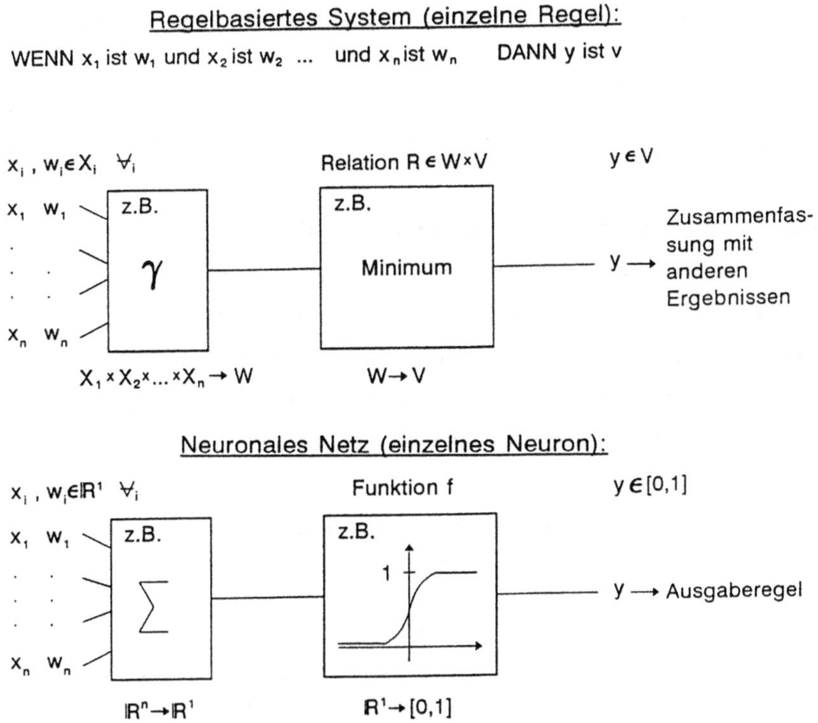

Abb. 2: Vergleich der Informationsverarbeitungsprozesse in einer Regel und in einem Neuron.

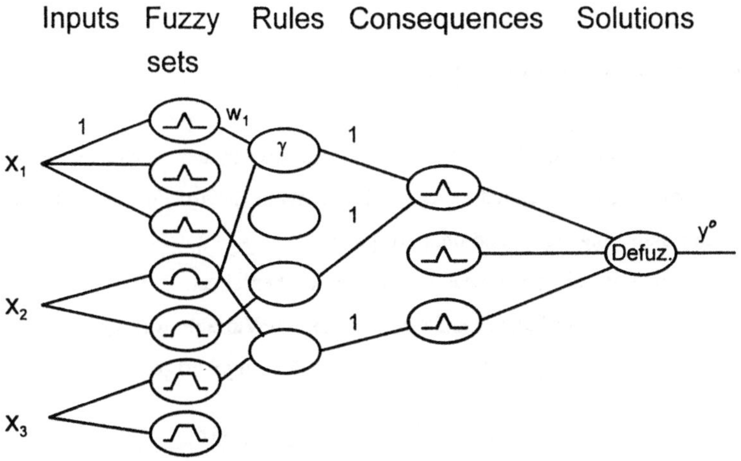

Abb. 3: Fuzzy Controller als neuronales Netz.

In diesem neuro-fuzzy System gibt es jeweils vereinfachte Neuronen für die Zugehörigkeitsfunktionen der Eingangsvariablen, für die Regeln, für die Zugehörigkeitsfunktionen der Ausgabevariablen und für die Defuzzifikation. Im nächsten Kapitel wird gezeigt wie bei geschickter Wahl der Neuronen bei diesem Netz der Backpropagationalgorithmus [Rumelhart et al., 1986] angewendet werden kann.

3. Vorstellung eines speziellen neuronalen Lernverfahrens

Die Funktionsweise dieses verallgemeinerten Mamdani-Controller kann wie folgt erläutert werden [Mamdani und Assilian, 1975; Lee, 1990a, b].

Die Regel R^i hat folgende Form: R^i: x_1 is A^i_1, x_2 is A^i_2, ..., x_n is A^i_n $\rightarrow$ y^i is B^i

wobei die A^i_j fuzzy Variablen sind und y^i der Output der i-ten Regel ist, welcher durch die fuzzy Variable B^i bestimmt wird. Für die Zugehörigkeitsfunktionen der Eingabe gibt es jeweils ein Neuron, welches als Eingabe ein x^0_j besitzt und den entsprechenden Zugehörigkeitswert $A^i_j(x^0_j)$ als Ausgabe weitergibt.

Bei gegebenen Eingangsdaten x^0_1 - x^0_n wird der Erfüllltheitsgrad der Prämisse der i-ten Regel wie folgt berechnet:

$$w^i = \left(\prod_{j=1}^{n} A^i_j(x^0_j)\right)^{(1-\gamma)} \left(1-\prod_{j=1}^{n} (1-A^i_j(x^0_j))\right)^{\gamma}$$

wobei: $0 \leq \gamma \leq 1$

Dies entspricht der Verknüpfung der Prämissen mit dem γ-Operator. Dieser Operator ist wegen seiner Allgemeinheit gewählt worden. Abhängig von γ kann es sich um eine T-Norm, eine S-Norm oder einen mittelnden Operator handeln ($\gamma=0 \rightarrow$ algebraic product, $\gamma=1 \rightarrow$ algebraic sum). Für jede Verknüpfung der Prämissen gibt es ein Neuron mit den Eingängen $A^i_j(x^0_j)$ und den Ausgang w^i für jede Regel R^i.

Im Prinzip könnten auch andere Operatoren verwendet werden.

Der Output der Regel R^i wird als Produkt berechnet:

$$y^i = w^i B^i(y)$$

Dies entspricht der Max-Prod Inferenz Methode. Ohne Beschränkung der Allgemeinheit kann hier ebenso die Max-Min Inferenz Methode benutzt werden [Zimmermann, 1990]. Für jeden Output existiert wiederum ein Neuron mit dem Eingang w^i und dem Ausgang y^i.

Der Output y^0 aller m Regeln wird als gewichtete Summe der F^is berechnet:

$$y^0 = \frac{\sum_{i=1}^{m} F^i \bar{y}^i}{\sum_{i=1}^{m} F^i}$$

wobei: $\bar{y}^i$: Schwerpunkt der unscharfen Variable $w^i B^i(y)$

F^i: gewichtete Fläche unter der Zugehörigkeitsfunktion

$$F^i = \int_{-\infty}^{+\infty} (w^i B^i(y))^{\alpha}\, dy \; ; \quad \alpha \geq 0$$

Diese Methode der Defuzzifikation entspricht der Center of Gravity Methode, wobei hier nicht direkt mit den Flächen gearbeitet wird sondern mit gewichteten Flächen. Damit erhält man eine parametrisierte Defuzzifikationsmethode ähnlich der Methode von Filev und Yager [Filev und Yager, 1991]. Diese wird durch ein einziges Neuron mit den Eingängen y^i und dem Ausgang y^0 in dem fuzzy-neuro Netz repräsentiert.

Bei gegebenen Beispieldaten kann der Entwurf des fuzzy-neuronalen Netzes wie folgt ablaufen. Zunächst wird der Typ, die Anzahl und die Vernetzung der Neuronen festgelegt. Anschließend kann auf dieses Netzwerk der Backpropagationalgorithmus angewendet werden, um die einzelnen Parameter zu lernen. Voraussetzung für die Anwendung dieses Lernverfahrens ist die Differenzierbarkeit der einzelnen Neuronen. Diese könnten zum Beispiel wie folgt aussehen:

1. Zugehörigkeitsfunktionen der Eingabe: Gaußfunktionen
2. Aggegration der Eingaben: γ-Operator (s.o.)
3. Verknüpfung der Erfülltheitsgrad der Prämissen mit der Ausgabe: Produkt (s.o.)
4. Zugehörigkeitsfunktionen der Ausgabe: Gaußfunktionen
5. Defuzzifikation: parametrisierte Center of Gravity Methode (s.o.)

Mit Hilfe folgender Substitution in 2. und 5. kann der Backpropagationalgorithmus ohne Einschränkungen verwendet werden. Damit können die Parameter a und b ohne weitere Beschränkungen gelernt werden, wobei die Bedingungen $\alpha \geq 0$ und $\gamma \in [0, 1]$ erfüllt werden [Krishnapuram und Lee, 1989].

$$\alpha = a^2 \quad \text{und} \quad \gamma = \frac{b^2}{b^2 + 1} \quad \text{wobei } a, b \in \mathbf{R}$$

Bemerkungen:

- Bei Benutzung des generalisierten Gradienten (Funktionen mit Knicken können behandelt werden) [Kiwiel, 1985 und Lemarechal, 1978] kann auch die Center of Area Methode bei der Defuzzifikation und der Minimum-Operator bei der Verknüpfung der Eingangsgrößen ohne Einschränkungen verwendet werden.
- Ohne Einschränkung der Allgemeinheit können auch Gewichte für jede Regel und für jeden Eingang eingeführt werden, welche ebenfalls gelernt werden können.
- Dieses Verfahren läßt sich auch auf den Fall des unsupervised Learning übertragen. Dies kann zum Beispiel durch die Addition einer stochastischen Größe zum Output des Reglers geschehen. Je nachdem welche Auswirkungen diese Größe auf die Bewertungsfunktion hat, können Lernschritte durchgeführt werden.
- Wenn γ gegen 0 oder 1 konvergiert, kann man sogar den γ Operator durch einen anderen Operator ersetzen [Krishnapuram und Lee, 1989]. Dafür geeignet sind z.B. die parametrisierten s- und t-Normen von Yager [Zimmermann, 1993, S.28].

4. Abschließende Bemerkung

Anstelle des Backpropagationalgorithmus können auch andere Lernverfahren wie zum Beispiel die Hebb'sche Lernregel verwendet werden [Hebb, 1988]. Weiterhin sind natürlich auch noch andere Definitionen eines fuzzy-neuronalen Netzwerkes denkbar. Das beschriebene Netzwerk steht hier also nur beispielhaft für eine neue Klasse von möglichen Varianten.

Numerische Ergebnisse für den obigen Fall sind zur Zeit in Vorbereitung.

5. Literaturverzeichnis

Buckley, J.J., Hayashi, Y. und Czogala, E., On the Equivalence of Neutral Nets and Fuzzy Expert Systems, in: FSS 53, 1993, S. 129-134.

Filev, D.P. und Yager, R.R., A Generalized Defuzzification Method via Bad Distributions, in: Int. J. of Intelligent Systems 6, 1991, S. 687-697.

Hebb, D.O., The Organization of Behavior, in: Neurocomputing, Hrsg. Anderson, J.A. und Rosenfeld, E., 3. Aufl., Cambridge, Massachusetts u.a., 1988, S. 60-78.

Kawamura, A., Watanabe, N., Okada, H. und Asakawa, K., A Prototype of Neuro-Fuzzy Cooperation System, in: IEEE International Conference on Fuzzy Systems, San Diego, 1992.

Kiwiel, K.C., Methods of Descent for Nondifferentiable Optimization, Lecture Notes in Mathematics 1133, Springer, 1985.

Kong, S.-G. und Kosko, B., Comparison of Fuzzy and Neural Truck Backer-Upper Control Systems, in: Kosko, 1992, S. 339-361.

Kosko, B., Neural Networks and Fuzzy Systems: A Dynamical Approach to Machine Intelligence, USA, 1992.

Krishnapuram, R. und Lee, J., Fuzzy-Connective-Based Hierarchical Aggegration Networks for Decision Making, in: FSS, 46, 1989, S. 11-27.

Lee, C.C., Fuzzy Logic in Control Systems: Fuzzy Logic Controller - Part I, IEEE Transactions on Systems, Man, and Cybernetics, Vol. 20, No. 2, S. 404-418, 1990a.

Lee, C.C., Fuzzy Logic in Control Systems: Fuzzy Logic Controller - Part II, IEEE Transactions on Systems, Man, and Cybernetics, Vol. 20, No. 2, S. 419-435, 1990b.

Lemarechal, C., Nonsmooth Optimization and Descent Methods, Institute for Applied Systems Analysis Laxenburg, RR-78-4, Austria, 1978.

Mamdani, E.H. und Assilian, S., An Experiment in Linguistic Synthesis with a Fuzzy Logic Controller, in: Int. J. Man Maschine Studies 7, 1975, S. 1-13.

Rumelhart, D.E., Hinton, G.E. und Williams, R.J., Learning Representations by Back-Propagating Errors, in: Nature 323, 1986, S. 533-536.

Schaefer, J., Neuronale Netze: Grundlagen und Modellkonzepte, Arbeitsbericht 92/02, Lehrstuhl für Unternehmensforschung, RWTH Aachen, 1992.

Wassermann, P.D., Neural Computing: Theory and Practise, New York, 1989.

Zadeh, L.A., Fuzzy Sets, in: Information and Control 8, 1965, S. 338-353.

Zimmermann, H.-J., Fuzzy Set Theory and Its Applications, 2. Ed., Boston, Dordrecht, London, 1990.

Zimmermann, H.-J., Fuzzy Technologien - Prinzipien, Werkzeuge, Potentiale, VDI Verlag, Düsseldorf, 1993.

EXPERTENSYSTEM FÜR KONFIGURIERUNGSAUFGABEN

Hans-Jürgen Sebastian, RWTH Aachen
Karin Müller, ELITE Foundation

Konfigurierung ist eine spezielle Form des Designs (des Entwerfens) von komplexen aus elementaren Strukturen. Konfigurierungsprobleme treten vor allem in stark strukturierten Domänen, die für Technik und Wirtschaft charakteristisch sind, in vielfältiger Weise und großer Zahl auf. Sie sind in der Regel, bedingt durch die Komponentenstruktur, durch einen großen Lösungsraum charakterisiert. Das Wissen über die zu konfigurierenden Systeme und deren Komponenten läßt sich in Hierarchien, Relationen und Restriktionen modellieren. Damit bieten sich Konzepte und Techniken aus der Expertensystemtechnologie an, um wissensbasiertes Konfigurieren durchzuführen.

Folgende Methoden kennzeichnen Systeme zur wissenbasierten Konfigurierung:
- eine objektorientierte Repräsentation der Konfigurierungsobjekte,
- die Verwaltung der Relationen und Randbedingungen mit Constraint-Systemen,
- ein Top-Down-Entwurf, der sich an der Komponentenstruktur orientiert.

Möchte man in Erweiterung der Möglichkeiten bekannter Konfigurierungs-Expertensysteme auch innovatives Design realisieren, benötigt man eine erweiterte Modellierung des Anforderungsbegriffes. Anforderungen

- an die Funktionalität des zu konfigurierenden Systems,
- in Form von Constraints oder
- als Zielkriterien

sind teilweise nur linguistisch sinnvoll formulierbar und können häufig scharf oder unscharf modelliert werden. Beispielsweise kann eine Anforderung wie "gutes Design" sinnvollerweise nur mit Fuzzy-Methoden erfaßt werden. Deshalb haben wir ein Modell zum optimierungsbasierten Konfigurieren unter Einbeziehung unscharfen Wissens entwickelt und Kontrollstrategien entworfen und erprobt.

Ein Baustein "Fuzzy-MADM" ist für solche Konfigurierungsaufgaben, bei denen aus einer endlichen Alternativenmenge auszuwählen ist, besonders geeignet. Wir demonstrieren die Funktionalität dieses Moduls "Fuzzy-MADM" innerhalb eines Konfigurierungs-Expertensystemes an einem Beispiel aus der Technik. Die Implementierung ist durch Integration eines Konfigurierungs-Shells mit einem Datenbanksystem, einem 3D-CAD-System und einem Fuzzy-Tool gekennzeichnet.

Humankapitaltheoretische Fundierung von Personalplanungsmodellen bei terminologischer und relationaler Unschärfe

Thomas Spengler, Frankfurt am Main, FRG

Zusammenfassung: In der vorliegenden Arbeit wird gezeigt, wie entscheidungsabhängige Lohnformen und Lohnsätze in Personalplanungsmodelle integriert werden können. Darüberhinaus werden Effekte von Schulungsangeboten des Unternehmens in Ansatz gebracht. Dabei gehen wir durchgängig von relationaler und terminonologischer Unschärfe aus.

Summary: In the present paper it is shown, that decisions about wage rates can be made by the use of personnel planning models. Furthermore we integrate possible effects of schooling in these models. The models consider terminological and relational vagueness.

1. Problemstellung

In der Literatur werden verschiedentlich Defizite der mikroökonomischen Fundierung der Personalwirtschaftslehre konstatiert /1/, /5/. Im vorliegenden Beitrag wollen wir uns mit der Humankapitaltheorie einen einschlägigen mikroökonomischen Ansatz herausgreifen und untersuchen, wie sich dessen Basiselemente in Personalplanungsmodelle integrieren lassen. Zu den wichtigsten Problemstellungen der Humankapitaltheorie zählen Fragen nach den Trägern (Financiers) beruflicher Schulungsprozesse, nach den Wirkungen der Finanzierungsart auf die Lohnsätze und auf das betriebliche Rekrutierungspotential, nach dem Nutzen potentieller Qualifizierungsgänge für die von ihnen betroffenen Parteien und nach der Dauer alternativer Bildungsmaßnahmen /2/.

Wir werden in Abschnitt 2 zunächst ein einfaches Grundmodell der sog. reinen Personalbereitstellungsplanung /3/ formulieren, über das u.a. entschieden werden kann, ob (und in welcher Höhe) die Arbeitskräfte anforderungs- oder qualifikationsabhängig entlohnt werden sollen. Anschließend werden einige Erweiterungsmodule vorgestellt, über die beispielsweise die Interdependenzen zwischen Lohnsatz und Schulungspotential, Leistungsbereitschaft oder Fluktuationsverhalten des Personals im Modell abgebildet werden können. Da die Entscheidungsträger die benötigten Daten in vielen Fällen lediglich größenordnungsmäßig angeben können und der mit der Modellösung erzielbare Nutzen möglicherweise steigt, wenn Restriktionen nur *ungefähr* eingehalten werden, wollen wir durchgängig von terminologischer und relationaler Unschärfe ausgehen. Bezüglich der prozeduralen Aspekte der Modellösung verweisen wir aus Platzgründen auf die einschlägige Literatur [z.B. /4/, /6/, /7/].

2. Ein Grundmodell

Das Grundmodell der reinen Personalbereitstellungsplanung /3/ sieht vor, daß die Entscheidungen bezüglich Lohnart und -höhe bereits getroffen wurden. Wir werden dieses Modell nun dahingehend erweitern, daß wir die Höhe der Lohnsätze und die Bemessungsgrundlage der Entlohnung zum Gegenstand der Entscheidung erheben. Wir betrachten folgende Entscheidungssituation:

(1) Es sind in den nächsten T Perioden Q Tätigkeitsarten zu erledigen. Dazu können Mitarbeiter aus R Arbeitskräftekategorien herangezogen werden.

(2) Die tätigkeitsspezifischen Personalbedarfe liegen in Form von LR-Fuzzy Intervallen, die Entlassungsobergrenzen und Rekrutierungspotentiale in Form von LR-Fuzzy Zahlen vor.

(3) Der Personalplaner muß entscheiden, welche Arbeitskräfte(kategorien) anforderungsabhängig und welche qualifikationsabhängig entlohnt werden. Außerdem stehen verschiedene Lohnsätze zur Auswahl.

(4) Die Rekrutierungs- und die Entlassungsobergrenzen sind abhängig von der gewählten Lohnart und vom Lohnsatz.

(5) Gesucht ist jene Personalausstattung und derjenige Personaleinsatzplan, die den Personalbedarf kostenminimal (Lohn-, Einstellungs-, Entlassungskosten) decken.

Symbolverzeichnis:

Indizes und Indexmengen:

$\underline{R} := \{r | r = 1,2,...,R\}$ Menge der Arbeitskräftekategorien

$\underline{Q} := \{q | q = 1,2,...,Q\}$ Menge der Bedarfskategorien

$\underline{T} := \{t | t = 1,2,...,T\}$ Menge der Teilperioden

$G_r := \{g | g = 1,...,\overline{G}_r\}$ Lohnsatzspektrum von Arbeitskräften der Art r (qualifikationsabhängige Entlohnung)

$L_q := \{l | l = 1,...,\overline{L}_q\}$ Lohnsatzspektrum bei Erledigung von Tätigkeiten der Art q (anforderungsabhängige Entlohnung)

R_q := Bereitstellungsspektrum für Tätigkeiten der Art q

Q_r := Verwendungsspektrum von Arbeitskräften der Art r

Q_l := Menge aller Tätigkeitsarten q, für die der Lohnsatz l gezahlt wird

L_r := Menge aller Lohnsätze $l \in \bigcup_{q \in Q_r} L_q$, die für Arbeitskräfte der Art r in Betracht kommen

Daten:

ϕ^l_{qt} (ϕ^g_{rt}) := Ausprägung des Lohnsatzes $l \in L_q$ ($g \in G_r$) in Periode t

ϕ^h_{rt} (ϕ^f_{rt}) := Einstellungs- (Entlassungs-)Kosten für eine Arbeitskraft der Art r in Periode t

ϕ^n_{rt} := Grundlohn einer in Periode t nicht eingesetzten Arbeitskraft der Art r

$\widetilde{PB}_{qt}$:= unscharfer Personalbedarf zur Erledigung von Tätigkeiten der Art q in Periode t

$\tilde{H}^{gt}_r(\max)$ [$\tilde{H}^{lt}_r(\max)$] := Rekrutierungsobergrenze in Periode t von Arbeitskräften der Art r, die für Lohnsatz g [l] arbeiten würden (LR-Fuzzy Zahl)

$\tilde{F}_r^{gt}(\max)[\tilde{F}_r^{lt}(\max)]$:= Entlassungsobergrenze in Periode t von Arbeitskräften der Art r, die für Lohnsatz g [l] arbeiten (LR-Fuzzy Zahl)

M:= hinreichend große positive Zahl

Entscheidungsvariable:

PA_r^{lt} (PA_r^{gt}):= Ausstattung mit Arbeitskräften der Art r, die in Periode t Lohnsatz l (g) erhalten

h_r^{lt} (h_r^{gt}):= Anzahl in Periode t eingestellter Arbeitskräfte der Art r, die Lohnsatz l (g) erhalten

f_r^{lt} (f_r^{gt}):= Anzahl in Periode t entlassener Arbeitskräfte der Art r, die Lohnsatz l (g) erhalten

PE_{rq}^{lt} (PE_{rq}^{gt}):= Einsatz von Arbeitskräften der Art r, die für Tätigkeit der Art q in Periode t Lohnsatz l (g) erhalten

PEN_r^{lt}:= Anzahl der in Periode t nicht eingesetzten Arbeitskräfte der Art r, die im Einsatzfall Lohnsatz l erhalten würden

$z_r, z_{rg^*t}, z_{rql^*t}$:= (binäre) Hilfsvariable

Mit diesen Symbolen läßt sich das Basismodell wie folgt formulieren:

Zielfunktion:

$$\sum_t \sum_r \left(\sum_{q \in Q_r} \sum_{l \in L_q} \phi_{qt}^l \cdot PE_{rq}^{lt} + \sum_{g \in G_r} \left[\phi_{rt}^g \cdot PA_r^{gt} + \phi_{rt}^h \cdot h_r^{gt} + \phi_{rt}^f \cdot f_r^{gt} \right] + \right.$$

$$\left. + \sum_{l \in L_r} \left[\phi_{rt}^n \cdot PEN_r^{lt} + \phi_{rt}^h \cdot h_r^{lt} + \phi_{rt}^f \cdot f_r^{lt} \right] \right) \stackrel{!}{=} \min \qquad [1]$$

u.d.N.:

Abstimmung Personalbedarf - Personaleinsatz:

$$\tilde{PB}_{qt} \underset{\sim}{=} \sum_{r \in R_q} \left(\sum_{l \in L_q} PE_{rq}^{lt} + \sum_{g \in G_r} PE_{rq}^{gt} \right) \quad \forall q,t \qquad [2]$$

Abstimmung Personaleinsatz - Personalausstattung:

[3.1]

$$\sum_{q \in Q_r} PE_{rq}^{gt} \leq PA_r^{gt} \quad \forall t,r,g \in G_r \qquad [3.2]$$

Rekrutierungsobergrenzen:

$$h_r^{gt} \underset{\sim}{\leq} \tilde{H}_r^{gt}(\max) \quad \forall t,r,g \in G_r \qquad [4.1]$$

$$h_r^{lt} \underset{\sim}{\leq} \tilde{H}_r^{lt}(\max) \quad \forall t,r,l \in L_r \qquad [4.2]$$

Entlassungsobergrenzen:

$$f_r^{gt} \underset{\sim}{\leq} \tilde{F}_r^{gt}(\max) \quad \forall t,r,g \in G_r \qquad [5.1]$$

$$f_r^{lt} \leq \tilde{F}_r^{lt}(\max) \quad \forall t, r, l \in L_r$$ [5.2]

Personalausstattungsfortschreibungen:

$$PA_r^{gt} = PA_r^{g,t-1} + h_r^{g,t} - f_r^{g,t} \quad \forall t, r, g \in G_r$$ [6.1]

$$PA_r^{lt} = PA_r^{l,t-1} + h_r^{l,t} - f_r^{l,t} \quad \forall t, r, l \in L_r$$ [6.2]

Ausschlußbedingungen für die Lohnsätze:
(a) Potentiallohn:

$$PA_r^{g^*t} \leq z_{rg^*t} \cdot M \quad \forall t, r, g^* \in G_r$$ [7.1]

$$\sum_{g \in G_r \setminus \{g^*\}} PA_r^{gt} \leq (1 - z_{rg^*t}) \cdot M \quad \forall t, r, g^* \in G_r$$ [7.2]

(b) Anforderungslohn:

$$PE_{rq}^{l^*t} \leq z_{rql^*t} \cdot M \quad \forall t, q, r \in R_q, l^* \in L_q$$ [7.3]

$$\sum_{l \in L_q \setminus \{l^*\}} PE_{rq}^{lt} \leq (1 - z_{rql^*t}) \cdot M \quad \forall t, q, r \in R_q, l^* \in L_q$$ [7.4]

Ausschlußbedingungen für die Lohnformen:

$$\sum_{g \in G_r} PA_r^{gt} \leq z_r \cdot M \quad \forall r, t$$ [8.1]

$$\sum_{q \in Q_r} \sum_{l \in L_q} PE_{rq}^{lt} \leq (1 - z_r) \cdot M \quad \forall t, r$$ [8.2]

Variablenbedingungen:

$$PA_r^{gt}, PA_r^{lt}, h_r^{gt}, h_r^{lt}, f_r^{gt}, f_r^{lt}, PE_{rq}^{lt}, PE_{rq}^{gt}, PEN_r^{lt} \geq 0 \quad \forall rel.\ r, q, t, l, g$$ [9.1]

$$z_r, z_{rg^*t}, z_{rql^*t} \in \{0,1\} \quad \forall\ relevanten\ r, q, t, g^*, l^*$$ [9.2]

Die Zielfunktion [1] sieht die Minimierung der gesamten Anforderungslohn-, Qualifikationslohn-, Einstellungs- und Entlassungskosten vor. Über die Restriktionen vom Typ [2] wird die *ungefähre* Deckung der unscharfen Personalbedarfe gewährleistet. Die Nebenbedingungen vom Typ [3] sehen vor, daß nicht mehr Arbeitskräfte eingesetzt werden können, als dem Betrieb zur Verfügung stehen. Die Restriktionen [4] und [5] bilden die humankapitaltheoretische These der Lohnabhängigkeit von Rekrutierungs- und Entlassungspotentialen im Modell ab. Die Restriktionenteilsysteme [7] und [8] verarbeiten bestimmte Lohngleichheitspostulate.

3. Modellerweiterungen

Da das Absentismus- und Fluktuationsverhalten sowie die Leistungsbereitschaft der Arbeitskräfte (auch) von der Bemessungsgrundlage und der Höhe der Entlohnung abhängen, werden wir nun skizzieren, wie sich entsprechende (auch nach Lohnform und Lohnsatz differenzierte) unscharfe Anwesenheits- und Verbleibensquoten sowie Leistungsfaktoren in das Grundmodell integrieren lassen. Dazu benötigen wir folgende Symbole:

$\tilde{\beta}_r^g \left(\tilde{\beta}_r^l\right)$:= (gegebene) unscharfe Verbleibensrate von Arbeitskräften der Art r, sofern diese den Lohnsatz g (l) erhalten [LR-Fuzzy Intervall]

$\tilde{\delta}_r^g \left(\tilde{\delta}_r^l\right)$:= (gegebene) unscharfe Anwesenheitsrate von Arbeitskräften der Art r, sofern diese den Lohnsatz g (l) erhalten [LR-Fuzzy Intervall]

$\tilde{\alpha}_r^g \left(\tilde{\alpha}_r^l\right)$:= (gegebener) unscharfer Leistungsfaktor von Arbeitskräften der Art r, sofern diese den Lohnsatz g (l) erhalten [LR-Fuzzy Intervall; definiert als Ist-/Standardarbeitsproduktivität]

Für die Berücksichtigung der Fluktuationseffekte sind die Personalausstattungsfortschreibungsgleichungen [6.1] und [6.2] zu ersetzen durch die Gleichungen [6.3] und [6.4]:

$$PA_r^{gt} \underset{\sim}{=} \tilde{\beta}_r^g \cdot PA_r^{g,t-1} + h_r^{g,t} - f_r^{g,t} \quad \forall t,r,g \in G_r \qquad [6.3]$$

$$PA_r^{lt} \underset{\sim}{=} \tilde{\beta}_r^l \cdot PA_r^{l,t-1} + h_r^{l,t} - f_r^{l,t} \quad \forall t,r,l \in L_r \qquad [6.4]$$

Die oben definierten unscharfen Anwesenheitsraten und Leistungsfaktoren können als Koeffizienten der entsprechenden Personaleinsatzvariablen in Restriktion [2] integriert werden. Wir zeigen dies exemplarisch anhand Restriktion [2.1] (Berücksichtigung von Absentismus):

$$\widetilde{PB}_{qt} \underset{\sim}{=} \sum_{r \in R_q} \left(\sum_{l \in L_q} \tilde{\delta}_r^l \cdot PE_{rq}^{lt} + \sum_{g \in G_r} \tilde{\delta}_r^g \cdot PE_{rq}^{gt} \right) \quad \forall q,t \qquad [2.1]$$

Je *günstiger* die gewährten Lohnsätze (g bzw. l) im Urteil der Arbeitskräfte ausgeprägt sind, desto höhere Werte werden die jeweiligen Leistungsfaktoren, Anwesenheits- und Verbleibensquoten annehmen. Dies kann dazu führen, daß über das Modell ein Optimum ermittelt wird, in dem die Arbeitskräfte (im Vergleich zum Grundmodell) zwar höhere Löhne erhalten, der Zielfunktionswert jedoch insgesamt kleiner ist, da geringere Einstellungs- und Entlassungskosten anfallen und der Personalbedarf kostengünstiger gedeckt wird.

Im Fokus der Humankapitaltheorie steht (auch) das Entscheidungsproblem, in welcher Höhe Arbeitskräfte und Betriebe Weiterbildungsinvestitionen tragen sollen. Die Arbeitskräfte werden diese Entscheidungen (u.a.) von der Erhöhung ihres Restlebenseinkommens abhängig machen -was im Modell jedoch nicht (differenziert) erfaßt wird-, die Unternehmen werden nur dann in Weiterbildungsmaßnahmen investieren, wenn dadurch die Personalbereitstellung künftig kostengünstiger gestaltet werden kann. Wir wollen im folgenden skizzieren, wie sich Schulungsprozesse und vor allem die möglicherweise mit dem Angebot von Schulungsmaßnahmen einhergehenden Veränderungen der Rekrutierungspotentiale in das Modell integrieren lassen. Wir definieren folgende Symbole:

Indizes und Indexmengen:

R_r^*:= Menge aller Zielqualifikationen r*, die von Qualifikation r aus erreicht werden können

$R_r^{'}$:= Menge aller Basisqualifikationen r', von denen aus die Qualifikation r erreicht werden kann

Daten:

$\tau_{r,r^*}(\tau_{r',r})$:= Dauer eines Schulungsganges von der Basisqualifikation r (r') zur Zielqualifikation r* (r)

ϕ^{lt}_{r,r^*} (ϕ^{gt}_{r,r^*}):= Kosten einer Ausbildung von Qualifikation r nach r* in Periode t, sofern die Arbeitskraft den Lohnsatz l (g) erhält

$\widetilde{HS}^{gt}_{r}(\max)[\widetilde{HS}^{lt}_{r}(\max)]$:= Rekrutierungsobergrenze in Periode t von Arbeitskräften der Art r, die geschult werden sollen und für Lohnsatz g [l] arbeiten würden (LR-Fuzzy Zahl)

ε:= hinreichend kleine positive Zahl

Entscheidungsvariable:

PES^{lt}_{r,r^*} (PES^{gt}_{r,r^*}):= Anzahl von Arbeitskräften der Art r, die Lohnsatz l (g) erhalten und in Periode t eine Ausbildung nach r* beginnen

$z^{'}_{rgt}, z^{*}_{rlt}$:= (binäre) Hilfsvariable

Das Grundmodell ist nun dahingehend zu variieren, daß in die Restriktionen [3.1] und [3.2] die Terme

$$\sum_{r^* \in R^*_r} \sum_{t'=t-\tau_{r,r^*}+1}^{t} PES^{lt'}_{r,r^*} \quad [3.1.a] \text{ und } \sum_{r^* \in R^*_r} \sum_{t'=t-\tau_{r,r^*}+1}^{t} PES^{gt'}_{r,r^*} \quad [3.2.a]$$

additiv eingestellt werden, die die Anzahl aller in der aktuellen Periode in Schulung befindlichen Arbeitskräfte der entsprechenden Kategorie repräsentieren. Die Personalausstattungsfortschreibungsgleichungen [6.1] und [6.2] sind um die jeweiligen Zu- und Abgänge aus Schulungsmaßnahmen ([6.1.a] und [6.2.a]) zu ergänzen:

$$+\sum_{r' \in R^{'}_r} PES^{g,t-\tau_{r',r}}_{r',r} - \sum_{r^* \in R^*_r} PES^{g,t-\tau_{r,r^*}}_{r,r^*} \quad [6.1.a]$$

$$+\sum_{r' \in R^{'}_r} PES^{l,t-\tau_{r',r}}_{r',r} - \sum_{r^* \in R^*_r} PES^{l,t-\tau_{r,r^*}}_{r,r^*} \quad [6.2.a]$$

Die Zielfunktion [1] muß um den Term [1.a] erweitert werden:

$$+\sum_{r^* \in R^*_r} \left[\sum_{l \in L_r} \phi^{lt}_{r,r^*} \cdot PES^{lt}_{r,r^*} + \sum_{g \in G_r} \phi^{gt}_{r,r^*} \cdot PES^{gt}_{r,r^*} \right] \quad [1.a]$$

Die Veränderungen der Rekrutierungspotentiale durch Schulungsangebote lassen sich durch den folgenden Restriktionensatz [10] in Ansatz bringen:

$$\sum_{r^* \in R^*_r} PES^{gt}_{r,r^*} \le z^{'}_{rgt} \cdot M \quad \forall t,r,g \in G_r \quad [10.1]$$

$$\sum_{r^* \in R_r^*} PES_{r,r^*}^{gt} \geq z'_{rgt} \cdot \varepsilon \quad \forall t,r,g \in G_r \tag{10.2}$$

$$\sum_{r^* \in R_r^*} PES_{r,r^*}^{lt} \leq z_{rlt}^* \cdot M \quad \forall t,r,l \in L_r \tag{10.3}$$

$$\sum_{r^* \in R_r^*} PES_{r,r^*}^{lt} \geq z_{rlt}^* \cdot \varepsilon \quad \forall t,r,l \in L_r \tag{10.4}$$

$$h_r^{gt} \underset{\sim}{\leq} z'_{rgt} \cdot \widetilde{HS}_r^{gt}(\max) + \left(1 - z'_{rgt}\right) \cdot \tilde{H}_r^{gt}(\max) \quad \forall t,r,g \in G_r \tag{10.5}$$

$$h_r^{lt} \underset{\sim}{\leq} z_{rlt}^* \cdot \widetilde{HS}_r^{lt}(\max) + \left(1 - z_{rlt}^*\right) \cdot \tilde{H}_r^{lt}(\max) \quad \forall t,r,l \in L_r \tag{10.6}$$

Das Restriktionensystem [10] bewirkt, daß die z-Variablen nur dann den Wert 1 annehmen können, wenn entsprechende Schulungen stattfinden und daß auch nur dann die korrespondierenden Rekrutierungspotentiale gelten. Für die Schulungsvariablen müssen noch Nichtnegativitätsbedingungen formuliert werden und die z-Variablen sind als 0/1-Variable zu definieren.

Literatur

/1/ Hax, Herbert: Theorie der Unternehmung - Information, Anreize und Vertragsgestaltung. Betriebswirtschaftslehre und ökonomische Theorie, hrsg. v. D. Ordelheide et al. Stuttgart: Poeschel (1991), S. 51-72.

/2/ Joll, Caroline/McKenna, Chris./McNabb, Robert/Shorey, John: Developments in Labour Market Analysis. London, Boston, Sydney: George Allen & Unwin (1983).

/3/ Kossbiel, Hugo: Personalbereitstellung und Personalführung. Allgemeine Betriebswirtschaftslehre, hrsg. v. H. Jacob, 5. überarb. Aufl. Wiesbaden: Gabler (1988), S. 1045-1257.

/4/ Rommelfanger, Heinrich: Entscheiden bei Unschärfe. Fuzzy-Decision Support-Systeme. Berlin u.a.: Springer (1988).

/5/ Sadowski, Dieter: Humankapital und Organisationskapital - Zwei Grundkategorien einer ökonomischen Theorie der Personalpolitik im Unternehmen. Betriebswirtschaftslehre und ökonomische Theorie, hrsg. v. D. Ordelheide et al. Stuttgart: Poeschel (1991), S. 127-141.

/6/ Spengler, Thomas: Fuzzy-Entscheidungsmodelle für die Planung der Personalbereitstellung. Operations Research Proceedings 1991, hrsg. v. W. Gaul et al. Berlin u.a.: Springer (1992), S. 501-508.

/7/ Spengler, Thomas: Lineare Entscheidungsmodelle zur Organisations- und Personalplanung. Heidelberg: Physica (1993).

UNTERSTÜTZUNG DER STRATEGISCHEN TECHNOLOGIEPLANUNG DURCH WISSENSBASIERTE SYSTEME

Brigitte Werners, Ruhr-Universität Bochum

Zusammenfassung: Die Möglichkeiten der Unterstützung strategischer Technologieplanung durch wissensbasierte Systeme werden überblicksartig dargestellt und durch einen eigenen Vorschlag erweitert. In den bisher verwendeten Systemen sind vage, unsichere oder verbal ausgedrückte Informationen häufig nur unzureichend berücksichtigt. Da diesen gerade bei der strategischen Planung besondere Bedeutung zukommt, wurde auf ihre angemessene Behandlung ausdrücklicher Wert gelegt. Auf dieser Grundlage wird das wissensbasierte System StraTeP zur strategischen Technologieplanung vorgeschlagen, welches bei der Analyse der gegenwärtigen Wettbewerbs- und technologischen Situation unterstützt, mögliche zukünftige Entwicklungen berücksichtigt und Empfehlungen für Technologieentscheidungen gibt.

Summary: The possibilities of supporting the strategic planning of technologies through knowledge-based systems are presented in a brief survey. Then they are extended through an own suggestion. Vague, uncertain, or verbally expressed parts of information are often treated insufficiently in the systems used up to now. As this kind of information is of major importance in the context of strategic planning, the appropriate representation is stressed explicitly. On this basis, the knowledge-based system StraTeP is suggested for the strategic planning of technologies. It supports in analyzing the current competitive and technological situation, allows to consider possible future developments and recommends decisions concerning the technologies.

Einführung

Die für die strategische Unternehmensplanung vorgeschlagenen und entwickelten Methoden und Instrumente beinhalten einerseits zumindest in Teilaspekten technologische Gesichtspunkte und können andererseits, teilweise direkt, für die Anwendung im Bereich der strategischen Technologieplanung übertragen werden. Wie die strategische Planung allgemein hat auch die strategische Technologieplanung zum Ziel, langfristige Erfolgspotentiale zu schaffen und zu sichern. Dazu ist der Einsatz von Technologien unter Berücksichtigung der Abhängigkeiten zwischen Produkt, Markt und Technologie zu planen und zu steuern. Zur Unterstützung der strategischen Technologieplanung werden in der Literatur verschiedene Instrumente und Modelle vorgeschlagen, einen Überblick gibt Werners (1993). Einige dieser Vorschläge betonen den technologischen Aspekt in allgemein anwendbaren Ansätzen. Auch hier wurden wissensbasierte Systeme zur Unterstützung entwickelt wie zum Beispiel EXSTRABS (Lelke, Werners 1991). Zu den bekanntesten Instrumenten zur Unterstützung der strategische Planung gehören die unterschiedlichen Portfoliomodelle, von denen im folgenden einige verbreitete Ansätze insbesondere unter technologischen Gesichtspunkten kurz behandelt werden.

Die Grundidee der Portfolio-Ansätze besteht darin, einzelne Elemente eines Betrachtungsgegenstandes geeignet zusammenzustellen, um so zu einer ausgewogenen Gesamtheit zu gelangen. Im Zusammenhang mit der strategischen Planung werden dazu die Betrachtungselemente hinsichtlich verschiedener Kriterien bewertet, welche meist zu zwei Dimensionen, den unternehmensinternen Fähigkeiten und den Umweltbedingungen, aggregiert werden. Durch die zweidimensionale Betrachtung ist eine visuelle Präsentation der Analyseergebnisse in einer Portfoliomatrix eindrucksvoll möglich. Die Positionierung in der Matrix wird i.a. mit dem Vorschlag von Normstrategien verbunden. Allgemein lassen sich die im Zusammenhang der technologischen Planung relevanten Portfolio-Ansätze unterteilen in die allgemeinen Produkt/Marktbezogenen Ansätzen, in Ansätze, die aus Produkt/Marktportfolios technologische bzw. FuE-Bereichsstrategien ableiten, und in Technologieportfolios.

Technologieportfoliomatrizen

Eines der bekanntesten Technologieportfoliomodelle (Pfeiffer et al. 1987), welches auf einem integrierten Produktlebenszykluskonzept basiert, wird im folgenden detaillierter dargestellt. Betrachtungsobjekt sind die gegenwärtig in dem Unternehmen vorkommenden Technologien, und zwar getrennt nach Produkt- und nach Prozeßtechnologie. Die Produkt- und die Prozeßtechnologieportfoliomatrizen besitzen übereinstimmende Dimensionen, jeweils zum einen die Umweltsituation im Technologiebereich, welche mit Technologieattraktivität bezeichnet wird, und zum anderen die unternehmensinterne Stärke zur Entwicklung von Technologien, zusammengefaßt unter dem Begriff Ressourcenstärke. Die Achsen der Matrizen werden in je drei Bereiche mit den möglichen Ausprägungen der Dimensionen "niedrig", "mittel" und "hoch" eingeteilt. Insgesamt ergeben sich die Bewertungen durch Aggregation über folgende Hierarchien:

Technologieattraktivität
- Technologiepotentialrelevanz
    - Weiterentwickelbarkeit
    - Zeitbedarf bis zur nächsten Entwicklungsstufe
- Technologiebedarfs-Relevanz
    - Anwendungsumfang, Anwendungsarten
    - Diffusionsverlauf über die Zeit

Ressourcenstärke
- Finanzielle Ressourcen
    - Budgethöhe
    - Kontinuität des Budgets
- Know-how-Stärke
    - Know-how-Stand
    - Stabilität des Know-how

Insbesondere die Zusammenfassung der Bewertungen bez. der untergeordneten Faktoren zu der Ausprägung der Bewertung eines höheren Faktors gibt aufgrund ihrer Kontextabhängigkeit Ansatzmöglichkeiten für den Einsatz von Expertensystemen. Den verschiedenen Feldern des resultierenden Technologie-Portfolios werden nun Handlungsempfehlungen für die Zuteilung von Ressourcen zugeordnet mit den grundsätzlichen Strategieempfehlungen Investieren,

Selektieren und Desinvestieren. Dabei sind jedoch, wie auch Pfeiffer et al. (1987) angeben, die Strategievorschläge für die unterschiedlichen Technologien aufeinander abzustimmen. In der Erweiterung dieses Ansatzes (Pfeiffer, Dögl 1990) werden die Komponententechnologien eines Produktes zum einen separat in dem Technologieportfolio abgebildet, zum anderen werden die Ergebnisse aggregiert dargestellt. Die Modifikation einer Komponententechnologiebeurteilung hat damit möglicherweise Auswirkungen auf mehrere Produkttechnologiebewertungen. Da bei diesem Ansatz auch Komplementärtechnologien als zukünftige Potentiale mit einbezogen werden sollen, ergeben sich damit eine Reihe von Ansatzpunkten, durch wissensbasierte Systeme strategische Technologieentscheidungen zu unterstützen. Mehrere Systeme sind bereits entwickelt worden, von denen einige, welche insbesondere auch Portfolioansätze unterstützen, im folgenden kurz vorgestellt werden.

Ausgewählte wissensbasierte Systeme zur strategischen Planung

STRAPLAN (Gabriel et al. 1990) ist ein prototypisches wissensbasiertes System zur Unterstützung der Strategischen Planung, speziell der Portfolio-Analyse, in der Versicherungsbranche. Als Wissensbasis wird das Konzept der Marktattraktivitäts-Wettbewerbspositionsmatrix verwendet, welches um branchenspezifische Kenntnisse eines Experten erweitert ist. Über die Analyse hinaus werden in Abhängigkeit von der konkreten Position einer strategischen Geschäftseinheit branchenbezogene Strategien vorgeschlagen. Die Bewertung der strategischen Faktoren erfolgt durch qualitative Angaben. Dabei sind für jeden Faktor drei Ausprägungen wie "hoch", "mittel" oder "niedrig" vorgesehen. Quantitativen Ausprägungen wird durch entsprechende Regeln ein Term zugeordnet, dies geschieht über die Vorgabe von strikten Intervallen auf einer Basisvariable. Die Umsetzung erfolgte mittels des Expertensystemtools K-EXPERT, welches ausschließlich WENN-DANN-Regeln zur Wissensrepräsentation zuläßt (Gabriel et al. 1990). Eine explizite Unsicherheitsmodellierung erfolgt nicht.

CASA, Computer-assistiertes Strategie Audit (Krallmann 1990) ist ein System zur wissensbasierten Strategieberatung von mittelständischen Unternehmen, welches insbesondere die Informationsgewinnung und die Analysephase der strategischen Planung unterstützen soll. Das System enthält die Teile Segmentierung in strategische Geschäftseinheiten, Ermittlungen der Ausprägungen der Unternehmenskultur, der Marktattraktivität und der relativen Wettbewerbsposition der strategischen Geschäftseinheiten und eine strategische Kosten- und Ergebnisanalyse. Für die einzelnen strategischen Geschäftseinheiten werden Sollstrategien ermittelt und den Iststrategien gegenübergestellt (Müller-Wünsch, Woltering 1990). Das diesem System zugrundeliegende Wissen stammt sowohl aus Lehrbüchern als auch von Beratern mittelständischer Unternehmen. Zur Implementierung wurde die produktionsregel-orientierte Programmierumgebung KES PS verwendet in Kombination mit weiteren Programmen wie EXCEL. Auch hier wird keine explizite Modellierung von Unsicherheiten durchgeführt.

STRATEX (Plattfaut 1988) ist ein Expertensystem, bei dem die Wissensdomäne die kombinierte Anwendung eines Produkt/Marktportfolios mit einem Technologieportfolio ist. Die Zuordnung von Normstrategien erfolgt gemäß der üblichen Vorschläge für die beiden Einzelportfolios, wobei aufgrund der speziellen Kenntnisse in dem anwendenden Unternehmen Modifikationen vorgenommen wurden. Der Benutzer führt eine Bewertung der strategischen Geschäftseinheiten bez. der Indikatoren vermittels der Terme "niedrig", "mittel", "hoch" durch. Diesen werden Zahlen der Skala -5 bis +5 zugeordnet, mit denen die Aggregationen dann durchgeführt werden, anschließend erfolgt die erneute Zuordnung von Termen. Zusätzlich erfolgt die Bewertung der Unsicherheit durch Konfidenzzahlen, die gleichermaßen wie die Originalbewertungen aggregiert werden. Die Umsetzung erfolgte zunächst mit HEXE und mit einer PROLOG2-Umgebung.

Von Reminger (1990) ist ein Expertensystem zur strategischen Technologieplanung entwickelt worden, welches auf dem Technologieportfoliokonzept von Pfeiffer basiert. Zur Bestimmung der Dimensionen Technologieattraktivität und Ressourcenstärke wird ähnlich vorgegangen wie bei dem System STRATEX, jedoch werden feiner untergliederte Fragebäume mit weniger starrer Gliederung der einzelnen Bewertungsaspekte verwendet. Das Wissen wird in Form von Regeln in mehreren miteinander verknüpften Wissensbasen repräsentiert. Es kommen unterschiedliche Inferenzstrategien zur Anwendung. Die Implementierung erfolgte mittels XiPLUS. Für die Repräsentation von Unsicherheiten wird eine dreiwertige Plausibilitätskennzahl verwendet, deren Aggregation mittels einer Verknüpfungstabelle durchgeführt wird.

Wie auch die kurze Beschreibung der wissensbasierten Systeme zeigt, verwenden die anwendungsorientierten Portfolioansätze sowohl zur Beschreibung der Einteilung der Dimensionen als auch zur Erfassung der Ausprägungen der strategischen Erfolgsfaktoren überwiegend verbale Ausdrücke wie beispielsweise "hoch", "mittel", "gering". Zur Erfassung und Verarbeitung der Bewertungen kritischer Erfolgsfaktoren werden den verbalen Begriffen häufig Werte zugeordnet, mit denen dann weitergerechnet wird. Der Verzicht auf die Berücksichtigung des Skalenniveaus und die relativ willkürliche, strikte Zuordnung von Werten ist dabei kritisch zu sehen. Ein weiterer Kritikpunkt an der Anwendung von Portfoliomatrizen zur strategischen Planung besteht darin, daß die Unsicherheit, welche mit der Beurteilung der Ausprägungen von Faktoren für eine strategische Geschäftseinheit auf reellwertiger Basis verbunden ist, nicht ausreichend berücksichtigt wird. Statt dessen werden genau anzugebende Punktbewertungen bez. der verschiedenen Faktoren verlangt, und deren Aggregation zu den Dimensionen einer Matrix führt zu einer Punktpositionierung der strategischen Geschäftseinheit. Meist ist es nicht möglich, derart detaillierte, genaue Informationen vom Anwender zu erhalten, sondern allenfalls Bereichsangaben, die häufig auf intuitiven Einschätzungen beruhen (Ansoff et al. 1981, S. 965). Daher wird von diesen Autoren eine methodische Weiterentwicklung von Portfoliomodellen vorgeschlagen, um die bei der Beurteilung strategischer Geschäftseinheiten auftretenden Unsicherheiten durch eine Unschärfepositionierung sichtbar zu machen. Besser geeignet erscheint die Berücksichtigung dieser nicht nur stochastisch beschreibbaren Unsicherheit durch die Verwendung linguistischer Variablen.

Linguistische Variablen zur Bewertung

Eine linguistische Variable ist definiert als Quintupel (X, T, U, G, M), wobei X der Name der Variable, T der Wertebereich dieser Variable, auch Termmenge genannt, U der Definitionsbereich der Basisvariable, G die Syntax zur Konstruktion neuer Werte und M die Semantik ist. Durch die Definition der linguistischen Variable ist die Möglichkeit gegeben, jeden Term durch eine Zugehörigkeitsfunktion auf der Basisvariable zu beschreiben. Grundsätzlich sind viele verschiedene Funktionen zur Modellierung von Termen geeignet. Für praktische Anwendungen empfiehlt sich die Verwendung von Funktionen mit einfacher mathematischer Struktur, die aufgrund weniger Parameter bestimmbar sind und damit für den Anwender des Konzepts plausibel erscheinen und mit denen effizient weitere Berechnungen durchführbar sind. Besonders einfach für den Anwender und für spätere Berechnungen ist die Verwendung von speziellen unscharfen Intervallen, die durch wenige Parameter zu charakterisieren sind und einen stückweise linearen Verlauf der Zugehörigkeitsfunktionen aufweisen. Hier werden insbesondere die trapezförmigen Zugehörigkeitsfunktionen vorgeschlagen, wobei ein unscharfes Intervall auf einer Basisvariable U dann charakterisiert wird durch ein Quadrupel (a,b,c,d). Die Werte b und c grenzen den Bereich vollständiger Zugehörigkeit ein, a und d sind äußerstenfalls noch als zugehörig zu betrachten.

Für die Anwendung in einem wissensbasierten System wird vorgeschlagen, das allgemeine Konzept einer linguistischen Variable als Objekt mit den Slots Termmenge, Basisvariable und Semantik zu repräsentieren. Damit können allgemeine Eigenschaften, wie die Gestalt der Zugehörigkeitsfunktionen, vererbt werden. Das subjektive Wissen eines Branchenexperten über die Bedeutung bestimmter Terme läßt sich auf diese Weise strukturiert, effizient und nachvollziehbar verwenden und dokumentieren. So sind beispielsweise die Aussagen eines Planers eines pharmazeutischen Unternehmens, der objektiv feststellbare Marktanteilswerte, also die Basisvariable, mit "stark" in einem Bereich von etwa 10 bis 20 %, "haltbar" in einem Bereich von etwa 2 bis 5 % beurteilt, als Terme beschreibbar über die Quadrupel (8,10,20,22) für "stark" und (1.8, 2, 5, 5.5) für "haltbar".

Damit sind auch Aggregationen linguistischer Variablen, welche die inhärente Unsicherheit adäquat beibehalten, über verschiedene Aggregationsstufen möglich. Dazu wird auf Grundlage des Erweiterungsprinzips die Aggregation über verschiedene Hierarchieebenen durchgeführt. Bei der Aggregation lassen sich verschiedene Operatoren verwenden, insbesondere auch solche, die Kompensationsmöglichkeiten der unterschiedlichen Kriterien zulassen. Dazu wurde eine Erweiterung des "unscharfen und" und des "unscharfen oder" (Werners 1984) entwickelt, welche auch verschiedene Gewichtungen der Faktoren erlaubt (Werners 1993). Die verbale Unsicherheit bei der Beurteilung auf der Faktorebene wird entsprechend in dem Ergebnis, also einer unscharfen Menge auf der Basis der Ergebnisdimension, ausgedrückt. Die abschließende Positionierung der Betrachtungsobjekte, also der strategischen Geschäftseinheiten oder der Technologien, erfolgt einschließlich der Darstellung der Unschärfe des Ergebnisses, indem sowohl ein inneres Rechteck mit Zugehörigkeitsgrad 1 der Positionierung als auch ein äußeres Rechteck mit fallenden Zugehörigkeitsgraden bis 0 abgebildet wird. Bei dem von Zimmermann

(1989) vorgestellten prototypischen System ESP, Expert System for Strategic Planning, ist ein ähnlicher Ansatz implementiert.

Schlußfolgern mit Termen

Liegen über mathematische Strukturen von Termmengen keine Informationen vor oder ist aus anderen Gründen kein geschlossener funktionaler Ausdruck für die Aggregation untergeordneter Kriterien zu einem übergeordneten Kriterium bekannt, kann die explizite Angabe des Zusammenhangs über eine Relation, z.B. einer Implikationsrelation, auf einer Termmenge eine adäquate Verknüpfung darstellen. So läßt sich jede einzelnen Regel eines regelbasierten Systems mathematisch als Relation auffassen. Beispielsweise eine Aussage der Form

WENN A DANN B

mit der Bedeutung "Aus A folgt B" ist durch das Tupel (A,B) beschreibbar. Sind X und Y Variablen, die Werte aus den Termmengen T1 bzw. T2 annehmen, und ist A ein Term aus T1 und B ein Term aus T2, dann kann folgende Regel

WENN X ist A DANN Y ist B

als Relation R auf T1xT2 aufgefaßt werden. Diese Relation wird mächtiger, falls mehrere Regeln in Abhängigkeit von unterschiedlichen Ausprägungen der Variablen X und Y formuliert werden können. Der Wissenserwerb bez. derartiger Abhängigkeiten zwischen Ausprägungen von X und Y entspricht damit einer Vervollständigung des fragmentarischen Wissens, bis im Extrem durch Erfassung aller Ausprägungen, die durch T1 für X möglich sind, und Angabe der zugehörigen Ausprägungen aus T2 für Y die Relation auf T1xT2 eine Funktion von T1 nach T2 darstellt. Es sind weiter Relationen möglich, in denen einer Voraussetzungsausprägung A mehrere Konklusionen zugeordnet sind. Auf Grundlage des generalisierten Modus ponens lassen sich bei Kombination mehrerer Regeln, d.h. einer mächtigeren Relation, auch approximative Inferenzmechanismen anwenden. Detailliertere Erörterungen sind Werners (1993) zu entnehmen. Das führt dazu, daß selbst bei nur vage bekannten Zusammenhängen aus recht genauen Voraussetzungen recht präzise Ergebnisse gefolgert werden können, wie sich ähnlich auch beim Einsatz von Fuzzy Control in der Praxis zeigt (Zimmermann 1993).

StraTeP

Im folgenden werden die wesentlichen Bestandteile des wissensbasierten Systems zur strategischen Technologieplanung StraTeP dargestellt. Um eine schnellere Verarbeitung und eine bessere Wartungs- und Entwicklungsmöglichkeit des Systems zu erzielen, wird das Wissen in mehreren Wissensbasen abgelegt. Eine wesentliche Wissensbasis enthält die verschiedenen Portfoliokonzepte, die dem Anwender zur Verfügung gestellt werden. Das mit diesen Konzepten verbundene Wissen umfaßt neben den jeweiligen Beschreibungen auch die Aggregationshierarchien zur Ermittlung der Dimensionen der Portfoliomatrizen und die Normstrategievorschläge. In einer zweiten Wissensbasis werden alle bereits festgelegten linguistischen Variablen erfaßt. Das Expertenwissen über die Durchführung einer Portfolioanalyse ist in einer übergeordneten Wissensbasis enthalten, die auf die untergeordneten Wissensbasen bei Bedarf zugreift. Hiermit wird die Analyseabwicklung gesteuert und Teiluntersuchungen initiiert. Ergebnis einer Sitzung ist die detaillierte Darstellung der Ausprägungen der kritischen Erfolgsfaktoren jedes Betrachtungsobjektes und die Portfoliodarstellung unter Berücksichtigung der Unsicherheiten, die mit den Beurteilungen verbunden sind.

In weiteren Analyseschritten besteht nun die Möglichkeit, Zusammenhänge zwischen Betrachtungsobjekten herzustellen und zu beurteilen. Dies kann etwa durch Berücksichtigung zukünftiger unterschiedlicher Entwicklungen im Sinne von Szenariountersuchungen geschehen oder durch gemeinsame, vergleichende Auswertung unterschiedlicher Portfoliokonzepte wie Produkt-Markt-Portfolio gemeinsam mit einem Technologieportfolio. Grundlage ist die entsprechende Erfassung der Zusammenhänge. Zur Herstellung des Zusammenhangs zwischen Produkt-Markt- und Technologieportfolio werden die zu den strategischen Geschäftsfeldern gehörenden Technologien ermittelt und in ihrer Bedeutung für das strategische Geschäftsfeld erfaßt.

Literatur:

Ansoff, I., Kirsch, W., Roventa, P. (1981): Unschärfenpositionierung in der strategischen Portfolio-Analyse, in: ZfB, 51. Jg., H. 10 (1981) S. 963-988

Gabriel, G., Gräff, H., Borgard, A. (1990): Entwicklung des wissensbasierten Systems STRAPLAN zur Strategischen Planung in der Versicherungsbranche, in: Zeitschrift für Planung, Bd. 1, Nr. 3 (1990) S. 245-263

Krallmann, H. (1990): Wissensbasierte Strategiefindung, in: Zahn, E. (Hrsg.): Europa nach 1992, Wettbewerbsstrategien auf dem Prüfstand, Stuttgart 1990, S. 159-174

Lelke, B., Werners, B. (1991): Modellierung und Implementierung von EXSTRABS: Ein Expertensystem zur Branchenstrukturanalyse der strategischen Planung, in: Wirtschaftsinformatik, 33. Jg., H. 4 (1991) S. 316-324

Müller-Wünsch, M., Woltering, A. (1990): Computergestützte Strategieberatung für den Mittelstand, in: Ehrenberg et al. (Hrsg.): Wissensbasierte Systeme in der Betriebswirtschaft, Grundlagen, Entwicklung, Anwendungen, Berlin 1990, S. 523-544

Pfeiffer, W., Dögl, R. (1990): Das Technologie-Portfolio-Konzept zur Beherrschung der Schnittstelle Technik und Unternehmensstrategie, in: Hahn, D., Taylor, B. (Hrsg.): Strategische Unternehmensplanung - Strategische Unternehmungsführung, Stand und Entwicklungstendenzen, 5., neu bearb. u. erw. Aufl., Heidelberg 1990, S. 254-282

Pfeiffer, W., Metze, G., Schneider, W., Amler, R. (1987): Technologie-Portfolio zum Management strategischer Zukunftsgeschäftsfelder, 4., unveränderte Aufl., Göttingen 1987

Plattfaut, E. (1988): DV-Unterstützung strategischer Unternehmensplanung, Beispiele und Expertensystemansatz, Berlin, Heidelberg, N.Y. 1988

Reminger, B. (1990): Expertensystem zur Unterstützung der strategischen Technologieplanung, Berlin 1990

Werners, B. (1984): Interaktive Entscheidungsunterstützung durch ein flexibles mathematisches Programmierungssystem, München 1984

Werners, B. (1993): Unterstützung der strategischen Technologieplanung durch wissensbasierte Systeme, Aachen 1993

Zimmermann, H.-J. (1989): Strategic planning, operations research and knowledge based systems, in: Verdegay, J.-L., Delgado, M. (Hrsg.): The interface between artificial intelligence and operations research in fuzzy environment, Köln 1989, S. 253-274

Zimmermann, H.-J. (Hrsg.): Fuzzy Technologien - Prinzipien, Werkzeuge, Potentiale, Düsseldorf 1993

DAS VERFAHREN ZUR KONSTRUKTION EINER UNSCHARFEN PRÄVALENZMATRIX IM ENTSCHEIDUNGSUNTERSTÜTZUNGSSYSTEM (DSS) FRED (FUZZY PREFERENCE DECISION SUPPORT SYSTEM)

Martina Wiemers
Ruhr-Universität Bochum

Auf der Grundlage der verstärkten Beschäftigung mit Entscheidungssituationen mit mehreren Zielen der letzten Jahre wird ein Verfahren zur Modellierung des Entscheidungsprozesses vorgestellt. Hauptaugenmerk ist hierbei die Konstruktion einer Prävalenzmatrix unter Einbeziehung von unscharfen Relationen. Das Verfahren gliedert sich in die Methodenklasse der Outrankingverfahren (im Rahmen der Multiattributiven Entscheidungstheorie bzw. Multikriteria Systeme mit endlicher Alternativmenge, (vgl. Electre, Promethee)) ein, bei denen keine vollständige Information über die Alternativen bzw. Attributsausprägungen bzw. Präferenzvorstellungen des Entscheidungsträgers (ET) vorliegen.
Wesentlicher Ansatz ist hier die Betrachtung der unscharfen Bewertungen der Alterantiven bzw. einzelner Kriterien bzw. deren Unvergleichbarkeit, sowie die graduelle Kompensationsbereitschaft bzw. - abneigung des ET bzgl. einzelner Alternativausprägungen.
Prävalenzrelationen erweitern hier die klassischen Präferenzrelationen dadurch, daß zwischen den Begriffen der Unvergleichbarkeit, Präferenz und Dominanz graduelle Übergänge ermöglicht werden. Auf der Grundlage eines axiomatischen Ansatzes zur Darstellung der Präferenzvorstellung des ET für quantitative Meßgrößen (z. B. Preis in DM) mit eingeschränkter Kompensationsmöglichkeit zwischen den Zielen wird eine Erweiterung auf qualitative Bewertungen (im Sinne von Linguistischen Termen) vorgenommen. Die vorgestellte Methode findet Anwendung in dem DSS FRED (s.o.).
Das System verfügt weiterhin über einen wissensbasierten Ansatz (unscharfe Regelwerke) zur Problemmodellierung. Zur Ermittlung eines Entscheidungsvorschlages werden Verallgemeinerungen von Electre, bzw. Promethee verwendet. Als Fallstudie ist die Prüfung von Lebensversicherungsanträgen vorgesehen, bei denen ein Antragsteller mit einer Gruppe von Bezugspersonen des gleichen Typs bzw. gleicher Kategorie verglichen wird. Dabei werden vertrags-, kunden-, und unternehmensspezifische Kriterien zur Beurteilung des Antrages berücksichtigt. Durch die schrittweise Rangordnung des potentiellen Kundens im Vergleich zu den Bezugskunden werden die Stärken und Schwächen des Antragstellers für die Versicherung deutlich.

FUZZY SET AND PATTERN RECOGNITION AND DATA ANALYSIS

Hans-Jürgen Zimmermann, RWTH Aachen

Fuzzy technology has become widely accepted as a tool to solve problems for which traditional methods are not quite applicable. In the past most of the applications that became known, were in the area of control engineering. These were knowledge-based approaches to control problems. In the recent past an area has emerged which seems to be of even higher importance than fuzzy control, that of fuzzy data analysis. By contrast to fuzzy control in this area algorithmic fuzzy approaches are being used as well as knowledge-based approaches. The main goal is to reduce complexity and to recognize structures.

This reduction of complexity achieves in a number of steps:

1. In modelling the problem or describing the process the first abstraction takes place.

2. In feature analysis the number of features used to describe objects is further reduced.

3. On the basis of these features classes (classifiers) are designed, which eventually reduce the complexity to a degree acceptable for human perception.

4. These classes are then used in the area of classification to structure observed masses of data.

Traditionally methods such as structured modelling, discriminant - and factor - analysis, cluster analysis etc. are used to achieve the goals.

THE EVALUATION OF INFORMATION UNDER RISK AND AMBIGUITY

Markus Berg, Universität Kiel
Roselies Eisenberger, Universität Mannheim

In many real world decision problems the set of opportunities the decision maker has, contains gathering further information which reduces the uncertainty involved. As gathering information may involve costs the determination of the optimal level of information is an important decision problem.

In our talk we will present the results of an experimental study on the willingness to pay for information that resolves uncertainty. In the experiment 157 students were asked to give information values for two-state-lottery decision problems with lotteries defined on three different risky or ambiguous events.

These experimental results will be contrasted with theoretical values of information derived from the Subjective Expected Utility Model and the Choquet Expected Utility Model. We find that neither model describes the observed information buying behaviour well. There are strong systematic violations of even simplest predictions.

We will conclude the talk with some comments on special features of information evaluation processes that might explain the results. We will also comment on some more general problems of determining values of information in models that allow for violations of the independence axiom like the Choquet Expected Utility Model.

INHERITANCE SETTLEMENT PROCEDURES AND INCENTIVES FOR MISREPRESENTATION

Thomas Braun, Eberhard-Karls-Universität Tübingen

Summary: The article considers the settlement of an estate consisting of an indivisible object that has to be divided among several heirs. Allocating the object to the one who values it most requires to ask the heirs for the cash amount which would render them indifferent between owner- and non-ownership of the whole object. The problem in doing so is that there is no guarantee that the heirs quote their actual cash equivalents. The paper examines the conditions for three different settlement schemes to be incentive compatible at least. In case of two heirs agreeing on the absence of systematic biases in valuation and mutual conjectures regarding valuation a settlement scheme which may best be described by the property of fair division is found to be truth inducing.

1. The problem

Inheritance settlement schemes have to perform two distinct funktions: Property rights must be allocated to one of the entitled heirs (allocation function) and those heirs who have to transfer their shares must be compensated (compensation function).

Appropriate procedures should be ex-post efficient requiring that the object is allocated to the one heir that values the object most. This, in turn, requires to know all the heirs' valuations which can only be experienced by asking them to quote the cash amount which would render them indifferent between owner- and non-ownership (reservation price) of the object as a whole. The problem with this approach is that the heirs might be induced to under- or overstate. In general the inclination to misrepresent is a matter of personality as well as the "rules of the game". Absolute risk averters not knowing the others' quotations at the time they have to quote, for example, will not misrepresent whenever they face the risk of paying too much for the others' shares or receiving too little for their own share.

The most desirable solution to the problem would be a scheme for which it is a dominant strategy to quote the actual cash equivalent. From the theory of auctions it is well known that for a sealed bid second price auction [1, 2] it is a dominant strategy for a potential buyer to bid his reservation price. Finding an answer to the question whether this property pertains to the problem of inheritance settlement as well demands a closer look at the causes that lead to the elimination of any incentive to misrepresent when participating in such an auction. Focusing on the incentive to understate first it proves that understating is detrimental in terms of expected profit because it puts a potential buyer at risk to miss the opportunity of buying the object at prices which may be considerably below his/her cash equivalent without reducing the price he/she has to pay if this risk doesn't materialize. The latter is true because the price the buyer has to pay is exclusively determined by the second highest bid. Overstating, without any compensating advantage, on the other hand only exposes the buyer to the risk of paying more than his/her cash equivalent and thus is detrimental in terms of expected profit too.

Unfortunately the sealed bid second price auction scheme doesn't eliminate incentives for misrepresentation when it comes to solve the problem at hand. The reason being that in the situation discussed here instead of having the choice between a buy or no buy the alternatives are either to buy or to sell. Being a potential seller as well as a potential buyer the incentive to overstate could only be ruled out if the compensation received for foregoing his/her property rights was totally independent of a buyer's own quote. This, however, contradicts the fact that the second highest quote determines the compensations implying everybody's chance to influence the compensation he/she will eventually receive with positive probability.

Upon reflection of the above reasoning it is hard to imagine the existence of a scheme for which to quote one's actual cash equivalent is a dominant strategy. Thus it seems approriate to examine whether common schemes are at least incentive compatible. Incentive compatibility requires that there must be no incentive for anybody to be dishonest given the assumption that the remaining heirs are honest.

2. Analysis of incentive compatibility

2.1. Basic assumptions of formal analysis

In order to highlight any incentive to misrepresent the following formal analysis of three different settlement procedures generally assumes risk neutral behaviour. There is a total of n heirs each of which being entitled to an equal share of the estate. Every single heir has identical beliefs about the cash equivalents of each of the remaining n-1 heirs. These beliefs may be depicted by independent random variables X which are equally distributed over $[0, g]$. The respective distribution und density functions are $F(x)$ and $f(x)$. ΔW refers to the increase in wealth realized by a decision maker, e to his actual cash equivalent and q to his quote. Each procedure allocates the object to the heir who cites the highest cash equivalent. This permits the following distinction: C^- is the compensation to be paid by a decision maker if he quotes the highest cash equivalent and C^+ is the compensation to be received in all other cases. The remaining part of this article examines the conditions under which different specifications of C^- and C^+ are incentive compatible. The next paragraph deals with a scheme that binds the heir that is awarded the object to pay a compensation exclusively based on his quote, in paragraph 2.3. it is the lowest quote that takes this role and in paragraph 2.4. the compensations are calculated according to the notion of fairly dividing the surplus resulting from differing quotes.

2.2. Maximum-quote-based compensation

If the highest (truthfully) quoted cash equivalent exclusively determines the compensations a decision maker faces the following conditional compensations

$$C^- = \frac{n-1}{n} q$$

and

$$C^+ = \frac{1}{n} x_{max} .$$

For

$$F_{max}(x, n) := F(x)^{n-1}$$

and

$$f_{max}(x, n) := (n-1) F(x)^{n-2} f(x)$$

the objective function reads

$$E(\Delta W) = (e - C^-(q)) F_{max}(q, n) + \int_q^g C^+(x) f_{max}(x, n) dx \rightarrow \max_q$$

It is straightforward to show that for the optimal quote q* the following condition

$$e - q^* = \frac{1}{n} \frac{F(q^*)}{f(q^*)} \tag{1}$$

must hold. (1) implies that a necessary condition for not to understate is F(e) = 0 which means that a decision maker has to be shure that it doesn't make sense to understate because there is nobody to undercut. For X being equally distributed over the interval [0, g] this requires e = 0. In general it is sufficient to assume that it is common knowlegde that nobody would ever state less than e. Except for this unrealistic case Equation (1) reveals that basing the compensations exclusivly on the highest quote causes the heirs to understate. This is easily understood by breaking down the effect of a marginal increase in q into the following two parts: The first part which will be referred to as *compensation effect* (CE) hereafter does only include the effects on C^- and C^+ holding the corresponding probabilities constant and the second part which will be referred to as *allocation effect* (AE) in what follows covers the effect on the probabilities holding C^- and C^+ constant. For the procedure at hand we have

$$\frac{\partial C^-}{\partial q} = \frac{n-1}{n} \text{ and } \frac{\partial C^+}{\partial q} = 0$$

and

$$CE = -\frac{\partial C^-}{\partial q} F_{max}(q, n) + \frac{\partial C^+}{\partial q}(1 - F_{max}(q, n)) = -\frac{\partial C^-}{\partial q} F_{max}(q, n) \leq 0$$

which means that as long as there is a positive probability of being awarded the object the compensation effect is always negative. Thus for the necessary condition to hold the allocation effect must be positive. This, again, demands to understate the actual cash equivalent as increasing q always raises the chance of an award which is beneficial if and only if q < e as is shown by

$$AE = (e - C^-(q)) f_{max}(q, n) - C^+(q) f_{max}(q, n) = (e - q) f_{max}(q, n) .$$

2.3. Minimum-qote-based compensation

If the compensation paid to those who have to transfer their shares is exclusively based on the lowest quote the objective function for n = 2, 3 runs

$$E(\Delta W)) = (n-1)(n-2) \int_0^q (e - E(C^- | X \leq x)) (F(q) - F(x))^{n-3} F(x) f(x) dx + \left(\frac{q}{n}\right) (1-F(q))^{n-1}$$

$$+ (n-1) F(q) (1-F(q))^{n-2} E(C^+ | X \leq q) \rightarrow \max_q$$

(For $n \geq 4$ a fourth term, namely

$$\sum_{i=2}^{n-2} \binom{n-1}{i} i (i-1) \int_0^q E(C^+ | X \leq x) (F(q) - F(x))^{i-2} F(x) f(x) dx$$

has to be added). For a total of two respectively three heirs the optimal statement requires

$$(q^* - e) f(q^*) = \left(\frac{1}{2}\right) (1 - F(q^*)) \tag{2a}$$

respectively

$$(q^* - e) 2 F(q^*) f(q^*) = \left(\frac{1}{3}\right) (1 - 2F(q^*) + 4 F(q^*)^2) \tag{2b}$$

to hold. (2b) implies that if there are three heirs the incentive to overstate never disappears whereas according to (2a) with only two heirs it disappears if and only if e = g. The basic reasoning behind this result is that as long as there is some chance that somebody states a higher cash equivalent the compensation effect is positive. Therefore the optimal quote requires the allocation effect to be negative thus making an overstatement of one's true cash equivalent a necessary condition. Adding a third heir makes for a qualitative difference for the following reason. If there is only one competing heir whose quote will never exceed g there is no incentive to overstate if e = g because even in the limit case of both quotes amounting to g it doesn't matter whom the object is allocated to so that nothing can be gained from bidding further up. This, however, does not pertain to the case of two out of three heirs quoting g because the minimum quote which is the basis for compensation could be considerably lower at the same time. Under such circumstances a tie could mean a severe loss to one of the two heirs quoting g.

2.4 Sharing-the-surplus-based compensation

The last settlement procedure to be discussed here complies to the notion of fairness in that all compensations are based on the individual quotes and the total surplus resulting from differing quotes ($\bar{q}$ represents the arithmetic mean of all quotes)

$$S = q_{max} - \bar{q}$$

is equally shared among the heirs [3]. This amounts to the following (conditional) compensations

$$C^- = \frac{n-1}{n} q - \frac{1}{n} S = \frac{n-1}{n} q - \frac{1}{n}(q - \bar{q})$$

and

$$C^+ = \frac{1}{n} q + \frac{1}{n} S = \frac{1}{n} q + \frac{1}{n}(q_{max} - \bar{q})$$

For

$$E(C^- | q) = \frac{n-1}{n} q - \frac{1}{n} E(S | X \leq q) = \frac{n-1}{n} q - \frac{1}{n}\left(q - \frac{q + (n-1)\, E(X | X \leq q)}{n}\right)$$

and

$$E(C^+ | x) = \frac{1}{n} q + \frac{1}{n} E(S | X \leq x) = \frac{1}{n} q + \frac{1}{n}\left(x - \frac{x + q + (n-2)\, E(X | X \leq x)}{n}\right)$$

the objective function reads

$$E(\Delta W) = (e - E(C^- | q))\, F_{max}(q, n) + \int_q^g E(C^+ | x)\, f_{max}(x, n)\, dx \rightarrow \max_q$$

Making proper use of characteristic features of equally distributed random variables (as has been made before) it may be shown that the optimal quote for this scheme requires

$$(e - q^*)\, f_{max}(q^*, n) = \left(\frac{n-1}{n}\right)\left(F_{max}(q^*, n) - \frac{1}{n}\right) \tag{3}$$

to hold. Equation (3) immediately reveals that quoting one's actual cash equivalent is optimal if and only if

$$F(e) = \left(\frac{1}{n}\right)^{\frac{1}{n-1}} \tag{4}$$

In conjunction with

$$\lim_{n\to\infty} F(e) = \lim_{n\to\infty} \exp(\ \ln(F(e))\ ) = \exp\left(\lim_{n\to\infty} \frac{-1}{n}\right) = 1$$

Equation (4) implies that for every $n \geq 2$ there are combinations (e, g) of cash equivalents and beliefs about anybody else's cash equivalents that will render the scheme under discussion incentive compatible.

The question, however, is to explain how such beliefs could arise. It will be shown below that in case of only two heirs there is a plausible explanation for such combinations to arise. For n = 2 Equation (4) implies

$$F_{max}(q, 2) = F(q) = \frac{1}{2}$$

so that

$$q^* = F^{-1}\left(\frac{1}{2}\right) = E(X) = e$$

becomes an optimal quote. Once more the economic background may easily be captured by separating the allocation from the compensation effect. For n = 2 we have

$$C^- = \frac{1}{2}(q - S) = \frac{1}{2}\left(q - \frac{q - X}{2}\right) = \frac{1}{2}\left(\frac{q + X}{2}\right)$$

and

$$C^+ = \frac{1}{2}(q + S) = \frac{1}{2}\left(q + \frac{X - q}{2}\right) = \frac{1}{2}\left(\frac{q + X}{2}\right)$$

rendering

$$E(C^-) = \frac{1}{2}\left(\frac{q + E(X \mid X \leq q)}{2}\right) \tag{5a}$$

and

$$E(C^+) = \frac{1}{2}\left(\frac{q + E(X \mid X > q)}{2}\right) \tag{5b}$$

Together with

$$E(X \mid X \leq q) = \frac{q}{2} \text{ and } E(X \mid X > q) = \frac{1}{2}(g + q)$$

(5a) and (5b) imply

$$\frac{\partial E(C^-)}{\partial q} = \frac{\partial E(C^+)}{\partial q}$$

so that for $q^* = E(X)$ we have

$$AE = -\frac{\partial E(C^-)}{\partial q} F(q^*) + \frac{\partial E(C^+)}{\partial q}(1-F(q^*)) = 0$$

and

$$CE = f(q^*)\,(\,e - E(C^- \mid q^*) - E(C^+ \mid q^*)\,) = f(q^*)\,(\,e - E(X)\,)$$

rendering $q^* = E(X) = e$ an optimal quote.

Thus splitting the surplus is not only fair but truth inducing when the absence of systematic biases in valuation (implying $E(X) = e$) is common knowledge (impliying each heir being convinced that the other one has unbiased expectations concerning his/her own cash equivalent).

3. Conclusion

The evident proposal of sharing the surplus resulting from differing quotes among the heirs implies that the compensations are calculated on the basis of weighted averages of all the heirs' quotes instead of just one of these. It is shown that this provides for a better balance of incentives to misrepresent in one or the other direction at least. For only two heirs the problem of misrepresentation can be avoided by splitting the surplus if they mutually agree on the absence of systematic biases in valuation. This seems to be quite plausible in the absence of specific information pointing at the opposite. In general, however, further research covering more sophisticated schemes is needed to solve the problem of misrepresentation.

References

[1] *Vickrey, W.* (1961), Counterspeculation, auctions and competive sealed tenders, JoF, 8 - 37
[2] *Riley, J. G. and W. F. Samuelson* (1981), Optimal auctions, AER, 381 - 392
[3]*Samuelson, W.* (1980), The object distribution problem revisited, QJoE, 85 - 98

DECISION MAKING WITH IMPRECISE PROBABILITIES

Frank P.A. Coolen
Eindhoven University of Technology

Summary In many decision problems the only information available about a random event is expert opinion. The theory of imprecise probabilities, a generalization of standard subjective probability, allows us to deal with such information. In this paper the use of imprecise probabilities is discussed, with emphasis on elicitation and combination of opinions and decision making, and some recent results are briefly mentioned.

1. Imprecise probabilities

We are concerned with an event A. We quantify a person's opinions about the event in terms of betting behavior. A bet is defined such that the owner of the bet receives 1 if A occurs, and 0 otherwise (expressed in some linear unit of utility, where money may not be suitable; see Walley, 1991; sect. 2.2). Your lower probability for event A, denoted by $\underline{\mathbb{P}}(A)$, is defined as the supremum of all prices against which you are willing to buy the bet; your upper probability for event A, denoted by $\overline{\mathbb{P}}(A)$, is defined as the infimum of all prices against which you are willing to sell the bet. Here it is assumed that you are only willing to buy or sell a bet if you expect profit. In the classical theory of probability, with a subjective interpretation in terms of betting behavior (De Finetti, 1974), $\underline{\mathbb{P}}(A) = \overline{\mathbb{P}}(A)$ is assumed, implying that there exists only one price at which you are indifferent between selling or buying the bet, and you do not have the freedom to accept no bet for more prices. A thorough fundamental theory of imprecise probabilities is built only on the assumption of coherence, meaning that one is never willing to accept combinations of bets that incur sure loss (Walley, 1991).

An immediate consequence of the assumption of precision is outlined by the following example, due to Gärdenfors and Sahlin (1982). Consider Miss Julie who is invited to bet on the outcome of two different tennis matches. As regards match I she is very well-informed about the two players; she knows everything about the results of their earlier matches, she has watched them play several times, she is familiar with their present physical condition and the setting of the match, etc. Given all this information, Miss Julie predicts that it will be a very even match and that a mere chance will determine the winner. In match II, she knows nothing whatsoever about the relative strength of the contestants (she has not even heard their names before) and she has no other information that is relevant for predicting the winner of the match.

If *pressed* to evaluate precise probabilities of the various possible outcomes of the matches, Miss Julie would say that in both matches, given the information she has, each of the players has a 50% chance of winning. In this situation a strict Bayesian would say that Miss Julie should be willing to bet at equal odds on one of the players winning in one of the matches if and only if she is willing to place a similar bet in the other match. It seems, however, perfectly rational if Miss Julie decides to bet on match I, but not on II.

This difference in her opinions about both matches should also be represented by the quantified expression of her thoughts, which is not possible through the use of precise probabilities. If probability is used as a language between a Decision Maker (DM) and consulted experts (Cooke, 1991; French, 1986), this restriction to precise probabilities leads to loss of essential information, since it is obvious that a DM interested in betting on the outcome of the tennis matches I and II, who consults Miss Julie, should be aware of the difference in her knowledge about these matches. When using the concept of imprecise probabilities to describe her betting behavior, and thereby quantifying her opinions, Miss Julie could represent the difference between both matches in a logical way, and the idea behind it is that she would be free to bet if she was offered the possibility, and was not pressed.

With *imprecision* defined by $\Delta(A) = \overline{P}(A) - \underline{P}(A)$, it seems quite reasonable to assume that $\Delta(A) = 1$ if and only if one has no information at all about A, and $\Delta(A) = 0$ if and only if one has perfect information about A, where of course A may still be a random event. This makes clear that two kinds of uncertainty play a role, stochastic uncertainty and uncertainty by lack of knowledge. Theories of precise probability do not distinguish between these fundamentally different kinds of uncertainty, for example, information measures related to entropy (such as Shannon's) are always related to stochastic uncertainty. In a theory of imprecise probabilities the amount of information can be related to $\Delta(A)$.

Although this relation is very obvious, in literature no attention has been paid to it so far, except by Walley (1991; sect. 5.3), who proposes, on informal grounds, a measure of the amount of information about A: $I(A) = \Delta(A)^{-1} - 1$. Coolen (1994) analyzes this relation further and provides a theoretical foundation for Walley's $I(A)$, and proposes a theory of updating of imprecise probabilities in the light of new information in which the relation between imprecision and information, expressed through $I(A)$, plays a central role.

Theories using imprecise probabilities are also known in robust Bayesian analysis (Berger, 1990). Robust Bayes is fundamentally different to our concept since the lower and upper probabilities do not have an interpretation as betting behavior, and probability is assumed precise with bounds used only as means for sensitivity analysis. Of course this can be a sensible way to perform sensitivity analysis, but it lacks the essential and necessary relation between imprecision and information.

So far we know of only one case-study using imprecise probabilities as main concept, so not as a tool to perform sensitivity analysis, presented by Walley and Campello de Souza (1990) in a decision analysis about the economic viability of solar heating systems in Brazil. In this case study imprecise probabilities are used to quantify opinions of several experts at one moment in time, without attention to updating in the light of new information.

2. Bayesian decision theory with imprecise prior probabilities

Many decision problems of practical interest are too complex to ask an expert to directly state the best decision, and this is the point where mathematical modeling enters the framework. Lindley (1990) gives an overview of the Bayesian ideology of decision making in a brilliant discussion, leaving to us only the task to present the generalization of allowing imprecise prior probabilities in decision making. An overview of important contributions and interesting ideas on decision making with imprecise probabilities is presented by Walley (1991a; sect. 1.8).

Decision problems (Lindley, 1990) can, with a good deal of generality, be described as follows. Let X be the sample space of points x and Θ the parameter space of points θ. These are connected by a probability density of X for a given θ. Let D be the decision space of points d, and $L(d,\theta)$ the *loss* in selecting d when θ obtains (we assume $L(d,\theta) \in R$ for all d and θ). The Bayesian approach uses a prior probability density $\pi(\theta)$ over parameter space Θ and chooses as the optimum decision that d which minimizes the expected loss (if such d exists):

$$EL(d|x) = \int_{\Theta} L(d,\theta)\pi(\theta|x)\, d\theta \,, \tag{1}$$

where $\pi(\theta|x)$ is the posterior probability density of θ, given x, obtained by Bayes' theorem. If no data x are available, the decision can be based on the prior, in which case the optimum decision is that d which minimizes the expected loss:

$$EL(d) = \int_{\Theta} L(d,\theta)\pi(\theta)d\theta \,. \tag{2}$$

Although in literature notation in terms of posteriors is often preferred (e.g. Moreno and Pericchi, 1992), without loss of generality we use the form 2, since only the fact that there is some pdf for θ is of interest here, and the posterior at one stage is just the prior for the next.
Generalizing the standard Bayesian concept to allow imprecise probabilities, the assumption of a single pdf $\pi(\theta)$ is replaced by assuming a class Π of pdf's for θ. The expected loss for $\pi \in \Pi$ and $d \in D$ is:

$$EL(d,\pi) = \int_{\Theta} L(d,\theta)\pi(\theta)\, d\theta \,. \tag{3}$$

For each decision $d \in D$ this leads to a set values of expected loss, denoted by:

$$\mathcal{EL}(d,\Pi) = \{EL(d,\pi)|\pi \in \Pi\} \,. \tag{4}$$

To choose an optimal decision an additional criterion is needed, by which these sets $\mathcal{EL}(d,\Pi)$ can be compared for all $d \in D$. In this direction, two useful characteristics of $\mathcal{EL}(d,\Pi)$ are the lower expected loss $\underline{\mathcal{EL}}(d,\Pi)$ and the upper expected loss $\overline{\mathcal{EL}}(d,\Pi)$:

$$\underline{\mathcal{EL}}(d,\Pi) = \inf \mathcal{EL}(d,\Pi) \; ; \tag{5}$$

$$\overline{\mathcal{EL}}(d,\Pi) = \sup \mathcal{EL}(d,\Pi) \; . \tag{6}$$

Since the main goal is to minimize expected loss, $d_1 \in D$ is obviously preferred to $d_2 \in D$ if $\overline{\mathcal{EL}}(d_1,\Pi) < \underline{\mathcal{EL}}(d_2,\Pi)$. For other situations the preference is less clear, although it seems reasonable that d_1 is also preferred to d_2 if both $\underline{\mathcal{EL}}(d_1,\Pi) < \underline{\mathcal{EL}}(d_2,\Pi)$ and $\overline{\mathcal{EL}}(d_1,\Pi) < \overline{\mathcal{EL}}(d_2,\Pi)$. Walley (1991a; sect. 3.9.7) briefly proposes a minimax rule as possible additional criterion, leading to the choice of a decision d_0 with $\overline{\mathcal{EL}}(d_0,\Pi) \leq \overline{\mathcal{EL}}(d,\Pi)$ for all $d \in D$, but Walley immediately remarks that there does not seem to be any good reason to prefer this rule to some other possibilities. In fact, it appears to be attractive to confront the DM with the consequences of the lack of perfect information by presenting a set of possible values of expected loss corresponding to each possible decision, with no further distinction possible based on the current amount of information. For practical acceptance of a quantitative decision theory it is better that lack of perfect information is reported by means of imprecise expected loss than that an optimal decision is presented, based on precise loss as the result of a classical Bayesian analysis, especially if one has little information.

In this paper the loss function $L(d,\theta)$ is assumed to be precisely known. If this cannot be assumed, e.g. because some cost figures are not precisely known, but it is known that $\underline{L}(d,\theta) \leq L(d,\theta) \leq \overline{L}(d,\theta)$, then generalization of a decision theory based on comparison of possible decisions by $\underline{\mathcal{EL}}(d,\Pi)$ and $\underline{\mathcal{EL}}(d,\Pi)$ is straightforward by determining $\underline{\mathcal{EL}}(d,\Pi)$ corresponding to loss function $\underline{L}(d,\theta)$ and $\overline{\mathcal{EL}}(d,\Pi)$ to $\overline{L}(d,\theta)$.

A mathematically attractive form for the class Π of pdf's is the intervals of measures class (DeRobertis and Hartigan, 1981), also called the density ratio class in the literature on robust Bayesian analysis (Berger, 1990), consisting of all pdf's that are proportional to functions g with $l(\theta) \leq g(\theta) \leq u(\theta)$ for all θ, where l and u are nonnegative functions. This class is attractive for imprecise prior probabilities in a Bayesian model for two reasons: Updating becomes very simple (Perricchi and Walley, 1991) and $\underline{\mathcal{EL}}(d,\Pi)$ and $\overline{\mathcal{EL}}(d,\Pi)$ can be derived in a few steps (Coolen, 1994) without calculation problems other than those that appear in deriving $EL(d,\pi)$ for a single prior distribution π (this has got much attention in Bayesian literature lately).

3. Implementation

For practical application of a theory of decision making based on expert opinions, elicitation and combination of such opinions is of the utmost importance. A natural problem for research in this area is that conclusions can only be drawn after real world applications, for which much attention is necessary by a multi disciplinary team of researchers, since not only mathematical but also psychological and communicational aspects play a very important role. We have not yet been able to perform such a study, but as a first step towards implementation of decision theory using imprecise probabilities we discuss some powerful theoretical arguments about elicitation and combination, in favor of this concept when compared to a theory using precise probabilities (Cooke, 1991). We conjecture that a decision theory using imprecise probabilities has more appeal to decision makers and consulted experts, and can play an important role in convincing managers of the power of quantitative methods as tools to assist them in making decisions under uncertainty.

Let us first regard combination of imprecise probabilities of $K \in \mathbb{N}$ experts. Suppose for event A expert k ($k = 1, .., n$) has assigned $\underline{\mathbb{P}}_k(A) \leq \overline{\mathbb{P}}_k(A)$. From the interpretation of imprecise probabilities in terms of betting behavior, two rules for combination follow immediately (Walley, 1991; ch.4), with clear interpretation in terms of group betting behavior:

$$I \qquad \underline{\mathbb{P}}^I(A) = \min_k \underline{\mathbb{P}}_k(A) \; ; \qquad \overline{\mathbb{P}}^I(A) = \max_k \overline{\mathbb{P}}_k(A) \; ;$$

$$II \qquad \underline{\mathbb{P}}^{II}(A) = \max_k \underline{\mathbb{P}}_k(A) \; ; \qquad \overline{\mathbb{P}}^{II}(A) = \min_k \overline{\mathbb{P}}_k(A) \; .$$

Combination by rule I implies that the group only accepts to buy or sell the bet on A if each member of the group accepts it individually, so these numbers indicate consensus. Combination by rule II implies that the group accepts to buy or sell the bet on A if at least one member accepts this. It is easy to see that rule I leads to coherent betting behavior, but rule II can lead to incoherence since $\underline{\mathbb{P}}^{II}(A) > \overline{\mathbb{P}}^{II}(A)$ is possible: This indicates that there are two members within the group who would be willing to exchange a bet on A at prices between $\overline{\mathbb{P}}^{II}(A)$ and $\underline{\mathbb{P}}^{II}(A)$, and thus there is obvious conflict between the thoughts of these experts.
The results of combination by rules I and II represent facts of the opinions of all members of the group that are important for a decision maker. Remark that restriction to precise probabilities always leads to the situation that two experts are willing to exchange bets, except if their probabilities are identical. In literature much attention has been paid to combination of precise probabilities (Cooke, 1991), but rules like (I) and (II) following from the interpretations of probabilities do not exist. Coolen (1994) proposes a third rule for combining imprecise probabilities by means of weighted averages, which also leads to coherent group betting behavior with quite logical interpretation, especially from the decision maker's point of view. Suppose the decision maker assigns weight w_k to expert k, with $0 < w_k < 1$ and $\sum_{k=1}^{K} w_k = 1$:

$$III \qquad \underline{\mathbb{P}}^{III}(A) = \sum_{k=1}^{K} w_k \underline{\mathbb{P}}_k(A) \; ; \qquad \overline{\mathbb{P}}^{III}(A) = \sum_{k=1}^{K} w_k \overline{\mathbb{P}}_k(A) \; .$$

Whereas rules I and II report important information, for practical decision making rule III seems to be more promising and attractive, since rule I can lead to very much imprecision and rule II to very little imprecision, or even incoherent group betting behavior, and it seems in accordance with intuition to express confidence in the expertise of different experts by weights and combine their opinions linearly. To be able to use rule III in practice, research is needed on derivation of sensible weights w_k, especially if new information becomes available, which has not yet been presented in literature. Perhaps some ideas on scoring rules from precise probability theory (Cooke, 1991) could be generalized, and it would also be of interest to look for non-linear combinations of weighted expert opinions.

Next we briefly discuss some relevant aspects of elicitation. Crucial for success of elicitation is that the consulted expert should know for which goal his knowledge is elicited, and must be willing to cooperate. Further, he should not be asked to state his opinion about some phenomenon that is unfamiliar to him, such as unobservable parameters in a mathematical model. Therefore, we suggest to restrict discussion to observables. Because no testcases have been performed, it is very hard to say anything more about the elicitation process, but Walley and Campello de Souza (1990) suggest to ask imprecise cumulative distribution functions for the observable random variable of interest, where the experts should be taught the betting interpretation of these imprecise probabilities. This is an interesting area for research, with open problems that need to be solved before the concept can be applied on large scales.

If imprecise probabilities for some events of interest are gathered, these can be combined, if necessary, and then we suggest to fit a Bayesian model such that the imprecise predictive probabilities based on the model and the imprecise priors fit well to the experts probabilities (Coolen, 1994). We believe that this is the job of the mathematician, that is important since final decisions depend on the chosen model. The choice of a parametrical model, that is necessary within the Bayesian concept, should get thorough attention, and should be based on knowledge of the process which could also come from the experts and Decision Maker.

A theory of imprecise probabilities might, for example, be useful in reliability analysis, since quite often in reliability problems a lack of historical or experimental data forces us to use expert opinion. Coolen and Newby (1993) have introduced the above concept into this area, where often the variable of interest is lifetime of equipment, whereas suitable parametrical models such as the Weibull or lognormal distributions have parameters that are hard to interpret, so indeed it seems best to concentrate on elicitation methods through questions about the lifetime.

An example of possible application of the concept to an age-replacement model for technical equipment is briefly discussed by Coolen (1994), but only after real-world applications a new quantitative concept for assisting on decision making can be evaluated.

Acknowledgement

The author is grateful to Martin Newby for useful cooperation, and to Paul van der Laan, Peter Sander and Emiel van Berkum for discussion and remarks during the research.

References

Berger, J.O. (1990) Robust Bayesian analysis: Sensitivity to the prior, *Journal of Statistical Planning and Inference* 25, 303-328.

Cooke, R.M. (1991) *Experts in Uncertainty* (Oxford University Press).

Coolen, F.P.A. (1994) Statistical Modeling of Expert Opinions using Imprecise Probabilities (PhD-Thesis Eindhoven University of Technology, to appear).

Coolen, F.P.A. and Newby, M.J. (1993) Bayesian reliability analysis with imprecise prior probabilities, *Reliability Engineering and System Safety*, to appear.

De Finetti, B. (1974) Theory of probability (2 volumes) (Wiley, London).

DeRobertis, L. and Hartigan, J.A. (1981) Bayesian inference using intervals of measures, *Annals of Statistics* 9, pp. 235-244.

French, S. (1986) *Decision theory: An introduction to the mathematics of rationality* (Ellis Horwood, Chichester).

Gärdenfors, P. and Sahlin, N.E. (1982) Unreliable probabilities, risk taking and decision making, *Synthese* 53, pp. 361-386.

Lindley, D.V. (1990) The present position in Bayesian statistics (1988 Wald memorial lectures, with discussion), *Statistical Science* 5, pp. 44-89.

Moreno, E. and Pericchi, L.R. (1992) Bands of probability measures: A robust Bayesian analysis, in: J.M Bernardo, J.O. Berger, A.P. Dawid and A.F.M. Smith (eds.) *Bayesian Statistics* 4, pp. 707-713 (Oxford University Press).

Pericchi, L.R. and Walley, P. (1991) Robust Bayesian credible intervals and prior ignorance, *International Statistical Review* 58, pp. 1-23.

Walley, P. (1991) Statistical Reasoning with Imprecise Probabilities (Chapman and Hall, London).

Walley, P. and Campello de Souza, F.M. (1990) Uncertainty and indeterminacy in assessing the economic viability of energy options: A case study of solar heating systems in Brazil, *Energy Systems and Policy* 14, pp. 281-304.

REACHABILITY OF NASH EQUILIBRIA IN COMPETITIVE LOCATION MODELS

H.A. Eiselt, University of New Brunswick, Canada
J. Bhadury, University of New Brunswick, Canada

In this paper we present a competitive location model. Models of this type have been used since Hotelling (1929) in his seminal paper "Stability in Competition" suggested a scenario in which duopolists compete on a linear market, i.e. a line segment, in prices and locations. His conclusion was that an equilibrium exists in which both competitors locate at the center of the market. This result was dubbed the "principle of minimum differentiation" and has been used by various authors to explain why existing products and political platforms are so similar to each other.

It was not until 1979, fifty years after the original paper was published, that d'Aspremont *et al.* discovered that the Hotelling model did, in fact, not possess an equilibrium at all. The underlying reason is that while the possibility of capturing additional markets between the two facilities forces them together, intense price competition à la Bertrand forces them apart. The authors also showed the volatility of the model: replacing the original linear transportation costs by quadratic ones results in an equilibrium in which competitors locate as far apart as possible, i.e. maximum, rather than minimum, differentiation.

What followed was a plethora of different models, most of them slight variations on the original theme. Even if only the major contributions are counted, Eiselt *et al.* (1993) have classified more than hundred of them. Some of the models possess equilibria while others don't, emphasizing the extreme sensitivity of the model. The approach in this paper focuses on a different aspect of equilibria rather than their existence. In short, we consider a model in which duopolists have fixed prices and compete only in locations. The original locations of the two competitors are arbitrary. Each competitor will now relocate considering his opponent's location as fixed. In this locational game we distinguish between two cases, a game with sequential moves and one in which competitors relocate simultaneously. Assuming that a locational equilibrium exists, our question is whether or not the equilibrium will always be reached in such a relocation game.

In the model under consideration, the market is in the shape of a tree with customers, whose demand for the homogeneous good is fixed, located at the vertices. The two competitors charged fixed, but not necessarily equal, mill prices, so that customers are responsible for the transport cost. Customers buy from the competitor with the lowest delivered price, i.e. mill price plus transport cost.

The case of equal prices is easy. Locational equilibria always exist and are always reached. If prices are unequal, it is shown that existing equilibria may actually be missed. However, we demonstrate that with an appropriate tie-breaking rule, existing equilibria will always be found.

A graphical user interface for interactive searching strategies in discrete vector optimization

Walter Habenicht, Universität Hohenheim
Ralph Scheubrein, Universität Hohenheim

Interactive approaches are very powerfull tools in vector optimization problems. The system *ENUQUAD* is devoted to discrete vector optimization, namely to those problems with a large number of alternatives and a limited number of criteria. Problems of this type often occur in multicriteria combinatorial problems. *ENUQUAD* determines the set of efficient solutions of the problem and provides the user with a set of searching options like

- Reference point approach,
- Aspiration levels,
- Neighborhood scanning,
- Filtering approaches.

The aim of this paper is to describe the graphical user interface of the system, which is a user oriented one, too. It provides the user with a set of options concerning the presentation of the relevant information about the efficient set.

The whole system can be interpreted as a tool-box, which can be used by a decisionmaker to build up his individual searching strategy using graphical representations according to his choice.

AN OBJECT-ORIENTED DECISION SUPPORT SYSTEM FOR MCDM

Thomas Hanne, FernUniversität Hagen

Zusammenfassung: In diesem Beitrag werden Konzept und Implementierung eines objektorientierten, entscheidungsunterstützenden Systems (DSS) vorgestellt, welches insbesondere für den Fall der Entscheidungsfindung unter mehrfacher Zielsetzung (MCDM) dient. Ansätze zur Integration unterschiedlicher MCDM-Methoden in dieses System werden vorgestellt. Eine Möglichkeit besteht in der interaktiven Anwendung einer einheitlichen Benutzeroberfläche. Ein anderer Ansatz basiert auf neuronalen Netzen und ähnlichen Konzepten. Ein wichtiger Aspekt hierbei liegt im Lernvermögen neuronaler Netze, das in entsprechend konzipierten entscheidungsunterstützenden Systemen das Lernen aus historischen Entscheidungsprozessen ermöglichen soll. In einem entsprechend konzipierten DSS können Fragen hinsichtlich der Auswahl einer Methode, der Bestimmung von Parametern und der Verwendung mehrerer Methoden angegangen werden.

Im objektorientierten Implementierungskonzept werden Entscheidungsprobleme und -methoden wie z. B. neuronale Netze oder MCDM-Verfahren als Objekte repräsentiert. Eine Klassenhierarchie von Problemen und Methoden wird vorgestellt. Jedem Problem kann eine Methode zugeordnet werden, die es löst. Eine Methode kann auf eine andere Methode (Metamethode) verweisen, die ein Lernverfahren bereitstellt. Lernverfahren wenden die Methoden auf Probleme an und vergleichen die Ergebnisse mit Referenzergebnissen. Die Beziehung von Problemen, Methoden und Lernverfahren wird an einem Anwendungsbeispiel veranschaulicht.

Summary: This paper deals with the basic ideas and implementation of an object-oriented decision support system (DSS), especially for multiple criteria decision making (MCDM). There are various ways of integrating different MCDM methods in a DSS. One way is the interactive application of different methods by means of an integrative user interface. Integration can also be achieved by utilizing neural networks and similar concepts. Neural network-based DSS are able to combine different methods and to learn from historical decisions. They are able to answer questions like: Which method should be used? How can the parameters be adjusted? Is it useful to apply different methods and to aggregate their results to a compromise solution? And how should this be done?

In the object-oriented implementation concept decision making problems and methods like neural networks and MCDM procedures are represented by objects. A class hierarchy of different problems and methods is presented. Each problem can be linked with a method which can solve it. A method can be linked with another method (metamethod) which provides a learning strategy. Learning strategies work by applying the methods on some input problems and by comparing the computed results with some reference results. The relationship between problems, methods, and learning methods is illustrated by an example.

1. Introduction: The Concept of the DSS

In the recent decades a large number of multiple criteria decision making (MCDM) methods has been elaborated. To facilitate the application of the methods, it is useful to develop decision support systems (DSS) for MCDM (Zeleny (1982), p.488-493). At the last International Conference on MCDM, Petrovsky (1992) has given a survey of implemented multiple criteria DSS (MCDSS). Most of the software available for MCDM only provides one or few methods. The integration of different MCDM methods is a goal of the implemented MCDSS.

There are two alternative approaches to the application of the (MC)DSS. ("MC" is put in parantheses if the according statement is not specific for multiple criteria applications.) One of them is based on an interactive user interface. The user chooses a method and assigns parameters. After the solution is calculated he or she may change the parameters or can try another method until satisfactory results are achieved. This approach utilizes the knowledge accumulated in the experiences of a decision maker.

The other approach is to use learning methods which should determine the choice of a method or the assignment of parameters. Historical data of (possibly non-method-based) decisions should be utilized to automate these processes in order to reduce costs and analyze the former decision making with the aim to improve it. Here, the experiences expressed in former decisions are utilized. Both approaches can possibly be combined. For instance, first, the (MC)DSS learns with the help of historical data and proposes a solution. Then the decision maker can interactively use the system to obtain a more satisfying result.

There are different concepts of the integration of methods to achieve these goals. Beyond the pure supply of different methods in such a system one could combine methods. For instance: A problem is solved by various methods and the results are further analysed and aggregated to form a compromise solution. Another possibility is to construct new methods by recombining parts of other methods. (For an example see Hanne (1993).)

A more sophisticated way of integration is the use of neural networks and other network-like structures. Neural networks are based on the idea of achieving the functionality of nervous systems by rebuilding its structure which consists of many simple interacting elements (neurons). Neural networks can be regarded as a simulation of MCDM or a corresponding method (where the neural network has learned to behave like the method) or as a new, individual MCDM concept (or *the* MCDM method in general). Neural networks are, however, successful in many subjects of decision making. So far, their importance for MCDM has not yet been explored sufficiently. Neural network concepts also consider the above mentioned aspect of learning. Many types of neural networks possess learning strategies like the well-known error back propagation in layered feed-forward networks (perceptrons) which is based on a gradient method (see, for instance, Müller and Reinhardt (1990)). Other learning methods like the evolution strategies are universal in the sense that they are applicable in a broad range of situations, e.g. for the improvement of more general networks (see Hanne (1993)) which could be used for the combination of "classical" MCDM methods. The nodes of this kind of network calculate their results not like neurons but like the underlying MCDM methods. With the combined application of neurons and other nodes neural MCDM networks can be constructed.

2. The Implementation

2.1 The Object-Oriented Concept

Prior to presenting the implemented (MC)DSS the main concepts of object-orientation are briefly summarized (see Mühlhoff (1990)): In the object-oriented paradigm data and appropriate operations are combined and called objects. Objects respond to messages they receive by executing suitable operations. As a result the internal state of the object (for example: its size) may change. (In fact, in purely object-oriented systems this is the only way to change the state of an object.) In object-oriented terminology the operations of an object are called methods. The structure of an object is described in its class. In the class the methods of its objects and their relationships to messages are defined. The decision which method should be executed when a specific message is received is made at the run-time of a program by performing a method search (dynamic binding).

Inheritance is the key object-oriented concept. It is the means by which the specialization of functionality can be achieved. A derived class inherits the data and the methods from its base class. The derived class may contain additional data and methods which possibly redefine the data and the methods of the base class.

For the implementation of the (MC)DSS the object-oriented programming language C++ has been chosen. C++ is an extension of the widespread high-level programming language C which also allows efficient assembler-like programming. C and C++ are also highly portable and well standardized by ANSI (Ellis and Stroustrup (1991)). This facilitates the usage of the program source code on different computers and operating systems. C++ surpasses C by providing object-oriented and other extensions. Especially, C++ permits strict data typing and allows the construction of user-defined "first class" data types.

2.2 Implementation Concept of the (MC)DSS

Decision support systems handle problems and provide methods to solve them. Apparently, a straightforward object-oriented implementation could represent problems as objects whose (object-oriented) methods correspond with the required methods for problem solving. This implementation approach has some major problems: First, many methods need problem specific information for their application. For example: The simple additive weighting methods use weights, the PROMETHEE methods require criterion specific preference functions (see Brans, Mareschal, and Vincke (1984)), neural networks need connection strengths etc. Second, problems can be solved by applying different methods. Different problems (possibly also of different classes) can occasionally be solved by using the same methods. Especially in an interactive DSS it is useful to handle the problem method links in a dynamic way.

To meet these requirements (problem solving) methods are turned into objects. These new method objects combine the procedures for problem solving (and some others) with the method-specific data. The objectization of methods has some other advantages. Methods can be treated as normal objects. This facilitates the implementation and application in various aspects. Methods and problems can be linked in a dynamic way (see Figure 2). A class hierarchy of methods can be constructed (see Figure 1).

In the (MC)DSS there is a base class *elemproblem* containing problem objects or – more precisely – appropriate data structures for the representation of decision problems. Typical data structures which are derived classes of *elemproblem* are *vector, matrix, namedmatrix* and *madm*; *vector* and *matrix* contain representations of vectors and matrices; *namedmatrix* serves to represent tables, matrices with names for the rows and columns; *madm* contains representations of MADM (multiple attribute decision making) problems. Between these classes there are also derived class relationships, e. g. *namedmatrix* is a derived class of *matrix*.

The class *method*, respectively its derived classes, represent problem solving methods. These methods are for example MADM methods, neurons, and neural networks (like the perceptron or associative memory). Each problem object possesses a pointer *solver* at a method which can solve it. This pointer is NULL as long as no problem solving method is specified. A problem can be solved by an (object-oriented) method *solve* which applies the specified (problem solving) method to the problem. A method solves a problem by executing a method *calculate* and returns the solution which is also an object of the class *problem.*

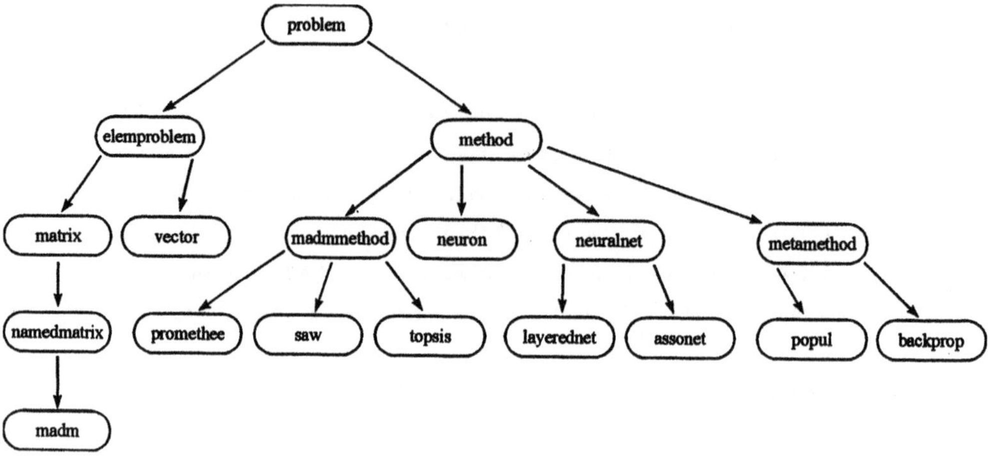

Figure 1: Part of the Class Hierarchy

Besides problem solving, learning is an important feature of the (MC)DSS. Learning is the modification of methods according to a specific functionality. For example: The weights (connection strengths) of a neural network shall be adjusted in a way so that the network calculates the desired outputs. This is done by special methods – metamethods – which can be applied to other methods. Each method, like an elementary problem, possesses a pointer *solver* to allow a link to a learning method.

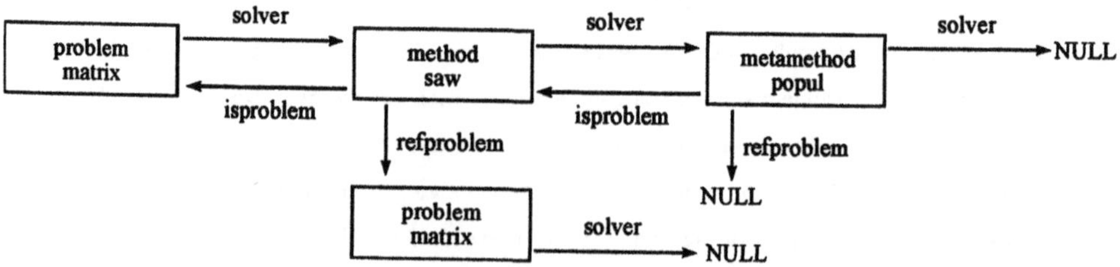

Figure 2: Relationship between Problems, Methods, and Metamethods: The Case of the Determination of Weights for Simple Additive Weighting with an Evolution Strategy as a Learning Method. (For an explanation of the pointers *isproblem* and *refproblem* see below.)

Learning is very similar to problem solving: The appropriate modification of a method can be regarded as a problem that can be solved by the application of learning methods to the method. Because of this, the methods (the class *method*) and elementary problems (the class *elemproblem*) are integrated in a joint base class *problem.*

3. Examples of Applications

So far, the (MC)DSS has not yet been used for real-world applications as it is still in the process of implementation.

As an important universal learning and optimization algorithm the evolution strategy has been implemented. Evolution strategies, like genetic algorithms, neural networks, and simulated annealing, follow the concept of "learning from nature". The evolution strategies are based on the abstract imitation of the principles of evolution which are replication, mutation, recombination, and selection. With another approach of application, they have been used for MCDM by Kursawe (1992).

In the (MC)DSS evolution strategies (ES) (an object of the class *popul*) build up a population of the method which should be improved. First, the $\mu, \mu \in N$, methods of the parental generation are replicated. The data of the $\lambda, \lambda \in N$, children methods become mutated. It is also possible to recombine the data so that a children method inherits its information from different parents. After that, the μ children methods with the best fitness are selected as the parents of the next generation. In the so-called $(\mu + \lambda)$-ES the children and their parents are considered in the selection process whereas in (μ, λ)-ES only the children can survive.

The fitness of a method which is the criterion to be improved during the learning process is calculated on the basis of l_p metrics. Therefore, the method is applied on its problem (linked with the method via a pointer *isproblem*) and the result's distance from a reference result which is linked with the method via a pointer *refproblem* is computed.

In an example the evolution strategy has been used to adjust the weights of a simple additive weighting (SAW) method. The example (fighter aircraft problem) is taken from Hwang / Yoon (1981), p. 100f. First, the fighter aircraft problem has been solved with a SAW method applying the weights $w = (0.2, 0.1, 0.1, 0.1, 0.2, 0.3)$ as in Hwang / Yoon (1981) to obtain reference results for the later learning application. Then a SAW method with random weights has been defined. For this method, a (2+6)-ES without recombination has been used as learning strategy. Several runs with variants of the ES have been executed. Some of the results are summarized in the table below.

Generations	Run 1	Run 2	Run 3	Run 4
1	0.116887	0.112302	0.116887	0.116887
3	0.063874	0.092139	0.082315	0.063874
5	0.032956	0.049142	0.057879	0.032956
10	0.023083	0.026247	0.032912	0.023136
20	0.014161	0.014392	0.015499	0.011930
200	0.000739	0.003620	0.002279	0.002079

Table 1: Values of Fitness of SAW Methods Improved by an ES

In Run 1, a (2+6)-ES with a special rule for the adaptation of mutations but without recombination has been used. The same method but with recombination was applied in Run 2. Run 3 uses the ES of Run 1 without the adaptation of mutations. In Run 4, a (2,6)-ES with the adaptation of mutations and without recombination has been used.

The surprising results of the not very successful ES using recombination (Run 2 compared with Run 1) is possibly effected by a too strong use of this feature. Note that it has been shown (see Schwefel (1977)) that, usually, recombination leads to better results in optimization. As expected, the lack of the adaptation of mutations leads to a slower approximation of the optimum. The use of a (2,6)-ES leads (temporarily) to a faster approximation. The reasons for this apparently surprising result are explained in Schwefel (1977).

Further tests applying the ES as a learning method have been made to improve the TOPSIS method (see Hwang and Yoon (1981)). Compared with the SAW case, similar results have been obtained.

In another example the evolution strategy has been used for the determination of the weights of 0-1-neurons which should compute the comparison operator ">". For these kind of neurons it is not possible to apply error back propagation or any other learning strategy using derivations. The process was started with random weight neurons. There was a small data sample of 6 inputs and the corresponding outputs to fit the weights. Two sample runs of the evolution process have been made, one with, the other without recombination. In the first case it took 31 generations to find out weights to obtain the correct results in all the six sample cases. Without recombination, only 21 generations have been necessary.

4. Conclusions

In this paper the basic ideas and the implementation concept of a decision support system which is especially suitable for MCDM have been presented. First experiences with the system have been elucidated. Besides the completion of the implementation and the expansion of the system, further tests and real-world applications are necessary to rate the usability of the system for decision making.

References

Brans, J. P., Mareschal, B., Vincke, P., PROMETHEE: A New Family in Outranking Methods in Multicriteria Analysis, in: Brans, J. P. (ed.), *Operational Research 84*, Amsterdam, 1984, p. 477-490.

Ellis, M. A., Stroustrup, B., *The Annotated C++ Reference Manual*, Reading (Massachusetts), 1991.

Hanne, T., Die Integration von Multikriteria-Verfahren insbesondere mittels neuronaler Netze, to appear.

Hwang, C.-L., Yoon, K., *Multiple Attribute Decision Making, Methods and Applications, A State-of-the-Art Survey*, Berlin, 1981.

Kursawe, F., Evolution Strategies for Vector Optimization, in: *Proceedings of the Tenth International Conference on Multiple Criteria Decision Making*, Taipei, 1992, Vol. III, p. 187-193.

Mühlhoff, H., *Entwurf und Implementierung einer Objektbanksprache als Basis für die Integration von Anwendungsprogrammen der Systemanalyse*, Diplomarbeit, Dortmund, 1990.

Müller, B., Reinhardt, J.: *Neural Networks - An Introduction*, Berlin, 1990.

Petrovsky, A. B., Computer Support for Human Decision Making, in: *Proceedings of the Tenth International Conference on Multiple Criteria Decision Making*, Taipei, 1992, Vol. II, p. 303-313.

Schwefel, H.-P., *Numerische Optimierung von Computer-Modellen mittels der Evolutionsstrategie*, Basel, 1977.

Zeleny, M., *Multiple Criteria Decision Making*, New York, 1982.

ENTSCHEIDUNGSUNTERSTÜTZUNG DES MANAGEMENTS IN ORGANISATIONSFRAGEN MIT PETRI-NETZEN
-beispielhaft demonstriert an der Organisation eines überbetrieblichen Controllings-

Ralf Helbig, Rheinische Friedrich-Wilhelm-Universität Bonn

In einer Zeit stetiger Verkürzung der Innovationszyklen, muß ein erfolgsorientiertes Unternehmen auch in seiner Organisation sehr flexibel sein. Entsprechend wird das Management häufiger mit damit verbundenen komplexen Fragestellungen konfrontiert, weswegen Entscheidungsunterstützung auch in diesem Gebiet gefordert ist. Bei Annahme einer durch sich verkürzende Innovationszyklen bedingten schnellen Änderung der Organisationsumgebung wäre eine Art Organisations-Controlling, das die Organisation beständig auf Optimalität in Bezug auf die Anforderungen an sie überprüft und sie entsprechend ändert, wünschenswert.

Im Hinblick darauf wurde ein Petri-Netz-Modell der Organisation eines überbetrieblichen Controllings, wie es in der landwirtschaftlichen Beratung praktiziert wird, aufgebaut. Anhand dieses Modells wird demonstriert, welche organisatorischen Fragestellungen mit Petri-Netzen bearbeitet werden können. Das Spektrum reicht von Optimierung des personellen Einsatzes, des Betriebsmitteleinsatzes, der Technologie bis hin zur Optimierung der Organisationstruktur, jeweils auch unter Berücksichtigung der Kosten.

Durch die flexibel gestaltete Eingabe der Umgebungsvariablen kann eine Organisation mit ähnlicher Aufgabenstellung spezifiziert werden und durch das Modell analysiert, simuliert und optimiert werden. Ebenso wird ein Organisations-Controlling möglich, da Änderungen in der Organisatorischen Umgebung sofort eingegeben und Konsequenzen sowie mögliche notwendige Reaktionen ersichtlich, aber auch getestet werden können.

Im Vortrag wird auch die gestaltete Animationsoberfläche, wodurch die Petri-Netze für den Manager in den Hintergrund treten, vorgestellt. Er kann anhand der Animation Ergebnisse verfolgen, Entscheidungen am Modell testen und sie entsprechend optimieren. Für einen Organisator oder Systemanalytiker besteht aber weiterhin die Möglichkeit durch einen direkten Zugriff auf das dahinterliegende Petri-Netz individuelle Anpassungen oder zu testende Strukturänderungen vorzunehmen.

THE APPLICATION OF MIXED INTEGER PROGRAMMING TO THE "BUSINESS SYSTEMS PLANNING"-PROBLEM

Gerhard F. Knolmayer, University of Berne

Summary: Business Systems Planning (BSP) is a method for determining different subsystems of an information system. The paper discusses several mixed integer programming models for solving BSP problems and presents computational experiences with MPSX-MIP and OSL in solving these problems.

Zusammenfassung: Business Systems Planning (BSP) ist eine Methode zur Bestimmung von Subsystemen eines Informationssystems. Dieser Beitrag beschreibt verschiedene Modelle der gemischt-ganzzahligen Programmierung zur Lösung von BSP-Problemen und berichtet über rechentechnische Erfahrungen, die bei Lösung dieser Probleme mit MPSX-MIP und OSL gewonnen wurden.

1. Introduction

Strategic planning of information systems is proposed as a first step in the information engineering concept [cf. 12] and supported by CASE workbenches e.g. in "Planning Workstations" [10]. Strategic planning methods primarily use matrices in which associations between relevant objects are represented. The best-known method in this field is IBM's Business Systems Planning (BSP) which uses processes and data classes as objects [4;5]. Fig. 1 gives an example of a MxN-dimensional BSP-matrix [11, p. 192]. Non-null elements of the BSP-matrix show the symbol C for processes creating and U for those using certain data classes. BSF is concerned with determining subsystems in such a way that almost all symbols are positioned within the SubMatrices at the Main Diagonal (SMMD) of a restructured BSP-matrix. The resulting information subsystems and the data flows between them are called an Information Systems Architecture [3;14] which provides several benefits [1;9]. This paper, however, concentrates on formal aspects of restructuring BSP-matrices.

	D1	D2	D3	D4	D5	D6	D7	D8	D9	D10
P1				U	C					
P2					U			U	U	U
P3		U	U	U						
P4	C			C						
P5					U	C	C			
P6		C	C	C						
P7				C		U	U			
P8	C			C						
P9					U	C	C			
P10		C	C	C						
P11				C		U	U			
P12								C		
P13										U
P14								U	U	U
P15					U				U	
P16									U	C
P17									C	

	D9	D10	D8	D6	D7	D5	D1	D4	D2	D3
P17	C									
P16	U	C								
P13		U								
P12			C							
P2	U	U	U			U				
P14	U	U	U							
P9				C	C	U				
P5				C	C	U				
P1						C		U		
P15	U					U				
P4							C	C		
P6								C	C	C
P10								C	C	C
P7				U	U			C		
P8							C	C		
P11				U	U			C		
P3								U	U	U

Fig. 1: Original and restructured BSP matrix

Early publications [e.g. 5; 11] provide examples of restructuring without describing the underlying methods. Later on, methods like ISMOD appeared which define an affinity index to determine an appropriate sequence of processes for a given sequence of data classes [3; 6; 14]; however, until recently no procedures for two-dimensional restructuring of BSP-matrices were available. Therefore we decided to develop methods for this task which may be modelled as a (modified) quadratic assignment problem. One technique for solving this type of problem is to linearize it and to apply linear mixed-integer programming (MIP). The determination of MIP-solutions for BSP-problems seems interesting

- to provide a benchmark for a comparison with solutions obtained by heuristic methods
- to obtain experiences about the power of today's systems in solving quadratic assignment problems and
- for computational comparisons with specialized branch&bound methods.

2. Decisions in modelling the BSP-problem as a MIP

2.1. Decisions affecting the resulting solutions

As a basis for formal planning, the information architect has to decide which properties the resulting matrix should have:

- Can and should all C entries of the BSP-matrix be positioned in a SMMD?
- Which number R of SMMD should be established?
- Is there a minimal number K of rows in each submatrix?
- Is there a minimal number L of columns in each submatrix?
- How should the objective function be defined? Should one minimize the number of U entries outside the SMMD or the number of null-values inside the SMMD? Which parameter values should be used for attaining these goals?

2.2. Decisions affecting only the necessary computational resources

Quadratic assignment problems can be linearized by defining binary variables x_{ijrs} and then solved by applying MIP-algorithms. If $x_{ijrs}=1$, row i is assigned to submatrix r and column j to submatrix s; if $x_{ijrs}=0$, either row i is not assigned to r and/or column j is not assigned to s. Defining x_{ijrs} explicitly as binary leads to coordination problems between feasible x_{ijrs}-values and results in a very large number of binary variables. This can be avoided by introducing explicit binary variables u_{ir} resp. v_{js} which equal 1 if row i is assigned to submatrix r resp. column j to submatrix s and 0 otherwise. Appropriate constraints make all formally continuous variables x_{ijrs} binary. The objective function becomes

$$\Sigma_{ijr} \; c_{ijrr} \, x_{ijrr} \; \text{->} \; \max! \qquad (1)$$

We use capital indices I and J if a special pair of indices is referenced. The parameters c_{IJrr} are 1 for every U element at row I and column J and -p for every null-value in the resulting SMMD r. These parameter values avoid the generation of very large SMMD which necessarily contain almost all U-entries. Each process i has to be assigned to exactly one submatrix r

$$\Sigma_r \quad u_{ir} \quad = 1 \quad \forall i \tag{2}$$

and each data class has to be assigned to exactly one submatrix s:

$$\Sigma_s \quad v_{js} \quad = 1 \quad \forall j \tag{3}$$

The u- resp. v- and x-variables are coordinated by constraints which can be differently formulated. If $u_{IR}=0$, all variables x_{IjRs} must also be zero; if $u_{IR}=1$, there must be as many positive x_{IjRs} variables as there are columns:

$$\Sigma_{js} \quad x_{ijrs} - N\, u_{ir} \quad = 0 \quad \forall i,r \tag{4a}$$

An equivalent, but tighter formulation is

$$\Sigma_s \quad x_{ijrs} - \quad u_{ir} \quad = 0 \quad \forall i,j,r \tag{4b}$$

The analogous equations for the data classes are:

$$\Sigma_{ir} \quad x_{ijrs} - M\, v_{js} \quad = 0 \quad \forall j,s \tag{5a}$$

$$\Sigma_r \quad x_{ijrs} - \quad v_{js} \quad = 0 \quad \forall i,j,s \tag{5b}$$

One may consider using both (4a) and (4b) resp. (5a) and (5b) in parallel. We assume that every C-entry (in row I and column J) can be assigned to a SMMD. This can either be expressed as

$$\Sigma_r \quad x_{IJrr} \quad = 1 \quad \forall I,J \tag{6a}$$

or as

$$u_{Ir} - \quad v_{Jr} \quad = 0 \quad \forall I,J,r \tag{6b}$$

Again one may consider using both (6a) and (6b) in parallel. Of course, only one combination of r and s is allowed for each combination of i and j:

$$\Sigma_{rs} \quad x_{ijrs} \quad = 1 \quad \forall i,j \tag{7}$$

However, (7) may also be cancelled without affecting the resulting solution. If there is a minimal number K of rows in each submatrix, constraints

$$\Sigma_i \quad u_{ir} \quad \geq K \quad \forall r \tag{8a}$$

may be added. An equivalent formulation introduces additional integer variables g_r as slack variables:

$$\Sigma_i \quad u_{ir} - \quad g_r \quad = K \quad \forall r \tag{8.1b}$$

$$\Sigma_r \quad g_r \quad = M - K R \tag{8.2b}$$

Analogous properties of the data classes may be modelled as

$$\Sigma_j \quad v_{js} \quad \geq L \quad \forall s \tag{9a}$$

or

$$\Sigma_j \quad v_{js} - \quad g_s \quad = L \quad \forall s \tag{9.1b}$$

$$\Sigma_s \quad g_s \quad = N - L R \tag{9.2b}$$

Withous loss of generality it is possible to assign an arbitrary C-element to the first SMMD. This assignment may force further processes and/or data classes also to the first SMMD. Assume that in the BSP-matrix shown in Fig. 1 we let $u_{41}=1$; this implies

$$u_{61}=u_{71}=u_{81}=u_{10,1}=v_{11}=v_{21}=v_{31}=1 \quad (10)$$

Such data dependent assignments may allow to fix many binary and continuous variables; null variables and the corresponding constraints (2)-(5) may also be physically removed.

Another possibility which takes advantage of specific data is to explicitly formulate that columns (rows) with C-elements in the same row (column) must be assigned to the same SMMD; e.g. there are y=6 C-entries in column 4 of the original BSP-matrix shown in Fig. 1 which lead to y(y-1)/2 constraints

$$u_{ir} - u_{Ir} = 0 \qquad \text{for } i,I=4,6,7,8,10,11;i<I \quad (11a)$$

However, it is also sufficient to define a "master entry", e.g. u_{4r}, and define y-1 equations

$$u_{ir} - u_{4r} = 0 \qquad \text{for } i=6,7,8,10,11 \quad (11b)$$

Finally there exist nonnegativity and integer constraints

$$x_{ijrs} \geq 0;\ u_{ir}, v_{js} = \{0,1\};\ g_r = \{0,1,2,...,M\text{-}KR\};\ g_s = \{0,1,2,...,N\text{-}LR\} \quad (12)$$

3. Computational experiences

The modelling options described above may be selected independently. It is probably not worthwhile to study all modelling variants with a large number of test cases to establish a (probably partial) ranking. Instead of investigating this large number of equivalent models, some formulations based on ceteris paribus changes of modelling options were solved. A matrix generator was developed in Turbo-PASCAL which generates the MPS format from a BSP-matrix specified in a spreadsheet via an ASCII-file [cf. 8]. Several BSP-matrices described in the literature were optimized using MPSX-MIP, Release 2, on an IBM 3090-38J resp. OSL [2; 16], Release 2, on an IBM RS6000-320. These packages provide several preprocessing facilities which, however, make formulation methodologies not obsolete. Due to sophisticated techniques employed by these packages, the number of iterations is not always well suited as an indicator for the solution effort needed; furthermore, there are also some well-known problems in comparing computational efforts by regarding CPU-times.

3.1. Experiences with MPSX/MIP

3.1.1. Experiences with a 17x10 matrix

We started with formulating several modelling variants for the rather easy test problem from [11, p. 192] shown in Fig. 1, R=3, K=L=2 and p=0.3. The maximal objective function value of all LP-relaxations regarded is 21; the value of the optimal integer solution (visualized in the right part of Fig. 1) is 6.8. The computational results obtained confirm the recommendation that a MIP should be formulated as tight as possible in a continuous sense [15]: After (4a) and (5a) were replaced by (4b) and (5b), the CPU-time changed from 25.38 to 1.89 CPU-minutes and the number of iterations from

61615 to 11911. Using (6b) instead of (6a) further reduced the number of iterations to 8756 and CPU-minutes to 1.61. The test runs provided no evidence that it is worthwhile to use logically equivalent constraints like (4a) and (4b) in parallel. Cancelling (7) showed no clear effects on the solution time in several model variants. Substituting (8a) and (9a) by (8.1b),(8.2b),(9.1b) and (9.2b) as well as adding (11a) or (11b) had insignificant effects. However, the solution effort was reduced to 0.35 CPU-minutes and 2167 iterations if (10) was used to fix the variables and to 0.26 CPU-minutes and 1915 iterations if the fixed variables and the thus redundant constraints were physically removed. From this rather limited computational experience we decided to use (4b), (5b), (6b), no (7), (8b), (9b) and a physical elimination based on (10) for investigating larger BSP-matrices.

3.1.2. Experiences with a 19x12 matrix

For the BSP-matrix given in [7] only 1 data class and 3 processes can immediately be assigned to the first SMMD analogeous to (10). The solutions found had far better objective function values than the one given in [7]. The computational effort varied considerably for different values of K, L and R: In most cases the optimality of the solution was proved in less than 5 CPU-minutes. However, e.g. solving the problem for K=2, L=1 and R=6 needed more than 151000 iterations and took about 48 CPU-minutes.

3.1.3. Experiences with a 37x24 matrix

The rather large BSP-matrix regarded in [12, Vol. 2, p. 175] resulted for R=5, K=L=3 in a MIP-model with 277 binary variables, almost 18000 continuous variables und more than 8200 rows. On using analogeous assignments as in (10), the definition of 16 binary variables for assigning data classes and 12 for assigning processes was avoided. MPSX-MIP found the continuous solution after approx. 1 CPU-hour but no integer solution after more than 8 CPU-hours. One reason for this is certainly the high dual degeneracy which results in large numbers of iterations without improving the objective function value.

3.2. Experiences with OSL

Release 2 of the OSL was used on an IBM RS 6000-320H. Computational comparisons between MPSX and OSL are given e.g. in [2]. With appropriate formulations we needed 1567 iterations and 0.44 CPU-minutes for optimization of the 17x10 BSP-model. For the 37x24 model we obtained the 6 integer solutions shown in Fig. 2; 5 of them were found after approximately 5.5 hours. However, after 5 days of dedicated use and more than 3 million iterations no proof of optimality was obtained.

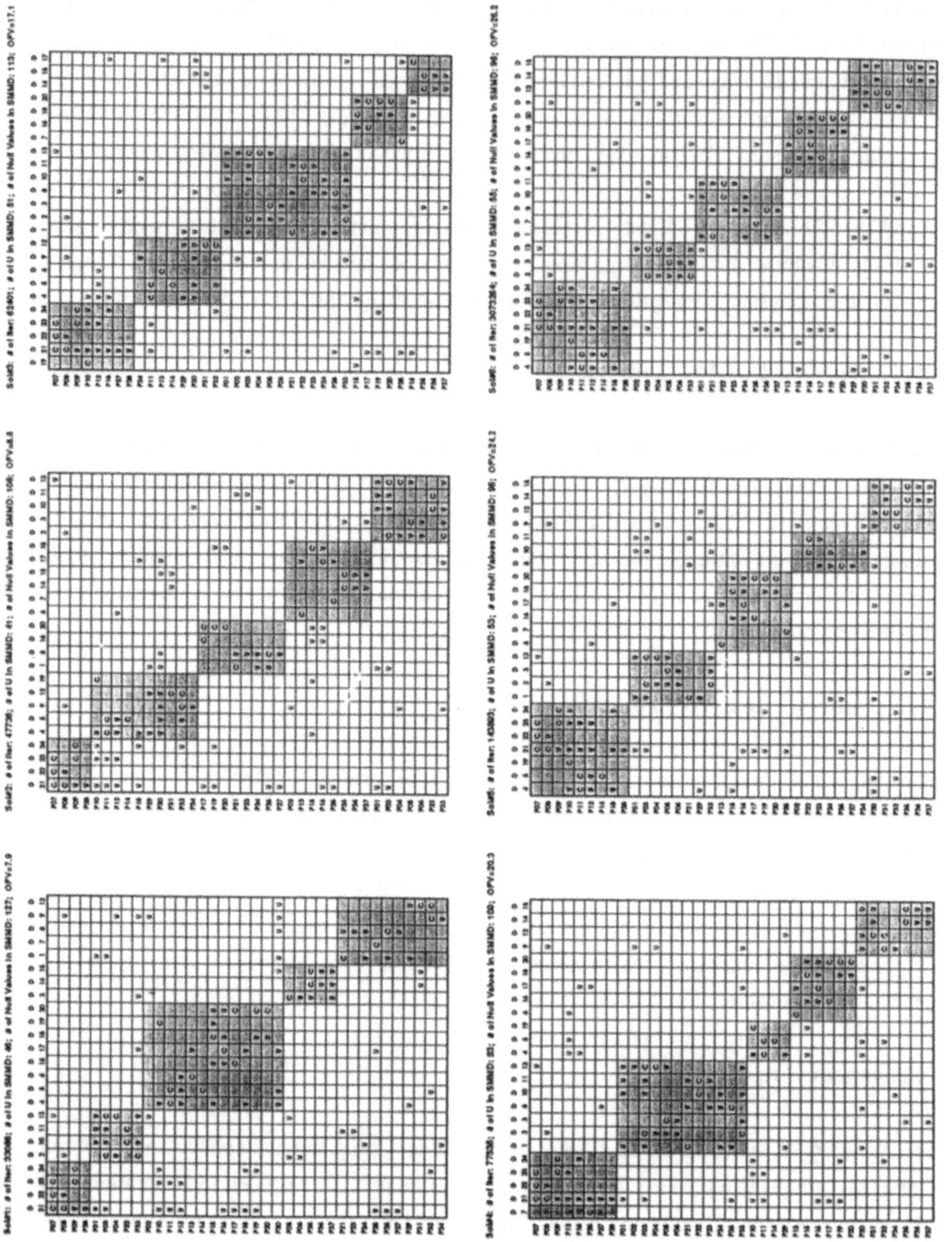

Fig. 2: A comparison of restructured BSP-matrices

4. Summary and Outlook

The results above show that the BSP-problem is in principle tractable as MIP and that OSL is an attractive option for solving large MIPs. However, the optimization of large quadratic assignment problems via MIP may consume remarkable computational resources and inhibit the generation of a variety of parametric solutions. Therefore the development of algorithms which specialize in the structure of the BSP-problem is relevant. Experiences with specialized branch&bound-algorithms for the BSP-problem are presented in [9; 13].

References

[1] Brancheau, J.C., Schuster, L., March, S.T., Building and Implementing an Information Architecture, in: Data Base 20 (1989) 2, 9-17.

[2] Ciriani, T.A., Model Structures and Optimization Strategies, in: T.A. Ciriani/R.C. Leachman (Eds.), Optimization in Industry, Chichester: Wiley 1993, 263-272.

[3] Hein, K.P., Information System Model and Architecture Generator, in: IBM Systems Journal 24 (1985), 213-235.

[4] Heinrich, L.J., Informationsmanagement, 4. Aufl., München-Wien: Oldenbourg 1992.

[5] IBM (Hrsg.), Business Systems Planning: Handbuch zur Planung von Informationssystemen, IBM-Form GE 12-1400-1, o.O.: IBM Deutschland 1980.

[6] IBM (Ed.), Informations System Model and Architecture Generator, Operation Guide, Release 1.3, IBM-Form SB 11-5989-2, 1990.

[7] Kiewiet, D.J., Steegwe, R.A., Conceptual Modeling and Cluster Analysis: Design Strategies for Information Architectures, in: J.I. DeGross et al. (Eds.), Proceedings of the 12th International Conference on Information Systems, Baltimore: ACM 1991, 315-326.

[8] Knolmayer, G., Spreadsheet-orientierte Matrixgeneratoren, in: H. Isermann et al. (Eds.), Operations Research Proceedings 1986, Berlin et al.: Springer 1987, 634-641.

[9] Knolmayer, G., Spahni, D., Darstellung und Vergleich ausgewählter Methoden zur Bestimmung von IS-Architekturen, in: H. Reichel (Hrsg.), Informatik - Wirtschaft - Gesellschaft, Berlin et al.: Springer 1993, S. 99 - 104.

[10] KnowledgeWare (Ed.), Application Development Workbench, ADW/Planning Workstation User Guide, Atlanta 1992.

[11] Martin, J., Strategic Data-Planning Methodologies, Englewood Cliffs: Prentice-Hall 1982.

[12] Martin, J., Information Engineering, 3 Vols., Englewood Cliffs: Prentice-Hall 1989/90.

[13] Spahni, D., Solving the "Business System Planning"-Problem Using Specialized Algorithms, Paper presented at the 18th Symposium über Operations Research, Cologne 1993.

[14] Vetter, M., Strategie der Anwendungssoftware-Entwicklung - Planung, Prinzipien, Konzepte, Stuttgart: Teubner 1988.

[15] Williams, H.P., Model Building in Mathematical Programming, 2nd ed., Chichester et al.: Wiley 1985.

[16] Wilson, D.G., Ruding, B.D., Introduction to the IBM Optimization Subroutine Library, in: IBM Systems Journal 31 (1992), 4-10.

DEVELOPMENT OF A SYSTEMIC EVOLUTIONARY AND BIOCYBERNETIC METHODOLOGY USING THE CHAOS AND CATASTROPHES THEORIES TO HANDLE COMPLEX DECISION SITUATIONS IN MANAGEMENT

Veneta Marzen, University of Southampton, UK

A critical review of many existing methodologies shows that these could not cope with the apparent complexity of the real world. In this sense, the limits of analytical approaches (e.g. MS / OR, MCDM) have been analyzed in relation to the recognition of the chaos theory which questioned the validity of the Newtonian deterministic paradigm.
However complexity of a decision situation could not only be related to different states of an open system. Rather complexity is concerning with the whole human social system. In this sense complexity involves the degree of inter connectivity of all hierarchical levels in the decision process, their cross-over links, feedbacks and also consequences for the decision field.
The phases of the proposed decision-aid/planning methodology for complex decision situations are summarised as follows:

Phase 1: Take a block of system's variables covering all aspects of our life. Reduce the complexity of the system by taking only relevant key factors able to create a rough picture of its reality.

Phase 2: Find out the inter dependencies of all relevant systemic variables.

Phase 3: Determine the active, passive, critical and buffer elements.

Phase 4: Elaborate time aspects. Follow future development paths and simulate if possible all scenarios in the form of systems dynamic or causal-loops diagrams.

Phase 5: Build a controlling model and show the controllable and non-controllable elements. Build a policy test.

Phase 6: Use the biocybernetic checklist. Determine controlling actions based on cybernetic and biocybernetic rules.

These methodology was used to analyze the EC project MARTA which aims to introduce telematics for control of future traffic flows in Europe.

MHZE - STRUCTURING OF OBJECTIVES

Sinan Perin, Universität zu Köln
Rüdiger von Nitzsch, Universität zu Köln

There are a lot of multicriteria decision support methods. Most of the methods assist the decision maker by choosing among different alternatives on the basis of a *known* system of objectives. Just a few approaches provide a support in structuring or defining objectives.

As a basis for an evaluation of alternatives a system of objectives has to fulfill the following five essential conditions. Primarily it has to be guaranteed that all objectives in the system are *fundamental*. An objective is fundamental if the fulfillment of the goal for itself represents a value for the decision maker. An objective is not fundamental if there is only a positive influence to another (fundamental) objective. Furthermore the system has to comprise all aspects which are of importance for the decision maker (*complete*), no aspect should be found in more than one objective (*not redundant*), it must be possible to value the outcomes in each objective independent of the other objectives (*decomposable*) and the objectives must be *operationable*.

We introduce a new computer-based approach *MHZE* (*m*ulti*h*euristische *Z*iel*e*rmittlung) which is designed to aid the decision maker by defining a system of objectives that fulfills the five conditions. *MHZE* integrates some procedures to test respectively improve the fulfillment of the conditions. An interactive question modul and two preference dependencies tests verifies whether the objectives are fundamental or not. Test of preference dependencies are also used to examine whether the goals are decomposable. In order to attain an complete system of objectives creativity tables and a cluster algorithm are embodied. Correlations between outcomes of alternatives are calculated to find hints to redundancies. It is possible to develop a simple influence model for the construction of artificial scales for objectives which can not be measured on a natural scale. Structuring the objectives as a hierarchy is a tool to ease the examination of all conditions in the whole approach.

KOSTENRECHNUNG UND NEOINSTITUTIONELLE BETRACHTUNGSWEISE: ÖKONOMISCHE BEGRÜNDUNGSANSÄTZE VON FIX- UND GEMEINKOSTENALLOKATIONEN

Dieter Pfaff, Johann Wolfgang Goethe-Universität Frankfurt a.M.

Gegenstand des Übersichtsvortrags ist die ökonomische Begründung von Verfahren der Fix- und Gemeinkostenallokation. Einerseits werden in der "traditionellen" Theorie häufig Verfahren der Vollkostenrechnung abgelehnt. Andererseits konnten sich Teilkostensysteme in der Praxis nicht hinreichend durchsetzen. Diese Diskrepanz läßt sich insbesondere damit begründen, daß die in der Theorie gewonnenen Erkenntnisse auf der Annahme eines einzigen (in der Regel risikoneutralen) Entscheidungsträgers beruhen, der über alle entscheidungsrelevanten Informationen verfügt. In realen Organisationen dagegen herrschen dezentrale Entscheidungssysteme vor, in denen Informationen asymmetrisch verteilt sind und die einzelnen Entscheidungsträger ihren Informationsvorsprung im eigenen Interesse und nicht im Sinne der vorgesetzten Instanzen nutzen. Sie geben u.a. auch Fehlinformationen, um die Entscheidungen dieser Instanzen im eigenen Interesse zu beeinflussen.

Erkenntnisse, die in einer "idealen" Welt eines einzelnen wohlinformierten Entscheidungsträgers gewonnen werden, sind damit nicht ohne weiteres auf Organisationen übertragbar. Die Mehrzahl der in der Fachliteratur diskutierten ökonomischen Begründungsansätze von Fix- und Gemeinkostenallokationen läßt sich unter vier Forschungsschwerpunkten zusammenfassen: Präferenz, Dekomposition, Anreiz und Koordination. Im ersten Fall hängt die Entscheidungsrelevanz von Fixkosten nur von der Gestalt der subjektiven Präferenzen des Entscheidungsträgers - üblicherweise seiner Risikoeinstellung - ab. Alle Präferenzmodelle betrachten einen einzigen Entscheider mit lediglich einem Entscheidungsproblem in einem Unternehmen. Demgegenüber gibt es bei den Dekompositionsansätzen zum Teil mehrere Entscheidungsträger, zumindest aber mehr als ein Entscheidungsproblem. Dabei wird die Fähigkeit von Gemeinkostenallokationen analysiert, die regelmäßig auftretenden Interdependenzen zwischen den einzelnen Entscheidungsproblemen zu approximieren. Unterstellt man zusätzlich zu den Annahmen der Dekompositionsmodelle asymmetrische Informationsverteilung und opportunistisches Verhalten, kommt man zu Anreiz- und Koordinationsansätzen. Erstere untersuchen, ob die Zurechnnung von Gemein- oder Fixkosten das Verhalten eines Agents so beeinflussen kann, daß sich der Nutzen des Principals erhöht. Letztere behandeln die Allokationswirkungen auf mindestens zwei Agents, wodurch im Unterschied zum einfachen Anreizmodell zusätzlich Abstimmungs- und Risikoverteilungsprobleme zwischen den Agents relevant werden können.

EFFIZIENTE GESTALTUNG VON LIEFERVERTRÄGEN BEI UNVOLLSTÄNDIGER INFORMATION

Heike Y. Schenk-Mathes, Universität Frankfurt am Main

Zusammenfassung: In diesem Beitrag wird die Gestaltung von Lieferverträgen für den Fall untersucht, daß der Lieferant unvollständig über leistungsbestimmende Merkmale eines bereits feststehenden Abnehmers informiert ist. Vertragsparameter sind Liefermenge und Zahlung des Abnehmers an den Lieferanten. Der Lieferant muß sein aus mehreren Verträgen bestehendes Angebot so gestalten, daß jeder als möglich erachtete Abnehmertyp zu kooperieren bereit ist. Es wird gezeigt, in welcher Weise das Vertragsangebot von der First-Best-Lösung abweicht und welchen Einfluß eine vor Vertragsangebot stattfindende Produktionsentscheidung des Lieferanten hat.

Summary: In this paper we analyse the design of supply contracts in the case of incomplete information about the type of buyer. It is assumed that the buyer whose characteristics are not known with certainty has already been chosen. Contract parameters are the quantity delivered and the payment the supplier receives from the buyer. The supplier's offer which can be composed of more than one contract must be designed so that each possible type of buyer will be willing to cooperate. The deviation of the menu of contracts from the First-Best solution is shown and the supplier's decision about the quantity to produce is considered.

1 Problemstellung

In dem Beitrag wird die Gestaltung von bilateralen Lieferverträgen bei unvollständiger Information über den Abnehmer betrachtet. Durch den Lieferanten erfolgt ein Angebot an den bereits feststehenden Abnehmer in Form eines Vertrages oder auch mehrerer Verträge, zwischen denen der Abnehmer wählen kann. Ein Vertrag besteht aus einem Liefermengenangebot und der dieser Menge entsprechenden Zahlung des Abnehmers an den Lieferanten. Die übrigen Vertragsbedingungen, wie z.B. Qualitätsvorgaben sowie Angaben zu deren Überprüfung und weitere Lieferungs- und Zahlungsbedingungen, sind vorgegeben und werden im folgenden nicht berücksichtigt. Der vom Abnehmer gewählte Vertrag wird gemäß den Vereinbarungen abgewickelt, d.h. Probleme der Nichterfüllung des Vertrages sind ausgeschlossen.

Zum Zeitpunkt der Festlegung des Vertragsangebotes, das im folgenden als Vertragsmenu bezeichnet wird, sind dem Lieferanten die leistungsbestimmenden Merkmale des Abnehmers nicht mit Sicherheit bekannt. Man spricht hier auch von unvollständiger Information über den Abnehmertyp. Im folgenden werden zwei Modelle vorgestellt, die sich nur hinsichtlich der Relevanz der Produktionsmengenentscheidung unterscheiden.

Das Vertragsmenu muß für jeden als möglich erachteten Abnehmertyp mindestens einen Vertrag enthalten, den dieser zu akzeptieren bereit ist. Dahinter kann sich z.B. die Aussicht auf Exklusivverträge verbergen, oder es sind Vorverhandlungen mit dem betreffenden Abnehmer geführt worden. Die Vertragsparteien verhalten sich risikoneutral.

Die behandelten Ansätze sind den Self-Selection-Modellen zuzurechnen [vgl. insbesondere 1,7 und als Anwendungsfall für Koordinationsprobleme innerhalb der Unternehmung 8]. Man kann von Screening-Modellen sprechen, da der Lieferant bei unvollständiger Information über seinen

Vertragspartner als erster agiert, indem er ein Vertragsmenu anbietet, an welches er gebunden ist [zu Vertragsmodellen vgl. z.B. 2,3,5 und zu Lieferverträgen zwischen zwei Unternehmungen bei unvollständiger Information auf beiden Vertragsseiten z.B. 6].

2 Modell I: Optimierung des Vertragsmenus ohne Berücksichtigung der Produktionsmengenentscheidung

2.1 Herleitung des Optimierungsansatzes

Der Gewinn des Lieferanten sei gegeben durch

(1) $G_L(q,Z) = Z - C(q),$

wobei q die Liefer- bzw. Produktionsmenge und Z die Zahlung des Abnehmers an den Lieferanten bezeichnet. Die Kostenfunktion C(q) sei zweimal stetig differenzierbar, streng konvex und verlaufe monoton steigend.

Die Liefermenge bildet den Faktorinput für den Abnehmer. Er verfügt über eine lineare Technologie mit einem Produktionskoeffizienten von eins; der resultierende Output wird zu einem konstanten Stückpreis abgesetzt (vollständige Konkurrenz auf dem Absatzmarkt). Aus Sicht des Lieferanten kann der Abnehmer zu einem von zwei Typen gehören. Die Abnehmertypen unterscheiden sich hinsichtlich ihrer Deckungsbeiträge und ihrer Fixkosten. Geht man von Fixkosten in Höhe von F_t und von einem Deckungsbeitrag d_t ohne Berücksichtigung der Zahlung Z an den Lieferanten aus, so ergibt sich für den Abnehmer des Typs t ein Gewinn von

(2) $G_A^t(q,Z) = d_t \cdot q - F_t - Z \geq G^m.$

Wir setzen $d_1 < d_2$, d.h. Abnehmertyp 1 weist höhere variable Stückkosten auf als Abnehmertyp 2. Der Abnehmer fordert einen Mindestgewinn G^m, der unabhängig von seinem Typ ist. Bei Indifferenz wählt der Abnehmer jenen Vertrag, den der Lieferant präferieren würde.

Wäre dem Lieferanten der Abnehmertyp zum Zeitpunkt des Vertragsangebotes bekannt, so könnte er einen Vertrag wählen, der dem betreffenden Typ gerade den Mindestgewinn bietet. Dieser optimale Vertrag bei vollständiger Information sei als First-Best-Vertrag, (q_t^F, Z_t^F), des betreffenden Typs t bezeichnet. Es gilt

(3) $d_t = C_q(q_t^F)$ und $Z_t^F = d_t \cdot q_t^F - F_t - G^m$, wobei im folgenden $q_1^F > 0$ vorausgesetzt wird.

Bei unvollständiger Information über den Abnehmertyp läßt sich zeigen, daß der Lieferant seine Position durch das Angebot zweier Verträge verbessern kann. Mit $\pi_t \in (0,1)$, t=1,2, sei die Wahrscheinlichkeit für das Vorliegen des Typs t bezeichnet. Das Optimierungsproblem lautet nun

(4) $\underset{(q_1,q_2,Z_1,Z_2)}{\text{Max}} \; E[G_L(q_1,q_2,Z_1,Z_2)] = \pi_1 \cdot (Z_1 - C(q_1)) + \pi_2 \cdot (Z_2 - C(q_2))$

(5) $d_1 \cdot q_1 - Z_1 \geq d_1 \cdot q_2 - Z_2$ und $d_2 \cdot q_2 - Z_2 \geq d_2 \cdot q_1 - Z_1,$

(6) $d_1 \cdot q_1 - Z_1 \geq M_1 = F_1 + G^m$ und $d_2 \cdot q_2 - Z_2 \geq M_2 = F_2 + G^m,$

(7) $q_1 \geq 0$ und $q_2 \geq 0.$

Durch (5) wird garantiert, daß jeder Abnehmertyp sich bei Wahl des für ihn vorgesehenen Vertrages zumindest nicht schlechter stellt als bei Wahl des anderen Vertrages [zu diesem Revelationsprinzip vgl. z.B. 4]. Die Restriktionen (6) stellen sicher, daß den einzelnen Abnehmertypen wenigstens der Mindestgewinn geboten wird.

2.2 Eigenschaften der optimalen Lösung

Mit Hilfe der Lagrangefunktion

$$\begin{aligned}L(q_1,q_2,Z_1,Z_2,\lambda_1,\lambda_2,\mu_1,\mu_2) &= \pi_1\cdot(Z_1-C(q_1))+\pi_2\cdot(Z_2-C(q_2))\\ &+\lambda_1\cdot(d_1\cdot q_1-d_1\cdot q_2-Z_1+Z_2)+\lambda_2\cdot(d_2\cdot q_2-d_2\cdot q_1-Z_2+Z_1)\\ &+\mu_1\cdot(d_1\cdot q_1-Z_1-M_1)+\mu_2\cdot(d_2\cdot q_2-Z_2-M_2)\end{aligned} \tag{8}$$

lassen sich neben den Restriktionen (5)-(7) folgende Bedingungen mit $\lambda_1,\lambda_2,\mu_1,\mu_2\geq 0$ für ein Optimum herleiten:

$$\frac{\partial L(\cdot)}{\partial q_1}=-\pi_1\cdot C_q(q_1)+\lambda_1\cdot d_1-\lambda_2\cdot d_2+\mu_1\cdot d_1\leq 0, \tag{9}$$

$$\frac{\partial L(\cdot)}{\partial q_2}=-\pi_2\cdot C_q(q_2)-\lambda_1\cdot d_1+\lambda_2\cdot d_2+\mu_2\cdot d_2\leq 0,$$

$$\frac{\partial L(\cdot)}{\partial Z_1}=\pi_1-\lambda_1+\lambda_2-\mu_1=0 \quad \text{und} \quad \frac{\partial L(\cdot)}{\partial Z_2}=\pi_2-\lambda_2+\lambda_1-\mu_2=0, \tag{10}$$

$$\frac{\partial L(\cdot)}{\partial q_t}\cdot q_t=0,\quad \frac{\partial L(\cdot)}{\partial \lambda_t}\cdot\lambda_t=0,\quad \frac{\partial L(\cdot)}{\partial \mu_t}\cdot\mu_t=0 \quad \text{für } t=1,2. \tag{11}$$

Proposition 1:

a) Es existiert eine optimale Lösung, für die $q_1\leq q_2$ und $Z_1\leq Z_2$ gilt. Diese Bedingungen sind sogar für jede zulässige Lösung erfüllt.

b) Aus $q_1=q_2$ folgt $Z_1=Z_2$ und umgekehrt.

♦ Beweisskizze: Der Lösungsraum ist offensichtlich nicht leer. Eine zulässige Lösung ist $q_1=q_2=0$ und $Z_1=Z_2=-\mathrm{Max}\{M_1,M_2\}$. Sei $q_1>q_2$ optimal. Die Self-Selection-Bedingung (5) für Typ 2 läßt sich umformen zu $Z_1-Z_2\geq d_2\cdot(q_1-q_2)$. Da $d_1<d_2$ gilt, muß $Z_1-Z_2>d_1\cdot(q_1-q_2)$ erfüllt sein. Diese Bedingung verletzt jedoch die Self-Selection-Bedingung für Typ 1. $q_1>q_2$ ist nicht zulässig und damit nicht optimal. Auch $Z_1>Z_2$ ist nicht zulässig, da die Self-Selection-Bedingung $Z_1-Z_2\leq d_1\cdot(q_1-q_2)$ für Typ 1 verletzt ist. Für den Beweis von Teil b) verwendet man beide Self-Selection-Bedingungen. ♦

Mit $(\hat{q},\hat{Z})$ sei jener Vertrag bezeichnet, bei dem beide Kooperationsbedingungen als Gleichungen erfüllt sind, d.h. es gilt $\hat{q}=\frac{M_2-M_1}{d_2-d_1}=\frac{F_2-F_1}{d_2-d_1}$. $\hat{q}$ ist also jene Menge, die die Abnehmertypen bei gegebener Zahlung an den Lieferanten zu gleichen Kosten herstellen können. Ein optimales Vertragsmenu $\{(q_1^*,Z_1^*),(q_2^*,Z_2^*)\}$ besitzt dann folgende Eigenschaften:

Proposition 2:

a) Wenn $q_1^F>\hat{q}$ und $q_2^F>\hat{q}$ gilt, dann ist ein Vertragsmenu $\{(q_1^*<q_1^F,Z_1^*<Z_1^F),(q_2^F,Z_2^*\leq Z_2^F)\}$ optimal.

b) Wenn $q_1^F \le \hat{q} \le q_2^F$ gilt, dann ist das Vertragsmenu $\{(q_1^F, Z_1^F), (q_2^F, Z_2^F)\}$ optimal.

c) Wenn $q_1^F < \hat{q}$ und $q_2^F < \hat{q}$ gilt, dann ist ein Vertragsmenu $\{(q_1^F, Z_1^* \le Z_1^F), (q_2^* > q_2^F, Z_2^* > Z_2^F)\}$ optimal.

♦ Beweisskizze: Nach Einsetzen von (10) in (9) erhält man

(12) $\pi_1 \cdot (d_1 - C_q(q_1)) + \lambda_2 \cdot (d_1 - d_2) \le 0$ und $\pi_2 \cdot (d_2 - C_q(q_2)) + \lambda_1 \cdot (d_2 - d_1) \le 0$.

Wegen $q_1^F > 0$ ist $q_1 = q_2 = 0$ und wegen Proposition 1a) ist $q_1 > 0$ und $q_2 = 0$ nicht optimal. Zu betrachten sind die Fälle $q_1, q_2 > 0$ und $q_1 = 0, q_2 > 0$.

Wegen (11) müssen für den Fall $q_1, q_2 > 0$ die Bedingungen (12) als Gleichungen erfüllt sein.

Wenn $\lambda_1 = \lambda_2 = 0$ gilt, dann folgt aus (12) $d_1 = C_q(q_1^*)$ bzw. $d_2 = C_q(q_2^*)$ und $q_1^* = q_1^F$ bzw. $q_2^* = q_2^F$. Wegen (10) und $\pi_1 \in (0,1)$ muß $\mu_1 > 0$ und $\mu_2 > 0$ sein. Zusammen mit (11) folgt, daß die Kooperationsbedingungen (6) als Gleichungen erfüllt sein müssen, d.h. $Z_1^* = Z_1^F$ und $Z_2^* = Z_2^F$. Setzt man die optimalen Werte in die Self-Selection-Bedingungen (5) ein, so erhält man $M_2 - M_1 \le q_2^F \cdot (d_2 - d_1)$ und $M_2 - M_1 \ge q_1^F \cdot (d_2 - d_1)$ und damit $q_1^F \le \hat{q} \le q_2^F$.

Wenn $\lambda_1 = 0$ und $\lambda_2 > 0$ gilt, dann folgt aus (12) $d_1 > C_q(q_1^*)$ bzw. $d_2 = C_q(q_2^*)$ und $q_1^* < q_1^F$ bzw. $q_2^* = q_2^F$. Wegen (10) und $\pi_1 \in (0,1)$ muß $\mu_1 = \pi_1 + \lambda_2 > 0$ sein und damit $Z_1^* = d_1 \cdot q_1^* - M_1 < d_1 \cdot q_1^F - M_1 = Z_1^F$. Wegen $\lambda_2 > 0$ und (11) muß zudem die Self-Selection-Bedingung (5) für Typ 2 als Gleichung erfüllt sein, so daß $Z_2^* = d_2 \cdot (q_2^F - q_1^*) + Z_1^*$ $= d_2 \cdot (q_2^F - q_1^*) + d_1 \cdot q_1^* - M_1$. Setzt man die optimalen Werte in die Kooperationsbedingung (6) für Typ 2 ein, so erhält man $q_1^* \ge \hat{q}$. Für $q_1^* > \hat{q}$ gilt $Z_2^* < Z_2^F$ und für $q_1^* = \hat{q}$ gilt $Z_2^* = Z_2^F$. Wegen $q_1^* < q_1^F$ und Proposition 1a) ergibt sich $q_1^F > \hat{q}$ und $q_2^F > \hat{q}$. Die Self-Selection-Bedingung (5) für Typ 1 ist erfüllt. q_1^* läßt sich folgendermaßen ermitteln: Wegen (10) gilt $\mu_1 + \mu_2 = 1$. Für $\mu_2 = 0$ folgt $\lambda_2 = \pi_2$ und aus (12) $\pi_1 \cdot (d_1 - C_q(q_1^*)) + \pi_2 \cdot (d_1 - d_2) = 0$. Wenn $\mu_2 > 0$ gilt, so folgt $q_1^* = \hat{q}$.

Für $\lambda_1 > 0$ und $\lambda_2 = 0$ kann man analog zeigen, daß $q_1^* = q_1^F$ und $q_2^* > q_2^F$ sowie $q_1^F < \hat{q}$ und $q_2^F < \hat{q}$ gilt. Für $\lambda_1 > 0$ und $\lambda_2 > 0$ folgt nach Einsetzen in (5) $q_1^* = q_2^* = \hat{q}$, was mit (12) nicht vereinbar ist.

Betrachten wir nun den Fall $q_1^* = 0$, $q_2^* > 0$, d.h. nur die zweite Bedingung (12) ist als Gleichung erfüllt. Für $\lambda_2 = 0$ ist diese Lösung nicht optimal, denn gemäß (12) müßte $d_1 \le C_q(0)$ erfüllt sein, was wegen $q_1^F > 0$ zum Widerspruch führt. Für $\lambda_1 = 0$ und $\lambda_2 > 0$ ist die Beweisführung analog zum ersten Fall. Für $\lambda_1 > 0$ und $\lambda_2 > 0$ folgt wieder $q_1^* = q_2^* = \hat{q} = 0$; diese Konstellation kann wegen $q_1^F > 0$ nicht optimal sein.

Da eine vollständige und separierbare Fallunterscheidung durchgeführt wurde, ist die Umkehrung der Aussagen möglich. ♦

3 Modell II: Relevanz der Produktionsmengenentscheidung

Bisher haben wir zeitliche Probleme bei der Produktion der vereinbarten Liefermenge außer acht gelassen. Es soll nun davon ausgegangen werden, daß die Produktionsmenge festgelegt werden muß, bevor die Liefermenge vereinbart wird. Zur Vereinfachung gehen wir weiterhin davon aus, daß das betrachtete Gut nicht lagerfähig ist und zuviel produzierte Einheiten vernichtet werden müssen. Dabei entstehen allerdings keine Kosten.

Wenn der Abnehmertyp bei Festlegung des Vertragsmenus nicht bekannt ist, dann muß der Lieferant wieder ein Vertragsmenu anbieten, das mindestens einen die Kooperationsbedingung erfüllenden Vertrag für jeden Typ enthält. Es soll nun untersucht werden, ob in diesem Fall ein Menu mit zwei Verträgen zu besseren Ergebnissen führen kann als das Angebot eines einzelnen Vertrages.

Werden zwei Verträge angeboten, dann muß die Produktionsmenge beim Lieferanten so groß sein wie die größte angebotene Menge. Ist die Produktionsmenge kleiner als die größte angebotene Menge, so kann ein Vertrag gegebenenfalls nicht erfüllt werden, was unzulässig ist. Im anderen Fall entstehen vermeidbare Kosten. Gemäß Proposition 1a) muß bezüglich der optimalen Liefermengen $q_1 \leq q_2$ gelten. Die Behauptung wird unter Zuhilfenahme der Self-Selection-Bedingungen bewiesen, die weiterhin erfüllt sein müssen. In der Zielfunktion (4) muß dann lediglich die Menge q_2 als Argument der Kostenfunktion bei beiden Typen eingesetzt werden. Die Restriktionen (5)-(7) gelten weiterhin. Modifiziert man auch die Lagrangefunktion (8) entsprechend, so muß die notwendige Bedingung (9) durch

$$\text{(13)} \quad \frac{\partial L(\cdot)}{\partial q_1} = \lambda_1 \cdot d_1 - \lambda_2 \cdot d_2 + \mu_1 \cdot d_1 \leq 0 \text{ und}$$

$$\frac{\partial L(\cdot)}{\partial q_2} = -C_q(q_2) - \lambda_1 \cdot d_1 + \lambda_2 \cdot d_2 + \mu_2 \cdot d_2 \leq 0$$

ersetzt werden.

Proposition 3:

a) Wenn $q_1^F \geq \hat{q}$ und $q_2^F > \hat{q}$ gilt, ist entweder $\{(q_1^F, Z_1^F)\}$ oder $\{(\hat{q}, \hat{Z}), (q_2^{**}, Z_2^{**})\}$ optimal, wobei q_2^{**} implizit gegeben ist durch $C_q(q_2^{**}) = \pi_2 \cdot d_2$ und $Z_2^{**} = d_2 \cdot q_2^{**} - M_2$ gilt.

b) Wenn $q_1^F < \hat{q} < q_2^F$ gilt, ist entweder $\{(\hat{q}, \hat{Z})\}$ oder $\{(\hat{q}, \hat{Z}), (q_2^{**}, Z_2^{**})\}$ optimal.

c) Wenn $q_1^F < \hat{q}$ und $q_2^F \leq \hat{q}$ gilt, ist $\{(q_2^F, Z_2^F)\}$ optimal.

♦ Beweisskizze: Durch Einsetzen von (10) in (13) erhält man

$$\text{(14)} \quad \pi_1 \cdot d_1 + \lambda_2 \cdot (d_1 - d_2) \leq 0 \quad \pi_2 \cdot d_2 - C_q(q_2) + \lambda_1 \cdot (d_2 - d_1) \leq 0.$$

Wegen $\pi_1 \in (0,1)$ kann die erste Ungleichung nur erfüllt sein, wenn $\lambda_2 > 0$ gilt.

Wenn zugleich $\lambda_1 > 0$ gilt, so müssen die Self-Selection-Bedingungen (5) strikt erfüllt sein, was wiederum zu $q_1^* = q_2^* = q^*$ und $Z_1^* = Z_2^* = Z^*$ führt. Wegen $q_1^F > 0$ ist $q^* = 0$ nicht optimal. Für den Fall, daß auch $\mu_1 > 0$ und $\mu_2 > 0$ gilt, erhält man $(q^*, Z^*) = (\hat{q}, \hat{Z})$. Für (13) kann man schreiben $\lambda_2 \cdot d_2 - \lambda_1 \cdot d_1 = \mu_1 \cdot d_1$ und damit $C_q(\hat{q}) = \mu_1 \cdot d_1 + \mu_2 \cdot d_2$. Wegen $\mu_1 + \mu_2 = 1$ (aus

(10)) folgt $q_1^F < \hat{q} < q_2^F$. Ist nur ein μ_t positiv, so daß $\mu_t = 1$ gilt, so erhält man gemäß (13) $q^* = q_t^F$. Außerdem ist die entsprechende Kooperationsbedingung (6) für Typ t strikt erfüllt, womit Z^* in Abhängigkeit von $q^* = q_t^F$ ermittelt werden kann. Setzt man die Zahlung in die jeweils andere Kooperationsbedingung ein, so erhält man $q_1^F \geq \hat{q}$ und damit auch $q_2^F > \hat{q}$ für $t = 1$ bzw. $q_1^F < \hat{q}$ und $q_2^F \leq \hat{q}$ für $t = 2$. $\mu_1 = \mu_2 = 0$ ist nicht optimal.

Wenn $\lambda_1 = 0$ gilt, so folgt aus (14) $C_q(q_2^*) = \pi_2 \cdot d_2$ und damit $q_2^* = q_2^{**}$. Wegen $\mu_1 = \pi_1 + \lambda_2 > 0$ (aus (10)) ist die Kooperationsbedingung für Typ 1 als Gleichung erfüllt und wegen $\lambda_2 > 0$ ist dies auch der Fall für die Self-Selection-Bedingung für Typ 2. Für den Fall, daß $\mu_2 = 0$ und damit $\mu_1 = 1$ ist, erhält man unendlich viele Optima, wobei gemäß (13) $d_1 = \pi_2 \cdot d_2$ und $q_1^F \geq \hat{q}$ sowie $q_2^F > \hat{q}$ erfüllt sein müssen. Das einelementige Vertragsmenu $\{(q_1^F, Z_1^F)\}$ ist eine optimale Lösung. Für den Fall, daß $\mu_1 > 0$ und $\mu_2 > 0$ gilt, erhält man nach Einsetzen der Zahlungen, die man über die Kooperationsbedingungen bestimmt, in die Self-Selection-Bedingung für Typ 2 den Ausdruck $M_2 - M_1 + q_1^* \cdot (d_1 - d_2) = 0$, d.h. $q_1^* = \hat{q}$. Die Self-Selection-Bedingung für Typ 1 ergibt, daß diese Lösung nur dann optimal sein kann, wenn $q_2^{**} \geq \hat{q}$ und damit auch $q_2^F > \hat{q}$ gilt. ♦

4 Interpretation und Beispiel

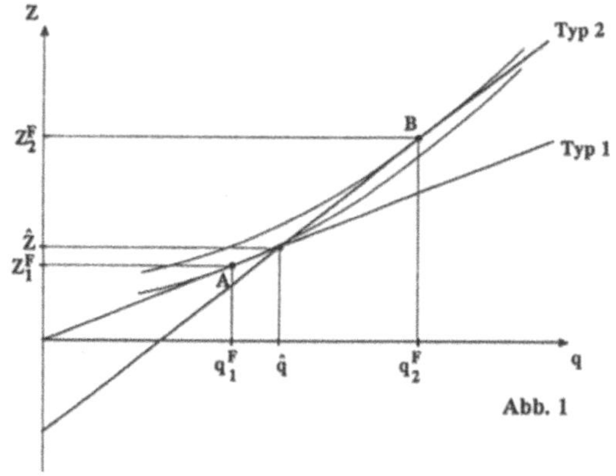

Abb. 1

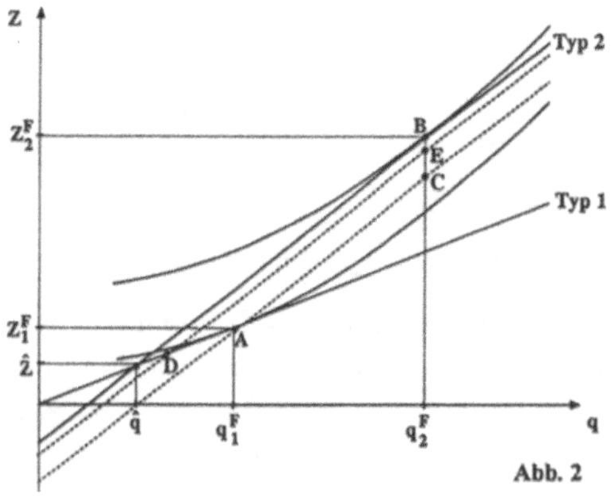

Abb. 2

Die Ergebnisse sollen anhand numerischer Beispiele verdeutlicht werden.

Beispiel 1:

$C(q) = \frac{1}{2}q^2, d_1 = 1, d_2 = 2, F_1 = 0, F_2 = \frac{5}{4}$, Mindestgewinn=0. Gemäß (3) lauten die First-Best-Verträge $(q_1^F, Z_1^F) = (1,1)$ und $(q_2^F, Z_2^F) = (2, \frac{11}{4})$.

Beispiel 2:

$F_2 = \frac{1}{2}$, sonst wie Beispiel 1. Der First-Best-Vertrag des Typs 2 lautet nun $(q_2^F, Z_2^F) = (2, \frac{7}{2})$.

In Abbildung 1 und 2 sind lineare Indifferenzkurven der Abnehmertypen eingezeichnet; diese führen jeweils zu dem Mindestgewinn. Je weiter rechts unten die Indifferenzkurven verlaufen, um so höher ist der ihnen zugeordnete Gewinn. Die Indifferenzkurven des Lieferanten verlaufen streng konvex und repräsentieren einen um so höheren Gewinn, je weiter links oben sie sich befinden. Die Punkte A und B stellen die First-Best-Verträge dar, da es sich hier um die Tangentialpunkte der Mindestgewinn-Indifferenzkurven der Abnehmertypen mit den Indifferenzkurven des Lieferanten handelt. In Beispiel 1 gilt $q_1^F < \hat{q} = \frac{5}{4} < q_2^F$. Gemäß Proposition 2 ist für das Modell I ein Vertragsmenu bestehend aus den First-Best-Verträgen optimal (vgl. Abbildung 1). Der Abnehmer gibt seinen wahren Typ preis, ohne daß der Lieferant für einen der Typen mehr als den Mindestgewinn bieten muß. In Beispiel 2, für das $q_1^F > \hat{q}$ und $q_2^F > \hat{q}$ gilt (vgl. Abbildung 2), führt dieses Vertragsmenu dazu, daß der Abnehmer unabhängig von seinem Typ den Vertrag A wählt. Der Lieferant stellt sich bereits durch das Angebot des Vertragsmenus {A,C} besser. Je nach Wahrscheinlichkeitsverteilung über die Abnehmertypen ergibt sich für diesen Fall ein optimales Vertragsmenu bestehend aus zwei Verträgen mit folgenden Eigenschaften: Das Menu enthält einen die Kooperationsbedingung für Typ 1 erfüllenden Vertrag, jedoch erfolgt eine Abweichung vom First-Best-Vertrag. Die Verträge des Menus liegen auf einer linearen Indifferenzkurve des Abnehmertyps 2, der in der Regel mehr als den Mindestgewinn erzielt bei einer Liefermenge, die seiner First-Best-Menge entspricht. Während in Beispiel 1 die First-Best-Verträge das optimale Menu bilden, ist es nun möglich, daß dem Abnehmertyp, der dem Lieferanten den höheren First-Best-Gewinn bietet, ein sogenannter Informationsgewinn gezahlt werden muß. So ergibt sich z.B. für $\pi_1 = 0{,}75$ als optimales Vertragsmenu $\{(\frac{2}{3},\frac{2}{3}),(2,\frac{10}{3})\}$ (vgl. das Vertragsmenu {D,E} in Abbildung 2). Bei Indifferenz wählt der Abnehmer annahmegemäß jenen Vertrag, den der Lieferant präferiert; für Abnehmertyp 2 ist dies der Vertrag $(2,\frac{10}{3})$. Er erzielt damit einen Gewinn, der um $\frac{1}{6}$ höher als sein Mindestgewinn ist.

Für das Modell II erhält man gemäß Proposition 3 als optimales Vertragsmenu für Beispiel 1 $\{(\hat{q},\hat{Z}) = (\frac{5}{4},\frac{5}{4})\}$ für $\pi_2 \in (0,\frac{5}{8}]$ und $\{(\frac{5}{4},\frac{5}{4}),(q_2^{**},Z_2^{**}) = (2\pi_2, 4\pi_2 - \frac{5}{4})\}$ für $\pi_2 \in (\frac{5}{8},1)$ und für Beispiel 2 $\{(q_1^F, Z_1^F) = (1,1)\}$ für $\pi_2 \in (0,\frac{1}{2}]$ und $\{(\hat{q},\hat{Z}) = (\frac{1}{2},\frac{1}{2}),(q_2^{**},Z_2^{**}) = (2\pi_2, 4\pi_2 - \frac{1}{2})\}$ für $\pi_2 \in (\frac{1}{2},1)$. Die Kooperationsbedingungen sind stets strikt erfüllt, der Abnehmer erhält also keinen Informationsgewinn, aber für bestimmte Wahrscheinlichkeitsbereiche sind zweielementige Vertragsmenus optimal.

Literatur

[1] Harris, M. und R.M. Townsend: Resource Allocation under Asymmetric Information, *Econometrica* 49 (1981), S.33-64.

[2] Hart, O. und B. Holmström: The Theory of Contracts, *Advances in Economic Theory*, T.F. Bewley (Hrsg.), S.71-155.

[3] Milgrom, P. und J. Roberts: Economics, Organization and Management, Englewood Cliffs 1992.

[4] Myerson, R.B.: Incentive Compatibility and the Bargaining Problem, *Econometrica* 47 (1979), S.61-73.

[5] Rasmusen, E.: Games and Information, Cambridge 1991.

[6] Riordan, M.H.: Uncertainty, Asymmetric Information and Bilateral Contracts, *Review of Economic Studies* 51 (1984), S.83-93.

[7] Sappington, D.: Limited Liability Contracts between Principal and Agent, *Journal of Economic Theory* 29 (1983), S.1-21.

[8] Wagenhofer, A.: Verrechnungspreise zur Koordination bei Informationsasymmetrie, *Controlling*, K. Spremann und A. Zur (Hrsg.), Wiesbaden 1992.

Konzipierung von Cafeteria-Systemen als neuere Form der Anreizgestaltung

Thomas Spengler und Matthias Vieth, Frankfurt am Main, FRG

Zusammenfassung: Cafeteria-Systeme dienen der Individualisierung der Anreizgewährung. In der vorliegenden Arbeit werden drei Entscheidungsmodelle zur Konstruktion und Evaluation von Cafeteria-Systemen skizziert: Das erste (zweite) Modell nimmt die Optimierung aus Sicht des Unternehmens (der Arbeitskräfte) vor, das dritte Modell dient der Generierung der optimalen Kompromißlösung.

Summary: Cafeteria-systems are used to individualize gratifications. The present paper considers three decision-models for building and evaluating cafeteria-systems. The first (second) model optimizes in the view of the firm (staff), the third model generates the optimal compromise solution.

1. Problemstellung

Unternehmen gewähren ihren Arbeitskräften eine Vergütung, und zwar nicht nur um ihrer gesetzlichen und vertraglichen Pflicht nachzukommen, sondern auch und vor allem um sie zu mehr oder minder exakt determinierten Verhaltensweisen zu bewegen (Motivationsaspekt). Sofern die Vergütung der Befriedigung individueller Bedürfnisse dienlich ist, verkörpert sie ein Anregungspotential (Konzept eigeninteressengeleiteten Handelns). Zur Bedürfnisbefriedigung fragen die Arbeitskräfte bestimmte Mengen nutzenstiftender Güter am Markt nach. Mit Hilfe eines Cafeteria-Systems der Vergütung beabsichtigt das Unternehmen, das Konzept marktlicher Transaktion für eine Effizienzsteigerung des betrieblichen Anreizsystems nutzbar zu machen. Charakteristikum *marktlicher* Gütertransaktionen ist die Zusammenstellung eines Güterbündels durch die Arbeitskraft unter Beachtung ihres Lohnbudgets, ihrer individuellen Präferenzen sowie des Güterangebots. Im Rahmen eines *Cafeteria-Systems* werden den Arbeitskräften von seiten des Unternehmens verschiedene Anreizarten als mögliche Entgeltbestandteile angeboten, aus denen sie sich das für sie 'optimale Menü' zusammenstellen können [Individualisierung der Anreizgewährung]. Die Optionen der Mitarbeiter werden durch ihren Nettolohn und durch das Cafeteria-Angebot beschränkt /1/, /7/.
Die Verlagerung marktlicher Transaktionen in das Unternehmen in der Form eines Cafeteria-Systems der Vergütung mag u.U. prima facie nicht vorteilhaft erscheinen: Zum einen entstehen dem Unternehmen durch die Installation eines internen Gütermarktes Kosten (z.B. Anreizbereitstellungs-, Verwaltungskosten), die jedoch über die gewählten Entgeltbestandteile als Lohnäquivalente an die Arbeitskräfte weitergegeben werden können. Zum anderen haben die Arbeitskräfte die Alternative, eine Vielzahl von Gütern direkt am externen Gütermarkt nachzufragen. Die Einrichtung des internen Marktes führt somit nur dann zu einer Annäherung an eine Pareto-effiziente Position, wenn die Arbeitskräfte durch das Cafeteria-System im Vergleich zum externen Markt Einsparungen erzielen und das Cafeteria-System für das Unternehmen zu keinen zusätzlichen Anreizkosten (-überhängen) führt /5/. Während das Unternehmen mit der Installation eines Cafeteria-Systems die Realisierung ökonomisch rationaler Leistungsmotivationseffekte

anstrebt, kann sich die Vorteilhaftigkeit der innerbetrieblichen Gütertransaktionen für die Arbeitskräfte aus drei Gründen ergeben:

(1) Das Unternehmen fragt die präferierten Güter am externen Gütermarkt nach und erzielt, im Vergleich zu einer einzelnen Arbeitskraft Kosteneinsparungen durch die Realisierung von Größendegressionseffekten ("economies of large scale"). In diesem Fall müssen diese günstigen Konditionen der Beschaffung am externen Markt an die Arbeitskräfte, zumindest zu einem Teil, weitergegeben werden. Sie erzielen dann einen Nutzenzuwachs durch das Cafeteria-System, indem sie das gleiche Gut am internen Markt für einen geringeren Preis erstehen können. Beispiel: Einem Unternehmen bietet sich die Möglichkeit, einen Firmenwagen zu günstigen Konditionen zu mieten. Diesen Vorteil gibt das Unternehmen an seine Arbeitskräfte weiter; die einzelne Arbeitskraft erzielt bei Wahrnehmung dieses Angebotes einen monetären Vorteil in Höhe der Differenz zwischen dem Preis des Gutes am internen Markt und dem Preis des Gutes am externen Markt.

(2) Das Unternehmen stellt die am externen Gütermarkt angebotenen Güter selbst her und bietet sie den Arbeitskräften zu günstigeren Konditionen als am externen Gütermarkt an. Beispiel: Ein Unternehmen produziert Kraftfahrzeuge, die es seinen Arbeitskräften unter dem Marktpreis anbietet.

(3) Das Unternehmen fertigt nicht marktlich angebotene Güter und bietet sie im Rahmen des Cafeteria-Systems an. Mit diesen Gütern ist aus Sicht der Arbeitskräfte ein Nutzenzuwachs verbunden. Beispiel: Die Arbeitskräfte können in Eigenfertigung erzeugte Kraftfahrzeuge vom Unternehmen erstehen, die über eine Sonderausstattung verfügen und nicht am externen Markt gehandelt werden.

In den folgenden Abschnitten formulieren wir drei Entscheidungsmodelle zur Konstruktion optimaler Cafeteria-Systeme. Im Gegensatz zu den meisten agency-theoretischen Untersuchungen /3/ betrachten wir jedoch nicht einzelne Arbeitnehmer, sondern - wie in der Personalplanung üblich - (homogene) Gruppen von Arbeitskräften (mit jeweils ähnlichen Fähigkeiten). Dabei machen wir es uns insofern einfach, als wir lediglich den deterministischen Fall thematisieren (bzw. die relevanten Daten als Erwartungswerte interpretieren): Modell A dient der simultanen Optimierung des Personalplans und des Cafeteria-Systems, wobei über die Zielfunktion lediglich Unternehmensziele extremiert werden. Modell B optimiert das Cafeteria-System aus der Perspektive der Arbeitskräfte, während Modell C einen integrativen Ansatz darstellt, der das Auffinden des optimalen Kompromisses zwischen Arbeitgeber- und Arbeitnehmerinteressen ermöglichen soll.

2. Die Gestaltung der Cafeteria-Systeme aus Unternehmensperspektive (Modell A)

Das Unternehmen befreit sich durch die Einrichtung eines Cafeteria-Systems von der rigiden Festlegung der Anreizarten und -ausprägungen für die einzelnen Arbeitskräfte(-gruppen) (bounded self-selection). Allerdings muß das Unternehmen für jede Arbeitskräftekategorie eine Entscheidung zwischen den alternativen Cafeteria-Systemen treffen. Ein Problem bei dieser Entscheidung stellt die Antizipation der Wirkungen alternativer Cafeteria-Systeme auf das Leistungsverhalten der einzelnen Arbeitskräfte(-gruppen) dar. Wir betrachten folgende Entscheidungssituation:

(1) Das Unternehmen strebt die vollständige Deckung ihres nach verschiedenen Tätigkeitsarten (q) differenzierten (und als Datum bekannten) Personalbedarfs an.

(2) Zur Deckung der Teilpersonalbedarfe können in begrenztem Umfang Arbeitskräfte verschiedener, nach Qualifikationen differenzierter, Kategorien (r) bereitgestellt werden.

(3) Das Unternehmen prüft, welche Cafeteria-Systeme (j) den Arbeitskräften angeboten werden sollen. Es ist bekannt, welche Cafeteria-Systeme von den verschiedenen Arbeitskräftekategorien akzeptiert werden ($\underline{J}_r$). Die Cafeteria-Systeme unterscheiden sich hinsichtlich der wählbaren Anreizarten[1] und Anreizausprägungen[2] *sowie* hinsichtlich des Cafeteria-Budgets.

(4) Die Arbeitskräfte erhalten einen qualifikationsabhängigen (Grund-)Lohn und - sofern sie im Leistungsprozeß eingesetzt werden - das Angebot, an einem Cafeteria-System j zu partizipieren.

(5) Die Arbeitskräfte weisen einen qualifikationsspezifischen Leistungsfaktor auf, der zusätzlich davon abhängt, welches Cafeteria-System ihnen letztendlich angeboten wird und welche Tätigkeit sie (nach Maßgabe des zu erstellenden Personaleinsatzplans) ausführen (sollen). Die Leistungsfaktoren ergeben sich aus dem Verhältnis von tatsächlich zu erwartender Arbeitsproduktivität und Standardarbeitsproduktivität und werden i.d.R. Werte um Eins annehmen.

(6) Das Unternehmen strebt die Minimierung der Lohn- und der Cafeteria-System-Kosten an.

Zur Formulierung des Modells werden die folgenden Symbole benötigt:

<u>Indizes und Indexmengen:</u>

$\underline{J}$:= {j|j ist ein Cafeteria-System; j = 1,...,J}

$\underline{R}$:= {r|r ist eine Arbeitskräftekategorie; r = 1,...,R}

$\underline{Q}$:= {q|q ist eine Tätigkeitsart; q = 1,...,Q}

$\underline{R}_q$:= {r|r ist eine Arbeitskräftekategorie, die Tätigkeitsart q erledigen kann}

$\underline{Q}_r$:= {q|q ist eine Tätigkeitsart, die von Arbeitskräftekategorie r erledigt werden kann}

$\underline{J}_r$:= {j|j ist ein Cafeteria-System, das Arbeitskräftekategorie r angeboten wird}

<u>Daten:</u>

$\overline{PA}_r^{max}$:= maximale Anzahl der verfügbaren Arbeitskräfte der Arbeitskräftekategorie r

$\overline{PB}_q$:= Anzahl benötigter Arbeitskräfte zur Erledigung der Tätigkeitsart q

M := hinreichend große Zahl

α_{rq}^j := Leistungsfaktor der Arbeitskräftekategorie r bei Cafeteria – System j und Erledigung der Tätigkeitsart q

Φ_r^G := qualifikationsabhängiger Grundlohnsatz für eine Arbeitskraft der Art r

Φ_j^S := Kosten des Cafeteria-Systems j

<u>Entscheidungsvariable:</u>

PE_{rq}^j := Anzahl der eingesetzten Arbeitskräfte der Kategorie r bei Tätigkeit der Art q und Geltung des Cafeteria – Systems j

PA_r := Anzahl der verfügbaren Arbeitskräfte der Kategorie r

$x_{jr} := \begin{cases} 1, & \text{wenn Cafeteria - System j für Arbeitskräfte der Kategorie r gilt} \\ 0, & \text{sonst} \end{cases}$

<u>Zielfunktion</u>

$$\sum_{r=1}^{R} \left[\Phi_r^G \cdot PA_r + \sum_{q \in \underline{Q}_r} \sum_{j \in \underline{J}_r} \Phi_j^S \cdot PE_{rq}^j \right] \overset{!}{=} \min$$

---

[1] z.B. Werkswohnungen, Kinderbetreuungsmöglichkeiten, Versicherungen, Dienstwagen

[2] z.B. 50 m^2 Wohnfläche, 50000 Dienstwagen-Kilometer

u.d.N.:

Abstimmung zwischen Personalbedarf und Personaleinsatz

$$\overline{PB}_q \leq \sum_{r \in \underline{R}_q} \sum_{j \in \underline{J}_r} \alpha_{rq}^j \cdot PE_{rq}^j \qquad \forall\, q \qquad (A.1)$$

Abstimmung zwischen Personaleinsatz und Personalausstattung

$$\sum_{j \in \underline{J}_r} \sum_{q \in \underline{Q}_r} PE_{rq}^j \leq PA_r \qquad \forall\, r \qquad (A.2)$$

Abstimmung zwischen der Personalausstattung und der Personalausstattungsobergrenze

$$PA_r \leq \overline{PA}_r^{max} \qquad \forall\, r \qquad (A.3)$$

Vollständigkeits- und Ausschlußbedingungen

$$\sum_{j \in \underline{J}_r} x_{jr} \leq 1 \qquad \forall\, r \qquad (A.4.1)$$

$$\sum_{q \in \underline{Q}_r} PE_{rq}^{j*} \leq M \cdot x_{j*r} \qquad \forall\, r, j^* \in \underline{J}_r \qquad (A.4.2)$$

$$\sum_{q \in \underline{Q}_r} \sum_{j \in \underline{J}_r \setminus \{j^*\}} PE_{rq}^j \leq M \cdot (1 - x_{j*r}) \qquad \forall\, r, j^* \in \underline{J}_r \qquad (A.4.3)$$

Nichtnegativitätsbedingungen, Ganzzahligkeitsbedingungen

$$x_{jr} \in \{0,1\} \qquad \forall\, r, j \in \underline{J}_r \qquad (A.5.1)$$

$$PE_{rq}^j, PA_r \geq 0 \qquad \text{für alle relevanten } r, q \in \underline{Q}_r, j \in \underline{J}_r \qquad (A.5.2)$$

In den ersten drei Restriktionstypen werden die drei Problembereiche der Personalplanung miteinander abgestimmt (Modell der reinen Personalbereitstellungsplanung /2/). Teilrestriktionssystem (A.4.1),(A.4.2) und (A.4.3) gewährleistet, daß jeder eingesetzten Arbeitskraft (bzw. Arbeitskräftekategorie) genau ein Cafeteria-System angeboten und der entsprechende Leistungsfaktor 'realisiert' wird.

3. Die Gestaltung der Cafeteria-Systeme aus Arbeitnehmerperspektive (Modell B)

Das Unternehmen stellt den Arbeitskräften alternative Cafeteria-Systeme zur Auswahl. Die Arbeitskräfte stehen vor dem Entscheidungsproblem, eines der angebotenen Cafeteria-Systeme zu wählen. Sie beziehen in ihre Überlegungen zum einen die potentiellen Optionen der verschiedenen Cafeteria-Systeme hinsichtlich der jeweiligen Anreizarten und deren Ausprägungen ein (z.B. flexible-benefit-systems oder flexible-benefit-plan /7/). Zum anderen wird die Auswahl des Cafeteria-Systems durch die aus der von der einzelnen Arbeitskraft übernommenen Tätigkeit resultierenden Arbeitsfreude (bzw. Arbeitsleid) beeinflußt. Die Entscheidungssituation der Arbeitskräfte läßt sich wie folgt beschreiben:

(1) Den Arbeitskräften ist bekannt, welche Cafeteria-Systeme $j \in \underline{J}_r$ prinzipiell für sie in Betracht kommen und wie hoch die jeweiligen Budgets sind.

(2) Sie stehen vor dem Problem, sich für eines dieser Cafeteria-Systeme entscheiden zu müssen und die für sie optimalen Ausprägungen der relevanten Anreizarten (i) zu wählen.

(3) Die Arbeitskräfte kennen die Preise, die sie für je eine Einheit der Anreizarten am externen Markt zu zahlen haben.

(4) Jede Arbeitskraft kann den für sie geltenden - vom zugewiesenen Cafeteria-System (j) und der auszuführenden Tätigkeit (q) abhängenden - Arbeitsleid-/Arbeitsfreudekoeffizienten angeben. Diese Koeffizienten sind auf das Intervall [-1;1] normiert und nehmen bei vollkommenem Arbeitsleid (vollkommener Arbeitsfreude) den Wert -1 (1), ansonsten dazwischenliegende Werte an. Zur Verein-

fachung der Notation gehen wir davon aus, daß alle Arbeitskräfte derselben Kategorie (r) auch dieselben Koeffizienten aufweisen.

(5) Die Arbeitskräfte streben die Maximierung der resultierenden Arbeitsfreude an. Darüberhinaus möchten sie das Ausmaß an Einsparungen beim Erwerb von Bedürfnisbefriedigungsmitteln maximieren.

(6) Die Arbeitskräfte können angeben, wieviele Anreizarten mindestens in einem Cafeteria-System enthalten sein sollen, welches Mindestniveau für die Anreizausprägungen gelten sollen und welche Anreizarten wechselseitig inkompatibel sind.

Die bereits erklärten Symbole werden um folgende ergänzt:

Indizes und Indexmengen:

$\underline{I}$:= {i|i ist eine Anreizart; i = 1,...,I}

$\underline{I}_j$:= {i|i ist eine Anreizart, die im Cafeteria-System j enthalten ist}

$\underline{I}_i^*$:= {i*|i* ist eine Anreizart, die nicht gemeinsam mit Anreizart i in einem Cafeteria-System vorkommt}

Daten:

$\overline{B}_j$:= Budgetobergrenze des Cafeteria-Systems j

ε := hinreichend kleine Zahl

β_{rq}^j := Arbeitsleid – / Arbeitsfreudekoeffizient der Arbeitskräftekategorie r bei Cafeteria – System j und Erledigung der Tätigkeitsart q

$\underline{u}$:= von der Arbeitskraft geforderte Mindestzahl an zu gewährenden Anreizarten

$\underline{M}_i$:= von der Arbeitskraft gefordertes Mindestniveau für die Ausprägung der Anreizart i

Φ_i^U := Preis einer Einheit der Anreizart i im Unternehmen

Φ_i^M := Preis einer Einheit der Anreizart i am externen Markt

Entscheidungsvariable:

z_{ij} := Anzahl der Einheiten von Anreizart i bei Cafeteria-System j, die von der Arbeitskraft gewählt werden

$$x_{jr} := \begin{cases} 1, \text{wenn das Cafeteria – System } j \text{ von Arbeitskräften der Art } r \text{ gewählt wird} \\ 0, \text{sonst} \end{cases}$$

$$u_{ij} := \begin{cases} 1, \text{wenn Anreizart } i \text{ bei Cafeteria – System } j \text{ von Arbeitskräften gewählt wird} \\ 0, \text{sonst} \end{cases}$$

Zielfunktionsvektor

$$\begin{pmatrix} \sum\limits_{j \in \underline{J}_r} \sum\limits_{i \in \underline{I}_j} \left(\Phi_i^M - \Phi_i^U\right) \cdot z_{ij} \\ \sum\limits_{j \in \underline{J}_r} \sum\limits_{q \in \underline{Q}_r} \beta_{rq}^j \cdot x_{jr} \end{pmatrix} \stackrel{!}{=} \max$$

u.d.N.:

Budgetrestriktion

$$\sum_{i \in \underline{I}_j} \Phi_i^U \cdot z_{ij} \le \overline{B}_j \cdot x_{jr} \qquad \forall\ j \in \underline{J}_r \qquad \text{(B.1)}$$

Vollständigkeits- und Ausschlußbedingung

$$\sum_{j \in \underline{J}_r} x_{jr} \le 1 \qquad \text{(B.2)}$$

Anreizvielfaltsbedingung

$$z_{ij} \le M \cdot u_{ij} \qquad \forall\ j \in \underline{J}_r, i \in \underline{I}_j \qquad \text{(B.3.1)}$$

$$z_{ij} \geq \varepsilon \cdot u_{ij} \qquad \forall\ j \in \underline{J}_r, i \in \underline{I}_j \qquad \text{(B.3.2)}$$

[3]

$$\sum_{j=1}^{J} \sum_{i \in \underline{I}_j} u_{ij} \geq \underline{u} \qquad \text{(B.3.3)}$$

Anreizverträglichkeitsbedingung

$$\sum_{i^* \in \underline{I}_i^*} z_{i^*j} \leq M \cdot \left(1 - u_{ij}\right) \qquad \forall\ j \in \underline{J}_r, i \in \underline{I}_j \qquad \text{(B.4)}$$

Mindestniveau der einzelnen Anreizarten

$$\underline{M}_i \cdot u_{ij} \leq z_{ij} \qquad \forall\ j \in \underline{J}_r, i \in \underline{I}_j \qquad \text{(B.5)}$$

Nichtnegativitätsbedingungen, Ganzzahligkeitsbedingungen

$$z_{ij} \geq 0 \qquad \forall\ j \in \underline{J}_r, i \in \underline{I}_j \qquad \text{(B.6.1)}$$

$$u_{ij}, x_{jr} \in \{0,1\} \qquad \text{für alle relevanten } j \in \underline{J}_r, i \in \underline{I}_j \qquad \text{(B.6.2)}$$

Jede Arbeitskraft/Arbeitskräftegruppe (r) erstellt diesen Entscheidungskalkül. Restriktionstyp (B.1) beschränkt die Wahlmöglichkeiten der betrachteten Arbeitskraft/Arbeitskräftegruppe durch das verfügbare Budget des gewählten Cafeteria-Systems nach oben, während Restriktionstyp (B.2) die Wahl nur eines Cafeteria-Systems zuläßt. Beschränkungen der Anreizwahlmöglichkeiten innerhalb eines Cafeteria-Systems werden durch das Restriktionsteilsystem (B.3.1) bis (B.5.2) zum Ausdruck gebracht.

4. Kompromißmodell (Modell C)

Während Modell A die Einführung eines Cafeteria-Systems unter (fast) ausschließlicher Berücksichtigung der Unternehmensinteressen betrachtet, fokussiert Modell B die Arbeitnehmerinteressen. Eine sukzessive Anwendung der beiden Entscheidungskalküle wird i.d.R. zu unzulässigen bzw. zu 'second-best' Lösungen führen, da z.B. die über Modell B eruierten Vorstellungen der Arbeitskräfte zum Personaleinsatz nicht notwendigerweise mit dem aus Sicht des Unternehmens optimalen Personaleinsatzplan (Modell A) und die Vorstellungen des Betriebes über die Cafeteria-Angebote nicht mit den Präferenzen der Arbeitskräfte übereinstimmen müssen. In dem folgenden Mehrziel-Optimierungsansatz[4] werden beide Entscheidungskalküle in ein Planungsmodell integriert, mithin ein optimaler Kompromiß angestrebt. Das Symbolverzeichnis wird um folgende Entscheidungsvariable ergänzt:

$$x_{rq}^{j} := \begin{cases} 1, & \text{wenn Arbeitskräftekategorie r bei Erledigung der Tätigkeitsart q Cafeteria-System j erhält} \\ 0, & \text{sonst} \end{cases}$$

Zielfunktionsvektor

$$\begin{pmatrix} -\sum_{r=1}^{R} \left[\Phi_r^G \cdot PA_r + \sum_{q \in \underline{Q}_r} \sum_{j \in \underline{J}_r} \Phi_j^S \cdot PE_{rq}^j \right] \\ \sum_{j=1}^{J} \sum_{i \in \underline{I}_j} \left(\Phi_i^M - \Phi_i^U\right) \cdot z_{ij} \\ \sum_{r=1}^{R} \sum_{j \in \underline{J}_r} \sum_{q \in \underline{Q}_r} \beta_{rq}^j \cdot x_{rq}^j \end{pmatrix} \overset{!}{=} \max$$

u.d.N: (A.1),(A.2),(A.3),(B.3.1),(B.3.2),(B.3.3),(B.4),(B.5) sowie

---

3 häufig redundant wegen (B.5)

4 Aus Platzgründen können wir hier lediglich auf die gängigen Verfahren zur Lösung von Vektormaximummodellen verweisen /8/, empfehlen jedoch die Verwendung eines fuzzy set-basierten interaktiven Nutzenmodells /4/, /6/.

Budgetrestriktion

$$\sum_{i \in \underline{I}_j} \Phi_i^U \cdot z_{ij} \le \overline{B}_j \cdot x_{rq}^j \qquad \forall\ r,q \in \underline{Q}_r, j \in \underline{J}_r \qquad (C.1)$$

Ausschlußbedingungen

$$\sum_{j \in \underline{J}_r} x_{rq}^j \le 1 \qquad \forall\ r,q \in \underline{Q}_r \qquad (C.2.1)$$

$$PE_{rq}^{j*} \le M \cdot x_{rq}^{j*} \qquad \forall\ r,q \in \underline{Q}_r, j^* \in \underline{J}_r \qquad (C.2.2)$$

$$x_{rq}^{j*} \le M \cdot PE_{rq}^{j*} \qquad \forall\ r,q \in \underline{Q}_r, j^* \in \underline{J}_r \qquad (C.2.3)$$

$$\sum_{j \in \underline{J}_r \setminus \{j^*\}} PE_{rq}^j \le M \cdot \left(1 - x_{rq}^{j*}\right) \qquad \forall\ r,q \in \underline{Q}_r, j^* \in \underline{J}_r \qquad (C.2.4)$$

Nichtnegativitätsbedingungen, Ganzzahligkeitsbedingungen

$$x_{rq}^j, u_{ij} \in \{0,1\} \qquad \text{für alle relevanten } r,q \in \underline{Q}_r, j \in \underline{J}_r, i \in \underline{I}_j \qquad (C.3.1)$$

$$PE_{rq}^j, PA_r, z_{ij} \ge 0 \qquad \text{für alle relevanten } r,q \in \underline{Q}_r, j \in \underline{J}_r, i \in \underline{I}_j \qquad (C.3.2)$$

Literatur

/1/ Dycke, Axel/Schulte, Christof: Cafeteria-Systeme.
Die Betriebswirtschaft, 46. Jg. (1986), S. 577-589.

/2/ Kossbiel, Hugo: Personalbereitstellungsplanung bei Arbeitszeitflexibilisierung.
Zeitschrift für Betriebswirtschaft, 62. Jg. (1992), S. 175-198.

/3/ Laux, Helmut/Schenk-Mathes, Heike Y.: Lineare und nichtlineare Anreizsysteme.
Heidelberg: Physica (1992).

/4/ Rommelfanger, Heinrich: Entscheiden bei Unschärfe. Fuzzy Decision Support-Systeme.
Berlin, Heidelberg, New York: Springer (1988).

/5/ Sadowski, Dieter: Der Handel mit Sozialleistungen - Zur Ökonomie und Organisation der betrieblichen Sozialpolitik.
Die Betriebswirtschaft, 44 Jg. (1984), S.579-590.

/6/ Spengler, Thomas: Lineare Entscheidungsmodelle zur Organisations- und Personalplanung.
Heidelberg: Physica (1993).

/7/ Wagner, Dieter: Anreizpotentiale und Gestaltungsmöglichkeiten von Cafeteria-Modellen.
Handbuch Anreizsysteme in Wirtschaft und Verwaltung, hrsg. von Günther Schanz.
Stuttgart: Poeschel (1991), S. 91-109.

/8/ Zimmermann, Hans-Jürgen/Gutsche, Lothar: Multi-Criteria-Analyse.
Berlin u.a.: Springer (1991).

MODELLING AND ALGORITHMS FOR LOCATION PROBLEMS

Christiane Tammer, Martin-Luther-Universität Halle-Wittenberg

This paper deals with several kinds of location problems, necessary optimality conditions and algorithms based on the optimality conditions.

We will study some generalizations of Fermat-Weber location problem, at first the following convex vector-valued L_p-Approximation problem:

$$(P1) \qquad f(u) = \begin{pmatrix} \| \sum_{j=1}^{n} u_j f_j^1 - f_o^1 \| \\ \vdots \\ \| \sum_{j=1}^{n} u_j f_j^m - f_o^m \| \end{pmatrix} \rightarrow \underset{u \in U}{\text{v-min}}$$

with $f_j^i \in L_p(S)$, $f_o^i \in L_p(S)$, $1 \leq p \leq \infty$, [$i = 1, \ldots, m$], $S \subset R^q$ compact, $U = R^n$.

Since the vector optimization problem (P1) is convex, we can use a scalarization with linear continuous functionals. Applying a variational principle we get necessary conditions for approximately efficient points of (P1) by using the subdifferential calculus. This necessary conditions we will use in order to derive a proximal point algorithm for the vector optimization problem (P1).

Further, we will discuss a population model, which can be described as a nonconvex vector-valued location problem. So we can't apply the results of convex analysis. We will show relations between the efficient solutions of the vector optimization problem and a surrogate scalar optimization problem. By using the directional derivative of the objective functions and a variational principle we derive necessary conditions for approximately efficient solutions.

TRANSFER PRICING UNDER ASYMMETRIC INFORMATION - AN EVALUATION OF ALTERNATIVE METHODS

Alfred Wagenhofer, University of Graz, Austria

This paper analyzes optimal transfer prices in a simple setting with two divisions. One division receives a special order. In order to fulfill it an upstream division has to supply an intermediate good. The production costs of the divisions are their respective private information. The objective of headquarters is to induce the divisions to take decisions which maximize firm value. It uses transfer prices to coordinate these decisions.

Cost-based, market-based, and negotiated transfer prices are compared whether they achieve optimal coordination, and whether they are undominated by the other methods. Market-based transfer prices fail to achieve optimal coordination among the divisions if there are sufficiently large production economies. If this is the case, the other methods may do better. Cost-based prices are shown to belong to the class of marginal cost plus a profit margin. The profit margin essentially gives the low cost divisions an incentive not to pretend to have high costs. Negotiation is discussed for two stylized settings, one in which one division is assigned all the bargaining power, and the other with equal divisional bargaining power.

Two main results are shown. (i) Cost-based prices always perform better than negotiated transfer prices, essentially because conflicting interests may lead divisions to depart from achieving the trade which is best from the view of the firm. (ii) Otherwise, the transfer pricing methods cannot be ranked with respect to achieving the maximization of firm value. Which method does better depends on the cost information held by the divisions, and the market prices for the intermediate good. The validity of these results under more general settings is discussed.

AHP IN DER PRAXIS

Karl Weber, Justus-Liebig-Universität Gießen

Unter den multikriteriellen Entscheidungsmethoden erweist sich der AHP-Ansatz (Analytic Hierarchy Process) nicht nur als theoretisch bedeutsam, sondern auch als praxisrelevant.
Nach einer Kurzdarstellung der AHP-Grundkonzepte und der zur computergestützten Modellbearbeitung verfügbaren Softwarepakete (AutoMan, Criterium 1.1, Expert Choice 8.0, Newtech Expert Choice) werden Beispiele aus den verschiedensten Wissensbereichen behandelt, wobei auch Besonderheiten der benutzten Softwarepekete verdeutlicht werden.
Die meisten Beispiele stammen aus dem Bereich der Wirtschaftswissenschaften. Aus der Betriebswirtschaftslehre werden AHP-Anwendungen im Absatz-, Produktions- und Verwaltungsbereich behandelt; dabei gelangen der AHP-Grundansatz und Varianten (Festwert-, Festmodell-Verfahren) zum Einsatz.
Die Modelle aus dem Gebiet der Volkswirtschaftslehre verdeutlichen den AHP-Einsatz unter Beizug unterschiedlicher Expertengruppen sowie Möglichkeiten zur Differenzierung zwischen Kosten-/Nutzenanalysen (Verkehrswirtschaft) und zur stufenweisen Problembearbeitung unter Verwendung des Supermatrix-Verfahrens (Wirtschaftsprognostik).
Der AHP-Einsatz auf dem Gebiet der Wirtschaftsinformatik führt zur AHP-gestützten Evaluation von AHP-Softwarepaketen. Weiterhin kann AHP als Entscheidungshilfe zur Selektion von OR-Software benutzt werden, was ebenfalls anhand eines praktischen Beispiels gezeigt wird.
Schließlich wird auch ein AHP-Beispiel aus der Politik (US Präsidentschaftswahlen) diskutiert und auf andere - in der Literatur erwähnte - Einsatzbereiche (Hauswirtschaft, Bildung/Berufswahl, Medizin, etc.) verwiesen.

Literatur:

Weber, Karl: Multikriterielle Entscheidungen. München: Oldenbourg, 1993.

THE PARTIAL SUM CRITERION FOR STEINER TREES IN GRAPHS AND SHORTEST PATHS

Cees W. Duin and Anton Volgenant
University of Amsterdam, Amsterdam

Many optimization problems are defined in a weighted graph. Generally, the criterion is min sum, i.e., the total weight of all edges in the solution; here we consider a different criterion.

The partial sum criterion with parameter p sums the largest p edge weights in the solution, giving the criterion value of the considered solution.

For $p = 1$ the criterion is known as bottleneck or minmax criterion. For the Steiner Tree Problem in Graphs - see for this problem Duin and Voss (1994) - with minmax criterion we describe in Duin and Volgenant (1993) an $O(|E|)$ algorithm with E the set of edges in the problem. The algorithm can be extended easily to the min-max Steiner Forest Problem with a restriction on the number of components - see Duin and Volgenant (1987).

For special choices of K (the set of special nodes in this problem) the algorithm solves the bottleneck shortest path problem corresponding with $|K| = 2$, Punnen (1991), and the bottleneck spanning tree problem corresponding with $|K| = n$, Camerini (1978); thereby it unifies these two existing algorithms.

We show for larger values of p that the partial sum Steiner Trees in Graphs problem is NP-hard; this result holds both for the *connected* case, where the p chosen edges must form a subtree as well as for the case, where the set of these edges need not be connected.

For the shortest path problem we consider the criterion for arbitrary values of p, defining it for solutions with less than p edges as the total sum. The connected version of the problem appears to be NP-hard, while for the free case a polynomial algorithm is derived, that is related to the algorithm for the p-sum Linear Assignment Problem as given by Grygiel (1981).

Applications are present in communication networks. Indirect applications arise within other algorithms, e.g., the minmax path algorithm can be used as a subroutine within a minmax Linear Assignment Problem. The partial sum criterion can be useful for instance in problems in location theory or as a criterion in fitting a line to given data points, minimizing the 'distance' of p outlyers to the line through the data points.

Steiner Tree Heuristics - A Survey

Cees Duin, Amsterdam

Stefan Voß, Darmstadt

Abstract

Given a graph with weights on its edges Steiner's problem in graphs is to determine a minimum cost subgraph spanning a set of specified vertices. Real world problems arising in the layout of connection structures in networks as, e.g., in topological network design or VLSI design may often be decomposed into a number of well known combinatorial optimization problems. Steiner's problem in graphs is included within this context. According to its complexity one is interested in developing efficient heuristic algorithms to find good approximate solutions. In this paper we give a survey on recent results on heuristic methods for the problem.

Zusammenfassung

Das Steiner-Problem in Graphen besteht darin, zu einem gegebenen kantenbewerteten Graphen einen Teilgraphen zu bestimmen, der eine vorgegebene Teilmenge der Knotenmenge kostenminimal aufspannt. Praktische Probleme bei der Konstruktion von Netzwerken oder im Bereich des VLSI-Design lassen sich oftmals in bekannte kombinatorische Optimierungsprobleme einschließlich des Steiner-Problems in Graphen dekomponieren. Aufgrund der Komplexität des Problems ist man an der Entwicklung effizienter Heuristiken interessiert. In dieser Arbeit geben wir einen Überblick über aktuelle Ergebnisse bei derartigen Verfahren zur Lösung des Problems.

1 Introduction

Consider an undirected connected graph $G = (V, E)$ with vertex set V, edge set E, and nonnegative weights associated with the edges. Given a set $Q \subseteq V$ of basic vertices *Steiner's problem in graphs* (**SP**) is to find a minimum cost subgraph of G such that there exists a path in the subgraph between every pair of basic vertices. In order to achieve this minimum subgraph additional vertices from the set $S := V - Q$, so-called Steiner vertices, may be included. Since all edge weights are assumed to be nonnegative there is an optimal solution which is a tree, a so-called *Steiner tree.*

Correspondingly, *Steiner's problem in directed graphs* (**SPD**) is to find a minimum cost directed subgraph of a given graph that contains a directed path between a root node and every basic vertex. Note, that any instance of SP may easily be transformed into an instance of SPD by replacing each undirected edge (i, j) with two arcs $[i, j]$ and $[j, i]$ directed opposite each other both having the same weight as the original edge.

Two well known special cases of SP are polynomially solvable. If $|Q| = 2$ the problem reduces to a shortest path problem and in the case where $Q = V$ SP reduces to the *minimum spanning tree* (**MST**) problem. However, in the general case SP is an NP-hard problem. For a comprehensive survey on SP the reader is referred to [17, 18, 42, 49].

According to SP's complexity one is interested in developing efficient heuristic algorithms to find good approximate solutions. Up to now there is a large amount of references with respect to heuristics for both problems, SP as well as SPD. For a comprehensive survey including extensive numerical results the reader is referred to Voß [43]. Nevertheless, there is still much effort going on for improved ideas in Steiner tree heuristics.

In the first part of this paper (see Section 2) we present a survey on existing methods including reduction techniques, deterministic exchange algorithms (compare also Section 3) and worst case analysis results.

The main drawback of algorithms such as deterministic exchange procedures is their inability to continue the search upon becoming trapped in a local optimum. This suggests consideration of recent techniques for guiding known heuristics to overcome local optimality. Following this theme, in the second part of the paper we investigate the application of local search techniques such as the tabu search metastrategy for solving Steiner's problem in graphs (Section 3). Finally, some conclusions are drawn (Section 4).

2 Finding Initial Feasible Solutions

In the sequel we survey existing heuristics as well as some new ideas for SP. First a general frame is presented together with two representatives having gained considerable interest within the literature. Then, by reviewing ideas of heuristic measure a unified approach is obtained (Section 2.2). Different aspects of comparability are treated in Sections 2.3 and 2.4. Further enhancements show the potential of additional topics within Steiner tree heuristics (Section 2.5). Finally, we consider SPD.

2.1 Build by Shortest Path Insertion

Before describing some heuristics for SP as well as SPD in detail we consider a classification scheme that helps to establish some common principles as well as some dissimilarities within different algorithms (cf. [43]).

A central theme to all heuristics is the use of some principles known from algorithms to find minimum spanning trees together with the computation of shortest paths. The general idea is to build up a feasible solution by insertion of shortest paths. Two methods may be considered:

- **Single COMPONENT EXTENSION**
  Start with a *partial solution* $T = (\{w\}, \emptyset)$ consisting of a single basic vertex w, the so-called root chosen arbitrarily from Q, if not stated otherwise. Extend T to a feasible solution by successive insertion of at most $|Q|$ shortest paths to basic vertices. This method may correspond to the proceeding of Prim's MST algorithm [35].

- **COMPONENT CONNECTING**
  Start with a *partial solution* $T = (Q, \emptyset)$ consisting of $|Q|$ singleton components. Expand T to a feasible solution by repeatedly selecting components which are connected by shortest paths. When in each step two nearest components are connected this method corresponds to the proceeding of Kruskal's MST algorithm [25].

By distinguishing the two main topics COMPONENT EXTENSION and COMPONENT CONNECTING some kind of classification of the algorithms is given. The first may be referred to as that special case of the second one where all components except one must remain singleton. Besides a specific partial solution to start with different heuristics arise according to the number of build up steps used to obtain a feasible solution T. Most frequently one of the following rules is applied:

- **1BASIC:** Build by single component extension (e.g. cheapest insertion).

- **2BASIC:** Given two components (a single basic vertex not yet part of a partial solution may be viewed as a component) connect them by a shortest path.

- **3BASIC:** Given three components connect them by use of some shortest paths.

Of course, one may connect more than three components at a time, too. Furthermore, heuristics differ in the amount of knowledge used in the computation for calculating a feasible solution, e.g., using accurate or approximated shortest path distances. Together with the specific implementation using COMPONENT EXTENSION or COMPONENT CONNECTING and including one of the building blocks described above we get different algorithms. In the sequel we present two well-known algorithms, namely a cheapest insertion method (CHINS) and the shortest distance graph heuristic (SDISTG). The first approach gives special emphasis to the basic vertices as well as the partial solution under development whereas the second approach uses only the basic vertices as its knowledge base. Takahashi and Matsuyama [41] where the first to describe a cheapest insertion method whereas SDISTG has been independently developed by several authors [5, 24, 31].

The following notation will be used: d_{ij} is the weight of a shortest path $P(i,j)$ between nodes i and j in G. Furthermore, $d(i,T) = \min\{d_{ij} | j \in V_T\}$ where V_T denotes the vertex set of T. A vertex of degree 1 in T is called leaf. Whenever ties occur they are broken arbitrarily if not stated otherwise.

CHINS: *Cheapest insertion*

(1) Start with $T = (\{w\}, \emptyset)$ consisting of a single basic vertex w denoted as the root.

Repeat

(2) Find nearest vertices $v* \in V_T$ and $p*$ being a basic vertex not in V_T,
i.e. $d_{v*p*} = min\{d_{vp} | v \in V_T, p \in Q - V_T\}$.

(3) Add the vertices and edges of $P(v*, p*)$ to T.

until T contains all basic vertices.

One of the most interesting aspects of CHINS is the influence of how the root w is chosen. A repetitive approach results in the following modified algorithm called **CHINS-Q**: *Cheapest insertion (all roots)*. CHINS is applied for every basic vertex defined as root w and the cheapest of all evaluated solutions gives the overall solution. Correspondingly, in **CHINS-2Q** CHINS is applied for every pair p, q of basic vertices where the root component is initialized as $P(p,q)$ and again the overall solution is the cheapest of all solutions obtained. In addition in **CHINS-V** all basic vertices and all Steiner vertices are assumed to be root vertices within CHINS once. Further modifications including additional repetitive applications of CHINS as well as incorporating 3BASIC into the framework of CHINS are possible and can be found in [50] and [43], respectively (compare Section 2.5).

SDISTG: *Shortest distance graph*

(1) Construct the complete graph with vertex set Q and each edge having the weight of a shortest path between the corresponding vertices in G.

(2) Construct an MST of this (shortest distance) graph.

(3) Replace each edge of the tree by its corresponding shortest path in G.

(4) While maintaining feasibility (i.e. there exists a path between every pair of basic vertices) delete edges and Steiner vertices such that no cycles exist and all leaves are basic vertices.

If $T = (V_T, E_T)$ is the resulting graph of Step (3) then Step (4) of SDISTG may be further specified. This indeed is the first improvement procedure that comes into mind when considering SP, see e.g. [36].

(4.1) Construct an MST $T' = (V'_T, E'_T)$ of the subgraph of G induced by V_T.

(4.2) **While** there exists a leaf of T' being a Steiner vertex **do** delete that leaf and its incident edge.

SDISTG has received considerable attention in the literature with respect to an efficient implementation using sophisticated data structures (cf. [23, 28, 48, 52]). When constructing the complete graph for vertex set $Q \cup S'$ for all subsets $S' \subseteq S$ with $0 \leq |S'| \leq k$ in Step (1) we get another heuristic method (cf. [31, 40]). For $k = |Q| - 2$ this results in an enumerative approach to solve SP exactly (with running time at least exponential). For $|Q| = 3$ and $k = 1$ we get a polynomial algorithm 3BASIC, which is useful as a subroutine in a component connecting approach: Determine with 3BASIC for every triple of components in the partial solution the cost to connect these three components. Execute the connection of lowest cost by corresponding path insertion (decreasing the number of components by two) and repeat these two steps until a feasible solution is obtained. (A somewhat different description of 3BASIC is given in [3]).

2.2 Heuristic Measure: A Unified Approach

If the vertex set V_T of an optimal solution is known for given data of SP the remaining problem consists of polynomially computing an MST of the subgraph of G induced by V_T. The set $V_T - Q$

of Steiner vertices may be determined approximately by means of a heuristic function f (see e.g. [12, 36]). The value $f(v)$ is used to express numerically the attractiveness of a vertex v to be included into a solution (i.e. acting as a Steiner vertex for $v \in S$). A general description of a corresponding heuristic is as follows.

HEUM: *Heuristic measure*

(1) Start with $T = (Q, \emptyset)$ comprising $|Q|$ basic vertices (subtrees of the final subgraph).
While T is not connected **do**
(2) Choose a vertex v using a heuristic function f.
(3) Unite the two components of T which are nearest to v by combining them with v via shortest paths (the vertices and edges of these paths are added to T).
endwhile

Generally the heuristic function f will be minimized where all vertices may come into consideration independently of either being an element of V_T or not. By choosing a suitable heuristic function f, e.g. f_1 below, SDISTG may be viewed as a special form of HEUM. Likewise CHINS is obtained by measure f_2. Note that in f_2, opposite to f_1, the non-basic vertices in T_0 are active.

A more general function that has been proposed incorporating the COMPONENT CONNECTING approach is as indicated by f_3 (cf. [12, 36, 43]). Intuitively $f(v)$ gives a measure of proximity of v to elements of a subset of vertices in T, or the least average distance of v to a set of vertices in T, respectively.

$$f_1(v) := \min_{1 \leq i \leq \sigma} \{d(v, T_0 \cap Q) + d(v, T_i \cap Q) | T_0, ..., T_\sigma \text{ are the components of } T\}$$

$$f_2(v) := \min_{1 \leq i \leq \sigma} \{d(v, T_0) + d(v, T_i) | T_0, ..., T_\sigma \text{ are the components of } T \text{ s.t. } w \in T_0\}$$

$$f_3(v) := \min_{1 \leq t \leq \sigma} \{\frac{1}{t} \sum_{i=0}^{t} d(v, T_i) | T_0, ..., T_\sigma \text{ are the components of } T \text{ s.t. } d(v, T_i) \leq d(v, T_j), 0 \leq i < j \leq \sigma\}$$

2.3 Numerical Results

Recently, it turned out that a specific set of test problems available through e-mail (see [2]) has been considered by several authors. Although these problems tend to be very sparse graphs we will adopt them here for reason of comparability. There are currently four sets of problems available with different numbers of vertices and edges ranging from 50 to 2500 nodes and 63 up to 65000 edges with integer edge weights as well as various numbers of basic vertices (see the corresponding values for two of the four data sets in Tables 1 - 2). For all problems but one (E-18 of the fourth data set) the optimal solution values are known (and given in columns Z_{opt}, see e.g. [4, 26]). For an algorithm computing problem instances with known optimal solutions see [22].

Numerical results on all different Steiner tree heuristics are a dime a dozen. For a comprehensive comparison of recent heuristics see [43, 50, 36]. Summarizing the numerical results in these sources with respect to the heuristics described above SDISTG is clearly outperformed by CHINS and HEUM with the function f_3, for instance. Furthermore, various modifications and repetitive applications of CHINS and HEUM as well as the dual ascent procedure of Wong [51] (see Section 2.6 and [43]) turn out to produce even more competitive results with respect to solution quality.

In Tables 1 - 2 we present numerical results e.g. for CHINS-Q indicating that the results are quite close to optimality, especially for small graphs. We added some results for CHINS-Q as well as HEUM (with f_3) combined with a simple posterior improvement procedure (as in Step (4) of SDISTG, see Section 3). Note that the time complexity of HEUM with f_3, $O(|V|^3)$, is not competitive with that of CHINS-Q if $|Q| \ll |V|$ (see Section 2.5). In each table the results are summarized by giving the average and maximum deviation, respectively, as well as the number of optimal solutions obtained. The programs are coded either in FORTRAN or PASCAL and run on a 486 PC. (Additional entries of the tables will be described below.)

problem	| V |	| E |	| Q |	Z_{opt}	CHINS-Q	CHINS-Q-I	HEUM-I	CHINS-LS	CHINS - REM
B - 1	50	63	9	82	82	82	82	82	85 - opt (1)
B - 2			13	83	84	83	83	83	84 - opt (6)
B - 3			25	138	138	138	138	138	138 - opt (0)
B - 4	50	100	9	59	59	59	62	59	59 - opt (0)
B - 5			13	61	62	61	61	61	62 - opt (1)
B - 6			25	122	125	124	124	124	127 - opt (9)
B - 7	75	94	13	111	111	111	111	111	111 - opt (0)
B - 8			19	104	104	104	104	104	104 - opt (0)
B - 9			38	220	220	220	220	220	222 - opt (8)
B - 10	75	150	13	86	86	86	90	86	90 - opt (1)
B - 11			19	88	90	90	90	90	92 - opt (6)
B - 12			38	174	174	174	174	174	174 - opt (0)
B - 13	100	125	17	165	168	165	172	174	178 - 170 (27)
B - 14			25	235	238	235	236	235	239 - opt (4)
B - 15			50	318	318	318	318	318	322 - opt (1)
B - 16	100	200	17	127	130	127	129	127	134 - opt (13)
B - 17			25	131	131	131	132	132	131 - opt (0)
B - 18			50	218	220	219	219	220	222 - 219 (2)
average deviation (in %)				0	0.78	0.24	1.17	0.61	0.19
maximum deviation (in %)				0	2.5	2.2	4.7	5.5	3.03
optimal solutions				18	10	15	9	13	16

Table 1: Numerical results for data set B

problem	| V |	| E |	| Q |	Z_{opt}	CHINS-Q	CHINS-Q-I	HEUM-I	CHINS-LS	CHINS - REM
C - 1	500	625	5	85	85	85	85	87	85 - opt (0)
C - 2			10	144	144	144	144	144	144 - opt (0)
C - 3			83	754	762	761	759	755	762 - opt (3)
C - 4			125	1079	1090	1084	1080	1083	1099 - 1082 (120)
C - 5			250	1579	1588	1581	1580	1580	1592 - opt (3)
C - 6	500	1000	5	55	55	55	55	55	55 - opt (0)
C - 7			10	102	102	102	108	102	102 - opt (0)
C - 8			83	509	519	513	511	511	520 - 510 (11)
C - 9			125	707	720	713	713	710	724 - 710 (163)
C - 10			250	1093	1108	1098	1099	1096	1112 - opt (11)
C - 11	500	2500	5	32	34	34	34	33	34 - 33 (1)
C - 12			10	46	46	46	48	46	48 - 47 (1)
C - 13			83	258	266	263	261	264	269 - 260 (26)
C - 14			125	323	331	328	324	326	333 - 324 (71)
C - 15			250	556	562	558	559	556	566 - opt (11)
C - 16	500	12500	5	11	11	11	12	12	13 - 12 (3)
C - 17			10	18	19	19	18	19	19 - 19 (0)
C - 18			83	113	125	122	123	124	123 - 120 (1)
C - 19			125	146	159	157	156	152	159 - 154 (3)
C - 20			250	267	268	268	292	268	269 - 269 (0)
average deviation (in %)				0	2.13	1.58	2.76	1.96	1.75
maximum deviation (in %)				0	13.3	10.6	9.3	9.7	9.1
optimal solutions				20	6	6	4	5	8

Table 2: Numerical results for data set C; $\mid V \mid = 500$

2.4 Comparability and Worst Case Analysis Results

Comparing different Steiner tree heuristics has been part of experimental as well as theoretical research for a long time. In this chapter we survey some of the results in this area including the recent breakthrough of finding a Steiner tree heuristic with a worst case bound strictly smaller than 2.

Let Z_i and Z_j denote the objective function values derived by algorithm A_i and A_j, respectively. A_i and A_j are called *incomparable* if instances with $Z_i < Z_j$ as well as instances with $Z_i > Z_j$ exist. If A_i and A_j are incomparable independently of how ties are broken we call them *strongly incomparable*. For the algorithms above, namely SDISTG and CHINS we get (cf. [48]) that they are strongly incomparable. However, this definition of incomparability does not give any idea on which algorithms could be preferred against others because even the simplest ones **SPATH** (i.e. computing the shortest path from a root w to any basic vertex) and **MST+P** (i.e. computing a minimum spanning tree on G and then successively pruning any Steiner vertex with degree one, cf. Step (4.2) of SDISTG) have to be added to this list of strongly incomparable algorithms (cf. [42]). For further results the reader is referred to [33, 34].

Alternatively algorithms can be compared with respect to their worst case performance. Unfortunately, nearly all of the heuristics given in the literature (including CHINS and SDISTG) have a worst case bound of 2, i.e., $Z_T/Z_{opt} \leq 2(1-1/|Q|)$, where Z_T and Z_{opt} denote the objective function values of a feasible solution and an optimal solution, respectively (see e.g. [5, 24, 31, 32, 33, 43, 47]).

Combining 3BASIC with SDISTG might lead to the following algorithm.[1]It successively applies 3BASIC to every three-element subset of Q (or later a modified set of basic vertices) and then it effectively contracts the most promising of such subsets, to eventually obtain a new Steiner vertex after each contraction.

SDISTG-3: *Shortest distance graph with 3BASIC*

(1) Start with $T = (Q, \emptyset)$ and denote by $T_0, ..., T_\sigma$ the components of T; $S' := \emptyset, r := 1$.

While ($\sigma \geq 2$ **and** $r > 0$) **do**

(2) **For all** three element subsets $\{i, j, k\}$ of indices from $\{0, \ldots, \sigma\}$ **do**
Apply 3BASIC to obtain a vertex v_{ijk} minimizing the heuristic function $f_4(v) := \sum_{h\in\{i,j,k\}} d(v, T_h \cap Q)$ and obtain T_{ijk} from T by uniting T_i, T_j and T_k (by combining them with each other and v_{ijk} through 3BASIC). Let $Z_{MST}(T)$ be the length of an MST of the complete graph with the vertex set corresponding to the components of T and edge weights $d(T_p, T_q) := \min\{d_{ij} | i \in T_p \cap Q, j \in T_q \cap Q\}$. $Z_{MST}(T_{ijk})$ is given, correspondingly.

(3) Define $r := \max\{Z_{MST}(T) - Z_{MST}(T_{ijk}) - f_4(v_{ijk}) | \forall \{i, j, k\} \subseteq \{0, \ldots, \sigma\}\}$ and $(ijk)*$ as *argmax* of subsets $\{i, j, k\}$ leading to r.
If $r > 0$ **then** $T := T_{(ijk)*}, S' := S' \cup \{v_{(ijk)*}\}$.

endwhile

(4) Apply SDISTG to the problem with modified set $Q \cup S'$ of basic vertices.

The main idea of SDISTG-3 is to compute, based on some heuristic measure, a sequence of promising Steiner vertices that have to be included into the solution and then to apply SDISTG on this modified set of basic vertices. Recently, Zelikovsky [53] proved that this algorithm has a worst case bound of 11/6 instead of 2. For further references on recent worst case results see [7, 54].

2.5 Enhancements and Further Algorithms

In this section we briefly give some ideas on additional topics that have been treated previously or are part of an additional study still under way. This includes mentioning some reduction techniques as well as further developments mainly based on the principle idea of CHINS-Q and HEUM.

[1]Due to the numerical results obtained by SDISTG, without considering these recent theoretical achievements it is not quite clear why SDISTG has received so much attention in the literature. Note, that a combination of 3BASIC with different algorithms for SP, especially HEUM and CHINS, has already been proved to be a reasonable approach. The heuristic function f_4 used in SDISTG-3 is a somehow restricted version of a function used in [3, 43].

Reduction Techniques

In a large number of investigations (see [1, 8, 9, 42, 50]) it has been shown that a very important step towards solving SP is to perform some preprocessing routines. Among these routines are the following: Any vertex $v \in S$ of degree 1 can be removed and any vertex $v \in S$ of degree 2 can be replaced by a single edge. Any edge with its weight larger than a shortest path between its endvertices can be removed. If an edge between two basic vertices appears in an MST on the entire graph then it belongs to an optimal solution and the graph can be contracted along this edge.

Though quite simple, the above mentioned rules may already be effective. Applying them together with some more sophisticated rules described in [8, 9] we obtain, e.g., the following results: All 18 problems of data set B (cf. Table 1) can be solved to optimality solely by applying reduction techniques.

Running Time Improvements

Considering naive implementations of the above mentioned algorithms we obtain the following complexities: $O(|Q| \cdot |V|^2)$ for CHINS, $O(|Q|^2 \cdot |V|^2)$ for CHINS-Q and $O(|Q|^3 \cdot |V|^2)$ for CHINS-2Q as well as $O(|Q| \cdot |V|^2)$ for SDISTG and $O(|Q|^3 \cdot |V| + |V|^3)$ for SDISTG-3.

With respect to clever implementations using sophisticated data structures these running times can be reduced (see e.g. [28, 54] for SDISTG and SDISTG-3). An overall idea in this respect is to use efficient shortest path and MST computations as e.g. presented by Gabow et al. [13]. With respect to the repetitive applications of CHINS, however, we can reduce the complexity in a different setting by simply rearranging some calculations. We explain this for CHINS-Q (the same principle applies to CHINS-2Q).

First, observe that only shortest paths between vertices of V and Q are necessary during Step (2) of CHINS. For CHINS-Q we can calculate all such shortest paths once at forehand. With all such paths available CHINS consumes only $O(|Q| \cdot |V|)$ time (cf. [11]). Therefore, for CHINS-Q (and also for the better variant CHINS-V) a time complexity equal to that of a single pass of CHINS follows.

Pilot Heuristic

To our opinion one of the weaker aspects of CHINS and similar heuristics is the fact that new paths are selected only on the basis of shortest distances without evaluating the corresponding paths themselves. The algorithms fail to judge all intermediary nodes on these paths as junction nodes. Only after inclusion one hopes to exploit these connective qualities.

In this respect HEUM with function f_3 is a first improvement. Another idea is to incorporate some sort of tie breaking rule whenever more than one shortest path of the same length exists (cf. [11, 21]). However, it seems to be desirable that an evaluation function accounts for overlaps between paths and calculates in advance some connective consequences of including the paths associated with the evaluated node. This can be done by a look ahead method or a so-called pilot heuristic (see Duin and Voß [11]), i.e., the impact of the vertices will be evaluated by additional Steiner tree calculations before inserting them. At any time within a component connecting approach, the partial solution $T = (V_T, E_T)$ induces an alternative smaller SP: assume the remaining components as basic vertices, while the non-basic vertex set is $V - V_T$. In this SP we can tentatively run, as a heuristic measure, CHINS for various root nodes. When this is done for all possible root nodes while applying 2BASIC the complexity order is $O(|Q|^2 \cdot |V|^2)$.

With respect to numerical results different faster variants of the pilot heuristic proved to be successful. For instance, one variant which still retains the time complexity of $O(|Q| \cdot |V|^2)$ solves 35 out of 38 of the B-data and C-data to optimality (with a maximum deviation of 0.9% for problems C-3, C-13 and C-18).

2.6 Finding Steiner Arborescences

Obviously, finding Steiner arborescences, i.e. feasible solutions for SPD, is a more general problem than SP having additional real world applications. Nevertheless, it has not yet received the same amount of interest as its undirected version. Because, according to e.g. Chopra et al. [4] and [42], the directed version of a mathematical formulation has some advantages over its undirected counterpart it is believed that SPD will achieve more interest in the future.

Wong [51], as SPD's most influencing source describes a dual ascent algorithm based on a multicommodity network flow formulation of SPD. The dual ascent procedure produces as a side result reduced arc costs. The arcs with reduced costs zero define a ***support graph***, a subgraph G' of G of promising arcs. Good upper bounds are obtained by running heuristics previously mentioned in this support graph. (Even naive algorithms like MST+P are then efficient; cf. [43]).

Subsequently different ideas including reduction techniques and generalizations of algorithmic ideas from SP have been developed (see [10, 27, 42]). In [42, 45] the worst case performance of several heuristics for SPD including directed versions of CHINS, SDISTG and SPATH is investigated. With respect to this criterion it is shown that none of the described algorithms for SPD dominates any other. Contrary to SP even one of the simplest algorithms has the same worst case error ratio as the more sophisticated ones. Indeed, the best bound which can be given for any of the heuristics considered is $Z_T/Z_{opt} \leq |Q|$, which is quite disencouraging. The proof idea is simply that the length of a shortest path from the root w to any basic vertex is at most Z_{opt}, i.e. $Z_T \leq |Q| \cdot Z_{opt}$. Furthermore, examples can be found to show that this bound is tight. Finally, it should be noted that all of these algorithms are strongly incomparible.

3 Improvement Procedures

Recently some improvement procedures have been investigated within the framework of SP. The first idea is some kind of pruning routine as in **MST+P** (cf. Step (4.2) of SDISTG):

(1) Let $T = (V_T, E_T)$ be a feasible solution of SP. Construct an MST T' of the subgraph of G induced by V_T.

(2) **While** there exists a leaf of T' being a Steiner vertex **do** delete that leaf and its incident edge.

Tables 1 - 2 show that with this simple approach considerable improvements are possible, i.e., column **CHINS-Q-I** gives the best results of CHINS-Q with the pruning routine applied to all $|Q|$ solutions encountered (for further results see [36, 43]). Following ideas from Section 2.5 we may incorporate deterministic path exchanges into insertion methods. In the sequel we follow a slightly different concept in the sense that posterior exchanges will be performed with respect to a given feasible solution. First two different neighbourhood definitions are distinguished and then their application within the framework of recent metastrategies is reported.

3.1 Neighbourhood Definition

With respect to a proper neighbourhood definition for SP two main ideas may be distinguished:

Edge-oriented Transformation

(1) Given a feasible solution $T = (V_T, E_T)$ randomly select an edge $(i, j) \in E_T$ and delete it from E_T.

(2) **While** there exists a leaf of T being a Steiner vertex **do** delete that leaf and its incident edge.

(3) Let T_i and T_j be the two disconnected components of T. Find nearest vertices $i*$ and $j*$ belonging to T_i and T_j, respectively, and add the vertices and edges of $P(i*, j*)$ to T (eventually disallowing to use (i, j)).

Node-oriented Transformation

(1) Given a feasible solution $T = (V_T, E_T)$. Let $(x_1, \ldots, x_{|S|})$ be a binary vector with x_i denoting that i is included ($x_i = 1$) or excluded ($x_i = 0$) from the solution.

(2) Choose a vertex $i \in S$ and define $x_i := 1 - x_i$. Recompute an MST of the graph induced by $Q \cup \{i \in S | x_i = 1\}$.

Within the literature usually an edge-oriented transformation is described (see e.g. [19, 37]). In [11] an edge oriented transformation is performed during the build up steps of a cheapest insertion method with $O(|V|\log|V| + |E| + |Q| \cdot |V|)$ time complexity. In Tables 1, 2 the results are reported in columns **CHINS-LS**.

An exception is the approach of Minoux [29] who successively adds vertices as long as they give an improvement to the solution under consideration. Also related to the node-oriented transformation is the approach within SDISTG-3, i.e., choosing additional Steiner vertices as long as shortcuts seem to be possible. Comparing the effort of evaluating all possible exchanges within either of the neighbourhoods shows some advantage for the node-oriented approach. It should be noted, however, that the latter approach allows for some infeasibilities whenever the induced graph in Step (2) is not connected (in that case the objective function value should be ∞).

3.2 Recent Metastrategies

Given a feasible solution for SP or SPD both transformation approaches may be applied in a first-fit or best-fit manner. The main drawback of such procedures is their inability to continue the search upon becoming trapped in a local optimum. This suggests consideration of recent techniques for guiding these heuristics to overcome local optimality. Following this theme we sketch some applications of local search techniques such as simulated annealing, genetic algorithms and tabu search for SP.

Simulated Annealing

Applications of simulated annealing with respect to SP can be found in Dowsland [6] as well as Schiemangk [37]. For an adaptation to SPD see [30].

In Dowsland [6] the usual Boltzmann distribution and a geometric cooling schedule were used together with an edge-oriented transformation approach. She achieved best results with a starting temperature of 0.3 which was reduced by a factor of $\frac{3}{4}$ after a prescribed number of 500 to 1000 iterations (depending on the temperature decrease and allowing an overall maximum of 5684 iterations). The test data were randomly generated graphs with 100 vertices, real valued edge weights, a density (i.e. ratio of actual edges to the maximum number of possible edges) of $\frac{1}{2}, \frac{3}{4}$, and 1 as well as $|Q| \in \{30, 50, 80\}$. The simulated annealing algorithm is tested while initializing either randomly or by applying CHINS. The algorithms are compared with two local search methods exchanging one or two paths at a time (cf. the edge-oriented transformation) as well as HEUM. HEUM shows an average deviation of $1.1 - 1.3\%$ compared to the results obtained by simulated annealing indicating that the simulated annealing approach can be effective with respect to solution quality.

Genetic Algorithms

Genetic algorithms are a class of adaptive search procedures based on the principles derived from the dynamics of natural population genetics. In the sequel we report the results of Kapsalis et al. [20]. For further references on genetic algorithms for SP see [15, 16].

SP is represented by a bit string of size $|S|$ where the entry in position i refers to Steiner vertex i either included or excluded from the solution. The fitness of any solution is computed by Prim's algorithm [35] restricted to the graph induced by the basic vertices plus all included Steiner vertices. For a population size of ten the authors define a breeding cycle as follows. Based on our above mentioned idea of a node-oriented transformation a new generation is obtained by successively choosing parents and replacing them by new solutions determined by applying the genetic operators crossover and mutation with probabilities 0.9 and 0.02, respectively. The best results have been obtained by initializing the gene pool either randomly or MST-seeded and applying the well-known roulette strategy for selecting solutions to join the gene, or parent pool. Numerical results of Kapsalis et al. for the B-data of Table 1 show that using one or another of the parameter setting the genetic algorithm finds optimal solutions in all of the test problems with at most 4000 seconds for each single run on an AppleMacIIfx.

Tabu Search

Following a steepest descent mildest ascent approach, any transformation may either result in a best possible improvement or a least deterioration of the objective function value. To prevent the search from cycling between the same solutions endlessly, in tabu search some exchanges will be made tabu, i.e. forbidden for a specific number of iterations. In the sequel we refer to a tabu search approach based on the node-oriented neighbourhood definition (cf. [44]).

In the reverse elimination method (REM) a dynamic tabu strategy is performed that assigns any exchange a tabu status for at most one iteration at a time. The aim is to prevent any solution from being revisited in future iterations by elaborating a necessary and sufficient condition. With respect to SP any transformation or move is evaluated under the MST+P routine. For data sets B and C (cf. Tables 1, 2) we applied this method for at most 200 iterations starting from a solution provided by CHINS. Each entry in columns CHINS - REM gives the starting value and the objective function value obtained after a specific number of iterations given in parenthesis. The results show that this method gives reasonable results without having to think about too many parameters (only the number of iterations and the move evaluation). However, additional topics like some intensification as well as diversification strategies have to be observed to get further improvements (cf. [44]).

4 Conclusions

In this paper we have surveyed Steiner tree heuristics including some new ideas especially relevant in a cheapest insertion approach. Recent developments show that there is considerable impact in different areas of research as, for instance, the investigation of worst case analysis results. Approximating shortest paths might help to reduce computation times without sacrifycing solution quality.

Further research going beyond the topics mentioned could be to perform some *target analysis*, i.e. a careful analysis of differences from feasible solutions obtained with either algorithm and the outcome after performing some algorithmic enhancements or the optimal solution (whenever available), respectively. In this sense problems with known optimal solutions are of great help (see Tables 1 - 2 as well as [22]). Furthermore, the development of efficient heuristics for special cases could be useful and trying to project the findings onto SP and SPD. An example might be the consideration of directed acyclic graphs with rectilinear distances where even the problem's complexity is still open (see e.g. [46]), or the general case of undirected rectilinear graphs (see e.g. [38, 39]).

Another area worth to be investigated is the recalculation of MSTs when only slight modification are made (e.g. in- or excluding a vertex and its incident edges, see e.g. [29, 14] for some ideas). This might help especially in speeding up any of the metastrategies when based on the node-oriented transformation.

References

[1] A. Balakrishnan and N.R. Patel, Problem reduction methods and a tree generation algorithm for the Steiner network problem, *Networks* **17** (1987) 65-85.

[2] J.E. Beasley, OR-Library: distributing test problems by electronic mail, *Journal of the Operational Research Society* **41** (1990) 1069-1072.

[3] N.P. Chen, New algorithms for Steiner tree on graphs, in: *Proceedings of the IEEE International Symposium on Circuits and Systems* (1983) 1217-1219.

[4] S. Chopra, E.R. Gorres and M.R. Rao, Solving the Steiner tree problem on a graph using branch and cut, *ORSA Journal on Computing* **4** (1992) 320-335.

[5] E.A. Choukhmane, Une heuristique pour le probleme de l'arbre de Steiner, *R.A.I.R.O. Recherche Opérationelle* **12** (1978) 207-212.

[6] K.A. Dowsland, Hill-climbing, simulated annealing and the Steiner problem in graphs, *Engineering Optimization* **17** (1991) 91-107.

[7] D.-Z. Du and Y. Zhang, On better heuristics for Steiner minimum trees, *Mathematical Programming* **57** (1992) 193-202.

[8] C.W. Duin and A. Volgenant, An edge elimination test for the Steiner problem in graphs, *Operations Research Letters* **8** (1989) 79-83.

[9] C.W. Duin and A. Volgenant, Reduction tests for the Steiner problem in graphs, *Networks* **19** (1989) 549-567.

[10] C.W. Duin and A. Volgenant, Reducing the hierarchical network design problem, *European Journal of Operational Research* **39** (1989) 332-344.

[11] C.W. Duin and S. Voß, On the use of shortest paths in Steiner tree heuristics, Working Paper, Univ. Amsterdam (1991).

[12] L.R. Foulds and V.J. Rayward-Smith, Steiner problems in graphs: algorithms and applications, *Engineering Optimization* **7** (1983) 7-16.

[13] H.N. Gabow, Z. Galil, T. Spencer and R.E. Tarjan, Efficient algorithms for finding minimum spanning trees in undirected and directed graphs, *Combinatorica* **6** (1986) 109-122.

[14] F. Glover, D. Klingman, R. Krishnan and R. Padman, An in-depth empirical investigation of non-greedy approaches for the minimum spanning tree problem, *European Journal of Operational Research* **56** (1992) 343-356.

[15] J. Hesser, R. Männer and O. Stucky, Optimization of Steiner trees using genetical algorithms, in: J.D. Schaffer (ed.) *Proceedings of the Third International Conference on Genetic Algorithms*, Morgan Kaufmann, San Mateo (1989) 231-236.

[16] J. Hesser, R. Männer and O. Stucky, On Steiner trees and genetic algorithms, in: J.D. Becker, I. Eisele and F.W. Mündemann (eds.) *Parallelism, Learning, Evolution*, Lecture Notes in Computer Science **565**, Springer, Berlin (1991) 509-525.

[17] F.K. Hwang and D.S. Richards, Steiner tree problems, *Networks* **22** (1992) 55-89.

[18] F.K. Hwang, D.S. Richards and P. Winter, *The Steiner Tree Problem*, North-Holland, Amsterdam (1992).

[19] Y.-B. Ji and M.-L. Liu, A new scheme for the Steiner problem in graphs, in: *Proceedings of the IEEE International Symposium on Circuits and Systems* (1988) 1839-1842.

[20] A. Kapsalis, V.J. Rayward-Smith and G.D. Smith, Solving the graphical Steiner tree problem using genetic algorithms, *Journal of the Operational Research Society* **44** (1993) 397-406.

[21] B.N. Khoury and P.M. Pardalos, A heuristic for the Steiner problem in graphs, Working Paper, Univ. of Florida, Gainesville (1993).

[22] B.N. Khoury, P.M. Pardalos and D.-Z. Du, A test problem generator for the Steiner problem in graphs, Working Paper, Univ. of Florida, Gainesville (1993).

[23] L.T. Kou, On efficient implementation of an approximation algorithm for the Steiner tree problem, *Acta Informatica* **27** (1990) 369-380.

[24] L. Kou, G. Markowsky and L. Berman, A fast algorithm for Steiner trees, *Acta Informatica* **15** (1981) 141-145.

[25] J.B. Kruskal, On the shortest spanning subtree of a graph and the travelling salesman problem, *Proc. Amer. Math. Soc.* **7** (1956) 48-50.

[26] A. Lucena, Steiner problem in graphs: Lagrangean relaxation and cutting-planes, *Bulletin of the Committee on Algorithms* **21** (1992) 2-7.

[27] N. Maculan, P. Souza and A. Candia Vejar, An approach for the Steiner problem in directed graphs, *Annals of Operations Research* **33** (1991), 471-480.

[28] K. Mehlhorn, A faster approximation algorithm for the Steiner problem in graphs, *Information Processing Letters* **27** (1988) 1255-128.

[29] M. Minoux, Efficient greedy heuristics for Steiner tree problems using reoptimization and supermodularity, *INFOR* **28** (1990) 221-233.

[30] L.J. Osborne and B.E. Gillett, A comparison of two simulated annealing algorithms applied to the directed Steiner problem on networks, *ORSA Journal on Computing* **3** (1991) 213-225.

[31] J. Plesnik, A bound for the Steiner tree problem in graphs, *Math. Slovaca* **31** (1981) 155-163.

[32] J. Plesnik, Worst-case relative performance of heuristics for the Steiner problem in graphs, *Acta Math. Univ. Comenianae* **60** (1991) 269-284.

[33] J. Plesnik, Heuristics for the Steiner problem in graphs, *Discrete Applied Mathematics* **37/38** (1992) 451-463.

[34] J. Plesnik, On some heuristics for the Steiner problem in graphs, in: J. Nesetril and M. Fiedler (eds.), *Fourth Czechoslowakian Symposium on Combinatorics, Graphs and Complexity*, Elsevier, Amsterdam (1992) 255-257.

[35] R.C. Prim, Shortest connection networks and some generalizations, *Bell Syst. Techn. J.* **36** (1957) 1389-1401.

[36] V.J. Rayward-Smith and A. Clare, On finding Steiner vertices, *Networks* **16** (1986) 283-294.

[37] C. Schiemangk, Thermodynamically motivated simulation for solving the Steiner tree problem and the optimization of interacting path systems, in: A. Iwainsky (ed.) *Optimization of Connection Structures in Graphs*, Central Institute of Cybernetics and Information Processes, Berlin/GDR (1985) 91-120.

[38] M. Servit, Heuristic algorithms for rectilinear Steiner trees, *Digital Processes* **7** (1981) 21-32.

[39] C.C. de Souza and C.C. Ribeiro, Heuristics for the minimum rectilinear Steiner tree problem, Working Paper G-90-32, GERAD, Montreal (1990).

[40] G.F. Sullivan, Approximation algorithms for Steiner tree problems, Technical Report 249, Department of Computer Science, Yale Univ., New Haven, CT (1982).

[41] H. Takahashi and A. Matsuyama, An approximate solution for the Steiner problem in graphs, *Math. Japonica* **24** (1980) 573-577.

[42] S. Voß, *Steiner-Probleme in Graphen*, Hain, Frankfurt/Main (1990).

[43] S. Voß, Steiner's problem in graphs: heuristic methods, *Discrete Applied Mathematics* **40** (1992) 45-72.

[44] S. Voß, A node-oriented tabu search approach for Steiner's problem in graphs, Working Paper, TH Darmstadt (1992).

[45] S. Voß, Worst case performance of some heuristics for Steiner's problem in directed graphs, *Information Processing Letters*, to appear.

[46] S. Voß and C.W. Duin, Heuristic methods for the rectilinear Steiner arborescence problem, *Engineering Optimization* **21** (1993) 121-145.

[47] B.M. Waxman and M. Imase, Worst-case performance of Rayward-Smith's Steiner tree heuristic, *Information Processing Letters* **29** (1988) 283-287.

[48] P. Widmayer, Fast approximation algorithms for Steiner's problem in graphs, Inst. für Angewandte Informatik und Formale Beschreibungsverfahren, Univ. Karlsruhe (1986).

[49] P. Winter, Steiner problem in networks: a survey, *Networks* **17** (1987) 129-167.

[50] P. Winter and J. MacGregor Smith, Path-distance heuristics for the Steiner problem in undirected networks, *Algorithmica* **7** (1992) 309-327.

[51] R.T. Wong, A dual ascent approach for Steiner tree problems on a directed graph, *Mathematical Programming* **28** (1984) 271-287.

[52] Y.F. Wu, P. Widmayer and C.K. Wong, A faster approximation algorithm for the Steiner problem in graphs, *Acta Informatica* **23** (1986) 223-229.

[53] A.Z. Zelikovsky, An $\frac{11}{6}$-approximation algorithm for the Steiner problem on graphs, in: J. Nesetril and M. Fiedler (eds.), *Fourth Czechoslowakian Symposium on Combinatorics, Graphs and Complexity*, Elsevier, Amsterdam (1992) 351-354.

[54] A.Z. Zelikovsky, A faster approximation algorithm for the Steiner tree problem in graphs, *Information Processing Letters* **46** (1993) 79-83.

GREEDY ALGORITHMS IN COMBINATORIAL OPTIMIZATION

Ulrich Faigle, Twente University of Technology, Enschede
The Netherlands

The *greedy algorithm* is perhaps the intuitively most natural optimization principle: take in each step the locally best decision, where "best" is measured by an objective function that is evaluated locally. The question, then, arises under what conditions such a local strategy leads to a globally optimal solution. Particular attention has therefore been payed to combinatorial structures for which the greedy algorithm works provably optimally (at least relative to certain types of objective functions): for example, matroids, polymatroids and their generalizations (e.g., greedoids).

The greedy approach is often quite successfully employed not as a full algorithm but as a fast subroutine within other algorithmic frameworks (e.g., pruning rule in branch-and-bound procedures or local optimizer in a Lagrange relaxation). In general, the greedy algorithm may be seen as a direct attempt to find local optima in search procedures. The greedy strategy appears to be of particular importance in the design of on-line algorithms (e.g., coloring, partitioning, scheduling), where it is often known as the *first fit* principle.

Recent interest in probabilistic aspects of optimization has exhibited the greedy strategy to be quite powerful in the design of deterministic algorithms based on the evaluation of conditional probabilities (derandomization techniques) or in the analysis of average case behavior (TSP, graph coloring).

This survey lecture cannot give a full treatment of all aspects of greedy principles in combinatorial optimization but tries to sketch some of the results and recent developments in the aforementioned areas.

A COMPARISON BETWEEN SIMULATED ANNEALING AND TABU SEARCH WITH AN EXAMPLE FROM THE PRODUCTION PLANNING

Albrecht Heinrici FernUniversität Hagen

Zusammenfassung: In diesem Beitrag werden die Gemeinsamkeiten und die Leistungsfähigkeit von Simulated Annealing und Tabu Search untersucht und dargestellt. Verschiedene moderne Suchverfahren für kombinatorische Probleme weisen eine gemeinsame Grundstruktur auf. Am Beispiel des Bandabgleichproblems werden die Verfahren konkretisiert und die Leistungsfähigkeit verglichen, wobei Lösungen für Standardprobleme aus der Literatur dargestellt werden. Tabu Search zeigt sich dabei als erfolgreich und anderen Verfahren überlegen.

Abstract: In this paper we investigate and depict the similarities and the performance of simulated annealing and tabu search. Different modern search methods which are used for combinatorial problems have a common structure. The realisation of both methods is shown for the assembly line balancing problem and the performance is compared. The problem set consists of well known literature problems. In this comparison tabu search is successful and superior to other methods.

1. Introduction

Simulated annealing (SA) and tabu search (TS) are search methods which can be used for several kinds of combinatorial problems. Starting from an initial solution other solutions are produced and this way a part of all admissible solutions is scanned. For producing a new solution of another solution a set of transitions (or moves) is defined. All solutions x' which can be reached by a transition from the solution x are called neighbours of x. The steering of the search process among these neighbour solutions is different in the search methods. A well known version of the search in this way is a general descent method which stops when a local optimum is found and which is normally run many times from different initial solutions.

SA, which was developed by Kirkpatrik et al. [9] and by others in similar versions, uses a stochastic way of steering while TS can be used fully deterministically. TS was developed by Glover [3] and similarly by Hansen [5]. Both methods do not guarantee an optimal solution in finite time and can be classified as heuristics.

The example used here is the simple assembly line balancing problem (SALBP). It occurs in the production planning when the total work which is to be done on an assembly line has to be assigned to the work stations at the line. The object is to find a solution where the work time is best utilised .

In the next pages at first the two methods are shown. After depicting the line balancing problem the realisation of both of the methods is shown. A set of literature problems is solved and the results are given. Concludingly some hints on new developments and other methods are given.

2. Description of the search methods

Normal descent methods have the problem that they can not overcome local optima. They stop when no further improvement of the value of the objective function is possible.

Both of the methods which are to be presented here work in a similar way as descent methods but do not have this problem. Like the descent methods they start at any initial solution and proceed by constructing new solutions. These new solutions belong to a defined neighbourhood of the original solution. In the search process a neighbour solution is chosen which fulfills specific criteria. Transitions which lead to solutions with better objective function values are preferred but other transitions can also be accepted. With this strategy the search methods normally do not get stuck in local optima.

The philosophy of the search is extremely different in both methods. SA was developed from a correspondence between physical processes and combinatorial problems while TS can be seen as connected with artificial intelligence [3].

2.1. Simulated annealing

Statistical mechanics describes the behaviour of sets of atoms. Because it is impossible to describe the behaviour of all of the single atoms their behaviour is described by their most probable behaviour. The probability of each configuration of the atoms in a set is weighted by its Boltzmann probability factor. There is a simple algorithm which simulates the probability distribution of a given collection of atoms by small random displacements of atoms in the set. By lowering the temperature of the system the configurations with small energy become more probable. For finding a low-energy state of a material in practice you first have to heat the material and then lower the temperature very slowly.

The value of the objective function of a combinatorial problem can be seen as the energy of a state of material. This way the global minimum of the value of the objective function is relative to the state of a material with minimal energy.

The SA algorithm which can be used to solve combinatorial problems works as follows:

1. Start with any initial solution x and chose a temperature parameter t.
2. Chose a transition randomly and construct the corresponding solution x'.
3. Calculate the value of the objective function and the difference d in the value between x and x'.
4. If x' is better than x accept x' as the current solution x. If not accept the solution x' with a probability $p(d, t)$.
5. Lower the temperature t
6. If $t > 0$ GOTO 2.

2.2. Tabu search [4]

TS is more similar to descent methods than SA. It follows the rule that generally the best neighbour is chosen. Because this policy leads into cycles at local optima, transitions which lead back to known solutions are forbidden.

To find the best transition from any solution, you normally must examine all neighbours. In neighbourhoods with many neighbours this might cost much time. To avoid this a small candidate subset of the complete neighbourhood is to be defined, so that only this subset has to be examined. If there is no admissible or no good enough solution in it you can extend the subset. The structure which is used to store information about the transitions which are forbidden (or tabu) is called tabu list. In this list special attributes of the recently used transitions are stored and all transitions which undo the attributes in the list are tabu. The length of the tabu list is limited to n so that only the reversion of the last n transitions is tabu. This length n is normally set between 7 and 20 depending on the problem class and dimension.

Because it is possible that this way transitions are tabu which lead to good solutions aspiration criteria are introduced. If a transition which is tabu leads to a solution which fulfills a criterion the tabu status is overridden.

There exist many extensions to this standard form of TS but they are not included in this study.

3. Realisation at the SALBP and computational comparison

The SALBP is an operational problem which occurs when the tasks of an assembly system have to be assigned to the stations of an assembly line. The tasks which have to be executed can not be done in any succession. There exist precedence relations which say that particular tasks have to be done before others. Every task has a time which is needed to perform it. The sum of the task times assigned to a station is called work content of the station. The cycle time is the maximal work content of a station and after the cycle time all the products at the line are moved to the next station. The time of a station which is not used for work is called idle time. The object is to minimise the idle time of the workers at the stations [1].

This can be done in two different ways: The first is to minimise the number of stations for a given cycle time (type-1) and the second is to minimise the cycle time for a given number of stations (type-2). Both types of problems are NP-complete [1]. Here the type-2-problem is examined.

In the following illustration an example problem of Jackson 1956 is shown. On the left side a precedence network is depicted where every node represents a task. The above number represents the number of the task and the below represents the task time. The arrows represent the precedence relations so that every task on the left side of an arrow is an immediate predecessor of the task at the right side of the arrow. The illustration on the right side shows a balanced line with the assignment of all tasks, the work content of the stations, and the cycle time. The idle time of the stations is grey coloured.

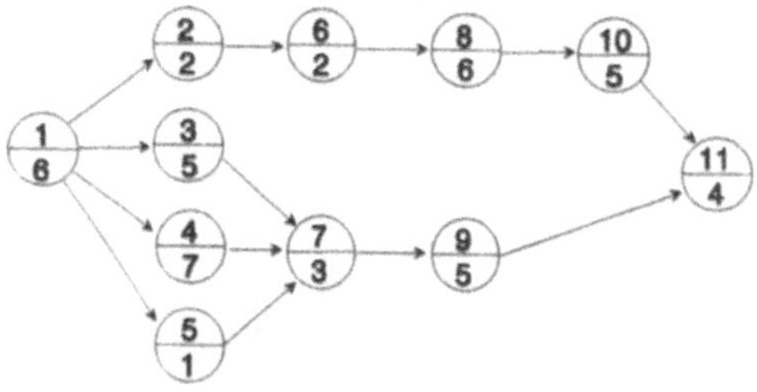

Figure 1 Precedence Network

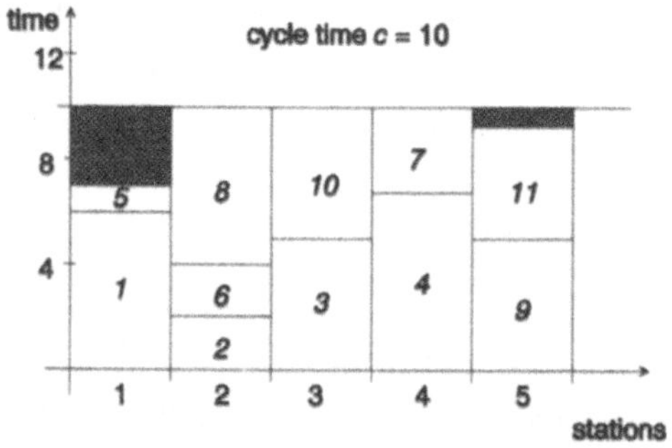

Figure 2 Assignment Chart

Type-1-problems are solvable optimally by the algorithm of Johnson [8] or Hoffmann [7]. The algorithms are fast but not effective for type-2-problems and an effective exact algorithm for the type-2-problem is not known. This is the reason why further on heuristics are interesting for this problem.

The initial solutions for both methods can be produced by any heuristic. An often used and fast heuristic for this problem is the ranked positional weight technique (RPWT) [6]. Finally the search methods are to be compared with the RPWT. Therefore it would be unfair to start the search methods with the solutions of the RPWT. To produce initial solutions a "retrograde" RPWT was created which is as fast as the original procedure but produces worse results. The procedure has to be started several times with decreasing cycle times because it is originally designed to solve type-1-problems.

In general there are many different possibilities for transitions between solutions. The most obvious transitions for sequencing problems are to shift an element or to swap to elements. Here that means to shift one task to another station or to swap two tasks between stations. Only such transitions are allowed which produce admissible solutions. In the next illustrations examples for both of the transitions are given.

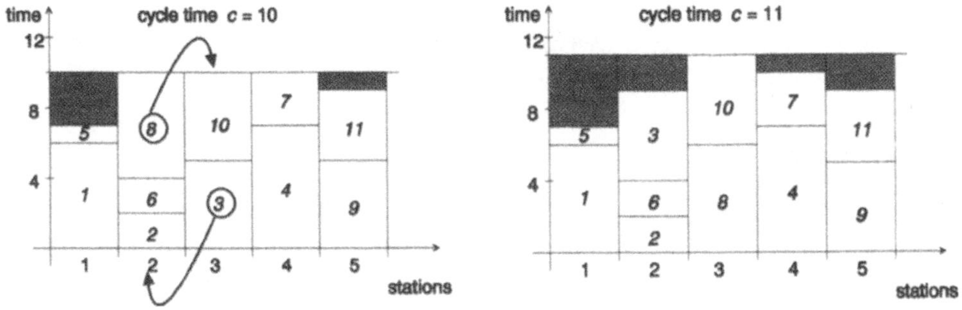

Figure 3 Swapping Tasks

A result of research at SA is that the transitions should produce solutions with similar objective function values. This result is confirmed by this study. Experiments showed that swapping is the best transition for SA because it produces neighbours which differ little from the preceding solution in the values of the objective function. SA starts with a high temperature which is let down exponentially. The quality of the solutions was found to be nearly independent from the number of iterations on the same temperature if the overall number of iterations was fixed.

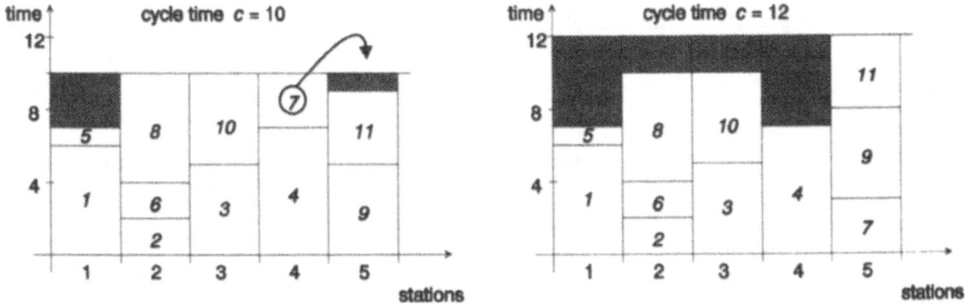

Figure 4 Shifting a Task

The candidate set of transitions for TS consists only of shifts of tasks out of the station with the greatest work content because this set is very small and can be examined fast. If there is no admissible solution found the neighbourhood is extended to all possible shifts and swaps.

The attribute which is stored in the tabu list is the task time of the moved task and the number of the station where the task came from. No task with the same task time can be shifted back to the station while the tabu list contains this attribute. The length of the tabu list is chosen as "5 + 'average number of tasks per station' ". The aspiration criterion of TS is fulfilled if a transition leads to a better solution than it led to when it became forbidden.

In the next illustrations typical courses of both of the searches are depicted. Every bar in the chart describes the best and the worse solution of a set of iterations. Both charts show the search at the example of Kilbridge et al. 1961 for ten stations. The optimal solution has a cycle time of 56 which is found by TS in the 198 th iteration.

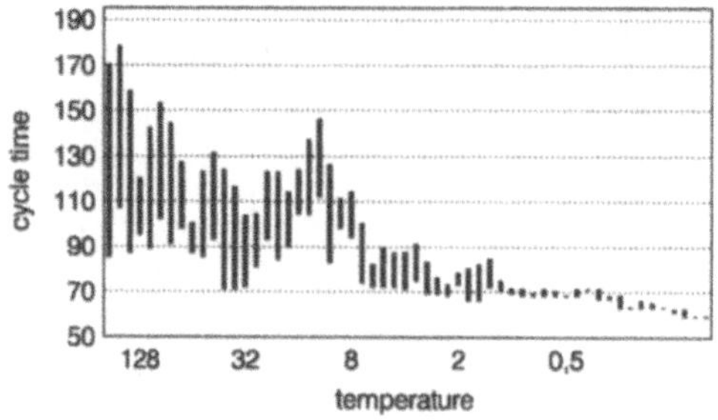

Figure 5 Course of SA

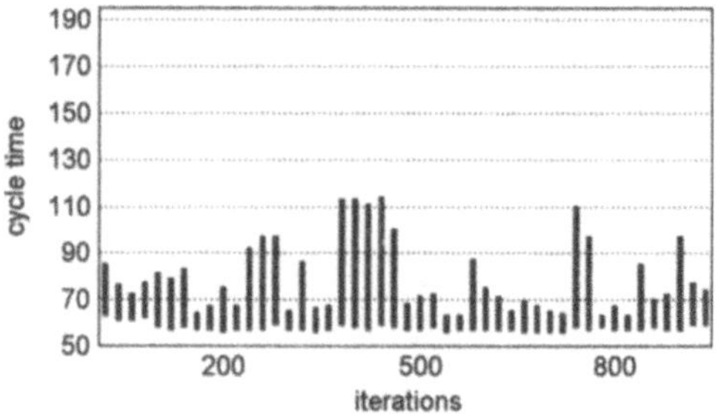

Figure 6 Course of TS

Table 1 Computational Results

no. of stations	RPWT	SA	TS	bound
Bowman 9				
4	25	22	22	19
5	18	17	17	15
6	17	17	17	13
Dar-El 11				
2	97	93	93	93
3	64	62	64 z	62
4	52	48	48	47
Mertens 7				
3	10	10	10	10
4	9	9	9	8
5	7	7	7	6
Jaeschke 9				
4	13	12	12	10
5	11	10	10	8
6	9	9	9	7
Jackson 11				
4	13	12	12	12
5	11	10	10	10
6	9	9	9	8
Mitchell 21				
3	36	35	35	35
5	22	21	21	21
8	14	15	14	14
Heskiaoff 28				
5	205	206	205	205
6	173	175	172	171
7	148	150	148	147

no. of stations	RPWT	SA	TS	bound
Sawyer 30				
8	43	44	41	41
9	40	39	38	36
10	35	35	34	33
11	32	34	32	30
12	30	34	31 z	27
Kilbridge 45				
3	186	184	184	184
4	139	138	138	138
6	93	93	92	92
7	81	81	79	79
10	57	59	56	56
Tonge 70				
7	507	507	503	502
8	447	459	452 z	439
9	399	411	392	390
10	359	358	353	351
15	240	251	239	234
Arcus 83				
10	7827	7698	7651	7571
11	7113	7452	7091	6883
12	6546	7091	6518	6309
13	6076	6035	5942	5824
14	5610	5604	5467	5408
Arcus 111				
15	10159	10454	10100	10034
16	9546	9806	9435	9407
17	8965	9424	8958	8853
18	8621	9210	8517	8362
19	8027	8642	8165	7921

A reason for the better performance of TS in comparison to SA seems to be that its course of search spends more time in the "interesting" part of the solution space.

In the next table a list of the solutions of all computed problems and algorithms is presented. The number behind each name of the problem is the number of tasks in the problem. In the column RPWT the results of the original procedure and not the initial solutions are given. The shown solutions of SA are average values. The CPU time for SA and TS was set equal. The bound is the theoretical minimum cycle time. The solutions where cycling occurred are marked with z. The table shows that TS produced equal or superior results than SA in most of the cases.

4. Conclusions

In this study a SA algorithm and a TS algorithm for the simple assembly line balancing problem are developed. A special neighborhood is constructed for both of the methods. The performance of the algorithms is compared at a set of standard literature problems. Over most of the problems TS performed equal or better than SA and as a standard heuristic.

In difference to branch-and-bound-like algorithms the algorithms shown here are extendible to operational restrictions without difficulties. Both methods are convenient to use but TS needs more special knowledge about the problem and more code in the implementation.

New general optimisation algorithms which use a similar neighbourhood of solutions are presented by Dueck [2]. The dependencies between all these algorithms including genetic algorithms are not totally understood and should be a future field of research.

6. References

[1] Baybars, I. (1986) "A Survey of Exact Algorithms for the Simple Assembly Line Balancing Problem", Management Science, 32, 8, 909-932.

[2] Dueck, G. (1993) "New Optimization Heuristics", Journal of Computational Physics, 104, 86-92

[3] Gover, F. (1986) "Future Path for Integer Programming and Links to Artificial Intelligence", Computers and Operations Research, 5, 533-549.

[4] Glover, F., M. Laguna (1992) "Tabu Search", preprint from: Modern Heuristic Techniques for Combinatorial Problems.

[5] Hansen, P. (1986) "The Steepest Ascent Mildest Descent Heuristic for Combinatorial Programming", presented at the Congress on Numerical Methods in Combinatorial Optimization, Capri, Italy.

[6] Helgeson, W.B., D.P. Birnie (1961) "Assembly Line Balancing Using Ranked Positional Weight Technique", The Journal of Industrial Engeneering, 7, 6, 394-398

[7] Hoffmann, T.R. (1992) "EUREKA: A Hybrid System for Assembly Line Balancing", Management Science, 38, 1, 39-47.

[8] Johnson, R.V. (1988) "Optimally Balancing Large Assembly Lines with 'FABLE' ", Management Science, 34, 2, 240-253.

[9] Kirkpatrik, S., C.D. Gelatt, M.P. Vecchi, (1983) "Optimization by Simulated Annealing", Science, 220, 671-680.

[10l] Talbot, F.B., J.H. Patterson, W.V. Gehrlein, (1986) "A Comparative Evaluation of Heuristic Line Balancing Techniques", Management Science, 32, 4, 430-454

For complete references to the examples used see [10].

COMPLEX SEQUENCING PROBLEMS

Johann Hurink, Universität Osnabrück

A complex sequencing problem may be formulated as follows. Let X be a subset of all permutations of a finite set. Associated with each permutation π in X there is an optimization problem $OPT(\pi)$. Let $f(\pi)$ be the optimal solution value of $OPT(\pi)$. Find a permutation π^* which optimizes f over all permutations in X. Usually $OPT(\pi)$ is a problem which can be solved efficiently (e.g. a linear program, a shortest path problem etc.)

We will present some practical problems which may be formulated as complex sequencing problems. Among them there are machine scheduling problems, railway scheduling problems, the problem of finding placement sequences and bin locations for robots, tool switching problems on flexible machines, etc. Most of these problems arise in connection with flexible manufacturing.

In general complex sequencing problems are NP–hard although each optimization problem $P(\pi)$ is efficiently solvable. Therefore we used local search techniques to calculate heuristic solutions for complex sequencing problems. These techniques yields very good results for a wide class of problems. The best known local search techniques are iterative improvement, simulated annealing and tabu search. The most important step when designing a local search method is the definition of a neighborhood. A neighborhood is given be a set of moves which allows to calculate for each permutation π a set $N(\pi)$ of neighbors. The iterative improvement method replaces the current permutation by its best neighbor and iterates this process until there is no neighbor which improves the current solution. The disadvantage of this method is the possibility to be trapped in a local minimum which is arbitrarily bad. Simulated annealing avoids this by with certain probability allowing moves which do not improve the current solution . Unfortunately simulated annealing is a very slow method. A better choice is the tabu search method.

We report on experience with local search applied to complex sequencing problems.

SEMI-GREEDY HEURISTICS FOR ACADEMIC COURSE SCHEDULING

Jan Juretzka, Christian-Albrechts-Universität Kiel
Andreas Drexl, Christian-Albrechts-Universität Kiel

We consider an academic course scheduling problem, which is based on the following **assumptions**: Let W denote the planning horizon (e.g. one week with W = 5 days). w = 1,...,W represents macro-periods (days). Each macro-period w is subdivided into p_w micro-periods (e.g. hours). Thus we have t = 1,...,T periods (time slots) with $T = \Sigma p_w$. r = 1,...,R categories of rooms are given, with λ_{rt} being the number of rooms of category r available in time slot t. There are z = 1,...,Z zones, the rooms belong to; H_z denotes the set of rooms in zone z. Changing from a room in zone z to a room in zone z', $z \neq z'$, in subsequent time slots is feasible only, if there is enough travel (setup) time. Let j = 1,...,J denote assignments of room categories to time slots ($J = R \cdot T$), in short room-periods. i = 1,...,I subjects have to be scheduled, where subject i has a duration of τ_i micro-periods. Let $J_i \in \{1,...,J\}$ denote the set of feasible assignments of subject i to room-periods. c_{ij} denotes the "cost" (teachers preferences) of assigning subject i to room-period $j \in J_i$. V_i denotes the set of subjects i' which have to scheduled in parallel with subject i. There are d = 1,...,D teachers (professors) available; in addition, there are s = 1,...,S groups of students (classes) attending the subjects (more than one class may belong to one subject). Θ_{dt} (Γ_{st}) denotes the "availability level" (e.g. 0 or 1) of teacher d (class s) in time slot t.

The problem under consideration is modelled as an **integer programming model** with binary variables with the following structure: Maximize the sum of assignment preferences s.t. the following constraints: subject (job) completion, subject parallelism, room/class/professors availability, travel time restrictions. In addition, an upper bound on the maximal number of subjects scheduled per professor, class, and day has to be observed. Note that, in contrast to other course scheduling models (cf. e.g. [1]), we already have an assignment of teachers to subjects.

It should be noticed, that the course scheduling problem considered here is closely related to preemptive multi-mode resource-constrained project scheduling where the **resources** are **partially renewable**.

The problem is solved to suboptimality with the aid of specially designed semi-greedy heuristics based on **randomized regrets** [2]. In addition, benchmark solutions are provided via LP-based heuristics.

References

[1] Dinkel, J.J., Mote, J., Venkataramanan, M.A., "An efficient decision support system for academic course scheduling", *Operations Research,* Vol. 37 (1989), pp. 853-864.

[2] Drexl, A., "Scheduling of project networks by job assignment", *Management Science,* Vol. 37 (1991), pp. 1590-1602.

EXACT AND APPROXIMATION ALGORITHMS FOR THE TACTICAL FIXED INTERVAL SCHEDULING PROBLEM

Leo G. Kroon, Erasmus University Rotterdam, NL
Marc Salomon, Erasmus University Rotterdam, NL
Luk N. Van Wassenhove, INSEAD Fontainebleau Cedex, France

The Tactical Fixed Interval Scheduling Problem (TFISP) is the problem of determining the ***minimum number*** of parallel non-identical machines, such that a feasible schedule exists for a given set of jobs. In TFISP each job must be carried out in a prespecified time interval and belongs to a specific job class. The problem is complicated by the restrictions

(i) that each machine can handle only one job at a time,

(ii) that each machine can handle only jobs from a subset of the job classes, and

(iii) that preemption is not allowed.

In this presentation we discuss the occurrence of TFISP in practice, we analyze the computational complexity of TFISP, and we present exact and approximation algorithms (with error bounds) for solving TFISP. The paper is concluded with a computational study.

Keywords:
Fixed interval scheduling, heuristics, integer programming, Lagrangean relaxation, non-preemptive scheduling, upper and lower bounds.

A GENERALIZED PROBABILISTIC ANALYSIS OF A POLYHEDRON'S NUMBER OF SHADOW VERTICES

Karl-Heinz Küfer, Universität Kaiserslautern

Introduction: Let $A := \{a_1, \ldots, a_m\}$, $a_i \in \mathbb{R}^n$, $m \geq n \geq 2$, where vectors a_i are considered i.i.d.–vectors under an arbitrary rotationally symmetric distribution in $\mathbb{R}^n$. The random event A may be identified with a polyhedron X_A by $X_A := \{x \in \mathbb{R}^n \,|\, a^T x \leq 1, a \in A\}$. Furthermore, let L_k, $L_k := \text{span}(u_1, \ldots, u_k)$, $u_i \in S^{n-1}$, be an arbitrary $\mathbb{R}^n$–subspace of dimension $k \in \{1, \ldots, n\}$. We call a vertex of X_A "L_k-shadow vertex" if it remains a vertex of X_A's image under an orthogonal projection into L_k. The number of X_A's L_k-shadow vertices will be denoted by $S_{L_k}(X_A)$. Knowledge about a polyhedron's number of shadow vertices is of interest within the probabilistic analysis of algorithms, where the the cardinality of a polyhedron's vertex subset is the main benchmark of complexity. A famous example is the "shadow vertex algorithm" — a variant of the simplex algorithm — in case of planary projection spaces (k=2), cf. [1,2].

Results: We study expectation value and variance of the random variable $S_{L_k}(X_A)$ in two different ways, the first leading to asymptotic results for fixed dimension n, $m \to \infty$, and special distributions, the latter overestimating moments independently from the underlying distribution. In both cases we concentrate on the increasing rate in m and give only qualitative results about the growth in n, k is always assumed to be fixed.

In case of expectation values a typical asymptotic result among many others is the following:

$$\text{(1)} \qquad \mathrm{E}(S_{L_k}) = C_{n,k}\, m^{(k-1)/(n+1)}(1 + o(1)),\ m \to \infty,$$

in case of a_i being uniformly distributed in the unit ball of $\mathbb{R}^n$, $n \geq 2$. Here, the constant $C_{n,k}$ is polynomial in n for fixed k. Independently from the underlying rotationally invariant distribution holds:

$$\text{(2)} \qquad \mathrm{E}(S_{L_k}) \leq C_{n,k}\, m^{(k-1)/(n-1)},\ m \geq n \geq 2,$$

$C_{n,k}$ being polynomial in n for fixed k again. Corresponding results for variances of S_{L_k} are

$$\text{(3)} \qquad \frac{\mathrm{Var}(S_{L_k})}{\mathrm{E}^2(S_{L_k})} = \Theta(m^{-(k-1)/(n+1)}),\ m \to \infty.$$

for uniformly distributed a_i in the n-dimensional unit ball and

$$\text{(4)} \qquad \frac{\mathrm{Var}(S_{L_k})}{\mathrm{E}^2(S_{L_k})} \leq C_{n,k},$$

which holds independently from the underlying distribution. In (3) and (4) the constants $C_{n,k}$ are polynomial in n for fixed k. By the aid of Chebychev's inequality (3) implies that in case of large m and fixed n even small relative deviations from the mean are very rare. This observation is true for all rotationally invariant distributions with compact support, while there are distributions with non-compact support, where the investigated quotient in (3) and (4) is positively bounded from below.

References:

[1] Borgwardt, K. H. (1982) *The Average Number of Pivot Steps Required by The Simplex Method is Polynomial.* ZOR 26: 157-177.

[2] Küfer, K. H. (1992) *On the Variance of the Number of Pivot Steps Required by the Simplex Algorithm.* To appear in ZOR, Preprint Universität Kaiserslautern.

ANWENDUNG EINES GENETISCHEN ALGORITHMUS AUF DAS SWAPPING-PROBLEM

Carl-Heinz Meyer, FernUniversität Hagen

Bei der Steuerung von Roboter-Greifarmen tritt das folgende Problem auf:

Auf einer Ebene seien *n* Gegenstände an bestimmten Punkten aufgestellt. Zu jedem Gegenstand gibt es eine gewünschte Endposition, in die dieser mit Hilfe eines Roboter-Greifarmes bewegt werden soll. Leider kann der Greifarm lediglich *einen* Gegenstand aufnehmen und zu einem neuen Punkt transportieren. Das Problem besteht nun darin, einen kürzesten Weg für den Greifarm zu ermitteln, sodaß alle Gegenstände ohne zwischenzeitliches Absetzen an ihre gewünschte Positionen überführt werden. (Man stelle sich ein gerade benutztes Schachbrett vor, dessen Figuren wieder in ihre Ausgangsposition gebracht werden sollen)

Genetische Algorithmen gehören zur Klasse der stochastischen parallelen Verfahren, mit denen sich auch für große NP-harte Probleme oftmals gute suboptimale Lösungen auffinden lassen. Die Problemstellung wird hierbei durch Mengen (Generationen) von binären Strings mit einer festen Länge repräsentiert. Aus diesen Generationen werden einige Strings für die Auswahl von Nachfolgegenerationen ausgewählt, wobei die ausgewählten Strings nach bestimmmten Regeln verändert werden. Sowohl die Auswahl als auch die Veränderungen orientieren sich nach Prinzipien, die auch in der Natur Verwendung finden, wie z.B. "Darwinian Survival of the Fittest" bzw. "Mutationen" und "Crossover"

Dieser Artikel beschreibt die Anwendung sowie einige Simulationsergebnisse von Genetischen Algorithmen bei der Lösung des Swapping-Problems.

SOLVING THE PALLET LOADING PROBLEM MORE EFFICIENTLY

Josef Nelißen, RWTH Aachen

This paper looks at the packing problem called the 'pallet loading problem' (PLP for short) – that of finding an optimal layout for identical rectangular boxes on a rectangular pallet (a problem of type (3, B, O, C) according to the typology of Dyckhoff). Practical considerations usually require that the boxes must be placed orthogonally (i.e. with the box edges parallel to the pallet edges), and in layers in which the vertical orientation of the boxes is fixed. These restrictions mean that the problem is reduced to the two–dimensional problem of packing a large containing rectangle, orthogonally, with identical copies of a small 'contained' rectangle (type (2, B, O, C)). This problem is commonly referred to as pallet loading problem as well.

Since 1979 many heuristics have been developed to solve the PLP, mainly based on the placing of blocks, rectangles which contain a number of packing rectangles with equal orientation. The optimality of a solution can be proven by upper bounds, for whose computation elaborate procedures have been developed. If the heuristics fail to reach the upper bound, exact algorithms with an exponentially increasing running time have to be applied.

After an introduction the talk first presents a new method to compute upper bounds, which tries to reduce the problem to another problem the exact solution of which is known. Despite its simplicity, this method has been proven to be quite useful in cases where other procedures fail.

In the second part the notion of partitions, a very important concept for the PLP, is introduced. Extending work done by Dowsland and Naujoks, this can on the one hand be used to characterize a set of equivalent instances of the PLP, which can be indexed by an open interval of the real line. On the other hand the partitions can be used to define a linear programming problem, whose solution with different cost functions leads to a deeper insight into the structure of an optimal solution. By means of a one–dimensional minimizer the first property can be used to find an instance of an equivalence class which has almost minimal waste. Structure analysis of this instance, based on the second property, often leads to a decreased upper bound.

In the third part a new heuristic is described, which reduces the given problem instance by placing L-shapes, formed of three blocks, in the lower left corner of the large rectangle, and solves the resulting instance either by block heuristics or an iteration of this L-reduction. This approach, which can be regarded as an extension of the diagonal heuristic of Exeler, often yields optimal solutions where block heuristics fail.

The last part describes an exact Branch&Bound algorithm, which doesn't operate on the given instance of the problem, but on the above mentioned equivalent problem with minimal waste, which often results in a vast reduction of run time. By use of a grid technique, preprocessing, upper bounds on the number of vertical/horizontal placed packing rectangles, and upper and lower bounds on the frequency of partitions, the algorithm is able to solve medium sized problems in reasonable time.

ANALYSING $E_k \mid E_r \mid c$ QUEUES

Ivo J.B.F. Adan, Eindhoven University of Technology, NL

The $E_k \mid E_r \mid c$ queueing system is a typical example of a problem the analysis of which is more complex than the simple formulation suggests. The equilibrium equations become almost intractable if one attempts to solve these equations via multi-dimensional generating function techniques.

In this talk we show that the equilibrium distribution can be represented by a linear combination of geometric terms by using a direct approach (without generating functions). This approach is based on the separation of variables technique, familiar from the theory of partial differential equations, but which is less well known for partial difference equations. The essential idea is first to construct a sufficiently rich base of separable solutions of the equations in the interior of the state space and then to use this base to construct a linear combination that also satisfies the conditions at the boundaries of the state space. The separable solutions appear to be geometric in each of the state variables.

The representation of the equilibium distribution provides insight in the behaviour of the processes and it can directly be used for numerical calculations.

Bottapprox: Eine Engpaßanalyse für geschlossene Warteschlangennetze auf der Basis der Summationsmethode

Gunter Bolch und Martin Fischer, Universität Erlangen-Nürnberg

Zusammenfassung

In dieser Arbeit wird das Verfahren Bottapprox zur Analyse geschlossener Warteschlangennetze vorgestellt. Bottapprox basiert auf der Summationsmethode, verwendet aber ein schnelleres Iterationsverfahren, das mit einer Engpaßanalyse startet. Es wird gezeigt, daß Bottapprox weniger Rechenzeit benötigt, genauere Ergebnisse als die Summationsmethode liefert, und daß die Genauigkeit für die meisten Anwendungsfälle ausreichend ist.

Summary

In this paper we present the new method Bottapprox for analysing closed queueing networks. Bottapprox is based on the summation method , but uses a faster iteration algorithm, which starts with a bottleneck analysis. It is shown that Bottapprox needs less computing time and is more accurate than the summation method and that the accuracy is sufficient for most applications.

1. Einleitung

Zur Analyse von geschlossenen Warteschlangennetzen werden sehr häufig approximative analytische Verfahren angewendet. Diese benötigen gegenüber der Simulation weit weniger Rechenzeit und sind auf eine größere Anzahl von Knotentypen anwendbar als exakte analytische Verfahren wie z.B. die Mittelwertanalyse. Ein solches approximatives Verfahren ist das hier vorgestellte Verfahren Bottapprox.

2. Notation

In dieser Arbeit werden nur geschlossene Warteschlangennetze mit einer Auftragsklasse betrachtet. Die formale Beschreibung für diese Netze sieht wie folgt aus (mit den verwendeten Symbolen):

- N Anzahl der Knoten
- K Anzahl der Aufträge
- $(k_1, \ldots, k_n)$ Zustand des Warteschlangennetzes
- k_i Anzahl der Aufträge im i-ten Knoten Für geschlossene Netze gilt:

$$\sum_{i=1}^{N} \bar{k}_i = K$$

- m_i Anzahl der parallelen Bedieneinheiten im i-ten Knoten
- μ_i Mittlere Bedienrate von Aufträgen im i-ten Knoten
- μ_i^{-1} Mittlere Bedienzeit eines Auftrages im i-ten Knoten
- p_{ij} Übergangswahrscheinlichkeit von Knoten i zu Knoten j, d.h., daß ein bedienter Auftrag vom Knoten i zum Knoten j überwechselt
- λ_i Gesamte mittlere Ankunftsrate von Aufträgen bei Knoten i
- e_i Anstelle der Übergangswahrscheinlichkeiten kann man auch die Besuchshäufigkeit (auch relative Ankunftsrate genannt) angeben.
  Für geschlossene Netze gilt dabei folgender Zusammenhang:

$$e_i = \sum_{j=1}^{N} e_j \cdot p_{ji}$$

Weiterhin werden nur Netze mit Knoten von folgenden Typen (Kendall'sche Notation) berücksichtigt: M/M/1-FCFS (Typ1), M/M/m-FCFS (Typ2), M/G/1-PS (Typ3), M/G/0-IS(Infinite Server) (Typ4), M/G/1-LCFS PRE (Typ5) ,-/G/1-FCFS (Typ6) und -/G/m-FCFS (Typ7). Diese Einschränkung wurde gemacht, um den Umfang dieser Arbeit einzugrenzen. Prinzipiell ist es aber möglich, den Algorithmus der Engpaßanalyse auch auf andere Knotentypen zu erweitern.

In nachfolgender Auflistung stehen wichtige Leistungsgrößen mit den in dieser Arbeit verwendeten Symbolen:

λ_i	Durchsatz im i-ten Knoten
λ	Durchsatz im gesamten Netz
ρ_i	Auslastung des i-ten Knotens
$\bar{k}_i$	Mittlere Anzahl von Aufträgen im i-ten Knoten
p(k)	Wahrscheinlichkeit, daß sich k Aufträge im Wartesystem befinden
P_m	Wahrscheinlichkeit, daß ein ankommender Auftrag nicht sofort bedient werden kann

Der Knoten i im Netz, dessen Auslastung ρ_i am höchsten ist, heißt Engpaßknoten. Die Leistungsgrößen des Engpaßknoten werden immer durch den Index bott gekennzeichnet (bottle-neck = Flaschenhals).

3. Konzept des Verfahren Bottapprox

Bei Bottapprox wird, wie bei der Summationsmethode, für jeden Knoten des geschlossenen Warteschlangennetzes die mittlere Anzahl von Aufträgen $\bar{k}_i$ in Abhängigkeit des Durchsatzes λ des Gesamtnetzes berechnet.

Dieser Zusammenhang kann funktional wie folgt dargestellt werden:

$$\bar{k}_i = f_i(\lambda_i) = \bar{f}_i(\lambda) = \hat{f}_i(\rho_i)$$

Die Funktionen f_i, $\bar{f}_i$ und $\hat{f}_i$ unterscheiden sich nur durch die Wahl der Leistungsgröße; λ, λ_i und ρ_i hängen direkt voneinander ab und können untereinander wie folgt ausgetauscht werden:

$$\lambda = \frac{\lambda_i}{e_i} \text{ und } \lambda_i = m_i \cdot \mu_i \cdot \rho_i \quad \forall (i \in \{1, \ldots, N\})$$

Die Funktion f_i (und damit auch $\bar{f}_i$ bzw. $\hat{f}_i$) besitzt für Knoten mit lastunabhängigen Bedienraten folgende Eigenschaften:

- die Funktion $f_i(\lambda)$ ist streng monoton steigend

  $$f_i(\lambda_i) < f_i(\lambda_i + \varepsilon) \quad \forall (\varepsilon > 0)$$

- wenn keine Aufträge im Knoten ankommen, d.h. die Ankunftsrate gleich 0 ist, sind auch keine Aufträge im Knoten:

  $$f_i(0) = 0$$

- da für die Auslastung ρ_i in jedem Knoten i die Ungleichung

  $$0 \le \rho_i \le 1 \quad \forall (i \in \{1, \ldots, N\})$$

  gilt, ergibt sich für den Definitionsbereich von λ_i:

  $$\lambda_i \in [0; (m_i \cdot \mu_i)]$$

  Für Infinite-Server-Knoten und Knoten mit $m_i > K$ Bedieneinheiten ergibt sich der Definitionsbereich:

  $$\lambda_i \in [0; (K \cdot \mu_i)],$$

  da sich dann maximal K Aufträge im Knoten aufhalten können.

Unter Berücksichtigung dieser Eigenschaften und der Einführung eines Korrekturfaktors in die Formel (G1) bis (G4) für einen einzelnen Knoten

(G1) $$\bar{k} = \frac{\rho}{1-\rho}$$ für Typ1, 3, 5

(G2) $$\bar{k} = m \cdot \rho + \frac{\rho}{1-\rho} \cdot P_m$$ für Typ2

(G3) $$\bar{k} = \rho + \rho^2 \cdot \frac{1 + c_B^2}{2 \cdot (1-\rho)}$$ für Typ6

(G4) $$\bar{k} = m \cdot \rho + \rho \cdot \frac{1 + c_B^2}{2 \cdot (1-\rho)} \cdot P_m$$ für Typ7 kommt man nach [Bolch89] S.160ff und S.210f zu folgender Formel für $f_i(\lambda_i)$:

$$
\text{(F1)} \quad \bar{k}_i = \begin{cases}
\dfrac{\rho_i}{1-(\frac{K-1}{K}\cdot\rho_i)} & \text{für Typ1, 3, 5} \\[2ex]
m_i\cdot\rho_i + \dfrac{\rho_i}{1-\left(\frac{K-m_i-1}{K-m_i}\cdot\rho_i\right)}\cdot P_{m_i} & \text{für Typ2} \\[2ex]
\dfrac{\lambda_i}{\mu_i} & \text{für Typ4} \\[2ex]
\rho_i + \dfrac{\rho_i^2\cdot a}{1-(\frac{K-1-a}{K-1}\cdot\rho_i)} & \text{für Typ6} \\[2ex]
m_i\cdot\rho_i + \dfrac{\rho_i}{1-\left(\frac{K-m_i-a}{K-m_i}\cdot\rho_i\right)}\cdot a\cdot P_{m_i} & \text{für Typ7}
\end{cases}
$$

mit $a = \frac{c_B^2+1}{2}$ und c_B^2 ist quadrierter Variationskoeffizient des Bedienprozesses.

ρ_i läßt sich dabei nach [Bolch89] wie folgt in Abhängigkeit von λ berechnen:

$$\rho_i = \frac{\lambda_i}{m_i\cdot\mu_i} = \frac{\lambda\cdot e_i}{m_i\cdot\mu_i}.$$

P_{m_i} ist die Wahrscheinlichkeit , daß ein ankommender Auftrag nicht sofort bedient werden kann. Für P_{m_i} gilt folgender Zusammenhang:

$$\text{(G5)} \quad P_{m_i} = P(k_i \geq m_i) = \frac{(m_i\cdot\rho_i)^{m_i}}{m_i!\,(1-\rho_i)}\cdot p_i(0)\,,$$

wobei $p_i(0)$ die Wahrscheinlichkeit angibt, daß sich im Knoten i keine Aufträge befinden.
Für $p_i(0)$ gilt:

$$\text{(G6)} \quad p_i(0) = \left[\sum_{k=0}^{m_i-1}\left(\frac{m_i\cdot\rho_i}{k!}\right)^k + \left(\frac{(m_i\cdot\rho_i)^{m_i}}{m_i!\cdot(1-\rho_i)}\right)\right]^{-1}.$$

Da man im Fall $\rho_i = 1$ davon ausgehen muß, daß sämtliche Aufträge im Knoten i sind, wird der Korrekturfaktor so gewählt, daß wenn ρ_i gegen 1 konvergiert $\bar{k}_i$ gegen K konvergiert:

$$\lim_{\rho_i\to 1}\bar{k}_i \to K$$

Die Wahl des Korrekturfaktors für die Funktion $f_i(\lambda)$ ist nicht eindeutig, und man kann deshalb auch andere Funktionen zur Berechnung der mittleren Auftragszahl $\bar{k}_i$ verwenden (Beispiel siehe [Fisch92]). In der Praxis hat sich jedoch die Formel (F1) bewährt und durchgesetzt.

Da für die Auslastung ρ_i in jedem Knoten i die Ungleichung

$$0 \leq \rho_i \leq 1 \quad \forall (i \in \{1, \ldots, N\})$$

gilt, und für die Ankunftsrate λ_i im i-ten Knoten die Formel

$$\lambda_i = \mu_i\cdot m_i\cdot\rho_i \quad \forall (i \in \{1, \ldots, N\})$$

gilt, kann man für die Ankunftsrate λ_i folgende Einschränkung machen:

$$0 \leq \lambda_i \leq (\mu_i\cdot m_i) \quad \forall (i \in \{1, \ldots, N\})$$

Da nun aber für den Durchsatz λ des Gesamtnetzes der Zusammenhang

$$\lambda = \frac{\lambda_i}{e_i} \quad \forall (i \in \{1, \ldots, N\})$$

besteht, gilt für den Durchsatz λ auch die Einschränkung

$$0 \le \lambda \le min_{i=1}^{N} \left(\frac{\mu_i \cdot m_i}{e_i} \right) \quad \forall (i \in \{1, \ldots, N\}).$$

Initialisierung des Startwerts:

Für den Durchsatz λ des Netzes gilt die Beziehung:

$$\lambda = \frac{\mu_i \cdot m_i}{e_i} \cdot \rho_i \quad \forall (i \in \{1, \ldots, N\}) \text{ mit } 0 \le \rho_i \le 1 \quad \forall (i \in \{1, \ldots, N\}).$$

Bei der Engpaßanalyse nimmt man an, daß die Auslastung ρ_{bott} im Engpaßknoten ziemlich nahe bei 1 liegt und wählt für den Durchsatz λ folgenden Startwert:

$$\lambda^{(0)} = min_{i=1}^{N} \left(\frac{\mu_i \cdot m_i}{e_i} \right) = \frac{\mu_{bott} \cdot m_{bott}}{e_{bott}},$$

wobei der Minimalwert im Engpaßknoten liegt.

Iteration:

Bottapprox ist ein iteratives Verfahren, das in jeder Iteration für jeden Knoten i die mittlere Auftragszahl $\bar{k}_i$ in Abhängigkeit des gewählten Durchsatzes $\lambda^{(k)}$ berechnet:

$$\bar{k}_i = f_i(\lambda_i) = \tilde{f}_i(\lambda).$$

Für die Funktion f_i wird die approximative Funktion (F1) gewählt.
Danach werden alle mittleren Auftragszahlen $\bar{k}_i$ aufsummiert:

$$g(\lambda) = \sum_{i=1}^{N} \bar{k}_i,$$

die Summe $g(\lambda)$ wird mit der tatsächlichen Auftragszahl K verglichen und es wird ein Korrekturfaktor kor

$$kor = \frac{K}{g(\lambda)}$$

eingeführt.
Es wird nun angenommen, daß in jedem Knoten i der Fehler bei der Berechnung der mittleren Auftragszahlen $\bar{k}_i$ gleich groß ist, d.h. in jedem Knoten müßte

$$\overline{k_i^{(k+1)}} = \overline{k_i^{(k)}} \cdot \frac{K}{g(\lambda)}$$

gelten.
Für den Engpaßknoten bott berechnet man nun diese neue mittlere Auftragszahl $\overline{k_{bott}^{(k+1)}}$:

$$\overline{k_{bott}^{(k+1)}} = \overline{k_{bott}^{k}} \cdot \frac{K}{g(\lambda)}$$

und ermittelt die neue Auslastung $\rho_{bott}^{(k+1)}$ in Abhängigkeit der neuen mittleren Auftragszahl $\overline{k_{bott}^{(k+1)}}$ im Engpaßknoten

$$\rho_{bott}^{(k+1)} = h(\overline{k_{bott}^{(k+1)}}),$$

wobei h die Umkehrfunktion zu $\hat{f}_i$ ist.
In Abhängigkeit der neuen Auslastung im Engpaßknoten wird nun ein neuer Wert für den Durchsatz λ errechnet:

$$\lambda^{(k+1)} = \frac{\rho_{bott}^{(k+1)} \cdot m_{bott} \cdot \mu_{bott}}{e_{bott}}.$$

Wenn keines der folgenden Abbruchkriterien erfüllt ist, wird eine neue Iteration gestartet.

Abbruchkriterien:

(1) Wenn der Korrekturfaktor kor im Intervall $[(1-\varepsilon);(1+\varepsilon)]$ liegt, wird die Iteration abgebrochen, da ein guter Näherungswert für den Durchsatz λ vorliegt.
Versuche haben ergeben, daß für $\varepsilon = 0.025$ das Verfahren gute Ergebnisse erzielt.
Wählt man ε kleiner, tritt keine Ergebnisverbesserung mehr ein, sondern es erhöht sich nur die Anzahl der Iterationsschritte.

(2) Wenn bei der Berechnung von $\rho_{bott} = h(\overline{k_{bott}})$ ein Wert größer als 1 auftritt, wird $\rho_{bott} = 1$ gesetzt und das Verfahren abgebrochen, da man sonst in eine Endlosschleife kommen würde.

(3) Um Endlosschleifen ganz ausschließen zu können, vereinbart man eine maximale Anzahl an Iterationsschritten; wird diese überschritten, wird das Verfahren abgebrochen.

Die Funktion h wird wie folgt definiert:

(1) Für die Knotentypen M/M/1-FCFS, M/G/1-PS und M/G/1-LCFS-PR kann man ganz leicht die Umkehrfunktion zu f_i bilden:

$$h(\overline{k_{bott}}) = \frac{\overline{k_{bott}}}{1+\frac{K-1}{K}\cdot\overline{k_{bott}}}$$

(2) Für den Knotentyp M/G/0-IS lautet die Umkehrfunktion zu f_i:

$$h(\overline{k_{bott}}) = \frac{\overline{k_{bott}}}{K}$$

(3) Für die Knotentypen M/G/1-FCFS und G/G/1-FCFS ergibt sich die Funktion

$$h(\bar{k}_{bott}) = \frac{1}{2\cdot(a-b)}\cdot\left(-(1+b\cdot\bar{k}_{bott})+\sqrt{(1+b\cdot\bar{k}_{bott})^2+4\cdot\bar{k}_{bot}\cdot(a-b)}\right)$$

mit

$$a = \frac{c_B^2+1}{2} \text{ und } b = \frac{K-1-a}{K-1}.$$

(4) Für die Knotentypen M/M/m-FCFS und -/G/1-FCFS kann man die Umkehrfunktion zu f_i nicht mehr bilden. Deshalb wählt man eine geeignete Approximation:

$$h(\bar{k}_i) = \frac{b-\sqrt{c}}{2\cdot f\cdot m_i}$$

mit

$$f = \frac{K-m_i-a}{K-m_i},\ a = \frac{1+c_B^2}{2},\ b = \bar{k}_i\cdot f+m_i+P_{m_i}\cdot a \text{ und } c = b^2-(4\cdot m_i\cdot\bar{k}_i\cdot f).$$

Da P_{m_i} in Formel (F1) nach Gleichung (G5) in Abhängigkeit der Auslastung ρ_i berechnet wird, ist die Funktion f_i zu komplex, um sie nach ρ_i auflösen zu können, d.h. es nicht möglich die Umkehrfunktion zu f_i zu bilden.
Da die Engpaßanalyse ein Iterationsverfahren ist, kann man die Ergebnisse schrittweise verbessern. Im (k+1)-ten Iterationsschritt kann man P_{m_i} mit dem Wert für ρ_i aus dem k-ten Iterationsschritt berechnen und in die Gleichung (F1) einsetzen. Damit bekommt man dann die Gleichung

$$k_i^{(k+1)} = m_i\cdot\rho_i^{(k+1)}+\frac{\rho_i^{(k+1)}}{1-\left(\frac{K-m_i-a}{K-m_i}\cdot\rho_i^{(k+1)}\right)}\cdot a\cdot P_{m_i}(\rho_i^{(k)})$$

Diese Gleichung kann man nach $\rho_i^{(k+1)}$ auflösen und man kommt so zu der oben angegebenen Funktion h.

In C-ähnlicher Notation kann man den Algorithmus wie folgt darstellen:

Initialisierung:

$$\lambda = min_{i=1}^{N}\left(\frac{\mu_i \cdot m_i}{e_i}\right);$$

$bott = i;$

Berechnungsvorschrift:

do{

1. Berechnung der Leistungsgrößen:

for (i = 1; i <= N; i++) {

$$\lambda_i = \lambda \cdot e_i; \qquad \rho_i = \frac{\lambda_i}{m_i \cdot \mu_i}; \qquad \bar{k}_i = f_i(\lambda); \}$$

2. Berechnung des neuen Durchsatzes λ:

$$g(\lambda) = \sum_{i=1}^{N} \bar{k}_i; \qquad \bar{k}_{bott} = \bar{k}_{bott} \cdot \frac{K}{g(\lambda)}; \qquad \rho_{bott} = h(\bar{k}_{bott});$$

$$\lambda = \frac{\rho_{bott} \cdot m_{bott} \cdot \mu_{bott}}{e_{bott}};$$

} while (($\left|\frac{K}{g(\lambda)} - 1\right| > \varepsilon$));

Letzte Korrektur:

for (i = 0; i < K; ++i)

$$\bar{k}_i = \bar{k}_i \cdot \frac{K}{g(\lambda)}$$

Abbildung 1: Algorithmus Bottapprox in C-ähnlicher Notation

4. Vergleich und Bewertung

Um das Verfahren Bottapprox beurteilen zu können, wurden 64 Produktformnetze und 96 Nichtproduktformnetze untersucht und mit den Ergebnissen des approximativen Verfahren Summationsmethode verglichen. Als Bewertungsgrundlage für die Produktformnetze dienen die Ergebnisse der Mittelwertanalyse, für die Nichtproduktformnetze die der Simulation mit einen 2,5%-igen Konfidenzintervall.

Als Testgrößen werden der Durchsatz λ und die mittlere Verweilzeit mvz im Gesamtnetz, sowie die mittlere Anzahl der Iterationsschritte betrachtet.

Für λ und mvz werden jeweils die relativen Abweichungen vom exakten Wert genommen. Für die relative Abweichung gilt dabei:

$$relative \text{ Abweichung} = \left|\frac{Wert_{exaktes \text{ Verfahren}} - Wert_{approximatives \text{ Verfahren}}}{Wert_{exaktes \text{ Verfahren}}}\right| \cdot 100$$

Da der Rechenaufwand pro Iterationsschritt für beide Verfahren nahezu gleich ist, kann man die Anzahl der Iterationsschritte als geeignetes Maß für die Rechengeschwindigkeit nehmen.

Für die Ergebnisse werden jeweils der minimale Wert (min), der Mittelwert (mittel) und der maximale Wert (max) der relativen Abweichung von λ und mvz, sowie die Anzahl der Iterationsschritte genommen. Die Beschreibung der Testnetze findet man in [Fisch92].

160 getestete Netze ergaben eine mittlere relative Abweichung von ca. 3% für den Gesamtdurchsatz und ca. 5% für die mittlere Antwortzeit gegenüber den exakten Verfahren Mittelwertanalyse bzw. Simulation. Dies ist eine Genauigkeit, die für die meisten Anwendungsfälle völlig ausreichend ist. Im Vergleich dazu liefert das approximative Verfahren Summationsmethode für die

gleichen Netze eine Abweichung von knapp 5% für den Gesamtdurchsatz und knapp 7% für die mittlere Antwortzeit. Zudem benötigt das Verfahren Bottapprox im Durchschnitt nur ein Drittel der Anzahl der Iterationsschritte des Verfahrens Summationsmethode, bei annähernd gleicher Anzahl von Rechenoperationen pro Iterationsschritt.

Tabellarisch kann man die Ergebnisse wie folgt darstellen:

	λ		mvz		# Iterationen	
	sum	bott	sum	bott	sum	bott
min	0	0	0	0	1	1
mittel	4.46	3.09	6.72	5.27	8.91	2.26
max	36.05	21.66	128.84	128.44	67	7

Abbildung 2: Ergebnistabelle

5. Zusammenfassung und Ausblick

Grundlage des Verfahrens Bottapprox ist eine Approximationsformel mit deren Hilfe die mittlere Anzahl von Aufträgen in den Knoten in Abhängigkeit vom Gesamtdurchsatzes des Netzes berechnet werden kann. Zusammen mit der Bedingung, daß die Gesamtanzahl der Aufträge im Netz konstant ist, ergibt sich hieraus die sog. Systemgleichung, aus der sich der Gesamtdurchsatz iterativ bestimmen läßt und damit auch andere Leistungsgrößen wie Antwortzeit, Auslastungen oder Warteschlangenlängen. Der Gesamtdurchsatz wird bei jedem Iterationsschritt neu in Abhängigkeit von der Auslastung des sog. Engpaßknotens, d.h. des Knotens mit der größten Auslastung bestimmt. Je größer die Auslastung des Engpaßknotens ist, desto geringer ist die Zahl der notwendigen Iterationsschritte, die Genauigkeit hängt nicht davon ab.

Bottapprox ist ein gutes approximatives Verfahren zur Analyse von geschlossenen Warteschlangennetzen. Gegenüber den exakten Verfahren 'Mittelwertanalyse' und 'Simulation' hat Bottapprox den Vorteil des geringeren Speicherplatzbedarfes und des wesentlich geringeren Rechenaufwandes. Gegenüber den approximativen Verfahren 'Summationsmethode' hat sie den Vorteil der etwas genaueren Resultate und des geringeren Rechenaufwandes bei gleichen Speicherplatzbedarf.

Um eine größere Vielfalt an geschlossenen Warteschlangennetzen untersuchen zu können, wäre es sinnvoll den Algorithmus der Engpaßanalyse noch auf *Mehrklassennetze* und andere *Knotentypen* zu erweitern. Die Erweiterung auf Mehrklassennetze, d.h auf Netze mit verschiedenen Auftragsklassen könnte dabei ähnlich wie bei der Summationsmethode durch [Hütter89] erfolgen. Durch Anwenden der Erkenntnisse von [Hütter89] und [Hahn90] könnte man analog zur Summationsmethode Bottapprox um Knoten mit statischen Prioritäten (sowohl mit, als auch ohne Verdrängung) erweitern. Ebenso könnte man nach den Erkenntnissen von [Brag90] die Engpaßanalyse auf asymetrische Warteräume ausdehnen.

6. Literaturverzeichnis

[Beck88] Beckmann J.: "Die Summationsmethode zur Analyse geschlossener Warteschlangennetze", Studienarbeit am IMMD 4 der FAU Erlangen-Nürnberg, 1988.

[Bolch89] Bolch G.: "Leistungsbewertung von Rechensystemen mittels analytischer Warteschlangenmodelle", B.G. Teubner, Stuttgart, 1989.

[Brag91] Brag W.: "Analytische Methoden zur Leistungsbewertung von Warteschlangennetzen", Diplomarbeit am IMMD 4 der FAU Erlangen-Nürnberg, 1991.

[Fisch92] Fischer M.: "Engpaßanalyse von Warteschlangennetzen", Studienarbeit am IMMD 4 der FAU Erlangen-Nürnberg, 1992.

[Hahn90] Hahn T.: "Erweiterung und Validierung der Summationsmethode", Diplomarbeit am IMMD 4 der FAU Erlangen-Nürnberg, 1990.

[Hütter89] Hütter T.: "Erweiterung der Summationsmethode zur Leistungsbewertung von Rechensystemen", Studienarbeit am IMMD 4 der FAU Erlangen-Nürnberg, 1988.

[PEPSY92] G. Bolch, M. Kirschnik, "PEPSY Performance Evaluation and Prediction SYstem", Interner Bericht am IMMD 4 der FAU Erlangen-Nürnberg, 1992.

OPTIMIZATION OF FIXED TIME POLLING SCHEMES

Sem Borst und Onno Boxma, CWI Amsterdam, NL
Jeroen Harink und George Huitema, PTT Groningen, NL

Recently PTT Telecom has introduced itemized telephone billing. A mediation system has been implemented to manage the extensive data transmission that is involved in itemized telephone billing. The mediation system may be modeled as a polling system. A polling system basically consists of several queues cyclically attended to by a single server. Technological restrictions in the mediation system that is currently implemented require a fixed time polling scheme. A fixed time polling scheme specifies not only the order of the visits to the queues but also the starting times.

We propose an approach to the problem of constructing efficient fixed time polling schemes, i.e., fixed time polling schemes that minimize a weighted sum of the mean waiting times at the various queues. As fixed time polling schemes are not likely to be amenable to an exact analysis, we resort to approximations for the mean waiting times at the various queues. Subsequently we formulate the problem as a mathematical programming problem. As it is NP-hard, we solve the problem heuristically by dividing it into subproblems that appear easier to handle, viz., successively:

1. Determination of the visit numbers;
2. Determination of the visit order;
3. Determination of the visit times.

The approach that we propose is tested by numerically investigating the fixed time polling schemes obtained, for realistic as well as for fictive system configurations, the latter constructed by systematic variation of the system parameters. The fixed time polling schemes obtained appear to be efficient compared with some other natural fixed time polling schemes.

Product forms for stochastic Petri nets

Richard J. Boucherie, CWI, Amsterdam

In the literature on queueing networks in equilibrium product form distributions play a vital role, and many product form queueing networks have been discovered. For example Jackson networks, and their generalisation to BCMP-networks play a fundamental role in the queueing network literature. In contrast, for stochastic Petri nets product form results seem to be less common. An immediate explanation of the discrepancy between queueing networks and Petri nets is that resource sharing and competition over resources cannot be modelled by queueing networks. It are these properties of stochastic Petri nets that destroy product form equilibrium distributions. An interesting exception are the Petri nets described in Lazar and Robertazzi, where conditions are given under which a Petri net has a product form equilibrium distribution. These conditions basically come down to independence conditions on 'firing sequences' in subnets of the Petri net. As a consequence, also if competition over resources occurs a product form distribution is obtained. It must be noted here, that the results of this reference are valid for Petri nets containing linear firing sequences only, and that in each place of the Petri net at most one token is allowed. It is the purpose of this talk to show that the results of Lazar and Robertazzi can immediately be generalised to Petri nets consisting of arbitrary subnets if an 'independence condition' similar to the condition given in Lazar and Robertazzi is satisfied.

The framework used in this talk is that of a collection of Markov chains. For this collection the product process is introduced: let S_k, q_k, $k = 1, \ldots, K$, be the state spaces resp. transition rates of the Markov chains in the collection, then $S = S_1 \times \cdots \times S_K$ is the state space of the product process, and the transition rates of the product process in dimension k are given by q_k. It is assumed in this framework that in each transition of the product process the state in one dimension changes only, that is in each transition of the product process one of the underlying Markov chains changes its state only. Competition over resources can then be modelled as exclusion of a part of the state space S, say the product process cannot enter $A \subset S$. The independence condition guaranteeing product form roughly states that *if the product process is in state* $\bar{n} = (\bar{n}_1, \ldots, \bar{n}_k, \ldots, \bar{n}_K)$, *where* $\bar{n}_i \in S_i$, *and* $\bar{n}' \in A$, *where* $\bar{n}' = \bar{n}$ *except for component* $\bar{n}_k$ *(i.e.* $\bar{n}'_k \neq \bar{n}_k$, $\bar{n}_i = \bar{n}'_i$, $i \neq k$*) then Markov chain* k *cannot change its state.* Under this condition the product process has a product form equilibrium distribution at $S \setminus A$ given by $\pi(\bar{n}) = B \prod_{k=1}^{K} \pi_k(\bar{n}_k)$, where π_k is the equilibrium distribution of Markov chain k at state space S_k, and B is a normalising constant determined by the exact form of A.

Lazar, A.A. and Robertazzi, T.G. (1991) Markovian Petri net protocols with product form solution, *Performance Evaluation*, **12**, 67-77.

THE USE OF ADJUSTED TRANSFER RATES IN QUEUEING NETWORKS WITH BACKGROUND LOAD

Hans Daduna, Universität Hamburg
Rolf Schassberger, TU Braunschweig

Product form networks are well suited to model the behaviour of complex production systems if quick and easy to apply analytical methods of performance analysis are requested for. When studying only a part of the network it is common to isolate this part and modeling the load, which is put on the subnetwork from external modes by independent Poisson processes. It has been observed in some special models of telecommunication theory that these "background load processes" may be substituted by carefully adjusting the service rate ("transfer rate") of the nodes.

This could be done in a way that the first-order quantities as throughput, mean transfer time, mean waiting time, etc., of the first system are truly reproduced by the system with "adjusted transfer rates". We first generalize this "method of adjusted transfer rates" and then show that the assumptions which are put on the system to reproduce exactly first-order performance characteristics are sufficient for reproducing transfer time characteristics of higher order, even transfer time quantiles - as long as these are accessable by analytical methods in considering overtake-free paths. We present examples which show that such paths appear naturally e.g. in modeling job-shop like queueing systems.

MAINTENANCE MANAGEMENT AND AVAILABILITY ASSESSMENT DURING THE LAST DECADE

Rommert Dekker, Erasmus University Rotterdam, NL
Marcel A.J. Smith, Erasmus University Rotterdam, NL
Frank A. Van der Duyn Schouten, Tilburg University, NL

One of the most important developments in the area of maintenance mangement and availability assessment during the last ten years is undoubtedly the increasing interest in industry for a scientific approach of these management areas. Apart from economic reasons also the shifting attention in the scientific world towards more realistic and applicable models contributed to this development to a large extent. In this lecture the major achievements of the last ten years will be reviewed.

First we present an overview of the new mathematical models on multi-component maintenance, which are strongly related with queueing models. Most multi-component models are based on the possibility to acquire economies of scale by grouping maintenance activities together. However, also situations where simultaneous maintenance on similar units should be avoided as much as possible are considered.

Two other directions in which maintenance modelling has been developed during the last few years are opportunity maintenance, and the integration of maintenance within production. In opportunity maintenance models the limited availability of maintenance possibilities is explicitly taken into account. Integration of maintenance and production into one model is based on the idea that by fine-tuning of maintenance and production considerable improvements of overall performance can be achieved.

As far as the techniques of mathematical analysis of maintenance problems is concerned attention will be given to new or revived techniques, like marginal cost analysis, delay time analysis, aggregation and disaggregation and the so-called method of stages.

Finally we will mention some software tools, developed either to support managerial decision making in practice or to compare different approaches (in particular simulation versus approximative analysis).

JOINT PREVENTIVE REPLACEMENT IN AN OPERATIONAL PLANNING PHASE

Rommert Dekker, Erasmus University Rotterdm, NL
Ralph Wildeman, Erasmus University Rotterdam, NL

In this paper we address the problem of combining preventive replacements in a multi-component maintenance system in an operational planning phase. Given an infinite or a finite time horizon, decisions concerning replacement of components have to be made at discrete time epochs within this horizon. We present a heuristic procedure that specifies an action for a given epoch, based on the current state of the system and on an expected behaviour of the system in the near future.

Firstly, for each component an expected initial replacement moment is determined, according to a control limit that has been obtained for each component separately, using a Markov decision model. These initial moments induce a finite planning horizon that starts at the current epoch, and in which the moments may be shifted so as to allow joint replacement, which saves set-up costs. However, these shifts usually imply a deviation from the initial plan, which may involve penalty costs. The objective is to find an optimal grouping of the replacements planned within the current planning horizon. The combination of components jointly replaced at the current epoch, corresponds to the action to be chosen at this epoch. This is repeated for each epoch. Thus we apply a 'rolling horizon' within the given infinite or finite time horizon. With numerical examples and a comparison with the optimal strategy, both for an infinite and a finite time horizon, we aim to illustrate and to justify our approach.

"WHERE WOULD WE BE WITHOUT TRAINS?"

A queueing modelling to determine train delays

Nico M. van Dijk, University van Amsterdam

Abstract: *For strategic planning purposes of required Railway Network Capacities on a long term basis, the Dutch Railway Company aims to obtain insights in where scheduling conflicts may arise at railway stations. A queueing model is developed to compute conflict probabilities at railway stations. An analytic solution is provided under an unrealistic loss assumption. This solution, however, still reflects realistic qualitative aspects and provides an indication of orders of magnitude.*

The results are applied to a real-world railway station supported by a graphical software program. The underlying assumptions are tested by simulation. These results indicate a practical usefulness of the model and its analytic solution.

1. INTRODUCTION

1.1 Background

The Dutch Railway Company (Nederlandse Spoorwegen) aims to reach high quality service standards for their train users despite the continuing if not explosive growth in public transportation usage (From 1980 through 1990 the average number of train users has nearly to doubled). Particularly, a large scale program is set up under the name of RAIL 21 to achieve such goals for beyond the year 2000. This program roughly focuses at three aspects:

1. To provide a sufficient **global infrastructure** (new routes for TGV's and transport, international connections etc.).
2. To achieve a high degree of **customer service** (comfortable trains, accurate information, good train connections etc.).

And last but not least:

3. To determine an "acceptable degree" of **capacities** and resources (railway stations, platforms, rail sections, junctions etc.).

Here the word "**acceptable**" is to be interpreted for two parties:
(i) For the customers as not leading to *unacceptable* delays and conflicts
(ii) For the Railway Company to realize a *proper scheduling* of all the trains.

The first aspect is one of a more macro-economic and political flavour, while the second one of a marketing and customer service nature. The third, however, is one of a more technical detailed level that actually requires quantitative insights (thus by some performance measurements) in capacity aspects and their consequences for train users.

1.2 Capacity planning

More precisely, the third aspect can be reformulated at three levels:

- At **strategic** level one needs to obtain insights in bottlenecks particularly at railway stations during 2000-2010.
- At **operational** level this means that detailed trainschedules can be made without running into substantial conflicts.
- At **performance and customer** level this translates to **delays** that trains (and thus customers) may experience due to the fact that there is (temporarily) insufficient capacity of common resources (infraelements such as rail sections, platforms, and junctions).

An example (Arnhem station)

For example, for the railway station Arnhem as in the figure below, at first glance it is clear that the east-side functions as a global bottleneck. But how strongly? Does it really lead to conflicts in making a schedule and how tight will this schedule be given that no more than approximately 60 % of time the railsections at the east-side are used (follow-up times included). These conflicts also depend on the availabilities of free platforms, rail sections and junctions, the different routes of the trains (15-20 routes per hour) and their interdependencies (e.g. which trains require the same platform). It thus depends on the total railway station as a whole. Last but not least, to what extend do these conflicts actually lead to train delays (just incidentally for a couple of minutes, or once every so many times for a specific train, or rather consistently for several trains at the same time).

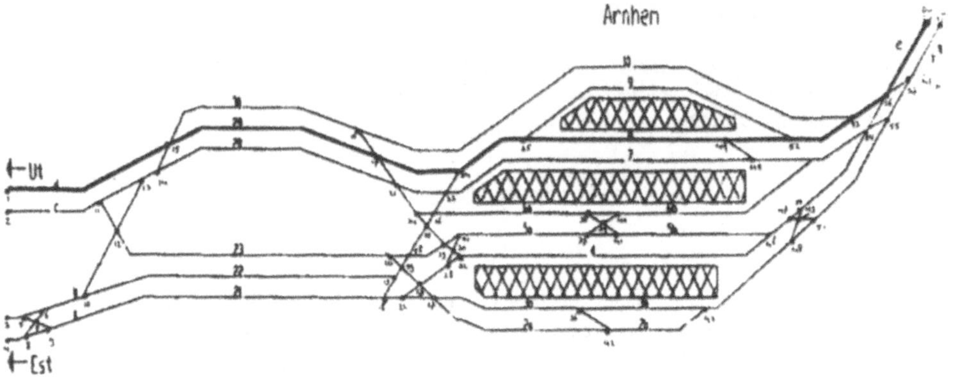

1.3 Why a problem with detailed schedules?

Here one might question why at all conflicts in trainschedules and at stations may arise given that the fractions of time that capacities are used are still far below 100 % (where a passing train is usually calculated as a 2-3 minutes occupation time) and given that trains can be scheduled deterministically. For example, for a simple junction or station with 2 trains on the hour east-west and 2 north-south, one could simply schedule them on the half-hour and a quarter apart, so that there will never be any conflict. This is indeed true, for a single junction or even station.

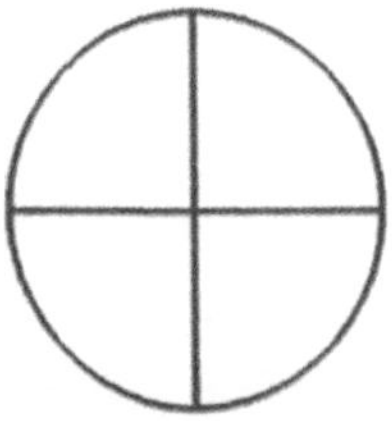

But for one thing by fixing an appropriate detailed schedule at one station one also fixes various trains for the neighbouring stations. One thus necessarily also need to take into account the neighbouring schedules etc. Consequently, one necessarily needs to make a total schedule for all stations simultaneously. This is a highly complex task that despite costly computerization still requires a couple of month time (at present still 3-4 month) for actual realization and fine-tuning. (Further software development to this end is under intensive study but principally it will remain a costly and laborious job).

Secondly, a fixed schedule just plans and assumes one (ideal) situation but it does not directly provide qualitative insights in for example the effects of an alternate route, of changes in the train frequencies or of disturbances by a variety of possible causes. Particularly one would like to know whether the infrastructure is sufficiently robust for such changes or whether and where it is highly sensitive.

Thirdly, a detailed schedule requires a detailed input of train frequencies and routes. However, the underlying motivation is the investigation of infrastructural capacities at a long-term basis, that is for beyond the year 2000, for which only very rough prognoses of such inputs are available.

1.4. Global objective

On the other hand, in order to investigate the question of sufficient infra-element capacities on a long-term basis, one is merely interested in a global performance indication and not in detailed schedules. In other words:

A global performance evaluating tool would be sufficient.

One common tool to such ends would be the technique of simulation.
However, in large the various points raised above also apply to simulation as summarized by:

- It is still costly and time-consuming
- It requires detailed input and technical assumptions
- It is primarily directed to quantitative results for a fixed situation and much less appropriate for a qualitative comparison of different situations or assumptions, other than by a large number of runs.

This is not to say that simulation is out of the picture as a highly supportive tool (as will also be illustrated later on). But for the purpose of prime interest:

To obtain indications of qualitative results for changes as mentioned above,

also a simple evaluation tool (for example one that could be run by a PC in a couple of minutes) would be of practical interest. The development of such a tool was the objective of our study. More precisely, that is:

1.5 Detailed objective

To develop a simple tool to obtain quick insights in the performance of a railway station in terms of train conflict or delay probabilities, particularly to get an indication of the effect of changes and of the (in)sensitivity to input assumptions. (train-frequencies, passing times, routes, etc.)

Conflictscore method
A first natural thought here is that of a conflictscore method which calculates *a cónflictscore* proportional to the number of possible conflicting trains. For example, in the earlier simple situation **x** trains North-South and **y** trains West-East would lead to a conflictscore **xy**, as if these trains are totally randomly distributed over the hour (e.g. with follow-up times of 3 minutes per train). Such a method however, does not seem to give a realistic indication of quality. For example, the two situations below both give a conflictscore 20, but they are quite incomparable in quality and realization.

$$\begin{Bmatrix} x = 5 \\ \\ y = 4 \end{Bmatrix} \qquad \begin{Bmatrix} x = 20 \\ \\ y = 1 \end{Bmatrix}$$

Essentially, this discrepancy arises as this method does not take into account:

- Dependencies on the hour (like fixed interarrival times)
- Dependencies of all infra-elements along a train route in stations.

In response to our objective we thus have to develop a model that can be analyzed rather easily on the one hand but that does take into account these dependencies on the other. As measure of interest we still study the **conflict probability** of arriving trains.

2. QUEUEING NETWORK MODEL

We can regard a railway station as a simple two-station queueing network as follows.

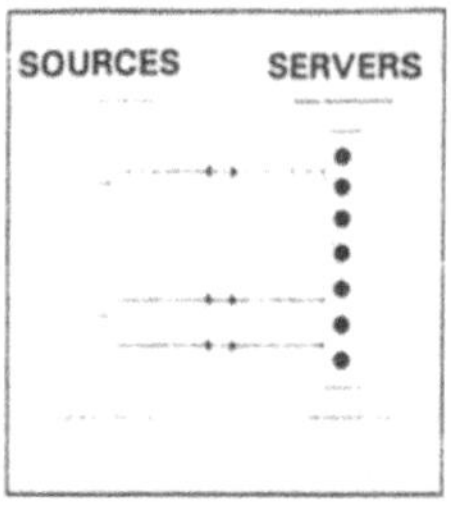

Let each different trainroute in the station be identified as a *source (or user)* that from time to time (realistically at fixed intervals) generates a *job* (representing the passing through the station by a train on that route). This job simultaneously requires a couple of *servers* (which represent the various infra-elements as rail sections, junctions, crossings and the platform that are to be reserved as a whole for letting the train enter a station). When upon a generation (= train arrival) not **all** required servers (= all infra elements along its path) are free, the job (= train) is blocked and realistically it has to wait until its path is given free (say in some priority order, e.g. first-come, first-served).

We thus have a fixed **number of sources** (the possible routes), say labelled by $h=1,\ldots,M$, a fixed **number of servers**, (the various infra-elements), say labelled by $s=1,\ldots,S$ and for each source h a **set of servers**, $S_h \subset \{1,\ldots,S\}$ which are required simultaneously.

The intergeneration (= interarrival) and service (= passing through) times are assumed to have an intensity of λ_h and μ_h per hour for source h, (e.g. $\lambda_h=2$ and $\mu_h=15$ mean that source h generates 2 trains per hour each with an average passing through time of 1/15 hour = 4 minutes). The distributions of these times are assumed to be arbitrary but typically may be thought of as being rather deterministic. (Fixed interarrival times and passing through times with a small normal random fluctuation).

No-analytic solution (literature)
The present description has direct similarities with circuit switch telecommunication systems in which a telephone call requires a total path of links to be free during the whole duration of a Call (e.g. [4]). When blocked call requests are delayed no analytic solution such as for the blocking probabilities has been reported in the corresponding telecommunication literature. In the queueing literature systems with simultaneous servicing can be classified in systems in which simultaneous servers are given free independently (the *independent* case) and all at the same time (the **concurrent** case). In essence, the description presented here pertains to the concurrent case. For the independent case some analytical and approximative results have been reported (cf. [1], [2], [3], [5]). For the concurrent case with delays no explicit formulae at all are available other than an approximate expression in [6] under the assumption of indistinguishable sources. In the present setting however, the distinction of different sources is quite essential.

3. ANALYTIC SOLUTION

In order to establish an analytic solution for the system described, where the source (= train) distinctions and interdependencies are considered as crucial aspects that cannot be disregarded, the following assumptions are made.

1. Jobs are generated as by a *finite source input* (that is, when a source has generated a job, it cannot generate any other job before its service is completed).

2. When a job is blocked to enter as its path is not free, it is lost. In queueing terms that is, we study a *loss* rather than a *delay problem.*

While the first assumption is a technical minor one which will turn out later to have a relatively small effect, the second one is most unrealistic in the setting of trains. However, as our prime interest is not so much to obtain accurate estimates of blocking (conflict) probabilities but much more to obtain qualitative insights in effects, among other things also reflected in conflict probabilities, of changes and alternatives, the assumptions might still lead to results that have a qualitative or indicative value. This will be supported later on.

To present our result let a state $H \subset \{1,\ldots,M\}$ represent the set of sources that are active (that is, of which a job is in service) and note that the possible states are restricted to the set of admissible states:

(3.1) $$C = \{H \mid S(h_1) \cap S(H_2) = \varnothing \text{ for all } h_1, h_2 \in H\} \; .$$

Result 3.1

Under the assumptions 1 and 2 and with c a normalizing constant with respect to the set of admissible states C, the steady state distribution $\pi(H)$ has the form:

(3.2.) $$\boxed{\pi(H) = C \prod_{h \in H} \left[\frac{\lambda_h}{\mu_h}\right] \quad (H \in C)} \; .$$

Proof: See [7] for the general case. For the (unrealistic) exponential case (that is, with both the interarrival and passing through times assumed to be exponential) the proof can be given rather directly as follows. Let $H+h = H \cup \{h\}$ and $H-h = H/\{h\}$.
The global balance equations in state H then become:

(3.3.) $$\left\{\begin{array}{c} \pi(h)\ \Sigma_{h \notin H}\ \lambda_h\ 1(H+h \in C) \\ + \\ \pi(H)\ \Sigma_{h \in H}\ \mu_h \end{array}\right\} = \left\{\begin{array}{c} \Sigma_{h \notin H}\ 1(H+h \in C)\ \pi(H+h)\ \mu_h \\ + \\ \Sigma_{h \in H}\ \pi(H-h)\ \lambda_h \end{array}\right\} .$$

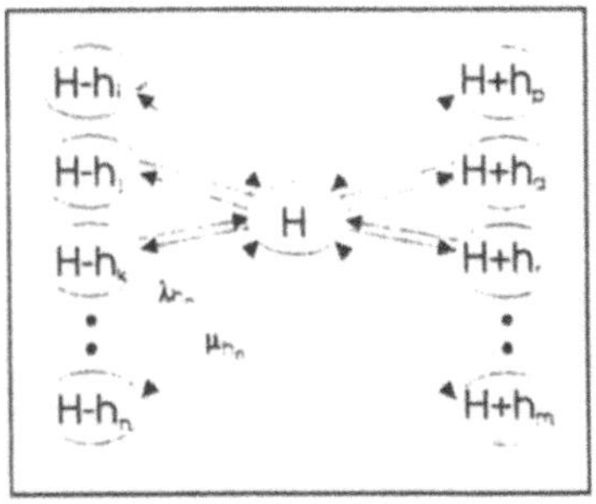

These in turn are directly reduced and verified by substituting (3.2) by the more detailed balançes:

(3.4) $$\pi(H)\ \mu_h = \pi(H-h)\ \lambda_h$$

for all $H \in C$ and $h \in H$. □

Remark 3.2 (Insensitivity)

Note that expression (3.2) depends on the interarrival and passing through times only through their *intensity values* λ_h and μ_h. The actual form of randomness that these times may experience is thus irrelevant. In queueing literature this is generally referred to as *insensitivity* (e.g. [9]). Most notably, *deterministic* interarrival and passing times are thus included. (In a pure deterministic case one may only speak of *a* rather than *the* steady state distribution. This, however, is merely theoretical as randomness, even though minor is realistically involved).

Remark 3.3 (Conflict probabilities)

From expression (3.2) one can easily compute various measures of interest such as the conflict probability of a train h as:

$$(3.5)\quad B^h = \Sigma_{H\in C}\, \pi(H)\; 1(H+h \notin C)\,,$$

or the fractions of time that certain groups of infra-elements are occupied. This will be illustrated graphically below.

4. AN APPLICATION

The results of section 3 have been applied to the "real-world railway station of Arnhem" given in figure 2 for their daily *peak-period*. In this period their are 15 different routes, 10 of which with a train frequency 1 and 5 of which with a train frequency of 2 per hour. All infra-elements (rail sections, junctions, platforms and endpoints were numbered 1...85. In figure 3 below these are set out horizontally. Each route is represented by a row in which all dark elements are the infra-elements used along that route. The different grey/dark tints (in real colours) represent the different *time fractions* that these elements are occupied (loads). The grey/dark tints (in real colours) on the left for the 15 routes represent the *conflict probabilities* for these routes. For the size of this realistic complex situation this figure can be computed in a minutes time.

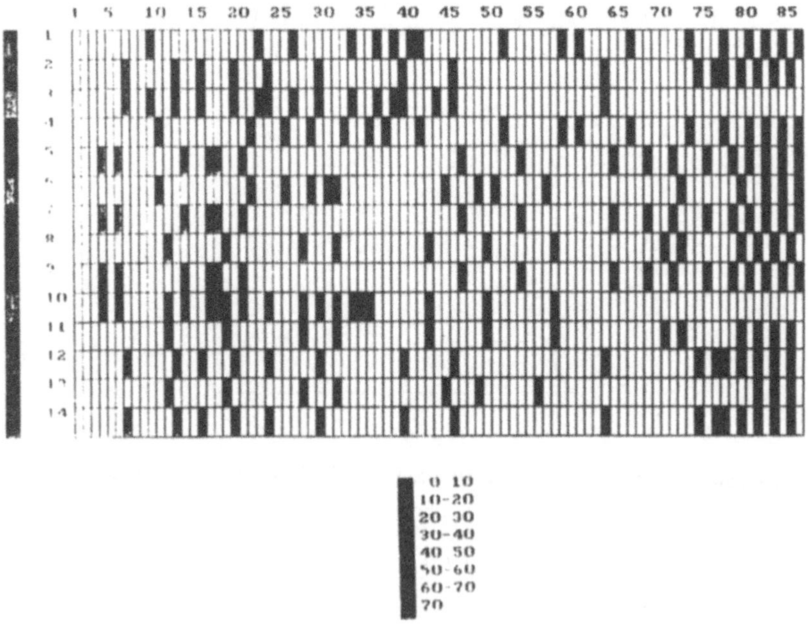

The results of section 3 in combination with this graphical support (which can also be made in a real graphical station version rather than in a matrix) directly enable one to detect bottlenecks and to see effects of various sorts of changes such as an alternate routing of a specific train.

Check of assumptions

To check the usefulness of the results for qualitative purposes as well as the underlying assumptions 1 and 2, a simulation program was written. Some numbers are given below which represents the experienced loads of specific crucial infra-elements. The first column gives the results as based on the analytic solution (3.2), the second column provides the simulation results under deterministic arrival times under assumption 1, the third column without assumption 1 and the last column also without assumption 2. The first assumption thus appears quite reasonable. The second gives rise to a consistent factor of approximately 1.5 (as could have been expected beforehand) but does not have a significant effect on qualitative results. (For example, the ratios for elements 1 and 3 etc.).

Element	Analytic	Simulation	Assumption 1	Assumption 2
1	28.63	28.08	28.49	45.44
2	28.01	26.72	25.69	44.41
3	12.16	12.40	10.90	19.87
4	20.01	20.72	19.31	30.24
12	7.87	7.44	5.16	20.27
21	31.24	30.56	31.24	50.01
25	3.00	2.32	2.4	4.28

Evaluation

Based on an analytic solution a simple performance tool has been provided to obtain qualitative insights in loads and conflict probabilities at railway stations. This tool may provide quick indications of where capacity expansions are needed, where routing changes would be preferred or where extra loads can be placed.

Acknowledgement

The author wishes to acknowledge the Dutch Railway Company (NS), in casu drs. J. van der Berg, for the approval of this paper and the support provided for its underlying research. He is also grateful to the Centre of Applied Mathematics of the Rabo-bank (CTW) for their assistance in the simulation.

References

[1] Federgruen, A. and Green, L.: *"An M/G/c queue in which the number of servers required is random"*, Columbia Bus. Sch. Res. Paper 504A, Columbia University, (1983).

[2] Green, L.: *"A queueing system in which customers require a random number of servers"*, perations Res. **28**, 1335-1346, (1980).

[3] Green, L.: *"Comparing operating characteristics of queues in which customers require a random number of servers"*, Management Sci, 27, 65-74, (1981).

[4] Shwartz, M.:*"Telecommunication Networks*, Addison Wesley, (1987).

[5] Seila, A.F.: *"On waiting times for a queue in which customers require simultaneous service from a random number of servers"*, Operations Res. 32, 1181-1184, (1984).

[6] Tijms, H.C.: *"Stochastic Modelling and Analysis"*, Wiley New York, (1986).

[7] Van Dijk, N.M.: *"A Queueing Model for Railway Stations: An Analytic Solution"*, Technical Report in possesion of the Dutch Railway Company (1993).

[8] Van Dijk, N.M. and Tijms H.C.: *"Insensitivity in two-node blocking models with applications"*, Teletraffic Analysis and Computer Performance Evaluation, North-Holland, Amsterdam, 329-340, (1986).

A MATRIX-GEOMETRIC ANALYSIS OF DISCRETE TIME QUEUEING SYSTEMS WITH PERIODIC SERVICE INTERRUPTIONS

Michel van Eenige, Eindhoven University of Technology, NL

In production systems as well as real time computer systems, the servicing of customers can be interrupted. These interruptions may occur fairly random (e.g., server breakdowns, rush orders, deficiencies of material or information) or may follow a more periodic (cyclic) pattern (e.g., periodic production schemes, safety maintenance, clocked software schedules).

In this talk, we consider a discrete time queueing system with scheduled interruptions, each of them taking a fixed amount of time. At this system, customers arrive according to a non-homogeneous Bernoulli process with a discrete phase type service time distribution.

By applying the matrix-geometric approach, we obtain the equilibrium distribution of the number of customers in the system. From this equilibrium distribution, we can compute the stationary sojourn time distribution and study the effects of interruptions on the probability a customer receives service before some specific due date. Moreover, examples show the influence on this probability of balancing interruptions over a cycle (e.g., by splitting an interruption into smaller ones). Finally, we consider some generalizations of this model.

[1] Eenige, M.J.A. van, J.A.C. Resing, J. van der Wal, A Matrix-Geometric Analysis of Queueing Systems with Periodic Service Interruptions, Eindhoven University of Technology, Department of Mathematics and Computing Science, Memorandum COSOR 93-32, November 1993

THE SYMMETRIC LONGEST QUEUE MODEL

Geert-Jan van Houtum, Eindhoven University of Technology, NL

Suppose we have a number of identical production lines, where at the end the finished products are packed by one server. Each production line has its own finite buffer in which finished products are stored before being handled by the server. If a production line finishes a product and the buffer is full, then the line will be blocked until space becomes available in the buffer. To prevent blocking as much as possible the server always picks a new product from the buffer containing the largest number of finished products.

If all production times are assumed to be exponential and if the production lines are assumed to have always raw materials available to work on, then the above production system may be modeled as an exponential, symmetric longest queue model. To compute the fraction of time that each production line is blocked, we derive an upper and lower bound model which both can be analyzed by using the matrix-geometric approach of Neuts. This leads to an efficient algorithm for the computation of the minimal buffer size required for meeting some target level for the fraction of time that each production line works.

ON A REPAIRMAN ASSIGNMENT PROBLEM

Ger Koole
CWI

Consider a model with m parallel queues, each with its own server, and the same finite capacity B. All service times are exponentially distributed with the same rate. Arriving customers have to be assigned to one of the non-empty queues. It is well known that, if the decision is based on the current queue lengths, assigning to the shortest queue is optimal, for a large class of reward functions. Also the arrivals can be quite general, even allowing the arrival stream to depend in some sense on the queue lengths.

Let us now assume that we do not have the current queue lengths available, but only the queue lengths just after the previous arrival. Thus, there is a delay in the arrival of information on the system. It can be proven that with this information delay it is optimal to assign to the queue which was smallest at the previous arrival instant. Note that this queue need not be the smallest anymore (although it is, if we look at it stochastically).

This result, and some variants of it, can be used to analyze several repairman problems. Indeed, consider a system consisting of m groups of parallel components, each with at most B functioning components. The completion of a repair by the repairman corresponds to the arrival of a customer in the routing model described above. It is natural to assume that the repairman is assigned to a non-functioning component at the beginning of the repair period, representing repair on the spot. This is where the delay comes in; the repairman is assigned to a failed component at the beginning of its repair period, but the repair naturally occurs at the end. If we assume furthermore that in each group at most one component is subject to failure at a time, then this repairman model is completely equivalent with the routing model, and therefore the repairman should be assigned to the group with the smallest number of functioning components. Suitable reward functions are the total number of functioning components, or the probability that each group has at least one functioning component.

Some variants of this model lead to the same result. In the first all functioning components are subject to failure, representing a warm stand-by model. In the second we define the system to be up if each group has at least one functioning component. Only if the system is up components can fail, one at a group. The expected time that the system is up in this cold stand-by model is again maximized by assigning to the group with the smallest number of functioning components.

INSPECTION AND REPAIR DECISIONS FOR HYDRAULIC STRUCTURES UNDER ISOTROPIC DETERIORATION

Jan M. van Noortwijk, Delft University of Technology, NL
Roger M. Cooke, Delft University of Technology, NL
Matthijs Kok, Delft Hydraulics, Emmeloord, NL

The subject of study is minimising the cost of inspection and repair of a hydraulic structure. The optimisation is based on isotropic properties of the underlying physical deterioration process. Probabilities of preventive repair and failure are obtained conditional on the only available (subjective) information, i.e. the average damage with regard to a finite or an infinite time-horizon. Hereby it is assumed that the damages per time-unit are exchangeable and that the decision-maker is cognitively indifferent to the way information about the average damage is gathered. By introducing a prior we can account for uncertainty in the decision problem. Advantages of our Bayesian approach are that we base our mathematical models on an observable quantity, namely the damage, and that our results can be expressed explicitly conditional on the average damage. For more details see van Noortwijk [1] and van Noortwijk, Cooke & Kok [2].

One illustration from the field of hydraulic engineering is studied: determining a cost-optimal periodic frequency of inspecting the concrete-weighted erosion mats of the Eastern Scheldt barrier. The mats are under water and must ensure the stability of the barrier. Failure occurs when a scour hole arises and its depth exceeds a certain safety level.

References

[1] Jan M. van Noortwijk. Inspection and repair decisions for hydraulic structures under symmetric deterioration. Technical Report ESRC 92-17, University of California at Berkeley, U.S.A., June 1992.

[2] Jan M. van Noortwijk, Roger M. Cooke, and Matthijs Kok. A Bayesian failure model based on isotropic deterioration. *European Journal of Operational Research (to appear)*, 1994.

ALGORITHMISCHE UNTERSUCHUNGEN VON BMAP/G/1 BEDIENUNGSMODELLEN MIT PAUSENZEITEN

Helmut Schellhaas, Technische Hochschule Darmstadt

Bedienungsmodelle mit Pausenzeiten haben im Hinblick auf wichtige Anwendungen (etwa) in der Informatik (Rechnerverkehrstheorie) reges Interesse gefunden. Wir betrachten Einbedienermodelle mit zwei verschiedenen Pausenpolitiken, nämlich der T-Politik und der k-Politik. In beiden Modellen legt der Bediener Pausen ein, sobald das System leer wird. Während bei der T-Politik die Bedienung wieder beginnt, sobald bei einem Pausenende mindestens ein Kunde im System ist, beginnt bei der k-Politik die Bedienung wieder, sobald mindestens k Kunden im System sind. Ankunftsprozeß ist der von Lucantoni (1991) eingeführte Markovsche Ankunftsprozeß mit Gruppen (BMAP), der eine ganze Reihe bekannter Ankunftsprozesse als Spezialfälle umfaßt (Poisson-Prozeß, Phasen-Erneuerungsprozeß, Markov-modulierter Poisson-Prozeß, Neuts' versatiler Markovscher Punktprozeß).

Ziel der Untersuchungen sind Darstellungen der stationären Verteilung der Kundenzahl (und gewisser Hilfsvariabler) in numerisch brauchbarer Form. Dazu werden zunächst geeignete eingebettete Markov-Ketten mit den modernen auf M.F. Neuts zurückgehenden matrix-analytischen Methoden untersucht. Es wird gezeigt, daß bekannte algorithmische Techniken für Matrix-M/G/1- Modelle Verwendung finden können. Darauf aufbauend werden mit Hilfe der Theorie der semi-regenerativen Prozesse algorithmische Lösungen für die zeitstationäre Verteilung der Kundenzahl entwickelt. Die Berechnung dieser Verteilung kann rekursiv erfolgen basierend auf der stationären Verteilung der eingebetteten Markov-Kette, für die algorithmische Lösungen vorliegen. Schließlich werden die Bedienungsmodelle mit einer Bewertungsstruktur verknüpft und die Kostenrate bestimmt.

Eine ausführliche Darstellung findet sich in:

H. Schellhaas: Single server queues with a batch Markovian arrival process and server vacations. Preprint-Nr. 1566, Fachbereich Mathematik, Technische Hochschule Darmstadt, 1993. Erscheint in: Operations Research Spektrum.

ASSIGNING IDENTICAL OPERATORS TO DIFFERENT MACHINES

Jan van der Wal, Eindhoven University of Technology, NL

Consider a system with M different machines and s identical operators. No machine can work without the constant attention of an operator and all operators can serve all machines. At machine m jobs arrive according to a Poisson arrival process with rate λ_m and service time distribution F_m. A job arriving at machine m can be served at machine m only. Whenever an operator finishes a job a decision has to be taken about the machine where he should go to next.

There are various ways in which this could be done. The operator could choose at random one of the machines that are waiting. He could also choose at random but using weights depending on queuelength. Or he could use a criterion such as FCFS.

The performance of the system, as in the shortest queue problem, is not as good as in the s-server system because it may happen that an operator is idle while there is still a job waiting for a machine which is already busy.

For the simplest case with all jobs exponentially distributed with a mean which is independent of the machine, Durlinger and Geurts [1] give a good approximation by describing the system as a two-dimensional Markov process. One of the dimensions is the number of jobs in the system, the other dimension counts the number of empty queues.

Another good approximation is obtained by looking at the asymptotic behaviour for large m and s. As we will show this approach leads to very good approximations for the case of non exponential jobs as well (cf. [2]).

[1] Durlinger, P.P.J. and J.H.J. Geurts, Determining work order throughput times in dual constrained production departments, Eindhoven University of Technology, Department of Industrial Engineering, Working paper BDK/KBS/89-13, December 1989.

[2] Kok, A.G. de, and J. van der Wal, Assigning identical operators to different machines, Eindhoven University of Technology, Department of Mathematics and Computing Science, Memorandum COSOR 93-31, November 1993.

Rationalisation of building maintenance through Markov decision models

Chris van Winden and Rommert Dekker
Econometric Institute, Erasmus University Rotterdam

In this paper we present Markov decision models for rationalisation of building maintenance. The models were developed for the Dutch State Building Agency, which is responsible for about 4000 state owned buildings with a value of some 20 billion guilders and an annual maintenance budget of some 90 million guilders. The models are intended to increase insight into strategic maintenance decision making and are in a preliminary phase only. Each building is divided into several parts, which state is determined each year by inspection and the result classified on a 6 point scale (varying from excellent condition to a very bad condition). Assuming residence times for each state, a semi-Markov decision model (SMDM) can be formulated with several maintenance actions in each state. From norms on the maximal discounted frequencies of certain states a minimum discounted cost policy can be determined by formulating the SMDM as a linear programming model. To incorporate the initial state of buildings and yearly budget constraints, however, the SMDM has to be transformed into an equivalent Markov decision model (MDM) with the residence time as second state variable. Modelling deterioration dependencies between e.g. protective parts (painting) and protected parts, however,leads to very large states spaces. Furthermore, a discounted cost analysis does not allow the control over the long term average state frequencies. Finally, by averaging over all building parts the MDM also allows the construction of a so-called trade-off curve between the average condition of buildings and the long term average costs required to keep the buildings in that condition.

PARALLEL ITERATIVE SOLUTION METHODS FOR MARKOV DECISION PROCESSES

Thomas W. Archibald, University of Edinburgh, Scotland
Kenneth I.M. McKinnon, University of Edinburgh, Scotland
Lyn C. Thomas, University of Edinburgh, Scotland

Applications of Markov decision processes arise in many problem areas including maintenance and replacement, inventory control and queuing models. However the extent to which Markov decision processes are used in practice is limited because the solution of realistic problems on serial machines is often impractical due to their large memory and processing time requirements. This talk will consider the use of parallel processors in the solution of Markov decision processes — an approach which has long been considered a potential solution to the computational intractability of these problems on serial machines.

Pre-Jacobi, one variant of value iteration, is acknowledged as one of the best serial solution methods for Markov decision processes. The fast convergence of pre-Jacobi is due to the use of an extrapolation which exploits the stochastic property of the iteration matrices. Without this extrapolation pre-Jacobi is not competitive with other value iteration methods such as Gauss-Seidel. To preserve the good convergence properties and hence the effectiveness of pre-Jacobi, a parallel algorithm must be synchronous.

We will describe the factors which influence the efficiency of synchronous parallel versions of the pre-Jacobi algorithm on distributed memory multiple instruction multiple data (MIMD) computers. These factors, which include the allocation of data to processors, the relative speed of calculation and communication, the degree of conflict for memory resource between calculation and communication and the cost of starting up a communication, also have a bearing on the efficiency of synchronous and asynchronous parallel algorithms for other iterative methods. Such methods include other iterative methods for Markov decision processes (for example policy iteration, reward revision and other variants of value iteration) and iterative methods for dynamic programming and the solution of linear and differential equations.

By comparing a number of approaches to the problem of allocating data to processors, we will show that for value iteration row-wise partitioning of the transition matrices often leads to a reduction in the amount of data which must be passed between processors during an iteration. Our analysis of this communication requirement will suggest that a multinode broadcast is appropriate and our analysis of the scheduling of calculation and communication on each processor will suggest a phased pipeline approach. For 60000 state sparse Markov decision processes using a Meiko Computing Surface with 121 transputers our phased pipeline algorithm gives 60 fold speed-ups compared to the best sparse serial algorithm.

Results from a timing model will be used to show that the effectiveness of the phased pipeline algorithm does not depend critically on the hardware characteristics of the Meiko Computing Surface. These results also show that the optimal solution time is largely determined by the rate at which data can be passed between processors and not the speed of the central processing unit (CPU).

SOME SEQUENTIAL DESIGNS FOR BERNOULLI EXPERIMENTS

Hans-Joachim Girlich and Siegmar Pohlenz
Universität Leipzig

We consider a special class of stochastic decision processes which may be applied for modelling optimal sequential allocation of a single processor to n tasks that do not evolve when they are not beeing processed, or of n possible treatments with unknown success probabilities to a sequence of patients where every patient gets only one treatment. The aim of our note is to study good allocation rules for a multi-armed Bernoulli bandit. In general these designs depart from an optimal design which maximizes the total expected reward from a sequence of experiments. We investigate other designs than Gittins'index policy (cf. [1]) because it depends on the discount factor, the prior, and an infinite sequence of trials which has not yet been widely accepted in practice.
In the case of Bellman's discounted one-armed bandit problem where the unknown parameter has an a priori beta distribution it is a well-known fact that the one-step look-ahead policy (or myopic policy) is optimal for small discounts. This policy is based on comparing the conditional expectation of the winning probability given previous successes and failures of the unknown arm with the known probability of success. We present a method extending results developed by Kumar/Seidman (1982), Glazebrook/Jones (1983) and Berry/Fristedt (1988) to recognize the optimality of the myopic policy also for bigger discounts.
In the second part of this note we propose a method to improve asymptotic optimal allocation rules over short sequence of trials. We use monotonicity properties of an optimal policy and a suitable partial ordering for reaching our objective. The performance of improvement is investigated by a numerical case study using Bather's well-known randomized allocation rule (cf.[2]).

References

[1] Gittins, J.C.: Multi-armed bandit allocation indices. New York etc.: John Wiley & Sons. (1989)
[2] Girlich, H.-J., and S. Pohlenz: On the improvement of allocation rules for a multi-armed bandit problem. Optimization (1993), (submitted).

OPTIMIZATION OF DISCOUNTED RENEWAL REWARD PROCESSES

Søren Glud Johansen, University of Aarhus, Denmark

Summary: We study renewal reward processes which are semi-Markov decision processes with denumerable state space and denumerable action space. Focus on renewals enables us to unify the discounted and undiscounted cases and we present a powerful technique to obtain, simultaneously for both cases, results about the form of the infinite-horizon optimal policy. The technique is illustrated by a model for transfer pricing of a service department. We conclude that the optimal transfer price is (*i*) increasing as a function of the work backlog and (*ii*) increasing and convex as a function of the service time.

1. Introduction

It is well-known that the theory of renewal reward processes provides results about undiscounted semi-Markov decision processes (SMDPs) [3]. Results about discounted SMDPs are normally derived differently but we show in this paper[1] that focus on renewals enables us to unify the discounted and undiscounted cases. We present a powerful technique to obtain, simultaneously for the discounted and undiscounted cases, results about the form of the optimal policy for infinite-horizon SMDPs. The technique is illustrated by a model for transfer pricing of a service department.

2. Discounted infinite-horizon SMDP

Let α denote the continuously-compounded interest rate. Consider a renewal reward process which is a SMDP with denumerable state space $\mathcal{S}$ and denumerable action space $\mathcal{A}$. The regeneration state is $r \in \mathcal{S}$ and the SMDP has the following unichain property: For each policy, state r can be reached from any other state. Denote by $R_{\alpha,a}(s)$ the expected discounted rewards earned until the next decision point from a decision point in which the state is s and the action a is chosen. Let S_n and a_n be respectively the $n+1^{st}$ state of the SMDP and the $n+1^{st}$ action chosen (i.e., $n=0$ refers to the initial decision point). Also, let T_n be the time to the $n+1^{st}$ decision point (i.e., $T_0=0$) and define $N := min\{n | S_{n+1} = r\}$. Hence, T_{N+1} is the time to the first transition to state r.

If $\alpha>0$ then, for any reasonable policy π and initial state s, the expected discounted rewards earned over infinite horizon is well-defined as

$$V_{\alpha,\pi}(s) := E_\pi[\textstyle\sum_{n=0}^{\infty} exp(-\alpha\cdot T_n)\cdot R_{\alpha,a_n}(S_n)|S_0=s],\ s\in\mathcal{S}, \tag{2.1}$$

where E_π represents the conditional expectation given that policy π is employed. The expected discounted first-passage rewards earned over time T_{N+1} is

$$FPR_{\alpha,\pi}(s) := E_\pi[\textstyle\sum_{n=0}^{N} exp(-\alpha\cdot T_n)\cdot R_{\alpha,a_n}(S_n)|S_0=s],\ s\in\mathcal{S}. \tag{2.2}$$

[1]The paper is partly based on joint work with Jan van der Wal, Eindhoven University of Technology, The Netherlands.

Define the expected α-adjusted first passage time as

$$FPT_{\alpha,\pi}(s) := E_\pi[\int_0^{T_{N+1}} exp(-\alpha\cdot t)dt|S_0=s],\ s\in\mathcal{S}, \tag{2.3}$$

and note that the expected discount factor for the first passage time is

$$E_\pi[exp(-\alpha\cdot T_{N+1})|S_0=s] = 1-\alpha\cdot FPT_{\alpha,\pi}(s),\ s\in\mathcal{S}. \tag{2.4}$$

Suppose that policy π is stationary. Then the unichain property ensures that the process regenerates itself each time state r is visited. This observation implies

$$V_{\alpha,\pi}(s) = FPR_{\alpha,\pi}(s)+(1-\alpha\cdot FPT_{\alpha,\pi}(s))\cdot V_{\alpha,\pi}(r),\ s\in\mathcal{S}. \tag{2.5}$$

The relative value of starting the process in state s rather than in state r is $v_{\alpha,\pi}(s):=V_{\alpha,\pi}(s)-V_{\alpha,\pi}(r)$ and we obtain from (2.5)

$$v_{\alpha,\pi}(s) = FPR_{\alpha,\pi}(s)-g_{\alpha,\pi}(r)\cdot FPT_{\alpha,\pi}(s),\ s\in\mathcal{S}, \tag{2.6}$$

where

$$g_{\alpha,\pi}(r) := \frac{FPR_{\alpha,\pi}(r)}{FPT_{\alpha,\pi}(r)}. \tag{2.7}$$

We interpret $g_{\alpha,\pi}(r)$ as policy π's gain rate for state r with interest rate α. Note that the long-run average returns earned per unit time by employing policy π is specified by (2.7) with $\alpha=0$. Also, the relative values specified by (2.6) with $\alpha=0$ are in accordance with the usual specification for the undiscounted case [4, p. 207]. Hence, (2.6) and (2.7) are applicable for all $\alpha\geq 0$.
The objective for the discounted case ($\alpha>0$) is to maximize the expected discounted rewards earned over infinite horizon. For the undiscounted case ($\alpha=0$), the objective is to maximize the long-run average returns earned per unit time. In the sequel we study both cases simultaneously.

3. First-passage process with revised rewards

Consider a decision point in which the state is s and the action a is chosen. Let $Y_a(s)$ be distributed as the state in the next decision point and define, analogous to (2.3), $T_{\alpha,a}(s,y)$ as the expected α-adjusted time to the next dicision point, conditional on the event that its state is y. Furthermore, define $T_{\alpha,a}(s):=E[T_{\alpha,a}(s,Y_a(s))]$ and note, analogous to (2.4), that $1-\alpha\cdot T_{\alpha,a}(s,y)$ is the expected discount factor for the time to the next decision point, conditional on the event that its state is y. For scalar g, define the expected g-revised discounted rewards as

$$RR_{\alpha,g,a}(s) := R_{\alpha,a}(s)-g\cdot T_{\alpha,a}(s),\ s\in\mathcal{S} \text{ and } a\in\mathcal{A}, \tag{3.1}$$

and let f be a real-valued function on $\mathcal{S}$ with the property that $(U_{\alpha,g,a}\circ f)(s)$ is well-defined by

$$(U_{\alpha,g,a}\circ f)(s)$$
$$:= RR_{\alpha,g,a}(s)+E\Big[\Big(1-\alpha\cdot T_{\alpha,a}(s,Y_a(s))\Big)\cdot 1_{\mathcal{S}\setminus\{r\}}(Y_a(s))\cdot f(Y_a(s))\Big],$$
$$s\in\mathcal{S} \text{ and } a\in\mathcal{A}, \tag{3.2}$$

where $1_{\mathcal{S}\setminus\{r\}}(y)$ is 1 if $y\in\mathcal{S}\setminus\{r\}$ and 0 if $y=r$.

For policy π and initial state s, the expected g-revised discounted first-passage rewards earned over time T_{N+1} is

$$RFPR_{\alpha,g,\pi}(s) := E_\pi[\sum_{n=0}^{N} exp(-\alpha\cdot T_n)\cdot RR_{\alpha,g,a_n}(S_n)|S_0{=}s],\ s\in\mathcal{S}. \quad (3.3)$$

If π is a stationary policy prescribing the action $\pi(s)$ in state s then, by conditioning on the state in the next decision point after the initial one and using (2.3) and (2.4), we obtain

$$RFPR_{\alpha,g,\pi}(s) = (U_{\alpha,g,\pi(s)}\circ RFPR_{\alpha,g,\pi})(s),\ s\in\mathcal{S}. \quad (3.4)$$

Note that the expected $g_{\alpha,\pi}(r)$-revised discounted first-passage rewards are the relative values specified by (2.6) and we have

$$v_{\alpha,\pi}(s) = (U_{\alpha,g_{\alpha,\pi}(r),\pi(s)}\circ v_{\alpha,\pi})(s),\ s\in\mathcal{S}. \quad (3.5)$$

For scalar g, let $\mathcal{U}_{\alpha,g}$ be a complete metric space of functions f on $\mathcal{S}\backslash\{r\}$ with the properties that $U_{\alpha,g}\circ f$ is well-defined by (3.6) and $U_{\alpha,g}$ is a contraction mapping of $\mathcal{U}_{\alpha,g}$ into itself.

$$(U_{\alpha,g}\circ f)(s) := sup_{a\in\mathcal{A}}(U_{\alpha,g,a}\circ f)(s),\ s\in\mathcal{S}\backslash\{r\} \text{ and } f\in\mathcal{U}_{\alpha,g}. \quad (3.6)$$

Often contaction can only be established with a clever definition of the metric [1,5] or the decision points. For the model considered in Section 4, $U_{\alpha,g}$ is a contractive in the metric induced by the usual supremum norm because we have defined the decision points appropriately.

Suppose that $\mathcal{U}_{\alpha,g}$ is nonempty, which, by the fixed point theorem for contraction mappings [3], implies that $U_{\alpha,g}$ has a unique fixed point $f_{\alpha,g}\in\mathcal{U}_{\alpha,g}$. Furthermore, under reasonable assumptions [1]

$$f_{\alpha,g}(r) := sup_{a\in\mathcal{A}}(U_{\alpha,g,a}\circ f_{\alpha,g})(r) \quad (3.7)$$

is strictly decreasing and continuous as a function of g in some interval $\mathcal{G}_\alpha$ within which $g_\alpha(r)$ is uniquely determined by the condition $f_{\alpha,g_\alpha(r)}(r){=}0$. If, for $g{=}g_\alpha(r)$ and $f{=}f_{\alpha,g_\alpha(r)}$, the suprema in (3.6) and (3.7) are attained as maxima for actions $\pi_\alpha(s)\in\mathcal{A}$, $s\in\mathcal{S}$, then the stationary policy π_α that prescribes those actions is optimal [1]. Moreover, the optimal relative value of state s is

$$v_\alpha(s) = f_{\alpha,g_\alpha(r)}(s),\ s\in\mathcal{S}. \quad (3.8)$$

Our technique to obtain results about the form of the optimal policy is, for any scalar $g\in\mathcal{G}_\alpha$ and some complete metric[2] space $\mathcal{F}_{\alpha,g}$ of functions on $\mathcal{S}\backslash\{r\}$, to define a contraction mapping $F_{\alpha,g}$ on $\mathcal{F}_{\alpha,g}$ into itself with fixed point $f_{\alpha,g}$ and other desired properties. By the above observations we can then conclude that $f_{\alpha,g_\alpha(r)}$ has the desired properties and this function specifies the relative values of an optimal policy if such one exists.

4. Optimal transfer pricing of a service department

We illustrate our technique by considering a model for transfer pricing of a service department (SD) [2]. SD is an $M/G/1$ queue with input being jobs

[2]The metric can be defined differently from the metric of $\mathcal{U}_{\alpha,g}$.

submitted by other departments within the organization to which SD belongs. Requests for service of jobs are Poisson with rate λ and, for each job the service and delivery times are observable upon request. The jobs' service times are mutually independent and distributed as the generic variable W.
The input control is exercised by announcing the transfer price which should be charged - upon delivery - for serving each submitted job. The department in possession of a job submits it to SD when the department by doing so earns a nonnegative net-benefit. The net-benefit is the difference between the department's benefit of having the job served by SD (rather than having it served elsewhere) and the transfer price. The benefit, observable upon request, is earned upon delivery and is distributed as the generic random variable $B(u,w)$ which depends on SD's work backlog u and on the job's service time w. The work backlog is the sum of the remaining service times of the jobs submitted, but yet not delivered by SD. SD serves jobs in the order they are submitted and, consequently, the delivery time is $u+w$. We let the term, decreasing, include invariability and make the following assumption.
Assumption about benefits. $Pr\{B(u,w)\geq b\}$ is left-continuous and decreasing as a function of the work backlog u for any service time w and any real b.
The left-continuity ensures that we below can restrict our attention to spaces of functions for which the left-hand derivative is continuous to the left. The monotonicity is satisfied when short delivery times are prefered.
Interpret the benefits earned by SD's service of jobs as rewards. The objective for the undiscounted case ($\alpha=0$) is to maximize the long-run average benefits earned per unit time. For the discounted case ($\alpha>0$), the objective is to maximize the expected discounted benefits earned over infinite horizon. The decision points are the epochs in which *either* request for service of a job occurs and it has just been decided whether the job is submitted to SD *or* the work backlog of SD drops to 0. At each decision point the state, defined as the work backlog including the service time of the job possibly just submitted, is observed and an action is chosen. We assume that $UL<\infty$ is an upper limit for acceptable delivery times and specify the state space as $\mathcal{S}:=[0,UL]$. The regeneration state is $r=0$ and can be reached from any other state. The action prescribes the transfer price which should be charged - upon delivery - for serving the job, if any, for which request for service occurs just before the next decision point. The price is specified as a function $a(t,u,w)$ in which the arguments are

t: the time from the decision point to request for service occurs
u: the work backlog upon request for service
w: the service time of the job for which request for service occurs.

Each of these arguments are nonnegative reals and the action space $\mathcal{A}$ consists of all thinkable price functions, i.e., no restrictions are imposed on the choice of price function.
Suppose that $S_n=s\in\mathcal{S}\backslash\{0\}$, $a_n=a\in\mathcal{A}$ and $T_{n+1}-T_n=t<s$ are respectively the work backlog, the action chosen and the time to the next decision point on the $n+1^{st}$ decision point and that request for service of a job with service time $W_n=w$ occurs when the work backlog is $u:=s-t$ just before the next decision point. Hence, the transfer price charged - upon delivery - for serving

this job is $b := a(t,u,w)$. The job is submitted to SD with the probability $Pr\{B(u,w) \geq b\}$ and we define for any scalar c

$$\tilde{R}_b(u,w,c) := Pr\{B(u,w) \geq b\} \cdot (b-c) + E[B(u,w)-b]^+. \tag{4.1}$$

With c interpreted as SD's cost of having the job submitted, we can interpret the two terms in (4.1) as SD's expected profit and the expected surplus earned by the department who possibly submits the job, respectively. Note that [2]

$$\tilde{R}(u,w,c) := sup_{b \in \mathbb{R}} \tilde{R}_b(u,w,c) = \tilde{R}_c(u,w,c). \tag{4.2}$$

Conditional on $S_n = s \in \mathcal{S} \backslash \{0\}$ and $a_n = a \in \mathcal{A}$, the expected discounted benefit earned on the $n+1^{st}$ decision point is

$$R_{\alpha,a}(s) = \int_0^s E[exp(-\alpha \cdot (s+W)) \cdot \tilde{R}_{a(t,s-t,W)}(s-t,W,0)] \cdot \lambda \cdot exp(-\lambda \cdot t) dt \tag{4.3}$$

and the expected α-adjusted time between the $n+1^{st}$ and the $n+2^{nd}$ decision point is, indepently of a,

$$T_\alpha(s) = \int_0^s exp(-(\alpha+\lambda) \cdot t) dt. \tag{4.4}$$

For real-valued functions f on $\mathcal{S} \backslash \{0\}$, define

$$\|f\| := sup_{s \in \mathcal{S} \backslash \{0\}} |f(s)| \tag{4.5}$$

and, with the metric induced by this norm, define $\mathcal{U}$ as the complete metric space of continuous functions f on $\mathcal{S} \backslash \{0\}$ with the properties that $lim_{s \downarrow 0} f(s) = 0$ and that the left-hand derivative

$$f'(s) := lim_{\Delta \downarrow 0} \frac{f(s)-f(s-\Delta)}{\Delta}, \; s \in \mathcal{S} \backslash \{0\}, \tag{4.6}$$

is continuous to the left. For any scalar g, use (4.2) to conclude that the mapping defined by (3.6) with $\mathcal{U}_{\alpha,g} = \mathcal{U}$ is specified in the following way

$$(U_{\alpha,g} \circ f)(s)$$
$$= \int_0^s \Big(\lambda \cdot f(s-t) - g + \lambda \cdot exp(-\alpha \cdot (s-t)) \cdot (\Phi_\alpha \circ f)(s-t) \Big) \cdot exp(-(\alpha+\lambda) \cdot t) dt, \tag{4.7}$$

where $f(s-s) := 0$ and

$$(\Phi_\alpha \circ f)(u) := E[exp(-\alpha \cdot W) \cdot \tilde{R}(u, W, exp(\alpha \cdot (u+W)) \cdot \Big(f(u) - f(u+W) \Big))], \; u \in \mathcal{S}. \tag{4.8}$$

Our assumption about benefits ensures that $U_{\alpha,g}$ maps $\mathcal{U}$ into itself. Furthermore, for any functions f and ϕ belonging to $\mathcal{U}$, it can be concluded

$$\| U_{\alpha,g} \circ f - U_{\alpha,g} \circ \phi \| \leq \Big(1 - exp(-\lambda \cdot UL) \Big) \cdot \|f - \phi\|, \tag{4.9}$$

showing that $U_{\alpha,g}$ is a contraction mapping. Hence, there exists a unique $f_{\alpha,g} \in \mathcal{U}$ such that

$$f_{\alpha,g}(s) = (U_{\alpha,g} \circ f_{\alpha,g})(s), \; s \in \mathcal{S} \backslash \{0\}, \tag{4.10}$$

and this fixed-point property implies that $f_{\alpha,g}$'s left-hand derivative is

$$f_{\alpha,g}{}'(s) = -g + \lambda \cdot exp(-\alpha \cdot s) \cdot (\Phi_\alpha \circ f_{\alpha,g})(s) - \alpha \cdot f_{\alpha,g}(s), \; s \in \mathcal{S} \backslash \{0\}. \tag{4.11}$$

Note that (4.11) can be rewritten as

$$\Big(exp(\alpha \cdot s) \cdot f_{\alpha,g}(s) \Big)' = -exp(\alpha \cdot s) \cdot g + \lambda \cdot (\Phi_\alpha \circ f_{\alpha,g})(s), \; s \in \mathcal{S} \backslash \{0\}, \tag{4.12}$$

which yields

$$f_{\alpha,g}(s) = exp(-\alpha\cdot s)\cdot\int_0^s\Big(-exp(\alpha\cdot u)\cdot g+\lambda\cdot(\Phi_\alpha\circ f_{\alpha,g})(u)\Big)du,\ s\in\mathcal{S}\backslash\{0\}. \quad (4.13)$$

In order to investigate $f_{\alpha,g}$ further, we specify $\mathcal{F}_{\alpha,g}$ as the space of continuous functions f on $s\in\mathcal{S}\backslash\{0\}$ with the properties

(i) $f(UL)=f_{\alpha,g}(UL)$,
(ii) the left-hand derivative f' is continuous to the left and,
(iii) $exp(\alpha\cdot s)\cdot f'(s)$ is decreasing in s.

We equip this space with the metric defined as

$$\rho_\alpha(f,\phi) := sup_{s\in\mathcal{S}\backslash\{0\}} exp((\alpha+\lambda)\cdot s)\cdot|f'(s)-\phi'(s)|,\ f\in\mathcal{F}_{\alpha,g} \text{ and } \phi\in\mathcal{F}_{\alpha,g}, \quad (4.14)$$

and define for $f\in\mathcal{F}_{\alpha,g}$ and $s\in\mathcal{S}\backslash\{0\}$

$$(F_{\alpha,g}\circ f)(s)$$
$$:= exp(-\alpha\cdot s)\cdot\Big(exp(\alpha\cdot UL)\cdot f_{\alpha,g}(UL)-\int_s^{UL}[exp(\alpha\cdot u)\cdot g+\lambda\cdot(\Phi_\alpha\circ f)(u)]du\Big). \quad (4.15)$$

Note that $(F_{\alpha,g}\circ f)(UL)=f_{\alpha,g}(UL)$. For any f and ϕ belonging to $\mathcal{F}_{\alpha,g}$ and $s\in\mathcal{S}\backslash\{0\}$, it is easily verified that

$$|(\Phi_\alpha\circ f)(s)-(\Phi_\alpha\circ\phi)(s)| \leq exp(\alpha\cdot s)\cdot\int_s^{UL}|f'(u)-\phi'(u)|du \quad (4.16)$$

and that

$$(F_{\alpha,g}\circ f)'(s) = exp(-\alpha\cdot s)\cdot\Big(-\tilde{g}+\lambda\cdot[(\Phi_\alpha\circ f)(s)+\alpha\cdot\int_s^{UL}(\Phi_\alpha\circ f)(u)du]\Big), \quad (4.17)$$

where

$$\tilde{g} := exp(\alpha\cdot UL)\cdot(g+\alpha\cdot f_{\alpha,g}(UL)). \quad (4.18)$$

Use (4.17) to conclude that $(F_{\alpha,g}\circ f)'$ is continuous to the left and that $exp(\alpha\cdot s)\cdot(F_{\alpha,g}\circ f)'(s)$ is decreasing in s (since $(\Phi_\alpha\circ f)(s)$ is decreasing in s and nonnegative [2]). Hence, $F_{\alpha,g}$ maps $\mathcal{F}_{\alpha,g}$ into itself. Furthermore, use (4.16) to conclude [2]

$$\rho_\alpha(F_{\alpha,g}\circ f,F_{\alpha,g}\circ\phi) \leq \Big(1-exp(-\lambda\cdot UL)\Big)\cdot\rho_\alpha(f,\phi),\ f\in\mathcal{F}_{\alpha,g} \text{ and } \phi\in\mathcal{F}_{\alpha,g}, \quad (4.19)$$

showing that $F_{\alpha,g}$ is a contraction mapping. Moreover, since we have defined $F_{\alpha,g}$ such that $F_{\alpha,g}\circ f_{\alpha,g}=f_{\alpha,g}$, we conclude that $f_{\alpha,g}$ is the unque fixed point of $F_{\alpha,g}$ and belongs to $\mathcal{F}_{\alpha,g}$. An important implication of this conclusion is that

$$exp(\alpha\cdot s)\cdot f_{\alpha,g}{}'(s) \text{ is decreasing in } s\in\mathcal{S}\backslash\{0\} \text{ for any scalar } g. \quad (4.20)$$

Consider a decision point in which the state is $s=0$. The α-adjusted time to the next decision point is, indepently of the action chosen, $T_\alpha(0)=1/(\alpha+\lambda)$. We use (4.2) to conclude that $f_{\alpha,g}(0)$, defined by (3.7), is specified in the following way

$$f_{\alpha,g}(0) = \frac{\lambda\cdot(\Phi_\alpha\circ f_{\alpha,g})(0)-g}{\alpha+\lambda},\ g\in\mathbb{R}. \quad (4.21)$$

The optimal gain rate $g_\alpha(0)$ for state 0 is uniquely determined by the condition $f_{\alpha,g_\alpha(0)}(0)=0$ and, this condition is equivalent to

$$g_\alpha(0) = \lambda\cdot(\Phi_\alpha\circ f_{\alpha,g_\alpha(0)})(0). \quad (4.22)$$

The following theorem states our result about SD's optimal transfer pricing.

Theorem. For any interest rate $\alpha \geq 0$, the stationary policy with transfer prices prescribed by

$$a_\alpha(t,u,w) := exp(\alpha\cdot(u+w))\cdot\Big(f_{\alpha,g_\alpha(0)}(u)-f_{\alpha,g_\alpha(0)}(u+w)\Big) \tag{4.23}$$

is optimal. The optimal price $a_\alpha(t,u,w)$ is nonnegative and does not depend on the time t from the decision point to the request for service occurs. The price is (i) increasing in the work backlog u and (ii) increasing and convex in the service time w.

Proof. The stationary policy with actions prescribed by (4.23) is optimal since

$$f_{\alpha,g_\alpha(0)}(s) = (U_{\alpha,g_\alpha(0),a_\alpha}\circ f_{\alpha,g_\alpha(0)})(s),\ s\in\mathcal{S}. \tag{4.24}$$

Use first (4.11) with $g=g_\alpha(0)$ and then (4.22) to conclude

$$f_{\alpha,g_\alpha(0)}{}'(0) := lim_{s\downarrow 0}f_{\alpha,g_\alpha(0)}{}'(s) \leq -g_\alpha(0)+\lambda\cdot(\Phi_\alpha\circ f_{\alpha,g_\alpha(0)})(0) = 0. \tag{4.25}$$

Hence, from (4.20) we have that $exp(\alpha\cdot s)\cdot f_{\alpha,g_\alpha(0)}{}'(s)$ is nonpositive and decreasing. Since

$$a_\alpha(t,u,w) = \int_u^{u+w} exp(\alpha\cdot(u+w-s))\cdot\Big(-exp(\alpha\cdot s)\cdot f_{\alpha,g_\alpha(0)}{}'(s)\Big)ds, \tag{4.26}$$

we conclude that $a_\alpha(t,u,w)\geq 0$ and, with t deleted,

(i) $a_\alpha(u,w) \leq \int_u^{u+w} exp(\alpha\cdot(u+w-s))\cdot\Big(-exp(\alpha\cdot(s+\Delta))\cdot f_{\alpha,g_\alpha(0)}{}'(s+\Delta)\Big)ds$

$= a_\alpha(u+\Delta,w),\ 0\leq\Delta\leq UL-(u+w);$

(ii) $lim_{\Delta\downarrow 0}\dfrac{a_\alpha(u,w)-a_\alpha(u,w-\Delta)}{\Delta} = \alpha\cdot a_\alpha(u,w)-exp(\alpha\cdot(u+w))\cdot f_{\alpha,g_\alpha(0)}{}'(u+w)$

is nonnegative and increasing in w. □

References

1. Johansen, S. G., 1991. First Passage Processes Provide New Results on semi-Markov Decision Processes. Department of Operations Research, University of Aarhus, Denmark.
2. Johansen, S. G., 1993. Optimal Prices of an $M/G/1$ Job Shop. Accepted for publication in *Operations Research* 41.
3. Ross, S. M., 1970. *Applied Probability Models with Optimization Applications*, Holden-Day, San Francisco, CA.
4. Tijms, H. C., 1986. *Stochastic Modelling and Analysis.* John Wiley & Sons, Chichester.
5. Wessels, J., 1977. Markov Programming by Successive Approximations with respect to Weighted Supremum Norms. *Journal of Mathematical Analysis and Applications* 58.

On Solutions for a Multi-Location Ressource Model with Redistribution

Peter Köchel, TU Chemnitz-Zwickau

Summary:The matter of this paper are periodic review multi-location models with redistribution. We give an overview of results achieved at the technical university in Chemnitz. Analytical investigations using Markovian decision theory as well as simulation and a software tool are sketched.

1. Introduction

In mathematical inventory theory multi-location models with redistribution comprise continuous review models as well as periodic review models.Usually, the former models assume repairable items, Poisson demand, (S-1, S) orders (without optimality proof) and steady state conditions (see [1], [7]). For analysis mainly queueing theory and mathematical programming are used.

In the present paper we consider periodic review models. We give a short overview of results achieved at the technical university in Chemnitz. Our approach applies Markovian decision theory. The next section is devoted to a description of the inventory model with redistribution (IMR) and the corresponding Markovian decision model (MDM). Results on periodic stationary MDM's, relevant to our purposes, are given in Section 3. Section 4 describes a procedure to receive an analytical solution for the IMR and related models. The final Section 5 reports some further models and solutions.

2. The basic inventory problem and the MDM

We assume N locations which are subject to a stochastic demand for a given product. Let $\underline{s}=(\underline{s}_1,\dots,\underline{s}_N)$ denote the random demand vector with distribution function F and finite expectation $m_i=E(s_i)$ for $i=1(1)N$. Periodically each location orders a defined amount of the product by an order decision (OD). Thereby location capacities $Q_i<\infty$, $i=1(1)N$, are to be taken into account. A small span of time to the following OD the stock on hand can be redistributed by a transportation decision (TD). OD and TD occur immediately. For location i the parameters h_i, p_i, $k_i>0$ denote holding, shortage and order cost per product unit, respectively. Finally, let $c_{ij}>0$ denote transporting cost per unit, transported to location $j\neq i$. We assume $c_{ii}=0$ for $i=1(1)N$. Cost of period $t=1,2,\dots$ will be discounted by $\alpha^{t-1}, 0\le\alpha\le1$. We introduce following conditions (see [3] for an economic interpretation):

$$
\begin{aligned}
&a)\ h_i+p_j-c_{ij} > \alpha\ (k_i-k_j)\\
&b)\ c_{ij}+h_j-h_i > \alpha\ (k_j-k_i)\\
&c)\ c_{ir}+c_{rj} > c_{ij},\ i,r,j=1(1)N, i\neq r\neq j,\ \alpha\in[0,1].
\end{aligned}
\qquad (2.1)
$$

Unsatiesfied demand is backlogged.

The problem now is to choose such OD's and TD's which minimize the expected discounted cost over an infinite horizon.

Apparently, after a cycle of two decisions the decision process recurs in some sense. To analyse decision processes with such a property K-periodic stationary MDM's are an appropriate tool. In [2] a K-periodic stationary MDM is defined as a tupel $M^K=((X_n),(A_n),(D_n),(p_n),(q_n),\alpha)_{n=1(1)K}$ with following meaning: Let us consider the n-th period of a cycle, $n=1(1)K$. The process will be found in a state $x\in X_n$, where X_n is the corresponding state space. In state $x\in X_n$ an action $a\in D_n(x)$ should be choosen from a set of actions, admissible for state x. A_n is the action space; p_n and q_n denote transition law and cost function, respectively, both depending on x and a. Cost of future cycles will be discounted by a factor α, $0\le\alpha\le1$.

Obviously, for the defined inventory problem K=2 holds, and the characteristics describing the process for n=1 are related to the OD and for n=2 to the TD. Let $x=(x_1,\dots,x_N)$ represent the vector of stocks <u>before</u> an OD is taken, $a=(a_1,\dots,a_N)$ the inventory level <u>after</u> an OD and $y=a-s$ the vector of rest stock after occuring realization $s=(s_1,\dots,s_N)$ of the demand. Redistribution is organized in accordance with TD $b=(b_{ij})$ with $b_{ij}\ge0$ as the transported quantity from location i to location $j(i,j=1(1)N)$. It holds:

$$X_1 = \{ x = (x_1,\dots,x_N) : x_i \le Q_i,\ i=1(1)N \};$$

$$D_1(x) = \{ a = (a_1,\dots,a_N) : x_i \le a_i \le Q_i,\ i=1(1)N \},\ x \in X_1;$$

$$A_1 = \{ a = (a_1,\dots,a_N) : a_i \le Q_i,\ i=1(1)N \};$$

$$X_2 = \{ y = (y_1,\dots,y_N) : y_i \le Q_i,\ i=1(1)N \};$$

$$D_2(y) = \{b=(b_{ij}) : b_{ij} \ge 0, \sum_{j=1}^{N} b_{ij} = y_i^{+},\ y_i + \sum_{j=1}^{N} b_{ji} - \sum_{j=1}^{N} b_{ij} \le Q_i,\ i,j=1(1)N \},$$

where $c\pm = \max(0;\pm c)$ for real c;

$$A_2 = \{ b = (b_{ij}) : b_{ij} \ge 0, \sum_{j=1}^{N} b_{ij} \le Q_i,\ i,j=1(1)N \}.$$

The transition laws result from evident balance equations

$$y = a - s \qquad \text{and} \qquad x = (y_1 \oplus b, \dots, y_N \oplus b), \tag{2.2}$$

where for i=1(1)N operation $\oplus$ means $\quad y_i \oplus b = y_i + \sum_{j=1}^{N} b_{ji} - \sum_{j=1}^{N} b_{ij}$.

The cost functions are $\quad q_1(x,a) = \sum_{i=k}^{N} k_i (a_i - x_i),\ a \in D_1(x),\ x \in X_1$
and

$$q_2(y,b) = \sum_{i=1}^{N} [h_i (y_i \oplus b)^{+} + p_i (y_i \oplus b)^{-} + \sum_{j=1}^{N} c_{ij} b_{ij}],\ b \in D_2(y),\ y \in X_2.$$

3. A brief review of results on periodic stationary MDM's

We summarize basic results on periodic stationary MDM's. For mathematical details see [2].

A. There is a system of optimality equations

$$V_n(x) = \min_{a \in D_n(x)} \{ q_n(x,a) + \int_{X_{n+1}} p_n(dy|x,a)\, V_{n+1}(y) \},\ x \in X_n,\ n=1(1)K-1;$$

$$V_K(x) = \min_{a \in D_K(x)} \{ q_K(x,a) + \alpha \int_{X_1} p_K(dy|x,a)\, V_1(y) \},\ x \in X_K. \tag{3.1}$$

Function V_n represents the optimal cost function for a process which starts at period n in the first cycle, n=1(1)K.

B. It exists a K-periodic stationary policy $d=(f_1,...,f_K)^\infty$ which is optimal.

C. The principle of myopic policies in a natural way carries over to so-called <u>cycle-optimal policies.</u>

To explain the latter statement let v_n be a function defined on X_n by

$$v_n(x) = \min_{a\in D_n(x)} \{q_n(x,a) + \int_{X_{n+1}} p_n(dy|x,a)\, v_{n+1}(y)\},\ x\in X_n,\ n=1(1)K-1;$$

$$v_K(x) = \min_{a\in D_K(x)} q_K(x,a),\ x\in X_K. \tag{3.2}$$

Thus v_n represents the optimal cost function for a process with duration from the n-th period of a cycle to the end of that cycle, n=1(1)K.
Let f_n^* realize the infimum at the right-hand side of (3.2), n=1(1)K. Policy $d^*=(f_1^*,...,f_K^*)^\infty$ is called cycle-optimal.
A sufficient condition for d^* to be (discount) optimal in some sense is the following (see[3],[4]):
(C.1) There exists a set $X_1^*\subset X_1$ of "good" states such that starting from a state in X_1^* we get the optimum during a cycle, i.e.,

$$v_1(x) = \min_{x'\in X_1} v_1(x') =: q,\ x\in X_1^*$$

(C.2) After finishing the present cycle the process is again in set X_1^* if d^* is applied.

(C.1)and (C.2) are very general and not easy to prove . Therefore we introduce

(C.3) p_1 and q_1 depend on $a\in A_1$ only.

(C.3) results in the property $v_1(x) = \min_{a\in D_1(x)} g(a),\ x\in X_1,$

where

$$g(a)=q_1(a)+\int_{X_2} p_1\ (dx'|a)\ v_2(x')\ ,\ a\in A_1. \tag{3.3}$$

(C.4) $\exists a^* \in A_1 : g(a^*) \leq g(a),\ a \in A_1$.

(C.5) $X_1^* = \{x \in X_1 : a^* \in D_1(x)\} \neq \emptyset$.

For (C.2) - (C.5) we have

$$V_1(x)=g(a^*)/(1-\alpha),\ x\in X_1^*,\ \text{and}\ f_1^*(x)=a^*,\ x\in X_1^*.$$

4. Outline of an analytical solution for the IMR

To receive an analytical solution for the IMR the following procedure will be successful (cp. [3]):

1. We derive general results, i.e., the system of optimality equations and the optimality of a two-periodic stationary policy.
2. We prove the optimality of a cycle-optimal policy.

Remark 4.1: For the IMR as well as for many other inventory models condition (C.3) can be established along the lines of Veinott's [8] approach. The idea is to use balance equations like (2.2) for corresponding transformation of the initial cost functions.

3. For resulting function g, defined by (3.3), we define the minimum point a^*.

In the IMR function g has the form

$$g(a)=\sum_{i=1}^{N}\ [(1-\alpha)k_i a_i+\alpha k_i m_i+L_i(a_i)]\ -C(a),\ a\in A_1, \tag{4.1}$$

where $L_i(a_i)$ describes the expected holding and shortage cost in location i with initial stock level $a_i, i=1(1)N$, and C(a) is the expected system's gain from optimal redistribution if OD a was choosen.

For IMR, which cost parameters fulfil (2.1), the following results are obtained in [3]:

a) There exists a unique solution a^* of the convex problem $\min_{a\geq 0} g(a)$, where g is defined by (4.1).

b) a^* is an optimal OD. In detail, a (S, S)-policy with $S=a^*$ is optimal for $x\leq a^*$ in the case $\alpha<1$ and for all $x\in X_1$ in the case $\alpha=1$.

c) For any rest stock vector y the optimal TD $b^*=(b_{ij})$ is a solution of an open linear transporting problem with supplier set $I^+=\{i=1(1)N:y_i<0\}$, consumer set $I^-=\{i=1(1)N: y_i<0\}$ and coefficients $C_{ij}=h_i+p_j-c_{ij}+\alpha(k_j-k_i)>0$ for $i\in I^+$, $j\in I^-$.

Since $C(a) = E[C(a,\underline{s})]$ with $C(a,s) = \sum_{i \in I^+} \sum_{j \in I^-} b^*_{ij} C_{ij}$ for $s \geq (0,\ldots,0)$,

from result c) follows that an analytical tractable expression for function g is possible only if N=2 or if $C_{ij} = C > 0$ for i,j=1(1)N, i≠j(cp[3]). Approaches to this general case are outlined in the following section.

5. Further investigations

In the present section we sketch some further investigations carried out in Chemnitz.

a) Approximate solution.

As mentioned in Section 4 the necessity to solve an open linear transporting problem for any demand realization prevents an analytical solution for an IMR with general coefficients C_{ij}. However, if we define $\underline{C} = \min \{C_{ij}: i,j=1(1)N,\ i \neq j\}$ and if we replace C_{ij} by $\underline{C}$ we get a twofold effect. We have a lower bound $\underline{C}(a)$ for C(a), $a \in A_I$, and a general solution of the transporting problem is possible, i.e., $\underline{C}(a)$ is given in an analytical way for $a \in A_I$. For more information see [3].

b) IMR with contingents and export.

For a modified IMR (at the beginning of a period each location gets a fixed amount of product, a part of which can be selled out of the system) in [6] are investigated existence and properties of optimal export and redistribution decisions.

c) Two-echelon model.

In [4] the procedure, outlined in Section 4, is applied to a two-echelon model with periodic ordering. A central warehouse orders only at the beginning of a cycle of K periods. From the central warehouse the N basic warehouses will be supplied at the beginning of periods 2 to K. It is shown that under additional assumptions (S.S)-type order rules are optimal.

d) Adaptive control.

If the distribution function F in the IMR depends on an unknown parameter the following identification scheme from adaptive control theory can be applied. At each period t we use the history of demand realizations to form an estimate for the unknown parameter. Then we choose OD and TD, which are optimal for the estimated parameter value. Under several conditions this scheme leads to an average optimal policy. Simulation results show the practicability of that approach (see [5]).

e) Simulation.

Another approach to deal with a general IMR is to combine simulation and optimization. We propose the following procedure for defining nearly-optimal decisions.

1. By simulation we generate a sample of demand realizations.
2. For given $a \in A_I$ we compute the estimate $\hat{g}(a)$ for $g(a)$ defined in (4.1).
3. Since $\hat{g}(.)$ is convex for every sample we use one of the methods for minimizing a convex function to find OD $\hat{a}$ which realizes the minimum.
4. We take OD $\hat{a}$ as approximation for a^* and the value $\hat{g}(\hat{a})$ as approximation for $g(a^*)$.

An essential advantage of the described approach is the possibility to investigate IMR's with arbitrary distributed demand and even with dependencies of the demand between locations.

f) Software.

For IBM-compatible PC's we have a software tool for solving IMR's with at most 10 locations. The solution is found by the approach described in item e). The method to define $\hat{a}$ is an adaptive one: the probabilities for the direction of the search and the step width are defined in an adaptive way. Automatically the computed results are compared with the case of independent locations without redistribution.

References

[1] Axsäter, S.: Modelling Emergency Lateral Transshipments in Inventory Systems. Management Sci., 36 (1990), 1329-1338

[2] Köchel, P.: Periodisch stationäre Markovsche Entscheidungsmodelle. Elektron.Inf.u.Kybern., 16 (1980), 554-567

[3] Köchel, P.: Ein dynamisches Mehr-Lager-Modell mit Transportbeziehungen zwischen den Lagern. Math.Operationsforsch.Statist., Ser.Optimization, 13 (1982), 267-286

[4] Köchel, P.: Optimal Inventory Control for a Two-echelon Model with Periodic Ordering. Elektron.Inf.u.Kybern.,20(1984),448-461

[5] Köchel, P.: Optimal Adaptive Inventory Control for a Multi-location Model with Redistribution. Optimization,19(1988), 525-537

[6] Köchel, P.: A Dynamic Multi-location Inventory Model with Contingents and Redistribution of Rest Stock. Engineering Costs and Production Economics, 19(1990), 359-363

[7] Nahmias, S.: Managing Repairable Item Inventory Systems: A Review. in "Multi-Level- Production/Inventory Control Systems: Theory and Practice, Studies in Man.Sciences, 16(1981), 253-277

[8] Veinott, A.F.: On the Optimality of (s,S) Inventory Policies: New Conditions and a New Proof. J.SIAM Appl.Mathem., 14(1966), 1067-1083

ON A CLASS OF STRATEGY PAIRS IN MARKOV GAMES

Heinz-Uwe Küenle, Technische Universität Cottbus

This contribution deals with Nash equilibria in two-person Markov games. A special class of strategy pairs is considered, so-called *intimidation strategy pairs*. The idea of such strategy pairs is the following one: Both players have an agreement to play some deterministic strategy pair, but if one of the players violates this agreement then the other player chooses such a strategy which is most unfavourable for the disloyal player.

We give simple conditions for the existence of intimidation strategy pairs which are also equilibrium strategy pairs. Furthermore, it can be shown that every loss of deterministic equilibrium pairs is also the equilibrium loss of an intimidation strategy pair. Recursive relations for the set of all deterministic equilibrium losses are presented. For Markov games with complete information it holds also that all effective equilibrium losses are losses of intimidation strategy pairs.

As an example the following situation is treated: Two persons harvest periodically from the same bounded source but every of this persons has the possibility to ruin the existence of the other person. The problem is here to find a compromise between the both persons how many of the source is allowed to take away.

The results can be generalised to n-person Markov games.

AN ACCEPT AND REJECTION SIMULATION METHOD TO ESTIMATE THE RELIABILITY OF DYNAMIC SYSTEMS

Jun-Der Leu, Taoyuan, Taiwan, R.O.C.

For evaluating systems performance, system designers model many dynamic systems, such as telecommunication, computer, electric power, transportation, and water distribution systems into stochastic networks to estimate system reliability. In acquiring the system reliability, both the problems to find all minimal paths or cuts and to compute the system reliability with known minimal paths or cuts are NP-hard. Current simulation methods to estimate system reliability focus mainly on how to reduce the variance of estimate given that minimal paths or cuts are known in advance. However, only 2.NP-hard obstacle is solved. See Figure.1

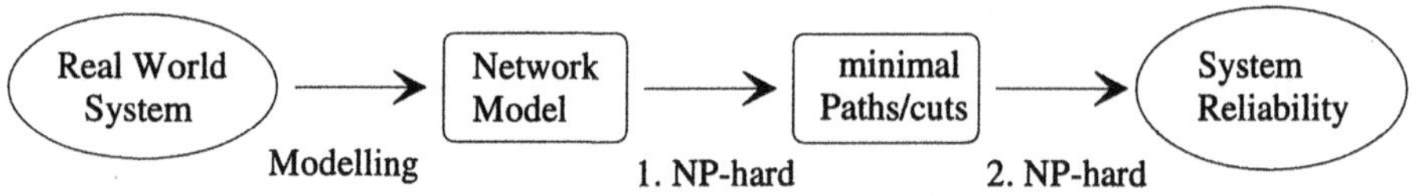

Figure. 1

In this paper, we derive first the distribution of numbers of failure components in system, and then, designing an Accept and Rejection Monte Carlo method to determine which components in system are failure, finally, applying a search algorithm to determine the system state. Repeating above simulation procedure for numerous replications, we have the system reliability estimate and its variance. As a result, no information about minimal paths or cuts is required to estimate system reliability so that both NP-hard obstacles are overcome in a time. See Figure. 2. A network example to compare the differences between the proposed method and the crude Monte Carlo method is presented

Figure. 2

ALGORITHMS FOR COMPUTING OPTIMAL POLICIES IN QUEUEING NETWORKS WITH PARTIAL INFORMATION

J.A. Loeve, University of Leiden, NL

In a Markov decision chain the optimal policy can be found by successive approximation. However, if it is not possible to observe all states of the system, the policy found in this way is not useful in practice. This is one of the reasons for looking at Markov decision chains with partial state information.

For example, this is the case in decentralized queueing networks. In such a queueing network there are several nodes or centres and each centre has its own decisionmaker who controls that centre. Each decisionmaker can observe only one centre, so the decisions are based on the information about that particular centre.

In general we can model this kind of situation by making a partition of the state space. Now the information on the state of the system available on each decision moment is the set of the partition in which the state is contained. So the exact state is not known and the decisions in all states in a set of the partition have to be the same.

Several new algorithms to obtain optimal policies in Markov decision chains with partial state information are designed. They can handle a larger state space than the well-known algorithms. However, there is no proof of convergence yet, although in practice it seems to work quite well. We will analyse these algorithms and discuss their applicability for finding decentralized optimal policies in some specific queueing models.

STOCHASTIC GAMES, AN OVERVIEW

O.J. Vrieze, Rijksuniversiteit Limburg, NL

Stochastic games can be viewed as an integration of two other classes of mathematical decision problems, namely Noncooperative Games and Markovian Decision Problems. With Noncooperative Games Stochastic Games have the same aspect in common, which concerns the decision situation in which two or more decision makers (players) have to take a decision and where these decision mutually influence each others outcome. With Markovian decision problems Stochastic Games have in common the aspect that the decision process proceeds dynamically along a stochastic process based on Markov chains.
In stochastic games usually the infinite horizon case is considered, in which the set of discrete decision moments coincides with the set of natural numbers.
The theory of Stochastic Games started in 1953 with an initiating paper of Shapley. The theory has been developed into different branches. We mention: two-person zero-sum models (one player has to pay the other), nonzero-sum models (each player has his own reward-function), discounted models (future rewards are discounted in order to intercompare them), limiting-average models (the overall evaluation of a stream of rewards is taken as some average).
The theory of zero-sum models is more or less completed. In 1980 Mertens and Neyman showed that the last open question, namely the existence of the value, can be answered affirmatively. For nonzero-sum models still many questions are open. The main problems arise from the fact that, unlike in the zero-sum case, the limiting average models cannot be treated as limits of discounted models. During this talk an overview of Stochastic Games will be presented. Specials emphasis will be given to evaluation rules like discounted criterium, limiting average criterium, total reward criterium and weighted reward criterium and their interrelation. For the nonzero-sum models the main difficulties will be explained with respect to existence proofs of equilibria.

INFLUENCE OF BREAKDOWN AND LIMITED REPAIR CAPACITY ON THE PERFORMANCE OF A CONTINUOUS PRODUCTION SYSTEM

Pieter Wartenhorst
McKinsey & Company, Amsterdam, NL

The influence of machine breakdown on the performance of continuous production- (or service-) systems is often not well understood or not taken into account at all. However, the performance of a system may be heavily affected by machine breakdown, and limited repair capacity. The overall production capacity will be affected, but also severe build-up of jobs may occur at a broken machine. Similar phenomena occur in the area of computer and communication systems, where failure and repair of processors have a major impact on the flow of jobs that have to be handled by those processors.
We present a queueing model that can be used to study such influences. The model consists of N single server queueing stations, each serving its own stream of customers (jobs). The servers (machines or processors) of the N stations are subject to breakdown. Broken servers are repaired by a joint repair facility consisting of K parallel repairmen. Whenever $K < N$, this repair facility is causing interference between the N stations.
We present both an exact (matrix-geometric) solution, and a simple approximation (employing stochastic decompostitions) of the distribution of the queue length at a particular station. With this model various design issues can be investigated such as the number of repairmen that is needed to maintain a pool of machines, or the number of machines that can be assigned to a certain crew of repairmen. Several numerical examples illustrate the approach.

Adressenverzeichnis

Adan 510
Dr. Ivo Adan, Department of Mathematics and Computing Science,
Eindhoven University of Technologie, P.O. Box 513, NL-5600 MB Eindhoven

Adner 42
Dr. Angelika Adner, Zentrales & Pharma-Controlling, Schering AG, Müllerstr. 170,
Postfach 650311, D-13342 Berlin

Altrogge VII
Prof. Dr. Günter Altrogge, Fachbereich Wirtschaftswissenschaften, Universität Hamburg,
Van-Melle-Park 5, D-20146 Hamburg

Ambrosi VII
Prof. Dr. Klaus Ambrosi, Institut für Betriebswirtschaftslehre, Universität Hildesheim,
Marienburger Platz 22, D-31141 Hildesheim

Archibald 537
Dr. Thomas Welsh Archibald, Department of Mathematics and Statistics,
The University of Edinburgh, Mayfield Road, Edinburgh EH9 3JZ, Scotland

Bachem 1
Prof. Dr. Achim Bachem, Mathematisches Institut, Universität zu Köln, D-50923 Köln

Bartsch 117
Thomas Bartsch, Lehrstuhl für Betriebliche Datenverarbeitung und Produktionswirtschaft,
Christian-Albrechts-Universität zu Kiel, Olshausenstraße 40, D-24118 Kiel

Bassan 323
Bruno Bassan, Politechnical Institute of Milano, Italien

Bastian VII
Prof. Dr. Michael Bastian, Institut für Wirtschaftsinformatik, RWTH Aachen,
Johanniterstraße 22-24, D-52056 Aachen

Behrens 23
Doris Anita Behrens, Mozartstr. 101/2, A-9020 Klagenfurt

Benz 324
Prof. Dr. Jochen Benz, Fachbereich Wirtschafts- und Sozialwissenschaften,
Fachhochschule Konstanz, Brauneggerstraße 55, D-72074 Konstanz

van den Berg 197
Ir. Jeroen van den Berg, Department of Mechanical Engineering, University of Twente,
P.O. Box 217, NL-7500 AE Enschede

Berg 432
Markus Berg, Institut für Betriebswirtschaftslehre, Christian-Albrechts-Universität zu Kiel,
Olshausenstraße 40, D-24118 Kiel

van Eenige 530
Drs. Michel van Eenige, Department of Mathematics and Computing Science, Technische Universiteit Eindhoven, P.O. Box 513, NL-5600 MB Eindhoven

Eiselt 447
Prof. Dr. H.A. Eiselt, Faculty of Administration, University of New Brunswick, Post Office Box 4400, Fredericton, E3B 5A3, N.B. Canada

Eisenberger 432
Dipl.-Vw. Roselies Eisenberger, Lehrstuhl für ABWL, Finanzwirtschaft, insb. Banken, L 5,2, D-68131 Mannheim

Elsebach 327
Dipl.-Math.oec Regina Elsebach, Fachbereich Wirtschafts- und Organisationswissenschaften, Universität der Bundeswehr Hamburg, D-22039 Hamburg

Erkens 222
Dipl.-Kfm. Elmar Erkens, Fachbereich 7 Wirtschaftswissenschaft, Universität Bremen, Postfach 330440, D-28359 Bremen

van Evert 77
Ir. Henk van Evert, CVI, P.O. Box 2233, NL-3500 GE Utrecht

Fahrion 367
Prof. Dr. Roland Fahrion, Alfred-Weber-Institut, Lehrstuhl für Wirtschaftsinformatik, Ruprecht-Karls-Universität Heidelberg, Grabengasse 14, D-69117 Heidelberg

Faigle 497
Prof. Dr. Ulrich Faigle, Faculty of Applied Mathematics, University of Twente, P.O. Box 217, NL-7500 AE Enschede

Fischer 85
Dr. Dietrich Fischer, GARONOR Planungs- und Betriebs-GmbH, Ritter-Hilprand-Str. 8, D-82024 Taufkirchen

Fischer 511
Martin Fischer, Institut für Mathematische Maschinen und Datenverarbeitung (Informatik) IV, Friedrich-Alexander-Universität Erlangen-Nürnberg, Martensstraße 1, D-91058 Erlangen

Fleisch 127
Mag. Elgar Fleisch, Institut für Höhere Studien, Stumpergasse 56, A-1060 Wien

Fleischmann 211
Prof. Dr. Bernhard Fleischmann, Lehrstuhl für Produktionswirtschaft, Universität Augsburg, Memminger Straße 14, D-86159 Augsburg

Fleischmann 45
Dipl.-Math. Sigund Fleischmann, CAP debis GEI GmbH, Oehleckerring 40, D-22404 Hamburg

Fleuren 209
Dr. Ir. Hein A. Fleuren, Centre for Quantitative Methods CQM B.V., P.O. Box 414, NL-5600 AK Eindhoven

Fochem 398
Dipl.-Ing. Manfred Fochem, Management Intelligenter Technologien GmbH, Korneliuscenter, Promenade 9, D-52076 Aachen

Furmans 151
Dr. Kai Furmans, Institut für Fördertechnik, Forschungsbereich Materialfluß, Universität Karlsruhe, Postfach 6980, D-76187 Karlsruhe

Gaalman 126
Gerard J.C. Gaalman, Faculty of Economics, University of Groningen, P.O. Box 800, NL-9700 AV Groningen

Gademann 197
Dr. Ir. Noud Gademann, Department of Mechanical Engineering, University of Twente, P.O. Box 217, NL-7500 AE Enschede

Gau 152
Thomas Gau, Institut für Wirtschaftswissenschaften, Operations Research, TU Braunschweig, Abt-Jerusalem Straße 4, Postfach 3329, D-38106 Braunschweig

Gaul 267
Prof. Dr. Wolfgang Gaul, Institut für Entscheidungstheorie und Unternehmensforschung, Universität Karlsruhe (TH), Postfach 69 80, D-76128 Karlsruhe

Geiger 128
Dr. Werner Geiger, Lehrstuhl für Betriebswirtschaftslehre V, Produktionswirtschaft, Universität Bayreuth, Postfach 101251, D-95440 Bayreuth

Giesberts 328
Paul Giesberts, Eindhoven University of Technology, P.O. Box 513, NL-5600 MB Eindhoven

Girlich 538
Prof. Dr. Hans-Joachim Girlich, Fachbereich Mathematik/Informatik, Universität Leipzig, Augustusplatz 10/Postfach 920, D-04009 Leipzig

Glud Johansen 539
Søren Glud Johansen, Department of Operations Research, University of Aarhaus, Build. 530, Ny Munkegade, DK-8000 Aarhus C

Golabi 86
Dr. Kamal Golabi, Optima Inc., 517 Washington Street, San Francisco, CA 94111, U.S.A.

Gronalt 153
Dr. Manfred Gronalt, Institut für Betriebswirtschaftslehre, Universität Wien, Brünner Straße 72, A-1210 Wien

Grünert 154
Dipl.-Ing. Tore Grünert, Institut für Wirtschaftswissenschaften, Operations Research, RWTH Aachen, Templergraben 64, D-52056 Aachen

Günther 153
Prof. Dr. Hans-Otto Günther, Institut für Betriebswirtschaftslehre, Universität Wien, Brünner Straße 72, A-1210 Wien

Haasis 297, 315, 322
Dr. habil. Hans-Dietrich Haasis, Institut für Industriebetriebslehre und Industrielle Produktion, Universität Karlsruhe (TH), Hertzstraße 16, Bau 06.33, D-76187 Karlsruhe

Habenicht 448
Prof. Dr. Walter Habenicht, Lehrstuhl für Industriebetriebslehre, Universität Hohenheim, Postfach 700562, D-70593 Stuttgart

Hagdorn - van der Meijden 212, 225
Drs. Lorike Hagdorn - van der Meijden, Faculty of Business Administration, Erasmus University Rotterdam, Burgemeester Oudlaan 50, P.O. Box 1738, NL-3000 DR Rotterdam

Van Hamme 87
L. Van hamme, Beyers Innovative Software, Michielssendreef 42, B-2930 Brasschaat

Hanne 449
Thomas Hanne, Fachbereich Wirtschaftswissenschaft/OR, Fernuniversität Hagen, Lütkenheider Str. 2, D-58084 Hagen

Harink 518
Drs. Jeroen H.A. Harink, PTT Research Tele-informatics, P.O. Box 15000, NL-9700 DC Groningen

Hauth 155
Dipl.-Kfm. Michael Hauth, Lehrstuhl für Unternehmungsforschung, Universität Mannheim, Schloss, D-68131 Mannheim

Hecking 46
Dr. Reinhard Hecking, Zentrale Absatzplanung (VM-31), Volkswagen AG, Postfach, D-38436 Wolfsburg

Heinrici 498
Dipl.-Kfm. Albrecht Heinrici, Lehrgebiet für BWL, insbes. Operations Research, Postfach 940, D-58097 Hagen

van der Heijden 209
Drs. Matthieu C. van der Heijden, Centre for Quantitative Methods CQM B.V., P.O. Box 414, NL-5600 AK Eindhoven

Helbig 456
Dipl. Ing. agr. Ralf Helbig, Institut für Landwirtschaftliche Betriebslehre, Rheinische-Friedrich-Wilhelms-Universität Bonn, Meckenheimer Allee 174, D-53115 Bonn

Hellberg 88
Man. Dir. Dr. Roland Hellberg, Sentraldistribusjon ANS, P.B. 113 Vindern, N-0319 Oslo

Hellwig 270
Prof. Dr. Klaus Hellwig, Abteilung Betriebswirtschaftslehre, Universität Ulm, Helmholtzstr. 18, D-89069 Ulm

Hörmann 329
Dr. Wolfgang Hörmann, Institut für Statistik und Datenverarbeitung, Wirtschaftsuniversität Wien, Augasse 2-6, A-1090 Wien

Juretzka 505
Jan Juretzka, Lehrstuhl für Betriebliche Datenverarbeitung und Produktionswirtschaft, Christian-Albrechts-Universität zu Kiel, Olshausenstraße 40, D-24118 Kiel

Kahn 399
Dipl. Forstw. Markus Kahn, Niedersächsische Forstländische Versuchsanstalt, Grätzelstr. 2, D-37079 Göttingen

Kallenberg VII
Dr. Lodewijk Kallenberg, Faculty of Applied Mathematics, Rijksuniversiteit Leiden, P.O. Box 9512, NL-2300 RA Leiden

Kallrath 44
Dr. Josef Kallrath, BASF AG, Abt. ZXA/ZC, Kaiser-Wilhelm-Str. 52, D-67056 Ludwigshafen

Karmarkar 2
Prof. Dr. Uday S. Karmarkar, University of Rochester, Graduate School of Business Administration, Rochester, New York 14627, U.S.A.

ten Kate 165
Drs. Hans A. ten Kate, Faculty of Management and Organization, University of Groningen, P.O. Box 800, NL-9700 AV Groningen

Kersten VII
Prof. Dr. Bert Kersten, Centrum voor Toegepaste Wiskunde, Rabobank Nederland, Postbus 17100, NL-3500 HG Utrecht

Kiewiet 374
Drs. Derk Jan Kiewiet, Faculteit Bedrijfskunde, Rijksuniversiteit Groningen, Postbus 800, NL-9700 AV Groningen

Kindt 221
Dipl. Wi.-Math. Andreas Kindt, Lehrstuhl für Betriebswirschaftslehre - Produktion und Logistik, Universität Augsburg, Memminger Straße 14, D-86159 Augsburg

Klein Haneveld 76
Prof. Dr. Willem K. Klein Haneveld, Department of Econometrics, Rijksuniversiteit Groningen, P.O. Box 800, NL-9700 AV Groningen

Knolmayer 457
Prof. Dr. Gerhard Knolmayer, Institut für Wirtschaftsinformatik, Universität Bern, Hallerstraße 6, CH-3012 Bern

Köchel 546
Prof. Dr. Peter Köchel, Technische Universität Chemnitz-Zwickau, Str. der Nationen 62, D-09009 Chemnitz

König 381
Dipl. Wi.-Ing. Hans-Jürgen König, Professur für BWL/Wirtschaftsinformatik, Justus-Liebig-Universität Giessen, Licher Str. 60, D-35394 Gießen

Kok 533
Matthijs Kok, Delft Hydraulics, P.O. Box 152, NL-8300 AD Emmeloord

de Kok 166
Prof. Dr. Ton G. de Kok, Technische Universiteit Eindhoven, Den Dolech 2, Paviljoen F4, Postbus 513, NL-5600 MB Eindhoven

Koole 532
Ger M. Koole, CWI, P.O. Box 4079, NL-1009 AB Amsterdam

Kopfer 222
Prof. Dr. Herbert Kopfer, Fachbereich 7 Wirtschaftswissenschaft, Universität Bremen, Bibliothekstraße, D-28359 Bremen

de Koster 96
Dr. Rene de Koster, Consultans and Engineers Ingenieursbureau, Groenewout BV, P.O. Box 3290, NL-4800 DG Breda

Kremer 271
Prof. Dr. Erhard Kremer, Institut für Mathematische Stochastik, Universität Hamburg, Bundesstraße 55, D-20146 Hamburg

Krishnapuram 406
Prof. Raghu Krishnapuram, ELITE Foundation, Promenade 9, D-52076 Aachen

Kroon 220, 506
Leo G. Kroon, Erasmus University Rotterdam, P.O. Box 1738, NL-3000 DR Rotterdam

Kruse 42, 60
Dr. Hermann-Josef Kruse, ZI-SEA, Schering Aktiengesellschaft, Müllerstr. 170, Postfach 650311, D-13342 Berlin

Küenle 553
Prof. Dr. Heinz-Uwe Küenle, Institut für Mathematik, Technische Universität Cottbus, Karl-Marx-Str. 17, PSF 101344, D-03013 Cottbus

Küfer 507
Dr. Karl-Heinz Küfer, Mathematik, Universität Kaiserslautern, Erwin-Schrödinger-Str., Postfach 3049, D-67663 Kaiserslautern

Kuhn 167
Dr. Heinrich Kuhn, Fachgebiet Produktionswirtschaft, TU Braunschweig, Pockelsstr. 14, D-38106 Braunschweig

Kulmann 382
Dipl.-Inform. Friedhelm Kulmann, FB Wiwi, AVZ II, Fernuniversität Hagen, Feithstr. 140, D-58097 Hagen

Lange 343
Tatjana Lange, Biesenbrower Straße 107, D-13057 Berlin

Leu 554
Jun-Der Leu, Department of General Course, Chung-Cheng Institute of Technology, Ta-Hsi, Tao-Yuan, Taiwan, Republic of China 33509

Meijboom 103
Dr. Ir. Bert R. Meijboom, Heineken Nederland B.V., Department of Physical Distribution, Postbus 530, NL-2382 PH Zoeterwoude

Meilijson 323
Prof. Isaac Meilijson, School of Mathematical Sciences, Tel Aviv University, Ramat Aviv, 69978 Tel Aviv, Israel

Melo 143
M. Teresa Melo, Econometric Institute, Erasmus University Rotterdam, Burg. Oudlaan 50, NL-3062 PA Rotterdam

Meyer 508
Dipl.-Wi.-Math. Carl-Heinz Meyer, Lehrgebiet für BWL, insbes. Operations Research, Fernuniversität Hagen, Postfach 940, D-58097 Hagen

van Montfort 109
Dr. Kees van Montfort, Free University, De Boelelaan 1105, NL-1081 HV Amsterdam

Mosler VII
Prof. Dr. Karl Clemens Mosler, Institut für Statistik und Quantitative Ökonomik II, Universität der Bundeswehr, Postfach 700822, D-22039 Hamburg

Muche 50
Dr. Gerhard Muche, Controlling Partner GmbH, Alte Landstraße 48, Postfach 1154, D-22949 Ammersbek

Müller 358
Prof. Dr. Johann-A. Müller, Fachbereich Informatik, Hochschule für Technik und Wirtschaft Dresden (FH), Friedrich-List-Platz 1, D-01069 Dresden

Müller 415
Dipl.-Inform. Karin Müller, ELITE Foundation, Promenade 9, D-52076 Aachen

Nass 224
Drs. Roel Nass, Econometric Institute, Erasmus University Rotterdam, P.O. Box 1738, NL-3000 DR Rotterdam

Natter 268
Mag. Martin Natter, Department of Business Administration and OR, Institute for Advanced Studies, A-1060 Wien

Nelißen 509
Dipl.-Inform. Josef Nelißen, Lehrstuhl für Angewandte Mathematik insbesondere Informatik, RWTH Aachen, Ahornstr. 55, D-52056 Aachen

Nicolai 296, 306
Dipl.-Ing. Marcus Nicolai, Deutsch-Französisches Institut für Umweltforschung (DFIU), Hertzstr. 16, D-76187 Karlsruhe

von Nitzsch 465
Dr. Rüdiger von Nitzsch, Lehrstuhl für Betriebswirtschaftslehre, Universität zu Köln, Albertus-Magnus-Platz, D-50923 Köln

van Noortwijk 533
Ir. Jan M. van Noortwijk, Faculty of Technical Mathematics & Informatics, Delft University of Technology, P.O. Box 5031, NL-2600 GA Delft

van Nunen . . . . VII, 212, 225
Prof. Dr. Ir. Jo A.E.E. van Nunen, Faculty of Business Administration, Erasmus University Rotterdam, P.O. Box 1738, NL-3000 DR Rotterdam

Oder 297
Dipl. Ing. Carsten Oder, Institut für Industriebetriebslehre und Industrielle Produktion, Universität Karlsruhe (TH), Hertzstraße 16, Bau 06.33, D-76187 Karlsruhe

Oehler 279
Dr. Andreas Oehler, Institut für Betriebswirtschaftslehre insbes. Bank- und Finanzwirtschaft, Fernuniversität Hagen, Feithstraße 152, D-58097 Hagen

Pankratz 222
cand. rer. pol. Giselher Pankratz, Deuzerstr. 12, D-57234 Wilnsdorf-Anzhausen

Peemöller 57
Dr. Jürgen Peemöller, c/o ORIGIN InformationTechnology Hamburg GmbH, Billstraße 80, D-20539 Hamburg

Perin 465
Dipl.-Inform. Sinan Perin, Lehrstuhl für Betriebswirtschaftslehre, Universität zu Köln, Albertus-Magnus-Platz, D-50923 Köln

Pesch 58
Christoph Pesch, Henkel KGaA, MMI/Systementwicklung, Postfach 101100, D-40191 Düsseldorf

Pesch 138
Dr. Erwin Pesch, Faculty of Economics and Business Administration, Rijksuniversiteit Limburg, P.O. Box 616, NL-6200 MD Maastricht

Peters 269
Dipl.-Kfm. Heinz-Theo Peters, Lehrstuhl für BWL, Universität zu Köln, Albertus-Magnus-Platz, D-50923 Köln

Pfaff 466
PD Dr. Dieter Pfaff, Fachbereich Wirtschaftswissenschaften, Johann Wolfgang Goethe-Universität Frankfurt am Main, Mertonstraße 17-25, D-60054 Frankfurt am Main

Piller 153
Mag. Ing. Franz Piller, Institut für Betriebswirtschaftslehre, Universität Wien, Brünner Straße 72, A-1210 Wien

Pohlenz 538
Dr. Siegmar Pohlenz, Florastr. 190, D-50733 Köln

Reese 298
Prof. Dr. Joachim Reese, Lehrstuhl für Produktion und Wirtschaftsinformatik, Universität Lüneburg, Lessingstraße 1, D-21332 Lüneburg

GPSR Compliance

The European Union's (EU) General Product Safety Regulation (GPSR) is a set of rules that requires consumer products to be safe and our obligations to ensure this.

If you have any concerns about our products, you can contact us on ProductSafety@springernature.com

In case Publisher is established outside the EU, the EU authorized representative is:

Springer Nature Customer Service Center GmbH
Europaplatz 3
69115 Heidelberg, Germany

Batch number: 10844827

Printed by Printforce, the Netherlands